Advances in Experimental Medicine and Biology

Volume 1200

More information about this series at http://www.springer.com/series/5584

Pierre Comizzoli • Janine L. Brown
William V. Holt
Editors

Reproductive Sciences in Animal Conservation

Second Edition

Springer

Editors
Pierre Comizzoli
Smithsonian Conservation Biology Institute
National Zoological Park
Washington, DC, USA

Janine L. Brown
Smithsonian Conservation Biology Institute
National Zoological Park
Front Royal, VA, USA

William V. Holt
Academic Unit of Reproductive
and Developmental Medicine
The University of Sheffield
Sheffield, UK

ISSN 0065-2598 ISSN 2214-8019 (electronic)
Advances in Experimental Medicine and Biology
ISBN 978-3-030-23632-8 ISBN 978-3-030-23633-5 (eBook)
https://doi.org/10.1007/978-3-030-23633-5

This Springer imprint is published by the registered company Springer Nature Switzerland AG
The registered company address is: Gewerbestrasse 11, 6330 Cham, Switzerland

Foreword

The recent United Nations IPBES (Intergovernmental Science-Policy Platform on Biodiversity and Ecosystem Services) assessment dramatically highlights the unprecedented pressures on biodiversity. Time is short, and the required commensurate increase in conservation effort worldwide requires a solid scientific underpinning: essentially an explosion of scientific innovation.

A vital segment of that is the very subject of this volume: reproductive sciences in animal conservation. Whether in the wild, captivity, marine, or terrestrial habitats, the ability of conservation to rise to the challenge will require a proliferation of new science. This volume represents a major contribution to that need.

When a field undergoes explosive growth as is now underway with reproductive science through the conservation lens, it is impossible for a single author (or a small number) to limn the topic. Only a multi-author volume is up to the task. This is that exciting and timely new book produced under the guidance of a troika of amazing editors.

It is interesting to see through this lens how reproductive sciences for conservation have progressed—rapidly and creatively in recent years. The field ranges from managing/encouraging reproduction in captivity—something those of us in the national capital area have seen through the lens of news bulletins from the national zoo about the giant pandas (followed by the public as avidly as if they emanated from Buckingham Palace) to struggles in the laboratory and the field with the amphibian extinctions driven by a lethal chytrid fungus.

At the other end of the spectrum is fertility control of exploded wildlife populations like white-tailed deer that destroy the habitat of ground-nesting birds in forests of Eastern North America because of the absence of original predators.

Modern zoos have long engaged in science to better understand and assist the reproduction of the species under their care. That led in cases like the golden lion tamarin to joint management of captive and wild populations. Given today's environmental challenges, that almost seems like an age of innocence.

Wild species, whether terrestrial, freshwater, or marine, are subject to molecules new to nature that affect their reproductive capability. They range from endocrine disruptors to plastics and their breakdown products which are virtually ubiquitous in the environment. Climate change adds another major component that can affect physiology or create ripples of change through food chains and in turn affect reproduction.

Conservation concern goes beyond single species to entire ecosystems. Tropical coral reefs are in that spotlight. They are buffeted by poor management of fisheries, which tip the balance in favor of macro algae, and by climate change and ocean acidification leading to coral bleaching and collapse of the reef ecosystem. That might seem impossible to address, but cryopreservation of coral sperm and larvae is showing great promise as means to create an afterlife for tropical corals.

Throughout the grand conservation challenge that confronts us and grows daily, there is an enormous need for new science. Science is needed to illuminate the impacts on biodiversity and possible ways to address them even as humanity wrestles with global change and charting a sustainable future for people and other forms of life. Reproductive science is a keystone in the necessary intellectual architecture as this pioneering volume demonstrates so well.

Thomas Lovejoy
University Professor of Environmental Science and Policy
George Mason University
Fairfax, VA, USA

Contents

Chapter 1
Reproductive Science as an Essential Component of Conservation Biology: New Edition

Pierre Comizzoli, Janine L. Brown, and William V. Holt

Abstract The previous edition of this book mainly provided a snapshot of the state of the art in terms of species-specific reproductive biology and emerging technologies. The influence of environmental changes on reproductive fitness was introduced but not fully explored. The objectives of this second edition were to (1) emphasize the need for holistic and global efforts to understand and sustain reproduction in a constantly changing environment and (2) provide more knowledge in the reproductive physiology of different taxa. The first section of the book is dedicated to survival and adaptation of species in a changing environment (including chapters on environmental impacts in different taxa, as well as the role of microbiomes). The second section focuses on progress in understanding, assisting or even suppressing reproduction in wild species, keeping in mind the influence of environmental factors as well. It contains chapters from the previous edition that were updated (reproduction in elephants, koalas, marsupials, amphibians, and corals), new chapters on species such as sharks and rays, and contributions about the increasing role of reproductive manipulations, such as assisted reproduction and contraception. While the present book emphasizes the overarching issue of environmental impacts on reproduction (resulting in infertility, subfecundity, or fitness), it also highlights the challenges of maintaining wild species in captivity, including those associated with ensuring good welfare. Captive environments can influence reproduction in a multitude of ways, some unexpected, such as the selection of unwanted genetic traits, an essential dimension to be considered to ensure the success of

---

P. Comizzoli (✉)
Smithsonian Conservation Biology Institute, National Zoological Park, Washington, DC, USA
e-mail: Comizzolip@si.edu

J. L. Brown
Smithsonian Conservation Biology Institute, National Zoological Park, Front Royal, VA, USA

W. V. Holt
Academic Unit of Reproductive and Developmental Medicine, The University of Sheffield, Sheffield, UK

P. Comizzoli et al. (eds.), *Reproductive Sciences in Animal Conservation*, Advances in Experimental Medicine and Biology 1200,
https://doi.org/10.1007/978-3-030-23633-5_1

conservation breeding programs. Lastly, new approaches, such as the use of allostatic load indexes and reproductive microbiome analyses also will be closely examined for the first time in rare and endangered species to address conservation issues.

1 Introduction

Conservation biology aims at understanding and sustaining thriving biodiversity because the disappearance of even a single animal or plant species can compromise functions of entire ecosystems. Preserving healthy wildlife populations is a complex task involving multiple disciplines spanning from social sciences to animal physiology, and where the aims are to prevent or mitigate the loss of species caused by human activities. These include major oil spills that devastate wildlife across large areas of coastline, increased atmospheric greenhouse gases, which are widely believed to cause global warming and acidification of the oceans; moreover, extreme weather events that result from such changes can wreak havoc on whole nations. On a more local scale, conservation biology is related to urbanization, industrialization, large-scale infrastructures (highways, pipelines), agriculture, forestry, mining, and other activities (Comizzoli and Holt 2019).

The core of animal conservation is the maintenance of genetic diversity within populations while ensuring that these are large enough to be sustainable. In addition to protecting viable populations in their natural habitat (*in situ* conservation), it is critical to maintain populations in captivity (*ex situ* conservation) for eventual reinforcements or reintroductions in the wild. Eventually, *in situ* and *ex situ* conservation breeding programs eventually enable habitats to be healthy and provide sustainable sources of clean water and food products to local human populations (Comizzoli and Holt 2019).

Reproduction is undeniably key to the survival of all species on earth, whether they live *in situ* or *ex situ*. Technology aside, the study of reproductive processes in animal species remains dauntingly complex, ranging from details of gametogenesis, fertilization and the subsequent processes of embryonic development, growth and sexual differentiation, endocrinology and aspects of behavior and brain function. As if this list was not broad enough, modern scientific advances have enabled us to drill down into the intricate details of gene expression, protein synthesis and the immune system as it affects each of the processes mentioned above. Reproductive sciences are key to understanding a wide range of problems, ranging from studying the consequences of pollutants on the development and survival of animals in the marine environment, to predicting the impact of climate changes on species survival across habitats.

The previous edition of our book mainly provided a snapshot of the state of the art in terms of species-specific reproductive biology and emerging technologies. The influence of environmental changes on reproductive fitness was introduced but not fully explored. This second edition emphasizes the need for holistic and global efforts to understand and sustain reproduction in a constantly changing environment.

Many recent studies illustrate this need. For instance, male mongooses that experience highly variable ecological conditions during development live longer and have greater lifetime fitness, while those that experience poor early-life conditions have reduced fertility (Marshall et al. 2017). In females, interestingly, there apparently is no such effect (Marshall et al. 2017). Thus, the first section of the book is dedicated to survival and adaptation of species in a changing environment (including chapters on environmental impacts in different taxa as well as the role of reproductive microbiomes). The second section focuses on progress in understanding, assisting and suppressing reproduction in wild species, keeping in mind the influence of environmental factors as well. It contains chapters from the previous edition that were updated (reproduction in elephants, koalas, marsupials, amphibians, and corals), new chapters on understudied species such as sharks and rays, and contributions about the increasing role of population control technologies like assisted reproduction and contraception.

While the present book emphasizes the overarching issue of environmental impacts on reproduction with regard to infertility, subfecundity, or fitness, it also highlights the challenges of maintaining wild species in captivity. Captive environments highly influence reproduction and can lead to the unwanted selection of genetic traits (Farquharson et al. 2018). This is an essential new dimension to consider to ensure the success of conservation breeding programs, especially if animals are be used for reintroduction. New approaches to assess welfare, such as the use of allostatic load indexes as a tool to measure the effects of accumulated stress over time, will be discussed in the context of how they can be used to determine what factors impact reproduction in rare and endangered species.

2 Survival and Adaptation of Species in a Changing Environment

Animals evolve and adapt to their environment to optimize fitness, and the science of understanding these interactions has led to the realization that phenomena such as temperature, photoperiod, diet and seasons have massive impacts on reproductive function. Environmental changes, both global and local, can affect the health and wellbeing of animals and humans alike during their entire life and even beyond (Gluckman et al. 2009). As the combination of climate change, epigenetics and adaptation provides an important and overarching context with links to most other aspects of reproductive sciences, we were keen to include a series of authoritative overviews of environmental changes and reproduction to understand some of its consequences.

In the opening chapter of that first section, Manlik mentions that survival and reproduction have similar impacts on population dynamics, since populations grow by one additional member regardless of whether a new individual is born, or an existing individual does not die. Interestingly, for slow-growing populations, previous analyses have shown that growth rates are more sensitive to changes in survival than to equal changes in reproduction, which is not necessarily accurate. The author is arguing that the role of reproduction in the conservation of slow-growing species is considerably greater than conventional sensitivity analyses would suggest. This is illustrated by case studies on wildlife populations that underscore the importance of reproduction for the conservation of slow-growing birds, ungulates, carnivores and cetaceans.

The next chapter from Kanda tackles the issue of exposure of wild species to endocrine disruptors in the environment. This is of particular concern given increasing wildlife exposures to cocktails of chemicals, which combined with other stressors, may play a greater role in reproductive disorders and survival than can be reproduced in laboratory experiments. Several approaches are proposed to mitigate the adverse effects of those compounds (Replacing—Recycling—Reusing). Importantly, growing cities and industries need to have a more responsible production and use of those detrimental compounds (concept of green chemistry).

The chapter by Watkins is about the epigenetic basis of adaptation and responses to environmental change. Most studies in that area have been conducted in human populations, or by laboratory studies of mice and rats. There is now general agreement that parental and grand-parental diets elicit significant epigenetically-based effects on body and organ size, and lifetime disease risks through transgenerational inheritance via the gametes and embryos (Watkins et al. 2018). There are many good lessons to be learned for wildlife on the long-term implications of parental environmental change for offspring health (including reproductive fitness), across a wide range of species. Epigenomics is an emerging discipline that will play a major role in conservation biology in the next decade.

The chapter of Edwards et al. highlights the development and use of new tools to measure environment impacts on animal welfare, which is inexorably linked to reproduction. Traditional methods of assessing stress have mostly relied on analyses of circulating or excreted glucocorticoids. Today, we recognize the need for more objective indicators that incorporate multiple physiological systems to identify both negative and positive welfare states. One technique measures allostatic load—a method by which multiple physiological markers are used to inform on factors related to morbidity and mortality risk in humans, which if applied to wildlife could be a powerful conservation tool. It is important to note that our goal is not to eliminate all stress—boredom is equally deleterious to welfare—but rather to understand how animals respond to challenges at individual and population levels, and identify mitigating strategies to maximize fitness. One important take-home message is that most welfare research is conducted in mammals, with little emphasis on other species, suggesting we may be underestimating the welfare needs of the majority of animals on the planet.

The chapter by Acevedo-Whitehouse focuses on marine mammals and presents evidence of how drivers of environmental change can exert effects at different levels of biological organization to influence the viability of individuals and populations, and discusses which aspects of environmental change could be the most likely to shift the physiological thresholds of different marine mammal species. Even minor environmental changes can have physiological impacts, especially if they occur during critical periods of growth and development. Marine mammals are particularly susceptible to the negative effects of human activities, with warming sea temperatures, melting sea ice, and acidification occurring at alarming rates. The chapter then concludes by identifying areas of further research in marine mammal conservation physiology in the current context of global environmental change.

The chapter of Mayer explores the role of reproductive sciences in the preservation and breeding of commercial and threatened teleost fishes. There is an urgent need to protect and conserve the huge genetic diversity offered by this diverse vertebrate group, not just as a source of genes for contemporary breeding and for protection against the consequences of climate change and disease, but also as part of national heritages. It is anticipated that the successful cryopreservation of fish germplasm will result in a range of economic, ecological and societal benefits. In partnership with emerging assisted reproductive technologies, the successful cryopreservation of fish germplasm will lead to more efficient reproduction in aquaculture, assist selective breeding programs, and be of crucial importance to future species conservation actions.

The closing chapter of that section by Comizzoli and Power reports the increasing need to understand and sustain reproductive microbiomes in wild animal species. Based on extensive research in humans, it is clear that microbiomes in the reproductive tract interact with reproductive cycles and pregnancy. However, those microbiomes are still poorly known in wild species, especially for their involvement in fertility or parental/offspring health. This emerging research area is highly relevant to conservation biology, from captive breeding management to successful reintroduction or maintenance of wild populations. Thus, reproductive microbiomes represent a novel dimension in conservation biology that will likely expand soon.

3 Progress in Understanding, Assisting, or Suppressing Reproduction in Wild Species

Recent review papers have reported inspiring breakthroughs in wildlife reproduction science that will set directions for future research and lead to more successes in conservation biology (Comizzoli and Holt 2019). Despite new tools and approaches allowing a better and faster understanding of key physiological mechanisms, we still know little about reproduction in endangered species. This is also true for the development of assisted reproductive techniques where species-specific protocols do not exist for most endangered species (Herrick 2019). Multiple factors still limit progress in that area (access to the animals, lack of training and funding, etc.).

From the outset, one of our major objectives in producing this new book has been to present a comprehensive progress report about various wildlife research programs that involve aspects of reproductive biology. Here we present a set of chapters that will inform the reader about the recent progress in that area.

The first chapter of that second section, by Brown, is an update on the comparative reproductive biology of elephants based on multiple years of research in different countries. In recent years, a number of significant advances have been made, primarily with respect to female elephants. Reproductive problems hamper captive propagation efforts, particularly those related to abnormal, or even lack of ovarian cyclicity. Recent large-scale, multi-institutional studies and use of epidemiological approaches have been important in identifying factors important for good welfare and reproduction; these include enrichment, feeding diversity, good elephant-keeper relations, social compatibility, exercise, and not being obese. Among wildlife, there is no match to the biological knowledge we have on elephants, which will only grow in the years to come.

The chapter of Martin-Wintle is an overview on giant panda populations that are managed through natural mating and artificial insemination. It is worth emphasising that the conservation status of the giant panda has recently been downlisted from "Endangered" to "Vulnerable", primarily through improved captive breeding based on knowledge of panda reproduction coupled with major investments made by Chinese scientists and their government. This chapter documents the development of successful giant panda managed breeding programs by focusing on three key areas: (1) the development of science-driven reproductive techniques to improve fecundity in a species where the mating system was poorly understood; (2) how targeted research and adaptive management of social settings surrounding estrus and breeding improved reproductive success; and (3) insights and solutions to challenges faced across the program history with future directions for research. As the giant panda program transitions to a maintenance phase for the captive population and focuses on producing individuals that will contribute to reintroduction efforts, it will be important to increase the proportion of animals that express natural behaviors, adaptability, and high cognitive abilities. Key to this effort will be increasing the proportion of individuals that mate naturally as well as maternally rear cubs, and being able to reliably transfer these traits to potential reintroduction candidates.

The next chapter by Rodger is a reflection on marsupial reproduction. Marsupials, like most wildlife, face a very uncertain future due to multiple and growing threats (from human activities to global changes). It therefore is time to better understand and learn to practically manipulate the reproduction of animals. However, the knowledge base of marsupial biology generally, and the absence of well-studied domestic or laboratory species so important in assisted reproductive technology development for eutherian mammals, exacerbates the challenge. The author has put forward a fairly strong case in his chapter to stop worrying about trying to maximize sperm motility, etc., after freezing, and just go ahead with freezing all sperm in the expectation that ICSI will be feasible in the future. Call for action: can we find more meaningful empathy, engagement and investment for our evolutionary relatives; iconic, attractive and unattractive, as they face this perfect storm of challenges to their survival due to humanities overt actions and ignorance?

The chapter from Johnston and Holt focuses on koala reproduction with rare and essential information on female anatomy and physiology. As with so many other species, koalas are threatened by climate change and habitat destruction, but also by disease and predation by dogs. It also is an excellent illustration of the integration of artificial insemination into a conservation program that has been ongoing for 25 years. The koala stands out as the one marsupial species where this technology has been useful, although it still is not a panacea for the damage caused by human activities. Challenges still remain, including developing effective semen cryopreservation techniques. Given the strong evidence of continued declines in koala population densities across the country, captive breeding is increasingly viewed as a conservation tool for this species. A major concern is loss of genetic diversity in this closed population, so assisted reproductive techniques may be needed to sustain the health of the population.

The chapter by Lanyon is about current methods and future approaches to examine reproductive biology in free-ranging marine mammals. This includes direct observations of reproductive behavior and output, endocrinology to determine reproductive state, and assisted reproductive technologies to enhance reproductive outcomes. It also addresses the most recent developments and innovations to reproductively sample marine mammals through two case studies (North Atlantic right whales and dugong). Climate disruption with increasingly severe coastal storms and flooding threaten the health of coastal seagrass beds, and consequently reproductive success and survival of this species. It is anticipated that climate change may have diverse and often serious effects on marine mammal reproduction in populations around the globe.

The chapter by Clulow and Clulow is about amphibian assisted reproductive technologies—Moving from technology to application. This is a very thorough review of successful technologies including non-invasive gamete collection, artificial fertilization and sperm cryopreservation that work well enough to be applied to many current conservation programs. Authors also consider new advances in technology (vitrification and laser warming) of cryopreservation of aquatic embryos of fish and some marine invertebrates that hold out hope of overcoming the block to amphibian oocyte and embryo cryopreservation. In addition, through two case studies (in New Guinea and Bolivia), they illustrate both the urgent need and the opportunity to implement immediately ARTs, cryopreservation and biobanking to amphibian conservation.

The chapter by Penfold et al. reports that rays and sharks are ecologically and reproductively diverse and that basic knowledge on their reproductive biology is still relatively unknown, including fundamental questions such as where birthing grounds are located for several large shark species (e.g. sand tiger shark, *Carcharius taurus*). Reproductive technologies used in shark and ray species include semen collection and artificial insemination. Elasmobranch spermatozoa are long-lived and cold resistant, lending itself well to overnight shipment to move genes rather than the animal from institution to institutions where offspring have been produced in several species. Classic steroid hormones (testosterone, progesterone and estradiol) control reproduction, and hormonal responses are influenced by environmental cues that are still not fully understood. These cues play an important role in initiating seasonal reproductive changes, including sperm quality, which renders elasmobranchs susceptible to environmental changes, such as ocean warming.

Current reproductive technologies will serve to generate data to mitigate anthropogenic changes that threaten the future of this vulnerable groups of fish.

The chapter from Hagedorn et al. is an outstanding review of the current knowledge on coral reproduction and successful attempts to use and integrate biotechnologies to save those species.

Cryopreservation is presented as a tool for restoration, and to facilitate the movement of genes across geographical barriers to avoid loss in genetic diversity. To date, 31 species of coral have been successfully cryopreserved and are stored in genome banks with the ultimate goal of using them to secure and restore reefs in threatened areas, such as the Great Barrier Reef. Challenges to coral survival revolve around warming and acidification of the oceans, which makes coral more susceptible to stress, bleaching and diseases. The greatest test will be whether efforts can be ramped up quickly enough to prevent diversity from diminishing further.

Lastly, the chapter by Cheryl Asa and Anneke Moresco is about fertility control in wildlife, which seems ironic given the threatened status of so many species, but there are numerous instances where effective population management requires some degree of reproductive control. This is especially true in well-protected areas where damage to ecosystems caused by uncontrolled population growth can negatively impact a multitude of species therein. A conservationist's conundrum—in some places there are too many animals—is particularly relevant for African elephants (http://www.earthisland.org/journal/index.php/articles/entry/a_conservationists_conundrum_in_some_places_there_are_too_many_elephants). Non-lethal forms of contraception can aid in mitigating human-animal conflict in high density areas, and control feral domestic or invasive species. Breeding programs in zoos also depend on highly effective and selective contraception to manage sustainable insurance populations. There are a number of effective fertility control methods in use today. The authors point out that while some have not lived up to past promises, others under development present hope for addressing remaining challenges in species management, especially in human-dominated landscapes.

4 Concluding Remarks

When compiling this book we were keen to emphasize and present the many remarkable dimensions of reproductive science and to show how they can and should inform the practice of wildlife conservation. Reproductive science is not only used to help with *ex-situ* animal breeding and management, but is able both to provide invaluable insights into the globally important consequences of environmental change and to mitigate some of the problems caused by human activities. We therefore chose to focus attention on topics that are timely, have shown exceptional progress, or that otherwise justify an update. This approach inevitably means that some fields of research were omitted or not expanded.

Regarding genomics, most efforts are still conducted in laboratory species to identify fertility genes, estimate the number of protein-coding genes essential for fertility of males and females, and predict the next major directions in the genetics of reproduction and fertility (Schimenti 2018). Even though more reference genomes of wild species have been published recently, the need for integration of genomic tools into conservation programs are still essential. This was highlighted in the first edition of the book and is still valid (Johnson and Koepfli 2014). Genomic and epigenomic information will only have value when solid phenotypical data are associated with them. Similar to the discipline of 'precision medicine' currently developed in humans, it is time to create and develop a concept of 'precision conservation' or 'precision conservation breeding'—a more customized/tailored approach to optimize the potential of each individual in a population (Comizzoli and Holt 2019).

Creating and developing Genome Resource Banks is still critical, even though we have not included a chapter dedicated to that topic, although it was touched on in several chapters dealing with specific taxa. Recent genetic analyses of the current black-footed ferret population, which was rescued from extinction by reproductive technologists, have demonstrated how the judicious use of frozen and banked semen dating back to the late 1980s has helped to mitigate the impact of inbreeding depression (Howard et al. 2016). We highlight through several chapters that much of the recent progress in reproductive sciences has been made in non-mammalian species, although not for the development of welfare tools as pointed out by Edwards et al. The cryopreservation of fish and amphibian oocytes and embryos, despite having been studied extensively, has remained stubbornly difficult to achieve, largely because they are so large and encased in impermeable vestments that prevent entry of cryoprotectants. Some novel approaches involving laser-warming of fish and coral embryos are showing promise; such sophisticated developments are driven mainly by the important potential value of fish models for biomedical research and the applications in commercial aquaculture rather than the applications in conservation biology.

We did not include a chapter on advanced reproductive technologies in birds, but we should nevertheless mention that new studies have demonstrated the success of artificial insemination (cranes; Songsasen et al. 2019) and encouraging results in primordial germ cell transplantations between species (in storks and in ibis; Tagami et al. 2017; Yakhkeshi et al. 2018). Specific research into the application of reproductive technologies in reptiles is not included either, but a recent review from Clulow and Clulow is still up to date (Clulow and Clulow 2016).

As implied by the title of this chapter, the editors firmly believe that it is extremely difficult to engage seriously in wildlife conservation without understanding how factors in the environment impact reproduction. Moreover, we hope to make the case that reproductive technologies can only be developed and improved by taking transdisciplinary approaches that benefit from expertise derived from many different fields. In keeping with this approach we are delighted with the breadth of topics we were able to include and would like to express our gratitude not only to the chapter authors, but also to the numerous reviewers for their willingness to help, their insights and their constructive comments.

References

Clulow J, Clulow S. Cryopreservation and other assisted reproductive technologies for the conservation of threatened amphibians and reptiles: bringing the ARTs up to speed. Reprod Fertil Dev. 2016;28:1116.

Comizzoli P, Holt WV. Breakthroughs and new horizons in reproductive biology of rare and endangered animal species. Biol Reprod. 2019; https://doi.org/10.1093/biolre/ioz031.

Farquharson KA, Hogg CJ, Grueber CE. A meta-analysis of birth-origin effects on reproduction in diverse captive environments. Nat Commun. 2018; https://doi.org/10.1038/s41467-018-03500-9.

Gluckman PD, Hanson MA, Bateson P, et al. Towards a new developmental synthesis: adaptive developmental plasticity and human disease. Lancet. 2009;373:1654–7.

Herrick JR. Assisted reproductive technologies for endangered species conservation: developing sophisticated protocols with limited access to animals with unique reproductive mechanisms. Biol Reprod. 2019; https://doi.org/10.1093/biolre/ioz025.

Howard JG, Lynch C, Santymire RM, Marinari PE, Wildt DE. Recovery of gene diversity using long-term cryopreserved spermatozoa and artificial insemination in the endangered black-footed ferret. Anim Conserv. 2016;19:102–11.

Johnson WE, Koepfli K. The role of genomics in conservation and reproductive sciences. Adv Exp Med Biol. 2014;753:71–96.

Marshall HH, Vitikainen EIK, Mwanguhya F, et al. Lifetime fitness consequences of early-life ecological hardship in a wild mammal population. Ecol Evol. 2017;7:1712–24.

Schimenti JC, Handel MA. Unpackaging the genetics of mammalian fertility: strategies to identify the "reproductive genome". Biol Reprod 2018; 99, 1119–1128.

Songsasen N, Converse SJ, Brown M. Reproduction and reproductive strategies relevant to management of whooping cranes ex situ. In: Whooping cranes: biology and conservation. Amsterdam: Elsevier; 2019. p. 373–87.

Tagami T, Miyahara D, Nakamura Y. Avian primordial germ cells. Adv Exp Med Biol. 2017;1001:1–18.

Watkins A, Dias I, Tsuro H, Allen D, Emes R, Moreton J, Wilson R, Ingram R, Sinclair K. Paternal diet programs offspring health through sperm- and seminal plasma-specific pathways in mice. Proc Natl Acad Sci U S A. 2018;115:10064.

Yakhkeshi S, Rahimi S, Sharafi M, Hassani S-N, Taleahmad S, Shahverdi A, Baharvand H. In vitro improvement of quail primordial germ cell expansion through activation of TGF-beta signaling pathway. J Cell Biochem. 2018;119:4309–19.

Part I
Survival and Adaptation of Species in a Changing Environment

Chapter 2
The Importance of Reproduction for the Conservation of Slow-Growing Animal Populations

Oliver Manlik

Abstract Both survival and reproduction are important fitness components, and thus critical to the viability of wildlife populations. Preventing one death (survival) or contributing one newborn (reproduction), has arguably the same effect on population dynamics—in each instance the population grows or is maintained by one additional member. However, for the conservation of slow-growing animal populations, the importance of reproduction is sometimes overlooked when evaluating wildlife management options. This has to do with the use of demographic sensitivity analyses, which quantify the relative contribution of vital rates to population growth. For slow-growing populations, the results of such analyses typically show that growth rates are more sensitive to changes in survival than to equal proportional changes in reproduction. Consequently, for slow-growing taxa, survival has been labelled a better fitness surrogate than reproduction. However, such a generalization, derived from conventional sensitivity analyses, is based on flawed approaches, such as omitting appropriate scaling of vital rates, and sometimes misinterpretations. In this chapter, I make the case that for the conservation of slow-growing species the role of reproduction is considerably greater than conventional sensitivity analyses would suggest. This is illustrated by case studies on wildlife populations that underscore the importance of reproduction for the conservation of slow-growing birds, ungulates, carnivores, and cetaceans.

Keywords Bears · Birds · Carnivores · Cetaceans · Elasticity analysis · Reproduction · Sensitivity analysis · Ungulates · Wildlife conservation

---

O. Manlik (✉)
Biology Department, College of Science, United Arab Emirates University, Al Ain, United Arab Emirates

Evolution and Ecology Research Centre, School of Biological, Earth and Environmental Sciences, University of New South Wales, Sydney, NSW, Australia
e-mail: oliver.manlik@uaeu.ac.ae

P. Comizzoli et al. (eds.), *Reproductive Sciences in Animal Conservation*, Advances in Experimental Medicine and Biology 1200,
https://doi.org/10.1007/978-3-030-23633-5_2

1 Introduction

Animal taxa can be grouped according to their life-history traits along a 'fast–slow' growing continuum (Stearns 1983; Gaillard et al. 1989; Harvey et al. 1989; Sæther and Bakke 2000; Bielby et al. 2007; Dobson and Oli 2007; Nilsen et al. 2009). Populations on the fast-end of this continuum reproduce early and often, during a relatively short lifespan, and typically have large litter (or clutch) sizes. On the other hand, slow-growing taxa commonly reproduce relatively late, have low fecundity, small litter (or clutch) sizes, and relatively long lifespans (Promislow and Harvey 1990; Lebreton and Clobert 1991). Slow-growing taxa encompass a variety of vertebrate classes, orders, and clades, including some birds (Lebreton and Clobert 1991) and mammals (Promislow and Harvey 1990), such as ungulates (Gaillard et al. 1998), carnivores (van de Kerk et al. 2013) and cetaceans (Mann et al. 2000), which are the focus of this chapter.

Given that for many threatened species, demographic and population-specific data are scarce, it is tempting to make generalizations based on shared life-history traits, which can then be applied to data-deficient populations (van de Kerk et al. 2013; Heppell et al. 2000; Crone 2001). One of these generalizations is that survival, especially adult survival, is often more important to the conservation of slow-growing taxa than reproduction. This generalization is derived from numerous studies, which appear to have shown that in slow-growing populations, the relative contribution of adult survival to population growth is greater than that of reproduction (Heppell et al. 2000; Crone 2001; Oli and Dobson 2003). In other words, in slow-growing taxa, adult survival typically has greater elasticity than reproduction.

These findings are based on elasticity analyses or other forms of demographic proportional sensitivity analyses. Such demographic sensitivity analyses are a set of analytical tools that try to assess the effect of vital rates on population growth, or other indicators for population viability (reviewed by Horvitz et al. 1997; Manlik et al. 2018a). This is typically done by manually changing vital rates and then assessing the potential effect of those changes on population growth. One of the most commonly used types of sensitivity analysis entails changing each vital rate by the same proportion. For example, both reproductive rates and survival rates of a given population might be increased by the same proportion of 1%, which in turn increases the forecast population growth of that population. If the 1% increase in survival leads to a greater increase in population growth compared to a 1% increase in reproduction, then survival is considered to have a relatively greater contribution to population growth than reproduction. Vice versa, reproduction is considered to have a larger contribution to population growth if the 1% increase in reproductive rates results in a greater population growth compared to changing survival rates by that same proportion. For slow-growing populations, the results of these proportional sensitivity analyses usually show that growth rates for such taxa are more sensitive to changes in survival than to changes in reproduction (Sæther and Bakke 2000; Lebreton and Clobert 1991; Gaillard et al. 1998; van de Kerk et al. 2013; Heppell et al. 2000; Crone 2001; Oli and Dobson 2003; Horvitz et al. 1997; Manlik et al. 2018a; Stahl and Oli 2006; Young and Keith 2011). A commonly made generalization, as shown in the examples above, is that conservation efforts targeting slow-growing animal populations should focus on survival and not reproduction.

The general pattern that adult survival typically has greater elasticity than reproduction has been shown for a wide range of taxa, including slow-growing bird species (Sæther and Bakke 2000; Lebreton and Clobert 1991; Stahl and Oli 2006), ungulates (Gaillard et al. 1998, 2000), carnivores (van de Kerk et al. 2013), and cetaceans (Brault and Caswell 1993; Young and Keith 2011). This has led to the misguided interpretation that adult survival is more important than reproduction for the viability of slow-growing animal populations, suggesting that reproduction only plays a minor role for the conservation of such taxa. For instance, Crone (2001) compared elasticity values of over 150 taxa. She found that, with few exceptions, reproduction had the lowest elasticity value, i.e. the relative smallest contribution to projected population growth in slow-growing taxa. In contrast, adult survival had consistently high elasticity values for long-lived taxa. Therefore, Crone concluded that "the single best fitness surrogate is annual adult survivorship, and the worst surrogate (from the suite of vital rates) is typically annual adult fecundity." Furthermore, she pointed out that "annual reproductive output is seldom the best indicator of differences in fitness" (p. 2611) (Crone 2001).

Other studies have stressed that wildlife management should focus on vital rates with the highest elasticity. For instance, Caswell (2000) generalized that "management efforts should focus on demographic parameters with the largest elasticity." Similarly, a review of 50 mammal populations relied on elasticity values when "classifying species for management purposes" (p. 654) (Heppell et al. 2000). Solely on the basis of elasticities, the authors concluded that for slow-growing mammal populations "fecundity and early offspring survival are less critical than juvenile survival to maturity" (p. 661) (Heppell et al. 2000). They argued that "for conservation purposes (…) growth rates of 'slow' mammals that mature late and have few offspring per year will respond better to improved adult or juvenile survival rates" (p. 662). The authors of this study portrayed the application of conventional elasticity analyses as "a shortcut for life history analysis that may be very useful for categorizing species for management" (p. 662). Heppell et al. (2000), however, also added words of caution, stating that "conservation biologists should apply elasticity analysis cautiously" (p. 662) and emphasized that elasticity analyses "cannot substitute for detailed, long-term studies of wild populations." Nonetheless, when comprehensive, long-term data on wildlife populations are lacking, elasticity analyses and generalizations based on shared life-history traits remain a tempting option to guide wildlife management.

On the surface, these generalizations are appealing and appear convincing. However, such studies often underestimate the importance of reproduction for slow-growing animals for two main reasons: (1) the way proportional sensitivity and elasticity analyses have been conventionally conducted are flawed; (2) the results of these analyses are often misinterpreted.

1.1 The Inherent Flaw of Conventional Sensitivity Analyses

The inherent flaw of conventional sensitivity analyses that aim to rank wildlife management options lies in the fact that the vital rates being compared are on different scales (Manlik et al. 2018a; Link and Doherty 2002). Proportional sensitivity analyses, including elasticity analyses, compare the relative influence on growth rate when vital rates are perturbed by equal proportions (reviewed by Manlik et al. 2018a). The problem of vital rates being on different scales can be explained by the following: Mortality and survival, which measure the same aspect of life history, are by definition, on opposite scales—if survival is 0.9, then mortality is 0.1. Therefore, increasing survival by a given proportion, instead of mortality, constitutes a much larger absolute change (see Table 2.1). In this example, increasing survival by 5% would effectively mean decreasing mortality by 45%, from 0.1 to 0.055 (Table 2.1)—not by 5%, as we would expect if mortality and survival rates were on an equal scale. Hence, comparing the effects of changing survival rates with changing reproductive rates at equal proportions, which is the convention for elasticity analyses, is akin to comparing apples and oranges. While it might be realistic to increase reproduction by 5%, increasing survival by 5%, which corresponds to a much higher proportional change in mortality for slow-growing species (i.e. 45% decrease in this example), may be impossible. Furthermore, as a consequence of this unequal scale, elasticity analyses on the basis of survival yield very different results compared to mortality-based elasticity analyses. For example, a mortality-based elasticity analysis on a killer whale (*Orcinus orca*) population showed that fertility had the greatest elasticity (Olesiuk et al. 1990). In contrast, a survival-based elasticity analysis on the same orca population, resulted in adult survival having much greater elasticity than fertility (Brault and Caswell 1993). In other words, the type of model used, mortality-based or survival-based, influences the results and conclusions—not only for cetaceans, but for a range of taxa, including birds (Mortensen and Reed 2016). Hence, the relative importance of reproduction, based on elasticity analyses, hinges on whether we choose survival-based or mortality-based analytical approaches. Survival-based approaches have since become the convention, although some have argued that mortality-based perturbations are biologically more meaningful (Gerber and Heppell 2004).

Table 2.1 Unequal consequences of changing survival versus mortality rates by equal proportions

Vital rate	Initial rate	Increase survival by 5%	% Change
Survival	0.90	0.90 → 0.945	**+5**
Mortality	0.10	0.10 → 0.0055	–45

Vital rate	Initial rate	Decrease mortality by 5%	% Change
Survival	0.90	0.90 → 0.905	+0.56
Mortality	0.10	0.10 → 0.095	**–5**

The top two rows show an increase in survival rates by 5% and its corresponding decrease in mortality rate. The lower two rows show a 5% decrease in mortality and its corresponding increase in survival rate.

Slow-growing animal populations tend to have relatively high adult survival rates, and comparatively low reproductive rates. This may partly explain the high survival-based elasticities for adult survival in such taxa. Hence the reported differences in elasticities between survival and reproduction, may at least partly, be a result of those vital rates being on different scales. Therefore, it is likely that results of conventional, survival-based elasticities underestimate the relative contribution of reproduction to population growth. Link and Doherty (2002) recognized this problem of scale in elasticity analyses and offered an alternative approach, called variance-stabilized sensitivity (VSS) analysis. VSS applies an arcsine square-root transformation in order to scale the vital rates (although there may be more suitable transformations than the arcsine for this purpose, e.g. Warton and Hui (2011)). Unfortunately, many studies still omit appropriate transformations to scale vital rates in elasticity analyses that are aimed at guiding wildlife management.

1.2 The Interpretation of Conventional Sensitivity Analyses

Wildlife management recommendations that make use of elasticity analyses, and other conventional sensitivity analyses hinge on the interpretation of the results. It is important to remind ourselves what elasticity analyses can and cannot tell us with regard to the importance of vital rates. Elasticity tells us how populations would respond to an arbitrary, incremental and proportionally equal change in a given vital rate. In other words, elasticity analyses measure the relative contribution of each vital rate to population growth, on the basis that each vital rate is altered by the same, infinitesimal proportion. However, elasticity analyses cannot tell us which vital rates are actually depressed, or changed the most, and are causing a given population to decline. Elasticities also cannot tell us whether a given vital rate is more or less amenable to conservation actions. Hence, the information gained from elasticity analyses is limited and needs to be interpreted with caution, especially when it is meant to guide wildlife conservation. Despite these limitations (and criticism) of conventional elasticity analyses, elasticity values are still sometimes used as the sole or primary indicator to guide wildlife management. For instance, recent studies on Yangtze river fish species (Wang et al. 2017) and on black oystercatchers (Meehan et al. 2018) based their management advice primarily on elasticities without vital rate scaling.

One of the pitfalls of elasticity analyses is that, unlike the underlying assumption of equal, infinitesimal change, vital rates rarely change at equal proportions in nature (Manlik et al. 2018a; Mills et al. 1999; Mills and Lindberg 2002; Mills 2013). Consequently, management actions may not be able to alter vital rates over the range typically used in proportional perturbations (de Kroon et al. 2000; Manlik et al. 2016). Vital rates that are near their maxima and that vary little over time may also be less amenable to management actions (Mortensen and Reed 2016; Manlik et al. 2016; Reed et al. 2009). In the case of slow-growing animal taxa, this is typically true for adult survival. In contrast, reproductive rates may be more malleable, as they display larger temporal variation in slow-growing taxa, and they may thus have a greater potential to be manipulated (Manlik et al. 2018a; Mortensen and Reed 2016; Reed et al. 2002). Furthermore, due to the greater natural variability of reproductive rates, reproduction is often the main determinant of realized change in population growth of slow-growing taxa. In other words, the large, natural fluctuation of reproduction often drive population dynamics—and thus determine whether populations increase or decrease. This has been observed in a range of slow-growing taxa, including birds (Cooch et al. 2001; Beissinger and Peery 2007), ungulates (Gaillard et al. 1998, 2000; Albon et al. 2000; Gaillard and Yoccoz 2003), carnivores (Beecham 1983; Mitchell et al. 2009; Beston 2011), and cetaceans (Manlik et al. 2016; Brough et al. 2016). The following case studies on birds, ungulates, carnivores, and cetaceans highlight the important role reproduction can play in conservation of slow-growing animal populations.

2 Case Studies

2.1 Birds

As for other taxa, bird species can be placed along a "slow-fast continuum," in which species that mature late, produce few offspring (mostly just one) and have long lifespans are at the slow-end of the continuum (Sæther et al. 1996). A study by Sæther and Bakke (2000) assessed elasticities and temporal variation of vital rates of 49 avian species across this slow-fast continuum. Consistent with other taxa, a general pattern emerged in that study: For slow-growing bird species, reproductive rates displayed greater fluctuations than survival rates, but had lower elasticities, especially compared to adult survival. In other words, infinitesimal changes in reproduction had a relatively small effect on population growth compared to the same proportional changes in adult survival. Another study by Stahl and Oli (2006), who conducted elasticity analyses on 159 bird populations, confirmed this pattern with respect to elasticity. They concluded that for slow-growing bird populations "survival parameters were most important." It is important to note that neither Sæther and Bakke (2000), nor Stahl and Oli (2006) applied appropriate scaling when comparing vital rates in their elasticity analyses, as suggested by Link and Doherty (2002). To be fair, Sæther and Bakke (2000) published their study before the Link and Doherty (2002) paper came out, and also incorporated temporal variation in their study. However, the lack of appropriate scaling of vital rates limits the conclusions with regard to the relative importance of vital rates of these bird populations. Nonetheless, Stahl and Oli (2006) relied on the elasticity values to offer wildlife management advice for slow-growing albatross populations, which had particularly high elasticities for survival rates: "These results suggest that the long-term persistence of these birds can be improved by increasing survival rates, which can be achieved by regulating long-line fishing" (p. 37) (Stahl and Oli 2006). The authors attributed the potential importance of reproduction only to species on the fast end of the continuum, with high fecundity elasticities: "An endangered species known to mature early and reproduce at a high rate would most likely benefit from efforts to increase the fertility rate of the species, perhaps by the protection of nesting sites" (p. 37) (Stahl and Oli 2006). It is important to keep in mind that elasticities only offer information about how population growth changes when a given vital rate is altered by an arbitrary, infinitesimal amount. Elasticity offers no indication as to which rates are actually driving realized population dynamics, or which rates may be amenable to management actions. Therefore, reproduction can play a much greater role for the conservation of slow-growing animal population than conventional elasticity analyses suggest. The following case study on a slow-growing seabird species, the marbled murrelet, and other examples, illustrate that reproduction can play a crucial role in the conservation of slow-growing birds.

2.1.1 Case Study: Marbled Murrelet

The marbled murrelet (*Brachyramphus marmoratus*; family Alcidae) (Fig. 2.1) is an endangered seabird species that primarily nests in old-growth and older-aged trees of the Pacific Northwest (United States and Canada) (Piatt and Ford 1993; Ralph et al. 1995; Burger 2002; Piatt et al. 2006; Barbaree et al. 2014). The species is IUCN Red Listed as endangered because of rapid population declines (BirdLife International 2018). A clutch size of only one egg, and one to two broods per year (Sealy 1974; Beissinger 1995), places the murrelet on the slower end of the 'fast-slow' continuum of avian life-history strategies. Previous studies (Beissinger 1995;

Fig. 2.1 Adult flying marbled murrelet (*Brachyramphus marmoratus*) at Kachemak Bay, Kenai Peninsula, Alaska, USA. Photo: Brian Sullivan

Peery et al. 2006) showed that population growth of marbled murrelet populations is more sensitive to adult survival than to fecundity. This is not surprising, given the common observation that population growth of slow-growing animal populations tends to be more sensitive to changes in adult survival than reproduction, e.g. Sæther and Bakke (2000), Crone (2001), Stahl and Oli (2006), and Mortensen and Reed (2016).

However, population growth being very sensitive to a given vital rate, does not necessarily mean that conservation efforts should focus on that vital rate. Another study by Beissinger and Peery (2007) nicely illustrates that point: In order to identify the cause of past population declines of a murrelet population in central California, the authors used museum specimens to reconstruct historic demographic parameters on the basis of age-ratios. The study revealed that during the late 1800s and early 1900s, when the population was stable, reproductive rates were 8–9 times larger compared to contemporary rates. In contrast, adult survival rates remained relatively constant over time. More importantly, the study showed that reproductive rates at historic levels would enable a reversal of contemporary population declines (Beissinger and Peery 2007). In this case, changes in reproduction and not survival, clearly drove the negative population trajectory of this slow-growing seabird.

The dependence of the murrelet on old-growth forests for nesting explains why reproduction, and thus nesting habitat, may be the key to the persistence of this unique seabird. The large population declines that have been reported for several populations have been strongly associated with loss of nesting sites (Burger 2002; Piatt et al. 2006; Burger and Waterhouse 2009). Deforestation in old-growth forest in California, Oregon, Washington, and British Columbia likely lowered reproductive output, caused population fragmentations and ultimately facilitated the decline of murrelet populations in these regions (BirdLife International 2018; Sealy and Carter 1984; Rodway et al. 1992; Stein and Miller 1992). A study by Long et al. (2011) estimated that suitable nesting habitat of marbled murrelet populations in British Columbia was reduced by 20–22% between 1978 and 2008. Consequently, current conservation strategies focus on identifying and preserving suitable nesting habitat to boost reproduction of the marbled murrelet (Burger et al. 2018).

2.1.2 Other Bird Studies

Other studies on slow-growing bird species further underscore the role of reproduction in avian conservation. For example, a study on the lesser snow goose (*Chen caerulescens caerulescens*) showed that realized population dynamics are more strongly influenced by past temporal changes in reproductive rates and juvenile survival than by adult survival, the rate with the highest elasticity (Cooch et al. 2001). A study by Norris and McCulloch (2003) showed that reproduction was crucial to the recovery of the endangered Seychelles magpie robin (*Copsychus sechellarum*). This long-lived bird species, endemic to the Seychelles archipelago, was nearly driven to extinction by 1970, with only a few individuals left on one of the islands (BirdLife International 2017). However, it has since partly recovered, especially due to substantial population growth in the early 1990s (population size doubled between 1990 and 1995) (Norris and McCulloch 2003). Significant increases in chick survival and reproduction were the primary drivers of the population recovery of this bird population. Despite its high elasticity, adult survival had little influence on the population dynamics of this population because those rates were already very high (>0.9) and could not be altered much (Norris and McCulloch 2003). Similar observations were made on a white-breasted thrasher (*Ramphocinclus brachyurus*) population, for which the capacity to alter survival rates was small because those rates were already relatively high (Mortensen and Reed 2016). These studies on slow-growing bird species indicate that reproduction can play a crucial role in the conservation of such taxa, especially because the highly variable reproductive rates often drive population dynamics, and because reproductive success is often more amenable to management actions.

2.2 *Ungulates*

Ungulates comprise a diverse group of hoofed mammals, and are usually characterized as even-toed (e.g. deer and camels) or odd-toed (e.g. horses and rhinoceroses). Several studies point to a general trend for ungulates with respect to variability and elasticity of vital rates: Reproductive rates typically vary more than survival rates, but survival rates, especially adult survival tend to have higher elasticity (Gaillard et al. 1998, 2000; Heppell et al. 2000; Gaillard and Yoccoz 2003; Escos et al. 1994; Walsh et al. 1995; Saether 1997; Raithel et al. 2007; DeCesare et al. 2011). A systematic review on more than 30 ungulate species across various biomes confirmed this pattern (Gaillard et al. 2000). Reproductive rates displayed large temporal variation, but had relatively low elasticity. This indicates that, although reproduction has a comparatively small, relative contribution to population growth, as measured by elasticity analyses, environmental stochasticity drives reproduction and ultimately population dynamics of many ungulate populations. The authors concluded: "Observed differences in temporal variation may be more important than estimated relative sensitivity or elasticity in determining the relative demographic impact of various fitness components (...) Consequently, temporal variation should play a prominent role in the population dynamics of large herbivores (Gaillard et al. 1998; Saether 1997), contrary to the simplistic approach that downplays environmental stochasticity for long-lived species" (Gaillard et al. 2000). In contrast, adult survival was shown to have the greatest elasticity and displayed little temporal variability, which may leave little room for management actions to alter adult survival, unless it is depressed below its typical range by anthropogenic mortality (e.g. hunting, road kill, etc.). The following case study on bighorn sheep illustrates that natural variability of reproduction can indeed play an important role in the conservation of ungulates.

2.2.1 Case Study: Bighorn Sheep

The bighorn sheep (*Ovis canadensis*; Family: Bovidae) (Fig. 2.2) is the largest sheep species in North America. It comprises five subspecies: the Sierra Nevada (*O. c. sierrae*), Rocky Mountain (*O. c. canadensis*), Nelson (*O. c. nelsoni*), Mexican (*O. c. mexicana*), and Peninsular (*O. c. cremnobates*) bighorn (Buchalski et al. 2016). The species' distribution covers a large area of mostly open terrain and mountainous regions of western North America. It ranges from northern Mexico, including Baja California, southwestern and Midwest USA, the Rocky Mountains, stretching to British Columbia and Alberta in Canada (Thorne et al. 1985; Sandoval 1985; Bleich et al. 1990; Rubin et al. 1998; Festa-Bianchet 2008). The IUCN currently lists the species as 'least concern' globally because of its wide distribution and mostly positive population trends (Festa-Bianchet 2008). However, subsequent to European settlements, there have been widespread local extinctions of bighorn sheep in the past, primarily due to overharvesting, livestock-transmitted diseases, and habitat loss, leading to fragmentation (Buchalski et al. 2016; Bleich et al. 1990; Seton 1929; Buechner 1960; Valdez and Krausman 1999). Among all subspecies, the Sierra

Fig. 2.2 Ewes and lamb Sierra Nevada bighorn sheep (*Ovis canadensis sierrae*), Wheeler Ridge, Sierra Nevada mountains, California, USA. Photo: Heather Johnson, California Department of Fish and Wildlife

Nevada bighorn sheep has the most restricted range and the lowest abundance in North America (U.S. Fish and Wildlife Service 2007).

Like other ungulates, bighorn sheep are slow reproducers, mature late, and have relatively long lifespans (Geist 1971). Females start to reproduce at about 2–4 years of age, which varies between populations and subspecies (Seton 1929; Turner and Hansen 1980; Geist 1971; Jorgenson et al. 1993; Shackleton et al. 1999). After a 6-month gestation period, bighorn sheep typically produce one, rarely two, offspring during distinct lambing seasons (Thompson and Turner 1982; Rubin et al. 2000). For example, in the Peninsular Ranges of California, between 1993 and 1996, 68 bighorn females were reported to give birth to 133 lamb, 87% of which were born between February and April (Rubin et al. 2000). The observed maximum lifespan for bighorn sheep is reported to be 19 years (Bérubé et al. 1999).

In order to identify vital rates that are crucial for the management of bighorn sheep populations, Johnson et al. (2010) used conventional elasticity analysis and other alternative methods, including a life-stage simulation analysis (LSA). LSA is a method that quantifies how much variation in population growth is explained by the actual range of vital rate fluctuations at certain times or localities (Mills and Lindberg 2002; Wisdom and Mills 1997; Wisdom et al. 2000). Johnson and colleagues performed LSA on three isolated Sierra Nevada bighorn sheep populations on the basis of vital rate variation (mean values, variances, and covariances) between 1980 and 2007. They detected large spatial and temporal fluctuations in vital rates, resulting in each vital rate having differing effects on population growths in different populations. In some instances, reproductive rates varied more than adult survival, and in other cases, adult survival displayed the largest fluctuations. This is in contrast to many other studies on ungulates, e.g. Gaillard et al. (1998, 2000) and Raithel et al. (2007), which report that reproductive rates generally display greater temporal variability than adult survival. It is noteworthy that the Sierra Nevada bighorn sheep populations examined by Johnson et al. (2010) represented small, and in some instances, declining populations. In contrast, many of the other ungulate studies were mostly on relatively large, stable populations. This may suggest that vital rate variability and its relative effect on population dynamics may depend on size and stability of the specific population. In any case, the observed population-specific variability of vital rates is consistent with other studies that show large spatial variations in several ungulates, including roe deer (*Capreolus capreolus*) (Nilsen et al. 2009), red deer (*Cervus elavus*) (Coulson et al. 1997), Soay sheep (*Ovis aries*) (Coulson et al. 1999), caribou (*Rangiver tarandus*) (Wittmer et al. 2007) and giraffes (*Giraffa camelopardalis*) (Lee et al. 2016). Although elasticity was consistently highest for adult survival in all three populations during any time period of the Johnson et al. (2010) study, the vital rate that best explained the realized population dynamics differed between populations. While population growth was driven by variation in adult survival in two of the populations, population dynamics for one of the populations was best explained by temporal variation in reproductive rates (Johnson et al. 2010). The authors emphasized that "Sierra Nevada bighorn sheep exhibited population-specific dynamics that did not follow theoretical expectations or those observed in other ungulates." Thus this case study questions the utility of making wildlife management inferences based on shared life histories, even for the same species. I concur with the conclusion of Johnson et al. (2010) that "effective management strategies for endangered species recovery may often need to be population-specific"—and not life-history specific.

Another study on red deer in Scotland showed that the relative importance of vital rates can vary within the same population over time (Albon et al. 2000). The authors of this study assessed the relative contribution of reproductive and survival rates to population growth in red deer between 1971 and 1997. While survival rates had the greatest relative effect on population growth between 1981 and 1997, variation in reproduction contributed the most to the observed change in growth rates during the first decade (1971–1980), at a time of sustained population growth (Albon et al. 2000). These studies show that the relative importance of reproduction versus survival depends on environmental stochasticity, as well as anthropogenic disturbances and threats.

2.3 *Carnivores*

For this section, 'carnivores' are defined as those species belonging to the order Carnivora, which is a large, diverse order of eutherian mammals, encompassing over 270 species (Eizirik et al. 2010). Having conducted a systematic review on 27 carnivore taxa from 35 studies, van de Kerk et al. (2013) proposed a "rule of thumb" for elasticity patterns with conservation implications. This rule of thumb for carnivores suggested that "for a slow reproducer [...] conservation strategies targeted on adult survival are expected to be more effective" because "the population growth rate of slow species generally has a high elasticity for adult survival" (p. 7) (van de Kerk et al. 2013). This generalization, which is solely based on elasticity values, implies that adult survival is generally a better target for the conservation of slow-growing carnivore populations than reproduction. Aside from the fact that the elasticity analyses of this study were conducted without appropriate scaling, such a generalization underestimates the role of reproduction. The following case studies on the American black bear, which were also included in the van de Kerk et al. (2013) study, demonstrate that reproduction can play a much greater role for the conservation of this slow-growing taxon than such generalizations imply.

2.3.1 Case Study: American Black Bear

The American black bear (*Ursus americanus* Family: Ursidae) (Fig. 2.3) is the most abundant and most widely distributed bear species in North America. It has a broad distribution across large areas of North America, ranging from northern Mexico through the United States of America (41 states) to Canada (12 providences and territories) (Mattson et al. 2005; Rojas-Martinez and Juarez-Casillas 2013; Sawaya et al. 2013; Obbard et al. 2014; Scheick and McCown 2014; Garshelis et al. 2016).

Fig. 2.3 Male American black bear (*Ursus americanus*) in the Pisgah National Forest. Photo: Roger Powell

This distribution encompasses a large portion of the historical range, and global population estimates are twice the number of all other bear species combined (Scheick and McCown 2014; Garshelis et al. 2016). Many of the populations in the U.S. have been expanding geographically and numerically, which is part of the reason why the IUCN currently lists the black bear as 'least concern' (Garshelis et al. 2016). However, habitat fragmentation and bear-human conflicts, including hunting and road-kill, pose a threat to the persistence of some local black bear populations, and have led to large population declines in the past (Beston 2011; Sawaya et al. 2013; Smith et al. 1991; Powell et al. 1996; Czetwertynski et al. 2007; Obbard and Howe 2008).

Black bears, which have been considered among the slowest reproducing terrestrial mammals (Jonkel and Cowan 1971; Bunnell and Tait 1981; Hristienko and McDonald 2007), are exemplary of a large, slow-growing, long-lived carnivore taxon. Black bear females typically start to reproduce at the age of 3 years, and produce a mean litter size of two to three cubs at a recruitment rate of approximately one cub per year (McDonald and Fuller 2001; Hristienko et al. 2004). They can successfully produce offspring until their mid-twenties (Hristienko and McDonald 2007). As is the case for most other slow-growing carnivores (van de Kerk et al. 2013), on the basis of elasticity analyses, population growth for black bears is most sensitive to changes in adult survival and relatively insensitive to proportional changes in reproduction (Mitchell et al. 2009; Beston 2011; Hebblewhite et al. 2003; Freedman et al. 2003). These studies on black bears were also included in the paper by van de Kerk et al. (2013) that formulated the "rule of thumb" for carnivores, which stipulates that "all else being equal (…), the most effective management targets adult survival for slowly reproducing species". Given the threats that many black bear populations face, adult survival is indeed often a suitable target for management action. For example, Hebblewhite et al. (2003) recommended to reduce road kills in order to elevate survival rates of the black bear population they studied in the Banff National Park, Canada's oldest national park. They used alternative sensitivity analyses to demonstrate that it is possible to stabilize the otherwise declining bear population ($\lambda = 0.953$) if human-induced adult female mortalities are decreased by 50%. This is explained by the observation that the majority of adult mortalities of radio-tracked individuals from this population were caused by humans (82%), 44% of which were due to road-kills (Hebblewhite et al. 2003). Therefore, for the persistence of this particular black bear population, reproduction does not appear to play an important role. Instead, adult survival seems to be key for the persistence of this population, apparently confirming the 'rule of thumb' for slow-growing carnivore populations.

However, studies on other black bear populations suggest that reproduction can play an important role for the conservation of this species—counter to the aforementioned rule of thumb for carnivores. Mitchell et al. (2009) used a combination of different approaches, including elasticity analyses and life table response experiments (LTRE), to evaluate the actual and potential effect of vital rates on population growth of black bears in the Pisgah Bear Sanctuary in the Appalachian Mountains of North Carolina. Unlike conventional elasticity analysis, LTRE assesses the sensitivity of population growth to a change of one standard deviation in the observed vital rates (Caswell 1989; Ehrlén and van Groenendael 1998). It thus allows us to project the response to past fluctuations and covariation of vital rates (Ehrlén and van Groenendael 1998). In their study, Mitchell et al. (2009) took advantage of a long-term dataset, encompassing 22 years of demographic data on bear populations in the Appalachian Mountains. Consistent with other studies on slow-growing taxa, they found that elasticity for adult survival was highest, and elasticity was comparatively low for reproductive rates. At the same time, reproductive rates of these black bears fluctuated more than any of the age-class specific survival rates. The LTRE analysis demonstrated that the relatively large, long-term fluctuations in reproduction had a greater impact on population growth rates than survival (Mitchell et al. 2009). The authors concluded that the influence of vital rates on population trajectories is likely population-specific and dependent on levels of protection. They hypothesized "that λ is most strongly influenced by recruitment in protected populations where adult survival is relatively high and constant, whereas adult survival will most influence λ for unprotected populations" (Mitchell et al. 2009). This illustrates that despite reproduction having lower elasticity, natural fluctuations in reproductive rates can drive population trajectories and thus the viability of slow-growing carnivores.

Another study by Beston (2011), who conducted a meta-analysis on 86 black bear populations, between 1959 and 2007, underscores the importance of reproduction. In her study, Beston compared the demography and population dynamics of black bear populations in the eastern and western part of North America. Overall, the populations in the east appeared more stable, with slightly more than half of the populations (55%) having positive population growth rates ($\lambda > 1$). In contrast, the western populations showed declining trends, with nearly two-thirds (66%) of them having negative population growth rates ($\lambda < 1$). Although the eastern populations were more stable, adult survival was slightly lower in the east (0.82) than in the west (0.88). Fecundity was the vital rate that seemed to best explain the difference in population growth between eastern and western populations. Mean fecundity of the more stable populations in the east (0.58) was substantially greater compared to the populations in the west (0.45). Adult survival had the greatest elasticity, and both adult survival and fecundity were positively correlated with population growth rates. However, for these bear populations, the differences in reproduction drove the differences in population trajectories. Beston suggested that, despite the high elasticity of adult survival, "the low spatial variation of adult survival decreases its importance for population growth rate." This is also consistent with the aforementioned study by Mitchell et al. (2009) and a decades-old study by Beecham (1983), who found that temporal fluctuations in reproductive rates were the main factor affecting population trajectories of a black bear population in Idaho between 1973 and 1977. These studies illustrate that elasticity values can mislead us about the importance of a vital rate with respect to population dynamics and ultimately the conservation of wildlife populations (Manlik et al. 2018a; Mills et al. 1999; Wisdom et al. 2000; Morris and Doak 2002). Consequently, this also undermines the usefulness and validity of 'rules of thumbs' for slow-growing taxa that are based on elasticity analyses and life-history generalizations. As is the case for the other case studies, adult survival is an appropriate target for conservation actions if there are factors that depress adult survival, e.g. by hunting or accidental road-kill. However, this is unrelated to the observation that elasticity is highest for adult survival. If there are no direct threats to adult survival, then reproductive rates are often the driver of population trends for black bears (Beecham 1983; Mitchell et al. 2009; Beston 2011) and other slow-growing animal taxa (Manlik et al. 2016; Beissinger and Peery 2007).

2.4 *Cetaceans*

One of the early demographic sensitivity analyses performed was on a killer whale (*Orcinus orca*) population (Brault and Caswell 1993; Olesiuk et al. 1990). This survival-based sensitivity analysis showed that adult and juvenile survival had the greatest relative effect on growth rates of this population (Brault and Caswell 1993). In contrast, population growth was relatively insensitive to proportional changes in fertility. The authors of this study concluded that "management plans for killer whales should pay attention to factors that might cause even small changes in survival probabilities, because these changes will have a large impact on population growth" (p. 1452) (Brault and Caswell 1993). Similar results and conclusions were obtained by a more recent study (Young and Keith 2011), which included elasticity analyses on five cetacean species (killer whales, *Orcinus orca*; sperm whales, *Physeter macrocephalus*; humpback whales, *Megaptera novaengliae*; North Atlantic right whale, *Eubalaena glacialis* and common bottlenose dolphins, *Tursiops truncatus*). Adult survival had the highest elasticity for all cetacean species examined, and, with the exception of the North Atlantic right whales, population growth was relatively insensitive to changes in fecundity (Young and Keith 2011). The authors offered the following generalization for slow-growing cetacean populations: "The implication is that adult female survival is the most important vital rate influencing the population growth or decline of these cetacean species (…) For cetacean management, then, any disturbance of this vital rate, anthropogenic, environmental, or otherwise, can greatly influence the health of the cetacean population" (p. 267) (Young and Keith 2011). These studies, and their interpretations of elasticity values, seem to suggest that, compared to adult survival, reproduction only plays a minor role in the conservation of cetaceans. The following case study on Indo-Pacific bottlenose dolphins shows that this is not the case—reproduction can be the key to the persistence of delphinid populations.

2.4.1 Case Study: Indo-Pacific Bottlenose Dolphin

Indo-Pacific bottlenose dolphins (*Tursiops aduncus*; family: Delphinidae) (Fig. 2.4) are broadly distributed in the temperate and tropical Indo-Pacific, ranging from South Africa, along the coasts of the Indian Ocean to southeast Asia and Australia (Connor et al. 2000; Wang et al. 2000; Wells and Scott 2002; Moller and Beteregaray

Fig. 2.4 Indo-Pacific bottlenose dolphin (*Tursiops aduncus*) mother and calf in Shark Bay, Western Australia. Photo: Madison Miketa

2001; Natoli et al. 2004; Allen et al. 2016). Abundance of this species is likely in the tens of thousands, but some local populations are threatened—primarily by anthropogenic sources of disturbance, such as coastal development, entanglement in fishing gear, and climate change, e.g. Manlik et al. (2016, 2018b), Reeves et al. (2003), Smith et al. (2016), Sprogis et al. (2016a, b), Bossley et al. (2017), Chabanne et al. (2017), and Wild et al. (2019). The IUCN currently lists this species as 'data-deficient' (Hammond et al. 2012). Like all cetaceans, bottlenose dolphins are large, long-lived, late-maturing, and slow-reproducing mammals. Generation time for Indo-Pacific bottlenose dolphins is estimated to be 21.1 years (Taylor et al. 2007), with a maximum lifespan of over 50 years—one adult female in Shark Bay is estimated to be in her 50s (personal communication, Janet Mann). Females typically start to reproduce between 8 and 12 years of age and produce one offspring at a time after a 12-month gestation period (Mann et al. 2000). Due to the relatively long weaning period of 3–4 years, the average inter-birth interval is approximately 4 years (Mann et al. 2000; Wallen et al. 2016; Karniski et al. 2018).

Bottlenose dolphins, thus, fit all the criteria for slow-growing taxa, and following conventional life-history generalizations, we would expect adult survival to have the relatively greatest effect on population trajectories. A study that my colleagues and I conducted on two coastal bottlenose dolphin populations in Western Australia, however, illustrates that the viability of these populations is more dependent upon reproduction than survival (Manlik et al. 2016). In this study, we conducted a population viability analysis to forecast population trajectories of the two populations. We also performed sensitivity analyses to identify the parameters that had the greatest relative effect on population growth and other indicators of population viability. These analyses included proportional sensitivity and elasticity analyses, in which each vital rate was changed by equal proportions, but we also assessed the effect of changes in vital rates based on observed, natural fluctuations over time. One of the two populations, Shark Bay, appeared stable, whereas the other, Bunbury, was forecast to sharply decline, unless supported by migration. Furthermore, the stable Shark Bay population displayed greater major histocompatibility complex (MHC) genetic diversity than the Bunbury population, which may suggest that it is better adapted to a changing environment (Manlik et al. 2019). Other genetic studies indicate that in the past, the Bunbury population was not a net recipient of immigrants, but instead was a source population, supporting other populations nearby (Manlik et al. 2018b). Consistent with other studies on long-lived, slow-growing taxa, including other cetacean populations (Brault and Caswell 1993; Young and Keith 2011), population growth for both populations was most sensitive to proportional changes in adult survival and least sensitive to changes in reproduction. Adult survival had the greatest elasticity, reproduction the lowest (Manlik et al. 2016). However, all other findings, based on alternative analyses, were in stark contrast to this, and underlined the importance of reproduction for the viability of the two populations: Over more than a decade, the reproductive rates of the stable Shark Bay population were never lower than that of the Bunbury population. Temporal fluctuations of reproduction had a much greater effect on population viability than changes in survival rates, which, unlike reproductive rates, remained relatively constant over time. Most importantly, the difference in viability between the two populations was due to the difference in reproduction—not survival. This was evident when we manually substituted the vital rates between the two populations. For example, applying the reproductive rates of the stable Shark Bay population to the Bunbury model resulted in a reversal of the forecast population decline. Hence, raising reproductive rates to the levels of the stable population would be an effective and realistic option to prevent or halt the population decline. In contrast, increasing survival rates, which were already at high levels, would be ineffective (Manlik et al. 2016). This case study illustrates that focusing on the vital rate with the highest elasticity, i.e. adult survival, would be ineffective in this case. Hence, the proposition that, for slow-growing animal taxa, survival is generally more important than reproduction is misleading.

Having identified reproduction as key to the persistence of the Bunbury dolphin population, how would we go about boosting reproductive output of wild dolphins? Although there may not be a clear-cut answer, it is probably best to try to identify factors that may have caused or contributed to depressing reproduction, and then find ways to eliminate or minimize these factors. This, of course, requires additional information on potential disturbances: The Bunbury dolphin population inhabits waters adjacent to the regional city center of Bunbury, one of the fastest growing industrial ports in Australia. Compared to the apparently stable Shark Bay population, the Bunbury population is exposed to relatively high anthropogenic activity, including port and marina development, dolphin provisioning, tourism, commercial and recreational fishing, as well as vessel traffic. Several studies (Lusseau 2005; Bejder et al. 2006a, b; Williams et al. 2006) have shown that the presence of boats can cause short-term behavioral alterations, which can lead to greater energetic expenditure in foraging and declines in relative abundance of coastal delphinids. There is some evidence that boat presence negatively impacts reproduction (Bejder 2005), although immediate behavioral responses to this disturbance may not necessarily alter vital rates in the long-term (New et al. 2013). It is certainly plausible that the sheer volume of vessel traffic in Bunbury waters, along with the associated increases in ambient noise levels and habitat modification, may have contributed to the relatively low reproductive output reported for the Bunbury dolphin population in recent years. Therefore, in order to positively influence mating behavior and, ultimately, reproductive success, Smith et al. (2016) recommended restrictions on vessel traffic, such as a vessel exclusion zone and speed limits that would mitigate the impact of vessel traffic. This management option could specifically target reproduction when implemented at sites and seasons that are crucial for mating and calving (Brough et al. 2016; Smith et al. 2013, 2016).

On a larger scale, establishing a Marine Protected Area (MPA) has proven effective. A recent study by Cheney et al. (2019) showed that an MPA in the Moray Firth of Scotland had a positive effect on bottlenose dolphin reproductive rates and calf survival rates. Since the establishment of the MPA, in the period between 2001 and 2016, the mean reproductive rate estimates have increased by more than 60% (Cheney et al. 2019). This increase in reproductive rates coincided with positive population trends of dolphins using the MPA. Mark-recapture analyses to estimate abundance of this dolphin population indicate that it has been stable throughout the time period that the MPA has been in place (Cheney et al. 2014).

These examples illustrate that reproduction is considerably more important for the conservation of some coastal bottlenose dolphin populations than conventional elasticity analyses or generalizations based on shared life-histories would imply. However, the importance of reproduction is not taxon-specific, or rooted in life-history—there is no indication that reproduction is always the driver of population dynamics of such taxa. In fact, there are numerous examples in which the viability of dolphin populations is threatened by factors influencing survival rather than reproduction. These include incidental capture in fisheries (Allen et al. 2009, 2017) or pathogen outbreaks, such a cetacean morbillivirus (Stephens et al. 2014; van Bressem et al. 2014). In other words, life-history traits or elasticity analyses, on their own, tell us little or nothing about which vital rate is most critical to conservation. Identifying which vital rate to target in order to achieve the optimum outcome of wildlife management actions requires a solid understanding of the species, demography and threats to specific wildlife populations.

3 Conclusion and Future Directions

I could be accused of cherry-picking case studies as examples that run counter to the rule of thumb that for the conservation of slow-growing animal populations (adult) survival is typically more important than reproduction. I readily admit that I carefully selected case studies that are good illustrations of why reproduction is more important than life-history theory generalizations would have us believe. However, these case studies are not some odd, rare exceptions to the rule. There are many such cases. More importantly, the generalization—that one or the other vital rates is relatively more important for certain taxa—is primarily based on proportional sensitivity analyses that omitted appropriate scaling of vital rates, e.g. Link and Doherty (2002). Moreover, for many, if not most slow-growing taxa, reproductive rates display much larger temporal fluctuations than survival rates, e.g. Gaillard et al. (1998), Manlik et al. (2016), Cooch et al. (2001), Mitchell et al. (2009), and Saether (1997) (but see Johnson et al. 2010). This has been interpreted as an evolutionary response, in which the vital rate with the highest elasticity—usually adult survival for slow-growing taxa—is buffered (Gaillard et al. 1998; Gaillard and Yoccoz 2003; Pfister 1998). Again, this interpretation is based mostly on studies that did not scale vital rates in a way that they would be comparable. Whether this buffering (or 'canalization') hypothesis holds up when appropriate vital rate scaling is applied remains to be seen. However, at least in one instance in which this was tested, it did not (Doherty et al. 2004). Regardless, for conservation actions to achieve the desired outcome, they need to act on factors that change vital rates on a proximate timescale—and not an evolutionary timescale.

When devising conservation strategies, I urge population modelers and wildlife managers to refrain from solely relying on elasticity analyses and to refrain from making broad generalizations based on shared life histories. Elasticity analyses can offer some, limited guidance to wildlife management, but only if appropriate scaling is applied and if interpreted with caution. I hope that the case studies that I presented in this chapter demonstrate that it is prudent to focus on specific threats or observed variability rather than on broad generalizations in order to guide wildlife management.

Future directions in the conservation of slow-growing animal populations should not neglect the important role of reproduction. This also calls for alternative approaches to assessing the importance of vital rates for population viability and conservation. Such alternatives include LTRE, which accounts for observed variability of reproduction and other vital rates and may thus better account for the influence of reproduction on realized population dynamics (Caswell 1989; Ehrlén and van Groenendael 1998). Another useful conservation tool, outlined in some of the case studies, is LSA (Mills and Lindberg 2002; Wisdom and Mills 1997; Wisdom et al. 2000). LSA allows population modelers to incorporate variation (and covariation) of vital rates during certain time periods and localities. It also enables us to simulate specific effects of different management options on population growth. Conservation practice would also benefit from using approaches that compare vital rates of stable versus declining populations, as well as evaluating the impact of specific threats and management options.

Perhaps most importantly, successful conservation actions require a thorough understanding of the demography, threats and biology of wildlife populations. The case studies presented in this chapter, show that this must also include a thorough understanding of the role of reproduction in animal conservation. For slow-growing, long-lived animals in particular, this requires time and resources, which are often lacking, especially for threatened taxa. However, unfortunately, there are no shortcuts to good conservation practice.

Acknowledgments I would like to thank everyone who has provided valuable comments when drafting this chapter, including Simon J Allen, Steve Beissinger, Heather Johnson, Janet Mann, Holly Raudino (nee Smith) and Kate Sprogis (in alphabetical order), but especially Julie Beston, Mike Mitchell and William B Sherwin who have provided thorough feedback on the entire chapter. My gratitude also goes to Madison Miketa, Roger Powell, Brian Sullivan and Heather Johnson for allowing me to use their photographs to be published. Correspondence and discussions with Robert C Lacy, Scott Mills and Michael Reed further contributed thoughtful ideas and inspirations for this chapter.

References

Albon SD, Coulson TD, Brown D, Guinness FE, Pemberton JM, Clutton-Brock TH. Temporal changes in key factors and key age groups influencing the population dynamics of female red deer. J Anim Ecol. 2000;69:1099–110.

Allen SJ, Tyne JA, Kobryn HT, Bejder L, Pollock KH, Loneragan NR. Patterns of dolphin bycatch in a north-western Australian trawl-fishery. PLoS One. 2009;9(4):e93178. https://doi.org/10.1371/journal.pone.0093178.

Allen SJ, Bryant KA, Kraus RH, Loneragan NR, Kopps AM, Brown AM, Gerber L, Kruetzen M. Genetic isolation between coastal and offshore, fishery-impacted bottlenose dolphin (*Tursiops* spp.) populations. Mol Ecol. 2016;25:2735–53.

Allen SJ, Pollock KH, Bouchet PJ, Kobryn HT, McElligott DB, Nicholson KE, Smith JN, Loneragan NR. Preliminary estimates of the abundance and fidelity of dolphins associating with a demersal trawl fishery. Sci Rep. 2017;7:4995. https://doi.org/10.1038/s41598-017-05189-0.

Barbaree BA, Nelson SK, Dugger BD, Roby DD, Carter HH, Whitworth DL, Newman SH. Nesting ecology of Marbled Murrelets at a remote mainland fjord in Southeast Alaska. Condor. 2014;116:173–84.

Beecham JJ. Population characteristics of black bears in west central Idaho. J Wildl Manage. 1983;47:405–12.

Beissinger SB. Population trends of the marbled murrelet projected from demographic analyses. In: Ralph CJ, Hunt GL, Raphael MG, Piatt JF, editors. Ecology and conservation of the marbled murrelet. General technical report PSW-GTR-152. Albany: Pacific Southwest Research Station, Forest Service, U.S. Department of Agriculture; 1995. p. 385–94.

Beissinger SB, Peery MZ. Reconstructing the historic demography of an endangered seabird. Ecology. 2007;88:296–305.

Bejder L. Linking short and long-term effects of nature-based tourism on cetaceans. PhD thesis, Dalhousie University, Halifax, Nova Scotia; 2005.

Bejder L, Samuels A, Whitehead H, Gales N, Mann J, Connor R, Heithaus M, Watson-Capps J, Flaherty C, Kruetzen M. Decline in relative abundance of bottlenose dolphins exposed to long-term disturbance. Conserv Biol. 2006a;20:1791–8.

Bejder L, Samuels A, Whitehead A, Gales N. Interpreting short-term behavioural responses to disturbance within a longitudinal perspective. Anim Behav. 2006b;72:1149–58.

Bérubé CH, Festa-Bianchet M, Jorgenson JT. Individual differences, longevity and reproductive senescence in bighorn ewes. Ecology. 1999;80:2555–65.

Beston JA. Variation in life history and demography of the American black bear. J Wildl Manage. 2011;75:1588–96.

Bielby J, Mace GM, Bininda-Emonds ORP, Cardillo M, Gittleman JL, Jones KE, Orme CD, Purvis A. The fast–slow continuum in mammalian life history: an empirical reevaluation. Am Nat. 2007;169:748–57.

BirdLife International. *Copsychus sechellarum* (amended version of 2016 assessment). The IUCN Red List of Threatened Species 2017: e.T22709978A111061160. 2017. https://doi.org/10.2305/IUCN.UK.2017-1.RLTS.T22709978A111061160.en.

BirdLife International. *Brachyramphus marmoratus*. The IUCN Red List of Threatened Species 2018; e.T22694870A132578946. 2018. https://doi.org/10.2305/IUCN.UK.2018-2.RLTS.T22694870A132578946.en.

Bleich VC, Wedhausen JD, Holl SA. Desert-dwelling mountain sheep: conservation implications of a naturally fragmented distribution. Conserv Biol. 1990;4:383–90. https://doi.org/10.1111/j.1523-1739.1990.tb00312.x.

Bossley MI, Steiner A, Rankin RW, Bejder L. A long-term study of bottlenose dolphins (*Tursiops aduncus*) in an Australian industrial estuary: increased sightings associated with environmental improvements. Mar Mamm Sci. 2017;33:277–90. https://doi.org/10.1111/mms.12368.

Brault S, Caswell H. Pod-specific demography of killer whales (*Orcinus orca*). Ecology. 1993;74:1444–54. https://doi.org/10.2307/1940073.

Brough TE, Henderson S, Guerra M, Dawson SM. Factors influencing heterogeneity in female reproductive success in a critically endangered population of bottlenose dolphins. Endanger Species Res. 2016;29:255–70. https://doi.org/10.3354/esr00715.

Buchalski MR, Sacks BN, Gille DA, Pendeo MC, Ernest HB, Morrison SA, Boyce WM. Phylogeographic and population genetic structure of bighorn sheep (*Ovis canadensis*) in North American deserts. J Mammal. 2016;97:823–38.

Buechner HK. The bighorn sheep in the United States, its past, present, and future. Wildl Monogr. 1960;4:1–174.

Bunnell FL, Tait DEN. Population dynamics of bears—implications. In: Fowler CW, Smith TD, editors. Dynamics of large mammal populations. New York, USA: John Wiley and Sons; 1981. p. 75–98.

Burger AE. Conservation assessment of Marbled Murrelets in British Columbia: review of the biology, populations, habitat associations, and conservation. Technical report series no. 387. Canadian Wildlife Service, Pacific and Yukon Region, Delta, B.C.; 2002.

Burger AE, Waterhouse FL. Relationships between habitat area, habitat quality, and populations of nesting Marbled Murrelets. J Ecosyst Manage. 2009;10:101–12.

Burger AE, Waterhouse FL, Deal J, Lank D, Donald DS. The reliability and application of methods used to predict suitable nesting habitat for Marbled Murrelets. J Ecosyst Manage. 2018;18:1–18.

Caswell H. Analysis of life table response experiments. Decomposition of effects on population growth rate. Ecol Model. 1989;46:221–37. https://doi.org/10.1016/0304-3800(89)90019-7.

Caswell H. Prospective and retrospective perturbation analyses: their roles in conservation biology. Ecology. 2000;81:619–27.

Chabanne DBH, Finn H, Bejder L. Identifying the relevant local population for environmental impact assessments of mobile marine fauna. Front Mar Sci. 2017;4:148.

Cheney BJ, Corkrey R, Durban JW, Grellier K, Hammond PS, Islas-Villanueva V, Janik VM, Lusseau SM, Parsons KM, Quick NJ, Wilson B, Thompsons PM. Long-term trends in the use of a protected area by small cetaceans in relation to changes in population status. Glob Ecol Conserv. 2014;2:118–28.

Cheney BJ, Thompson PM, Cordes LS. Increasing trends in fecundity and calf survival of bottlenose dolphins in a marine protected area. Sci Rep. 2019;9:1767. https://doi.org/10.1038/s41598-018-38278-9.

Connor RC, Wells R, Mann J, Read A. The bottlenose dolphin: social relationship in a fission-fusion society. In: Mann J, Connor RC, Tyack P, Whitehead H, editors. Cetacean societies: field studies of whales and dolphins. Chicago: University of Chicago Press; 2000. p. 91–126.

Cooch EG, Rockwell RF, Brault S. Retrospective analysis of demographic responses to environmental change: a lesser snow goose example. Ecol Monogr. 2001;71:377–400.

Coulson T, Albon SD, Guinness FE, Pemberton J, Clutton-Brock TH. Population substructure, local density, and calf winter survival in red deer. Ecology. 1997;78:852–63.

Coulson T, Albon SD, Pilkington J, Clutton-Brock TH. Small-scale spatial dynamics in a fluctuating ungulate population. J Anim Ecol. 1999;68:658–71.

Crone EE. Is survivorship a better fitness surrogate than fecundity? Ecology. 2001;55:2611–4.

Czetwertynski SM, Boyce MS, Schmiegelow FK. Effects of hunting on demographic parameters of American black bears. Ursus. 2007;18:1–18.

de Kroon HA, Plaisier A, van Groenendael J, Ehrlén J. Elasticities: a review of methods and model limitations. Ecology. 2000;81:607–18.

DeCesare NJ, Hebblewhite M, Bradley M, Smith KG, Hervieux D, Neufeld L. Estimating ungulate recruitment and growth rates using age ratios. Wildl Manage. 2011;76:144–53.

Dobson FS, Oli MK. Fast and slow life histories of mammals. Ecosci. 2007;14:292–9.

Doherty PF, Schreiber EA, Nichols JD, Hines JE, Link WA, et al. Testing life history predictions in a long-lived seabird: a population matrix approach with improved parameter estimation. Oikos. 2004;105:606–18. https://doi.org/10.1111/j.0030-1299.

Ehrlén J, van Groenendael J. Direct perturbation analysis for better conservation. Conserv Biol. 1998;12:470–4. https://doi.org/10.1046/j.1523-1739.1998.96420.x.

Eizirik E, Murphy WJ, Koepfli KP, Johnson WE, Dragoo JW, O'Brien SJ. Pattern and timing of the diversification of the mammalian order Carnivora inferred from multiple nuclear gene sequences. Mol Phylogenet Evol. 2010;56:49–63. https://doi.org/10.1016/j.ympev.2010.01.033.

Escos J, Alados CL, Emlen JM. Application of the stage-projection model with density-dependent fecundity to the population dynamics of Spanish ibex. Can J Zool. 1994;72:731–7.

Festa-Bianchet M. *Ovis canadensis*. The IUCN Red List of Threatened Species 2008: e.T15735A5075259. 2008. https://doi.org/10.2305/IUCN.UK.2008.RLTS.T15735A5075259.en

Freedman AH, Portier KM, Sunquist ME. Life history analysis for black bears (*Ursus americanus*) in a changing demographic landscape. Ecol Model. 2003;167:47–64.
Gaillard J-M, Yoccoz NG. Temporal variation in survival of mammals: a case of environmental canalization? Ecology. 2003;84:3294–306.
Gaillard J-M, Pontier D, Allainé D, Lebreton JD, Trouvilliez J, Clobert J. An analysis of demographic tactics in birds and mammals. Oikos. 1989;56:59–76.
Gaillard J-M, Festa-Bianchet M, Yoccoz NG. Population dynamics of large herbivores: variable recruitment with constant adult survival. Trends Ecol Evol. 1998;13:58–63. https://doi.org/10.1016/S0169-5347(97)01237-8.
Gaillard J-M, Festa-Bianchet M, Yoccoz NG, Loison A, Toigo C. Temporal variation in fitness components and population dynamics of large herbivores. Annu Rev Ecol Syst. 2000;31:367–93.
Garshelis DL, Scheick BK, Doan-Crider DL, Beecham JJ, Obbard ME. *Ursus americanus*. The IUCN Red List of Threatened Species: e.T41687A114251609. 2016.
Geist V. Mountain sheep: a study in behavior and evolution. Chicago: University of Chicago Press; 1971. p. 383.
Gerber LR, Heppell SS. The use of demographic sensitivity analysis in marine species conservation planning. Biol Conserv. 2004;120:121–8. https://doi.org/10.1016/j.biocon.2004.01.029.
Hammond PS, Bearzi G, Bjorge A, Forney KA, Karkzmarski L, Kasya T, Perrin WF, Scott M, Wang J, Wells RS, Wilson B. *Tursiops aduncus*. The IUCN Red List of Threatened Species: e.T41714A17600466. 2012. https://doi.org/10.2305/IUCN.UK.2012.RLTS.T41714A17600466.en.
Harvey PH, Read AF, Promislow EL. Life history variation in placental mammals: unifying the data with theory. In: Harvey PH, Partridge L, editors. Oxford surveys in evolutionary biology. Oxford: Oxford University Press; 1989. p. 13–31.
Hebblewhite M, Percy M, Serrouya R. Black bear (*Ursus americanus*) survival and demography in Bow Valley of Banff National Park, Alberta. Biol Conserv. 2003;112:415–25.
Heppell SS, Caswell H, Crowder LB. Life histories and elasticity patterns: perturbation analysis for species with minimal demographic data. Ecology. 2000;81:654–65.
Horvitz C, Schemske DW, Caswell H. The relative 'importance' of life-history stages to population growth: prospective and retrospective analyses. In: Tuljapurkar S, Caswell H, editors. Structured population models in marine, terrestrial and freshwater systems. New York: Chapman and Hall; 1997. p. 247–71.
Hristienko H, McDonald JE. Going into the 21st century: a perspective on trends and controversies in the management of the American black bear. Ursus. 2007;18:72–88.
Hristienko H, Pastuck D, Rebizant KJ, Knudsen B, Conner ML. Using reproductive data to model cub orphaning in Manitoba due to spring harvest of females. Ursus. 2004;15:23–34.
Johnson HE, Mills LS, Stephenson TR, Wehausen JD. Population-specific vital rate contributions influence management of an endangered ungulate. Ecol Appl. 2010;20:1753–65. https://doi.org/10.1890/09-1107.1.
Jonkel CJ, Cowan IM. The black bear in the spruce–fir forest. Wildl Mongr. 1971;27:57.
Jorgenson JT, Festa-Bianchet M, Lucherini M, Wishart WD. Effects of body size, population density, and maternal characteristics on age at first reproduction in bighorn ewes. Can J Zool. 1993;71:2509–17.
Karniski C, Krzyszczyk E, Mann J. Senescence impacts reproduction and maternal investment in bottlenose dolphins. Proc R Soc B. 2018;285:20181123. https://doi.org/10.1098/rspb.2018.1123.
Lebreton JD, Clobert J. Bird population dynamics, management and conservation: the role of mathematical modelling. In: Perrins CM, Lebreton JD, Hirons GJM, editors. Bird population studies: relevance to conservation and management. Oxford: Oxford University Press; 1991. p. 105–25.
Lee DE, Bond ML, Kissui BM, Kiwango YA, Bolger DT. Spatial variation in giraffe demography: a test of 2 paradigms. J Mammal. 2016;97:1015–25.

Link WA, Doherty PF. Scaling in sensitivity analysis. Ecology. 2002;83:3299–305. https://doi.org/10.2307/3072080.

Long JA, Hazlitt SL, Nelson TA, Laberee K. Estimating 30-year change in coastal old-growth habitat for a forest-nesting seabird in British Columbia, Canada. Endanger Species Res. 2011;14:49–59.

Lusseau D. Residency pattern of bottlenose dolphins *Tursiops* spp. in Milford Sound, New Zealand is related to boat traffic. Mar Ecol Prog Ser. 2005;295:265–72.

Manlik O, McDonald JA, Mann J, Raudino HC, Bejder L, Krützen M, Connor RC, Heithaus MR, Lacy RC, Sherwin WB. The relative importance of reproduction and survival for the conservation of two dolphin populations. Ecol Evol. 2016;6(11):3496–512. https://doi.org/10.1002/ece3.2130.

Manlik O, Lacy RC, Sherwin WB. Applicability and limitations of sensitivity analyses for wildlife management. J Appl Ecol. 2018a;55(3):1430–40. https://doi.org/10.1111/1365-2664.13044.

Manlik O, Chabanne D, Daniel C, Bejder L, Allen SJ, Sherwin WB. Demography and genetics suggest reversal of dolphin source-sink dynamics, with implications for conservation. Mar Mamm Sci. 2018b; https://doi.org/10.1111/mms.12555.

Manlik O, Krützen M, Kopps AM, Mann J, Bejder L, Allen SJ, Frère C, Connor RC, Sherwin WB. Is MHC diversity a better marker for conservation than neutral genetic diversity? A case study of two contrasting dolphin populations. Ecol Evol. 2019;00:1–13. https://doi.org/10.1002/ece3.5265.

Mann J, Connor RC, Barré LM, Heithaus MR. Female reproductive success in bottlenose dolphins (*Tursiops* sp.): life history, habitat, provisioning, and group size effects. Behav Ecol. 2000;11:210–9.

Mattson DJ, Herrero S, Merrill T. Are black bears a factor in the restoration of North American grizzly bear populations? Ursus. 2005;16:11–30.

McDonald JE, Fuller TK. Prediction of litter size in American black bears. Ursus. 2001;12:93–102.

Meehan TD, Harvey AL, Michel NL, Langham GM, Weinstein A. A population model exploring factors influencing black oystercatcher (*Haematopus bachmani*) population dynamics. Waterbirds. 2018;41:115–221.

Mills LS. Conservation of wildlife populations: demography, genetics and management. 2nd ed. Oxford: Blackwell; 2013.

Mills LS, Lindberg MS. Sensitivity analysis to evaluate the consequences of conservation action. In: Beissinger SR, McCullough DR, editors. Population viability analysis. Chicago: University of Chicago Press; 2002. p. 338–66.

Mills LS, Doak DF, Wisdom MJ. Reliability of conservation actions based on elasticity analysis of matrix models. Conserv Biol. 1999;13:815–29.

Mitchell MS, Pacifici LB, Grand JB, Powell RA. Contributions of vital rates to growth of a protected population of American black bears. Ursus. 2009;20:77–84.

Moller LM, Beteregaray LB. Coastal bottlenose dolphins from southeastern Australia are *Tursiops aduncus* according to sequences of the mitochondrial DNA control region. Mar Mamm Sci. 2001;17:249–63.

Morris WF, Doak DF. Quantitative conservation biology: theory and practice of population viability. Sunderland: Sinauer; 2002.

Mortensen JL, Reed MJ. Population viability and vital rate sensitivity of an endangered avian cooperative breeder, the white-breasted thrasher (*Ramphocinclus brachyurus*). PLoS One. 2016;11(2):e0148928. https://doi.org/10.1371/journal.pone.0148928.

Natoli A, Peddemors VM, Hoelzel AR. Population structure and speciation in the genus *Tursiops* based on microsatellite and mitochondrial DNA analyses. J Evol Biol. 2004;17:363–75.

New LF, Harwood J, Thomas L, Donovan C, Clark JS, Hastie G, Thompson PM, Cheney B, Scott-Hayward L, Lusseau D. Modelling the biological significance of behavioural change in coastal bottlenose dolphins in response to disturbance. Funct Ecol. 2013;27:314–22.

Nilsen EB, Gaillard J-M, Andersen R, Odden J, Delorme D, van Laere G, Linnell DC. A slow life in hell or a fast life in heaven: demographic analyses of contrasting roe deer populations. J Anim Ecol. 2009;78:585–94.

Norris K, McCulloch N. Demographic models and the management of endangered species: a case study of the critically endangered Seychelles magpie robin. J Appl Ecol. 2003;40:890–9. https://doi.org/10.1046/j.1365-2664.2003.00840.x.

Obbard ME, Howe EJ. Demography of black bears in hunted and unhunted areas of the boreal forest of Ontario. J Wildl Manage. 2008;72:869–80.

Obbard ME, Howe EJ, Wall LL, Allison B, Black R, Davis P, Dix-Gibson L, Gatt M, Hall MN. Relationships among food availability, harvest, and human–bear conflict at landscape scales in Ontario, Canada. Ursus. 2014;25:98–110.

Olesiuk PF, Bigg MA, Ellis GM. Life history and population dynamics of resident killer whales (*Orcinus orca*) in the coastal waters of British Columbia and Washington state. Rep Int Whaling Comm Spec Issue. 1990;12:209–43.

Oli MK, Dobson FS. The relative importance of life-history variables to population growth rate in mammals: Cole's prediction revisited. Am Nat. 2003;161:422–40.

Peery MZ, Becker BH, Beissinger SR. Combining demographic and count-based approaches to identify source-sink dynamics: an example using an endangered seabird. Ecol Appl. 2006;16:1516–28. https://doi.org/10.1890/10510761(2006)016[1516:CDACAT]2.0.CO;2.

Pfister CA. Patterns of variance in stage-structured populations: evolutionary predictions and ecological implications. Proc Natl Acad Sci U S A. 1998;95:213–8. https://doi.org/10.1073/pnas.95.1.213.

Piatt JF, Ford RG. Distribution and abundance of Marbled Murrelets in Alaska. Condor. 1993;95:662–9.

Piatt JF, Kuletz KJ, Burger AE, Hatch SA, Friesen VL, Birt TP, Arimitsu ML, Drew GS, Harding AMA, Bixler KS. Status review of the Marbled Murrelet (*Brachyramphus marmoratus*) in Alaska and British Columbia. Open-file report 2006-1387. U.S. Geological Survey; 2006.

Powell RA, Zimmerman JW, Seaman DE, Gilliam JF. Demographic analyses of a hunted black bear population with access to a refuge. Conserv Biol. 1996;10:224–34.

Promislow DEL, Harvey PH. Living fast and dying young: a comparative analysis of life history variation among mammals. J Zool. 1990;220:417–37.

Raithel JD, Kauffman MJ, Pletscher DH. Impact of spatial and temporal variation in calf survival on the growth of elk populations. J Wildl Manage. 2007;71:795–803.

Ralph CJ, Hunt GL, Raphael MG, Piatt JF. Ecology and conservation of the Marbled Murrelet. Albany: Pacific Southwest Research Station; 1995.

Reed MJ, Mills LS, Dunning JB, Menges ES, McKelvey KS, Frye R, Beissinger SR, Anstett M-C, Miller P. Emerging issues in population viability analysis. Conserv Biol. 2002;16:7–19.

Reed MJ, Fefferman N, Averill-Murray RC. Vital rate sensitivity analysis as a tool for assessing management actions for the desert tortoise. Biol Conserv. 2009;142:2710–7. https://doi.org/10.1016/j.biocon.2009.06.025.

Reeves RR, Smith BD, Crespo EA, Notarbartolo di Sciara G. Dolphins, whales and porpoises: 2002-2010 conservation action plan for the world's cetaceans. Gland and Cambridge: IUCN/SSC Cetacean Specialist Group; 2003.

Rodway MS, Carter HR, Sealy SG, Campbell RW. Status of the Marbled Murrelet in British Columbia. Proc West Found Vertebr Zool. 1992;5:17–41.

Rojas-Martinez AE, Juarez-Casillas LA. First record of American black bear (*Ursus americanus*) from Hidalgo, Mexico. Revista Mexicana de Biodiversidad. 2013;84:1018–21.

Rubin ES, Boyce WM, Jorgensen MC, Torres SG, Hayes CL, O'Brien CS, Jessup DA. Distribution and abundance of bighorn sheep in the Peninsular Ranges, California. Wildl Soc Bull. 1998;26:539–51.

Rubin ES, Boyce WM, Bleich VC. Reproductive strategies of desert bighorn sheep. J Mammal. 2000;81:769–86.

Saether B-E. Environmental stochasticity and population dynamics of large herbivores: a search for mechanisms. Trends Ecol Evol. 1997;12:143–9.

Sæther B-E, Bakke O. Avian life history variation and contribution of demographic traits to the population growth rate. Ecology. 2000;81:642–53.

Sæther B-E, Ringsby TH, Røskaft E. Life history variation, population processes and priorities in species conservation: towards a reunion of research paradigms. Oikos. 1996;77:217–26.

Sandoval AV. Status of bighorn sheep in the Republic of Mexico. In: Hoefs M, editor. Distribution, abundance, management and conservation of sheep of the world and closely related mountain ungulates. Yukon: Northern Wild Sheep and Goat Council, Whitehorse; 1985. p. 86–94.

Sawaya MA, Stetz JB, van Manen FT, Clark JD. Population monitoring options for American black bears in the northeastern United States and eastern Canada. A technical publication for the Northeast Black Bear Technical Committee 2013; 2013.

Scheick BK, McCown W. Geographic distribution of American black bears in North America. Ursus. 2014;25:24–33.

Sealy SG. Breeding phenology and clutch size in the Marbled Murrelet. Auk. 1974;91:10–23.

Sealy SG, Carter HR. At-sea distribution and nesting habitat of the Marbled Murrelet in British Columbia: problems in the conservation of a solitary nesting seabird. In: Croxall JP, Evans PGH, Schreiber RW, editors. Status and conservation of the world's seabirds. Cambridge: International Council for Bird Preservation; 1984. p. 737–56.

Seton ET. The bighorn. In: Seton ET, editor. Lives of the game animals. Garden City: Doubleday; 1929. p. 519–73.

Shackleton DM, Shank CC, Wikeem BM. Natural history of Rocky Mountain and California bighorn sheep. In: Valdez R, Krausman PR, editors. Mountain sheep of North America. Tucson: University of Arizona Press; 1999. p. 78–138.

Smith KG, Clark JD, Gipson PS. History of black bears in Arkansas: over-exploitation, near elimination, and successful reintroduction. Eastern Workshop Black Bear Res Manage. 1991;10:5–14.

Smith HC, Pollock K, Waples K, Bradley S, Bejder L. Use of the robust design to estimate seasonal abundance and demographic parameters of a coastal bottlenose dolphins (*Tursiops aduncus*) population. PLoS One. 2013;8 https://doi.org/10.1371/journal.pone.0076574.

Smith HC, Frère C, Kobryn H, Bejder L. Dolphin sociality, distribution and calving As important behavioural patterns informing management. Anim Conserv. 2016;19:462–71.

Sprogis KR, Pollock KH, Raudino HC, Allen SJ, Kopps AM, Manlik O, Tyne JA, Bejder L. Sex-specific patterns in abundance, temporary emigration and survival of Indo-Pacific bottlenose dolphins (*Tursiops aduncus*) in coastal and estuarine waters. Front Mar Sci. 2016a;3 https://doi.org/10.3389/fmars.2016.00012.

Sprogis KR, Raudino HC, Rankin R, MacLeod CD, Bejder L. Home range size of adult Indo-Pacific bottlenose dolphins (*Tursiops aduncus*) in a coastal and estuarine system is habitat and sex-specific. Mar Mamm Sci. 2016b;32:287–308. https://doi.org/10.1111/mms.12260.

Stahl JT, Oli MK. Relative importance of avian life-history variable to population growth rate. Ecol Model. 2006;198:23–39.

Stearns SC. The influence of size and phylogeny of covariation among life-history traits in the mammals. Oikos. 1983;41:173–87.

Stein JL, Miller GS. Endangered and threatened wildlife and plants: determination of threatened status for the Washington, Oregon, and California population of the Marbled Murrelet. Fed Regist. 1992;57(191):45328–37.

Stephens N, Duignan PJ, Wang J, Bingham J, Finn H, Bejder L, Patterson AP, Holyoake C. Cetacean morbillivirus in coastal Indo-Pacific bottlenose dolphins, Western Australia. Emerg Infect Dis. 2014;20:666–70.

Taylor LB, Chivers SJ, Larse J, Perrin WF. Generation length and percent mature estimates for IUCN assessments of cetaceans. Administrative report LJ-07-01. National Marine Fisheries Services, Southwest Fisheries Science Center, U.S.A.; 2007.

Thompson RW, Turner JC. Temporal geographic variation in the lambing season of bighorn sheep. Can J Zool. 1982;60:1781–93.

Thorne ET, Hickey WO, Stewart ST. Status of California and Rocky Mountain bighorn sheep in the United States. In: Hoefs M, editor. Distribution, abundance, management and conservation of sheep of the world and closely related mountain ungulates. Whitehorse: Northern Wild Sheep and Goat Council; 1985. p. 56–81.

Turner JC, Hansen CG. Reproduction. In: Monson SL, editor. The desert bighorn, its life history, ecology, and management. Tucson: University of Arizona Press; 1980. p. 145–51.
U.S. Fish and Wildlife Service. Recovery plan for the Sierra Nevada bighorn sheep. Sacramento: U.S. Fish and Wildlife Service; 2007.
Valdez R, Krausman PR. Description, distribution, and abundance of mountain sheep in North America. In: Valdez R, Krausman PR, editors. Mountain sheep of North America. Tucson: University of Arizona Press; 1999. p. 1–21.
van Bressem MF, Duignam PJ, Banyard A, Barbieri M, Colegrove KM, et al. Cetacean morbillivirus: current knowledge and future directions. Viruses. 2014;6:5145–81.
van de Kerk M, de Kroon H, Conde DA, Jongejans E. Carnivora population dynamics are as slow and as fast as those of other mammals: implications for their conservation. PLoS One. 2013;8:e70354.
Wallen M, Patterson E, Krzyszczyk E, Mann J. The ecological costs to females in a system with allied sexual coercion. Anim Behav. 2016;115:227–36.
Walsh NE, Griffith B, McCabe TR. Evaluating growth of the porcupine caribou herd using a stochastic model. J Wildl Manage. 1995;59:262–72.
Wang JY, Chou L-S, White BN. Differences in the external morphology of two sympatric species of bottlenose dolphins (genus *Tursiops*) in Chinese waters. J Mammal. 2000;81:1157–65.
Wang T, Gao G, Jakovlić I, Liu H-Z. Life tables and elasticity analyses of Yangtze River fish species with implications for conservation and management. Rev Fish Biol Fish. 2017;27:255–66. https://doi.org/10.1007/s11160-016-9464-8.
Warton DI, Hui FKC. The arcsine is asinine: the analysis of proportions in ecology. Ecology. 2011;92:3–10. https://doi.org/10.1890/10-0340.1.
Wells RS, Scott MD. Bottlenose dolphins *Tursiops truncatus* and *T. aduncus*. In: Perrin WF, Würsig B, Thewissen JGM, editors. Encyclopedia of marine mammals. London: Academic Press; 2002. p. 122–8.
Wild S, Krützen M, Rankin RW, Hoppitt WJE, Gerber L, Allen SJ. Long-term decline in reproduction and survival of dolphins following a marine heatwave. Curr Biol. 2019;29(7):R239–R240. https://doi.org/10.1016/j.cub.2019.02.047.
Williams RM, Lusseau D, Hammond PS. Estimating relative energetic costs of human disturbance to killer whales (*Orcinus orca*). Biol Conserv. 2006;133:301–11.
Wisdom MJ, Mills LS. Sensitivity analysis to guide population recovery: prairie chickens as an example. J Wildl Manage. 1997;61:302–12. https://doi.org/10.2307/3802585.
Wisdom MJ, Mills LS, Doak DF. Life stage simulation analysis: estimating vital-rate effects on population growth for conservation. Ecology. 2000;81:628–41. https://doi.org/10.1890/0012-9658(2000)081[0628:LSSAEV]2.0.CO;2.
Wittmer HU, McLellan BN, Serrouya R, Apps CD. Changes in landscape composition influence the decline of a threatened woodland caribou population. J Anim Ecol. 2007;76:568–79.
Young KE, Keith EO. A comparative analysis of cetacean vital rates using matrix population modeling analysis of cetacean vital rates. Int J Appl Sci Technol. 2011;1:261–77.

Chapter 3
Reproductive Impact of Environmental Chemicals on Animals

Rakesh Kanda

Abstract Wildlife is exposed to a diverse range of natural and man-made chemicals. Some environmental chemicals possess specific endocrine disrupting properties, which have the potential to disrupt reproductive and developmental process in certain animals. There is growing evidence that exposure to endocrine disrupting chemicals plays a key role in reproductive disorders in fish, amphibians, mammals, reptiles and invertebrates. This evidence comes from field-based observations and laboratory based exposure studies, which provide substantial evidence that environmental chemicals can cause adverse effects at environmentally relevant doses. There is particular concern about wildlife exposures to cocktails of biologically active chemicals, which combined with other stressors, may play an even greater role in reproductive disorders than can be reproduced in laboratory experiments. Regulation of chemicals affords some protection to animals of the adverse effects of exposure to legacy chemicals but there continues to be considerable debate on the regulation of emerging pollutants.

1 Introduction

Chemical pollution of the environment is inextricably linked with human activities such as the manufacture, handling, storage, use and disposal of chemicals. Regulation of chemicals affords some protection to the potential adverse effects of chemicals on human and animal health; nevertheless chemical pollution remains a major threat to animals and to ecosystem biodiversity (Malaj et al. 2014). Animal exposure to chemicals continues to grow and is closely linked to the massive growth in the global population which has surged in modern times and is forecast to reach 9.7 billion by 2050 (Melorose et al. 2015). Alongside population expansion, many countries have seen unprecedented economic and industrial growth resulting in a vast increase in the production volume and diversity of chemicals produced globally. Animal and human exposure to chemical pollutants has therefore become

R. Kanda (✉)
Institute of Environment, Health and Societies, Brunel University London, Uxbridge, UK
e-mail: rakesh.kanda@brunel.ac.uk

P. Comizzoli et al. (eds.), *Reproductive Sciences in Animal Conservation*, Advances in Experimental Medicine and Biology 1200,
https://doi.org/10.1007/978-3-030-23633-5_3

unavoidable (Guillette and Guillette 1996) and environmental management of chemicals during production, use and disposal has become a significant global challenge.

Of particular concern are the conservation implications of endocrine disrupting compounds (EDC), due to their potential to disrupt natural homeostasis, reproduction and developmental process in animals (Colborn et al. 1993). EDC's encompass a large and diverse range of compounds spanning multiple classes of chemicals and possessing a wide range of physicochemical properties, this group of non-discrete chemicals are best defined by their action, i.e. they have the potential to interfere or disrupt the production, binding action, metabolism and or elimination of natural hormones (Bergman et al. 2012) thus triggering unnatural responses in the body. Highly potent EDCs that mimic natural hormones i.e. estrogens, androgens and progesterone, and their antagonists have been the focus of most ecotoxicological studies to date. Evidence of the disruptive effect of a magnitude of EDCs is available from a number of laboratory studies conducted with single chemicals or mixtures of chemicals or by field-based evidence. In one of the earliest laboratory demonstration of hormonal disruption, researchers experimentally evidenced the effects of exposure of natural and synthetic hormones on platyfish, *Platypoecilus mallatus*. In this study, immature males were exposed to the natural hormone α-estradiol and immature females were exposed to the synthetic hormone pregneninolone alongside control fish. In the estrogen treated males, compared with the controls, the gonads were small, compact and appeared bipartite whilst the pregneninolone treated females had ovaries, compared to controls that were small and contained fewer oocytes (Cohen 1946). It should be noted that this and many other earlier studies linking chemicals with endocrine disruptive effects were conducted using high (in this case, milligram) exposure to certain chemicals and therefore did not reflect the environmental dose that animals would be exposed to naturally in the wild. More recent studies do however provide very compelling evidence that adverse effects, including reproductive affects also occur in wildlife with exposure to EDCs at environmentally relevant concentrations. For example, laboratory exposure studies, with Tributyltin (TBT) at only 5 ng/L have induced imposex on dog whelk, *Nucella lapillus*, a species that is particularly sensitive to TBT exposures (Gibbs et al. 1988). A number of field studies of harbours and marinas in Europe, North America and Asia prior to stringent TBT regulation (Antizar-Ladislao 2008) established that environmental concentrations in harbour and marinas substantially exceeded the laboratory exposure concentration. Such findings eventually resulted in the ban of TBT in antifouling paints in 2008.

A further example of a highly potent chemical having endocrine disrupting effects on the reproductive functionality of certain animals at low exposure concentrations has been evidenced for the highly potent endocrine disrupter, α-ethynylestradiol (EE2), the synthetic, active ingredient in many contraceptive pill formulations which is ubiquitous in surface waters receiving effluent discharges from municipal sewage treatment plants particularly within urbanised areas. In a laboratory study conducted by Parrott and Blunt, Fathead Minnows, *Pimephales promelas* were exposed to low concentrations of EE2 (0.32 and 0.96 ng/L), and at these sub ng/L concentrations endocrine disruption occurred in male fish, which showed decreased male sex characteristics and had lower egg fertilization success compared to non-exposed controls. However, female fish exposed at the same low concentration showed no changes in the gonadosomatic index (Parrott and Blunt 2005).

1.1 Wildlife Exposure and Endocrine Disruption

Early evidence linking EDC induced reproductive disorders in wildlife were found by numerous environmental studies in the Laurentian Great Lakes (Hamlin and Guillette 2010). The lakes during this period were highly contaminated with chemicals from industrial discharges from manufacturing plants and pesticide use. Of major significance was the contamination of the water body, the lake sediment and biota, with the now banned organochlorine (OCL) substances i.e. polychlorinated biphenyls (PCB) and the insecticide Dichlorodiphenyltrichloroethane (DDT). Contamination of the Great Lakes commenced with the large-scale production and use of these legacy chemicals in the early part of the last century. However, it was not until the late 1950s when the electron capture detector was developed that allowed Analytical Chemists to detect trace levels of "electron capturing compounds" including organohalogenated chemicals such as OCls and PCBs in water, sediment, biota and other environmental matrices, could the observed adverse effects in animals be linked with these chemicals (Clement et al. 2012). For example, wildlife observations in the 1970s by conservation specialists who had become increasingly concerned about the reproductive problems of Great Lake birds, including the bald eagle, *Haliaeetus leucocephalus*, noted that these birds had a number of reproductive problems including the failure to lay eggs or they produced eggs with thin shells or had chicks with deformities (Hamlin and Guillette 2010). An analysis of the failed eggs from these birds were found to contain high concentration of PCB (90 mg/g), DDT (16 mg/g) and 1-Dichloro-2,2-bis(p-chlorophenyl) ethylene (p,p-DDE), the degradation product of DDT. A reduction in eggshell thickness of up to 90% for some bird species was correlated with high concentration of p,p-DDE (Donaldson et al. 1999). The accumulation of PCB and DDT and its metabolites (p,p-DDE, p,p-DDD) and indeed any other endocrine disrupter in the eggs of birds, reptiles and other oviparous species is of significant concern because, unlike EDC placental exposures, where exposure to the chemical is dynamic and dependent on maternal exposure, the presence of contaminants in eggs will persist throughout development (Hamlin and Guillette 2010). A DDT PCB ban has resulted in a decrease in PCB concentrations in fish by a factor of 2 every 9–17 years and DDT in fish by a factor of 2 every 7–10 years (Hites and Holsen 2019). Consequently levels in bird eggs have also decreased, and the bald eagle has been removed from the federal list of endangered species as it no longer needs protection under the Endangered Species Act.

The North American Alligator, *Alligator mississippiensis*, is another animal reported to have significant vulnerability to endocrine disrupting compounds, specifically to OCL pesticides. It was particularly concerning when the alligator population declined dramatically in the period between 1980 and 1987. Chemical analysis of alligator eggs during this time showed elevated concentration of DDT metabolites, p,p-DDE and p,p-DDD in eggs, suggesting that DDT was responsible for these reproductive failures. Later reports suggest that other compounds, specifically dibromochloropropane and ethylene dibromide could also have played a role in the reproductive failure and population collapse (Semenza et al. 1997). Following a ban on DDT and PCBs, similar to the scenario with the bald eagle, the North American Alligator is no longer an endangered species, its population having recovered and now growing. This is either the result of a reduction in OCL exposure and/or as a result of a ban on hunting and from other conservation activity.

However both of the examples of endocrine disruption provided above are the result of gross pollution in highly contaminated environments with exposure occurring at relatively high concentration of multiple contaminants. These chemical exposures are atypical of the everyday exposures to wildlife in pristine environments, and exposures at these high concentrations would not be regarded as normal background in most urban environments. However, in certain environments, including many parts of the developing world, where stringent environmental control is not regulated, there may be wildlife exposure at levels that were observed in developed countries 50 years ago. Nevertheless, and regardless of dose, it is recognised that wildlife is exposed to a vast range of chemicals; however, the number and type of contaminants to which they are exposed remains uncertain. Any assessment of regional exposure to chemicals therefore requires detailed long-term observations and analytical monitoring to obtain both temporal and spatial data to describe normal chemical exposures. However, in general it can be assumed that wildlife and domestic animal exposure to most environmental chemicals would usually occur at low or ultra-low concentrations (parts per million to parts per trillion), but even at such low concentrations, micro-pollutants in the environment have shown adverse reproductive effects in some animals. An example is the feminization of male fish following exposure to treated sewage effluents containing low parts per trillion levels of natural and synthetic steroid estrogens (Jobling et al. 1998). There is concern that even lower concentration of parts per quadrillion (ppq) of some micropollutants present in water, air, soil or feedstuff can act additively or synergistically to elicit endocrine disrupting effects on different species. Furthermore, wildlife populations exposed to environmental chemicals may also be exposed to other environmental stressors e.g. climatic, physical, temperature; which in combination with chemicals may interact to enhance the adverse effects of chemical exposure (Hooper et al. 2013).

1.2 Mode of Action

Endocrine disrupters are defined as chemicals that interfere with an organism's natural endocrine functionality. The mechanistic mode of action involved in endocrine disruption has been reviewed in a number of excellent review articles, which provide detailed toxicological mechanisms involved in endocrine system interference (Yang et al. 2015, Wuttke et al. 2010; Shanle and Xu 2011). The review paper by Yang et al. looks at toxicological mechanisms of EDC's using molecular pathway analysis focusing on EDC mechanism pathways leading to obesity and cancer (Yang et al. 2015). The mini review by Wuttke et al. provides details on EDC reproductive mechanisms (Wuttke et al. 2010). Shanle and Xu provide details of the mechanism of action of EDCs on the estrogen receptors (Shanle and Xu 2011). The following section summarizes some of the mechanistic pathways and modes of action reported to be involved in environmental endocrine disruption that could be responsible for some reproductive disorders observed in wild animals.

Endocrine disrupters can impair the functionality of the endocrine system of an organism through various modes of action, including binding to steroid hormone receptors, interfering with the synthesis, signalling and other functionality of natural steroids. EDCs can be classified as agonists, antagonists and as mixed agonist-antagonists. EDC agonists are defined as chemicals that mimic, or partly mimic, naturally occurring hormones (estrogens, androgens, progestogen, thyroid hormones) and thus over stimulate the transcriptional or post-transcriptional mechanisms. Antagonists are compounds that block hormones from binding to their receptors, and include substances that bind to a receptor but do not activate it, and therefore block the activity of natural hormones thus inhibiting the transcriptional or post-transcriptional mechanisms. A mixed agonist/antagonist is used to refer to a compound which, under some conditions behaves as an agonist for a receptor while under other conditions, behaves as an antagonist. Thus some environmental EDCs, can behave as selective receptor modulators and therefore act as agonists for some receptor types and antagonist for others or agonist in some tissues while antagonist in others. Sohoni and Sumpter used an in vitro yeast-based assay to determine estrogenic, anti-estrogenic, androgenic and anti-androgenic activities in a variety of chemicals. They showed that many environmental estrogens also possess anti-androgenic activity, e.g. bisphenol A and butyl benzyl phthalate were both estrogenic and anti-androgenic, and that not all xenoestrogens were anti-androgenic (i.e. the estrogen nonylphenol was found to be a weak androgen agonist) (Sohoni and Sumpter 1998). A further consideration when determining the activity of environmental chemicals is to consider the activity of their degradation products, as environmental exposures to the parent and its degradation products will be concurrent dependent on the fate of the compound and time of exposure. For example, the pesticide DDT (principal component o,p-DDT) binds to the estrogen receptor and has anti-androgenic activity. However, the major and persistent degradation and metabolic products of DDT are o,p-DDE (a weak estrogen) and p,p-DDE which has a very low binding affinity to the estrogen receptor, but inhibits androgen binding to the androgen receptor, thus inhibiting androgen transcriptional activity in developing, pubertal and adult male rats. Thus abnormalities in male sex development induced by DDT in rats may be mediated by the androgen receptor via its metabolite p,p-DDE (Kelce et al. 1995).

2 Occurrence, Fate and Reproductive Toxicity

2.1 *Natural Sex Hormones*

Wildlife exposure to EDC's is most commonly associated with exposure to synthetic chemicals. Contamination of the environment from the direct discharge of human waste into the environment is assumed to be a third world problem and the concerns of municipal wastewater treatment plant operators discharging treated wastewater or disposing sewage sludge to land are mainly with sanitary and bulk parameters to ensure that discharges comply with regulatory standards. There is a growing regulatory focus at national and international levels to protect ecosystems from deleterious effects of toxicants, including micro-pollutants discharged into the environment. Furthermore, research has identified that even following extensive wastewater treatment, the effluents from municipal sewage treatment plants are the major cause of feminization observed in wild fish populations worldwide (Tyler and Jobling 2008). Research has identified that the most estrogenic substances in treated sewage effluents are the natural hormones estrone and 17β-estradiol, and the active ingredient of the birth control pill the synthetic hormone 17α-ethynylestradiol (Desbrow et al. 1998). The steroid load into municipal wastewater treatment plants is predominantly or wholly of human origin and the source of steroid input into receiving waters is primarily via municipal sewage treatment plants. There is also some, probably minor, contribution from effluents emanating from pharmaceutical plants manufacturing steroid based drugs and personal care products. Agricultural run-off from pig, cattle and sheep farming may be a major contributor of surface water contamination as these farm animals excrete much higher amounts of steroids than humans. In the case of dairy cows, they excrete two orders of magnitude more estrogens than humans (Johnson et al. 2006). Thus an understanding of steroidogenesis, metabolism, inactivation and catabolism of natural steroid hormones within humans and animals can provide insights into potential remediation strategies for their removal from human and animal waste.

The major classes of steroids in vertebrates, characterised by their structure and activity are estrogens (formed of C_{18}-estrane skeleton), androgens (C_{19}-androstane skeleton), progestagens (C_{21}-pregnane skeleton) and corticoids (C_{21} skeleton); their pathways of steroidogenesis and metabolism differ in different species. Active steroid hormones formed in the body are metabolized into their inactive forms in the kidneys and the liver and eliminated by secretion from the liver into bile and excreted via urine (major route) and faeces (generally the minor route). Although steroids can be eliminated in urine and faeces in their original or free form, elimination via urine involves converting the free active steroid into hydrophilic metabolites to increase water solubility, thus allowing solubilisation into biological fluids. The most usual and major catabolic pathway is the conjugation of active molecules to form hydrophilic sulphate and/or glucuronide derivatives. Steroids can also undergo other catabolic reactions including oxidation of the 17ß-hydroxyl group, or hydroxylation can occur at various ring positions. Structural changes can also involve the

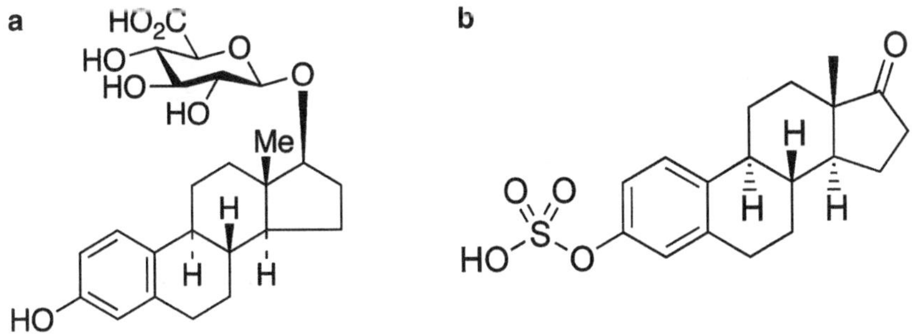

Fig. 3.1 Examples of metabolic conjugation and urinary excretion of natural steroids. (**a**) 17β-Estradiol glucuronide. (**b**) Estrone sulfate

reduction of a double bond at the C-4 position and reduction of an oxo(keto) group at C-3 to form a secondary alcoholic group (Steimer n.d.). Structural changes can be reversible and irreversible, e.g. estradiol is reversibly transformed to estrone and estrone sulphate and irreversibly to estriol. The metabolites from these transformations can be further hydroxylated and reduced and the resulting products can be conjugated to form sulphate and glucuronides (The Danish Environmental Protection Agency 2002) (Fig. 3.1).

Hence, steroid will be eliminated from the body in the original free form or metabolites and/or conjugates via urine or faeces. Thus a large number of structurally diverse molecules will be excreted in the conjugated or non-conjugated form in different ratios in urine and faeces (Eriksson and Gustafsson 1972). From a research perspective, the measurement of each of these components can be complex, as each metabolite will have different chemical properties, and for analytical methods to provide quantitative measurements they require pure analytical standards; moreover, with the exception of major metabolites, these standards are often not available. De-conjugation of samples using chemicals or enzymes allows analysts to determine the unconjugated and conjugated forms, but care is needed to ensure that de-conjugation does not lead to transformation of the original metabolite (Hauser et al. 2008). It is often not clear from literature reports when steroids are reported in the environment or in biota or human tissue whether the data provided is for free, conjugated or total steroids. Published data on the levels of all excreted individual metabolites from humans and animals is not readily available, however the literature does provide information on the excretion of total free and conjugated steroids in animals and humans reviewed (Adeel et al. 2017) in Table 3.1 below.

Furthermore, de-conjugation occurs naturally via microbial activity in sewers and in the wastewater treatment plants themselves. Wastewater treatment processes therefore require consideration of both the free and conjugated metabolites entering the plant. Microbiological degradation of steroids within treatment plants will also deconjugate steroid conjugates back to the free and active steroid and create further degradation products, which may also be active and therefore effluent discharges and sewage sludge may release a mixture of substances into the environment

Table 3.1 Steroid estrogen excretion by animals and humans per person (or animal) μg/day

Human/animal excretion	Faecal estrogen excretion μg/day			Urine estrogen excretion μg/day			Reference
	E1	E2	EEQ*	E1	E2	EEQ*	
Dairy cow	51	148	165	787	236	498	Johnson et al. (2006)
Cycling pig	12	9	13	82	ND	40	Johnson et al. (2006)
Pregnant pig	33	28	39	1400	ND	467	Johnson et al. (2006)
Human male	–	–	–	5.35	2.19	3.97	Lee (1967)
Human female	–	–	–	6.73	3.72	5.96	Lee (1967)
Human pregnant female	97.8**	202.8**	235.4	670§	170§	393	Adlercreutz and Martin (1976)

Where E1, estrone; E2, 17β-estradiol

*EEQ equals 17β-estradiol equivalents (calculated using EEQ = E2 + E1/3)

**Mean oestrogen concentrations of two 24-h pools of human pregnancy faeces

§Mean oestrogen concentrations in 13 pregnancy urine samples between the 36th and 40th week of gestation including the active free estrogens and a mixture of known and unknown degradation products (Baynes et al. 2012).

Based on a general assumption that about 20% of steroid estrogens will survive sewage treatment, Johnson and Williams (Johnson and Williams 2004) calculated that a combined 44 kg EEQ/year of estrogens from the human population of the UK would contaminate surface waters. In later work, the same authors reported, based on conservative estimates of 1% of estrogens waste generated by farm animals reaching rivers, that farms would input an estimated 8 kg EEQ/year of estrogens, which is about 15% of the calculated total (Johnson et al. 2006). Aquatic life is therefore exposed to a number of natural estrogens as well as progesterone and natural androgens. In the environment, these contaminants will undergo degradation resulting in transformation products, some of which may still be biologically potent or possess other types of activity. For example, a recent study has shown that progesterone, which has minimal androgenic activity, is transformed in river water and degrades to potent androgens (Ojoghoro et al. 2017). Thus exposure to surface waters containing natural steroids on their own and when combined with man-made contaminants may have adverse endocrine disrupting effects on aquatic biota.

2.2 *Synthetic Sex Hormones*

Synthetic steroids are of growing interest and significance as known and emerging environmental endocrine disrupters. A variety of synthetic steroids are in worldwide use in veterinary and human medicine, either prescribed or under illegal use. Synthetic steroids are designed to mimic natural hormones, to achieve desired therapeutic or contraceptive outcomes and they often have higher potency than their natural analogues. Furthermore, they often possess, by design, greater chemical, microbiological and photochemical stability, and thus they can pose a higher threat to wildlife compared to their natural analogues, even if present in the environment at considerably lower concentration.

Synthetic steroids can be broadly categorized into the following categories:

- *Androgenic-anabolic steroids (AAS):* structurally similar to testosterone with similar effects as the natural steroid, examples include oxandrolone, oxabolone, nandrolone
- *Glucocorticoids:* examples include prednisone, dexamethasone and triamcinolone
- *Estrogens:* mimic the natural steroids, E1, E2 and E3 and include diethylstilbestrol (DES), now banned and 17-α-ethynylestradiol the active ingredient of the contraceptive pill
- *Progestins:* mimic the action of progesterone and include norethindrone, 17-Hydroxyprogesterone caproate, danazol, medroxyprogesterone acetate

Ethynylestradiol (EE2) is a potent synthetic estrogen. Synthetic sex hormones (estrogens and progestins) have widespread use as oral contraceptives, administered by injection, taken orally, or used as skin patches. Formulations of oral contraceptives can contain progestin only or combined estrogen and progestin. Ethynylestradiol has been used as the synthetic estrogen for decades, initially at a dose of 100–150 μg EE2/tablet, later reduced to 20–35 μg of EE2/tablet to decrease the risk of thromboembolism (Laurenson et al. 2014). Human urine is considered as the main source of EE2 into the environment (Vethaak et al. 2005). The metabolism of EE2 in the body prior to excretion occurs through conjugation via glucuronidation and sulphation into the hormonally inert ethynylestradiol glucuronides (EE2-Gly) and ethynylestradiol sulphate (EE2-sulphate). Hydroxylation occurs in the liver to form 2-hydroxy-EE, 4-hydroxy-EE, 6α-hydroxy-EE, and 16β-hydroxy-EE. Unlike with the natural steroid E2, 16α-hydroxylation does not occur with EE2, owing to steric hindrance by its ethynyl group at C17α (Kuhl n.d.). Also the ethynylation of EE2 is largely irreversible and therefore EE2 cannot be metabolized into estradiol. The excretion of EE2 also differs from E1, E2; unlike natural steroids it has been reported that a higher proportion of EE2 is excreted in faeces compared to urine, 62% and 38% respectively. Also, as a result of bacterial action in the gut, where EE2 deconjugation occurs, faeces mainly contains free EE2, compared to urine where conjugates predominate. During sewerage and within the sewage treatment plant, EE2 is further deconjugated to its active free form through β-glucuronidase and sulphatase activity (Baynes et al. 2012; Lai et al. 2000).

The concentration of EE2 entering the influent of wastewater treatment plants is generally low compared to natural steroids (Fleming et al. 2016) and even with relatively poor removal, the concentration of EE2 in the effluent of sewage treatment is low (generally in the low ng/L range or lower) depending on the wastewater treatment plant design. Thus environmental concentrations of EE2 are expected to be low, and most studies have shown this to be the case. The same applies to EE2 concentrations in receiving waters (Williams et al. 2003; Tomšíková et al. 2012). However, the high chemical stability of EE2, its higher resistance to degradation and its higher Log P compared with E2 and E1 means EE2 has a greater potential to persist in the environment, accumulate in sediment and bioconcentrate in biota (Aris et al. 2014). Furthermore, human and fish studies have shown that EE2 has a high binding affinity to the estrogen receptor, which is up to twice that of E2 in humans and up to 20 times E2 in some fish species. Thus, the presence of EE2 in the environment, even at very low levels, poses a considerable risk to the sexual development of certain species of animals exposed to EE2.

Levonorgestrel (LNG) and Norethindrone (NOR) are synthetic steroids belonging to the progestin (also called gestagens, progestogens, or progestagens) category of drugs, used in a number of birth control approaches, including emergency birth control. Progestins have similar effects to those of the natural hormone progesterone, binding to the PR receptor. Progestins can also bind to a wide range of other receptors, including, estrogen, androgen, glucocorticoid, and mineralocorticoid receptors, and have anti-androgenic activity. Progestins are used in hormonal birth control (alone or in combination with estrogens) and menopausal hormone therapy and impact female reproductive systems, the breasts and the brain (Kuhl n.d.). They act as agonists of the progesterone receptor.

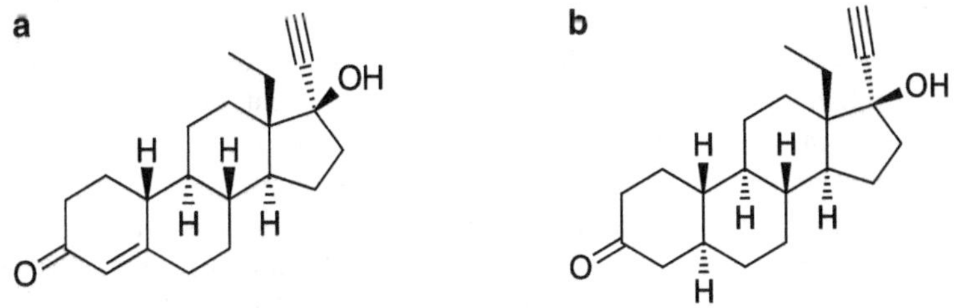

Fig. 3.2 Chemical structure of Levonorgestrel (LNG) a PR agonist and its active metabolite 5α-Dihydrolevonorgestrel (5α-LNG) a PR and AR agonist. (**a**) Levonorgestrel (LNG). (**b**) 5α-Dihydrolevonorgestrel (5α-LNG)

LNG acts as a progesterone receptor (PR) agonist. It is also a weak agonist of the androgen receptor (AR) and has little or no estrogenic or glucocorticoid, or anti-mineralocorticoid activity. Due to its progestogenic activity, LNG is able to suppress fertility and gonadal sex hormone production in both women and men. The metabolism of LNG prior to excretion includes hydroxylation and conjugation, with orally administered radiochemical studies, showing excretion in urine and faeces of 45% and 32% respectively based on ^{14}C activity (Stanczyk and Roy 1990). Chromatographic studies by the same researchers showed that the excretion products in urine were predominantly conjugated, mainly glucuronides. The major urinary metabolites of LNG identified were 3α,5α-tetrahydrolevonorgestrel glucuronide and minor amounts of LNG-glucuronide, 16-β-hydroxylevonorgestrel and 16-β-hydroxy-3α,5-β-tetrahydrolevonorgestrel. Lesser amounts of unconjugated LNG and its metabolites were also found, but in quantities of 0.1% or less of the administered dose (Stanczyk and Roy 1990). Later studies detected 5α-Dihydrolevonorgestrel (5α-LNG), which is an active metabolite of the levonorgestrel (Fig. 3.2), with in vitro studies determining that it has about one-third of the binding affinity of LNG for the progesterone receptor. However, 5α-LNG in contrast to LNG has a selective progesterone receptor modulator profile, having both progestogenic and antiprogestogenic activity as well as androgenic activity (Cabeza et al. 1995).

The human metabolism and excretory products of NOR differ significantly from LNG. Studies where women were administered NOR (a progestin) determined the presence of EE2 (an estrogen) in urine (Braselton et al. 1977). The predominant metabolites in urine are 3α,5-β-tetrahydronorethindrone glucuronide and sulfate conjugates. In addition, smaller amount of 3α, 5α-tetrahydronorethindrone and 3β, 5α-tetrahydronorethindrone have been detected (Stanczyk and Roy 1990).

Studies of the relative binding affinities of LNG and its metabolites to PR, AR and ER receptors by displacement analysis revealed that unchanged LNG binds with high affinity to PR and AR, but not to ER. 5α-dihydro LNG (5α-LNG) had a diminished binding with PR and an enhanced binding affinity for AR compared with LNG. The other major metabolite, 3α-5α-tetrahydro LNG (3α,5α-LNG) has no PR, AR or ER activity (Lemus et al. 1992). LNG and NOR are of ecotoxicological concern due to their ability to reduce egg production in fish at trace concentrations of 0.8–1.0 ng/L (Kumar et al. 2015). The binding affinity of a number of LNG and NOR metabolites may also have the same implications as the parent compounds, impacting on fish reproduction at low levels using different modes of action.

The major route of progestin entry into the environment is *via* sewage treatment plants through human excretion of progestins following use of oral contraceptives, which have a variety of formulations; some containing estrogens and progestins and some only containing progestins. The formulation of combined oral contraceptive pills (COCP) has been changing since the introduction of the pill in the early 1960s. The first generation of COCPs contained >50 μg EE2, in combination with progestins e.g. NOR, NOR acetate, noretynodrel or etynodiol acetate. The second generation COCPs released in the 1970s had a lower EE2 content (<50 μg) and lower progestin content, typically containing NOR and LNG. Third generation COCPs only contain progestins e.g. desogestrel, gestodene or norgestimate and fourth generation COCPs contain the progestin drospirenone, dienogest, or nomegestrol acetate (Speroff and Darney 2010). Thus the amount and type of progestins in the environment has been changing over a period of time since the introduction of the contraceptive pill. A critical review of progestins in the environment by Kumar et al (2015) reports that only a few of the progestins (of around 20 active compounds in use today) have been monitored in sewage effluents or river water and there have been limited ecotoxicological studies of progestins on aquatic organisms. Furthermore, the limited environmental occurrence data in the open literature shows considerable variability from one study to another (Kumar et al. 2015). The same authors report that the greatest concerns with progestins have been with exposure to LNG, NOR and gestodine, which have impaired female fathead minnow reproduction at 1 ng/L and propose that if the same effects could be replicated in cyprinid fish at the same low concentrations, this would signify a significant environmental problem (Kumar et al. 2015). Other potent synthetic progestins e.g. ethynodiol diacetate and desogestrel are also of concern, as exposure to ng/L or sub-ng/L of the active substance in surface waters provide fish a human therapeutic dose, but for other progestins much higher concentrations (tens or hundreds of ng/L) would be required to achieve the same dose (Kumar et al. 2015).

Thus, progestins have differing biological and physicochemical properties, with some compounds having the ability to impair female reproduction at very low concentrations and there is potential for their metabolites to have some effects. There is also growing concern about the cumulative effects of mixtures of synthetic progestins in the environment and these factors make synthetic progestins highly significant environmental EDCs.

2.3 *Pharmaceuticals and Personal Care Products (PPCP)*

PPCPs are a large and highly diverse range of chemicals which have received considerable attention as emerging environmental contaminants due to their ability to induce physiological effects in animals and humans at low doses (Ebele et al. 2017).

Table 3.2 Active Pharmaceutical Ingredients with potential endocrine disrupting ER, AR and PR agonistic or antagonistic potential

Hormone agonist			Hormone antagonist		
Estrogens	Androgens	Progestins	Anti-estrogens	Anti-androgens	Anti-progestogens
Estrone[a]	Mesterolone	Levonorgestrel	Clomifene	Bicalutamide	Mifepristone
Estradiol[a]	Testosterone[a]	Norethindrone	Tamoxifen	Cyproterone acetate	Lilopristone
Ethinylestradiol	Nandrolone	Megestrol	Danazol	Finasteride	Onapristone
	Oxabolone		Raloxifene	Flutamide	

Certain API's have multiple receptor binding affinities (not shown)
[a]Natural steroid also a prescribed API

Studies worldwide have confirmed the presence of PPCPs in the environment, which raises concerns about their exposure to, and subsequent effect on, wildlife. Due to these concerns the European Commission placed E2, EE2 and diclofenac on a Watch List of substances to be monitored in EU surface waters to determine the risk they pose to the aquatic environment and whether EU Environmental Quality Standards (EQS) should be set for them.

The first watch-list was recently reviewed by the EU's Joint Research Centre (JRC), who recommended withdrawing the pharmaceutical painkiller diclofenac and the sunscreen ingredient (2-ethylhexyl-4-methoxycinnamate) and advised on adding a number of new substances to a revised watch-list. These include some pharmaceuticals e.g. Erythromycin, Clarithromycin, Ciprofloxacin and Azithromycin. The estrogens, E1, E2 and EE2 are also on the updated watch-list (Barbosa et al. 2016).

There is particular interest in drugs that interact with the endocrine systems of humans and animals with respect to their potential as endocrine disrupting compounds. Examples of Active pharmaceutical ingredients falling under a number of categories of drugs that have potential endocrine disruption potential to wildlife are provided in Table 3.2. Synthetic drugs that mimic natural sex hormones were discussed in the previous section. These hormone agonists are used to treat certain conditions caused by low levels of testosterone, such as delayed puberty in boys. Hormone antagonists are used to treat a number of disorders such as abnormally high levels of sex hormones in the body. Both types of synthetic drugs are of interest due to their potential as endocrine disrupters. A number of pharmaceuticals, when metabolised in the body or degraded in the environment have the potential to generate metabolic or degradation by-products that have hormonal agonistic or antagonistic potential, as observed with natural Progesterone and discussed in the previous section (Ojoghoro et al. 2017).

Anti-androgen drugs function by blocking the natural function of androgens by binding to the AR and blocking androgen induced receptor activation. Cyproterone acetate is an anti-androgen used in the treatment of hypersexuality in males and in the treatment of advanced prostate cancer. Cyproterone acetate also has progestational activity. There is limited environmental occurrence and ecotoxicological data to date on anti-androgens including cyproterone acetate, although an ecotoxicologi-

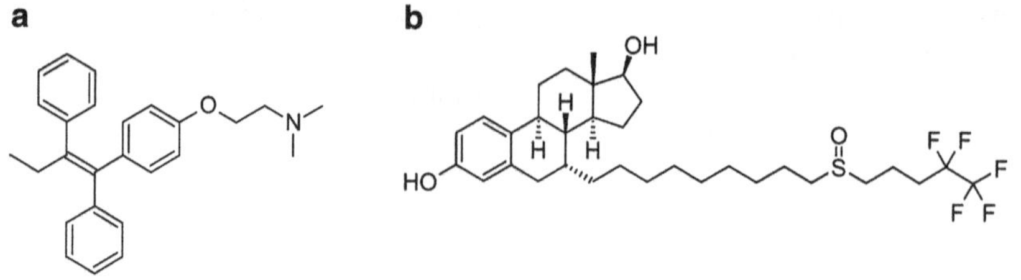

Fig. 3.3 Anti-estrogenic drugs used in the treatment of breast cancer. (**a**) Tamifen. (**b**) Fulvestrant

cal study has determined that cyproterone acetate alone or in combination with steroid estrogens did not impact sexual disruption in fish (Green et al. 2015).

Anti-estrogen drugs function by blocking the natural function of estrogens by binding to the ER and blocking estrogen-induced receptor activation. In a study to determine the sex differentiation and embryonic development of the domestic fowl, *Gallus gallus domesticus*, researchers (Jessl et al. 2018) examined the effects of *in ovo* exposure of the anti-estrogens fulvestrant and tamoxifen, injected into the yolk of fertilized eggs. An exposure to tamoxifen (0.1/1/10 μg/g egg) strongly impaired the differentiation of female gonads and led to a size reduction of the ovary and induced malformations of the ovarian cortex, while fulvestrant (0.1/1/10 μg/g egg) did not affect sexual differentiation (Jessl et al. 2018) (Fig. 3.3).

Anti-progestogens are drugs that function by blocking the natural function of progestogens like progesterone by binding or blocking the PR receptor or by inhibiting progestogen production. Examples of anti-progestogens include Mifepristone and Gestrinone. Mifepristone is used for emergency contraception or to abort a pregnancy, typically used with a prostaglandin such as misoprostol or gemeprost. There is limited information in the literature on environmental levels of these two drugs. Based on the approved dose of 200 mg of mifepristone and the frequency of use, it is probable that its occurrence in the environment and exposure to wildlife would be low. The ecotoxicological impact of anti-progestogens or indeed hormone agonists and antagonists in general present in the environment, albeit in most circumstances at much lower levels than the exposure levels of most laboratory ecotoxicology studies, may still pose a threat to sexual development of wild animals from exposure to these substances, either as discrete compounds, but most probably as mixtures of compounds acting in combination with other pharmaceuticals and anthropogenic chemicals.

2.4 Phthalates and Bisphenol-A

Bisphenols and phthalates have been extensively studied in the environment due to their established endocrine disrupting properties. Bisphenols constitute a group of chemicals containing dihydroxyphenyl functional groups and include Bisphenol-A (BPA), Bisphenol-F (BPF) and Bisphenol-S (BPS) with studies showing all three having endocrine disrupting properties (Structures Fig. 3.4). BPA is a high use

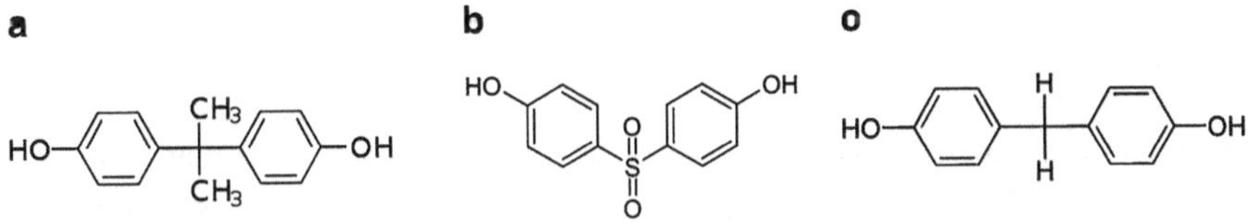

Fig. 3.4 Chemical structures of reported endocrine disrupting Bisphenols: Bisphenol-A, Bisphenol S and Bisphenol F. (**a**) Bisphenol-A. (**b**) Bisphenol-S. (**c**) Bisphenol-F

chemical (2012 world production was 4.4 million tonnes) used as a building block for polycarbonate plastics and in epoxy resins used to line food and beverage containers and mains water pipes, BPA has received worldwide attention regarding its reported human health implications (Rubin 2011).

BPA was considered a weak environmental estrogen based on its binding affinity to ER alpha and ER beta (the basis of the cell based bioassay); reported be 1000–10,000 times lower than E2 (Kuiper et al. 1998) however more recent studies provide evidence of higher potency and suggest the involvement of estrogen receptors located outside the nucleus in the cell membrane and mitochondria and possibly the cytosol (Ropero et al. 2006). BPA also has androgenic activity. In a recent review on Bisphenol A exposure, effects and policy Flint et al. (2012) conclude that aquatic wildlife exposure to BPA at or near environmentally relevant concentrations (0.08–12.5 μg/L) may have significant adverse on the development of a number of vertebrates, and a recent environmental risk assessment has proposed lowering the predicted no effect concentration of BPA from 100 to 0.06 μg/L (Wright-Walters et al. 2011).

Phthalic acid esters (commonly referred to as phthalates) are another group of chemicals that have received considerable scientific, media and public attention due to their reported endocrine disrupting effects. Phthalates are used as plasticizers in large quantities in the manufacture of plastics, such as polyvinylchloride (PVC), to give plastics their desired and useful properties such as flexibility and durability (Murphy 2001). Phthalate world consumption was nearly 5.7 million tonnes of in 2002 and forecasts suggest world production could exceed nine million tonnes by 2020 (Pritchard 2017). The most commonly used phthalates are diisononyl phthalate (DINP), diisodecyl phthalate (DIDP) and di(2-Propyl Heptyl) phthalate (DPHP), replacing some of the low mass derivatives such as di(2-ethylhexyl) phthalate (DEHP), diisodecyl phthalate (DIDP) and Dibutylphthalate (DBP), which have been restricted in Europe and other parts of the world in products such as toys. Plasticisers can migrate from materials and get released from plastics relatively easily by leaching into water over a period of time. This has implications for human health due to leaching of plasticisers into drinking water and food, and there is a risk to wildlife from the release of plasticisers from leaching from plastic pipes and their discharge into the environment from household and industry generated wastewater (Erythropel et al. 2014).

A number of studies have reported endocrine disrupting activity of phthalates. Using recombinant yeast screen bioassay, researchers (Harris et al. 1997) have found that phthalates have low estrogenic activity (potencies ranged from approximately 1×10^6 to 5×10^7 times less than E2 and relative potencies between variants decreased in the order butyl benzyl phthalate (BBP) > dibutylphthalate (DBP) > diisobutylphthalate (DIBP) > diethylphthalate (DEP) > diisiononylphthalate (DINP) (Harris et al. 1997). In a further study of the estrogenic and androgenic potential of phthalates, testing 15 phthalates, in vitro studies report that phthalates bind to estrogen receptors and have weak estrogenic and anti-androgenic potential (Kwon and Ji 2016).

Thus due to their high consumption and widespread use phthalates and bisphenol are compounds of concern due to their potential endocrine disruption potential on wildlife and there is increasing evidence that human reproductive health may also be implicated (Mathers et al. 2009).

2.5 *Pesticides*

There is significant and growing evidence that as well as legacy pesticides such as DDT, a number of currently approved pesticidal products have the potential to disrupt the endocrine system of animals. This is particularly concerning as worldwide usage of pesticides continues to grow, currently estimated usage is that over two million tonnes of pesticides are used worldwide (De et al. 2014) for a variety of purposes to destroy or control unwanted animals or plants; commonly grouped either into chemical families (e.g. organochlorine, organophosphorous, carbamates, phenoxyaliphatic acids, nitroanilines, bipyridyliums) or by the target organism (e.g. algaecides, bactericides, fungicides, herbicides, insecticides, molluscicides, and rodenticides). Over 80% of pesticides used in agricultural are herbicides, but numerous other pesticides are used in lower but significant quantities. A recent study to determine pesticide residues in agricultural soils across Europe found pesticide residues present in 83% of the tested soils and 58% of the soils contained multiple residues (Silva et al. 2019), thus animal exposure from the results of this study and other studies, suggest pesticide exposure to animals could be considerable.

Pesticides have potential to disrupt animal reproductive systems in a number of ways, for example as follows:

- through estrogenic effects: e.g. when a pesticide binds to and activates an estrogen receptor; examples include: alachlor (Klotz et al. 1996), methocarb (Raun Andersen et al. 2002), endosulfan and dieldrin (Wade et al. 1997)
- through anti-estrogenic effects: e.g. when a pesticide binds but does not activate an estrogen receptor; examples include: bifenthrin (Brander et al. 2012)
- through androgenic effects: e.g. when a pesticide binds to and activates an androgen receptor; examples include cyprodinil, pyrimethanil (Orton et al. 2011)
- through anti-androgenic effects: e.g. when a pesticide binds but does not activate an androgen receptor; examples include vinclozolin, procymidone and prochloraz (Müller et al. 2009)

3 Reproductive Effects on Wildlife

Information presented above supports the growing scientific evidence that certain man-made chemicals, as well as some natural substances of human, animal and plant origin, have the potential to disrupt the endocrine system of certain animals and may adversely interfere with the natural reproductive development processes of wild animals. Wildlife exposure to endocrine or hormone disrupting chemicals occurs following the introduction of these substances into the environment, through point and non-point sources from individual, industrial and agricultural use, resulting in pollutants entering water bodies, the atmosphere or on land. Exposure can occur via a number of routes including the ingestion of contaminated food and water or through inhalation of contaminated air and from skin contact. These exposures are described comprehensively in a landmark report of the "State of the Science of Endocrine Disrupting Chemicals" published in 2012 by the United Nations Environment Programme and the World Health Organisation, prepared by a group of experts to assess the evidence and implications of the exposure of EDCs on wildlife (Bergman et al. 2013). The report surmises that EDCs have the capacity to interfere with wildlife development and EDCs are a global threat that needs to be resolved, but concludes that "there is currently no widely agreed system for evaluating the strength of evidence of associations between exposures to chemicals and adverse health outcomes". Some of the evidence from this report and from the wider literature is summarised below with a focus on potential reproductive effects on wildlife (Jobling et al. 1998).

3.1 Effects in Fish

The development of adverse effects in fish as a result of exposure to endocrine disrupting chemicals has been a common feature observed in wild fish throughout Europe, North America and across the world since the 1980s. These adverse biological effects have been particularly pronounced in fish downstream of sewage treatment works. Effects observed in various wild fish include the occurrence within male and juvenile fish of the blood protein vitellogenin (VTG), an egg yolk protein precursor normally only present in mature females (Jobling and Tyler 2007). Field studies, now dating back a few decades, established through laboratory and field studies (e.g. in caged studies), showed that roach (*Rutilus rutilus*) exposed to treated sewage treatment works effluent in UK rivers had reduced reproductive capacity and fish exhibited intersex when exposed to treated effluents. Intersex is a condition in which fish possess both male and female characteristics within the same individual (Tetreault et al. 2011).

Treated sewage effluents are composed of a complex mixture of thousands of man-made and natural chemicals, including their metabolic and degradation products. The presence of this large cocktail of chemicals in effluents is due to their incomplete removal by wastewater treatment process. Studies to determine the causative agents within sewage that could be attributed to the biological effects have been investigated using effect-directed analysis, whereby large volumes of treated sewage effluent were extracted, concentrated and the resulting extract fractionated using high performance liquid chromatography, with each fraction being tested for estrogenic potency using the yeast estrogen screen to identify the most estrogenic fractions. Using gas chromatography mass spectrometry, the natural hormones, 17β-estradiol and estrone and the synthetic hormone 17 α-ethynylestradiol were identified as the most estrogenic components within sewage (Desbrow et al. 1998). Further studies have shown that chemicals present in other fractions, that have low estrogenic potency (or other endocrine disrupting properties) when tested singularly, may exert higher effects when present in combination with other chemicals. In an early study to test this hypothesis, researchers used the YES screen to study the effect of combinations of two weak environmental estrogens, such as dieldrin, endosulfan, or toxaphene and found they were 1000 times as potent as any chemical alone (Arnold et al. 1996). Despite the extensive work detailing single compound or mixture effects of estrogens in fish, it remains uncertain whether estrogenic effects are a threat to fish populations. In a recent study, researchers attempted to determine whether the exposure to domestic wastewater effluent including effluents containing potent steroid estrogens would harm fish populations by monitoring fish density data over 38 sites over several years, but no clear evidence between fish density and wastewater exposure could be found (Johnson and Chen 2017). Exposure of male fish to concentrations of androgenic compounds has also been reported to cause adverse reproductive effects. For example, exposure of environmental concentrations of Levonorgestrel, an androgenic progestin was shown to disrupt the seasonal breeding cycle in male three-spined stickleback (*Gasterosteus aculeatus*) (Svensson et al. 2014).

Endocrine disruption has also been evidenced in female fish. Exposure to environmentally relevant concentration of trenbolone for example, have resulted in reproductive behavioural change in females (Saaristo et al. 2013; Bertram et al. 2018). Other studies exposing sticklebacks to wastewater effluent showed no significant androgenic effects on the reproductive parameters (Björkblom et al. 2009) suggesting perhaps that concentrations of androgens in effluents are not high enough to induce androgenic effects.

3.2 Effects in Invertebrates

Although invertebrates are considered to be at risk from endocrine disruption at sub-lethal concentrations of chemicals in the environment, their studies may be limited as their hormone systems are not as well understood compared with vertebrates (Matthiessen et al. 2018). However, there is evidence that invertebrates are adversely effected by endocrine disrupting compounds, including insecticides e.g. endosulfan, tebufenozide, methoprene and DDT, and by herbicides e.g. diquat, atrazine and

MCPA (Oetken et al. 2004). The androgenic effects of tri-organotin compounds (e.g. Tributyltin (TBT), Triphenyl tin (TPT)) on molluscs is the best known example of endocrine disruption in invertebrates. TBT used in anti-fouling paints was responsible for the global decline the population of several molluscan species, through interference with reproduction. Molluscs exposed to TBT caused masculinization of females resulting in them developing male reproductive organs leading to infertility and a large decline in their populations worldwide (Alzieu 2000). Following a ban in using tributyltins within antifouling paints, introduced by the International Maritime Organisation in 2008, the population of molluscs began to recover (Matthiessen and Gibbs 1998).

3.3 *Effects in Amphibians*

Endocrine Disruption of reproduction in amphibians, which may be subject to exposure to chemicals in both the terrestrial and aquatic environments, are of particular interest due to the worldwide decline of many amphibian species (Norris 2011) and the possibility that this may be attributed to the exposure to EDCs at different stages of their metamorphosis. Amphibian metamorphosis is a post-embryonic process controlled by endogenous steroid hormones and any disruption in thyroid regulation at different stages of metamorphosis has the potential for reproductive and development impairment (Shi et al. 1996). Amphibians may be particularly sensitive to a wide variety of EDCs due to their penetration via their permeable integument, and they may be particularly susceptible to chemical exposures after hatching, as this is when their skin is highly permeable (Hamlin and Guillette 2010). Furthermore, the seasonal applications of pesticides in spring and early summer has received particular attention as this period coincides with the breeding seasons of a number of amphibian species. In a study exploring the amphibian exposome in an agricultural landscape using telemetry and passive sampling, researchers found that pesticide concentrations in frogs were greatest in spring when frogs were occupying wetlands, compared to late summer when frogs occupied terrestrial habitats (Swanson et al. 2018). A number of studies have focused on the ecotoxicological effects of pesticides on amphibians and researchers have suggested that exposure to pesticides may accelerate metamorphosis, which can negatively affect reproduction i.e. juveniles are likely to be undersized and under sized animals are less ecologically fit (Forson and Storfer 2006; Cheek et al. 1999; Cauble and Wagner 2005). There are far more reports suggesting that exposure to certain pesticides may delay metamorphosis and cause developmental retardation (Teplitsky et al. 2005; Mann et al. 2009). Anti-AChE pesticides e.g. organophosphorus and carbamate insecticides have been extensively reported with regard to environmental exposures leading to delayed metamorphosis. For example, malathion exposure of leopard frogs (*Rana pipiens*) in an outdoor mesocosom study led to large reductions in growth and development, and subsequent increased mortality; this was related to prolonging the larval period, thus increasing the risk of predation. In addition the environment dried out before frogs reached full maturity (Relyea and Diecks 2008).

Amphibian exposure to the anti-androgenic herbicide, linuron, has been shown to negatively impact reproductive physiology and fertility in *Xenopus tropicalis*. Exposure to μg/L levels of linuron resulted in demasculinisation of secondary sexual characteristics and reduced male fertility of test species (Orton et al. 2018). Atrazine, has been widely studied for its endocrine disrupting properties in amphibians. The scientific evidence presented in the literature of the endocrine disrupting effects of atrazine at environmentally relevant concentrations on amphibians is conflicting; e.g. a study on the effect of exposure to waterborne atrazine at environmental concentrations on wild leopard frogs (*Rana pipiens*) in different regions of the United States found that 10–92% of males exposed to atrazine showed gonadal abnormalities such as retarded development and hermaphroditism (Hayes et al. 2002). Further studies of atrazine exposure by the same group have reported sexual abnormalities in exposed male frogs, such as feminized laryngeal structures, decreased mature sperm, low fertility and incapability of competing for females. However, a critical review of the effects of atrazine on amphibians, as well as fish and aquatic reptiles, by Solomon et al. reports that the evidence of adverse effects of atrazine is not supported by the vast majority of observations (Solomon et al. 2008). A number of studies have documented immunosuppressive effects of pesticides, suggesting a role for environmental contaminants in increased pathogen virulence and disease rates when considering the cause of global amphibian decline (Hayes et al. 2010).

3.4 Effects in Mammals

Laboratory and field studies over the decades have shown that mammals, and in particular aquatic mammals, are susceptible when exposed to certain endocrine disrupting chemicals. Legacy organochlorine pesticides such as DDT and chlorinated compounds such as PCBs and TCDD can bioaccumulate in mammals and biomagnify in animals high in the food chain e.g. porpoise, seals, killer whale, mink and otter. High concentrations of organochlorine compounds in the bodies of these animals places them at risk (Fossi and Marsili 2007). For example, the presence of high levels of PCBs in Killer Whales (*Orcinus orca*), which persist to this day despite a near global ban in their manufacture over 30 years ago, is placing Killer Whale populations at risk as reported recently by Desforges et al. who used a model framework to predict that PCB mediated effects on the reproduction and immune function of Killer whales threaten the long-term viability of more than half of the world's killer whale populations (Desforges et al. 2018), raising further concern about PCBs in Whales as well as other species.

The potential impact of endocrine disrupting compounds on the reproduction and welfare of ruminants has been the subject of some recent studies. In a study to determine the potential reproductive effects of endocrine disrupting compounds on domestic animals, Boerjan et al. carried out a preliminary risk assessment which concluded that under certain circumstances animals will be exposed to steroid estrogens and other EDCs through drinking sewage contaminated water or from EDC contaminated herbage, but the implications of such exposures are largely unknown although they cannot be neglected (Boerjan et al. 2002). A review of the potential impact of EDCs on Farm animals (Rhind 2005) surmises from the small amount of data available at the time that ruminants may be exposed to EDCs, and may have significant effects on farm animals, although these effects were generally not apparent in practice. A further review by the same author at a later date (Rhind et al. 2010), reporting on the reproductive effects of environmental pollutants on ruminant farm animals exposed to a number of chemicals such as pesticides, herbicides, therapeutic drugs and chemicals from sludge-treatments, concludes that EDCs originating from these sources may disrupt ovarian follicle and oocyte development in mammals and that mixtures of EDCs are likely to have negative impacts on the developing testis in domestic animals. Furthermore, the authors state that while EDCs may affect all ages, the developing foetus was particularly vulnerable. In sheep, various organs may be affected by EDC exposure including the ovary, testis, and pituitary gland, but these changes are not reflected in changes in reproductive performance under the conditions tested. Nevertheless EDC exposure may pose a threat to both reproductive performance and welfare in the longer term (Rhind et al. 2010).

3.5 *Effects in Reptiles*

Like other vertebrates, reptiles are susceptible to adverse reproductive effects following exposure to EDCs, as observed in the 1980s following a major spill of dicofol into Lake Apopka in Florida causing a drastic fall in the population of the American alligators (*Alligator mississippiensis*) living in the lake (Semenza et al. 1997). The role of EDCs in the population decline of Alligators was subsequently investigated in the laboratory by Guillette et al. to determine the mode of action of certain pesticides, e.g. by exposing developing embryos to various pesticides; the authors described different reproductive effects on this species (Milnes and Guillette 2008).

There are various factors that make reptiles a favourable species for endocrine disrupter research such as their habitat, diet and their modes of sex determination (Bhandari et al. 2015). In many species of reptiles including crocodiles, lizards and turtles, sex is only determined after fertilization during the period of organogenesis (Boggs et al. 2011) and, in common with other classes of wild animals, studies have shown that EDC exposure during early life can significantly alter reproductive systems e.g. turtles exposed to PCBs during early life show altered sex ratios (Bergeron et al. 1994). Furthermore, most reptiles have a temperature dependent sex determination (Bull 1980) which may be over-ridden, resulting in reptiles producing females at otherwise male temperatures (Wibbels et al. 1993).

There are growing reports linking EDCs with various adverse effects in different reptiles; compounds which are ubiquitous in the aquatic and terrestrial environments, for example Atrazine, an organonitrogen triazine herbicide and one of the most frequently detected pesticides in groundwater and drinking water worldwide, has received significant attention for its reproductive endocrine disruption potential across various groups of animals and has been reported to induce feminisation and to demasculinize male gonads, producing testicular lesions associated with reduced germ cell numbers in reptiles (Hayes et al. 2011).

Another example of an environmental chemical suspected of causing endocrine disrupting effects is provided by Murray et al. who detected elevated levels of the synthetic androgen 17α-methyltestosterone in river water and in egg yolk and placenta of a highly male biased population of American crocodiles (*Crocodylus acutus*) in the Tempisque River (Murray et al. 2017). Furthermore, earlier studies by the same authors, involving the exposure of 17α-methyltestosterone *to American crocodile* embryos, led to the development of hermaphroditic primary sex organs and masculinization of the clitero-penis occurred (Murray et al. 2015). Thus there is growing evidence that some chemicals in the environment may play a significant role in the reproduction of certain species of reptiles,

4 Conclusion and Priorities for the Future

Wildlife is exposed to a diverse range of natural and man-made chemicals as well as to their metabolic and degradation products, which includes some highly potent endocrine disrupting chemicals. There is a growing body of evidence suggesting that exposure to these environmental chemicals plays a key role in reproductive disorders in some wildlife. This evidence comes from field based observations concerning wildlife reproductive abnormalities linked to the presence of environmental contaminants, as well as from a growing body of laboratory based exposure studies, which provide substantial supporting evidence that environmental chemicals can cause adverse effects at environmentally relevant doses. There is particular concern about wildlife exposures to cocktails of chemicals, which combined with other stressors, may play a greater role in reproductive disorders than can be reproduced in laboratory experiments.

However, there are still many gaps in our understanding of the risks posed by endocrine disrupting chemicals on humans and wildlife. For example, there is uncertainty on whether any adverse impacts observed in laboratory studies of a chemical on an individual are likely to lead to population level effects (Johnson and Sumpter 2016). There is considerable debate on whether regulation of EDCs should be based on a hazard or risk assessment of a substance. A risk assessment approach considers the likelihood of adverse effects and is derived from a combination of hazard and exposure, including the amount of exposure, the duration of exposure and the time period of exposure. This approach is concerned with aspects such as short-term exposure during critical windows of development, e.g. as a foetus when EDCs can cause delayed, but permanent, damage to an organism. In the United States, the endocrine disrupter screening programme has started a tier based EDC screening programme to aid a risk assessment of chemicals to which humans and wildlife are widely exposed. The EU proposes to regulate EDCs on hazard based criteria rather than a risk based approach. Furthermore, experts in the EU have proposed a clear distinction between hazard and risk assessment and the EDC State of the Art assessment report to the EU proposes that the socio-economic impacts of chemical regulation should not influence regulatory decisions (Faust and Kortenkamp 2009).

Thus, several approaches can be used to limit the adverse effects of endocrine disrupting chemicals in the environment. The first is to replace, ban or limit the production of the most potent substances. This is the most effective solution and has been highly effective with persistent legacy chemicals such as TBT, DDT and PCBs. However, as chemicals are banned, manufacturers have synthesised alternatives that can have similar or different consequences for wildlife. Therefore, when registering a new or replacement chemical or simply retaining registration for an existing chemical, it is essential that consideration is given to its environmental consequences. In this regard, the EU has introduced various legislation aimed at phasing out endocrine disruptors in industrial chemicals, plant protection products and biocides, e.g. in the REACH programme (Registration, Evaluation, Authorisation and restriction of Chemicals), endocrine disrupting chemicals are considered of similar regulatory concern as substances of very high concern, e.g. chemicals that may be persistent, bio-accumulative or toxic. In the US, although there is no specific regulation in relation to endocrine disrupting effects of chemicals, under the Toxic Substance Control Act (TSCA), the USEPA has authority to regulate all chemicals in commercial use that exhibit ED effects, except chemicals that are categorised as pesticides, food, drugs and cosmetics (Coady et al. 2016). Other means of chemical regulation include placing restrictions on use, i.e. chemicals that have substantial socioeconomic value can be restricted for only essential use and only where a safer option is not available. The third option is to reuse or recycle—this is highly relevant for many chemicals, i.e. the reuse or recycling of plastics would limit the discharge of endocrine disrupting plasticizers into the environment.

However all of these solutions ignore (1) the basic needs of modern society, which include the need for chemicals for the production of essential twenty-first century goods such as household consumer products as well as veterinary and human medicines where alternatives cannot be easily found, and (2) the problems with the growing human population and the rapid urbanisation of the world's population, which leads to increasing contamination of the environment resulting from the disposal of industrial and household waste into confined space. Municipal and industrial wastewater treatment plants and other forms of treatment do provide some protection to the receiving environment, including wildlife exposed to such discharges. However, there is growing evidence that current technology used for water treatment is not sufficient to safeguard against endocrine disruption of wildlife. Various technological solutions have been proposed to reduce the risk of such discharges, including options such as upgrading treatment plants with additional advanced end of pipe treatment. To support these improvements, legislation such as the Water Framework Directive (WFD) in the European Union has been established to ensure that the receiving environment is protected and that ultimately both the chemical and ecological status of the environment is acceptable in all member states. However, the implementation of such improvements has been difficult due to the high costs involved in implementing such changes to meet the new stringent requirements of the directive, and because of this, almost half of the rivers in Europe are still not compliant with WFD's water quality objectives to date. Therefore it is imperative that solutions other than end of pipe treatment are considered; for example, the production and retention of registration of chemicals that focus on using green and sustainable production methods should prioritise over cheaper alternatives, registration of new chemicals should be limited to green chemicals that are in themselves and including their metabolic and degradation products not persistent in the environment or toxic to wildlife, including in their endocrine disrupting potential. Green Chemistry can also play an important role in wastewater and solid waste treatment thus avoiding the use of more harmful chemicals. Stopping harmful chemicals getting into the environment and into the food chain, particularly those that have the potential to adversely impact wildlife is essential if we are to conserve the long-term biodiversity of our global ecosystems and protect the reproductive health of wildlife.

References

Adeel M, Song X, Wang Y, Francis D, Yang Y. Environmental impact of estrogens on human, animal and plant life: a critical review. Environ Int. 2017;99:107–19. https://www.sciencedirect.com/science/article/pii/S0160412016304494.

Adlercreutz H, Martin F. Oestrogen in human pregnancy faeces. Acta Endocrinol. 1976;83(2):410–9. http://www.ncbi.nlm.nih.gov/pubmed/989669.

Alzieu C. Impact of tributyltin on marine invertebrates. Ecotoxicology. 2000;9(1/2):71–6. https://doi.org/10.1023/A:1008968229409.

Antizar-Ladislao B. Environmental levels, toxicity and human exposure to tributyltin (TBT)-contaminated marine environment. A review. Environ Int. 2008;34(2):292–308. https://www.sciencedirect.com/science/article/pii/S0160412007001845.

Aris AZ, Shamsuddin AS, Praveena SM. Occurrence of 17α-ethynylestradiol (EE2) in the environment and effect on exposed biota: a review. Environ Int. 2014;69:104–19. https://www.sciencedirect.com/science/article/pii/S0160412014001226.

Arnold SF, Klotz DM, Collins BM, Vonier PM, Guillette LJ, McLachlan JA. Synergistic activation of estrogen receptor with combinations of environmental chemicals. Science. 1996;272(5267):1489–92. http://www.ncbi.nlm.nih.gov/pubmed/8633243.

Barbosa MO, Moreira NFF, Ribeiro AR, Pereira MFR, Silva AMT. Occurrence and removal of organic micropollutants: an overview of the watch list of EU decision 2015/495. Water Res. 2016;94:257–79. http://www.ncbi.nlm.nih.gov/pubmed/26967909.

Baynes A, Green C, Nicol E, Beresford N, Kanda R, Henshaw A, et al. Additional treatment of wastewater reduces endocrine disruption in wild fish. A comparative study of tertiary and advanced treatments. 2012. http://dbwt.us/cbnswwtp/A10-Additional-Treatment-of-Wastewater-Reduces-Endocrine-Disruption-in-Wild-Fish-A-Comparative-Study-of-Tertiary-and-Advanced-Treatments.pdf.

Bergeron JM, Crews D, McLachlan JA. PCBS as environmental estrogens—turtle sex determination as a biomarker of environmental contamination. Environ Health Perspect. 1994;102(9):780–1.

Bergman Å, Heindel J, Jobling S, Kidd K, Zoeller RT. State of the science of endocrine disrupting chemicals. Toxicol Lett. 2012;211:S3.

Bergman A, Jerrold HJ, Jobling S, Kidd KA, Zoeller TR. State of the science of endocrine disrupting chemicals—2012. In: WHO–UNEP; 2013.

Bertram MG, Saaristo M, Martin JM, Ecker TE, Michelangeli M, Johnstone CP, et al. Field-realistic exposure to the androgenic endocrine disruptor 17β-trenbolone alters ecologically important behaviours in female fish across multiple contexts. Environ Pollut. 2018;243:900–11. https://www.sciencedirect.com/science/article/pii/S0269749118318839.

Bhandari RK, Deem SL, Holliday DK, Jandegian CM, Kassotis CD, Nagel SC, et al. Effects of the environmental estrogenic contaminants bisphenol A and 17α-ethinyl estradiol on sexual development and adult behaviors in aquatic wildlife species. Gen Comp Endocrinol. 2015;214:195–219. https://www.sciencedirect.com/science/article/pii/S0016648014003712.

Björkblom C, Högfors E, Salste L, Bergelin E, Olsson P-E, Katsiadaki I, et al. Estrogenic and androgenic effects of municipal wastewater effluent on reproductive endpoint biomarkers in three-spined stickleback (gasterosteus aculeatus). Environ Toxicol Chem. 2009;28(5):1063–71. https://doi.org/10.1897/08-337.1.

Boerjan ML, Freijnagel S, Rhind SM, Meijer GAL. The potential reproductive effects of exposure of domestic ruminants to endocrine disrupting compounds. Anim Sci. 2002;74(1):3–12. https://www.cambridge.org/core/journals/animal-science/article/potential-reproductive-effects-of-exposure-of-domestic-ruminants-to-endocrine-disrupting-compounds/0FDF5FB744EE3E7ABF6D0ED965C7EF3E.

Boggs ASP, Botteri NL, Hamlin HJ, Guillette LJ. Endocrine disruption of reproduction in reptiles. Horm Reprod Vertebr. 2011:373–96. https://www.sciencedirect.com/science/article/pii/B9780123749307100147.

Brander SM, He G, Smalling KL, Denison MS, Cherr GN. The in vivo estrogenic and in vitro anti-estrogenic activity of permethrin and bifenthrin. Environ Toxicol Chem. 2012;31(12):2848–55. https://doi.org/10.1002/etc.2019.

Braselton WE, Lin TJ, Mills TM, Ellegood JO, Mahesh VB. Identification and measurement by gas chromatography-mass spectrometry of norethindrone and metabolites in human urine and blood. J Steroid Biochem. 1977;8(1):9–18. https://www.sciencedirect.com/science/article/pii/0022473177902102.

Bull JJ. Sex determination in reptiles. Q Rev Biol. 1980;55(1):3–21. https://doi.org/10.1086/411613.

Cabeza M, Vilchis F, Lemus AE, de León LD, Pérez-Palacios G. Molecular interactions of levonorgestrel and its 5α-reduced derivative with androgen receptors in hamster flanking organs. Steroids. 1995;60(9):630–5. https://www.sciencedirect.com/science/article/pii/0039128X95000752?via%3Dihub.

Cauble K, Wagner RS. Sublethal effects of the herbicide glyphosate on amphibian metamorphosis and development. Bull Environ Contam Toxicol. 2005;75(3):429–35.

Cheek AO, Ide CF, Bollinger JE, Rider CV, McLachlan JA. Alteration of leopard frog (Rana pipiens) metamorphosis by the herbicide acetochlor. Arch Environ Contam Toxicol. 1999;37(1):70–7.

Clement RE, Reiner EJ, Bhavsar SP. Organohalogen contaminants of emerging concern in Great Lakes fish: a review. Anal Bioanal Chem. 2012;404(9):2639–58. http://www.ncbi.nlm.nih.gov/pubmed/22926127.

Coady K, Matthiessen P, Zahner HM, Staveley J, Caldwell DJ, Levine SL, et al. Endocrine disruption: where are we with hazard and risk assessment? PeerJ. 2016;4:e2580v1.

Cohen H. Effect of sex hormones on the development of the platyfish, Platypoecilus mallatus. Zool Sci Contrib N Y Zool Soc. 1946;31(8–9):121–8. http://www.ncbi.nlm.nih.gov/pubmed/20278192.

Colborn T, Vom Saal FS, Soto AM. Developmental effects of endocrine-disrupting chemicals in wildlife and humans. Environ Health Perspect. 1993;101:378–84.

The Danish Environmental Protection Agency. Feminisation of fish. Danish Environmental Protection Agency. 2002. https://www2.mst.dk/udgiv/publications/2002/87-7972-305-5/html/default_eng.htm.

De A, Bose R, Kumar A, Mozumdar S. Worldwide pesticide use; 2014. p. 5–6. https://doi.org/10.1007/978-81-322-1689-6_2.
Desbrow C, Routledge EJ, Brighty GC, Sumpter JP, Waldock M. Identification of estrogenic chemicals in STW effluent. 1. Chemical fractionation and in vitro biological screening. Environ Sci Technol. 1998;32:1549–58. https://doi.org/10.1021/es9707973.
Desforges J-P, Hall A, McConnell B, Rosing-Asvid A, Barber JL, Brownlow A, et al. Predicting global killer whale population collapse from PCB pollution. Science. 2018;361(6409):1373–6. http://science.sciencemag.org/content/361/6409/1373.
Donaldson GM, Shutt JL, Hunter P. Organochlorine contamination in bald eagle eggs and nestlings from the Canadian Great Lakes. Arch Environ Contam Toxicol. 1999;36(1):70–80. http://www.ncbi.nlm.nih.gov/pubmed/9828264.
Ebele AJ, Abou-Elwafa Abdallah M, Harrad S. Pharmaceuticals and personal care products (PPCPs) in the freshwater aquatic environment. Emerg Contam. 2017;3(1):1–16. https://www.sciencedirect.com/science/article/pii/S2405665016300488.
Eriksson H, Gustafsson J-Å. Excretion of steroid hormones in adults steroids in urine from adults. Clin Chim Acta. 1972;41:79–90. https://www.sciencedirect.com/science/article/pii/0009898172904986.
Erythropel HC, Maric M, Nicell JA, Leask RL, Yargeau V. Leaching of the plasticizer di(2-ethylhexyl)phthalate (DEHP) from plastic containers and the question of human exposure. Appl Microbiol Biotechnol. 2014;98(24):9967–81. https://doi.org/10.1007/s00253-014-6183-8.
Faust M, Kortenkamp A. State of the art assessment of endocrine disrupters. 2009.
Fleming M, Achari G, Hassan QK. Modeling the loading and fate of estrogens in wastewater treatment plants. Cogent Environ Sci. 2016;2(1):1222690. https://doi.org/10.1080/23311843.2016.1222690.
Flint S, Markle T, Thompson S, Wallace E. Bisphenol A exposure, effects, and policy: a wildlife perspective. J Environ Manag. 2012;104:19–34. https://www.sciencedirect.com/science/article/pii/S0301479712001405.
Forson D, Storfer A. Effects of atrazine and iridovirus infection on survival and life history characteristics in long-toed salamanders, Ambystoma macrodactylum. Environ Toxicol Chem. 2006;25(1):168–73.
Fossi MC, Marsili L. Effects of endocrine disruptors in aquatic mammals. Pure Appl Chem. 2007;75(11–12):2235–47. https://www.degruyter.com/view/j/pac.2003.75.issue-11-12/pac200375112235/pac200375112235.xml.
Gibbs PE, Pascoe PL, Burt GR. Sex change in the female dog-whelk, Nucella Lapillus, induced by tributyltin from antifouling paints. J Mar Biol Assoc U K. 1988;68(4):715–31. http://www.journals.cambridge.org/abstract_S0025315400028824.
Green C, Brian J, Kanda R, Scholze M, Williams R, Jobling S. Environmental concentrations of anti-androgenic pharmaceuticals do not impact sexual disruption in fish alone or in combination with steroid oestrogens. Aquat Toxicol. 2015;160:117–27. http://www.ncbi.nlm.nih.gov/pubmed/25630045.
Guillette LJ, Guillette EA. Environmental contaminants and reproductive abnormalities in wildlife: implications for public health? Toxicol Ind Health. 1996;12:537–50.
Hamlin HJ, Guillette LJ. Birth defects in wildlife: the role of environmental contaminants as inducers of reproductive and developmental dysfunction. Syst Biol Reprod Med. 2010;56(2):113–21.
Harris CA, Henttu P, Parker MG, Sumpter JP. The estrogenic activity of phthalate esters in vitro. Environ Health Perspect. 1997;105(8):802–11. http://www.ncbi.nlm.nih.gov/pubmed/9347895.
Hauser B, Schulz D, Boesch C, Deschner T. Measuring urinary testosterone levels of the great apes—problems with enzymatic hydrolysis using Helix pomatia juice. Gen Comp Endocrinol. 2008;158(1):77–86. http://linkinghub.elsevier.com/retrieve/pii/S0016648008001962.
Hayes T, Haston K, Tsui M, Hoang A, Haeffele C, Vonk A. Herbicides: feminization of male frogs in the wild. Nature. 2002;419(6910):895–6. http://www.ncbi.nlm.nih.gov/pubmed/12410298.
Hayes TB, Falso P, Gallipeau S, Stice M. The cause of global amphibian declines: a developmental endocrinologist's perspective. J Exp Biol. 2010;213(6):921–33. http://www.ncbi.nlm.nih.gov/pubmed/20190117.

Hayes TB, Anderson LL, Beasley VR, de Solla SR, Iguchi T, Ingraham H, et al. Demasculinization and feminization of male gonads by atrazine: consistent effects across vertebrate classes. J Steroid Biochem Mol Biol. 2011;127(1–2):64–73. https://www.sciencedirect.com/science/article/abs/pii/S0960076011000665.

Hites RA, Holsen TM. Temporal trends of PCBs and DDTs in Great Lakes fish compared to those in air. Sci Total Environ. 2019;646:1413–8. https://www.sciencedirect.com/science/article/pii/S0048969718328468.

Hooper MJ, Ankley GT, Cristol DA, Maryoung LA, Noyes PD, Pinkerton KE. Interactions between chemical and climate stressors: a role for mechanistic toxicology in assessing climate change risks. Environ Toxicol Chem. 2013;32(1):32–48. http://www.ncbi.nlm.nih.gov/pubmed/23136056.

Jessl L, Lenz R, Massing FG, Scheider J, Oehlmann J. Effects of estrogens and antiestrogens on gonadal sex differentiation and embryonic development in the domestic fowl (Gallus gallus domesticus). PeerJ. 2018;6:e5094. http://www.ncbi.nlm.nih.gov/pubmed/30002959.

Jobling S, Tyler CR. Endocrine disruption in wild freshwater fish. Pure Appl Chem. 2007;75 (11–12):2219–34.

Jobling S, Nolan M, Tyler CR, Brighty G, Sumpter JP. Widespread sexual disruption in wild fish. 1998. https://doi.org/10.1021/es9710870

Johnson AC, Chen Y. Does exposure to domestic wastewater effluent (including steroid estrogens) harm fish populations in the UK? Sci Total Environ. 2017;589:89–96. https://www.sciencedirect.com/science/article/pii/S0048969717303911.

Johnson AC, Sumpter JP. Are we going about chemical risk assessment for the aquatic environment the wrong way? Environ Toxicol Chem. 2016;35(7):1609–16. https://doi.org/10.1002/etc.3441.

Johnson AC, Williams RJ. A model to estimate influent and effluent concentrations of estradiol, estrone, and ethinylestradiol at sewage treatment works. 2004. https://doi.org/10.1021/es035342u

Johnson AC, Williams RJ, Matthiessen P. The potential steroid hormone contribution of farm animals to freshwaters, the United Kingdom as a case study. Sci Total Environ. 2006;362(1–3):166–78. https://www.sciencedirect.com/science/article/pii/S0048969705004523.

Kelce WR, Stone CR, Laws SC, Gray LE, Kemppainen JA, Wilson EM. Persistent DDT metabolite p,p'–DDE is a potent androgen receptor antagonist. Nature. 1995;375(6532):581–5. https://doi.org/10.1038/375581a0.

Klotz DM, Beckman BS, Hill SM, McLachlan JA, Walters MR, Arnold SF. Identification of environmental chemicals with estrogenic activity using a combination of in vitro assays. Environ Health Perspect. 1996;104(10):1084–9. http://www.ncbi.nlm.nih.gov/pubmed/8930550.

Kuhl H. Pharmacology of estrogens and progestogens: influence of different routes of administration. n.d.. http://hormonebalance.org/images/documents/Kuhl05PharmEstroProgestClimacteric_1313155660.pdf.

Kuiper GGJM, Lemmen JG, Carlsson B, Corton JC, Safe SH, van der Saag PT, et al. Interaction of estrogenic chemicals and phytoestrogens with estrogen receptor β. Endocrinology. 1998;139(10):4252–63. https://doi.org/10.1210/endo.139.10.6216.

Kumar V, Johnson AC, Trubiroha A, Tumová J, Ihara M, Grabic R, et al. The challenge presented by progestins in ecotoxicological research: a critical review. Environ Sci Technol. 2015;49(5):2625–38. http://www.ncbi.nlm.nih.gov/pubmed/25611781.

Kwon B, Ji K. Estrogenic and androgenic potential of phthalates and their alternatives. Korean J Environ Health Sci. 2016;42(3):169–88. http://koreascience.or.kr/journal/view.jsp?kj=HROSBE&py=2016&vnc=v42n3&sp=169.

Lai KM, Johnson KL, Scrimshaw MD, Lester JN. Binding of waterborne steroid estrogens to solid phases in river and estuarine systems. 2000. https://doi.org/10.1021/es9912729

Laurenson JP, Bloom RA, Page S, Sadrieh N. Ethinyl estradiol and other human pharmaceutical estrogens in the aquatic environment: a review of recent risk assessment data. AAPS J. 2014;16(2):299–310. http://www.ncbi.nlm.nih.gov/pubmed/24470211.

Lee S. Determination of estrone, 17B-estradiol and estriol in urine of young adults. All Grad Theses Diss. 1967. https://digitalcommons.usu.edu/etd/4974.

Lemus AE, Vilchis F, Damsky R, Chávez BA, García GA, Grillasca I, et al. Mechanism of action of levonorgestrel: in vitro metabolism and specific interactions with steroid receptors in target organs. J Steroid Biochem Mol Biol. 1992;41(3–8):881–90. https://www.sciencedirect.com/science/article/abs/pii/096007609290442L.

Malaj E, von der Ohe PC, Grote M, Kühne R, Mondy CP, Usseglio-Polatera P, et al. Organic chemicals jeopardize the health of freshwater ecosystems on the continental scale. Proc Natl Acad Sci. 2014;111:9549–54.

Mann RM, Hyne RV, Choung CB, Wilson SP. Amphibians and agricultural chemicals: review of the risks in a complex environment. Environ Pollut. 2009;157(11):2903–27.

Mathers MJ, Sperling H, Rübben H, Roth S. The undescended testis: diagnosis, treatment and long-term consequences. Dtsch Arztebl Int. 2009;106(33):527–32. http://www.ncbi.nlm.nih.gov/pubmed/19738919.

Matthiessen P, Gibbs PE. Critical appraisal of the evidence for tributyltin-mediated endocrine disruption in mollusks. Environ Toxicol Chem. 1998;17(1):37–43. https://doi.org/10.1002/etc.5620170106.

Matthiessen P, Wheeler JR, Weltje L. A review of the evidence for endocrine disrupting effects of current-use chemicals on wildlife populations. Crit Rev Toxicol. 2018;48(3):195–216. http://www.ncbi.nlm.nih.gov/pubmed/29171327.

Melorose J, Perroy R, Careas S. World population prospects. New York: United Nations Departments of Economic and social Affairs; 2015.

Milnes MR, Guillette LJ. Alligator Tales: new lessons about environmental contaminants from a sentinel species. Bioscience. 2008;58(11):1027–36. https://doi.org/10.1641/B581106.

Müller AK, Bosgra S, van der Voet H, Nielsen E, Ladefoged O. Probabilistic cumulative risk assessment of anti-androgenic pesticides in food. Food Chem Toxicol. 2009;47(12):2951–62. https://www.sciencedirect.com/science/article/pii/S0278691509003779.

Murphy J. Additives for plastics handbooks. New York: Elsevier Science; 2001. 469 pp.

Murray CM, Easter M, Padilla S, Garrigós DB, Stone JA, Bolaños-Montero J, et al. Cohort-dependent sex ratio biases in the American crocodiles (Crocodylus acutus) of the Tempisque Basin. Copeia. 2015;103(3):541–5.

Murray CM, Merchant M, Easter M, Padilla S, Garrigós DB, Sasa Marin M, et al. Detection of a synthetic sex steroid in the American crocodile (Crocodylus acutus): evidence for a novel environmental androgen. Chemosphere. 2017;180:125–9.

Norris DO. Endocrine disruption of reproduction in amphibians. Horm Reprod Vertebr. 2011:203–11. https://www.sciencedirect.com/science/article/pii/B9780123749314100112.

Oetken M, Bachmann J, Schulte-Oehlmann U, Oehlmann J. Evidence for endocrine disruption in invertebrates. Int Rev Cytol. 2004:1–44. http://www.ncbi.nlm.nih.gov/pubmed/15261735.

Ojoghoro JO, Chaudhary AJ, Campo P, Sumpter JP, Scrimshaw MD. Progesterone potentially degrades to potent androgens in surface waters. Sci Total Environ. 2017;579:1876–84. http://www.ncbi.nlm.nih.gov/pubmed/27939196.

Orton F, Rosivatz E, Scholze M, Kortenkamp A. Widely used pesticides with previously unknown endocrine activity revealed as in vitro antiandrogens. Environ Health Perspect. 2011;119(6):794–800. http://www.ncbi.nlm.nih.gov/pubmed/21310686.

Orton F, Säfholm M, Jansson E, Carlsson Y, Eriksson A, Fick J, et al. Exposure to an anti-androgenic herbicide negatively impacts reproductive physiology and fertility in Xenopus tropicalis. Sci Rep. 2018;8(1):9124. http://www.nature.com/articles/s41598-018-27161-2.

Parrott JL, Blunt BR. Life-cycle exposure of fathead minnows (Pimephales promelas) to an ethinylestradiol concentration below 1 ng/L reduces egg fertilization success and demasculinizes males. Environ Toxicol. 2005;20(2):131–41. https://doi.org/10.1002/tox.20087.

Pritchard G. Global demand for plasticizers continues to rise. Addit Polym. 2017;2017(10):10–1. https://www.sciencedirect.com/science/article/pii/S0306374717301379.

Raun Andersen H, Vinggaard AM, Høj Rasmussen T, Gjermandsen IM, Cecilie Bonefeld-Jørgensen E. Effects of currently used pesticides in assays for estrogenicity, androgenicity, and aromatase activity in vitro. Toxicol Appl Pharmacol. 2002;179(1):1–12. http://linkinghub.elsevier.com/retrieve/pii/S0041008X01993470.

Relyea RA, Diecks N. An unforeseen chain of events: lethal effects of pesticides on frogs at sub-lethal concentrations. Ecol Appl. 2008;18(7):1728–42. https://doi.org/10.1890/08-0454.1.

Rhind S. Are endocrine disrupting compounds a threat to farm animal health, welfare and productivity? Reprod Domest Anim. 2005;40(4):282–90. https://doi.org/10.1111/j.1439-0531.2005.00594.x.

Rhind SM, Evans NP, Bellingham M, Sharpe RM, Cotinot C, Mandon-Pepin B, et al. Effects of environmental pollutants on the reproduction and welfare of ruminants. Animal. 2010;4(07):1227–39. http://www.journals.cambridge.org/abstract_S1751731110000595.

Ropero AB, Alonso-Magdalena P, Ripoll C, Fuentes E, Nadal A. Rapid endocrine disruption: environmental estrogen actions triggered outside the nucleus. J Steroid Biochem Mol Biol. 2006;102(1–5):163–9. http://linkinghub.elsevier.com/retrieve/pii/S0960076006002640.

Rubin BS. Bisphenol A: an endocrine disruptor with widespread exposure and multiple effects. J Steroid Biochem Mol Biol. 2011;127(1–2):27–34. http://www.ncbi.nlm.nih.gov/pubmed/21605673.

Saaristo M, Tomkins P, Allinson M, Allinson G, Wong BBM. An androgenic agricultural contaminant impairs female reproductive behaviour in a freshwater fish. PLoS One. 2013;8(5):e62782. https://doi.org/10.1371/journal.pone.0062782.

Semenza JC, Tolbert PE, Rubin CH, Guillette LJ, Jackson RJ, Jackson RJ. Reproductive toxins and alligator abnormalities at Lake Apopka, Florida. Environ Health Perspect. 1997;105(10):1030–2. http://www.ncbi.nlm.nih.gov/pubmed/9349835.

Shanle EK, Xu W. Endocrine disrupting chemicals targeting estrogen receptor signaling: identification and mechanisms of action. Chem Res Toxicol. 2011;24(1):6–19. http://www.ncbi.nlm.nih.gov/pubmed/21053929.

Shi Y-B, Wong J, Puzianowska-Kuznicka M, Stolow MA. Tadpole competence and tissue-specific temporal regulation of amphibian metamorphosis: roles of thyroid hormone and its receptors. BioEssays. 1996;18(5):391–9. https://doi.org/10.1002/bies.950180509.

Silva V, Mol HGJ, Zomer P, Tienstra M, Ritsema CJ, Geissen V. Pesticide residues in European agricultural soils—a hidden reality unfolded. Sci Total Environ. 2019;653:1532–45. https://www.sciencedirect.com/science/article/pii/S0048969718343420?via%3Dihub.

Sohoni P, Sumpter JP. Several environmental oestrogens are also anti-androgens. J Endocrinol. 1998;158(3):327–39. http://www.ncbi.nlm.nih.gov/pubmed/9846162.

Solomon KR, Carr JA, Du Preez LH, Giesy JP, Kendall RJ, Smith EE, et al. Effects of atrazine on fish, amphibians, and aquatic reptiles: a critical review. Crit Rev Toxicol. 2008;38(9):721–72. https://doi.org/10.1080/10408440802116496.

Speroff L, Darney PD. A clinical guide for contraception. 2010. https://books.google.co.uk/books/about/A_Clinical_Guide_for_Contraception.html?id=f5XJtYkiJ0YC&redir_esc=y.

Stanczyk FZ, Roy S. Metabolism of levonorgestrel, norethindrone, and structurally related contraceptive steroids. Contraception. 1990;42(1):67–96. http://www.ncbi.nlm.nih.gov/pubmed/2143719.

Steimer T. Steroid hormone metabolism. n.d.. https://www.gfmer.ch/Books/Reproductive_health/Steroid_hormone_metabolism.html.

Svensson J, Fick J, Brandt I, Brunström B. Environmental concentrations of an androgenic progestin disrupts the seasonal breeding cycle in male three-spined stickleback (Gasterosteus aculeatus). Aquat Toxicol. 2014;147:84–91. https://www.sciencedirect.com/science/article/pii/S0166445X13003548.

Swanson JE, Muths E, Pierce CL, Dinsmore SJ, Vandever MW, Hladik ML, et al. Exploring the amphibian exposome in an agricultural landscape using telemetry and passive sampling. Sci Rep. 2018;8(1):10045. http://www.nature.com/articles/s41598-018-28132-3.

Teplitsky C, Piha H, Laurila A, Merilä J. Common pesticide increases costs of antipredator defenses in Rana temporaria tadpoles. Environ Sci Technol. 2005;39(16):6079–85.

Tetreault GR, Bennett CJ, Shires K, Knight B, Servos MR, McMaster ME. Intersex and reproductive impairment of wild fish exposed to multiple municipal wastewater discharges. Aquat Toxicol. 2011;104(3–4):278–90. https://www.sciencedirect.com/science/article/pii/S0166445X11001378.

Tomšíková H, Aufartová J, Solich P, Nováková L, Sosa-Ferrera Z, Santana-Rodríguez JJ. High-sensitivity analysis of female-steroid hormones in environmental samples. TrAC Trends Anal Chem. 2012;34:35–58. https://www.sciencedirect.com/science/article/abs/pii/S0165993612000374.

Tyler CR, Jobling S. Roach, sex, and gender-bending chemicals: the feminization of wild fish in English rivers. Bioscience. 2008;58(11):1051–9. https://doi.org/10.1641/B581108.

Vethaak AD, Lahr J, Schrap SM, Belfroid AC, Rijs GBJ, Gerritsen A, et al. An integrated assessment of estrogenic contamination and biological effects in the aquatic environment of The Netherlands. Chemosphere. 2005;59(4):511–24. https://www.sciencedirect.com/science/article/pii/S0045653505000639.

Wade MG, Desaulniers D, Leingartner K, Warren GF. Interactions between endosulfan and dieldrin on estrogen-mediated processes in vitro and in vivo. Reprod Toxicol. 1997;11(6):791–8. https://www.sciencedirect.com/science/article/pii/S0890623897000622.

Wibbels T, Gideon P, Bull JJ, Crews D. Estrogen- and temperature-induced medullary cord regression during gonadal differentiation in a turtle. Differentiation. 1993;53(3):149–54. https://www.sciencedirect.com/science/article/pii/S0301468111600911.

Williams RJ, Johnson AC, Smith JJL, Kanda R. Steroid estrogens profiles along river stretches arising from sewage treatment works discharges. Environ Sci Technol. 2003;37(9):1744–50.

Wright-Walters M, Volz C, Talbott E, Davis D. An updated weight of evidence approach to the aquatic hazard assessment of bisphenol A and the derivation a new predicted no effect concentration (Pnec) using a non-parametric methodology. Sci Total Environ. 2011;409(4):676–85. http://www.ncbi.nlm.nih.gov/pubmed/21130487.

Wuttke W, Jarry H, Seidlova-Wuttke D. Definition, classification and mechanism of action of endocrine disrupting chemicals. Hormones (Athens). 2010;9(1):9–15. http://www.ncbi.nlm.nih.gov/pubmed/20363717.

Yang O, Kim HL, Weon J-I, Seo YR. Endocrine-disrupting chemicals: review of toxicological mechanisms using molecular pathway analysis. J Cancer Prev. 2015;20(1):12–24. http://www.ncbi.nlm.nih.gov/pubmed/25853100.

Chapter 4
Transgenerational Impact of Environmental Change

Hannah L. Morgan and Adam J. Watkins

Abstract The ability to adapt to changing environmental conditions is critical for any species to survive. Many environmental changes occur too rapidly for an organism's genome to adapt in time. Accordingly, being able to modify either its own phenotype, or the phenotype of its offspring to better suit future anticipated environmental conditions could afford an organism a significant advantage. However, a range of animal models and human epidemiological data sets are now showing that environmental factors such as changes in the quality or quantity of an individual's diet, temperature, stress or exposure to pollutants can all adversely affect the quality of parental gametes, the development of the preimplantation embryo and the health and wellbeing of offspring over multiple generations. This chapter will examine transgenerational effects of both maternal and paternal environmental factors on offspring development and wellbeing in both human and animal model studies. Changes in the epigenetic status of either parental or grand-parental gametes provide one candidate mechanism through which the impacts of environmental experience can be passed from one generation to another. This chapter will therefore also focus on the impact of parental and grand-parental diet on epigenetic transgenerational inheritance and offspring phenotype.

1 Introduction

In response to changing environmental conditions, many organisms are able to survive through strategies such as migration in order redistribute their populations, modifying their behaviour to accommodate the new conditions or through biological adaptations such as phenotypic plasticity. As the evolution of an organism's genome will not occur rapidly enough to accommodate immediate changes in

H. L. Morgan · A. J. Watkins (✉)
Division of Child Health, Obstetrics and Gynaecology, Faculty of Medicine, University of Nottingham, Nottingham, UK
e-mail: adam.watkins@nottingham.ac.uk

P. Comizzoli et al. (eds.), *Reproductive Sciences in Animal Conservation*, Advances in Experimental Medicine and Biology 1200,
https://doi.org/10.1007/978-3-030-23633-5_4

environmental conditions within one generation, phenotypic plasticity affords a mechanism by which organisms can adapt to rapidly-changing environments (Gienapp et al. 2008). Studies in both humans and animal models have shown that the impact of changing parental environmental factors can extend beyond the generation initially exposed, influencing the development and wellbeing of subsequent generations through a range of genetic and epigenetic mechanisms (Desai et al. 2015). Typically, these studies have focused on the quality and quantity of maternal nutrition during defined time frames around conception, during gestation and/or during early neonatal life (Sinclair and Watkins 2014). However, there is now a significant body of data showing that paternal lifestyle and environmental factors also affect the development and wellbeing of a male's offspring (Fleming et al. 2018).

The concept that our health as adults may be influenced by the wellbeing of our parents prior to our conception is now established, but the precise mechanisms linking changing parental environmental factors to impaired offspring health remain to be fully defined. The Developmental Origins of Health and Disease (DOHaD) hypothesis outlines how changes in parental environment impact on gamete maturation, early embryo development, fetal growth and adult offspring heath (Hanson and Gluckman 2014). These sensitive reproductive and developmental stages are characterised by significant changes in cell fate, epigenetic status and metabolism all of which can be modified by changes to 'normal' environmental cues, and which may be perpetuated into adult somatic or reproductive tissues (Fleming et al. 2015). The DOHaD hypothesis, and the biological processes underlying it, hence provide a mechanism by which changing environmental conditions for one generation may alter, either beneficially or negatively, the phenotype and wellbeing of subsequent generations.

Research in both humans and animal models has shown that even minor modifications of environmental set points—often within a normal range such that they do not impair the reproductive success of a species—are capable of driving adaptations in the developing offspring in an attempt to alter phenotypes to align with the potential future environment (Gluckman and Hanson 2004). These environmental influences can include a number of factors, such as diet (mostly excess or restriction of key nutrients), extreme temperature fluctuations, poor air quality (pollution, smoking etc.) and stress. Furthermore, these environmental influences can be intertwined with one another, for example with warmer more volatile climate events such as extended heatwaves, being increasingly more common (Christidis et al. 2015). Plant growth and thus food availability is being impacted, altering diets or restricting key nutrients for many species (Battisti and Naylor 2009).

In this review, we will highlight a range of human and animal model systems in which changing parental environmental factors impact on offspring development and wellbeing over multiple generations. Due to the exceedingly large number and variability of potential environmental insults, this chapter will focus predominantly on nutritional impairments. Studies exploring the impact of manipulating parental diet on offspring health have formed the foundation on which our understanding of DOHaD mechanisms has been explored. However, DOHaD studies have now expanded to include a range of different environmental conditions, providing additional insight into how parental wellbeing can influence offspring health over multiple generations (Hoffman et al. 2017). Initially, our review will explore the DOHaD hypothesis, outlining the state of our current understanding of the impact of changes in parental environment on offspring wellbeing across multiple generations. From here, we define how epigenetic mechanisms link parental environment to gamete quality and transgenerational offspring development. Finally, we outline future perspectives and priorities for offspring health and wellbeing in an ever changing global environment. The use of animal models has permitted detailed investigation into the underlying biological and molecular mechanisms (Heindel 2018). As such, a diverse range of non-human models have been developed with which the impact of environmental change on future generations can be defined. Therefore, throughout this review we will make reference to both human and non-human studies, providing a comprehensive overview of the transgenerational impact of environmental change.

2 Defining Transgenerational Programming

Prior to reviewing the literature regarding the transgenerational programming effects of altered parental environment, it is first appropriate to define our use of the term 'transgenerational programming'. For our review, we defined the original generation that was exposed to the environmental insult as the F0 (grand-parental) generation. We define the first generation of offspring born as the F1 generation and their offspring (the grand-offspring of the F0 generation) as the F2 generation. For some researchers, for a maternal environmental effect to be deemed to act transgenerationally, it must be transmitted into the F3 generation (Jirtle and Skinner 2007). The rationale here is that if the environment of the mother is altered during gestation, then the offspring of the F1 generation will be impacted directly as they develop *in utero*. However, the F2 generation cannot be correctly termed to have been influenced transgenerationally because the germ cells (which will eventually yield gametes) will be developing within the F1 fetus during gestation. The original maternal environmental challenge will potentially impact both F1 and F2 generations at the same time (see Fig. 4.1). In this scenario, it is only the F3 generation (great grand-offspring) and beyond that can be considered to be programmed transgenerationally. However, there are comparatively few studies that have investigated the impacts of parental F0 environmental factors into an F3 generation. Furthermore, if the mother is exposed to an environmental insult prior to conception, then it can be argued that changes in the F2 generation will have been programmed transgenerationally. In addition, this is also true for the environmental exposure of the father, as in this scenario only the F0 paternal gametes have been impacted, while the F1 generation has been exposed to the direct insult, the F2 generation has not. There is hence a greater body of phenotypic and mechanistic data defining the impacts of parental environmental manipulation of F2 generation phenotype.

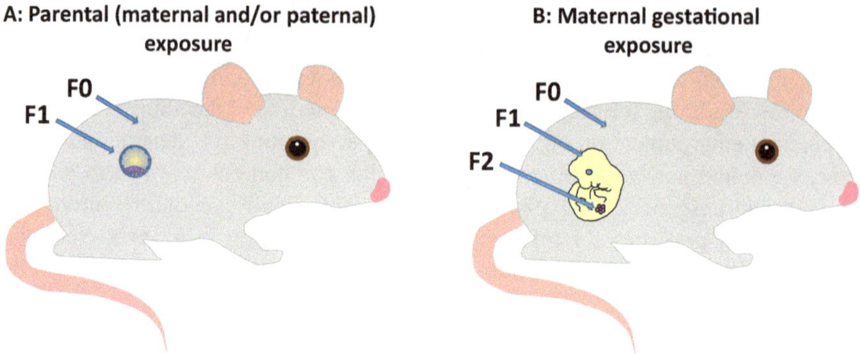

Fig. 4.1 Impacts of environmental change on offspring phenotype across multiple generations. (**a**) Poor parental environmental conditions which affect either the quality of the parental gametes or preimplantation embryo (pictured) will impact only on the immediate F1 generation they generate (intergenerational programming). Any phenotypic changes established within the F2 generation following manipulation of the F0 gametes or early embryo will have occurred via transgenerational inheritance. (**b**) In contrast, poor maternal (F0) environment during gestation can impact on both her gestational wellbeing and the development of her offspring (F1 generation) as well as the germ cells within those developing F1 offspring that will ultimately give rise to the F2 generation. As both the F1 offspring and their developing germ cells are exposed to the programming impact of the poor maternal environment, changes in F1 and/or F2 offspring phenotype are describes as being intergenerational. As the cells for the F3 generation are not yet present, any change in the phenotype of the F3 generation have occurred via transgenerational inheritance

3 State of the Art: Developmental Origins of Health and Disease (DOHaD)

3.1 Human Observational Studies

Over the past three decades, the DOHaD hypothesis has expanded greatly from its original focus on fetal growth restriction, weight at birth and the increased risks for coronary artery disease in later life (Barker et al. 1989). Original observations from Professor David Barker and colleagues identified strong associations between reduced weight at birth and an increasing risk of developing cardiovascular disease in adulthood (Barker et al. 1989; Barker and Osmond 1986). Such associations were validated further following analysis of epidemiological data sets from discrete periods of famine during the twentieth century such as the Dutch hunger winter (Painter et al. 2008). Here, during the winter of 1944–1945, individuals were exposed to a short, but significant period of nutritional deprivation. Offspring from women exposed to the famine during the first trimester of gestation displayed increased propensity to developing obesity, coronary heart disease and hyperlipidaemia in adulthood (Roseboom et al. 2000a, b). In contrast, offspring from women experiencing the famine during later gestation displayed impaired glucose tolerance as adults (Ravelli et al. 1998). Recently, links between paternal expose to the famine and elevated grand-offspring BMI have also been reported (Veenendaal et al. 2013).

Initially, support for the DOHaD hypothesis was limited due to the widely held belief that health in adulthood was predominantly a consequence of lifestyle in adulthood. The notion that our adult health could be programmed by factors occurring decades earlier was difficult for many to comprehend. However, since the reporting of the original epidemiological observations by Barker and colleagues, the field of DOHaD research has expanded dramatically to include a diverse range of human epidemiological data sets and animal models. Underlying the acceptance of the DOHaD hypothesis were some of the original animal studies into maternal gestational diet in models such as the rat (Langley and Jackson 1994), sheep (Osgerby et al. 2002) and baboon (Cox et al. 2006). These, and subsequent studies, have shown that the programming of offspring health can occur during many stages of the reproductive timeline, e.g. prior to, during and after conception, altering the developmental trajectory of the offspring and increasing the propensity later in life to developing non-communicable diseases such as obesity, cardiovascular disease and type 2 diabetes (Hanson and Gluckman 2008). As the DOHaD concept, that changes in the maternal gestational environment could significantly impact the development and wellbeing of her offspring, became established within mainstream research, questions regarding the persistence across multiple generations of such programming effects emerged (Hoffman et al. 2017). Such questions have been reinforced by analysis of data from the Överkalix study in which associations between grand-parental pre-pubertal nutrition and the health and wellbeing of their grand-children were identified. Here, an inverse relationship between the nutritional access of a grandfather and the mortality risk for heart disease in his grandson has been reported (Pembrey et al. 2014). By contrast, other large cohort data studies have not observed significant associations in BMI between fathers and offspring over three generations (Murrin et al. 2012), although strong maternal-offspring relationships were reported from the Lifeways data. These observations are in line with other reports in which adult grand-children of mothers exposed to the Dutch hunger winter have higher reported incidences of poor health from any cause (Painter et al. 2008). Other human cohort data sets also document effects of maternal undernutrition which can persist across multiple generations. These include cognitive impairment in grand-children of women exposed to famine in China between 1959–1963 (Li et al. 2015) and reduced birth weight of grandchildren whose grandmothers fasted for Ramadan during pregnancy (Alwasel et al. 2013).

3.2 *Animal Model Studies*

Animal models have become an essential tool with which the impacts of environmental change on offspring development and wellbeing can be studied (Segovia et al. 2015). Due to their short reproductive cycle and ease of use, rodent models have come to form a substantial base on which many programming studies are conducted (Aiken and Ozanne 2014). As in human epidemiological studies, the initial focus of many animal DOHaD studies centred on the impact of maternal undernutrition during specific stages of development and gestation. However, there is now a wide range of both maternal and paternal studies examining the impacts of factors such as over/undernutrition (Aiken and Ozanne 2014; Radford et al. 2014), stress (Nestler 2016) and age (Bloch Qazi et al. 2017). Furthermore, while rodent models still represent a majority of developmental programming studies, other models such as sheep (Sinclair et al. 2007), rabbit (Rousseau-Ralliard et al. 2019), seahorse (Otero-Ferrer et al. 2016) and even fruit flies (Ost et al. 2014) have helped to define the phenotypic impacts and underlying mechanisms linking parental diet with offspring ill-health.

The fact that diet of one generation can impact the wellbeing of future generations has major implications for farming and conservation (Murdoch et al. 2016; Feeney et al. 2014). However, due to the high cost and long generation times for domestic animal species, transgenerational studies in larger mammals are noticeably scarce. In pigs, one study has reported transgenerational inheritance of carcass traits following the feeding of F0 males diets high in B vitamins and supplements like folate (Braunschweig et al. 2012). Here, the authors observed significant reductions in F2 generation carcass fat content and increased shoulder muscle mass when compared to offspring from control diet fed males. For poultry, one review article has suggested that studies may demonstrate transgenerational inheritance of traits, though the precise mechanisms are often not defined (Berghof et al. 2013).

In rodents, transgenerational developmental programming, which is expressed as phenotypic changes within the F1 offspring, has been demonstrated to be impacted by both maternal and paternal diet (Benyshek et al. 2006; Frantz et al. 2011; Fullston et al. 2013). In response to maternal diet high in fat and sugar diet fed from before conception through to weaning, Saben et al., observed significant impairments in insulin signalling and mitochondrial function in offspring for up to three generations (Saben et al. 2016). Interestingly, in some mouse models, while F2 generation offspring of dams fed a high fat diet displayed reduced insulin sensitivity, F3 offspring appeared normal (Dunn and Bale 2011). Furthermore, paternal obesity in mice has been found to induce transgenerational changes in blood glucose and insulin levels of offspring as well as changes in the liver triglyceride and diaglyceride profiles when challenged with a Western diet for up to two generations (Cropley et al. 2016). In fruit flies, changes in dietary protein:carbohydrate ratios given to F0 males resulted in significantly increased longevity of offspring for up to three generations (Xia and de Belle 2016), while in fish (guppy), transgenerational programming of F2 offspring stress responses have been reported in response to environmental estrogenic contaminants (Volkova et al. 2015).

4 Sensitivity of Parental Gametes and Embryos

Changes to the quality of parental gametes provide one biological mechanism through which parental environmental factors may be transmitted to future generations. If these changes persist into the developing primordial germ cells of the F1 offspring during development, they may result in the transmission of programmed traits and phenotypes into the F2 generation and beyond.

Studies have shown that the maturing oocyte appears sensitive to changes in maternal nutritional status that can impact long-term health of offspring (Ashworth et al. 2009). In livestock (cows, sheep, pigs) poor maternal nutritional status results in reduced oocyte production rate and negatively impacts on quality (Ashworth et al. 2009). Moreover, in ewes, reduction in feed consumption for a brief period of 2 weeks resulted in oocyte gene expression changes, with proposed effects on embryo development through defects in various transporters (Pisani et al. 2008; Schmidt et al. 2009). Furthermore, environmental factors such as increased external temperature during El Niño periods have been correlated with poorer oocyte quality in South American cows (Bényei et al. 2003). In mice, maternal low protein diet fed to females exclusively during the window of oocyte maturation (3.5 days prior to mating) resulted in significant impairments in offspring cardiovascular function and altered behavioural profiles (Watkins et al. 2008).

It is becoming increasingly apparent that paternal diet and other environmental factors also affect sperm quality and influence offspring development and wellbeing similar in to that of maternal health (Fleming et al. 2018). In rodents, paternal protein restriction has been shown to blunt blastocyst metabolic gene expression, enhance fetal growth and skeletal formation, elevate adult adiposity and impair cardio-metabolic function in adult offspring (Watkins and Sinclair 2014; Watkins et al. 2017). Studies have also shown that F1 offspring of male mice fed a low folate diet had a higher incidence of skeletal malformations during development and significant changes in placental gene expression (Lambrot et al. 2013). Separately, rodent studies have also shown that cognitive function as well as anxiety- and depression-related behaviours in offspring can be programmed in response to either a folate-deficient diet (McCoy et al. 2018) or paternal obesity (Zhou et al. 2018). Paternal obesity has long been associated with reduced sperm quality and quantity and has been linked to fertility problems (Kort et al. 2006). However, more recently paternal obesity in mice has been found to induce transgenerational changes. Cropley et al. showed that a non-genetic inheritance influenced male F2 generation blood glucose and insulin levels as well as changes in the liver triglyceride and diglyceride profiles when challenged with a Western diet (Cropley et al. 2016). Interestingly, paternal dietary restriction programmed elevated adiposity and dyslipidaemia in adult offspring (McPherson et al. 2016). It therefore appears, at least in rodents, that paternal over- and undernutrition have similar phenotypic effects on offspring cardio-metabolic health. As discussed earlier, detailed studies of developmental programming in cattle and other domestic species are few, highlighting the need for additional studies into the impact of paternal diet on offspring development. However, some studies have already shown that sperm quality in pigs can be increased through the provision of n-3 and n-6 polyunsaturated fatty acids in the diet (Rooke et al. 2001). Similar results have also been shown in sperm of bulls (Moallem et al. 2015) and rams (Jafaroghli et al. 2014).

An additional focus on the role of a father in the programming of his offspring's health has come from the influence of the seminal plasma on the maternal uterine environment. The maternal uterine environment at the time of conception can influence the embryo's susceptibility to disease, its growth trajectory and metabolic phenotypes in later life (Bromfield et al. 2014). The father plays a key role in modulating the uterine environment via seminal fluid secretions (Peitz and Olds-Clarke 1986; Watkins et al. 2018). In this way the uterine environment is not just directly altered by the mother but can also receive information about the external environment from the father. Seminal fluid composition can be altered by male physiology (Camargo et al. 2018) and it plays a vital role in regulating sperm function and fertility (Caballero et al. 2012). In mice, absence of the seminal fluid at the time of mating impairs embryo development and blastocyst cell number as well as increasing adiposity and glucose intolerance in adult offspring. Our own mouse studies have also shown that offspring growth and metabolic health can be affected in a manner specific to sperm and seminal fluid. Here, we used artificial insemination in combination with vasectomised male mating to generate offspring derived from sperm and/or seminal plasma from males fed a control or low protein diet (Watkins et al. 2018). Interestingly, we observed the biggest impacts on offspring metabolic health occurred when the dietary background of the sperm mismatched that of the seminal plasma indicating that either the sperm or seminal fluid was sufficient to influence offspring health (Watkins et al. 2018). Apart from mice, seminal plasma has also been shown to have significant roles in regulating reproduction in domestic mammal species (Bromfield 2016) and humans (Robertson and Sharkey 2016). Interestingly, in women, there is some evidence that the duration of unprotected exposure to a partner's seminal plasma is inversely associated with risk of her developing preeclampsia during pregnancy (Klonoff-Cohen et al. 1989; Robillard et al. 1994).

5 Epigenetic Mechanisms

With evidence that transgenerational programming spans a wide range of species including mammals (Aiken et al. 2016) fish (Knecht et al. 2017), fruit flies (Brookheart and Duncan 2016) and nematode worms (Klosin et al. 2017), a universal mechanism for the transmission of environmental influences seems likely. Underlying transgenerational programming are mechanisms including insulin resistance, endoplasmic reticulum stress, altered tissue structure/function and immune system disruption. However, as the factors that can influence future generations of offspring health do not tend to alter DNA sequences, the majority exert their impact on the epigenome (Daxinger and Whitelaw 2010). Transgenerational epigenetic inheritance may provide one potential mechanism for the transfer of information from parents to their offspring regarding the environment they experienced. In other words, these mechanisms exist for organisms to "inform" their progeny about prevailing environmental conditions, thus preparing them for what they will experience in due course (Aiken and Ozanne 2014). The transgenerational inheritance of epigenetic marks was first realised with the viable yellow agouti gene in mice. This gene encodes a signalling factor that promotes the change in follicular melanocyte pigment production (from black eumelanin to yellow phaeomelanin). This results in patterns of coat colour that respond to the epigenetic status of the promoter of this gene (Duhl et al. 1994). This epigenetic status can be inherited, and furthermore altered by diet. Maternal dietary supplementation with methyl-donors, such as vitamin B12 and folate, can alter the coat-colour distribution in their offspring. This shows not only that diet can impact epigenetic status but that this it can impact the development of offspring (Cooney et al. 2002).

Epigenetics is the regulation of gene expression without changing the genomic DNA sequence. These modifiable signals include patterns of DNA methylation, non-coding RNAs, RNA methylation, histone variants and histone post translational modifications (Schübeler 2015; Buschbeck and Hake 2017; Bannister and Kouzarides 2011; Ponting et al. 2009). These mechanisms regulate gene expression and play key roles in developmental processes (Heard and Martienssen 2014). The epigenetic marks altered in response to the parental environment can be inherited by the offspring, persist into adulthood and be passed on to subsequent generations (Cavalli and Paro 1998; Lang-Mladek et al. 2010). In the context of reproduction, these changes can be passed to the offspring so that their development is altered to better fit the environment it will be born into (Gluckman and Hanson 2004). There is, however, a large body of evidence that disturbances to epigenetic regulators are linked to the development of various disorders (Dawson and Kouzarides 2012; Lee et al. 2014). Epigenetic inheritance is termed transgenerational if the epigenetic information originates in the parental germ line and any observable change across generations occurs in the absence of any direct exposure to the same environmental stimuli that caused the original modifications and is independent of the induced phenotype, such that the programmed phenotype is inherited (Nilsson et al. 2018).

DNA methylation is the most thoroughly studied of the epigenetic mechanisms to date, yet it is only one of several epigenetic mechanisms that can regulate chromatin state, gene expression profiles and tissue function. DNA methylation involves methyl groups being added to the DNA, usually at a cytosine base and predominantly at sites where a cytosine nucleotide is followed by a guanine nucleotide (termed CpG sites) (Nilsson et al. 2018; Bird 2002). DNA methylation status has been linked to nutritional status in many animals (Liu et al. 2003). Many animal studies have been able to manipulate this mechanism of epigenetic modification in dietary studies, with dietary supplementations or restrictions of methyl donor nutrients, such as choline, vitamin B12, folate and methionine (Sinclair et al. 2007; Cooney et al. 2002; Wolff et al. 1998).

The oocyte, sperm and embryo are subjected to DNA methylation erasure, that is a removal of the epigenetic marks, twice during early development; once just post-fertilisation and again in the primordial germ cells of the embryo (Heard and Martienssen 2014). This demethylation is followed by periods of remethylation and there is increasing evidence that there is a 'memory' of parental epigenetic marks and that these can be re-instated. Evidence for these retained epigenetic marks comes from studies on paternally imprinted genes, such as genes in specific histone modification areas, that are protected from the global demethylation and is thoroughly reviewed by Messerschmidt (2012) and Cantone and Fisher (2013).

Paternal environment can also influence the epigenetic status of the sperm and subsequently affect the development of the preimplantation embryo (Colaco and Sakkas 2018). The best explanation of how these detrimental offspring changes occur is that paternal epigenetic marks are altered by the male diet during spermatogenesis and gamete maturation and passed into the oocyte at fertilisation. These can include changes to the DNA methylation status of sperm (Watkins et al. 2018), RNA content (Chen et al. 2016a) and histone modifications (Hammoud et al. 2009). Even though sperm epigenetic markers undergo significant erasure during early embryo development, there is evidence of modifications persisting because sperm methylation status has been shown to be heritable (Molaro et al. 2011). The DNA within the sperm is bound mainly to protamines, instead of histones as in somatic cells, yet there is evidence that the retained 1–15% of histones can carry epigenetic marks which can be incorporated into the blastocyst at fertilisation (Donkin and Barres 2018). Sperm have been shown to contain a number of small non-coding RNAs (sncRNA), which are incorporated in the embryo, in addition to the nuclear genomic material (Gapp et al. 2014). Rodgers et al. showed that microRNA present in the sperm of paternal mice are increased in response to stress and that it was these modifications that drove changes observed in the offspring's hypothalamic–pituitary–adrenal responses (Rodgers et al. 2013, 2015). Another population of sncRNAs that can impact epigenetic transmission of information between generations is transfer RNAs (tRNAs) (Chen et al. 2016b). These tRNAs can be altered in response to agricultural fungicides, such as vinclozolin, which have been shown to induce disease phenotypes in F3 and F4 generations of rats (Schuster et al. 2016).

These environmentally-driven modifications that can be inherited through several generations has been hypothesised to be a mechanism to inform offspring about the environment they will encounter and thus prepare them with to respond appropriately. Therefore, it stands to reason that negative impacts of epigenetic inheritance can be increased when there is a mismatch between the perceived environment of the mother and father, for which there are epigenetically driven modifications, and the actuality of the environment encountered by offspring as adults (Godfrey et al. 2007). This has a tremendous importance in relation to the current climate change, which is occurring at alarming rates, exposing different generations to vastly different environments.

6 Priorities for the Future

There is now compelling data from a range of studies demonstrating that changes in metabolic and cardiovascular health of offspring can be transmitted across multiple generations (See Fig. 4.2). While these associations are now well-established, the focus of future studies will need to define the precise biological and molecular mechanisms underlying transgenerational programming of offspring health. While the use of rodent models has provided significant insight into the influence of environmental factors on parental gametes, embryos and adult offspring, observations

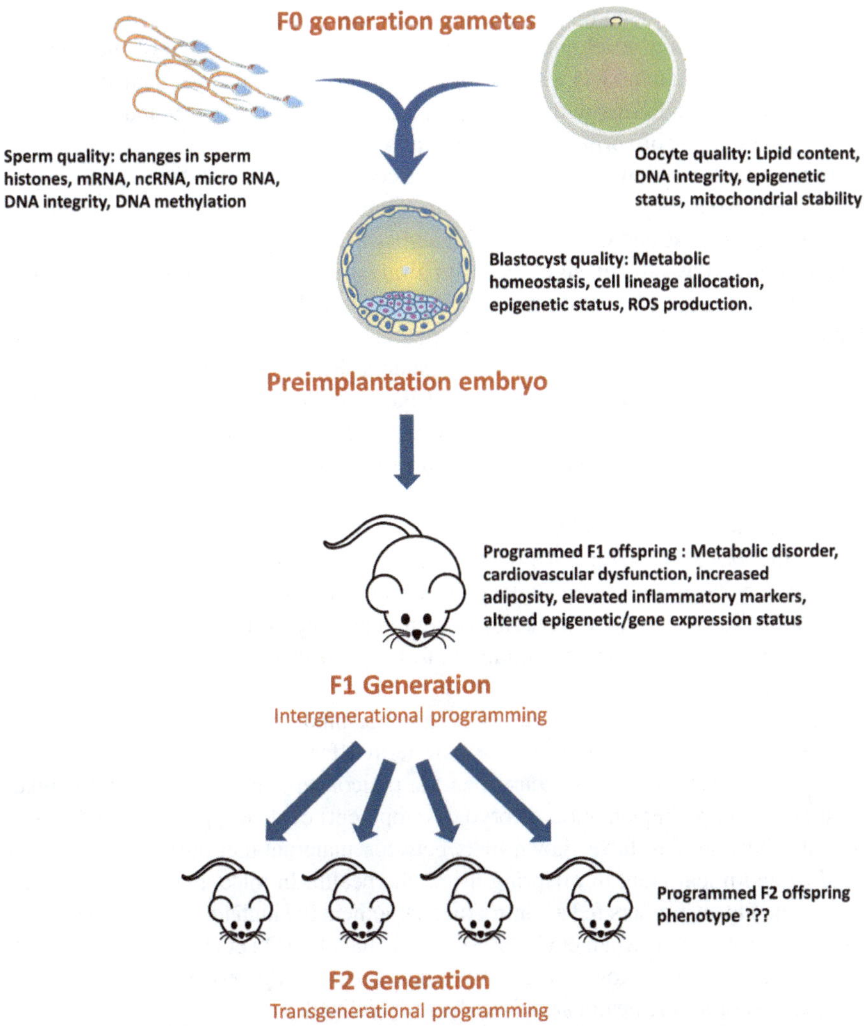

Fig. 4.2 Linking parental environmental change with gamete quality and transgenerational programming. Changes in F0 gamete quality and/or epigenetic status can influence the development of the preimplantation embryo including changes in metabolic, cell lineage and or epigenetic/gene expression status. Changes initiated within the preimplantation embryo, and persisting into post-implantation tissues can impact on adult offspring phenotype predisposing them to greater risk of developing cardio-metabolic disorders. Changes in F1 offspring phenotype are typically viewed as intergenerational effects, depending on whether the developing germ cells of the F1 offspring were exposed to the programming event (i.e. poor maternal diet during gestation) or not (paternal programming). Phenotypic traits may be passed onto an F2 generation, representing transgenerational programming

from rodent studies cannot always be extrapolated directly into other species. Therefore, exploring developmental programming in a wide range of mammalian and non-mammalian models will permit greater insight into the mechanisms linking environmental conditions with changes in parental gamete quality and (inter- and trans-generational) offspring development and wellbeing.

In this current review, while our focus has centred on parental diet, it is clear that there are a wealth of other environmental factors which can impact on parental reproductive fitness and offspring health across multiple generations. For example, several studies have shown significant impairment in both sperm and oocyte quality in response to environmental toxicants (Lea et al. 2016a, b; Maurice et al. 2018). Many environmental pollutants such as phthalates and bisphenol A (BPA) act as endocrine disruptors, affecting either the parents (Rattan et al. 2017; Rehman et al. 2018) or the offspring during gestation (Walker and Gore 2017; Zama and Uzumcu 2010). Similar to the effects of parental diet, studies have highlighted early gestational stages as being particularly susceptible to influences of environmental disruptors which can have major and far-reaching effects at extremely low doses and across multiple generations (Xin et al. 2015). Furthermore, as global rates of urbanisation increase, the effects of poor air quality on fundamental aspect of human reproduction has also become a new focus of research in recent years. Indeed, studies in both mice and humans have shown significant impairments in multiple aspects of fundamental fertility and reproduction following exposure to polluted air (Carre et al. 2017).

Another area of significant interest is in understanding the influence of environmental factors during perinatal life for long-term offspring health (Sarkar and Baffy 2018). While our review has focused on the periconception period (around gamete maturation and preimplantation embryo development) evidence, predominantly from animal model studies, have shown links between maternal diet during the perinatal window and impairment in offspring metabolic health. In mice, pups cross-fostered and weaned from mothers fed a control diet to mothers fed high fat diet develop symptoms of hypertension (Khan et al. 2005), type 2 diabetes (Du et al. 2015) and obesity (Tsuduki et al. 2013). However, the mechanism underlying perinatal environmental changes and offspring health are still to be fully defined.

A growing area of interest is the impact of paternal wellbeing on the health of his offspring. Understandably, the majority of DOHaD studies have focused on the influence a mother's wellbeing exerts on development and health of her offspring. However, over the past decade it has become increasingly evident that paternal lifestyle and physiology can significantly impact on semen quality, post-fertilisation development and long-term health of offspring (Lucas and Watkins 2017). Here, as in maternal programming studies, epigenetic factors appear central in linking poor paternal wellbeing with offspring ill-health. Similar to the oocyte, sperm quality appears sensitive to a range of paternal factors including diet (Schagdarsurengin and Steger 2016), pollutants (Mima et al. 2018), ozone (Sokol et al. 2006) and age (Jenkins et al. 2015), and it is able to program offspring phenotype across multiple generations (Pentinat et al. 2010). In addition, we are beginning to understand the complexity of paternal programming with studies showing both the sperm and the seminal plasma play significant roles in programming offspring health (Watkins et al. 2018; Robertson et al. 2015).

A final brief consideration should be given to the increased use of Assisted Reproductive Technologies (ART) in the field of human reproduction (Fleming et al. 2018) and animal conservation (Paulson and Comizzoli 2018). Here, the environment of the gametes and embryo during routine procedures such as in vitro fertilisation (IVF) differs considerably from that during natural conception. This exposure may have the potential to alter the blastocysts development trajectory and eventually lead to negative impacts on the health of the offspring in adult life (Rexhaj et al. 2013). Studies analysing mice conceived using IVF show that offspring display characteristics of hypertension and diabetes; with elevated blood pressure (Rexhaj et al. 2013), impaired glucose tolerance and insulin signalling (Chen et al. 2014) and also dysregulated gene expression in their liver, pancreas, muscles and adipose tissues (Feuer et al. 2014). Some studies also report similar changes in the growth (Dumoulin et al. 2010; Zandstra et al. 2018) and cardiometabolic health of ART children (Meister et al. 2018; Yeung and Druschel 2013), however the number of children analysed in such studies are comparatively low and further investigation on larger cohorts is essential.

Going forward, it is clear that reproductive fitness of both males and females are directly connected with and influenced by their immediate environment. As many species face rapid and significant environmental change, it should come as no surprise that a wealth of studies demonstrate that a change in the environment directly impacts sperm and oocyte quality and the wellbeing of the resultant offspring. It is therefore, essential that a focus on the long-term implications of parental environmental change for offspring health, across a wide range of species, be maintained to protect the health and well-being of future generations.

Acknowledgements Dr Watkins and his research are supported by the Biotechnology and Biological Sciences Research Council (BBSRC) under grant number BB/R003556/1.

References

Aiken CE, Ozanne SE. Transgenerational developmental programming. Hum Reprod Update. 2014;20(1):63–75.

Aiken CE, Tarry-Adkins JL, Ozanne SE. Transgenerational effects of maternal diet on metabolic and reproductive ageing. Mamm Genome. 2016;27(7–8):430–9.

Alwasel SH, Harrath A, Aljarallah JS, Abotalib Z, Osmond C, Al Omar SY, et al. Intergenerational effects of in utero exposure to Ramadan in Tunisia. Am J Hum Biol. 2013;25(3):341–3.

Ashworth CJ, Toma LM, Hunter MG. Nutritional effects on oocyte and embryo development in mammals: implications for reproductive efficiency and environmental sustainability. Philos Trans R Soc Lond B Biol Sci. 2009;364(1534):3351–61.

Bannister AJ, Kouzarides T. Regulation of chromatin by histone modifications. Cell Res. 2011; 21(3):381–95.

Barker DJ, Osmond C. Infant mortality, childhood nutrition, and ischaemic heart disease in England and Wales. Lancet. 1986;1(8489):1077–81.

Barker DJ, Osmond C, Golding J, Kuh D, Wadsworth ME. Growth in utero, blood pressure in childhood and adult life, and mortality from cardiovascular disease. BMJ. 1989;298(6673):564–7.
Battisti DS, Naylor RL. Historical warnings of future food insecurity with unprecedented seasonal heat. Science. 2009;323(5911):240–4.
Bényei B, Gáspárdy A, Cseh S. Effect of the El Niño phenomenon on the ovarian responsiveness and embryo production of donor cows. Acta Vet Hung. 2003;51(2):209–18.
Benyshek DC, Johnston CS, Martin JF. Glucose metabolism is altered in the adequately-nourished grand-offspring (F-3 generation) of rats malnourished during gestation and perinatal life. Diabetologia. 2006;49(5):1117–9.
Berghof TV, Parmentier HK, Lammers A. Transgenerational epigenetic effects on innate immunity in broilers: an underestimated field to be explored? Poult Sci. 2013;92(11):2904–13.
Bird A. DNA methylation patterns and epigenetic memory. Genes Dev. 2002;16(1):6–21.
Bloch Qazi MC, Miller PB, Poeschel PM, Phan MH, Thayer JL, Medrano CL. Transgenerational effects of maternal and grandmaternal age on offspring viability and performance in Drosophila melanogaster. J Insect Physiol. 2017;100:43–52.
Braunschweig M, Jagannathan V, Gutzwiller A, Bee G. Investigations on transgenerational epigenetic response down the male line in F2 pigs. PLoS One. 2012;7(2):e30583.
Bromfield JJ. A role for seminal plasma in modulating pregnancy outcomes in domestic species. Reproduction. 2016;152(6):R223–R32.
Bromfield JJ, Schjenken JE, Chin PY, Care AS, Jasper MJ, Robertson SA. Maternal tract factors contribute to paternal seminal fluid impact on metabolic phenotype in offspring. Proc Natl Acad Sci U S A. 2014;111(6):2200–5.
Brookheart RT, Duncan JG. Drosophila melanogaster: an emerging model of transgenerational effects of maternal obesity. Mol Cell Endocrinol. 2016;435:20–8.
Buschbeck M, Hake SB. Variants of core histones and their roles in cell fate decisions, development and cancer. Nat Rev Mol Cell Biol. 2017;18(5):299–314.
Caballero I, Parrilla I, Alminana C, del Olmo D, Roca J, Martinez EA, et al. Seminal plasma proteins as modulators of the sperm function and their application in sperm biotechnologies. Reprod Domest Anim. 2012;47(Suppl 3):12–21.
Camargo M, Intasqui P, Bertolla RP. Understanding the seminal plasma proteome and its role in male fertility. Basic Clin Androl. 2018;28:6.
Cantone I, Fisher AG. Epigenetic programming and reprogramming during development. Nat Struct Mol Biol. 2013;20(3):282–9.
Carre J, Gatimel N, Moreau J, Parinaud J, Leandri R. Influence of air quality on the results of in vitro fertilization attempts: a retrospective study. Eur J Obstet Gynecol Reprod Biol. 2017;210:116–22.
Cavalli G, Paro R. The Drosophila Fab-7 chromosomal element conveys epigenetic inheritance during mitosis and meiosis. Cell. 1998;93(4):505–18.
Chen M, Wu L, Zhao J, Wu F, Davies MJ, Wittert GA, et al. Altered glucose metabolism in mouse and humans conceived by IVF. Diabetes. 2014;63(10):3189–98.
Chen Q, Yan M, Cao Z, Li X, Zhang Y, Shi J, et al. Sperm tsRNAs contribute to intergenerational inheritance of an acquired metabolic disorder. Science. 2016a;351(6271):397–400.
Chen Q, Yan W, Duan E. Epigenetic inheritance of acquired traits through sperm RNAs and sperm RNA modifications. Nat Rev Genet. 2016b;17(12):733–43.
Christidis N, Jones GS, Stott PA. Dramatically increasing chance of extremely hot summers since the 2003 European heatwave. Nat Clim Change. 2015;5(1):46–50.
Colaco S, Sakkas D. Paternal factors contributing to embryo quality. J Assist Reprod Genet. 2018; 35:1–16.
Cooney CA, Dave AA, Wolff GL. Maternal methyl supplements in mice affect epigenetic variation and DNA methylation of offspring. J Nutr. 2002;132(8 Suppl):2393S–400S.
Cox LA, Nijland MJ, Gilbert JS, Schlabritz-Loutsevitch NE, Hubbard GB, McDonald TJ, et al. Effect of 30 per cent maternal nutrient restriction from 0.16 to 0.5 gestation on fetal baboon kidney gene expression. J Physiol. 2006;572(Pt 1):67–85.

Cropley JE, Eaton SA, Aiken A, Young PE, Giannoulatou E, Ho JWK, et al. Male-lineage transmission of an acquired metabolic phenotype induced by grand-paternal obesity. Mol Metab. 2016;5(8):699–708.

Dawson MA, Kouzarides T. Cancer epigenetics: from mechanism to therapy. Cell. 2012;150(1): 12–27.

Daxinger L, Whitelaw E. Transgenerational epigenetic inheritance: more questions than answers. Genome Res. 2010;20(12):1623–8.

Desai M, Jellyman JK, Ross MG. Epigenomics, gestational programming and risk of metabolic syndrome. Int J Obes (Lond). 2015;39(4):633–41.

Donkin I, Barres R. Sperm epigenetics and influence of environmental factors. Mol Metab. 2018;14:1–11.

Du Q, Hosoda H, Umekawa T, Kinouchi T, Ito N, Miyazato M, et al. Postnatal weight gain induced by overfeeding pups and maternal high-fat diet during the lactation period modulates glucose metabolism and the production of pancreatic and gastrointestinal peptides. Peptides. 2015;70:23–31.

Duhl DM, Vrieling H, Miller KA, Wolff GL, Barsh GS. Neomorphic agouti mutations in obese yellow mice. Nat Genet. 1994;8(1):59–65.

Dumoulin JC, Land JA, Van Montfoort AP, Nelissen EC, Coonen E, Derhaag JG, et al. Effect of in vitro culture of human embryos on birthweight of newborns. Hum Reprod. 2010;25(3):605–12.

Dunn GA, Bale TL. Maternal high-fat diet effects on third-generation female body size via the paternal lineage. Endocrinology. 2011;152(6):2228–36.

Feeney A, Nilsson E, Skinner MK. Epigenetics and transgenerational inheritance in domesticated farm animals. J Anim Sci Biotechnol. 2014;5(1):48.

Feuer SK, Liu X, Donjacour A, Lin W, Simbulan RK, Giritharan G, et al. Use of a mouse in vitro fertilization model to understand the developmental origins of health and disease hypothesis. Endocrinology. 2014;155(5):1956–69.

Fleming TP, Velazquez MA, Eckert JJ. Embryos, DOHaD and David Barker. J Dev Orig Health Dis. 2015;6(5):377–83.

Fleming TP, Watkins AJ, Velazquez MA, Mathers JC, Prentice AM, Stephenson J, et al. Origins of lifetime health around the time of conception: causes and consequences. Lancet. 2018;391(10132):1842–52.

Frantz ED, Aguila MB, Pinheiro-Mulder Ada R, Mandarim-de-Lacerda CA. Transgenerational endocrine pancreatic adaptation in mice from maternal protein restriction in utero. Mech Ageing Dev. 2011;132(3):110–6.

Fullston T, Ohlsson Teague EM, Palmer NO, DeBlasio MJ, Mitchell M, Corbett M, et al. Paternal obesity initiates metabolic disturbances in two generations of mice with incomplete penetrance to the F2 generation and alters the transcriptional profile of testis and sperm microRNA content. FASEB J. 2013;27(10):4226–43.

Gapp K, Jawaid A, Sarkies P, Bohacek J, Pelczar P, Prados J, et al. Implication of sperm RNAs in transgenerational inheritance of the effects of early trauma in mice. Nat Neurosci. 2014;17(5):667–9.

Gienapp P, Teplitsky C, Alho JS, Mills JA, Merila J. Climate change and evolution: disentangling environmental and genetic responses. Mol Ecol. 2008;17(1):167–78.

Gluckman PD, Hanson MA. Living with the past: evolution, development, and patterns of disease. Science. 2004;305(5691):1733–6.

Godfrey KM, Lillycrop KA, Burdge GC, Gluckman PD, Hanson MA. Epigenetic mechanisms and the mismatch concept of the developmental origins of health and disease. Pediatr Res. 2007;61(5 Pt 2):5R–10R.

Hammoud SS, Nix DA, Zhang H, Purwar J, Carrell DT, Cairns BR. Distinctive chromatin in human sperm packages genes for embryo development. Nature. 2009;460(7254):473–8.

Hanson MA, Gluckman PD. Developmental origins of health and disease: new insights. Basic Clin Pharmacol Toxicol. 2008;102(2):90–3.

Hanson MA, Gluckman PD. Early developmental conditioning of later health and disease: physiology or pathophysiology? Physiol Rev. 2014;94(4):1027–76.
Heard E, Martienssen RA. Transgenerational epigenetic inheritance: myths and mechanisms. Cell. 2014;157(1):95–109.
Heindel JJ. The developmental basis of disease: update on environmental exposures and animal models. Basic Clin Pharmacol Toxicol. 2018; https://doi.org/10.1111/bcpt.13118.
Hoffman DJ, Reynolds RM, Hardy DB. Developmental origins of health and disease: current knowledge and potential mechanisms. Nutr Rev. 2017;75(12):951–70.
Jafaroghli M, Abdi-Benemar H, Zamiri MJ, Khalili B, Farshad A, Shadparvar AA. Effects of dietary n-3 fatty acids and vitamin C on semen characteristics, lipid composition of sperm and blood metabolites in fat-tailed Moghani rams. Anim Reprod Sci. 2014;147(1-2):17–24.
Jenkins TG, Aston KI, Meyer T, Carrell DT. The sperm epigenome, male aging, and potential effects on the embryo. Adv Exp Med Biol. 2015;868:81–93.
Jirtle RL, Skinner MK. Environmental epigenomics and disease susceptibility. Nat Rev Genet. 2007;8(4):253–62.
Khan IY, Dekou V, Douglas G, Jensen R, Hanson MA, Poston L, et al. A high-fat diet during rat pregnancy or suckling induces cardiovascular dysfunction in adult offspring. Am J Physiol Regul Integr Comp Physiol. 2005;288(1):R127–33.
Klonoff-Cohen HS, Savitz DA, Cefalo RC, McCann MF. An epidemiologic study of contraception and preeclampsia. JAMA. 1989;262(22):3143–7.
Klosin A, Casas E, Hidalgo-Carcedo C, Vavouri T, Lehner B. Transgenerational transmission of environmental information in C. elegans. Science. 2017;356(6335):320–3.
Knecht AL, Truong L, Marvel SW, Reif DM, Garcia A, Lu C, et al. Transgenerational inheritance of neurobehavioral and physiological deficits from developmental exposure to benzo[a]pyrene in zebrafish. Toxicol Appl Pharmacol. 2017;329:148–57.
Kort HI, Massey JB, Elsner CW, Mitchell-Leef D, Shapiro DB, Witt MA, et al. Impact of body mass index values on sperm quantity and quality. J Androl. 2006;27(3):450–2.
Lambrot R, Xu C, Saint-Phar S, Chountalos G, Cohen T, Paquet M, et al. Low paternal dietary folate alters the mouse sperm epigenome and is associated with negative pregnancy outcomes. Nat Commun. 2013;4:2889.
Langley SC, Jackson AA. Increased systolic blood pressure in adult rats induced by fetal exposure to maternal low protein diets. Clin Sci (Lond). 1994;86(2):217–22. Discussion 121.
Lang-Mladek C, Popova O, Kiok K, Berlinger M, Rakic B, Aufsatz W, et al. Transgenerational inheritance and resetting of stress-induced loss of epigenetic gene silencing in Arabidopsis. Mol Plant. 2010;3(3):594–602.
Lea RG, Amezaga MR, Loup B, Mandon-Pepin B, Stefansdottir A, Filis P, et al. The fetal ovary exhibits temporal sensitivity to a 'real-life' mixture of environmental chemicals. Sci Rep. 2016a;6:22279.
Lea RG, Byers AS, Sumner RN, Rhind SM, Zhang Z, Freeman SL, et al. Environmental chemicals impact dog semen quality in vitro and may be associated with a temporal decline in sperm motility and increased cryptorchidism. Sci Rep. 2016b;6:31281.
Lee JH, Friso S, Choi SW. Epigenetic mechanisms underlying the link between non-alcoholic fatty liver diseases and nutrition. Nutrients. 2014;6(8):3303–25.
Li J, Na L, Ma H, Zhang Z, Li T, Lin L, et al. Multigenerational effects of parental prenatal exposure to famine on adult offspring cognitive function. Sci Rep. 2015;5:13792.
Liu L, Wylie RC, Andrews LG, Tollefsbol TO. Aging, cancer and nutrition: the DNA methylation connection. Mech Ageing Dev. 2003;124(10–12):989–98.
Lucas ES, Watkins AJ. The long-term effects of the periconceptional period on embryo epigenetic profile and phenotype; the paternal role and his contribution, and how males can affect offspring's phenotype/epigenetic profile. Adv Exp Med Biol. 2017;1014:137–54.
Maurice C, Kaczmarczyk M, Cote N, Tremblay Y, Kimmins S, Bailey JL. Prenatal exposure to an environmentally relevant mixture of Canadian Arctic contaminants decreases male reproductive function in an aging rat model. J Dev Orig Health Dis. 2018;9(5):511–8.

McCoy CR, Jackson NL, Brewer RL, Moughnyeh MM, Smith DL, Jr., Clinton SM. A paternal methyl donor depleted diet leads to increased anxiety- and depression-like behavior in adult rat offspring. Biosci Rep 2018;38(4).
McPherson NO, Fullston T, Kang WX, Sandeman LY, Corbett MA, Owens JA, et al. Paternal under-nutrition programs metabolic syndrome in offspring which can be reversed by antioxidant/vitamin food fortification in fathers. Sci Rep. 2016;6:27010.
Meister TA, Rimoldi SF, Soria R, von Arx R, Messerli FH, Sartori C, et al. Association of assisted reproductive technologies with arterial hypertension during adolescence. J Am Coll Cardiol. 2018;72(11):1267–74.
Messerschmidt DM. Should I stay or should I go: protection and maintenance of DNA methylation at imprinted genes. Epigenetics. 2012;7(9):969–75.
Mima M, Greenwald D, Ohlander S. Environmental toxins and male fertility. Curr Urol Rep. 2018;19(7):50.
Moallem U, Neta N, Zeron Y, Zachut M, Roth Z. Dietary alpha-linolenic acid from flaxseed oil or eicosapentaenoic and docosahexaenoic acids from fish oil differentially alter fatty acid composition and characteristics of fresh and frozen-thawed bull semen. Theriogenology. 2015; 83(7):1110–20.
Molaro A, Hodges E, Fang F, Song Q, McCombie WR, Hannon GJ, et al. Sperm methylation profiles reveal features of epigenetic inheritance and evolution in primates. Cell. 2011;146(6): 1029–41.
Murdoch BM, Murdoch GK, Greenwood S, McKay S. Nutritional influence on epigenetic marks and effect on livestock production. Front Genet. 2016;7:182.
Murrin CM, Kelly GE, Tremblay RE, Kelleher CC. Body mass index and height over three generations: evidence from the lifeways cross-generational cohort study. BMC Public Health. 2012;12:81.
Nestler EJ. Transgenerational epigenetic contributions to stress responses: fact or fiction? PLoS Biol. 2016;14(3):e1002426.
Nilsson EE, Sadler-Riggleman I, Skinner MK. Environmentally induced epigenetic transgenerational inheritance of disease. Environ Epigenet. 2018;4(2):dvy016.
Osgerby JC, Wathes DC, Howard D, Gadd TS. The effect of maternal undernutrition on ovine fetal growth. J Endocrinol. 2002;173(1):131–41.
Ost A, Lempradl A, Casas E, Weigert M, Tiko T, Deniz M, et al. Paternal diet defines offspring chromatin state and intergenerational obesity. Cell. 2014;159(6):1352–64.
Otero-Ferrer F, Izquierdo M, Fazeli A, Holt WV. Embryonic developmental plasticity in the long-snouted seahorse (Hippocampus reidi, Ginsburg 1933) in relation to parental preconception diet. Reprod Fertil Dev. 2016;28(7):1020–8.
Painter RC, Osmond C, Gluckman P, Hanson M, Phillips DI, Roseboom TJ. Transgenerational effects of prenatal exposure to the Dutch famine on neonatal adiposity and health in later life. BJOG. 2008;115(10):1243–9.
Paulson RJ, Comizzoli P. Addressing challenges in developing and implementing successful in vitro fertilization in endangered species: an opportunity for humanity to "give back". Fertil Steril. 2018;109(3):418–9.
Peitz B, Olds-Clarke P. Effects of seminal vesicle removal on fertility and uterine sperm motility in the house mouse. Biol Reprod. 1986;35(3):608–17.
Pembrey M, Saffery R, Bygren LO, Network in Epigenetic Epidemiology, Network in Epigenetic Epidemiology. Human transgenerational responses to early-life experience: potential impact on development, health and biomedical research. J Med Genet. 2014;51(9):563–72.
Pentinat T, Ramon-Krauel M, Cebria J, Diaz R, Jimenez-Chillaron JC. Transgenerational inheritance of glucose intolerance in a mouse model of neonatal overnutrition. Endocrinology. 2010;151(12):5617–23.
Pisani LF, Antonini S, Pocar P, Ferrari S, Brevini TA, Rhind SM, et al. Effects of pre-mating nutrition on mRNA levels of developmentally relevant genes in sheep oocytes and granulosa cells. Reproduction. 2008;136(3):303–12.

Ponting CP, Oliver PL, Reik W. Evolution and functions of long noncoding RNAs. Cell. 2009;136(4):629–41.
Radford EJ, Ito M, Shi H, Corish JA, Yamazawa K, Isganaitis E, et al. In utero effects. In utero undernourishment perturbs the adult sperm methylome and intergenerational metabolism. Science. 2014;345(6198):1255903.
Rattan S, Zhou C, Chiang C, Mahalingam S, Brehm E, Flaws JA. Exposure to endocrine disruptors during adulthood: consequences for female fertility. J Endocrinol. 2017;233(3):R109–R29.
Ravelli AC, van der Meulen JH, Michels RP, Osmond C, Barker DJ, Hales CN, et al. Glucose tolerance in adults after prenatal exposure to famine. Lancet. 1998;351(9097):173–7.
Rehman S, Usman Z, Rehman S, AlDraihem M, Rehman N, Rehman I, et al. Endocrine disrupting chemicals and impact on male reproductive health. Transl Androl Urol. 2018;7(3):490–503.
Rexhaj E, Paoloni-Giacobino A, Rimoldi SF, Fuster DG, Anderegg M, Somm E, et al. Mice generated by in vitro fertilization exhibit vascular dysfunction and shortened life span. J Clin Invest. 2013;123(12):5052–60.
Robertson SA, Sharkey DJ. Seminal fluid and fertility in women. Fertil Steril. 2016;106(3):511–9.
Robertson SA, Chin PY, Schjenken JE, Thompson JG. Female tract cytokines and developmental programming in embryos. Adv Exp Med Biol. 2015;843:173–213.
Robillard PY, Hulsey TC, Perianin J, Janky E, Miri EH, Papiernik E. Association of pregnancy-induced hypertension with duration of sexual cohabitation before conception. Lancet. 1994;344(8928):973–5.
Rodgers AB, Morgan CP, Bronson SL, Revello S, Bale TL. Paternal stress exposure alters sperm microRNA content and reprograms offspring HPA stress axis regulation. J Neurosci. 2013;33(21):9003–12.
Rodgers AB, Morgan CP, Leu NA, Bale TL. Transgenerational epigenetic programming via sperm microRNA recapitulates effects of paternal stress. Proc Natl Acad Sci U S A. 2015; 112(44):13699–704.
Rooke JA, Sinclair AG, Edwards SA. Feeding tuna oil to the sow at different times during pregnancy has different effects on piglet long-chain polyunsaturated fatty acid composition at birth and subsequent growth. Br J Nutr. 2001;86(1):21–30.
Roseboom TJ, van der Meulen JH, Osmond C, Barker DJ, Ravelli AC, Bleker OP. Plasma lipid profiles in adults after prenatal exposure to the Dutch famine. Am J Clin Nutr. 2000a;72(5):1101–6.
Roseboom TJ, van der Meulen JH, Osmond C, Barker DJ, Ravelli AC, Schroeder-Tanka JM, et al. Coronary heart disease after prenatal exposure to the Dutch famine, 1944-45. Heart. 2000b;84(6):595–8.
Rousseau-Ralliard D, Couturier-Tarrade A, Thieme R, Brat R, Rolland A, Boileau P, et al. A short periconceptional exposure to maternal type-1 diabetes is sufficient to disrupt the feto-placental phenotype in a rabbit model. Mol Cell Endocrinol. 2019;480:42–53.
Saben JL, Boudoures AL, Asghar Z, Thompson A, Drury A, Zhang W, et al. Maternal metabolic syndrome programs mitochondrial dysfunction via germline changes across three generations. Cell Rep. 2016;16(1):1–8.
Sarkar M, Baffy G. Perinatal programming of adolescent nonalcoholic fatty liver disease: a case for gender inequality? Hepatology. 2018;67(1):7–9.
Schagdarsurengin U, Steger K. Epigenetics in male reproduction: effect of paternal diet on sperm quality and offspring health. Nat Rev Urol. 2016;13(10):584–95.
Schmidt S, Hommel A, Gawlik V, Augustin R, Junicke N, Florian S, et al. Essential role of glucose transporter GLUT3 for post-implantation embryonic development. J Endocrinol. 2009;200(1):23–33.
Schübeler D. Function and information content of DNA methylation. Nature. 2015;517(7534): 321–6.
Schuster A, Skinner MK, Yan W. Ancestral vinclozolin exposure alters the epigenetic transgenerational inheritance of sperm small noncoding RNAs. Environ Epigenet. 2016;2(1)
Segovia SA, Vickers MH, Zhang XD, Gray C, Reynolds CM. Maternal supplementation with conjugated linoleic acid in the setting of diet-induced obesity normalises the inflammatory

phenotype in mothers and reverses metabolic dysfunction and impaired insulin sensitivity in offspring. J Nutr Biochem. 2015;26(12):1448–57.
Sinclair KD, Watkins AJ. Parental diet, pregnancy outcomes and offspring health: metabolic determinants in developing oocytes and embryos. Reprod Fertil Dev. 2014;26(1):99–114.
Sinclair KD, Allegrucci C, Singh R, Gardner DS, Sebastian S, Bispham J, et al. DNA methylation, insulin resistance, and blood pressure in offspring determined by maternal periconceptional B vitamin and methionine status. Proc Natl Acad Sci U S A. 2007;104(49):19351–6.
Sokol RZ, Kraft P, Fowler IM, Mamet R, Kim E, Berhane KT. Exposure to environmental ozone alters semen quality. Environ Health Perspect. 2006;114(3):360–5.
Tsuduki T, Kitano Y, Honma T, Kijima R, Ikeda I. High dietary fat intake during lactation promotes development of diet-induced obesity in male offspring of mice. J Nutr Sci Vitaminol (Tokyo). 2013;59(5):384–92.
Veenendaal MV, Painter RC, de Rooij SR, Bossuyt PM, van der Post JA, Gluckman PD, et al. Transgenerational effects of prenatal exposure to the 1944-45 Dutch famine. BJOG. 2013; 120(5):548–53.
Volkova K, Reyhanian Caspillo N, Porseryd T, Hallgren S, Dinnetz P, Olsen H, et al. Transgenerational effects of 17alpha-ethinyl estradiol on anxiety behavior in the guppy, Poecilia reticulata. Gen Comp Endocrinol. 2015;223:66–72.
Walker DM, Gore AC. Epigenetic impacts of endocrine disruptors in the brain. Front Neuroendocrinol. 2017;44:1–26.
Watkins AJ, Sinclair KD. Paternal low protein diet affects adult offspring cardiovascular and metabolic function in mice. Am J Physiol Heart Circ Physiol. 2014;306(10):H1444–52.
Watkins AJ, Wilkins A, Cunningham C, Perry VH, Seet MJ, Osmond C, et al. Low protein diet fed exclusively during mouse oocyte maturation leads to behavioural and cardiovascular abnormalities in offspring. J Physiol. 2008;586(8):2231–44.
Watkins AJ, Sirovica S, Stokes B, Isaacs M, Addison O, Martin RA. Paternal low protein diet programs preimplantation embryo gene expression, fetal growth and skeletal development in mice. Biochim Biophys Acta. 2017;1863(6):1371–81.
Watkins AJ, Dias I, Tsuro H, Allen D, Emes RD, Moreton J, et al. Paternal diet programs offspring health through sperm- and seminal plasma-specific pathways in mice. Proc Natl Acad Sci U S A. 2018;115(40):10064–9.
Wolff GL, Kodell RL, Moore SR, Cooney CA. Maternal epigenetics and methyl supplements affect agouti gene expression in Avy/a mice. FASEB J. 1998;12(11):949–57.
Xia B, de Belle JS. Transgenerational programming of longevity and reproduction by post-eclosion dietary manipulation in Drosophila. Aging (Albany NY). 2016;8(5):1115–34.
Xin F, Susiarjo M, Bartolomei MS. Multigenerational and transgenerational effects of endocrine disrupting chemicals: a role for altered epigenetic regulation? Semin Cell Dev Biol. 2015;43:66–75.
Yeung EH, Druschel C. Cardiometabolic health of children conceived by assisted reproductive technologies. Fertil Steril. 2013;99(2):318–26.
Zama AM, Uzumcu M. Epigenetic effects of endocrine-disrupting chemicals on female reproduction: an ovarian perspective. Front Neuroendocrinol. 2010;31(4):420–39.
Zandstra H, Brentjens L, Spauwen B, Touwslager RNH, Bons JAP, Mulder AL, et al. Association of culture medium with growth, weight and cardiovascular development of IVF children at the age of 9 years. Hum Reprod. 2018;33(9):1645–56.
Zhou Y, Zhu H, Wu HY, Jin LY, Chen B, Pang HY, et al. Diet-induced paternal obesity impairs cognitive function in offspring by mediating epigenetic modifications in spermatozoa. Obesity (Silver Spring). 2018;26(11):1749–57.

Chapter 5
Stress, Well-Being and Reproductive Success

Katie L. Edwards, Ashley N. Edes, and Janine L. Brown

Abstract To reverse the trend of declining wildlife populations globally, individuals must be provided with conditions that allow them to not just survive, but to thrive. It is no longer only the remit of captive breeding programs to ensure animal well-being; *in situ* conservation efforts also must consider how environmental and anthropogenic pressures impact wild populations, and how to mitigate them—especially with regards to reproduction and survival. Stress and welfare are complex concepts that necessitate an understanding of how stressors affect animals on both individual and population levels. There are species differences in how factors impact well-being, related in part to natural history, which also are shaped by individual perceptions and coping abilities. A multitude of stress-related responses then have the potential to disrupt fertility on many levels, and ultimately fitness. A major limitation to advancing welfare science is the lack of definitive tests to verify welfare status; i.e., is the animal happy or not? While analyses of circulating or excreted glucocorticoids have for decades been the primary method of assessing stress, today we recognize the need for more objective indicators that incorporate multiple physiological systems, including behavior, to assess both negative and positive welfare states. In this chapter, we discuss the potential for stress to disrupt, and sometimes facilitate reproduction, including the key role that glucocorticoids play. We then discuss a number of physiological biomarkers, which in addition to glucocorticoids, have the potential to assess well-being and the role of stress on reproduction. Finally, we discuss allostatic load, a method by which multiple physiological markers are used to inform on morbidity and mortality risk in humans, which if applied to wildlife, could be a powerful tool for conservation.

K. L. Edwards (✉) · A. N. Edes
Center for Species Survival, Smithsonian Conservation Biology Institute,
Front Royal, VA, USA
e-mail: edwardskl@si.edu

J. L. Brown
Smithsonian Conservation Biology Institute, National Zoological Park,
Front Royal, VA, USA

P. Comizzoli et al. (eds.), *Reproductive Sciences in Animal Conservation*,
Advances in Experimental Medicine and Biology 1200,
https://doi.org/10.1007/978-3-030-23633-5_5

Keywords Reproduction · Welfare · Well-being · Stress · Biomarkers · Allostatic load · Immune system · Glucocorticoids · Morbidity · Mortality · Wildlife

1 Introduction

A key focus of endangered species conservation efforts must be on understanding the fundamental elements that individuals and populations of a species need to not only survive, but to thrive. Modern zoos have an ethical responsibility to provide the best care for animals, and to increase our understanding of species' biology so that we can manage them in ways that meet both physical and psychological needs. Animals with optimal welfare are more likely to reproduce, maintain species-appropriate behaviors, and be in good health, all of which are central to successful conservation breeding programs. A lack of reproductive function can be an indicator of sub-optimal welfare (Mason 2010; Mason and Veasey 2010), and improvements in care can have positive effects on breeding success (Mellor et al. 2015). Good welfare is not only important for species under human care, however. *In situ* conservation typically focuses on populations, but understanding how changing environments and anthropogenic pressures affect individual members of a species can have important implications for ensuring long-term species persistence.

On a basic level, animal welfare science addresses how an individual copes with the conditions it faces. In reality, the concept of well-being is much more complex. A good representation of this complexity is the Five Domains Model, originally formulated in 1994 (Mellor and Reid 1994), which has more recently been expanded to incorporate positive welfare states (Mellor and Beausoleil 2015). This model outlines four physical/functional domains—nutrition, environment, physical health and behavior, and a fifth—the animals' mental state (Fig. 5.1) that act together to influence welfare. Unlike the Five Freedoms (Farm Animal Welfare Council 1979) upon which this model was based, these domains incorporate positive experiences, as opposed to merely freedom from negative states. It is widely accepted that affective states can be measured in two dimensions, valence (positive to negative) and arousal (low to high activation) (Mendl et al. 2010), with perhaps a third, potency/dominance, based on the animal's perception of how it will cope with the outcome (Scherer et al. 2006). An animal's mental state can therefore influence the extent to which positive or negative factors impact overall well-being. For that reason, cognitive bias testing, which measures how emotional states affect cognitive processes (i.e., glass half-full vs. half-empty), has emerged as an important tool for assessing welfare states in animals (see review, Clegg 2018). Although the focus of this review is not on behavioral measures of welfare *per se*, it is important to recognize that negative affective states can severely compromise quality of life (Scheier and Bridges 1995). In humans, depressed people are more likely to express negative judgments when confronted with ambiguous signals (Eysenck et al. 1991) and vice versa; the bias determines whether they are pessimistic or optimistic (Carver et al. 2010). In animals, those with better welfare make more optimistic choices (Clegg 2018; Mendl et al. 2009). Cognitive bias testing has been used with farm, laboratory, and companion species, but only recently has been applied to zoo animals, and

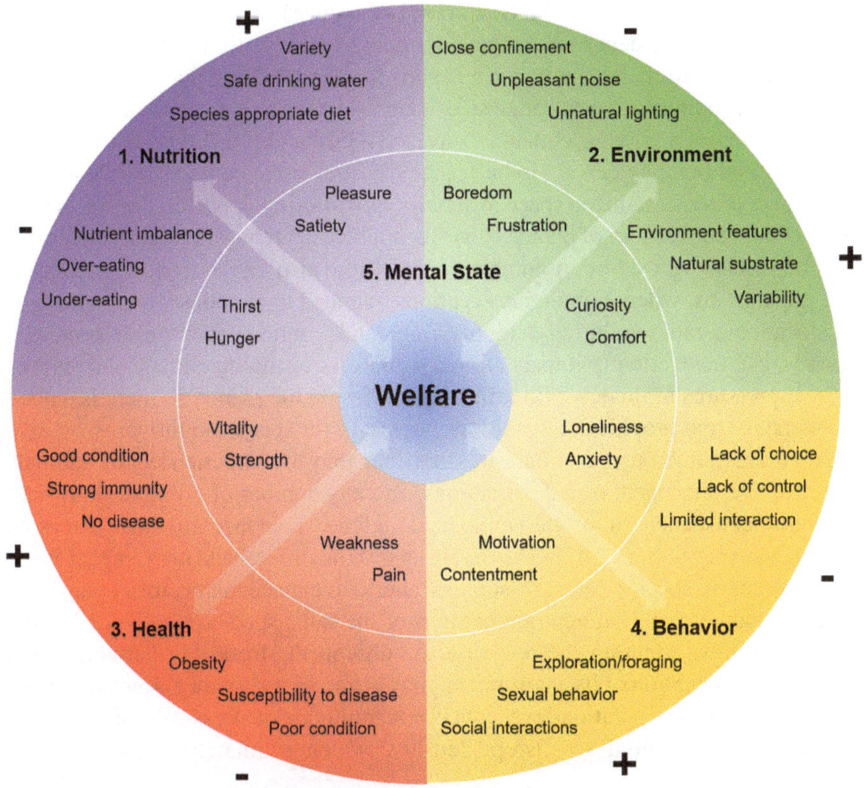

Fig. 5.1 The Five Domains Model incorporating factors within four physical/functional domains (nutrition, environment, health, and behavior) and associated affective states within a fifth mental domain that impact animal well-being. Factors can be either positive or negative, and their impact on welfare can be moderated by the mental state of the individual. Adapted from Mellor and Beausoleil (2015)

then only a few (Clegg et al. 2017; McGuire et al. 2017a, b). There are several types of biases that can occur: individuals in a negative affective state are more likely to make pessimistic judgements about ambiguous stimuli (judgement bias), pay more attention to negative stimuli (attention bias), or remember negative situations more readily (memory bias) than those in a more positive state, each of which can be tested empirically (Clegg 2018). Judgement bias tests are the most common, however, especially in animals. Developed by Harding et al. (2004), animals are trained to respond to a cue, which can be auditory, visual, spatial or olfactory, that is associated with a positive event (e.g., food), and to perform a different response to avoid a negative or less positive result (e.g., no food). Once the animals respond appropriately (i.e., preference for the positive cue, food), they are presented with an ambiguous cue or stimuli to determine if the animal responds negatively (pessimistic; lack of response to the cue) or positively (optimistic; approaches the cue). Those in a positive affective state exhibit more optimistic responses than those in a negative state. Thus, cognitive bias tests provide a non-invasive means of assessing welfare that, especially if combined with monitoring of biomarkers, can be used to inform decisions on management of animals under human care.

In a captive setting, factors that can negatively affect welfare may include unnatural social groupings or home-range size, the presence of people or predators, noise, or unnatural photoperiod or climate. Conversely, husbandry practices that facilitate positive social bonding can act to buffer the effect of some stressors, while enrichment can counteract reduced space availability by offering opportunities for mental stimulation, and freedom of choice can offer a mechanism by which individuals can moderate their own welfare. A number of challenges also face extensively managed wild populations, where human interventions intended to aid conservation, such as translocations, radio-collaring and tracking, and confinement within protected reserves, can inadvertently impact animal welfare. A changing climate and anthropogenic pressures ultimately will affect all species on the planet, so there is a need to understand and provide the conditions necessary for species to thrive.

There are numerous potential stressors that may be experienced by animals under captive, managed, or wild situations, and any number of possible short- and long-term consequences of the biological defense response to those stressors (Fig. 5.2). There may be species differences in how different factors impact well-being, related to species natural history or alternative mechanisms for coping with perceived challenges. Different populations of the same species may be under differing pressures, affecting the response to subsequent stress. Furthermore, each individuals' perception of its environment may differ, so well-being is an individual-based rather than a population-level concept. A multitude of stress-related responses then have the potential to disrupt fertility, which ultimately impacts fitness

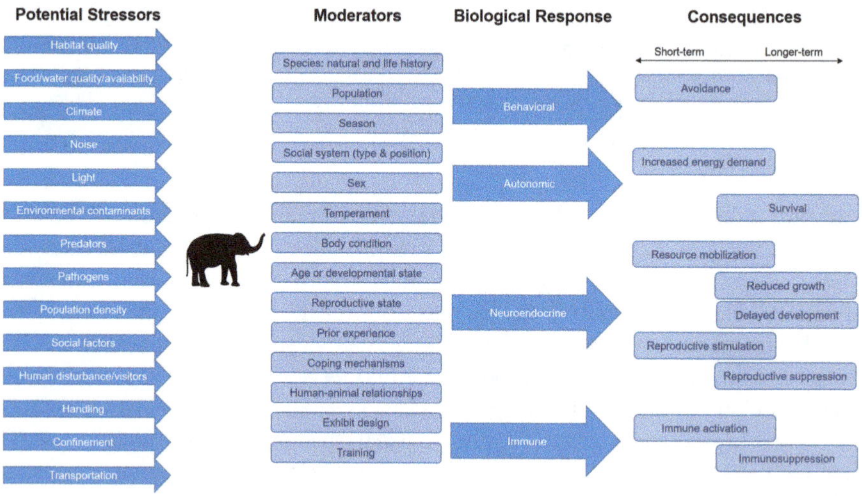

Fig. 5.2 Potential stressors that may be experienced by wildlife under captive, managed or wild situations, illustrated with moderators that may influence whether a stress response is initiated, and possible short- and long-term consequences of the biological defense response to stress on the organism. Adapted from Hofer and East (2012)

(Wingfield and Sapolsky 2003). Differences in the relative cost of disruption may influence how it impacts an individual's reproduction. For example, seasonal breeders that have only a finite window of opportunity to breed may be more resistant to acute natural stressors (Boonstra et al. 2001). Similarly, individuals with limited opportunity for breeding due to age (Boivin et al. 2006) or social status (Sapolsky 1985) may become more resistant to the effects of stress.

The challenge that conservation biologists face today is how to assess such a complex and subjective concept in a diverse array of taxa. A more holistic approach to understanding and assessing well-being will be key, because no one factor acts in isolation. In this chapter, we discuss the effect of stress on reproduction, both negative and positive, and highlight a number of physiological measures that could allow us to better assess animal well-being, including use of a multi-biomarker approach known as allostatic load that assesses the wear-and-tear on the body that accumulates as an individual is exposed to repeated or chronic stress. This approach has been used to inform on morbidity and mortality risk in humans, which if applied to wildlife, could be a powerful tool for conservation.

2 Stress and Reproduction

2.1 The Neuroendocrine Stress Response

When an animal experiences a stressor that could disrupt homeostasis, the resulting processes are organized into three stages: recognition of the stress by the central nervous system; the body mounting a biological defense response; and the consequences of that response (Moberg 2000). The biological defense response involves a combination of behavioral, autonomic nervous system, and neuroendocrine changes that can lead to altered biological function. If a behavioral response of avoidance is not feasible or sufficient, the autonomic nervous system may be activated, resulting in changes in catecholamines, heart rate, blood pressure, or gastrointestinal activity, to facilitate a 'fight or flight' response (Cannon 1929). The main neuroendocrine response to a stressor involves activation of the hypothalamic-pituitary-adrenal (HPA) axis (Matteri et al. 2000), or the hypothalamic-pituitary-interrenal axis in fish (Wendelaar Bonga 1997). Corticotropin-releasing hormone (CRH) is released from the hypothalamus, stimulating the anterior pituitary to produce adrenocorticotrophic hormone (ACTH), and finally glucocorticoids (GCs) such as cortisol and corticosterone from the adrenal cortex. GCs are regulated through positive and negative feedback mechanisms, and facilitate the mobilization of energy stores (Moberg 2000), allowing the body to respond to a stressor (McEwen and Wingfield 2003) and maintain homeostasis (Moberg and Mench 2000). Although both cortisol and corticosterone are secreted to some degree across taxa, cortisol is the primary GC in fish and most mammalian species, with the exception of murinae and rabbits where corticosterone is dominant (Palme 2019). Corticosterone is also the main GC in birds, amphibians and reptiles (Koren et al. 2012).

A wide range of positive and negative stimuli can elicit a stress response (Buwalda et al. 2012), including social (Creel 2001), physical (Neumann et al. 1998), physiological (Gasparotto et al. 2005), and psychological (Tamashiro et al. 2005) stressors. The body's response depends on a number of factors, including age, sex and genetics (Blas et al. 2006; Hovland et al. 2017; Kolbe et al. 2015; Koolhaas et al. 1999; Moberg 2000; Tilbrook et al. 2000), physiological state (Brann and Mahesh 1991b; Raouf et al. 2006), and how the stressor is perceived (Cooper and Baglioni 1988). Perception may, in turn, depend on prior experience (Mormède et al. 2007; Nisenbaum et al. 1991) and personality traits (Sapolsky 1994), as well as the predictability of the stressor itself (Morgan and Tromborg 2007) and availability of coping mechanisms (Goymann and Wingfield 2004). The stress response is primarily adaptive; however, negative consequences can result if the animal does not cope appropriately with a stressor that may occur repeatedly or is chronic. Processes that are non-essential to immediate survival may be down-regulated, and can result in the disruption of the immune response (Khansari et al. 1990), growth (Spencer et al. 2009), and reproduction (Dobson et al. 2003; Tilbrook et al. 2000). This potential for dysregulation has made investigations of stress physiology an important component of many research disciplines, including conservation biology (Busch and Hayward 2009; Cooke et al. 2013; Dantzer et al. 2014; Madliger and Love 2014; Sheriff et al. 2011; Wikelski and Cooke 2006). Here we focus on the potential impact stress has on reproduction and well-being, considering the complex interactions of the multiple biological processes that can be involved.

2.2 How Stress Can Disrupt Reproduction

The most common mechanism by which stress affects reproduction involves activation of the HPA axis, resulting in a disruption of the hypothalamic-pituitary-gonadal (HPG) axis (Uphouse 2011). This has been well documented in rodents (Brann and Mahesh 1991b; Rivier and Rivest 1991), domestic ungulates (Dobson et al. 2003; Dobson and Smith 2000; Tilbrook et al. 2002), primates (Olster and Ferin 1987; Tamashiro et al. 2005), humans (Chrousos et al. 1998; Schenker et al. 1992), and amphibians, reptiles, fish, and birds (Breuner 2011; Carr 2011; Fuzzen et al. 2011; Moore and Jessop 2003; Schreck 2010; Tokarz and Summers 2011). Physical (Dobson and Smith 2000), behavioral (Moberg 1991), social (Sapolsky 1985), nutritional (Wade and Schneider 1992), and environmental (Boonstra et al. 1998) stressors all have been shown to disrupt reproduction through activation of the HPA response.

Stress can disrupt reproductive function at many levels (Dobson and Smith 2000; Rivier and Rivest 1991). CRH and ACTH act primarily on the hypothalamus and higher brain centers, while GC effects are possible all along the HPG axis (Wingfield and Sapolsky 2003). Females are particularly sensitive to disruptions prior to ovulation, a time that relies on a delicate balance of hormonal signaling (Dobson and Smith 2000). During this period, ovulation can be disrupted via inhibition of pulsatile gonadotropin releasing hormone (GnRH) secretion from the hypothalamus, which influences both the frequency and amplitude of pulsatile luteinizing hormone (LH) secretion (Dobson et al. 2003), and is further impacted by direct inhibitory effects of GCs on the pituitary, making it less responsive to GnRH. Insufficient GnRH/LH frequency can prevent follicular development, leading to anestrous (Dobson and Smith 2000). If pulse frequency is sufficient for normal follicular development, but not for eliciting an ovulatory LH surge, cystic follicles can develop and render a female infertile (Dobson and Smith 2000; Ferin 1999). Finally, GCs act on the ovaries, decreasing sensitivity to LH through a reduction in receptor availability (Wingfield and Sapolsky 2003). Exposure to stress during the luteal phase can disrupt luteolysis through inhibition of luteolytic prostaglandins (Cooke and Benhaj 1989; Lee et al. 2007; Wang et al. 1993), resulting in abnormally long cycles. It can also interfere with uterine growth and implantation success (Kalantaridou et al. 2004, 2010; Rabin et al. 1990), perhaps due to a decline in progesterone required for uterine maturation, or via a stress-induced production of prolactin, which inhibits actions of progesterone in the uterus (Wingfield and Sapolsky 2003). Even if conception occurs, increased stress has been associated with early pregnancy failure (Einarsson et al. 1996; Parker and Douglas 2010). Finally, stress-induced GnRH and gonadotropin insufficiency can result in diminished biosynthesis of sex steroids (Kalantaridou et al. 2010), which impacts the expression of proceptive (Carter 1992) and receptive behaviors (Papargiris et al. 2011) important for breeding.

Activation of the HPA axis can similarly impact GnRH/LH pulsatility in males (Kalantaridou et al. 2010). Increases in GC concentration inhibit testosterone-biosynthetic enzyme activity (Orr et al. 1994), leading to a reduction in testosterone secretion (Hardy et al. 2005). There is also evidence that stress may have a direct impact on fertility through reduction in the number, as well as the function of Leydig cells through apoptosis (Hardy et al. 2005). Furthermore, severe psychosocial stress in humans has been associated with reduced sperm count, perhaps because of reduced testosterone, but possibly also due to direct effects on the seminiferous epithelium (Fenster et al. 1997). Cumulatively, these effects can lead to diminished libido and fertility (Phillips et al. 1989).

2.3 *The Role of Glucocorticoids in Normal Reproductive Function*

Although stress and increased HPA axis activity can disrupt reproduction in many ways, physiological concentrations of GCs also are a vital component of normal reproductive function (Brann and Mahesh 1991b). In the female, GCs play an important role in follicle maturation, ovulation, and luteinization (Hillier and Tetsuka 1998; Tetsuka 2007), and regulate production of, and sensitivity to, gonadotropins (Brann and Mahesh 1991b). During pregnancy, they can facilitate implantation (Keay et al. 2002), regulate immune processes (Whirledge and Cidlowski

2013), and mobilize maternal energy stores to accommodate a rapidly growing fetus (Atkinson and Waddell 1995). The presence of GC receptors within the ovary (Schreiber et al. 1982; Tetsuka et al. 1999), uterus (Korgun et al. 2003), placenta (Filiberto et al. 2011; Speeg and Harrison 1979), and testes (Evain et al. 1976; Haeussler and Claus 2007; Herrera-Luna et al. 2012; Schultz et al. 1993) also provide a means by which these hormones can regulate reproduction locally.

Early evidence for the importance of GCs to reproductive processes came from adrenalectomy studies in rodents, where delayed puberty (Macfarland and Mann 1977; Ramaley 1974), suppression of follicle stimulating hormone (FSH) production (Meijs-Roelofs and Kramer 1977), inhibition of follicular development and ovulation (Peppler and Jacobs 1976), and irregular cyclicity (Mann et al. 1975) were all reported as a result. By contrast, studies in rodents illustrate stimulatory effects of acute stress, ACTH, and GCs on LH and FSH release, ovulation, fertility, sexual maturation and sexual behavior (reviewed in Brann and Mahesh 1991b). Similarly, in non-rodent species, short-term administration of GCs resulted in initial stimulation of testosterone production in bulls (Thibier and Rolland 1976), enhanced LH release in response to exogenous GnRH in the boar (Liptrap and Raeside 1983), and increased basal LH concentrations in pre-pubertal gilts (Pearce et al. 1988). Transportation is a commonly used experimental stressor for domestic livestock, and is generally associated with inhibited reproduction (Dobson and Smith 1995, 2000). However, there are examples of stimulatory effects post-transport as well, such as attainment of puberty in gilts (Hughes 1982), estrous induction in domestic cats (Wildt 1980), and ovulation but not behavioral estrus in ewes and heifers (Braden and Moule 1964; Lamond 1962).

The endocrine milieu significantly influences the effect that HPA axis activity has on reproductive function. Estrogen priming in particular can have a profound impact on the response to stress, at least in females. Acute stress (Armario et al. 1987; Briski and Sylvester 1988; Higuchi et al. 1986) and acute administration of ACTH (Brann et al. 1991) and GCs (Brann et al. 1990, 1992) all have been shown to stimulate gonadotropin secretion in estrogen-primed female rats. Conversely, acute stress in ovariectomized rats results in inhibited LH secretion (Higuchi et al. 1986). Estradiol can alter the sensitivity of the anterior pituitary to GnRH (Brann and Mahesh 1991a), which may explain the different effects that stress has on LH in low versus high estrogen states.

Longitudinal monitoring of GCs has revealed a predictable pattern across the ovarian cycle in many species, although patterns tend to be species-specific. In women (Kirschbaum et al. 1999) and black rhino (K. Edwards, unpublished), GC concentrations are higher during the luteal than the follicular phase. Conversely, in Asian elephants (Fanson et al. 2014), giant pandas (Kersey et al. 2011), and ewes (Sosa et al. 2013), GCs are higher during the follicular phase in association with estrus/periestrus. Disruption of this cyclic release of GCs has been associated with ovarian dysfunction (Brann and Mahesh 1991b; Mahesh and Brann 1998; Tébar et al. 1995). Notably, there are reported differences in GC excretion between reproductively successful and unsuccessful giant pandas (Kersey et al. 2011) and Asian elephants (Fanson et al. 2014).

Adrenal activity plays a role throughout pregnancy and lactation. During early gestation, GCs regulate many of the processes required for implantation and development of the embryo and placenta (Michael and Papageorghiou 2008). Administration of GCs during this critical period has led to improved pregnancy rates in both human (Boomsma et al. 2007; Quenby et al. 2003, 2005) and animal (Duong et al. 2011) studies. In many species, the timing of parturition is also initiated by the maturation of the fetal HPA axis and a corresponding increase in cortisol (Ginther 1992; Knobil and Neill 1998). Periparturient cortisol surges have been noted in a number of wildlife species, such as black rhinoceros (K. Edwards, unpublished), elephants (Meyer et al. 2004), pygmy rabbits (Scarlata et al. 2012), pig-tailed macaques (Shields et al. 2016), and giant panda (Kersey et al. 2011). Mammary development, lactogenesis and milk secretion are also promoted by GCs (Casey and Plaut 2007; Whirledge and Cidlowski 2013).

Ultimately, there is a delicate balance between levels of GCs that promote or inhibit reproduction, which is important to consider when investigating the potential effects of acute or chronic stressors on reproductive function. Often the effects of GCs are species-specific, so it is equally important to first understand normal GC patterns and then determine how stress disrupts fertility on individual and population levels. This will require specialized approaches to consider species ecology and animal life histories, preferably coupled with longitudinal hormone monitoring.

2.4 *Measuring the Stress Response in Wildlife*

To date, the vast majority of stress-related research conducted in wildlife has focused on measuring GCs (Ganswindt et al. 2012), which can be a useful measure of stress physiology and the potential impacts of stress on animal welfare (Mormède et al. 2007; Palme 2012; Sheriff et al. 2011). Non-invasive fecal measures have predominated over the last 30 years because samples are relatively easy to obtain and collection does not disturb the animal. Other minimally-invasive techniques have been developed for a wide range of matrices, including urine (Anestis 2009; Bennett et al. 2008; Brown et al. 2010; Narayan et al. 2010, 2012), saliva (Ange-van Heugten et al. 2009; Behringer et al. 2013; Cross et al. 2004; Menargues et al. 2008, 2012), hair (Burnard et al. 2017; Heimbürge et al. 2018), feathers (Bortolotti et al. 2008; Freeman and Newman 2018), skin secretions (Santymire et al. 2018), scales (Aerts et al. 2015), shed skin (Berkvens et al. 2013), claws (Baxter-Gilbert et al. 2014; Matas et al. 2016), blubber (Kershaw and Hall 2016), baleen (Hunt et al. 2017), and blow (Burgess et al. 2018; Thompson et al. 2014). Each technique comes with advantages and disadvantages based on the ease of sample collection, need for extraction, storage requirements, hormone metabolism, routes of excretion, and the specific research questions of interest. Hormones appear in circulation within minutes of a stressor (Walker et al. 2005), and blood collection in itself can invoke a stress response (Cook et al. 2000). The delay between secretion and excretion/deposition for different sample matrices can be minutes (saliva), hours (urine), days (feces), or even months (hair), depending on species-specific factors such as gut-transit time and rate of hair or feather growth. In the case of whales, a single baleen plate provides a continuous, retrospective hormonal profile over a period of many years (Hunt et al. 2016). Some hormone production is episodic, complicating the interpretation of circulating concentrations and making the integrated nature of other sample types beneficial for assessing factors relating to welfare, as they represent an averaged concentration over time. However, when utilizing fecal and urinary measures in particular, native hormones (cortisol and corticosterone) are rarely present in large quantities. Instead, these hormones are metabolized by the liver and/or kidney to aid excretion (Palme 2005). In addition, the metabolites present vary by species and sometimes by sex (Touma et al. 2003). A recent comprehensive review (Palme 2019) summarized over 1200 fecal GC metabolite studies in domestic and wildlife species to provide an overview of GC metabolism and excretion, methodological considerations and potential applications. Central to the success of these studies is proper validation of techniques to ensure the predominant metabolites are suitably detected and an understanding of confounding factors that can influence results. When properly applied, non-invasive analyses of GCs and their metabolites can be a powerful tool for understanding stress physiology.

A wide range of factors can affect reproduction and welfare of wild populations. Anthropogenic stressors include land use activities such as logging, mining, military activity, and installation of turbines (Agnew et al. 2016; Barron et al. 2012; Leshyk et al. 2013; Łopucki et al. 2018; Mancera et al. 2017; Morgan et al. 2018), habitat fragmentation (Carlitz et al. 2016; Martinez-Mota et al. 2004), hunting pressure (Bryan et al. 2015; Burke et al. 2008; Gobush et al. 2008; Ordóñez-Gómez et al. 2016; Rimbach et al. 2013), environmental contaminants (Beck et al. 2014; Lattin et al. 2014; Meillère et al. 2016; Romero and Wikelski 2002; Tartu et al. 2014), and recreation and tourism (Creel et al. 2002; Piñeiro et al. 2012; Thiel et al. 2008; Zwijacz-Kozica et al. 2013). GCs have also been used to investigate habitat quality (Bryan et al. 2014; Caslini et al. 2016; Herring and Gawlik 2013; Lea et al. 2018b; Romero and Wikelski 2001; Woolley et al. 2009), the response of animal populations to a changing landscape (Baxter-Gilbert et al. 2014; Bechshøft et al. 2013; Brearley et al. 2012; French et al. 2008; Navarro-Castilla et al. 2014; Troïanowski et al. 2017), and to translocation or re-introduction (Drake et al. 2012; Franceschini et al. 2008; Hing et al. 2017; Ji et al. 2013; Lèche et al. 2016; Snyder et al. 2012; Viijoen et al. 2008). Last, and of particular interest to researchers, GC measures can inform on the degree of animal disturbance due to the presence of observers or use of population monitoring tools, like radiocollars (Moll et al. 2009; Pereia et al. 2009; Shutt et al. 2014; Sperry et al. 2009), which are important considerations for designing *in situ* studies.

Numerous studies demonstrate how anthropogenic stressors can negatively affect reproduction through increased GC production in wild populations. American kestrels nesting in high disturbance areas had higher corticosterone concentrations and were more likely to abandon nests than females in lower disturbance areas (Strasser and Heath 2013). Northern spotted owls close to noisy roads also had lower fledging rates than those situated by quieter roads (Hayward et al. 2011).

Similarly, female wood frogs produced more corticosterone in areas with high traffic noise, and also showed a reduction in appropriate breeding behavior in response to a male breeding chorus (Tennessen et al. 2014).

Ecotourism can contribute to conservation through revenue generation to support intensive protection (Robbins et al. 2011) and increasing awareness (Waylen et al. 2009), but human disturbance also can have negative consequences for breeding success and offspring survival if not properly regulated. Yellow-eyed penguins at sites with unregulated tourism had higher GCs, poorer breeding success and lower fledging weights, than those where human visitation was limited to population monitoring only (Ellenberg et al. 2007). Similarly, breeding performance of ospreys was reduced in areas with increased boat traffic compared to control areas (Monti et al. 2018). In that study, chicks also were directly impacted, with increased feather corticosterone concentrations in high-traffic areas. Although adult hoatzins were undisturbed by tourist presence, chicks at disturbed sites exhibited stronger stress-induced hormone responses, and had lower body mass and increased juvenile mortality than those at undisturbed sites (Müllner et al. 2004).

Ex situ management can also benefit from the use of GCs analyses. Examples include identifying the importance of environmental enrichment in felids (Carlstead et al. 1993; Moreira et al. 2007) and black footed ferrets (Poessel et al. 2011); how improving exhibits can reduce stress in black rhinos, pygmy rabbits, and clouded leopards (Carlstead and Brown 2005; Scarlata et al. 2013; Shepherdson et al. 2004; Wielebnowski et al. 2002a); and the positive effect of meeting social needs of primates, felids, and ursids (Frederick et al. 2013; Henkel et al. 2010; Machatschke et al. 2006; Smith et al. 2011; Smith and French 1997; Wielebnowski et al. 2002b). In addition, GCs have been used to assess the response to visitor presence and interaction in captive wildlife, such as spider monkeys (Davis et al. 2005), meerkats (Scott et al. 2017), and koalas (Webster et al. 2017), as well as investigating the efficacy of exhibit changes to mitigate visitor-associated stress in gorillas (Clark et al. 2011). These studies highlight the importance of taking natural history into consideration to inform on best management practices for each species, which can then lead to a better understanding of why individuals fail or thrive under human care.

Investigating how potential stressors in the captive environment affect reproduction is vital to the success of conservation breeding programs and long-term sustainability of endangered species. For example, in a group of captive douc langurs, a period of social change resulted in elevated fecal GC excretion and cycle irregularity (extended follicular phases, abbreviated luteal phases and/or anovulation), suggesting that social disruption has the potential to impair ovarian function in this species (Heistermann et al. 2004). Among wild male olive baboons, cortisol was higher in subordinates because aggression is used to achieve and maintain social rank, while in common marmosets, lower ranked females had lower cortisol than higher ranked females because social position is inherited and there is little aggression between group members (Abbott et al. 2003). Among female Barbary macaques, it is mid-ranking females that engage in the most agonistic interactions, and consequently have higher fecal GC concentrations than those both above and below them in the hierarchy (Edwards et al. 2013). Increased adrenal activity during the breeding season has also been associated with reduced conception rates and survival of subsequent litters in the pygmy rabbit (Scarlata et al. 2012). Additionally, although the exact role of stressors on reproductive function has yet to be fully understood, higher GC concentrations have been reported in non-cycling female cheetahs (Jurke et al. 1997), during periods of irregular cyclicity compared to when cycling normally in black rhinos (K. Edwards, unpublished), and higher GC variability in non-cycling white rhino females (Carlstead and Brown 2005).

As discussed above, GCs can be a useful marker of physiological state, especially when assessed non-invasively, but often there are confounding factors involved in their measurement. GC receptors are ubiquitous throughout body tissues (Ballard et al. 1974) and GCs play important roles in addition to the stress response, including the regulation of immunity (Webster et al. 2002), cardiovascular function (Schmid et al. 1967), metabolism (Munck and Koritz 1962; Olefsky and Kimmerling 1976), growth and development (Ballard and Ballard 1972), cognition (Mizoguchi et al. 2004), and behavior (Schiml and Rissman 1999). Thus, elevated concentrations do not always equate to poor welfare. Variation in concentrations can be representative of an animal coping appropriately with day to day challenges (Ralph and Tilbrook 2016), and may reflect other physiological factors, such as changes across the estrous cycle and pregnancy. Furthermore, substantial inter- and intra-individual differences in GC responses often make interpretations difficult (Bonier et al. 2009; Breuner et al. 2008; Busch and Hayward 2009; Cockrem 2013; Mostl and Palme 2002). For example, individuals may be more or less responsive to potential challenges due to differences in coping styles (Bonier et al. 2009; Koolhaas et al. 1999), physiological states (Raouf et al. 2006), or prior experiences (Ulrich-Lai and Herman 2009). This complexity of stress physiology relationships necessitates the incorporation of a variety of metrics to understand welfare implications within diverse biological systems. When an animal perceives a threat to its homeostasis, the biological response may be behavioral, autonomic, neuroendocrine, immunological, or more likely a combination of all four. Although we will not discuss behavioral indicators here, in the next section, we will introduce a number of complementary and/or alternative physiological biomarkers that have application to studies of animal welfare.

3 Alternative Methods to Assess Welfare

3.1 The Autonomic Nervous System

The autonomic nervous system (ANS) response to stress forms the basis of Cannon's 'fight or flight' response (Cannon 1929), that affects a number of biological systems including cardiovascular and gastrointestinal function (Moberg 2000). Changes in heart rate, blood pressure, respiratory rate, and body temperature occur as blood flow is redirected away from the extremities towards organs and musculature. These physiological changes are often used as indicators of an individual's ability to respond to a stressor. Heart rate, and perhaps more importantly, heart rate variability (HRV), have been used in animal research to investigate responses to psychological and environmental stressors, and how they are shaped by individual characteristics, such as temperament and coping strategies (for review, see Von Borell et al. 2007). Heart rate is controlled by non-additive effects and interactions between the two branches of the ANS: the sympathetic and the parasympathetic (Hainsworth 1995), leading to constant variability. HRV is the result of either an increase/decrease of sympathetic tone or an increase/decrease of parasympathetic tone, and may be a sensitive and useful indicator for determining the physiological and psychological flexibility of an animal and its ability to respond to stress (Friedman and Thayer 1998). Among animals, heart rate and HRV have been used to investigate social (de Jong et al. 2000; Gácsi et al. 2013), environmental (Accardi et al. 2017), and management-related stressors (Hagen et al. 2005; Parrott et al. 1998). For example in chimpanzees, exposure to conspecific threat, stress, and alarm vocalizations resulted in increased HRV (Berntson and Boysen 1989). Most studies conducted to-date have investigated cardiac responses to short-term stressors, but there is some evidence that chronic stress can have a cumulative effect on HRV (Park et al. 2017). Long-term monitoring of heart health can be facilitated by the use of implantable devices. Although not strictly non-invasive because they require implantation, heart rate monitors allow the assessment of physiological responses in free-ranging animals (Chaise et al. 2017), and for prolonged periods of time without additional contact (Laske et al. 2017); for example, the response of wild black (Ditmer et al. 2015, 2019; Laske et al. 2011) and brown (Støen et al. 2015) bears to human and environmental stressors. Similar devices have also been used in free-ranging Przewalski's horses to assess seasonal changes in stress and metabolic demand, with decreased HRV during the winter and spring compared to summer, and HRV negatively correlated with HR and body temperature (Pohlin et al. 2017).

A second ANS indicator, blood pressure, also has been used to investigate both mental (Hjortskov et al. 2004) and physical stress (Irvine et al. 1997). In one example, the effect of different housing conditions on rat welfare was assessed using heart rate, blood pressure, and body temperature (Krohn et al. 2003). Rats kept on grid or plastic flooring had significantly higher systolic and diastolic blood pressure, heart rate, and body temperature; blood pressure remained elevated even after the animals were returned to standard bedding. This suggests that changes in ANS function may persist even after the stressor has been removed. While there can be logistical challenges to overcome when training animals for blood pressure monitoring (e.g., gorillas: Murphy et al. 2018), efforts are ongoing at many zoos. Respiratory rate is a commonly used metric in studies of fish welfare. Increased ventilation rates have been described in response to stress (Barreto and Volpato 2011; Laitinen and Valtonen 1994) and pain (Ashley et al. 2009) in fish, and increased respiration rate and volume also occur in response to pain in humans (Kato et al. 2001), small mammals (Mayer 2007), and farm animals (Prunier and Leterrier 2014).

Changes in blood flow due to catecholamine (e.g., epinephrine, norepinephrine) and GC release can result in changes in body temperature. Infrared thermography has been used in a veterinary context for many years to assess injury (Church et al. 2009; Hilsberg 2002) and heat stress (Adamec et al. 1998; Kimmel et al. 1992). Re-direction of blood flow towards a site of injury and associated vasoconstriction or vasodilation can be detected using this approach. More recently, thermography has also been used to assess psychological stress. In these instances, so-called 'thermal windows', or areas of the body responsible for heat exchange (Weissenböck et al. 2010), can be used to assess arousal. Eye (Dai et al. 2015; Stewart et al. 2008b; Travain et al. 2016; Valera et al. 2012), nose (Ioannou et al. 2015; Kano et al. 2016; Kuraoka and Nakamura 2011; Nakayama et al. 2005; Proctor and Carder 2015), ear (Riemer et al. 2016), and comb temperature (Moe et al. 2012) all have been utilized for welfare assessments in a variety of species. A recent study on wild blue tits found correlations between eye thermal signatures and both body condition and circulating GC concentration, suggesting that thermography could integrate multiple aspects of physiological state (Jerem et al. 2018). These techniques also have the advantage of being non-invasive; however, valence can be difficult to establish (Travain et al. 2016). Temperature changes can be elicited during negative states such as fear and anxiety (Lecorps et al. 2016; Stewart et al. 2008a), but also during positive states (Travain et al. 2016), so data interpretation may not be straightforward. An additional caveat with ANS monitoring is that responses typically are acute, so alternative approaches may be required to assess long-term welfare.

In addition to these physiological assessments, there are several potential biomarkers that reflect ANS function. Catecholamines have been quantified in a wide range of taxa including domestic livestock (Hay and Mormède 1998), companion animals (Beerda et al. 2000), rodents (Niebylski et al. 2012), birds (Beuving and Blokhuis 1997), reptiles (Benn et al. 2019), primates (Dettling et al. 2002) and elephants (Dehnhard 2007), increasing in serum/plasma and urine in response to acute stress. In addition, salivary alpha-amylase and chromogranin-A have received much attention to-date, particularly in human psychological research.

Alpha-amylase is considered one of the most important enzymes in saliva, involved in the initiation of starch digestion in the oral cavity (Nater and Rohleder 2009), as well as having an important anti-bacterial function (Scannapieco et al. 1993). The salivary glands are under ANS control, and salivary alpha-amylase has been used to assess various psychological stress responses in humans, as reviewed by Nater and Rohleder (2009). Salivary alpha-amylase correlates with cardiovascular parameters (Bosch et al. 2003; Granger et al. 2006; Nater et al. 2006) and catecholamine release (Thoma et al. 2012), and increases have been reported in response to acute (Bosch et al. 1996) and accumulated stressors (Bosch et al. 1998), while both baseline (Vigil et al. 2010) and magnitudes of response (Nater et al. 2007) correlate with chronic stress. Takai et al. (2004, 2007) further demonstrated an increase in salivary alpha-amylase in response to a stressful paradigm and a decrease during relaxation. Additionally, the ratio of salivary alpha-amylase to cortisol has been linked to both physical (Filaire et al. 2013) and psychological (Booij et al. 2015) well-being and so may be a more useful marker of stress dysregulation in humans than either biomarker alone (Ali and Pruessner 2012). Measures of salivary alpha-amylase as an indicator of ANS activation during acute stress have been reported in a number of animal species, including rats (Matsuura et al. 2012), dogs (Contreras-Aguilar et al. 2017), pigs (Contreras-Aguilar et al. 2018; Fuentes et al. 2011), horses (Fuentes-Rubio et al. 2015), and non-human primates (Behringer et al. 2012, 2013; Higham et al. 2010; Mau et al. 2010; Smiley et al. 2010). However, species differences in alpha-amylase activity exist and may be related to diet, with greater levels present in species with a higher starch intake (Behringer et al. 2013). Potential species differences, as well as the impact of food intake and circadian rhythm on salivary biomarker production, should be explored further before applying it to wildlife species.

Chromogranin-A (CgA) is a member of the granin family of peptides secreted by the adrenal medulla (Cohn et al. 1982) and salivary glands (Saruta et al. 2005) and has also been utilized to assess ANS activity in relation to welfare. It is a prohormone for several biologically active peptides that influence endocrine, cardiovascular, and immune functions, and also affects calcium and glucose homeostasis (D'amico et al. 2014). In humans, circulating CgA is used as a diagnostic marker for pathologies including endocrine tumors (Duque et al. 2013; Giovinazzo et al. 2013; Massironi et al. 2010), heart failure and hypertension (Estensen et al. 2006; Jansson et al. 2008), inflammatory diseases (Di Comite et al. 2006; Zhang et al. 2009), and neurodegenerative and neuropsychiatric diseases (Bartolomucci et al. 2010; Conlon 2010; O'Toole et al. 2009). Salivary CgA, on the other hand, has been used primarily as an indicator of psychophysical stress due to its co-release with catecholamines (Kanno et al. 1998). In human studies, CgA has been associated with a variety of psychological (Kanamaru et al. 2006; Nakane et al. 1998; Takatsuji et al. 2008) and physical (Miyakawa et al. 2006; Yoshida et al. 2011) stressors. This biomarker has also been quantified in dogs (Akiyoshi et al. 2005) and rodents (Kanno et al. 1998), and in pigs, it has been investigated in studies exploring the effect of immobilization (Escribano et al. 2013), food deprivation (Ott et al. 2014), and social isolation (Escribano et al. 2015) on stress biomarkers, as well as the beneficial effect of environmental enrichment on improving well-being (Casal et al. 2017). It would be worth determining how these biomarkers are conserved across taxa, and which could be used to assess physical and/or psychological stress in wildlife species.

3.2 *Neuroendocrine*

As described above, GCs have been by far the most widely used physiological assessments of welfare to-date, and many excellent reviews have discussed their various uses in wildlife (Palme 2019; Sheriff et al. 2011). In addition to GCs, however, the adrenal gland also secretes dehydroepiandrosterone (DHEA), and the sulfated metabolite, DHEA-S, in response to ACTH stimulation (Kroboth et al. 1999). In humans and some primates, DHEA and DHEA-S act on multiple physiological systems (Kroboth et al. 1999). For example, DHEA-S acts as an antagonist to cortisol (McEwen and Seeman 1999), and has been shown to positively correlate with stressors such as acute psychosocial stress (Lennartsson et al. 2012; Shirotsuki et al. 2009), chronic physical and psychological stress (Bernton et al. 1995), and post-traumatic stress disorder (Morgan et al. 2004). Conversely, DHEA-S is negatively associated with age, and thought to be involved in somatic maintenance and healthy aging (Edes 2017). Decreased DHEA-S has also been described with several diseases, including renal (Zumoff et al. 1980) and liver disease (Floreani et al. 1991), diabetes mellitus (Yamauchi et al. 1996), coronary artery disease (Herrington et al. 1990), rheumatoid arthritis (Masi 1995), and Alzheimer's disease (Svec and Lopez 1989). Studies in other taxa suggest that acute and chronic stress may also increase DHEA-S concentrations in rhesus macaques (Maninger et al. 2010) and in birds (Monclús et al. 2018; Newman et al. 2008), while clinical pathology is associated with decreased DHEA in cattle (Almeida et al. 2008). Increased DHEA occurs following ACTH stimulation (Yon et al. 2007b) and during the physiological state of musth in Asian bull elephants (Yon et al. 2007a). It has been proposed that chronic stress and hyperactivity of the HPA axis could lead to a shift towards GC production at the expense of DHEA (Guilliams and Edwards 2010); accordingly the cortisol:DHEA ratio may be a useful metric in studies investigating the relationship between adrenal activity and stress. For example, Gundlach et al. (2018) found cortisol concentrations did not differ between diseased and healthy harbor and gray seals, but diseased animals had lower DHEA and a higher cortisol:DHEA ratio. DHEA-S can be measured non-invasively in feces (Takeshita et al. 2018) and hair (Peric et al. 2017), so it has the potential for wide application to wildlife studies. Another adrenal steroid, aldosterone, and related fecal metabolites were positively correlated with GCs in pregnant female and reproductively mature male North Atlantic right whales (Burgess et al. 2017), offering an additional marker of adrenal activity.

Another hormone that has been used to assess stress is prolactin, which is perhaps one of the most versatile hormones in the mammalian body, with over 300 known physiological roles described (as reviewed by Torner 2016). As well as key roles in pregnancy and lactation, prolactin also moderates the stress response through a variety of mechanisms, including but not limited to activation of the hypothalamic-pituitary-adrenal (HPA) axis. An increase in prolactin in association with increased GC secretion has been reported in fish under acute and chronic stress (Wendelaar Bonga 1997). However, in birds, prolactin levels often decrease in response to short-term stressors, such as handling and restraint (reviewed in Angelier and Chastel 2009). Thus, differences across taxa in the role that prolactin plays in the stress response should be understood before using this hormone as a potential welfare indicator in wildlife.

3.3 Immune Function

GCs play an important role in regulating immune function, exerting multiple effects on immune cells and their mediators to maintain homeostasis (reviewed in Webster et al. 2002). However, especially when prolonged, stress can have immunosuppressive effects (Siegel 1987), disrupting immune function in two broad ways. Hyper-activation of the HPA axis and GC production can lead one to be prone to infection, whereas suppressed HPA activity, as during chronic stress, can increase susceptibility to inflammation (Webster et al. 2002). Physiological and therapeutic GCs exert anti-inflammatory effects to prevent over-activation by down-regulating pro-inflammatory and up-regulating anti-inflammatory cytokines, promoting a shift from Th1 (cellular) to Th2 (humoral) immunity (Elenkov and Chrousos 1999), and suppressing the production of inflammatory mediators including prostaglandins (Newton et al. 1996) and nitric oxide (Korhonen et al. 2002). GCs also impact immune cells directly by suppressing maturation, differentiation, and proliferation, reducing leukocyte recruitment and adhesion to the site of injury, and promoting apoptosis of a variety of immune cells (Webster et al. 2002; Webster Marketon and Glaser 2008). Physiological levels of GCs regulate the fine balance between inflammatory and anti-inflammatory actions crucial to an appropriate immune response; but during stress, dysregulation can occur (reviewed in Webster Marketon and Glaser 2008). A wide variety of immune function biomarkers, including immune cells and protein mediators and products of the innate and adaptive immune response, have been used as indicators of welfare.

One of the earliest immune function tests in vertebrate species involved examination of leukocyte profiles in blood smears (reviewed in Davis et al. 2008). GCs increase the number and proportion of neutrophils (or heterophils in birds and reptiles) released from bone marrow, while sequestering lymphocytes, so the ratio of these two cell types (N:L or H:L) is a useful indicator of physiological state, with lower ratios indicating better health and welfare. Cellular changes have been observed in response to exogenous GC (Anderson et al. 1999; Kranendonk et al. 2005; Morici et al. 1997; Reidarson and McBain 1999; Wojtaszek et al. 2002), or ACTH (Bilandžić et al. 2006; McLeay 1973; Rossdale et al. 1982) administration, as well as during psychological disorders (Kronfol et al. 1984), transport stress (Frank et al. 2006; Groombridge et al. 2004; Kock et al. 1999; Noda et al. 2007; Parga et al. 2001), strenuous exercise (Cardinet et al. 1964; Rossdale et al. 1982), restraint stress (López-Olvera et al. 2007; Morrow-Tesch et al. 1993), social stress (Hickey et al. 2003), and exposure to environmental contaminants (Cabagna et al. 2005; Witeska 2005). In avian species, the H:L has also been used to predict susceptibility to disease (Al-Murrani et al. 2002, 2006), growth rates (Moreno et al. 2002) and survival to the next breeding season (Kilgas et al. 2006; Lobato et al. 2005). In box turtles, lower ratios were observed in individuals living under more enriched conditions (e.g., mulched floors, shredded paper and a box for shelter as opposed to newspaper flooring alone: Case et al. 2005); whereas in pigs, differences in humoral immune responses were reported for individuals raised in barren compared to environments enriched with straw bedding (Bolhuis et al. 2003). In sheep, outdoor access had beneficial effects on lymphocyte proliferation, while reduced space had a negative impact on humoral responses (Caroprese et al. 2009). Finally, rearing density was also correlated with H:L in farm-raised soft-shelled turtles (Chen et al. 2007).

Many of these cellular changes are also associated with changes in protein mediators, such as cytokines, chemokines, and acute phase proteins (APPs) that signal, attract, and moderate immune cell function and other inflammatory processes. Much of the research in this area has focused on the impact of psychological stress on immune biomarkers (reviewed by Hänsel et al. 2010). Low socioeconomic status correlates with both increased concentrations of the APP C-reactive protein (CRP) (Alley et al. 2006; Gimeno et al. 2007b; Lubbock et al. 2005; McDade et al. 2006; Rosvall et al. 2006) and pro-inflammatory cytokine interleukin (IL)-6 (Brydon et al. 2004; Gimeno et al. 2007a; Hemingway et al. 2003; Koster et al. 2006; Loucks et al. 2006). This same cytokine was reportedly elevated among caregivers with terminally ill relatives (Kiecolt-Glaser et al. 2003; Lutgendorf et al. 1999). High levels of job stress have been associated with increased concentrations of another pro-inflammatory cytokine, tumor necrosis factor-alpha (TNF-α) (Grossi et al. 2003; von Känel et al. 2008), as well as decreased anti-inflammatory cytokines (Mommersteeg et al. 2006; von Känel et al. 2008) and increased CRP (Toker et al. 2005). TNF-α and interferon-gamma (IFN-γ) were both shown to be associated with mood and psychosocial stress (O'Connor et al. 2009), and similar biomarkers have been used to investigate major depression (Lopresti et al. 2014), with increased concentrations of CRP (Howren et al. 2009; Valkanova et al. 2013), TNF-α, IL-1 and IL-6 being observed (Dowlati et al. 2010; Howren et al. 2009). Studies in poultry have also investigated the role of cytokines and chemokines during stress. Following the administration of corticosterone, cytokine and chemokine mRNA expression levels were altered in both lymphocytes and heterophils (Shini et al. 2010; Shini and Kaiser 2009). Pro-inflammatory proteins were initially up-regulated, presumably to facilitate the immune response, but then were suppressed following chronic GC treatment, with a shift from Th1 to Th2 dominance. Acute stress can also impact protein immune biomarkers. For example, transport stress is a commonly used experimental stressor in domestic livestock, and has been associated with increased APP concentrations in pigs (Pineiro et al. 2007), sheep (Piccione et al. 2012), and cattle (Arthington et al. 2003; Fazio et al. 2015; Lomborg et al. 2008). Although these biomarkers are generally measured in serum or plasma, there is evidence that some can also be quantified non-invasively in saliva. For example, acute immobilization stress in pigs was associated with an increase in salivary IL-18, a pro-inflammatory cytokine (Muneta et al. 2011), while increases in salivary CRP were observed in zoo-housed gorillas following social stress and aggression (Fuller and Allard 2018). Finally, physiological changes in immune function due to social rank may extend to the molecular level. For example, low social status in female rhesus macaques is associated with an up-regulation of the genes that are involved in inflammation, resulting in increased circulating levels of inflammatory markers (Snyder-Mackler et al. 2016). Conversely, in wild yellow baboon males, increased expression of genes involved in immune defense and inflammation are a precursor, rather than a consequence, of social status (i.e., males with up-regulated gene expression are more likely to have high social status: Lea et al. 2018a).

3.4 Biomarkers of Positive Affect

Affective states can be quantified in terms of level of arousal (low to high activation) and positive to negative valence (Mendl et al. 2010). Providing good welfare is not merely about the absence of negative experiences, but also about promoting positive ones. Incorporating measures of both positive and negative states allows us to evaluate welfare along a continuum, assessing factors that are good for an individual, as opposed to just not being bad. This may be the key to ensuring that individuals do not merely cope with their environment, but thrive. Indeed, many markers of ANS, neuroendocrine, and immune function can potentially inform about positive affects (see Boissy et al. 2007; Yeates and Main 2008; Whitham and Wielebnowski 2013 for reviews) as well as negative ones; although a challenge now is determining how to distinguish between the two. Several human studies have explored HRV as a potential indicator of positive affect, including its relation to positive mother:infant interactions (Fox 1989), self-induced positive emotional states (McCraty et al. 1995, 1998), and relaxation therapy that was correlated with pleasant and calm emotions (Lee et al. 2005). Furthermore, there is some evidence that the range of emotion, such as happiness or sadness, can be distinguished by the pattern of HRV (McCraty et al. 1995; Rainville et al. 2006; Tiller et al. 1996). The same may not be true for commonly used neuroendocrine indicators, however. Mice subject to either positive (sexual encounter) or negative (aggression) stimuli experienced similar GC responses (Buwalda et al. 2012); the same was true for horses restrained, exercised, or allowed to mate (Colborn et al. 1991), and pigs experiencing social success or defeat (Otten et al. 2002). This inability to clearly distinguish between positive versus negative responses has so far made GCs difficult to use for assessing positive states, at least in isolation.

One proposed biomarker of positive affect is immunoglobulin A (IgA), an antibody that plays an important role in the immune defense against pathogens (Kerr 1990). Produced at mucosal linings, secretory IgA is present in saliva, tears, bile, milk, and mucosal secretions of the reproductive, respiratory, and gastrointestinal systems (Pihl and Hau 2003), and acts as the first line of defense against pathogens including viruses and bacteria. A distinct circulatory form of IgA is produced by plasma cells in the bone marrow and acts as a secondary line of defense to eliminate pathogens that breach the mucosal surface (Woof and Kerr 2004). Secretory IgA has been shown to be negatively correlated with GCs, decreasing during times of stress, and increasing in response to positive stimuli. Physical stressors such as intensive exercise (Gleeson et al. 1995; Skandakumar et al. 1995), psychological challenges

(Deinzer and Schuller 1998; Ng et al. 1999), metabolic demand (Royo et al. 2005), and relocation to a new environment (Bundgaard et al. 2012) have all been associated with decreased secretory IgA. Conversely, positive stimuli such as relaxation and positive emotional states (Green et al. 1988), pleasant linguistic stimuli (Watanuki and Kim 2005), humor (Dillon et al. 1986; Lambert and Lambert 1995; McClelland and Cheriff 1997), and experiencing self-induced positive emotional states (McCraty et al. 1996) are associated with increased secretory IgA. Secretory IgA can be measured non-invasively in saliva (Kikkawa et al. 2003; Kvietkauskaite et al. 2014; Lucas et al. 2007), urine (Eriksson et al. 2004; Paramastri et al. 2007; Rehbinder and Hau 2006), and feces (Rehbinder and Hau 2006; Royo et al. 2004; Tress et al. 2006). However, unlike for secretory IgA, relatively little research has been conducted into the health and well-being implications of circulatory IgA (Leong and Ding 2014), although significant inter-individual variation has been reported (Edwards et al. 2019). This immunoglobulin appears to be well-conserved across the species studied to-date, including humans (Gleeson et al. 1995; Hucklebridge et al. 1998), cats (Gourkow et al. 2014), dogs (Kikkawa et al. 2003; Skandakumar et al. 1995; Svobodova et al. 2014), pigs (Muneta et al. 2010), cattle (Wilson et al. 2002), rodents (Bundgaard et al. 2012; Guhad and Hau 1996; Hau et al. 2001), non-human primates (Huang et al. 2014; Lantz et al. 2016; Paramastri et al. 2007), reindeer (Rehbinder and Hau 2006; Yin et al. 2015), and elephants (Edwards et al. 2019). Although the exact mechanism by which increases in IgA occur during positive affective states has not been fully elucidated, it has great potential as a marker of well-being in wildlife species. One important caveat is that just as with GCs, interpretation of IgA data may not always be straightforward. For example, rather than being associated with positive welfare, IgA and CGs were both elevated during an acute systemic illness in a zoo Asian elephant (Edwards et al. 2019), and others have shown both IgA and GCs increase in response to acute stressors of a non-immune nature (Jarillo-Luna et al. 2015; Tsujita and Morimoto 1999). These apparent divergent results may mean that IgA is a useful measure of psychological well-being, but in order to distinguish between positive affective states and physiological health may require longitudinal sample collection or additional measures used in combination.

Other potential biomarkers examined in the context of positive affective states include catecholamines, epinephrine and norepinephrine, which act to increase blood flow and cardiovascular activity in response to a variety of stimuli. Pressman and Cohen (2005) defined positive affect as either stable disposition-like (states), or relatively short-term bouts of positive emotions (traits). In both cases, positive affect is associated with lower levels of epinephrine and norepinephrine (Berk et al. 1989; Codispoti et al. 2003; Cohen et al. 2003). Neurohormones dopamine (Knutson et al. 2002), oxytocin (Liberzon et al. 1997; Mitsui et al. 2011), and serotonin (Flory et al. 2004), as well as endorphins (van Ree et al. 1999), have also been investigated in relation to positive affect. These substances are difficult to monitor in wildlife as they generally are measured centrally rather than peripherally. However, catecholamines, (epinephrine, norepinephrine and dopamine) have been measured in elephant urine, and show changes in response to stressful events (Dehnhard 2007).

Urinary neurohormone (dopamine, oxytocin and serotonin) differences have also been reported between African elephant females with normal and abnormal reproductive cycles (Prado et al. 2019). Finally, both alpha-amylase and CgA show promise as positive welfare indicators, with increased concentrations associated with pleasure/approach behavior (Fortunato et al. 2008) and laughter (Toda et al. 2007), respectively. Although much work is still needed to fully validate the use and biological relevance of some of these potential biomarkers, they provide exciting potential avenues for assessing positive emotion in animals that, when used in combination with other traditional physiological and/or behavioral measures, may provide a more holistic means of assessing animal well-being.

4 Multi-Biomarker Measures of Stress and Welfare

The examples above are just a few of the potential biological markers available today, and it is becoming clear that the sheer number of possibilities present a substantial challenge in determining which ones, or what combinations, are most effective for studying welfare under varying circumstances. Because multiple physiological systems are activated during the stress response, the most effective approach likely will rely on measurement of multiple analytes, which could differ depending on the origin and type of stressor. Stress affects the entire organism, with a multitude of effects occurring downstream of changes in the HPA axis (Everly and Lating 2013; Nelson 2011; Seaward 2006). It is this multi-systemic nature of stress responses and their effects that makes biomarkers reflective of ANS activation and immune function promising additions to GCs as physiological tools for assessing welfare and well-being. For that reason, allostatic load indices (ALIs) have received considerable attention as potential tools for assessing the welfare of animal collections in human care, as well as informing conservation efforts *in situ*. Frequently used in human research since 1997, ALIs have been translated to animals only recently through their application to zoo-housed gorillas (Edes et al. 2016a, b, 2018b).

ALIs are composites of biomarkers from multiple somatic systems that reflect allostatic load, or the physiological wear-and-tear that accumulates over the lifespan due to senescence and experiencing both acute and chronic stressors (McEwen and Stellar 1993). They consist of a combination of biomarkers representing the neuroendocrine, cardiovascular, metabolic, and immune systems, although the number and systems involved may vary by index. Originally designed to predict individuals at greatest risk for future poor health in sub-clinical human populations (Seeman et al. 1997), dozens of different ALIs have been applied since their first iteration (see Juster et al. 2010; Beckie 2012; Edes and Crews 2017). Measures of allostatic load have been shown to associate with a wide variety of outcomes, such as increased risk of poor mental and physical health, development of chronic disease, and shortened lifespans. Furthermore, research in humans has identified several factors that predict increased development of allostatic load, including early life adversity, lower socioeconomic position, and poor social connectedness. The flexibility with which allostatic load can be estimated and the overlap between relationships observed in humans and characteristics we share with other species have made the application of allostatic load theory and methodology appealing to numerous animal researchers. In this section, we first review the theory behind allostatic load and then demonstrate how the substantial body of work on humans can be applied to research on animal welfare and reproductive success.

4.1 Allostasis and Allostatic Load

Homeostasis has long served as the foundation of biological functioning (Bernard 1865; Cannon 1932) and is so widely accepted that deviations from defined physiological values often are used to diagnose and treat clinical conditions (Sterling 2012). This idea of internal constancy is accurate for some parameters, such as pH and oxygen tension (McEwen 1998a, b; Sterling 2012), but for many somatic systems, variability based on current need is best. For example, blood pressure and heart rate change in response to demand and, while we can calculate average values, it would be inefficient and maladaptive for the body to maintain an average value during times of exertion or rest (Sterling 2012; Sterling and Eyer 1988). This variation in physiological parameters to meet current needs is allostasis; Sterling and Eyer (1988) defined it simply as "stability through change". While homeostatic systems operate through local feedback, allostasis involves the entire soma and is directly regulated by the brain. This allows organisms to continuously re-evaluate their situation, as well as anticipate future needs and adjust (Sterling 2012; Sterling and Eyer 1988). Stress responses are highly integrated, multi-system allostatic responses designed to promote survival. Once stressors are controlled or eliminated, demand-based allostatic changes cease and systems return to baseline (McEwen 1998a, b). Allostasis promotes health by allowing organisms to adequately respond to challenges (Sterling 2012; Sterling and Eyer 1988), but these responses can be costly. Over time, as individuals experience repeated and/or chronic stressors, even from minor day-to-day events, damage to cells and tissues may occur, leading to the loss of physiological integration (McEwen and Stellar 1993). This systemic physiological impairment and dysregulation, or "wear-and-tear", is allostatic load (McEwen and Stellar 1993).

Assessment of allostatic load was first applied by Seeman et al. (1997), who used 10 biomarkers: systolic blood pressure, diastolic blood pressure, and high-density lipoprotein to total cholesterol ratio were used to assess cardiovascular health; waist-hip ratio and glycated hemoglobin (Hb_{A1c}) reflected the metabolic system; and cortisol, DHEA-S, epinephrine, and norepinephrine represented the neuroendocrine system. Allostatic load was estimated using the quartile method, which remains the most frequently used technique (Fig. 5.3) (Edes and Crews 2017; Edes et al. 2018a). With this approach, the sample distribution of each biomarker is divided into quartiles and one quartile is assigned as high-risk, depending on the biomarker (e.g., the highest/fourth quartile is high-risk for HbA_{1c}, but the lowest/first quartile is high-risk for DHEA-S). Quartiles were chosen as opposed to deciles or clinical thresholds because Seeman et al. (1997) wanted an index that could measure sub-clinical risk and thus predict those individuals who were at greatest risk for future health concerns, rather than simply those who were already known to be in poor health. Allostatic load scores were then determined by summing the number of biomarkers within high-risk quartiles for each individual; thus, scores could range from 0 (i.e., no biomarkers considered high-risk) to 10 (i.e., all biomarkers considered high-risk). Higher allostatic load scores were associated with poorer cognitive and physical functioning, larger declines in cognitive and physical functioning, and increased risk for cardiovascular disease (CVD; Seeman et al 1997). This research, and other studies since, demonstrated that multi-biomarker composites often are better at predicting future risk of morbidity and mortality than any one biomarker alone (Gruenewald et al. 2006; Karlamangla et al. 2002, 2006; Seeman et al. 1997, 2001, 2004a; Seplaki et al. 2004).

Seeman et al. (1997) also recognized that theirs was just one of many possible ALIs, paving the way for the wide variety in biomarkers and methodologies used today (Edes and Crews 2017; Juster et al. 2010). This flexibility has been essential to ongoing research, allowing the use of biomarkers from existing datasets to estimate allostatic load and address hypotheses through a variety of lenses. This precedence sets the stage for studying allostatic load in wildlife using a variety of biomarkers that have already been measured (Fig. 5.4). While some have criticized the lack of a "gold standard" for estimating allostatic load (Duong et al. 2017; Gallo et al. 2014; Mauss et al. 2015; Read and Grundy 2012), consistent results across the existing body of literature demonstrates that the concept and basic methodology are sound (Edes and Crews 2017; Prior et al. 2018). In fact, a standard set of biomarkers

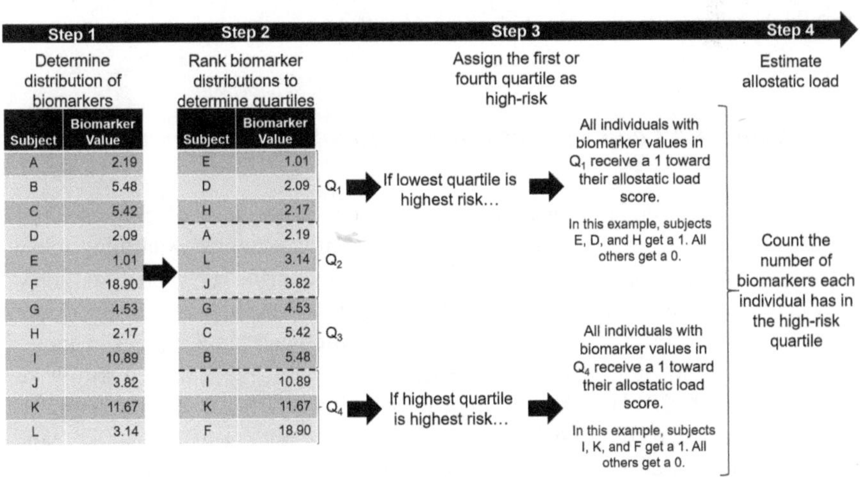

Fig. 5.3 Estimating allostatic load using one-tailed quartiles, following the original methodology in human research proposed by Seeman et al. (1997). Modified from Edes et al. (2018a)

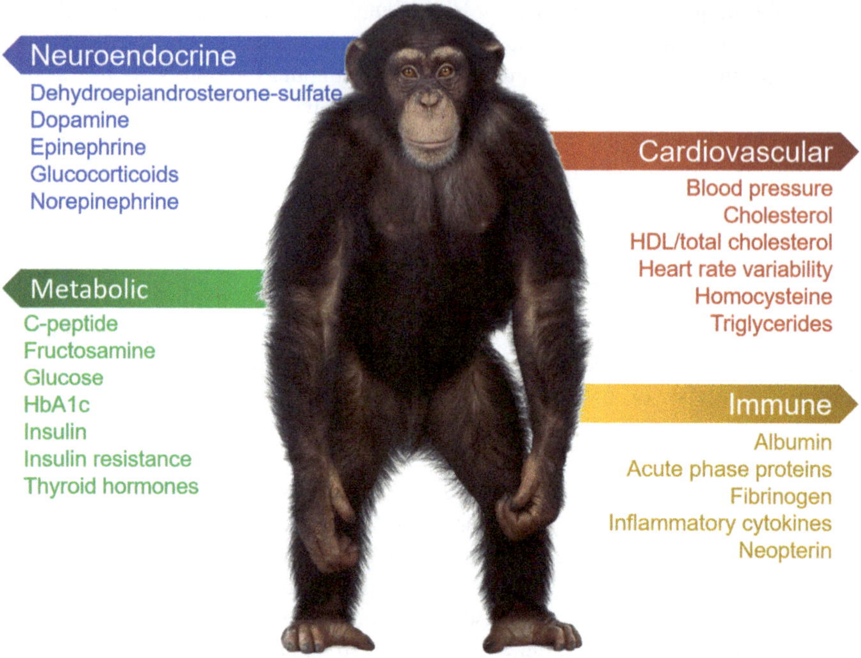

Fig. 5.4 Examples of biomarkers representing multiple body systems that have been or could be measured in wildlife species to estimate allostatic load

for estimating allostatic load may not be ideal. Rather, addressing different questions may require flexibility in the biomarkers chosen and methodologies employed (Delpierre et al. 2016) and applications to wildlife likely will require species-specific ALIs (Edes et al. 2018a).

4.2 Allostatic Load Theory and Methodology in Animals

Studies have used single biomarkers (e.g., glucocorticoids, testosterone) as simple proxies of allostatic load in some wild species. Examples include examining sociality, testosterone levels, and ornamentation in male mandrills (Setchell et al. 2008); body condition and group composition in dominant versus subordinate superb starlings (Rubenstein and Shen 2009); sibling competition and hunger stress in spotted hyena cubs (Benhaiem et al. 2013); energy storage in black-legged kittiwakes (Schultner et al. 2013); seasonal differences in sociality in female Assamese macaques (Fürtbauer et al. 2014); hierarchy stability among male bearded capuchins (Mendonça-Furtado et al. 2014); male response to female ovulation in white-faced capuchins (Schoof et al. 2014); effects of age and season on grey mouse lemurs (Hämäläinen et al. 2015); habitat degradation in common lizards (Josserand et al. 2017); seasonal variation in energetic demands in Przewalski's horses (Pohlin et al. 2017); group member replacement in blue-fronted Amazon parrots (Silva de Souza Matos et al. 2017); energetic expenditure in waterfowl (Johns et al. 2018); urbanization in yellow-bellied marmots (Price et al. 2018); chronic stress in European seabass and gilthead seabream (Samaras et al. 2018); and energetic demands in juvenile blue monkeys (Thompson et al. 2019). However, despite similarities in stress response physiology among vertebrates (Cavigelli and Caruso 2015; Nelson 2011), the application of multi-biomarker ALIs outside of humans has been limited to only one species of non-human primate: captive western lowland gorillas (Edes et al. 2016a, b, 2018b).

4.3 *From Human Research to Animal Welfare*

One of the most obvious applications for ALIs is to monitor the effects of aging on functional decline and to predict risk of poor health outcomes. Of interest is understanding why some zoo-housed species tend to outlive their free-ranging conspecifics, while others do not, making age-related patterns of decline and disease a concern for animal caretakers. Age is a well-established predictor of allostatic load in humans (Crimmins et al. 2003, 2009; Levine and Crimmins 2014; Rosenberg et al. 2014; Seeman et al. 2010, 2014; Taylor et al. 2018) and has been shown to positively associate with allostatic load in gorillas (Edes et al. 2016a, 2018b). In humans, allostatic load is associated with both chronic degenerative conditions such as CVD (Karlamangla et al. 2002; Mattei et al. 2010; Nelson et al. 2007; Sabbah et al. 2008; Seeman et al. 2002), hypertension (Mattei et al. 2010; Zota et al. 2013), diabetes (Mattei et al. 2010), arthritis (Mattei et al. 2010), and periodontal disease (Borrell and Crawford 2011; Sabbah et al. 2008), as well as increased mortality risk (Borrell et al. 2010; Castagné et al. 2018; Duru et al. 2012; Glei et al. 2014; Gruenewald et al. 2006; Hwang et al. 2014; Karlamangla et al. 2006; Levine and Crimmins 2014; Robertson et al. 2017; Seeman et al. 1997, 2001, 2004a). Many species of conservation concern also experience high rates of disease, particularly those in captivity.

4.3.1 Disease and Health Biomarkers

Although etiology can differ, allostatic load across taxa may be similarly associated with disease development and consequences. CVD is common among captive great apes (Lowenstine et al. 2016; McManamon and Lowenstine 2012; Murphy et al. 2018; Strong et al. 2016, 2018), giant pandas (Qiu and Mainka 1993), cheetahs (Munson et al. 1999), and marine mammals (Kreuder et al. 2005; Zabka et al. 2009). Similar chronic conditions occur in wildlife, arthritis in particular, with nearly all taxa affected; i.e., birds, reptiles, land and marine mammals (Miller and Fowler 2012; Nganvongpanit et al. 2017). Consistent with findings in humans, gorillas with at least one chronic degenerative condition (e.g., CVD, arthritis) have higher allostatic load (Edes, unpublished data). Infectious diseases, such as elephant endotheliotropic herpesvirus (EEHV) (Atkins et al. 2013; Garner et al. 2009; Kendall et al. 2016; Metzler et al. 1990; Richman et al. 2000a, b; Schaftenaar et al. 2010), encephalomyocarditis virus (Grobler et al. 1995; Jones et al. 2011; Lamglait et al. 2015; Masek-Hammerman et al. 2012; Reddacliff et al. 1997; Wells et al. 1989), malignant catarrhal fever (Hänichen et al. 1998; Li et al. 1999; Okeson et al. 2007; Uphoff Meteyer et al. 1989), salmonellosis (Adesiyun et al. 1998; Gopee et al. 2000; Jang et al. 2008; Onderka and Finlayson 1985), and tuberculosis (*Mycobacterium* spp.) (for review, see Montali et al. 2001; West 2006; Chambers 2009; Lécu and Ball 2011; Miller and Fowler 2012) also are observed in a variety of species. As stress is associated with immunosuppression (McEwen 1998a, b), higher allostatic load likely increases susceptibility to infectious diseases. Furthermore, the presence of disease may further contribute to allostatic load.

Allostatic load in animals also may associate with parasite load, which is a common proxy of health, immune status, and overall physical condition (Lim et al. 2008; Muehlenbein and Watts 2010; Robert and Schwanz 2013; Vidya and Sukumar 2002). In humans, individuals who tested positive for the *Toxoplasma gondii* parasite had 61% higher allostatic load than non-seropositive peers (Egorov et al. 2018). A number of studies have documented an association between habitat disturbance, such as forest fragmentation or urbanization, and increased fecal parasite richness and/or abundance in wild animals, including roe deer (Lutz and Kierdorf 1997), common fruit bats (Cottontail et al. 2009), agile antechinus and other forest marsupials (Fontúrbel et al. 2014; Johnstone et al. 2010), freshwater fishes (Lymbery et al. 2010), multiple lizard species (Italian wall lizards: Corti et al. 2009; ornate dragon lizards: Radwan et al. 2014; common wall lizards: Lazić et al. 2017), and several non-human primates (howler monkeys: Vitazkova and Wade 2007; grey mouse lemurs: Raharivololona and Ganzhorn 2009; lion-tailed macaques: Hussain et al. (2013); vervet monkeys: Thatcher et al. 2018; but see study on grey mouse lemurs and fat-tailed dwarf lemurs: Rakotoniaina et al. 2016). The ability for ALIs to predict future morbidity and mortality suggests this tool could be applied to assess the risk to wild populations under anthropogenic stress, to prioritize resources for intervention.

4.3.2 Diet and Nutrition

Proper diets are essential for ensuring health and well-being across species. For example, recent studies have shown that the diets of wild lorises and pottos are quite diverse and it has been suggested that long-term consumption of homogenous diets in captivity may contribute to their high morbidity and mortality rates (Fuller et al. 2014; Simpson et al. 2018). Monitored over the long-term, it may be possible to use ALIs to determine responses to dietary changes or quality both *in situ* and *ex situ*. In humans, lower quality diets, such as those lacking in necessary vitamins and minerals, are associated with higher allostatic load (Frei et al. 2015; Kusano et al. 2016;

Rodriguez et al. 2018). Understanding what animals eat in the wild has led to improved diets for many species in captivity, such as tapirs (Clauss et al. 2009), giraffe (Sullivan et al. 2010), western lowland gorillas (Smith et al. 2014), banteng (Rowden and Rose 2016), and lorises and pottos (Simpson et al. 2018). For some zoo-housed species, additional research on free-ranging conspecifics is needed to ensure dietary needs are being met (Maple and Bocian 2013; Simpson et al. 2018), and to understand how nutrition affects overall wellness in the long-term.

4.3.3 Physical Environment

Ensuring that animal habitats are appropriate has been a long-term animal welfare goal (Broom 1986; Fraser 2009; Howell and Cheyne 2019; Kagan et al. 2015; Markowitz 1982; Martin 2015; Poole 1998). In humans, it has been estimated that as much as 70% of an individual's allostatic load may be accounted for by the environment (Petrovic et al. 2016). With longitudinal approaches, ALIs may provide a tool for studying the appropriateness of captive environments, including husbandry and management practices, while in the wild, it could inform on effects of anthropogenic pressure or climate change. As appropriate environments should help moderate allostatic load, measuring allostatic load prior to major changes and again after animals have acclimated to those changes could provide quantitative feedback on how individuals are faring. For example, inactivity often is a sign of poor welfare and increased risk of poor health outcomes (Broom 1986; Burn 2017; Howell and Cheyne 2019; Maple and Bocian 2013). Spaces that encourage and permit animals to increase physical activity may successfully reduce allostatic load; in humans, those who engage in more physical activity have lower allostatic load than less active peers (Gay et al. 2015; Lipowicz et al. 2014; Petrovic et al. 2016; Rodriguez et al. 2018; Rosenberg et al. 2014; Schenk et al. 2018; Upchurch et al. 2015b). Simply providing hiding spaces was sufficient for increasing activity levels in pottos (Frederick and Fernandes 1996), decreasing GC concentrations in clouded leopards (Shepherdson et al. 2004), and reducing pacing and GC concentrations in leopard cats (Carlstead et al. 1993). Similarly, providing simultaneous access to indoor and outdoor spaces increased play behavior in polar bears (Ross 2006) and allowed western lowland gorillas to reduce proximity between group members, a species-typical strategy for reducing conflict (Kurtycz et al. 2014). While attention often is focused on spatial needs, reducing allostatic load may not require larger spaces *per se*, but rather it may be more important to have quality space that encourages natural behavior, as has been documented for zoo elephants (Brown et al. 2016; Cameron and Ryan 2016). However, it must be noted that the pathway between stress and poor health outcomes involves a long latency period (Taylor et al. 2018) and allostatic load reflects physiology over the long-term rather than acute responses. As such, sufficient time between longitudinal measures of allostatic load is required to document meaningful changes. Latency periods likely will vary by species based on life history factors.

4.3.4 Social Environment

The importance of species-typical social relationships has long been recognized by animal caretakers (e.g., dolphins: Waples and Gales 2002; elephants: Freeman et al. 2009; lorises and pottos: Fuller et al. 2013, 2014; manatees: Harper and Schulte 2005; for review, see Price and Stoinski 2007). Research on stress supports this as a priority for animal welfare. Social support is a moderator of physiological activity in humans (Hostinar et al. 2014; House et al. 1988; Levine et al. 1989; McEwen and Seeman 1999; Seeman et al. 1994; Seeman and McEwen 1996; Taylor 2006; Uchino 2006; Uchino et al. 1996), and is associated with improved health outcomes (Cohen and Wills 1985; Holt-Lunstad et al. 2010; Hostinar et al. 2014; House et al. 1988; Levine et al. 1989; Seeman 1996; Seeman et al. 1994; Seeman and McEwen 1996; Smith and Christakis 2008; Uchino 2006; Uchino et al. 1996). Additionally, those who report having less social support have higher allostatic load (Glei et al. 2007; Horan and Widom 2015; Seeman et al. 2002, 2004b; Weinstein et al. 2003). The effects of social support on physiological activity have also been well studied in non-human primates. For example, bonobos (Hohmann et al. 2009; Palagi and Norscia 2013; Sannen et al. 2004) and gorillas (Cordoni et al. 2006; Mallavarapu et al. 2006; Palagi et al. 2008) practice reconciliation and consolation, which may attenuate stress responses following conflict. In wild chimpanzees, GCs were only significantly elevated following intergroup encounters if their bond partner was absent (Wittig et al. 2015). Affiliative social interactions in rhesus macaques were associated with lower long-term HPA axis activity (Wooddell et al. 2017). In marmoset monkeys, the presence of their mate during stressful experiences attenuated HPA axis activity, such as lower cortisol reactivity (Cavanaugh et al. 2016). Moreover, a lack of social support can have very real consequences on health and lifespan. In wild female yellow baboons, higher levels of social connectedness with other adult females and adult males were associated with reduced risk of death and considerably longer lifespans (Archie et al. 2014). Given that social support likely moderates physiological activity across gregarious species, it is expected to be a potent mediator of allostatic load.

Among social animals, dominance hierarchies can be an important life strategy. In humans, 'dominance' is mediated by social position (e.g., socioeconomic status or educational attainment: Cummins 2005, 2016; Lea et al. 2018a; Shively and Day 2015) and has been shown to be inversely associated with allostatic load (Crimmins et al. 2009; Petrovic et al. 2016; Rosenberg et al. 2014; Seeman et al. 2004a, 2010, 2014; Taylor et al. 2018). While the traditional view of dominance hierarchies involves a pecking order where stronger, more dominant individuals continually harass lower ranked individuals who have few resources and coping mechanisms, many species have more complex, nuanced patterns of affiliation. The level of stress experienced by an individual may be more directly related to their social interactions, rather than dominance rank *per se* (Edwards et al. 2013), and even within classic dominance hierarchies, individuals can develop mechanisms for coping with stressful social interactions (Abbott et al. 2003; Goymann and Wingfield 2004; Maestripieri and Hoffman 2011). Socially mediated changes in HPA activity and immune function can vary across species, so associations of allostatic load with social status also likely will vary. While higher allostatic load due to rank may seem like something to avoid so as to mitigate potential health declines, hierarchies serve not only important socioecological roles but also help regulate physiology and are key to appropriate captive environments.

4.3.5 Life Events

When considering the impact of stressful events on allostatic load and health, it is important to remember that the effects actually begin *in utero* and early life sets the stage for adult physiology. Because early life experiences help program HPA axis responsiveness, individuals are particularly vulnerable during these stages (Lea et al. 2017; Lupien et al. 2006; McEwen 1998b, 2003; McEwen and Seeman 1999; Murray et al. 2018). Not surprisingly, early life experiences are associated with allostatic load during adulthood in humans (Barboza Solís et al. 2015; Berg et al. 2017; Christensen et al. 2018b; Horan and Widom 2015; McEwen 2000, 2001, 2004, 2012; Seeman et al. 2010; Thayer et al. 2017). For example, adults who experienced abuse or neglect as children have higher allostatic load (Danese and McEwen 2012; Horan and Widom 2015). While physiology in immature individuals is more likely to reflect changes due to growth and development than physiological dysregulation, once these physiological parameters stabilize at their adult levels of expression, allostatic load reflects early life experiences. Multiple studies support the hypothesis that early environments will affect allostatic load development across taxa. In common marmosets, parental deprivation early in life was associated with increases in GCs, epinephrine, and norepinephrine, and lower body weight; negative effects were also evident in their behavior as juveniles, with less locomotion, fewer social vocalizations, and less parental contact following reunion (Pryce et al. 2004). Being born during periods of ecological challenge was associated with faster reproductive senescence in red deer (Nussey et al. 2007) and Asian elephants (Mumby et al. 2015). In zebra finches and great tits, clutch size in adulthood was predicted by resource availability in early life (Haywood and Perrins 1992). Cumulative exposure to potential sources of early life adversity, such as low maternal dominance rank, drought, or the presence of a competitive sibling, was associated with shortened lifespans in wild yellow baboons (Tung et al. 2016). Furthermore, given the large impacts of social support and social position on physiology, it is unsurprising that variation in maternal social connectedness and rank can also moderate offspring outcomes. For example, nursery-reared rhesus macaque females had lower social status when entering adulthood than did mother-reared females (Dettmer et al. 2017) and stronger social bonds in female savannah baboons increased the likelihood of infant survival (Silk et al. 2003). Similarly, in humans, children born to parents with low socioeconomic position (Christensen et al. 2018a) or who experience instability (e.g., mothers separated or divorced during the first year of life: Christensen et al. 2018b) have higher allostatic load as adults. Likewise, wild-caught gorillas that experienced early life adversity when captured had significantly higher allostatic loads than zoo-born gorillas (e.g., mother- and nursery-reared: Edes et al. 2016b, 2018b).

By using ALIs, we also may be able to better identify what types of events represent major stressors to animals in human care. Not all stressors are negative and some activation of the HPA axis is necessary for healthy functioning (Barker 1992, 2003; Carlstead and Shepherdson 1994; Meehan and Mench 2007; Moodie and Chamove 1990; Peters et al. 2017; Whitham and Wielebnowski 2013; Wolf 2003), but repeated or chronic uncertainty contributes to allostatic load development (Peters et al. 2017). Such stressors in captive animals may include anesthesia (e.g., Hispaniolan Amazon parrots: Heatley et al. 2000; cheetahs and clouded leopards: Young et al. 2004; sable antelope: Loeding et al. 2011; southern three-banded armadillos: Howell-Stephens et al. 2012; red colobus monkeys: Wasserman et al. 2013); transport (e.g., tigers: Dembiec et al. 2004; long-tailed macaques: Honess et al. 2004; Asian elephants: Fanson et al. 2013; Canada lynx: Fanson and Wielebnowski 2013; Indian rhinos: Capiro et al. 2014); and introductions to new group members (e.g., African elephants: Burks et al. 2004; red-shanked douc langurs: Heistermann et al. 2004; sable antelope: Loeding et al. 2011; western lowland gorillas: Jacobs et al. 2014). Some conflict in gregarious species is typical and unavoidable, and experiences such as anesthesia and institutional transfers are necessary for maintaining the health and demographic and genetic viability of captive populations, but it is worth determining what the long-term effects of these stressors might be. It is important to note, however, that studies on the effects of cumulative stress on physiology or health are limited, particularly in long-lived wildlife species, and conclusions have not always been straightforward. For example, cumulative early life adversity predicted longevity in wild baboons (Tung et al. 2016) and number of transfers was positively correlated with stereotypy in zoo-housed elephants (Greco et al. 2016). In gorillas, counts of events such as anesthesia, agonistic interactions with wounding, and zoo transfers, as well as total counts of all stressful events, are associated with allostatic load in some collections, but not others (Edes et al. 2016a, 2018b). It may be that stressful events may need to be weighted by event type (Edes et al. 2018b), or the chosen biomarkers were not optimal for assessing the long-term consequences of those particular stressors. Further ALI testing and validation are needed to identify the most appropriate methods to use for the taxa being studied (Edes et al. 2018a) and the research questions being asked (Delpierre et al. 2016).

4.3.6 Cognitive Function, Temperament and Personality

Research in humans has revealed higher allostatic load is associated with reduced cognitive functioning (Booth et al. 2015; Karlamangla et al. 2002, 2014; Seeman et al. 1997, 2001), suggesting similar relationships may exist in other highly intelligent animals. In humans, performing mentally stimulating tasks helps maintain cognition and memory (Ball et al. 2002; Willis et al. 2006). Therefore, providing opportunities for animals to engage in enriching cognitive tasks may be sources of beneficial challenge (Meehan and Mench 2007) that help moderate allostatic load development or at least attenuate its associated declines in cognition and memory. At this point, most cognitive research in zoos is done in primates and bears (for review, see Hopper 2017), but other taxa likely would benefit as well, and depending on the type of study designed, there could be additional benefits beyond maintaining cognitive abilities. Participating in cognitive research has been shown to improve welfare in chimpanzees (Herrelko et al. 2012; Hopper et al. 2016), crested macaques (Whitehouse et al. 2013), and brown capuchins (Ruby and Buchanan-Smith 2015). Furthermore, when studying decision-making behavior in chimpanzees, Hopper et al. (2016) observed increased locomotion across the group during study participation, even in animals that elected not to participate. Thus, carefully designed cognitive research could have added benefits such as group-wide increases in activity levels, which, as discussed previously, should also moderate allostatic load.

The option to participate in cognitive studies also provides opportunities for zoo animals to exhibit agency and control. Allowing animals to make choices has long been recognized as being beneficial to animal welfare (Herrelko et al. 2012; Kagan et al. 2015; Markowitz 1982; Markowitz and Aday 1998; Poole 1998; Snowdon 1989). Moreover, perceptions of control are critical to moderating stress responses. How events are perceived shapes the physiological responses to those events and influences the development of allostatic load (Everly and Lating 2013; Juster et al. 2010; Maestripieri and Hoffman 2011; McEwen 1998a; Rosen and Schulkin 2004; Sterling 2004; Ulrich-Lai and Herman 2009). For example, higher allostatic load has been observed in people reporting higher levels of perceived stress (Goldman et al. 2005) and those who feel they lack purpose in life (Zilioli et al. 2015). Providing options and allowing animals to make decisions could reduce the likelihood of perceiving events as stressful and so moderate allostatic load. Because it promotes choice and control and reduces stress responses (Capiro et al. 2014; Lambeth et al. 2006; Laule and Desmond 1998; Maple and Bocian 2013; Poole 1998), positive reinforcement training (PRT) to allow animals to choose to participate in procedures such as routine health monitoring and receiving injections may be one way to reduce allostatic load in some species. As an added benefit, engaging in training sessions promotes welfare by serving as a form of enrichment (Carrasco et al. 2009; Fuller et al. 2013; Laule and Desmond 1998; Leeds et al. 2016; Pomerantz and Terkel 2009; Shepherdson 1998). In chimpanzees, PRT sessions decreased abnormal and stress-related behaviors and increased affiliative behaviors that lasted throughout the day, with effects that were particularly salient for lower-ranked individuals (Pomerantz and Terkel 2009). While beneficial for all individuals, PRT may be an additional way to help subordinate individuals cope with their low status and reduce its potential negative effects on health.

Other research has indicated that personality assessments may be useful tools to improve captive management (Herrelko et al. 2012; Schaefer and Steklis 2014; Watters and Powell 2012) and help animal management staff better predict how certain individuals will respond to change or which ones might be more likely to participate in research projects (Herrelko et al. 2012). While research in humans thus far is limited, it has been suggested that personality, via its effects on perception and coping strategies, affects the development of allostatic load (Beckie 2012; Christensen et al. 2018a; Goldman et al. 2005; Horan and Widom 2015; Seeman et al. 2010; Upchurch et al. 2015a). For example, people who are hostile (Kubzansky et al. 1999; Seeman et al. 2002) or have low anger control (Zilioli et al. 2017) have significantly higher allostatic load, while those with positive affect have lower allostatic load (Schenk et al. 2018). Additionally, humans with poor coping skills also have higher allostatic load (Fernandez et al. 2015; Glei et al. 2007). This research helps highlight the potential for differential morbidity and mortality based on personality and suggests animals who are frequently hostile or show high levels of aggression could have higher allostatic load. For example, the frequency of negative social interactions and aggression initiated by captive male Dorcas gazelles is correlated with GC levels measured in hair samples, suggesting high levels of aggression are driving increases in GCs (Salas et al. 2016), which could also translate into increases in allostatic load over longer time periods. On the other hand, chimpanzees rated higher in extraversion and lower in neuroticism were rated by keepers as being happier and having better welfare (Robinson et al. 2017), rhesus macaques rated higher in either confidence or anxiety had fewer injuries (Robinson et al. 2018), and personality traits that favor low rates of aggression and strong social bonds were associated with longer lifespans in chimpanzees (Altschul et al. 2018), suggesting lower allostatic load may be associated with these personality dimensions. Personality assessments may also inform housing decisions. For instance, given their single-male, multi-female social structures, there are more adult male gorillas than can be housed in breeding groups. All-male groups of gorillas exist in the wild but there is concern about the potential for escalated aggression and wounding when space is limited (Leeds et al. 2015; Stoinski et al. 2004). However, the formation of bachelor groups is one option for housing excess males and is preferable to solitary housing, so personality assessments may help determine who would do well in all-male groups (Schaefer and Steklis 2014; Stoinski et al. 2004). While personality assessments require investments of staff time, they are a cost-effective tool for helping captive management (Herrelko et al. 2012; Schaefer and Steklis 2014) and could provide an additional avenue with which to explore how personality shapes physiology, behavior, and health via allostatic load.

4.3.7 Allostatic Load and Reproduction

Gestation, parturition, lactation, and rearing offspring can be costly and stressful (Archie et al. 2014; Maestripieri and Georgiev 2015; Maestripieri and Hoffman 2011). For example, in wild yellow baboons, females were at the highest risk of injury when ovulating and healed more slowly from wounds while lactating (Archie et al. 2014). Similarly, in free-ranging rhesus macaques, females were more likely to die giving birth than during the mating season and had higher cortisol while lactating (Maestripieri and Georgiev 2015). Consistent with these costs, parous female gorillas had significantly higher allostatic load than nulliparous females (Edes et al. 2016a, 2018b). Given the costs of reproduction and that gestation, parturition, lactation, and rearing are stressful for females, allostatic load may be well poised to help address issues of poor fecundity in captivity.

Research examining the relationship between allostatic load and reproduction in women is growing and has focused thus far on associations with adverse pregnancy outcomes such as low birth weight, pre-term birth, small-for-gestational age infants, and pre-eclampsia. This research is complicated because estimating allostatic load is not recommended during pregnancy (Barrett et al. 2018; Morrison et al. 2013), as many biomarkers will reflect physiological changes due to gestation rather than stress-induced impairment and dysregulation. Although it has been suggested that estimating allostatic load in the early stages of pregnancy could be a valid approach (Hux et al. 2014; Hux and Roberts 2015), a better strategy might be to estimate allostatic load periodically over the lifespan and then examine whether allostatic load predicts reproductive success and outcomes. Zoos are in a good position to do this with animal collections by using cryopreserved samples collected during routine veterinary procedures prior to reproduction to estimate allostatic load and then continuing to monitor across the reproductive lifespan.

Surprisingly, the first study to examine associations of allostatic load with reproduction found no association with known risk factors such as birth weight, but higher allostatic load was inversely associated with gestational age (Wallace and Harville 2013). However, this study estimated allostatic load opportunistically in women undergoing glucose tolerance tests between 26–28 weeks of gestation. In studies that measured allostatic load during teenage years and then examined it in association with adult reproductive outcomes, higher allostatic load again was not associated with adverse birth outcomes such as low birth weight, pre-term birth, or small-for-gestational-age infants (Wallace et al. 2013a, b). While this research looked at allostatic load prior to reproduction, there is a long latency period between stress exposure and physiological dysregulation (Taylor et al. 2018). As such, measuring allostatic load at young ages in long-lived species may not allow sufficient time for allostatic load to accumulate.

Recognizing the issues associated with measuring allostatic load during advanced pregnancy or immature years, other researchers have taken different approaches. Researchers who estimated allostatic load using samples collected prior to 15 weeks of gestation demonstrated that allostatic load is higher in women with pre-eclampsia and that higher allostatic load is associated with increasing odds of developing pre-eclampsia, ultimately determining that allostatic load is a better predictor of pre-eclampsia than individual biomarkers or obesity (Hux and Roberts 2015). Some studies have also looked at allostatic load postpartum, suggesting that the differences in allostatic load between birth outcomes were likely present prior to birth as well (Hux et al. 2014). For example, Hux et al. (2014) found women with a history of low birth weight outcomes have higher allostatic load, while Accortt et al. (2017) observed higher allostatic load in women was associated with experiencing more than one adverse birth outcome. Further emphasizing the importance of meeting dietary needs for animal welfare, Accortt et al. (2017) also found vitamin D deficiency is an important biomarker to include in ALIs when interested in reproductive outcomes in women.

Only one study has measured allostatic load during adulthood prior to pregnancy thus far. Increasing allostatic load was associated with increased odds of pre-eclampsia, preterm birth, and low birthweight (Barrett et al. 2018). There was no association between allostatic load and fecundity, but this study was done in women receiving fertility treatments. Retrospective analysis of female reproductive records and estimates of allostatic load using previously banked samples would enable zoo researchers to examine whether allostatic load is associated with fecundity across a wide range of taxa. Such studies may also help us determine optimal reproduction levels. In humans, there is a U-shaped relationship between the number of children born and lifespan; women with two children having the lowest mortality risk compared to women with fewer than two or no children and women with more than two children (Wang et al. 2013). While few animals in zoos reach their reproductive potential, not breeding has been found to be detrimental to females across taxa, including several canid species, elephants, white rhinos, Seba's bats, wildebeest, stingrays, cheetahs, tigers, and naked mole rats, leading to reproductive failure (for review, see Penfold et al. 2014). As many zoo populations are currently not sustainable (Comizzoli and Holt 2019; Lees and Wilcken 2009; Penfold et al. 2014; Powell 2019), a better understanding of the relationship between allostatic load and fecundity could help guide species management efforts.

In women who became mothers as adults, complications during their own birth (Christensen et al. 2018b) or being born with low birth weight (Barboza Solís et al. 2015; Christensen et al. 2018b) is associated with higher allostatic load. Multi-generational work is needed across species to confirm that higher allostatic load in females results in low birth weight and other adverse pregnancy outcomes, and whether such pregnancy outcomes consequently predispose offspring to higher allostatic load in adulthood. If such relationships can be identified, allostatic load may be a pathway with which to trace risk of poor health outcomes in early life. For example, the rapid onset of EEHV in elephant calves, coupled with its high fatality rates, has had devastating consequences on captive populations (Atkins et al. 2013; Garner et al. 2009; Kendall et al. 2016; Metzler et al. 1990; Richman et al. 2000a; Schaftenaar et al. 2010). Stress may play a role in increasing susceptibility to EEHV infection (Kendall et al. 2016; Schaftenaar et al. 2010), however, given that it occurs at younger ages when measuring allostatic load is less reliable, it is unlikely to provide an early warning of susceptibility to EEHV infection. However, if there are intergenerational effects, the mother's allostatic load may help identify offspring at greatest risk. As such, allostatic load may provide an avenue with which to both explore how to improve reproduction in captive elephants, which already suffer low and unsustainable reproductive rates, as well as identify offspring who are most likely to be at risk of EEHV.

The relationship between sociality and reproduction has been well documented in a number of species. For example, wild elephants live in multi-generational herds with complex, hierarchical groups of related females (Freeman et al. 2004) and these relationships are critical for normal ovarian function and hormone status (Brown et al. 2016). Pituitary function in elephants also appears to be influenced by social factors. In a large welfare study in the U.S. (Brown et al. 2016), multi-variable modeling determined age (positive) and enrichment diversity (negative) are important factors for ovarian cyclicity, while age (positive), enrichment diversity (negative), alternate feeding methods (negative) and social group contact (positive) were predictors of hyperprolactinemia, a condition known to be associated with ovarian cycle problems in elephants and other species. Thus, providing elephants with a more enriched environment may support normal pituitary-ovarian function. By contrast, although elephants are highly social species, those that spent time in a greater number of different social groups had a greater risk of hyperprolactinemia (Brown et al. 2016). This strategy may be needed to mitigate problems with socially incompatible elephants, so dividing a herd into multiple groups and housing elephants in a variety of social configurations may yield a more demanding social environment to which they respond with elevated prolactin production and reduced ovarian cyclicity (Prado et al. 2019). In terms of temperament, consistently low prolactin was associated with a more fearful and apprehensive temperament, whereas a more popular and caring temperament was associated with elevated prolactin, again in elephants (N. Prado and J. Brown, unpublished). Finally, acyclic elephants are more often rated as herd peacekeepers (Freeman et al. 2004). Therefore, while increased levels of prolactin may help elephants cope with increased social demands, it could occur at the cost of normal reproductive function. Unfortunately, until recently, the importance of multi-generational herd structure and social compatibility to elephant reproduction and welfare had been largely overlooked (Brown et al. 2016; Freeman et al. 2004), so allostatic load measures to identify potential risk factors related to poor population sustainability should be explored.

5 Conclusions and Future Perspectives

Animal well-being is an important consideration for wildlife conservation, to ensure that individuals, populations, and species, continue to thrive. Stress is a vital part of life; indeed too little stress (i.e., lack of stimulation) can be equally problematic to well-being as too much. The stress response is an adaptive process that allows organisms to respond appropriately to challenges they face, whether positive or negative, and acute stress can have stimulatory effects on reproduction and immune function. However, excessive stress can have significant deleterious consequences. Delayed growth and development, immunosuppression and increased susceptibility to disease, and reduced reproduction and survival can all result, limiting viability at both the individual- and population-level. Wildlife under human care and in nature all experience events that have the potential to impact their well-being, and an individual's response to these challenges can be moderated by a variety of factors ranging from their own temperament or prior experience, to their social relationships and availability of coping mechanisms. To understand about underlying physiological responses to stress, potential avenues for reducing impact, and to predict how changing environments will affect different populations of wildlife, our ability to measure the response to stress is key.

Traditional methods of welfare assessment have considered elevated stress hormones, or consequences of negative stress such as the occurrence of abnormal behaviors, poor health and survival, and the lack of reproductive function. Although we have not discussed in detail the many behavioral measures of well-being here, it is important to consider that these are important moderators to the physiological response to stress, and whenever possible, behavioral changes should be monitored in addition to physiological ones. In addition to those, we have summarized many of the alternative or complimentary physiological biomarkers that have been used to assess welfare or well-being in humans or animals to-date. These include indicators of autonomic nervous, neuroendocrine, and immune activation, and cover an array of techniques that may allow us to assess valence as well as level of arousal, to distinguish things that are good for well-being as well as those that may be bad. However, due to the complexity of factors that can influence well-being, we conclude that the best welfare assessment tools will likely rely on a multidimensional measurement approach, incorporating neuroendocrine, immune, and metabolic markers, as well as putting these in context with behavior.

Moving forward, these multi-biomarker approaches such as allostatic load will likely be key to improving our understanding of individual differences in susceptibility to stress and to the efficacy of mitigation strategies. However, while allostatic load shows promise as a tool for assessing overall well-being and preventative health monitoring in animal species, it is important to remember that the relationship between allostatic load and a number of health conditions, such as disease or parasite loads, is likely reciprocal. Higher allostatic load may increase vulnerability to pathology, but the onset of disease or increasing parasite load may itself be stressful and result in additional increases in allostatic load, further perpetuating the physiological impairment and reducing the likelihood of positive health outcomes. Similarly, some of the potential biomarkers of positive well-being, such as IgA, play multiple roles that when measured in isolation may be difficult to distinguish. Measuring an array of biomarkers representing multiple biological systems will help put things into context. Furthermore, advancing technologies that facilitate proteomic approaches to determine what biomarkers are up- or down-regulated in response to specific types of stressors should help us ascertain whether there are species or situational differences, and what biomarkers will provide the best insight into an animal's state.

A further challenge for the future is how to conduct these multi-biomarker assessments non-invasively. Several of the biomarkers mentioned in this chapter have been analyzed in saliva and/or urine, occasionally feces, but as has been done for GC measures over the last three decades, validating these techniques for the diverse species of conservation concern, and determining the biological implications of detectable changes in minimally invasive matrices, will be vital to their use. Finally, most of the welfare monitoring techniques discussed here have been explored in relatively few species, with a significant taxa bias towards mammals, and lesser emphasis on birds, reptiles, amphibians, and fish. Even less is known about what constitutes good welfare in lower vertebrate and invertebrate species, many of which are commonly displayed in zoos and aquaria. Yet there is ample evidence that many more species meet the criteria for having consciousness, sentience, cognition and intelligence, including elasmobranchs (Schluessel 2015), cephalopods (Mather 2008), fish (Brown 2015) and even insects (Barron and Klein 2016), suggesting we may be underestimating the welfare needs of the majority of animals on the planet. To that end, we would be wise to take heed of the words of Henry Beston in how we consider animals:

> We need another and a wiser and perhaps a more mystical concept of animals. Remote from universal nature and living by complicated artifice, man in civilization surveys the creature through the glass of his knowledge and sees thereby a feather magnified and the whole image in distortion. We patronize them for their incompleteness, for their tragic fate for having taken form so far below ourselves. And therein do we err. For the animal shall not be measured by man. In a world older and more complete than ours, they move finished and complete, gifted with the extension of the senses we have lost or never attained, living by voices we shall never hear. They are not brethren, they are not underlings: they are other nations, caught with ourselves in the net of life and time, fellow prisoners of the splendour and travail of the earth.—The Outermost House: A Year of Life on the Great Beach of Cape Cod

Acknowledgments We thank the technical, veterinary, zoo and research staff who contributed to work described herein. Authors and associated work were funded or supplied with in-kind donations by the AZA Conservation Endowment Fund, Columbus Zoo and Aquarium, Department of Anthropology at The Ohio State University, FBB Capital Partners, Friends of the National Zoo, Institute of Museum and Library Services, International Elephant Foundation, Morris Animal Foundation (D15ZO-403), Siemens Medical Solutions Diagnostics, Shared Earth Foundation, Smithsonian Competitive Grants Program, Smithsonian Institution Fellowship and George E. Burch Fellowship Programs, Smithsonian Women's Committee, and the Society of Integrative and Comparative Biology.

References

Abbott DH, Keverne EB, Bercovitch FB, Shively CA, Medoza SP, Saltzman W, Snowdon CT, Ziegler TE, Banjevic M, Garland T, Sapolsky RM. Are subordinates always stressed? A comparative analysis of rank differences in cortisol levels among primates. Horm Behav. 2003;43:67–82.

Accardi M, Boulay E, Pouliot M, Maghezzi S, Forster R, Authier S. qEEG as a biomarker of animal welfare: assessing the influence of different caging. J Pharmacol Toxicol Methods. 2017;88:193–4.

Accortt EE, Mirocha J, Schetter CD, Hobel CJ. Adverse perinatal outcomes and postpartum multi-systemic dysregulation. Matern Child Health J. 2017;21:398–406.

Adamec T, Kunc P, Knížková I, Dolejš J, Toufar O. By evaporative cooling in summer session is possible to improve growth and feed conversion in pig fattening. In: 48th annual meeting of EAAP, Vienna, Austria; 1998. p. 219–20.

Adesiyun AA, Caesar K, Inder L. Prevalence of Salmonella and Campylobacter species in animals at Emperor Valley Zoo, Trinidad. J Zoo Wildl Med. 1998;29:237–9.

Aerts J, Metz JR, Ampe B, Decostere A, Flik G, De Saeger S. Scales tell a story on the stress history of fish. PLoS One. 2015;10:e0123411.

Agnew RC, Smith VJ, Fowkes RC. Wind turbines cause chronic stress in badgers (*Meles meles*) in Great Britain. J Wildl Dis. 2016;52:459–67.

Akiyoshi H, Aoki M, Shimada T, Noda K, Kumagai D, Saleh N, Sugii S, Ohashi F. Measurement of plasma chromogranin A concentrations for assessment of stress responses in dogs with insulin-induced hypoglycemia. Am J Vet Res. 2005;66:1830–5.

Ali N, Pruessner JC. The salivary alpha amylase over cortisol ratio as a marker to assess dysregulations of the stress systems. Physiol Behav. 2012;106:65–72.

Alley DE, Seeman TE, Kim JK, Karlamangla A, Hu P, Crimmins EM. Socioeconomic status and C-reactive protein levels in the US population: NHANES IV. Brain Behav Immun. 2006;20:498–504.

Almeida P, Weber P, Burton J, Zanella A. Depressed DHEA and increased sickness response behaviors in lame dairy cows with inflammatory foot lesions. Domest Anim Endocrinol. 2008;34:89–99.

Al-Murrani W, Al-Rawi I, Raof N. Genetic resistance to *Salmonella typhimurium* in two lines of chickens selected as resistant and sensitive on the basis of heterophil/lymphocyte ratio. Br Poultry Sci. 2002;43:501–7.

Al-Murrani W, Al-Rawi A, Al-Hadithi M, Al-Tikriti B. Association between heterophil/lymphocyte ratio, a marker of 'resistance'to stress, and some production and fitness traits in chickens. Br Poultry Sci. 2006;47:443–8.

Altschul DM, Hopkins WD, Herrelko ES, Inoue-Murayama M, Matsuzawa T, King JE, Ross SR, Weiss A. Personality links with lifespan in chimpanzees. Elife. 2018;7:e33781.

Anderson B, Watson D, Colditz I. The effect of dexamethasone on some immunological parameters in cattle. Vet Res Commun. 1999;23:399–413.

Anestis SF. Urinary cortisol responses to unusual events in captive chimpanzees (*Pan troglodytes*). Stress. 2009;12:49–57.

Angelier F, Chastel O. Stress, prolactin and parental investment in birds: a review. Gen Comp Endocrinol. 2009;163:142–8.

Ange-van Heugten KD, van Heugten E, Timmer S, Bosch G, Elias A, Whisnant S, Swarts HJM, Ferket P, Verstegen MWA. Fecal and salivary cortisol concentrations in woolly (*Lagothrix ssp.*) and spider monkeys (*Ateles spp.*). Int J Zool. 2009;2009:1–9.

Archie EA, Tung J, Clark M, Altmann J, Alberts SC. Social affiliation matters: both same-sex and opposite-sex relationships predict survival in wild female baboons. Proc R Soc Lond B. 2014;281:1–9.

Armario A, Restrepo C, Hidalgo J, Lopez-Calderon A. Differences in prolactin and LH responses to acute stress between peripuberal and adult male rats. J Endocrinol. 1987;112:9–13.

Arthington J, Eicher S, Kunkle W, Martin F. Effect of transportation and commingling on the acute-phase protein response, growth, and feed intake of newly weaned beef calves. J Anim Sci. 2003;81:1120–5.

Ashley PJ, Ringrose S, Edwards KL, Wallington E, McCrohan CR, Sneddon LU. Effect of noxious stimulation upon antipredator responses and dominance status in rainbow trout. Anim Behav. 2009;77:403–10.

Atkins L, Zong J-C, Tan J, Mejia A, Heaggans SY, Nofs SA, Stanton JJ, Flanagan JP, Howard L, Latimer E, Stevens MR, Hoffman DS, Hayward GS, Ling PD. Elephant endotheliotropic herpesvirus 5, a newly recognized elephant herpesvirus associated with clinical and subclinical infections in captive Asian elephants (*Elephas maximus*). J Zoo Wildl Med. 2013;44:136–43.

Atkinson HC, Waddell BJ. The hypothalamic-pituitary-adrenal axis in rat pregnancy and lactation: circadian variation and interrelationship of plasma adrenocorticotropin and corticosterone. Endocrinology. 1995;136:512–20.

Ball K, Berch DB, Helmers KF, Jobe JB, Leveck MD, Marsiske M, Morris JN, Rebok GW, Smith DM, Tennstedt SL, Unverzagt FW, Willis SL. Effects of cognitive training interventions with older adults. JAMA. 2002;288:2271–81.

Ballard PL, Ballard RA. Glucocorticoid receptors and the role of glucocorticoids in fetal lung development. Proc Natl Acad Sci U S A. 1972;69:2668–72.

Ballard PL, Baxter JD, Higgins SJ, Rousseau GG, Tomkins GM. General presence of glucocorticoid receptors in mammalian tissues. Endocrinology. 1974;94:998–1002.

Barboza Solís C, Kelly-Irving M, Fantin R, Darnaudéry M, Torrisani J, Lang T, Delpierre C. Adverse childhood experiences and physiological wear-and-tear in midlife: findings from the 1958 British birth cohort. Proc Natl Acad Sci U S A. 2015;112:E738–E46.

Barker DJP. The fetal and infant origins of adult disease. Br Med J. 1992;301:1111.

Barker DJP. The developmental origins of adult disease. Eur J Epidemiol. 2003;18:733–6.

Barreto RE, Volpato GL. Ventilation rates indicate stress-coping styles in Nile tilapia. J Biosci. 2011;36:851–5.

Barrett ES, Vitek W, Mbowe O, Thurston SW, Legro RS, Alvero R, Baker V, Bates GW, Casson P, Coutifaris C, Eisenberg E, Hansen K, Krawetz S, Robinson R, Rosen M, Usadi R, Zhang H, Santoro N, Diamond M. Allostatic load, a measure of chronic physiological stress, is associated with pregnancy outcomes, but not fertility, among women with unexplained infertility. Hum Reprod. 2018;33:1757–66.

Barron AB, Klein C. What insects can tell us about the origins of consciousness. Proc Natl Acad Sci U S A. 2016;113:4900–8.

Barron DG, Brawn JD, Butler LK, Romero LM, Weatherhead PJ. Effects of military activity on breeding birds. J Wildl Manage. 2012;76:911–8.

Bartolomucci A, Pasinetti G, Salton S. Granins as disease-biomarkers: translational potential for psychiatric and neurological disorders. Neuroscience. 2010;170:289–97.

Baxter-Gilbert JH, Riley JL, Mastromonaco GF, Litzgus JD, Lesbarrères D. A novel technique to measure chronic levels of corticosterone in turtles living around a major roadway. Conserv Physiol. 2014;2:cou036.

Bechshøft TØ, Sonne C, Rigét FF, Letcher RJ, Novak MA, Henchey E, Meyer JS, Eulaers I, Jaspers VL, Covaci A. Polar bear stress hormone cortisol fluctuates with the North Atlantic oscillation climate index. Polar Biol. 2013;36:1525–9.

Beck EM, Smits JE, St. Clair CC. Health of domestic mallards (*Anas platyrhynchos domestica*) following exposure to oil sands process-affected water. Environ Sci Technol. 2014;48:8847–54.

Beckie TM. A systematic review of allostatic load, health, and health disparities. Biol Res Nurs. 2012;14:311–46.

Beerda B, Schilder MBH, van Hooff J, de Vries HW, Mol JA. Behavioural and hormonal indicators of enduring environmental stress in dogs. Anim Welf. 2000;9:49–62.

Behringer V, Deschner T, Möstl E, Selzer D, Hohmann G. Stress affects salivary alpha-amylase activity in bonobos. Physiol Behav. 2012;105:476–82.

Behringer V, Borchers C, Deschner T, Möstl E, Selzer D, Hohmann G. Measurements of salivary alpha amylase and salivary cortisol in hominoid primates reveal within-species consistency and between-species differences. PLoS One. 2013;8:e60773.

Benhaiem S, Hofer H, Dehnhard M, Helms J, East ML. Sibling competition and hunger increase allostatic load in spotted hyaenas. Biol Lett. 2013;9:1–4.

Benn AL, McLelland DJ, Whittaker AL. A review of welfare assessment methods in reptiles, and preliminary application of the welfare quality(R) protocol to the pygmy blue-tongue skink, tiliqua adelaidensis, using animal-based measures. Animals. 2019;9:22.

Bennett C, Fripp D, Othen L, Jarsky T, French J, Loskutoff N. Urinary corticosteroid excretion patterns in the okapi (*Okapia johnstoni*). Zoo Biol. 2008;27:381–93.

Berg MT, Simons RL, Barr A, Beach SRH, Philibert RA. Childhood/adolescent stressors and allostatic load in adulthood: support for a calibration model. Soc Sci Med. 2017;193:130–9.

Berk LS, Tan SA, Fry WF, Napier BJ, Lee JW, Hubbard RW, Lewis JE, Eby WC. Neuroendocrine and stress hormone changes during mirthful laughter. Am J Med Sci. 1989;298:390–6.

Berkvens CN, Hyatt C, Gilman C, Pearl DL, Barker IK, Mastromonaco GF. Validation of a shed skin corticosterone enzyme immunoassay in the African house snake (*Lamprophis fuliginosus*) and its evaluation in the eastern massasauga rattlesnake (*Sistrurus catenatus catenatus*). Gen Comp Endocrinol. 2013;194:1–9.

Bernard C. An introduction to the study of experimental medicine. New York: Dover; 1865.

Bernton E, Hoover D, Galloway R, Popp K. Adaptation to chronic stress in military trainees: adrenal androgens, testosterone, glucocorticoids, IGF-1, and immune function. Ann N Y Acad Sci. 1995;774:217–31.

Berntson GG, Boysen ST. Specificity of the cardiac response to conspecific vocalizations in chimpanzees. Behav Neurosci. 1989;103:235–45.
Beuving G, Blokhuis HJ. Effect of novelty and restraint on catecholamines in plasma of laying hens. Br Poultry Sci. 1997;38:297–300.
Bilandžić N, Žurić M, Lojkić M, Šimić B, Milić D, Barač I. Cortisol and immune measures in boars exposed to three-day administration of exogenous adrenocorticotropic hormone. Vet Res Commun. 2006;30:433–44.
Blas J, Baos R, Bortolotti GR, Marchant TA, Hiraldo F. Age-related variation in the adrenocortical response to stress in nestling white storks (*Ciconia ciconia*) supports the developmental hypothesis. Gen Comp Endocrinol. 2006;148:172–80.
Boissy A, Manteuffel G, Jensen MB, Moe RO, Spruijt B, Keeling LJ, Winckler C, Forkman B, Dimitrov I, Langbein J. Assessment of positive emotions in animals to improve their welfare. Physiol Behav. 2007;92:375–97.
Boivin J, Sanders K, Schmidt L. Age and social position moderate the effect of stress on fertility. Evol Hum Behav. 2006;27:345–56.
Bolhuis JE, Parmentier HK, Schouten WG, Schrama JW, Wiegant VM. Effects of housing and individual coping characteristics on immune responses of pigs. Physiol Behav. 2003;79:289–96.
Bonier F, Martin PR, Moore IT, Wingfield JC. Do baseline glucocorticoids predict fitness? Trends Ecol Evol. 2009;24:634–42.
Booij SH, Bos EH, Bouwmans ME, van Faassen M, Kema IP, Oldehinkel AJ, de Jonge P. Cortisol and α-amylase secretion patterns between and within depressed and non-depressed individuals. PLoS One. 2015;10:e0131002.
Boomsma CM, Keay SD, Macklon NS. Peri-implantation glucocorticoid administration for assisted reproductive technology cycles. Cochrane Database Syst Rev. 2007;6:CD005996.
Boonstra R, Hik D, Singleton GR, Tinnikov A. The impact of predator-induced stress on the snowshoe hare cycle. Ecol Monogr. 1998;68:371–94.
Boonstra R, McColl CJ, Karels TJ. Reproduction at all costs: the adaptive stress response of male Arctic ground squirrels. Ecology. 2001;82:1930–46.
Booth T, Royle NA, Corley J, Gow AJ, Valdés Hernádez Mdel C, Maniega SM, Ritchie SJ, Bastin ME, Starr JM, Wardlaw JM, Deary IJ. Association of allostatic load with brain structure and cognitive ability in later life. Neurobiol Aging. 2015;36:1390–9.
Borrell LN, Crawford ND. Social disparities in periodontitis among US adults: the effect of allostatic load. J Epidemiol Community Health. 2011;65:144–9.
Borrell LN, Dallo FJ, Nguyen N. Racial/ethnic disparities in all-cause mortality in U.S. adults: the effect of allostatic load. Public Health Rep. 2010;125:810–6.
Bortolotti GR, Marchant TA, Blas J, German T. Corticosterone in feathers is a long-term, integrated measure of avian stress physiology. Funct Ecol. 2008;22:494–500.
Bosch JA, Brand HS, Ligtenberg TJ, Bermond B, Hoogstraten J, Amerongen AVN. Psychological stress as a determinant of protein levels and salivary-induced aggregation of *Streptococcus gordonii* in human whole saliva. Psychosom Med. 1996;58:374–82.
Bosch J, Brand H, Ligtenberg A, Bermond B, Hoogstraten J, Nieuw Amerongen A. The response of salivary protein levels and S-IgA to an academic examination are associated with daily stress. J Psychophysiol. 1998;12:384–91.
Bosch JA, de Geus EJ, Veerman EC, Hoogstraten J, Amerongen AVN. Innate secretory immunity in response to laboratory stressors that evoke distinct patterns of cardiac autonomic activity. Psychosom Med. 2003;65:245–58.
Braden A, Moule G. Effects of stress on ovarian morphology and oestrus cycles in ewes. Aust J Agr Res. 1964;15:937–49.
Brann DW, Mahesh VB. Regulation of gonadotropin secretion by steroid hormones. Front Neuroendocrinol. 1991a;12:165–207.
Brann DW, Mahesh VB. Role of corticosteroids in female reproduction. FASEB J. 1991b;5:2691–8.
Brann DW, Putnam CD, Mahesh VB. Corticosteroid regulation of gonadotropin and prolactin secretion in the rat. Endocrinology. 1990;126:159–66.

Brann DW, Putnam CD, Mahesh VB. Validation of the mechanisms proposed for the stimulatory and inhibitory effects of progesterone on gonadotropin secretion in the estrogenprimed rat: a possible role for adrenal steroids. Steroids. 1991;56:103–11.

Brann DW, Putnam CD, Mahesh VB. Regulation of follicle stimulating hormone secretion by natural and synthetic corticosteroids. In: Hunzicker-Dunn M, Schwartz NB, editors. Follicle stimulating hormone. New York: Springer; 1992. p. 303–9.

Brearley G, McAlpine C, Bell S, Bradley A. Influence of urban edges on stress in an arboreal mammal: a case study of squirrel gliders in southeast Queensland, Australia. Landsc Ecol. 2012;27:1407–19.

Breuner CW. Stress and reproduction in birds. In: David ON, Kristin HL, editors. Hormones and reproduction of vertebrates. London: Academic Press; 2011. p. 129–51.

Breuner CW, Patterson SH, Hahn TP. In search of relationships between the acute adrenocortical response and fitness. Gen Comp Endocrinol. 2008;157:288–95.

Briski KP, Sylvester PW. Effect of specific acute stressors on luteinizing hormone release in ovariectomized and ovariectomized estrogen-treated female rats. Neuroendocrinology. 1988;47:194–202.

Broom DM. Indicators of poor welfare. Br Vet J. 1986;142:524–6.

Brown C. Fish intelligence, sentience and ethics. Anim Cogn. 2015;18:1–17.

Brown JL, Kersey DC, Freeman EW, Wagener T. Assessment of diurnal urinary cortisol excretion in Asian and African elephants using different endocrine methods. Zoo Biol. 2010;29:274–83.

Brown JL, Paris S, Prado-Oviedo NA, Meehan CL, Hogan JN, Morfeld KA, Carlstead K. Reproductive health assessment of female elephants in North American zoos and association of husbandry practices with reproductive dysfunction in African elephants (*Loxodonta africana*). PLoS One. 2016;11:e0145673.

Bryan HM, Darimont CT, Paquet PC, Wynne-Edwards KE, Smits JE. Stress and reproductive hormones reflect inter-specific social and nutritional conditions mediated by resource availability in a bear–salmon system. Conserv Physiol. 2014;2:cou010.

Bryan HM, Smits JE, Koren L, Paquet PC, Wynne-Edwards KE, Musiani M. Heavily hunted wolves have higher stress and reproductive steroids than wolves with lower hunting pressure. Funct Ecol. 2015;29:347–56.

Brydon L, Edwards S, Mohamed-Ali V, Steptoe A. Socioeconomic status and stress-induced increases in interleukin-6. Brain Behav Immun. 2004;18:281–90.

Bundgaard CJ, Kalliokoski O, Abelson KS, Hau J. Acclimatization of mice to different cage types and social groupings with respect to fecal secretion of IgA and corticosterone metabolites. In Vivo. 2012;26:883–8.

Burgess EA, Hunt KE, Kraus SD, Rolland RM. Adrenal responses of large whales: integrating fecal aldosterone as a complementary biomarker to glucocorticoids. Gen Comp Endocrinol. 2017;252:103–10.

Burgess EA, Hunt KE, Kraus SD, Rolland RM. Quantifying hormones in exhaled breath for physiological assessment of large whales at sea. Sci Rep. 2018;8:10031.

Burke T, Page B, Van Dyk G, Millspaugh J, Slotow R. Risk and ethical concerns of hunting male elephant: behavioural and physiological assays of the remaining elephants. PLoS One. 2008;3:e2417.

Burks KD, Mellen JD, Miller GW, Lehnhardt J, Weiss A, Figueredo A, Maple TL. Comparison of two introduction methods for African elephants (*Loxodonta africana*). Zoo Biol. 2004;23:109–26.

Burn CC. Bestial boredom: a biological perspective on animal boredom and suggestions for its scientific investigation. Anim Behav. 2017;130:141–51.

Burnard C, Ralph C, Hynd P, Edwards JH, Tilbrook A. Hair cortisol and its potential value as a physiological measure of stress response in human and non-human animals. Anim Prod Sci. 2017;57:401–14.

Busch DS, Hayward LS. Stress in a conservation context: a discussion of glucocorticoid actions and how levels change with conservation-relevant variables. Biol Conserv. 2009;142:2844–53.

Buwalda B, Scholte J, de Boer SF, Coppens CM, Koolhaas JM. The acute glucocorticoid stress response does not differentiate between rewarding and aversive social stimuli in rats. Horm Behav. 2012;61:218–26.
Cabagna MC, Lajmanovich RC, Stringhini G, Sanchez-Hernandez JC, Peltzer PM. Hematological parameters of health status in the common toad *Bufo arenarum* in agroecosystems of Santa Fe Province, Argentina. Appl Herpetol. 2005;2:373–80.
Cameron EZ, Ryan SJ. Welfare at multiple scales: importance of zoo elephant population welfare in a world of declining wild populations. PLoS One. 2016;11:e0158701.
Cannon WB. Bodily changes in pain, hunger, fear and rage: an account of recent researches into the function of emotional excitement. New York: Appleton; 1929.
Cannon WB. The wisdom of the body. New York: W.W. Norton; 1932.
Capiro JM, Stoops MA, Freeman EW, Clawson D, Schook MW. Effects of management strategies on glucocorticoids and behavior in Indian rhinoceros (*Rhinoceros unicornis*): translocation and operant conditioning. Zoo Biol. 2014;33:131–43.
Cardinet G, Littrell J, Schalm O. Effects of sustained muscular activity upon blood morphology of the horse. Calif Vet. 1964;18:31–5.
Carlitz EH, Miller R, Kirschbaum C, Gao W, Hänni DC, van Schaik CP. Measuring hair cortisol concentrations to assess the effect of anthropogenic impacts on wild chimpanzees (*Pan troglodytes*). PLoS One. 2016;11:e0151870.
Carlstead K, Brown JL. Relationships between patterns of fecal corticoid excretion and behavior, reproduction and environmental factors in captive black (*Diceros bicornis*) and white (*Ceratotherium simum*) rhinoceros. Zoo Biol. 2005;24:215–32.
Carlstead K, Shepherdson D. Effects of environmental enrichment on reproduction. Zoo Biol. 1994;13:447–58.
Carlstead K, Brown J, Seidensticker J. Behavioral and adrenocortical responses to environmental changes in leopard cats (*Felis bengalensis*). Zoo Biol. 1993;12:321–31.
Caroprese M, Annicchiarico G, Schena L, Muscio A, Migliore R, Sevi A. Influence of space allowance and housing conditions on the welfare, immune response and production performance of dairy ewes. J Dairy Res. 2009;76:66–73.
Carr JA. Stress and reproduction in amphibians. In: David ON, Kristin HL, editors. Hormones and reproduction of vertebrates. London: Academic Press; 2011. p. 99–116.
Carrasco L, Colell M, Calvo M, Abelló MT, Velasco M, Posada S. Benefits of training/playing therapy in a group of captive lowland gorillas (*Gorilla gorilla gorilla*). Anim Welf. 2009;18:9–19.
Carter S. Neuroendocrinology of sexual behavior in the female. In: Becker J, Breedlove S, Crews D, editors. Behavioral endocrinology. Cambridge: MIT Press; 1992. p. 71–96.
Carver C, Scheier M, Segerstrom S. Optimism. Clin Psychol Rev. 2010;30:879–89.
Casal N, Manteca X, Escribano D, Cerón J, Fàbrega E. Effect of environmental enrichment and herbal compound supplementation on physiological stress indicators (chromogranin A, cortisol and tumour necrosis factor-α) in growing pigs. Animal. 2017;11:1228–36.
Case BC, Lewbart GA, Doerr PD. The physiological and behavioural impacts of and preference for an enriched environment in the eastern box turtle (*Terrapene carolina carolina*). Appl Anim Behav Sci. 2005;92:353–65.
Casey TM, Plaut K. The role of glucocorticoids in secretory activation and milk secretion, a historical perspective. J Mammary Gland Biol Neoplasia. 2007;12:293–304.
Caslini C, Comin A, Peric T, Prandi A, Pedrotti L, Mattiello S. Use of hair cortisol analysis for comparing population status in wild red deer (*Cervus elaphus*) living in areas with different characteristics. Eur J Wildl Res. 2016;62:713–23.
Castagné R, Garès V, Karimi M, Chadeau-Hyam M, Vineis P, Delpierre C, Kelly-Irving M. Allostatic load and subsequent all-cause mortality: which biological markers drive the relationship? Findings from a UK birth cohort. Eur J Epidemiol. 2018;33:441–58.
Cavanaugh J, Carp SB, Rock CM, French JA. Oxytocin modulates behavioral and physiological responses to a stressor in marmoset monkeys. Psychoneuroendocrinology. 2016;66:22–30.
Cavigelli SA, Caruso MJ. Sex, social status and physiological stress in primates: the importance of social and glucocorticoid dynamics. Philos Trans R Soc Lond B Biol Sci. 2015;370:20140103.

Chaise LL, Paterson W, Laske TG, Gallon SL, McCafferty DJ, Théry M, Ancel A, Gilbert C. Implantation of subcutaneous heart rate data loggers in southern elephant seals (*Mirounga leonina*). Polar Biol. 2017;40:2307–12.
Chambers MA. Review of the diagnosis and study of tuberculosis in non-bovine wildlife species using immunological methods. Transbound Emerg Dis. 2009;56:215–27.
Chen X, Niu C, Pu L. Effects of stocking density on growth and non-specific immune responses in juvenile soft-shelled turtle, *Pelodiscus sinensis*. Aquacult Res. 2007;38:1380–6.
Christensen DS, Flensborg-Madsen T, Garde E, Hansen ÅM, Pedersen JM, Mortensen EL. Early life predictors of midlife allostatic load: a prospective cohort study. PLoS One. 2018a;13:e0202395.
Christensen DS, Flensborg-Madsen T, Garde E, Hansen ÅM, Pedersen JM, Mortensen EL. Parental socioeconomic position and midlife allostatic load: a study of potential mediators. BMC Public Health. 2018b;18:1029.
Chrousos GP, Torpy DJ, Gold PW. Interactions between the hypothalamic-pituitary-adrenal axis and the female reproductive system: clinical implications. Ann Intern Med. 1998;129:229–40.
Church JS, Cook NJ, Schaefer AL. Recent applications of infrared thermography for animal welfare and veterinary research: everything from chicks to elephants. Proc Inframation. 2009;10:215–24.
Clark FE, Fitzpatrick M, Hartley A, King AJ, Lee T, Routh A, Walker SL, George K. Relationship between behavior, adrenal activity, and environment in zoo-housed western lowland gorillas *(Gorilla gorilla gorilla).* Zoo Biol. 2011;31:306–21.
Clauss M, Wilkins T, Hartley A, Hatt JM. Diet composition, food intake, body condition, and fecal consistency in captive tapirs (*Tapirus spp.*) in UK collections. Zoo Biol. 2009;28:279–91.
Clegg I. Cognitive bias in zoo animals: an optimistic outlook for welfare assessment. Animals. 2018;8:104.
Clegg IL, Rödel HG, Delfour F. Bottlenose dolphins engaging in more social affiliative behaviour judge ambiguous cues more optimistically. Behav Brain Res. 2017;322:115–22.
Cockrem JF. Individual variation in glucocorticoid stress responses in animals. Gen Comp Endocrinol. 2013;181:45–58.
Codispoti M, Gerra G, Montebarocci O, Zaimovic A, Augusta Raggi M, Baldaro B. Emotional perception and neuroendocrine changes. Psychophysiology. 2003;40:863–8.
Cohen S, Wills TA. Stress, social support, and the buffering hypothesis. Psychol Bull. 1985;98:310–57.
Cohen S, Doyle WJ, Turner RB, Alper CM, Skoner DP. Emotional style and susceptibility to the common cold. Psychosom Med. 2003;65:652–7.
Cohn D, Zangerle R, Fischer-Colbrie R, Chu L, Elting J, Hamilton J, Winkler H. Similarity of secretory protein I from parathyroid gland to chromogranin A from adrenal medulla. Proc Natl Acad Sci U S A. 1982;79:6056–9.
Colborn D, Thompson D, Roth T, Capehart J, White K. Responses of cortisol and prolactin to sexual excitement and stress in stallions and geldings. J Anim Sci. 1991;69:2556–62.
Comizzoli P, Holt WV. Breakthroughs and new horizons in reproductive biology of rare and endangered animal species. Biol Reprod. 2019:ioz031.
Conlon JM. Granin-derived peptides as diagnostic and prognostic markers for endocrine tumors. Regul Pept. 2010;165:5–11.
Contreras-Aguilar MD, Tecles F, Martínez-Subiela S, Escribano D, Bernal LJ, Cerón JJ. Detection and measurement of alpha-amylase in canine saliva and changes after an experimentally induced sympathetic activation. BMC Vet Res. 2017;13:266.
Contreras-Aguilar MD, Escribano D, Martínez-Subiela S, Martínez-Miró S, Cerón JJ, Tecles F. Changes in alpha-amylase activity, concentration and isoforms in pigs after an experimental acute stress model: an exploratory study. BMC Vet Res. 2018;14:256.
Cook C, Mellor D, Harris P, Ingram J, Matthews L. Hands-on and hands-off measurement of stress. In: Moberg G, Mench J, editors. The biology of animal stress: basic principles and implications for animal welfare. Wallingford: CABI Publishing; 2000. p. 123–46.
Cooke RG, Benhaj KM. Effects of ACTH and cortisol on luteolysis in the ewe. Anim Reprod Sci. 1989;20:201–11.

Cooke SJ, Sack L, Franklin CE, Farrell AP, Beardall J, Wikelski M, Chown SL. What is conservation physiology? Perspectives on an increasingly integrated and essential science. Conserv Physiol. 2013;1:cot001.
Cooper CL, Baglioni AJ Jr. A structural model approach toward the development of a theory of the link between stress and mental health. Br J Med Psychol. 1988;61:87–102.
Cordoni G, Palagi E, Tarli SB. Reconciliation and consolation in captive western gorillas. Int J Primatol. 2006;27:1365–82.
Corti C, Biaggini M, Berti R. Analysis of escape behaviour and ectoparasite load in *Podarcis sicula* (Lacertidate) populations in different agricultural habitats. Amphibia-Reptillia. 2009;30:453–61.
Cottontail VM, Wellinghausen N, Kalko EKV. Habitat fragmentation and haemoparasites in the common fruit bat, *Artibeus jamaicensi* (Phyllostomidae) in a tropical lowland forest in Panamá. Parasitology. 2009;136:1133–45.
Creel S. Social dominance and stress hormones. Trends Ecol Evol. 2001;16:491–7.
Creel S, Fox JE, Hardy A, Sands J, Garrott B, Peterson RO. Snowmobile activity and glucocorticoid stress responses in wolves and elk. Conserv Biol. 2002;16:809–14.
Crimmins EM, Johnston M, Hayward M, Seeman TE. Age differences in allostatic load: an index of physiological dysregulation. Exp Gerontol. 2003;38:731–4.
Crimmins EM, Kim JK, Seeman TE. Poverty and biological risk: the earlier "aging" of the poor. J Gerontol A Biol Sci Med Sci. 2009;64A:286–92.
Cross N, Pines MK, Rogers LJ. Saliva sampling to assess cortisol levels in unrestrained common marmosets and the effect of behavioral stress. Am J Primatol. 2004;62:107–14.
Cummins DD. Dominance, status, and social hierarchies. In: Buss DM, editor. The handbook of evolutionary psychology. Hoboken: Wiley; 2005. p. 676–97.
Cummins DD. Status and dominance hierarchies. In: Weekes-Shackelford V, Shackelford TK, Weekes-Shackelford VA, editors. Encyclopedia of evolutionary psychological science. Cham: Springer; 2016. p. 1–10.
D'amico MA, Ghinassi B, Izzicupo P, Manzoli L, Di Baldassarre A. Biological function and clinical relevance of chromogranin A and derived peptides. Endocr Connect. 2014;3:R45–54.
Dai F, Cogi NH, Heinzl EUL, Dalla Costa E, Canali E, Minero M. Validation of a fear test in sport horses using infrared thermography. J Vet Behav. 2015;10:128–36.
Danese A, McEwen BS. Adverse childhood experiences, allostasis, allostatic load, an age-related disease. Physiol Behav. 2012;106:29–39.
Dantzer B, Fletcher QE, Boonstra R, Sheriff MJ. Measures of physiological stress: a transparent or opaque window into the status, management and conservation of species? Conserv Physiol. 2014;2:cou023.
Davis N, Schaffner CM, Smith TE. Evidence that zoo visitors influence HPA activity in spider monkeys (*Ateles geoffroyii rufiventris*). Appl Anim Behav Sci. 2005;90:131–41.
Davis A, Maney D, Maerz J. The use of leukocyte profiles to measure stress in vertebrates: a review for ecologists. Funct Ecol. 2008;22:760–72.
de Jong I, Sgoifo A, Lambooij E, Korte SM, Blokhuis HJ, Koolhaas JM. Effects of social stress on heart rate and heart rate variability in growing pigs. Can J Anim Sci. 2000;80:273–80.
Dehnhard M. Characterisation of the sympathetic nervous system of Asian (*Elephas maximus*) and African (*Loxodonta africana*) elephants based on urinary catecholamine analyses. Gen Comp Endocrinol. 2007;151:274–84.
Deinzer R, Schuller N. Dynamics of stress-related decrease of salivary immunoglobulin A (sIgA): relationship to symptoms of the common cold and studying behavior. Behav Med. 1998;23:161–9.
Delpierre C, Barbosa-Solis C, Torrisani J, Darnaudery M, Bartley M, Blane D, Kelly-Irving M, Tomasdottir MO, Robertson T, Gustafsson PE. Origins of health inequalities: the case for allostatic load. Longit Life Course Stud. 2016;7:79–103.
Dembiec DP, Snider RJ, Zanella AJ. The effects of transport stress on tiger physiology and behavior. Zoo Biol. 2004;23:335–46.

Dettling AC, Feldon J, Pryce CR. Repeated parental deprivation in the infant common marmoset (*Callithrix jacchus*, primates) and analysis of its effects on early development. Biol Psychiatry. 2002;52:1037–46.
Dettmer AM, Wooddell LJ, Rosenberg KL, Kaburu SSK, Novak MA, Meyer JS, Suomi SJ. Associations between early life experience, chronic HPA axis activity, and adult social rank in rhesus macaques. Soc Neurosci. 2017;12:92–101.
Di Comite G, Marinosci A, Di Matteo P, Manfredi A, Rovere-Querini P, Baldissera E, Aiello P, Corti A, Sabbadini MG. Neuroendocrine modulation induced by selective blockade of TNF-α in rheumatoid arthritis. Ann N Y Acad Sci. 2006;1069:428–37.
Dillon KM, Minchoff B, Baker KH. Positive emotional states and enhancement of the immune system. Int J Psychiatry Med. 1986;15:13–8.
Ditmer MA, Vincent JB, Werden LK, Tanner JC, Laske TG, Iaizzo PA, Garshelis DL, Fieberg JR. Bears show a physiological but limited behavioral response to unmanned aerial vehicles. Curr Biol. 2015;25:2278–83.
Ditmer MA, Werden LK, Tanner JC, Vincent JB, Callahan P, Iaizzo PA, Laske TG, Garshelis DL. Bears habituate to the repeated exposure of a novel stimulus, unmanned aircraft systems. Conserv Physiol. 2019;7:coy067-coy.
Dobson H, Smith R. Stress and reproduction in farm animals. J Reprod Fertil. 1995;49:451–61.
Dobson H, Smith RF. What is stress, and how does it affect reproduction? Anim Reprod Sci. 2000;60–61:743–52.
Dobson H, Ghuman S, Prabhakar S, Smith R. A conceptual model of the influence of stress on female reproduction. Reproduction. 2003;125:151–63.
Dowlati Y, Herrmann N, Swardfager W, Liu H, Sham L, Reim EK, Lanctôt KL. A meta-analysis of cytokines in major depression. Biol Psychiatry. 2010;67:446–57.
Drake K, Nussear K, Esque T, Barber A, Vittum K, Medica P, Tracy C, Hunter K. Does translocation influence physiological stress in the desert tortoise? Anim Conserv. 2012;15:560–70.
Duong HT, Piotrowska-Tomala KK, Acosta TJ, Bah MM, Sinderewicz E, Majewska M, Jankowska K, Okuda K, Skarzynski DJ. Effects of cortisol on pregnancy rate and corpus luteum function in heifers: an in vivo study. J Reprod Dev. 2011:1112060422.
Duong MT, Bingham BA, Aldana PC, Chung ST, Sumner AE. Variation in the calculation of allostatic load scores: 21 examples from NHANES. J Racial Ethn Health Disparities. 2017;4:455–61.
Duque M, Modlin IM, Gupta A, Saif MW. Biomarkers in neuroendocrine tumors. JOP. 2013;14:372–6.
Duru OK, Harawa NT, Kermah D, Norris KC. Allostatic load burden and racial disparities in mortality. J Natl Med Assoc. 2012;104:89–95.
Edes AN. Dehydroepiandrosterone-sulfate (DHEA-S), sex, and age in zoo-housed western lowland gorillas (*Gorilla gorilla gorilla*). Primates. 2017;58:385–92.
Edes AN, Crews DE. Allostatic load and biological anthropology. Am J Phys Anthropol. 2017;162:44–70.
Edes AN, Wolfe BA, Crews DE. Assessing stress in zoo-housed western lowland gorillas (*Gorilla gorilla gorilla*) using allostatic load. Int J Primatol. 2016a;37:241–59.
Edes AN, Wolfe BA, Crews DE. Rearing history and allostatic load in adult western lowland gorillas (*Gorilla gorilla gorilla*) in human care. Zoo Biol. 2016b;35:167–73.
Edes AN, Wolfe BA, Crews DE. Evaluating allostatic load: a new approach to measuring long-term stress in wildlife. J Zoo Wildl Med. 2018a;49:272–82.
Edes AN, Wolfe BA, Crews DE. The first multi-zoo application of an allostatic load index to western lowland gorillas (*Gorilla gorilla gorilla*). Gen Comp Endocrinol. 2018b;266:135–49.
Edwards KL, Walker SL, Bodenham RF, Ritchie H, Shultz S. Associations between social behaviour and adrenal activity in female Barbary macaques: consequences of study design. Gen Comp Endocrinol. 2013;186:72–9.
Edwards KL, Bansiddhi P, Paris S, Galloway M, Brown JL. The development of an immunoassay to measure immunoglobulin A in Asian elephant feces, saliva, urine and serum as a potential biomarker of well-being. Conserv Physiol. 2019;7:coy077.

Egorov AI, Converse R, Griffin SM, Styles J, Klein E, Sams E, Hudgens E, Wade TJ. Environmental risk factors for *Toxoplasma gondii* infections and the impact of latent infections on allostatic load in residents of central North Carolina. BMC Infect Dis. 2018;18:421.

Einarsson S, Madej A, Tsuma V. The influence of stress on early pregnancy in the pig. Anim Reprod Sci. 1996;42:165–72.

Elenkov IJ, Chrousos GP. Stress hormones, Th1/Th2 patterns, pro/anti-inflammatory cytokines and susceptibility to disease. Trends Endocrinol Metab. 1999;10:359–68.

Ellenberg U, Setiawan AN, Cree A, Houston DM, Seddon PJ. Elevated hormonal stress response and reduced reproductive output in yellow-eyed penguins exposed to unregulated tourism. Gen Comp Endocrinol. 2007;152:54–63.

Eriksson E, Royo F, Lyberg K, Carlsson HE, Hau J. Effect of metabolic cage housing on immunoglobulin A and corticosterone excretion in faeces and urine of young male rats. Exp Physiol. 2004;89:427–33.

Escribano D, Soler L, Gutiérrez AM, Martínez-Subiela S, Cerón JJ. Measurement of chromogranin A in porcine saliva: validation of a time-resolved immunofluorometric assay and evaluation of its application as a marker of acute stress. Animal. 2013;7:640–7.

Escribano D, Tvarijonaviciute A, Tecles F, Cerón JJ. Serum paraoxonase type-1 activity in pigs: assay validation and evolution after an induced experimental inflammation. Vet Immunol Immunopathol. 2015;163:210–5.

Estensen ME, Hognestad A, Syversen U, Squire I, Ng L, Kjekshus J, Dickstein K, Omland T. Prognostic value of plasma chromogranin A levels in patients with complicated myocardial infarction. Am Heart J. 2006;152:927.e1.

Evain D, Morera AM, Saez JM. Glucocorticoid receptors in interstitial cells of the rat testis. J Steroid Biochem. 1976;7:1135–9.

Everly GS, Lating JM. The anatomy and physiology of the human stress response. A clinical guide to the treatment of the human stress response. 3rd ed. New York: Springer; 2013. p. 17–51.

Eysenck MW, Mogg K, May J, Richards A, Mathews A. Bias in interpretation of ambiguous sentences related to threat in anxiety. J Abnorm Psychol. 1991;100:144.

Fanson KV, Wielebnowski NC. Effect of housing and husbandry practices on adrenocortical activity in captive Canada lynx (*Lynx canadensis*). Anim Welf. 2013;22:159–65.

Fanson KV, Lynch M, Vogelnest L, Miller G, Keeley T. Response to long-distance relocation in Asian elephants (*Elephas maximus*): monitoring adrenocortical activity via serum, urine, and feces. Eur J Wildl Res. 2013;59:655–64.

Fanson KV, Keeley T, Fanson BG. Cyclic changes in cortisol across the estrous cycle in parous and nulliparous Asian elephants. Endocr Connect. 2014;3:57–66.

Farm Animal Welfare Council. Press release. 1979.

Fazio F, Ferrantelli V, Cicero A, Casella S, Piccione G. Utility of acute phase proteins as biomarkers of transport stress in ewes and beef cattle. Ital J Food Saf. 2015;4:4210.

Fenster L, Katz DF, Wyrobek AJ, Pieper C, Rempel DM, Oman D, Swan SH. Effects of psychological stress on human semen quality. J Androl. 1997;18:194–202.

Ferin M. Stress and the reproductive cycle. J Clin Endocrinol Metab. 1999;84:1768–74.

Fernandez CA, Loucks EB, Arheart KL, Hickson DA, Kohn R, Duka SL, Gjelsvik A. Evaluating the effects of coping style on allostatic load, by sex: the Jackson heart study, 2000–2004. Prev Chronic Dis. 2015;12:E165.

Filaire E, Ferreira JP, Oliveira M, Massart A. Diurnal patterns of salivary alpha-amylase and cortisol secretion in female adolescent tennis players after 16 weeks of training. Psychoneuroendocrinology. 2013;38:1122–32.

Filiberto AC, Maccani MA, Koestler DC, Wilhelm-Benartzi C, Avissar-Whiting M, Banister CE, Gagne LA, Marsit CJ. Birthweight is associated with DNA promoter methylation of the glucocorticoid receptor in human placenta. Epigenetics. 2011;6:566–72.

Floreani A, Titta M, Plebani M, Faggian D, Chiaramonte M, Naccarato R. Sex hormone changes in post-menopausal women with primary biliary cirrhosis (PBC) and with cryptogenic chronic liver disease. Clin Exp Obstet Gynecol. 1991;18:229–34.

Flory JD, Manuck SB, Matthews KA, Muldoon MF. Serotonergic function in the central nervous system is associated with daily ratings of positive mood. Psychiatry Res. 2004;129:11–9.

Fontúrbel FE, Candia AB, Salzar DA, Malebrán J, González-Browne C, Botto-Mahan C. How forest marsupials are affected by habitat degradation and fragmentation? A meta-analysis. Naturwissenchaften. 2014;101:599–602.

Fortunato CK, Dribin AE, Granger DA, Buss KA. Salivary alpha-amylase and cortisol in toddlers: differential relations to affective behavior. Dev Psychobiol. 2008;50:807–18.

Fox NA. Psychophysiological correlates of emotional reactivity during the first year of life. Dev Psychol. 1989;25:364.

Franceschini MD, Rubenstein DI, Low B, Romero LM. Fecal glucocorticoid metabolite analysis as an indicator of stress during translocation and acclimation in an endangered large mammal, the Grevy's zebra. Anim Conserv. 2008;11:263–9.

Frank D, Gauthier A, Bergeron R. Placebo-controlled double-blind clomipramine trial for the treatment of anxiety or fear in beagles during ground transport. Can Vet J. 2006;47:1102.

Fraser D. Assessing animal welfare: different philosophies, different scientific approaches. Zoo Biol. 2009;28:507–18.

Frederick C, Fernandes D. Behavioral changes in pottos (*Perodicticus potto*): effects of naturalizing an exhibit. Int J Primatol. 1996;17:389–99.

Frederick C, Hunt K, Kyes R, Collins D, Durrant B, Ha J, Wasser SK. Social influences on the estrous cycle of the captive sun bear (*Helarctos malayanus*). Zoo Biol. 2013;32:581–91.

Freeman NE, Newman AE. Quantifying corticosterone in feathers: validations for an emerging technique. Conserv Physiol. 2018;6:coy051.

Freeman EW, Weiss E, Brown JL. Examination of the interrelationships of behavior, dominance status, and ovarian activity in captive Asian and African elephants. Zoo Biol. 2004;23:431–48.

Freeman EW, Guagnano G, Olson D, Keele M, Brown JL. Social factors influence ovarian acyclicity in captive African elephants (*Loxodonta africana*). Zoo Biol. 2009;28:1–15.

Frei R, Haile SR, Mutsch M, Rohrmann S. Relationship of serum vitamin D concentrations and allostatic load as a measure of cumulative biological risk among the US population: a cross-sectional study. PLoS One. 2015;10:1–17.

French SS, Fokidis HB, Moore MC. Variation in stress and innate immunity in the tree lizard (*Urosaurus ornatus*) across an urban–rural gradient. J Comp Physiol B. 2008;178:997–1005.

Friedman BH, Thayer JF. Anxiety and autonomic flexibility: a cardiovascular approach. Biol Psychol. 1998;47:243–63.

Fuentes M, Tecles F, Gutiérrez A, Otal J, Martínez-Subiela S, Cerón JJ. Validation of an automated method for salivary alpha-amylase measurements in pigs (*Sus scrofa domesticus*) and its application as a stress biomarker. J Vet Diagn Invest. 2011;23:282–7.

Fuentes-Rubio M, Fuentes F, Otal J, Quiles A, Tecles F, Cerón JJ, Hevia ML. Measurements of salivary alpha-amylase in horse: comparison of 2 different assays. J Vet Behav. 2015;10:122–7.

Fuller G, Allard S. Preliminary data showing potential for salivary C-reactive protein as an indicator of welfare in western lowland gorillas (*Gorilla gorilla gorilla*). J Vet Behav. 2018;28:58–62.

Fuller G, Kuhar CW, Dennis PM, Lukas KE. A survey of husbandry practices for lorisid primates in North American zoos and related facilities. Zoo Biol. 2013;32:88–100.

Fuller G, Lukas KE, Kuhar C, Dennis PM. A retrospective review of mortality in lorises and pottos in North American zoos, 1980-2010. Endanger Species Res. 2014;23:205–17.

Fürtbauer I, Heistermann M, Schülke O, Ostner J. Low female stress hormone levels are predicted by same- or opposite-sex sociality depending on season in wild Assamese macaques. Psychoneuroendocrinology. 2014;48:19–28.

Fuzzen MLM, Bernier NJ, Glen Van Der K. Stress and reproduction. In: David ON, Kristin HL, editors. Hormones and reproduction of vertebrates. London: Academic Press; 2011. p. 103–17.

Gácsi M, Maros K, Sernkvist S, Faragó T, Miklosi A. Human analogue safe haven effect of the owner: behavioural and heart rate response to stressful social stimuli in dogs. PLoS One. 2013;8:e58475.

Gallo LC, Fortmann AL, Mattei J. Allostatic load and the assessment of cumulative biological risk in biobehavioral medicine: challenges and opportunities. Psychosom Med. 2014;76:478–80.

Ganswindt A, Brown JL, Freeman EW, Kouba AJ, Penfold LM, Santymire RM, Vick MM, Wielebnowski N, Willis EL, Milnes MR. International Society for Wildlife Endocrinology: the future of endocrine measures for reproductive science, animal welfare and conservation biology. Biol Lett. 2012;8:695–7.
Garner MM, Helmick K, Ochsenreiter J, Richman LK, latimer E, Wise AG, Maes RK, Kiupel M, Nordhausen RW, Zong JC, Hayward GS. Clinico-pathologic features of fatal disease attributed to new variances of endotheliotropic herpesvirus in two Asian elephants (*Elephas maximus*). Vet Pathol. 2009;46:97–104.
Gasparotto OC, Lopes DM, Carobrez SG. Pair housing affects anxiety-like behaviors induced by a social but not by a physiological stressor in male Swiss mice. Physiol Behav. 2005;85:603–12.
Gay JL, Salinas JJ, Buchner DM, Mirza S, Kohl Iii HW, Fisher-Hoch SP, McCormick JB. Meeting physical activity guidelines is associated with lower allostatic load and inflammation in Mexican Americans. J Immigr Minor Health. 2015;17:574–81.
Gimeno D, Brunner EJ, Lowe GD, Rumley A, Marmot MG, Ferrie JE. Adult socioeconomic position, C-reactive protein and interleukin-6 in the Whitehall II prospective study. Eur J Epidemiol. 2007a;22:675–83.
Gimeno D, Ferrie J, Elovainio M, Pulkki-Raback L, Keltikangas-Jarvinen L, Eklund C, Hurme M, Lehtimäki T, Marniemi J, Viikari J. When do social inequalities in C-reactive protein start? A life course perspective from conception to adulthood in the Cardiovascular Risk in Young Finns Study. Int J Epidemiol. 2007b;37:290–8.
Ginther OJ. Reproductive biology of the mare: basic and applied aspects. Cross Plains: Equiservices; 1992.
Giovinazzo F, Schimmack S, Svejda B, Alaimo D, Pfragner R, Modlin I, Kidd M. Chromogranin A and its fragments as regulators of small intestinal neuroendocrine neoplasm proliferation. PLoS One. 2013;8:e81111.
Gleeson M, McDonald WA, Cripps AW, Pyne DB, Clancy RL, Fricker PA, Wlodarczyk JH. Exercise, stress and mucosal immunity in elite swimmers. Adv Exp Med Biol. 1995;371A:571–4.
Glei DA, Goldman N, Chuang Y-L, Weinstein M. Do chronic stressors lead to physiological dysregulation? Testing the theory of allostatic load. Psychosom Med. 2007;69:769–76.
Glei DA, Goldman N, Rodríguez G, Weinstein M. Beyond self-reports: changes in biomarkers as predictors of mortality. Popul Dev Rev. 2014;40:331–60.
Gobush KS, Mutayoba BM, Wasser SK. Long-term impacts of poaching on relatedness, stress physiology, and reproductive output of adult female African elephants. Conserv Biol. 2008;22:1590–9.
Goldman N, Glei DA, Seplaki C, Liu IW, Weinstein M. Perceived stress and physiological dysregulation in older adults. Stress. 2005;8:95–105.
Gopee NV, Adesiyun AA, Caesar K. Retrospective and longitudinal study of salmonellosis in captive wildlife in Trinidad. J Wildl Dis. 2000;36:284–93.
Gourkow N, LaVoy A, Dean GA, Phillips CJC. Associations of behaviour with secretory immunoglobulin A and cortisol in domestic cats during their first week in an animal shelter. Appl Anim Behav Sci. 2014;150:55–64.
Goymann W, Wingfield JC. Allostatic load, social status and stress hormones: the costs of social status matter. Anim Behav. 2004;67:591–602.
Granger DA, Kivlighan KT, Blair C, El-Sheikh M, Mize J, Lisonbee JA, Buckhalt JA, Stroud LR, Handwerger K, Schwartz EB. Integrating the measurement of salivary α-amylase into studies of child health, development, and social relationships. J Soc Pers Relat. 2006;23:267–90.
Greco BJ, Meehan CL, Hogan JN, Leighty KA, Mellen J, Mason GJ, Mench JA. The days and nights of zoo elephants: using epidemiology to better understand stereotypic behavior of African elephants (*Loxodonta africana*) and Asian elephants (*Elephas maximus*) in North American zoos. PLoS One. 2016;11:e0144276.
Green ML, Green RG, Santoro W. Daily relaxation modifies serum and salivary immunoglobulins and psychophysiologic symptom severity. Appl Psychophysiol Biofeedback. 1988;13:187–99.
Grobler DG, Raath JP, Braack LEO, Keet DF, Gerdes GH, Barnard BJH, Kriek NPJ, Jardine J, Swanepoel R. An outbreak of encephalomyocarditis-virus infection in free-ranging African elephants in the Kruger National Park. Onderstepoort J Vet Res. 1995;62:97–108.

Groombridge JJ, Massey JG, Bruch JC, Malcolm TR, Brosius CN, Okada MM, Sparklin B. Evaluating stress in a Hawaiian honeycreeper, *Paroreomyza montana*, following translocation. J Field Ornithol. 2004;75:183–7.

Grossi G, Perski A, Evengård B, Blomkvist V, Orth-Gomér K. Physiological correlates of burnout among women. J Psychosom Res. 2003;55:309–16.

Gruenewald TL, Seeman TE, Ryff CD, Karlamangla AS, Singer BH. Combinations of biomarkers predictive of later life mortality. Proc Natl Acad Sci U S A. 2006;103:14158–63.

Guhad FA, Hau J. Salivary IgA as a marker of social stress in rats. Neurosci Lett. 1996;216:137–40.

Guilliams TG, Edwards L. Chronic stress and the HPA axis. Standard. 2010;9:1–12.

Gundlach NH, Schmicke M, Ludes-Wehrmeister E, Ulrich SA, Araujo MG, Siebert U. New approach to stress research in phocids—potential of dehydroepiandrosterone and cortisol/dehydroepiandrosterone ratio as markers for stress in harbor seals (*Phoca vitulina*) and gray seals (*Halichoerus grypus*). J Zoo Wildl Med. 2018;49:556–63.

Haeussler S, Claus R. Expression of the glucocorticoid receptor and 11β-hydroxysteroid dehydrogenase 2 in pig testes cells along fetal development. Reprod Fertil Dev. 2007;19:664–9.

Hagen K, Langbein J, Schmied C, Lexer D, Waiblinger S. Heart rate variability in dairy cows—influences of breed and milking system. Physiol Behav. 2005;85:195–204.

Hainsworth R. The control and physiological importance of heart rate. In: Malik M, Camm AJ, editors. Heart rate variability. New York: Futura Publishing; 1995. p. 3–19.

Hämäläinen A, Heistermann M, Kraus C. The stress of growing old: sex- and season-specific effects of age on allostatic load in wild grey mouse lemurs. Oecologia. 2015;178:1063–75.

Hänichen T, Reid HW, Wiesner H, Hermanns W. Malignant catarrhal fever in zoo ruminants. Tierarztl Prax Ausg G Grosstiere Nutztiere. 1998;26:294–300.

Hänsel A, Hong S, Camara RJ, Von Kaenel R. Inflammation as a psychophysiological biomarker in chronic psychosocial stress. Neurosci Biobehav Rev. 2010;35:115–21.

Harding EJ, Paul ES, Mendl M. Animal behaviour: cognitive bias and affective state. Nature. 2004;427:312.

Hardy MP, Gao HB, Dong Q, Ge RS, Wang Q, Chai WR, Feng X, Sottas C. Stress hormone and male reproductive function. Cell Tissue Res. 2005;322:147–53.

Harper JY, Schulte BA. Social interactions in captive female Florida manatees. Zoo Biol. 2005;24:135–44.

Hau J, Andersson E, Carlsson HE. Development and validation of a sensitive ELISA for quantification of secretory IgA in rat saliva and faeces. Lab Anim. 2001;35:301–6.

Hay M, Mormède P. Urinary excretion of catecholamines, cortisol and their metabolites in Meishan and large white sows: validation as a non-invasive and integrative assessment of adrenocortical and sympathoadrenal axis activity. Vet Res. 1998;29:119–28.

Hayward LS, Bowles AE, Ha JC, Wasser SK. Impacts of acute and long-term vehicle exposure on physiology and reproductive success of the northern spotted owl. Ecosphere. 2011;2:1–20.

Haywood S, Perrins CM. Is clutch size in birds affected by environmental conditions during growth? Proc R Soc Lond B. 1992;249:195–7.

Heatley JJ, Oliver JW, Hosgood G, Columbini S, Tully TN. Serum corticosterone concentrations in response to restraint, anesthesia, and skin testing in Hispaniolan Amazon parrots (*Amazona ventralis*). J Avian Med Surg. 2000;14:172–6.

Heimbürge S, Kanitz E, Otten W. The use of hair cortisol for the assessment of stress in animals. Gen Comp Endocrinol. 2018;270:10–7.

Heistermann M, Ademmer C, Kaumanns W. Ovarian cycle and effect of social changes on adrenal and ovarian function in *Pygathrix nemaeus*. Int J Primatol. 2004;25:689–708.

Hemingway H, Shipley M, Mullen MJ, Kumari M, Brunner E, Taylor M, Donald AE, Deanfield JE, Marmot M. Social and psychosocial influences on inflammatory markers and vascular function in civil servants (the Whitehall II study). Am J Cardiol. 2003;92:984–7.

Henkel S, Heistermann M, Fischer J. Infants as costly social tools in male Barbary macaque networks. Anim Behav. 2010;79:1199–204.

Herrelko ES, Vick S-J, Buchanan-Smith HM. Cognitive research in zoo-housed chimpanzees: influence of personality and impact on welfare. Am J Primatol. 2012;74:828–40.

Herrera-Luna C, Budik S, Aurich C. Gene expression of ACTH, glucocorticoid receptors, 11βHSD enzymes, LH-, FSH-, GH receptors and aromatase in equine epididymal and testicular tissue. Reprod Domest Anim. 2012;47:928–35.

Herring G, Gawlik DE. Differential physiological responses to prey availability by the great egret and white Ibis. J Wildl Manage. 2013;77:58–67.

Herrington DM, Gordon GB, Achuff SC, Trejo JF, Weisman HF, Kwiterovich PO, Pearson TA. Plasma dehydroepiandrosterone and dehydroepiandrosterone sulfate in patients undergoing diagnostic coronary angiography. J Am Coll Cardiol. 1990;16:862–70.

Hickey M-C, Drennan M, Earley B. The effect of abrupt weaning of suckler calves on the plasma concentrations of cortisol, catecholamines, leukocytes, acute-phase proteins and in vitro interferon-gamma production. J Anim Sci. 2003;81:2847–55.

Higham JP, Vitale AB, Rivera AM, Ayala JE, Maestripieri D. Measuring salivary analytes from free-ranging monkeys. Physiol Behav. 2010;101:601–7.

Higuchi T, Honda K, Negoro H. Influence of oestrogen and noradrenergic afferent neurones on the response of LH and oxytocin to immobilization stress. J Endocrinol. 1986;110:245–50.

Hillier SG, Tetsuka M. An anti-inflammatory role for glucocorticoids in the ovaries? J Reprod Immunol. 1998;39:21–7.

Hilsberg S. Clinical application of infrared-thermography in inflammation diagnosis in mega-herbivores. In: Proceedings of European Association of Zoo and Wildlife Veterinarians 4th scientific meeting; 2002. p. 315–20.

Hing S, Northover AS, Narayan EJ, Wayne AF, Jones KL, Keatley S, Thompson RA, Godfrey SS. Evaluating stress physiology and parasite infection parameters in the translocation of critically endangered woylies (*Bettongia penicillata*). Ecohealth. 2017;14:128–38.

Hjortskov N, Rissén D, Blangsted AK, Fallentin N, Lundberg U, Søgaard K. The effect of mental stress on heart rate variability and blood pressure during computer work. Eur J Appl Physiol. 2004;92:84–9.

Hofer H, East M. Stress and immunosuppression as factors in the decline and extinction of wildlife populations: concepts, evidence and challenges. In: Aguirre A, Ostfeld R, Daszak P, editors. New directions in conservation medicine: applied cases of ecological health. New York: Oxford University Press; 2012. p. 82–107.

Hohmann G, Mundry R, Deschner T. The relationship between socio-sexual behavior and salivary cortisol in bonobos: tests of the tension regulation hypothesis. Am J Primatol. 2009;71:223–32.

Holt-Lunstad J, Smith TB, Layton JB. Social relationships and mortality risk: a meta-analytic review. PLoS Med. 2010;7:1–20.

Honess PE, Johnson PJ, Wolfensohn SE. A study of behavioural responses of non-human primates to air transport and re-housing. Lab Anim. 2004;38:119–32.

Hopper LM. Cognitive research in zoos. Behav Sci. 2017;16:100–10.

Hopper LM, Shender MA, Ross SR. Behavioral research as physical enrichment for captive chimpanzees. Zoo Biol. 2016;35:293–7.

Horan JM, Widom CS. From childhood maltreatment to allostatic load in adulthood: the role of social support. Child Maltreat. 2015;20:229–39.

Hostinar CE, Sullivan RM, Gunnar MR. Psychobiological mechanisms underlying the social buffering of the HPA axis: a review of animal models and human studies across development. Psychol Bull. 2014;140:256–82.

House JS, Landis KR, Umberson D. Social relationships and health. Science. 1988;241:540–5.

Hovland AL, Rød AMS, Eriksen MS, Palme R, Nordgreen J, Mason GJ. Faecal cortisol metabolites as an indicator of adrenocortical activity in farmed silver foxes (*Vulpes vulpes*). Appl Anim Behav Sci. 2017;197:75–80.

Howell CP, Cheyne SM. Complexities of using wild versus captive activity budget comparisons for assessing captive primate welfare. J Appl Anim Welf Sci. 2019;22:78–96.

Howell-Stephens JA, Brown JS, Bernier D, Mulkerin D, Santymire RM. Characterizing adrenocortical activity in zoo-housed southern three-banded armadillos (*Tolypeutes matacus*). Gen Comp Endocrinol. 2012;178:64–74.

Howren MB, Lamkin DM, Suls J. Associations of depression with C-reactive protein, IL-1, and IL-6: a meta-analysis. Psychosom Med. 2009;71:171–86.

Huang S, Li L, Wu J, Li C, Bai J, Sun Y, Wang G. Seasonal variations in immunoreactive cortisol and fecal immunoglobulin levels in Sichuan golden monkey (*Rhinopithecus roxellana*). Turk J Zool. 2014;38:642–50.

Hucklebridge F, Clow A, Evans P. The relationship between salivary secretory immunoglobulin A and cortisol: neuroendocrine response to awakening and the diurnal cycle. Int J Psychophysiol. 1998;31:69–76.

Hughes P. Factors affecting the natural attainment of puberty in the gilt. In: Cole D, Foxcroft G, editors. Control of pig reproduction. London: Butterworths; 1982. p. 117–38.

Hunt KE, Lysiak NS, Moore MJ, Rolland RM. Longitudinal progesterone profiles in baleen from female North Atlantic right whales (*Eubalaena glacialis*) match known calving history. Conserv Physiol. 2016;4:cow014.

Hunt KE, Lysiak NS, Moore M, Rolland RM. Multi-year longitudinal profiles of cortisol and corticosterone recovered from baleen of North Atlantic right whales (*Eubalaena glacialis*). Gen Comp Endocrinol. 2017;254:50–9.

Hussain S, Ram MS, Kumar A, Shivaji S, Umapathy G. Human presence increases parasitic load in endangered lion-tailed macaques (*Macaca silenus*) in its fragmented rainforest habitats in southern India. PLoS One. 2013;8:e63685.

Hux VJ, Roberts JM. A potential role for allostatic load in preeclampsia. Matern Child Health J. 2015;19:591–7.

Hux VJ, Catov JM, Roberts JM. Allostatic load in women with a history of low birth weight infants: the national health and nutrition examination survey. J Womens Health. 2014;23:1039–45.

Hwang A-C, Peng L-N, Wen Y-W, Tsai Y-W, Chang L-C, Chiou S-T, Chen L-K. Predicting all-cause and cause-specific mortality by static and dynamic measurements of allostatic load: a 10-year population-based cohort study in Taiwan. J Am Med Dir Assoc. 2014;15:490–6.

Ioannou S, Chotard H, Davila-Ross M. No strings attached: physiological monitoring of rhesus monkeys (*Macaca mulatta*) with thermal imaging. Front Behav Neurosci. 2015;9:160.

Irvine RJ, White J, Chan R. The influence of restraint on blood pressure in the rat. J Pharmacol Toxicol Methods. 1997;38:157–62.

Jacobs RM, Ross SR, Wagner KE, Leahy M, Meiers ST, Santymire RM. Evaluating the physiological and behavioral response of a male and female gorilla (*Gorilla gorilla gorilla*) during an introduction. Zoo Biol. 2014;33:394–402.

Jang YH, Lee SJ, Lim JG, Lee HS, Kim TJ, Park JH, Chung BH, Choe NH. The rate of *Salmonella spp.* infection in zoo animals at Seoul Grand Park, Korea. J Vet Sci. 2008;9:177–81.

Jansson AM, Røsjø H, Omland T, Karlsson T, Hartford M, Flyvbjerg A, Caidahl K. Prognostic value of circulating chromogranin A levels in acute coronary syndromes. Eur Heart J. 2008;30:25–32.

Jarillo-Luna RA, Rivera-Aguilar V, Pacheco-Yépez J, Godínez-Victoria M, Oros-Pantoja R, Miliar-García A, Campos-Rodríguez R. Nasal IgA secretion in a murine model of acute stress. The possible role of catecholamines. J Neuroimmunol. 2015;278:223–31.

Jerem P, Jenni-Eiermann S, Herborn K, McKeegan D, McCafferty DJ, Nager RG. Eye region surface temperature reflects both energy reserves and circulating glucocorticoids in a wild bird. Sci Rep. 2018;8:1907.

Ji S, Yang L, Ge X, Wang B, Cao J, Hu D. Behavioural and physiological stress responses to transportation in a group of Przewalski's horses (*Equus ferus prezwalskii*). J Anim Plant Sci. 2013;23:1077–84.

Johns DW, Marchant TA, Fairhurst GD, Speakman JR, Clark RG. Biomarker of burden: feather corticosterone reflects energetic expenditure and allostatic overload in captive waterfowl. Funct Ecol. 2018;32:345–57.

Johnstone CP, Reina RD, Lill A. Impact of anthropogenic habitat fragmentation on population health in a small, carnivorous marsupial. J Mammal. 2010;91:1332–41.

Jones P, Cordonnier N, Mahamba C, Burt FJ, Rakotovao F, Swanepoel R, André C, Dauger S, Bakkali Kassimi L. Encephalomyocarditis virus mortality in semi-wild bonobos (*Pan paniscus*). J Med Primatol. 2011;40:157–63.

Josserand R, Dupoué A, Agostini S, Haussy C, Le Galliard J-F, Meylan S. Habitat degradation increases stress-hormone levels during the breeding season, and decreases survival and reproduction in adult common lizards. Oecologia. 2017;184:75–86.
Jurke MH, Czekala NM, Lindburg DG, Millard SE. Fecal corticoid metabolite measurement in the cheetah (*Acinonyx jubatus*). Zoo Biol. 1997;16:133–47.
Juster RP, McEwen BS, Lupien SJ. Allostatic load biomarkers of chronic stress and impact on health and cognition. Neurosci Biobehav Rev. 2010;35:2–16.
Kagan R, Carter S, Allard S. A universal animal welfare framework for zoos. J Appl Anim Welf Sci. 2015;18:S1–S10.
Kalantaridou SN, Makrigiannakis A, Zoumakis E, Chrousos GP. Stress and the female reproductive system. J Reprod Immunol. 2004;62:61–8.
Kalantaridou SN, Zoumakis E, Makrigiannakis A, Lavasidis LG, Vrekoussis T, Chrousos GP. Corticotropin-releasing hormone, stress and human reproduction: an update. J Reprod Immunol. 2010;85:33–9.
Kanamaru Y, Kikukawa A, Shimamura K. Salivary chromogranin-A as a marker of psychological stress during a cognitive test battery in humans. Stress. 2006;9:127–31.
Kanno T, Asada N, Yanase H, Iwanaga T, Nishikawa Y, Hoshino M, Yanaihara N. Autonomic control of submandibular chromogranin A secretion in the anaesthetized rat. Biomed Res. 1998;19:411–4.
Kano F, Hirata S, Deschner T, Behringer V, Call J. Nasal temperature drop in response to a playback of conspecific fights in chimpanzees: a thermo-imaging study. Physiol Behav. 2016;155:83–94.
Karlamangla AS, Singer BH, McEwen BS, Rowe JW, Seeman TE. Allostatic load as a predictor of functional decline: MacArthur studies of successful aging. J Clin Epidemiol. 2002;55:696–710.
Karlamangla AS, Singer BH, Seeman TE. Reduction in allostatic load in older adults is associated with lower all-cause mortality risk: Macarthur studies of successful aging. Psychosom Med. 2006;68:500–7.
Karlamangla AS, Miller-Martinez D, Lachman ME, Tun PA, Koretz BK, Seeman TE. Biological correlates of adult cognition: midlife in the United States (MIDUS). Neurobiol Aging. 2014;35:387–94.
Kato Y, Kowalski CJ, Stohler CS. Habituation of the early pain-specific respiratory response in sustained pain. Pain. 2001;91:57–63.
Keay SD, Harlow CR, Wood PJ, Jenkins JM, Cahill DJ. Higher cortisol:cortisone ratios in the preovulatory follicle of completely unstimulated IVF cycles indicate oocytes with increased pregnancy potential. Hum Reprod. 2002;17:2410–4.
Kendall R, Howard L, Masters N, Grant R. The impact of elephant endotheliotropic herpesvirus on the captive Asian elephant (*elephas maximus*) population of the United Kingdom and Ireland (1995-2013). J Zoo Wildl Med. 2016;47:405–18.
Kerr MA. The structure and function of human IgA. Biochem J. 1990;271:285–96.
Kersey DC, Wildt DE, Brown JL, Snyder RJ, Huang Y, Monfort SL. Rising fecal glucocorticoid concentrations track reproductive activity in the female giant panda (*Ailuropoda melanoleuca*). Gen Comp Endocrinol. 2011;173:364–70.
Kershaw JL, Hall AJ. Seasonal variation in harbour seal (*Phoca vitulina*) blubber cortisol-a novel indicator of physiological state? Sci Rep. 2016;6:21889.
Khansari DN, Murgo AJ, Faith RE. Effects of stress on the immune system. Immunol Today. 1990;11:170–5.
Kiecolt-Glaser JK, Preacher KJ, MacCallum RC, Atkinson C, Malarkey WB, Glaser R. Chronic stress and age-related increases in the proinflammatory cytokine IL-6. Proc Natl Acad Sci U S A. 2003;100:9090–5.
Kikkawa A, Uchida Y, Nakade T, Taguchi K. Salivary secretory IgA concentrations in beagle dogs. J Vet Med Sci. 2003;65:689–93.
Kilgas P, Tilgar V, Mänd R. Hematological health state indices predict local survival in a small passerine bird, the great tit (*Parus major*). Physiol Biochem Zool. 2006;79:565–72.
Kimmel E, Arkin H, Berman A. Evaporative cooling of cattle: transport phenomena and thermovision. St. Joseph: American Society of Agricultural Engineers; 1992. p. 14.

Kirschbaum C, Kudielka BM, Gaab J, Schommer NC, Hellhammer DH. Impact of gender, menstrual cycle phase, and oral contraceptives on the activity of the hypothalamus-pituitary-adrenal axis. Psychosom Med. 1999;61:154–62.

Knobil E, Neill J, editors. Encyclopedia of reproduction, vol. 1–4. New York: Academic Press; 1998.

Knutson B, Burgdorf J, Panksepp J. Ultrasonic vocalizations as indices of affective states in rats. Psychol Bull. 2002;128:961.

Kock RA, Mihok SR, Wambua J, Mwanzia J, Saigawa K. Effects of translocation on hematologic parameters of free-ranging black rhinoceros (*Diceros bicornis michaeli*) in Kenya. J Zoo Wildl Med. 1999;30:389–96.

Kolbe T, Palme R, Tichy A, Rülicke T. Lifetime dependent variation of stress hormone metabolites in feces of two laboratory mouse strains. PLoS One. 2015;10:e0136112.

Koolhaas JM, Korte SM, De Boer SF, Van Der Vegt BJ, Van Reenen CG, Hopster H, De Jong IC, Ruis MAW, Blokhuis HJ. Coping styles in animals: current status in behavior and stress-physiology. Neurosci Biobehav Rev. 1999;23:925–35.

Koren L, Whiteside D, Fahlman Å, Ruckstuhl K, Kutz S, Checkley S, Dumond M, Wynne-Edwards K. Cortisol and corticosterone independence in cortisol-dominant wildlife. Gen Comp Endocrinol. 2012;177:113–9.

Korgun E, Dohr G, Desoye G, Demir R, Kayisli U, Hahn T. Expression of insulin, insulin-like growth factor I and glucocorticoid receptor in rat uterus and embryo during decidualization, implantation and organogenesis. Reproduction. 2003;125:75–84.

Korhonen R, Lahti A, Hämäläinen M, Kankaanranta H, Moilanen E. Dexamethasone inhibits inducible nitric-oxide synthase expression and nitric oxide production by destabilizing mRNA in lipopolysaccharide-treated macrophages. Mol Pharmacol. 2002;62:698–704.

Koster A, Bosma H, Penninx BW, Newman AB, Harris TB, Van Eijk JTM, Kempen GI, Simonsick EM, Johnson KC, Rooks RN. Association of inflammatory markers with socioeconomic status. J Gerontol A Biol Sci Med Sci. 2006;61:284–90.

Kranendonk G, Hopster H, Eerdenburg FV, Reenen KV, Fillerup M, Groot JD, Korte M, Taverne M. Evaluation of oral administration of cortisol as a model for prenatal stress in pregnant sows. Am J Vet Res. 2005;66:780–90.

Kreuder C, Miller MA, Lowenstine LJ, Conrad PA, Carpenter TE, Jessup DA, Mazet JA. Evaluation of cardiac lesions and risk factors associated with myocarditis and dilated cardiomyopathy in southern sea otters (*Enhydra lutris nereis*). Am J Vet Res. 2005;66:289–99.

Kroboth PD, Salek FS, Pittenger AL, Fabian TJ, Frye RF. DHEA and DHEA-S: a review. J Clin Pharmacol. 1999;39:327–48.

Krohn TC, Hansen AK, Dragsted N. Telemetry as a method for measuring the impact of housing conditions on rats' welfare. Anim Welf. 2003;12:53–62.

Kronfol Z, Turner R, Nasrallah H, Winokur G. Leukocyte regulation in depression and schizophrenia. Psychiatry Res. 1984;13:13–8.

Kubzansky LD, Kawachi I, Sparrow D. Socioeconomic status, hostility, and risk factor clustering in the normative aging study: any help from the concept of allostatic load? Ann Behav Med. 1999;21:330–8.

Kuraoka K, Nakamura K. The use of nasal skin temperature measurements in studying emotion in macaque monkeys. Physiol Behav. 2011;102:347–55.

Kurtycz LM, Wagner KE, Ross SR. The choice to access outdoor areas affects the behavior of great apes. J Appl Anim Welf Sci. 2014;17:185–97.

Kusano Y, Crews DE, Iwamoto A, Sone Y, Aoyagi K, Maeda T, Leahy R. Allostatic load differs by sex and diet, but not age in older Japanese from the Goto Islands. Ann Hum Biol. 2016;43:34–41.

Kvietkauskaite R, Vaicaitiene R, Mauricas M. The change in the amount of immunoglobulins as a response to stress experienced by soldiers on a peacekeeping mission. Int Arch Occup Environ Health. 2014;87:615–22.

Laitinen M, Valtonen T. Cardiovascular, ventilatory and total activity responses of brown trout to handling stress. J Fish Biol. 1994;45:933–42.

Lambert RB, Lambert NK. The effects of humor on secretory immunoglobulin A levels in school-aged children. Pediatr Nurs. 1995;21:16–9.

Lambeth SP, Hau J, Perlman JE, Martino M, Schapiro SJ. Positive reinforcement training affects hematologic and serum chemistry values in captive chimpanzees (*Pan troglodytes*). Am J Primatol. 2006;68:245–56.

Lamglait B, Joris A, Romey A, Bakkali-Kassimi L, Lemberger K. Fatal encephalomyocarditis virus infection in an African savanna elephant (*Loxodonta africana*) in a French zoo. J Zoo Wildl Med. 2015;46:393–6.

Lamond D. Synchronization of oestrus and ovulation in beef heifers. Proc Aust Soc Anim Prod. 1962;4:72–8.

Lantz EL, Lonsdorf EV, Heintz MR, Murray CM, Lipende I, Travis DA, Santymire RM. Non-invasive quantification of immunoglobulin A in chimpanzees (*Pan troglodytes schweinfurthii*) at Gombe National Park, Tanzania. Am J Primatol. 2016;80:e22558.

Laske TG, Garshelis DL, Iaizzo PA. Monitoring the wild black bear's reaction to human and environmental stressors. BMC Physiol. 2011;11:13.

Laske TG, Iaizzo PA, Garshelis DL. Six years in the life of a mother bear-the longest continuous heart rate recordings from a free-ranging mammal. Sci Rep. 2017;7:40732.

Lattin CR, Ngai HM, Romero LM. Evaluating the stress response as a bioindicator of sub-lethal effects of crude oil exposure in wild house sparrows (*Passer domesticus*). PLoS One. 2014;9:e102106.

Laule G, Desmond T. Positive reinforcement training as an enrichment strategy. In: Shepherdson DJ, Mellen JD, Hutchins M, editors. Second nature: environmental enrichment for captive animals. Washington: Smithsonian Institution Press; 1998. p. 302–13.

Lazić MM, Carretero MA, Živković U, Crnobrnja-Isailović J. City life has fitness costs: reduced body condition and increased parasite load in urban common wall lizards, *Podarcis muralis*. Salamandra. 2017;53:10–7.

Lea AJ, Tung J, Archie EA, Alberts SC. Developmental plasticity: bridging research in evolution and human health. Evol Med Public Health. 2017;2017:162–75.

Lea AJ, Akinyi MY, Nyakundi R, Mareri P, Nyundo F, Kariuki T, Alberts SC, Archie EA, Tung J. Dominance rank-associated gene expression is widespread, sex-specific, and a precursor to high social status in wild male baboons. Proc Natl Acad Sci U S A. 2018a;115:E12163–E71.

Lea JMD, Walker SL, Kerley GIH, Jackson J, Matevich SC, Shultz S. Non-invasive physiological markers demonstrate link between habitat quality, adult sex ratio and poor population growth rate in a vulnerable species, the Cape mountain zebra. Funct Ecol. 2018b;32:300–12.

Lèche A, Cortez MV, Della Costa NS, Navarro JL, Marin RH, Martella MB. Stress response assessment during translocation of captive-bred greater rheas into the wild. J Ornithol. 2016;157:599–607.

Lecorps B, Rödel HG, Féron C. Assessment of anxiety in open field and elevated plus maze using infrared thermography. Physiol Behav. 2016;157:209–16.

Lécu A, Ball R. Mycobacterial infections in zoo animals: relevance, diagnosis and management. Int Zoo Yearb. 2011;45:183–202.

Lee MS, Rim YH, Jeong D-M, Kim MK, Joo MC, Shin SH. Nonlinear analysis of heart rate variability during Qi therapy (external Qigong). Am J Chin Med. 2005;33:579–88.

Lee HY, Acosta TJ, Tanikawa M, Sakumoto R, Komiyama J, Tasaki Y, Piskula M, Skarzynski DJ, Tetsuka M, Okuda K. The role of glucocorticoid in the regulation of prostaglandin biosynthesis in non-pregnant bovine endometrium. J Endocrinol. 2007;193:127–35.

Leeds A, Boyer D, Ross SR, Lukas KE. The effects of group type and young silverbacks on wounding rates in western lowland gorilla (*Gorilla gorilla gorilla*) groups in North American zoos. Zoo Biol. 2015;34:296–304.

Leeds A, Elsner R, Lukas KE. The effect of positive reinforcement training on an adult female western lowland gorilla's (*Gorilla gorilla gorilla*) rate of abnormal and aggressive behavior. Anim Behav Cogn. 2016;3:78–87.

Lees CM, Wilcken J. Sustaining the Ark: the challenges faced by zoos in maintaining viable populations. Int Zoo Yearb. 2009;43:6–18.

Lennartsson A-K, Kushnir MM, Bergquist J, Jonsdottir IH. DHEA and DHEA-S response to acute psychosocial stress in healthy men and women. Biol Psychol. 2012;90:143–9.
Leong KW, Ding JL. The unexplored roles of human serum IgA. DNA Cell Biol. 2014;33:823–9.
Leshyk R, Nol E, Chin EH, Burness G. Adult ovenbirds (*Seiurus aurocapilla*) show increased stress-responsiveness in logged forests. Gen Comp Endocrinol. 2013;194:295–9.
Levine ME, Crimmins EM. A comparison of methods for assessing mortality risk. Am J Hum Biol. 2014;26:768–76.
Levine S, Coe CL, Wiener SG. Psychoneuroendocrinology of stress: a psychobiological perspective. In: Brush FR, Levine S, editors. Psychoendocrinology. San Diego: Academic Press; 1989. p. 341–77.
Li H, Westover WC, Crawford TB. Sheep-associated malignant catarrhal fever in a petting zoo. J Zoo Wildl Med. 1999;30:408–12.
Liberzon I, Trujillo KA, Akil H, Young EA. Motivational properties of oxytocin in the conditioned place preference paradigm. Neuropsychopharmacology. 1997;17:353.
Lim YAL, Ngui R, Shukri J, Rohela M, Mat Naim HR. Intestinal parasites in various animals at a zoo in Malaysia. Vet Parasitol. 2008;157:154–9.
Lipowicz A, Szklarska A, Malina RM. Allostatic load and socioeconomic status in Polish adult men. J Biosoc Sci. 2014;46:155–67.
Liptrap R, Raeside J. Effect of cortisol on the response to gonadotrophin releasing hormone in the boar. J Endocrinol. 1983;97:75–81.
Lobato E, Moreno J, Merino S, Sanz JJ, Arriero E. Haematological variables are good predictors of recruitment in nestling pied flycatchers (*Ficedula hypoleuca*). Ecosci. 2005;12:27–34.
Loeding E, Thomas J, Bernier D, Santymire R. Using fecal hormonal and behavioral analyses to evaluate the introduction of two sable antelope at Lincoln Park Zoo. J Appl Anim Welf Sci. 2011;14:220–46.
Lomborg SR, Nielsen LR, Heegaard PM, Jacobsen S. Acute phase proteins in cattle after exposure to complex stress. Vet Res Commun. 2008;32:575–82.
López-Olvera JR, Marco I, Montané J, Casas-Díaz E, Lavín S. Effects of acepromazine on the stress response in southern chamois (*Rupicapra pyrenaica*) captured by means of drive-nets. Can J Vet Res. 2007;71:41.
Lopresti AL, Maker GL, Hood SD, Drummond PD. A review of peripheral biomarkers in major depression: the potential of inflammatory and oxidative stress biomarkers. Prog Neuropsychopharmacol Biol Psychiatry. 2014;48:102–11.
Łopucki R, Klich D, Ścibior A, Gołębiowska D, Perzanowski K. Living in habitats affected by wind turbines may result in an increase in corticosterone levels in ground dwelling animals. Ecol Indic. 2018;84:165–71.
Loucks EB, Sullivan LM, D'Agostino RB Sr, Larson MG, Berkman LF, Benjamin EJ. Social networks and inflammatory markers in the Framingham Heart Study. J Biosoc Sci. 2006;38:835–42.
Lowenstine LJ, McManamon R, Terio KA. Comparative pathology of aging great apes: bonobos, chimpanzees, gorillas, and orangutans. Vet Pathol. 2016;53:250–76.
Lubbock LA, Goh A, Ali S, Ritchie J, Whooley MA. Relation of low socioeconomic status to C-reactive protein in patients with coronary heart disease (from the heart and soul study). Am J Cardiol. 2005;96:1506–11.
Lucas RM, Ponsonby A-L, Dear K. Mid-life stress is associated with both up-and down-regulation of markers of humoral and cellular immunity. Stress. 2007;10:351–61.
Lupien SJ, Ouellet-Morin I, Hupbach A, Tu MT, Buss C, Walker D, Pruessner J, McEwen BS. Beyond the stress concept: allostatic load—a developmental biological and cognitive perspective. In: Ciccheti D, Cohen DJ, editors. Developmental psychopathology. 2nd ed. Hoboken: Wiley; 2006. p. 578–628.
Lutgendorf SK, Garand L, Buckwalter KC, Reimer TT, Hong S-Y, Lubaroff DM. Life stress, mood disturbance, and elevated interleukin-6 in healthy older women. J Gerontol A Biol Sci Med Sci. 1999;54:M434–M9.
Lutz W, Kierdorf H. Parasite loads of roe deer (*Capreolus capreolus L.*) from neighbouring habitats with different intensities of recreational traffic. Z Jagdwiss. 1997;43:251–8.

Lymbery AJ, Hassan M, Morgan DL, Beatty SJ, Doupé RG. Parasites of native and exotic freshwater fishes in south-western Australia. J Fish Biol. 2010;76:1770–85.
Macfarland LA, Mann DR. The inhibitory effects of ACTH and adrenalectomy on reproductive maturation in female rats. Biol Reprod. 1977;16:306–14.
Machatschke IH, Wallner B, Dittami J. Impact of social environment on female chimpanzee reproductive cycles. Horm Behav. 2006;50:126–31.
Madliger CL, Love OP. The need for a predictive, context-dependent approach to the application of stress hormones in conservation. Conserv Biol. 2014;28:283–7.
Maestripieri D, Georgiev AV. What cortisol can tell us about the costs of sociality and reproduction among free-ranging rhesus macaque females on Cayo Santiago. Am J Primatol. 2015;78:92–105.
Maestripieri D, Hoffman CL. Chronic stress, allostatic load, and aging in nonhuman primates. Dev Psychopathol. 2011;23:1187–95.
Mahesh VB, Brann DW. Regulation of the preovulatory gonadotropin surge by endogenous steroids. Steroids. 1998;63:616–29.
Mallavarapu S, Stoinski TS, Bloomsmith MA, Maple TL. Postconflict behavior in captive western lowland gorillas (*Gorilla gorilla gorilla*). Am J Primatol. 2006;68:789–801.
Mancera KF, Lisle A, Allavena R, Phillips CJ. The effects of mining machinery noise of different frequencies on the behaviour, faecal corticosterone and tissue morphology of wild mice (*Mus musculus*). Appl Anim Behav Sci. 2017;197:81–9.
Maninger N, Capitanio JP, Mason WA, Ruys JD, Mendoza SP. Acute and chronic stress increase DHEAS concentrations in rhesus monkeys. Psychoneuroendocrinology. 2010;35:1055–62.
Mann DR, Korowitz CD, Barraclough CA. Adrenal gland involvement in synchronizing the preovulatory release of LH in rats. Proc Soc Exp Biol Med. 1975;150:115–20.
Maple TL, Bocian D. Wellness as welfare. Zoo Biol. 2013;32:363–5.
Markowitz H. Behavioral enrichment in the zoo. New York: Van Norstrand Reinhold; 1982.
Markowitz H, Aday C. Power for captive animals: contingencies and nature. In: Shepherdson DJ, Mellen JD, Hutchins M, editors. Second nature: environmental enrichment for captive animals. Washington: Smithsonian Institution Press; 1998. p. 47–58.
Martin AL. The primatologist as a behavioral engineer. Am J Primatol. 2015;79:1–10.
Martinez-Mota R, Valdespino-Quevedo C, Sanchez-Ramos MA. Faecal cortisol levels as a measurement of stress due to habitat fragmentation in wild black howler monkeys *(Alouatta pigra)* in Mexico. Folia Primatol. 2004;75:299.
Masek-Hammerman K, Miller AD, Lin KC, MacKey J, Weissenböck H, Gierbolini L, Burgos A, Perez H, Mansfield KG. Epizootic myocarditis associated with encephalomyocarditis virus in a group of rhesus macaques (*Macaca mulatta*). Vet Pathol. 2012;49:386–92.
Masi A. Sex hormones and rheumatoid arthritis: cause or effect relationships in a complex pathophysiology? Clin Exp Rheumatol. 1995;13:227–40.
Mason GJ. Species differences in responses to captivity: stress, welfare and the comparative method. Trends Ecol Evol. 2010;25:713–21.
Mason GJ, Veasey JS. What do population-level welfare indices suggest about the well-being of zoo elephants? Zoo Biol. 2010;29:256–73.
Massironi S, Conte D, Sciola V, Spampatti MP, Ciafardini C, Valenti L, Rossi RE, Peracchi M. Plasma chromogranin A response to octreotide test: prognostic value for clinical outcome in endocrine digestive tumors. Am J Gastroenterol. 2010;105:2072.
Matas D, Keren-Rotem T, Koren L. A method to determine integrated steroid levels in wildlife claws. Gen Comp Endocrinol. 2016;230:26–8.
Mather JA. Cephalopod consciousness: behavioural evidence. Conscious Cogn. 2008;17:37–48.
Matsuura T, Takimura R, Yamaguchi M, Ichinose M. Estimation of restraint stress in rats using salivary amylase activity. J Physiol Sci. 2012;62:421–7.
Mattei J, Demissie S, Falcon LM, Ordovas JM, Tucker K. Allostatic load is associated with chronic conditions in the Boston Puerto Rican Health Study. Soc Sci Med. 2010;70:1988–96.
Matteri RL, Carroll JA, Dyer CJ. Neuroendocrine responses to stress. In: Moberg GP, Mench JA, editors. The biology of animal stress basic principles and implications for animal welfare. Oxon: CABI publishing; 2000. p. 43–76.

Mau M, Südekum KH, Johann A, Sliwa A, Kaiser T. Indication of higher salivary α-amylase expression in hamadryas baboons and geladas compared to chimpanzees and humans. J Med Primatol. 2010;39:187–90.

Mauss D, Li J, Schmidt B, Angerer P, Jarczok MN. Measuring allostatic load in the workforce: a systematic review. Ind Health. 2015;53:5–20.

Mayer J. Use of behavior analysis to recognize pain in small mammals. Lab Anim. 2007;36:43.

McClelland DC, Cheriff AD. The immunoenhancing effects of humor on secretory IgA and resistance to respiratory infections. Psychol Health. 1997;12:329–44.

McCraty R, Atkinson M, Tiller WA, Rein G, Watkins AD. The effects of emotions on short-term power spectrum analysis of heart rate variability. Am J Cardiol. 1995;76:1089–93.

McCraty R, Atkinson M, Rein G, Watkins AD. Music enhances the effect of positive emotional states on salivaryIgA. Stress Med. 1996;12:167–75.

McCraty R, Barrios-Choplin B, Rozman D, Atkinson M, Watkins AD. The impact of a new emotional self-management program on stress, emotions, heart rate variability, DHEA and cortisol. Integr Physiol Behav Sci. 1998;33:151–70.

McDade TW, Hawkley LC, Cacioppo JT. Psychosocial and behavioral predictors of inflammation in middle-aged and older adults: the Chicago health, aging, and social relations study. Psychosom Med. 2006;68:376–81.

McEwen BS. Protective and damaging effects of stress mediators. N Engl J Med. 1998a;338:171–9.

McEwen BS. Stress, adaptation, and disease. Allostasis and allostatic load. Ann N Y Acad Sci. 1998b;840:33–44.

McEwen BS. Allostasis and allostatic load: implications for neuropsychopharmacology. Neuropsychopharmacology. 2000;22:108–24.

McEwen BS. From molecules to mind: stress, individual differences, and the social environment. Ann N Y Acad Sci. 2001;935:42–9.

McEwen BS. Early life influences on life-long patterns of behavior and health. Ment Retard Dev Disabil Res Rev. 2003;9:149–54.

McEwen BS. Protective and damaging effects of the mediators of stress and adaptation: allostasis and allostatic load. In: Schulkin J, editor. Allostatis, homeostasis, and the costs of physiological adaptation. Cambridge: Cambridge University Press; 2004. p. 65–98.

McEwen BS. Brain on stress: how the social environment gets under the skin. Proc Natl Acad Sci U S A. 2012;109:17180–5.

McEwen BS, Seeman T. Protective and damaging effects of mediators of stress: elaborating and testing the concepts of allostasis and allostatic load. Ann N Y Acad Sci. 1999;896:30–47.

McEwen BS, Stellar E. Stress and the individual: mechanisms leading to disease. Arch Intern Med. 1993;153:2093–101.

McEwen BS, Wingfield JC. The concept of allostasis in biology and biomedicine. Horm Behav. 2003;43:2–15.

McGuire M, Vonk J, Fuller G, Allard S. Using an ambiguous cue paradigm to assess cognitive bias in gorillas (*Gorilla gorilla gorilla*) during a forage manipulation. Anim Behav Cogn. 2017a;4:70–83.

McGuire M, Vonk J, Johnson-Ulrich Z. Ambiguous results when using the ambiguous-cue paradigm to assess learning and cognitive bias in gorillas and a black bear. Behav Sci. 2017b;7:51.

McLeay D. Effects of ACTH on the pituitary-interrenal axis and abundance of white blood cell types in juvenile coho salmon, *Oncorhynchus kisutch*. Gen Comp Endocrinol. 1973;21:431–40.

McManamon R, Lowenstine L. Cardiovascular disease in great apes. In: Miller RE, Fowler ME, editors. Fowler's zoo and wild animal medicine: current therapy. St. Louis: Elsevier; 2012. p. 408–15.

Meehan CL, Mench JA. The challenge of challenge: can problem solving opportunities enhance animal welfare? Appl Anim Behav Sci. 2007;102:246–61.

Meijs-Roelofs H, Kramer P. Effects of adrenalectomy on the release of follicle-stimulating hormone and the onset of puberty in female rats. J Endocrinol. 1977;75:419–26.

Meillère A, Brischoux F, Bustamante P, Michaud B, Parenteau C, Marciau C, Angelier F. Corticosterone levels in relation to trace element contamination along an urbanization gradient in the common blackbird (*Turdus merula*). Sci Total Environ. 2016;566:93–101.

Mellor D, Beausoleil N. Extending the 'Five Domains' model for animal welfare assessment to incorporate positive welfare states. Anim Welf. 2015;24:241–53.
Mellor D, Reid C. Concepts of animal well-being and predicting the impact of procedures on experimental animals. In: Improving the well-being of animals in the research environment. Glen Osmond: Australian and New Zealand Council for the Care of Animals in Research and Teaching; 1994. p. 3–18.
Mellor DJ, Hunt S, Gusset M, editors. Caring for wildlife: the world zoo and aquarium animal welfare strategy. Gland: WAZA Executive Office; 2015.
Menargues A, Urios V, Mauri M. Welfare assessment of captive Asian elephants (*Elephas maximus*) and Indian rhinoceros (*Rhinoceros unicornis*) using salivary cortisol measurement. Anim Welf. 2008;17:305–12.
Menargues A, Urios V, Liminana R, Mauri M. Circadian rhythm of salivary cortisol in Asian elephants (*Elephas maximus*): a factor to consider during welfare assessment. J Appl Anim Welf Sci. 2012;15:383–90.
Mendl M, Burman OH, Parker RM, Paul ES. Cognitive bias as an indicator of animal emotion and welfare: emerging evidence and underlying mechanisms. Appl Anim Behav Sci. 2009;118:161–81.
Mendl M, Burman OH, Paul ES. An integrative and functional framework for the study of animal emotion and mood. Proc R Soc Lond B. 2010;277:2895–904.
Mendonça-Furtado O, Edaes M, Palme R, Rodrigues A, Siqueira JÉ, Izar PÍ. Does hierarchy stability influence testosterone and cortisol levels of bearded capuchin monkeys (*Sapajus libidinosus*) adult males? A comparison between two wild groups. Behav Processes. 2014;109:79–88.
Metzler AE, Ossent P, Guscetti F, R übel A, Lang EM. Serological evidence of herpesvirus infection in captive Asian elephants (*Elephas maximus*). J Wildl Dis. 1990;26:41–9.
Meyer JM, Walker SL, Freeman EW, Steinetz BG, Brown JL. Species and fetal gender effects on the endocrinology of pregnancy in elephants. Gen Comp Endocrinol. 2004;138:263–70.
Michael AE, Papageorghiou AT. Potential significance of physiological and pharmacological glucocorticoids in early pregnancy. Hum Reprod Update. 2008;14:497–517.
Miller RE, Fowler M, editors. Fowler's zoo and wild animal medicine current therapy, vol. 7. St. Louis: Elsevier Saunders; 2012.
Mitsui S, Yamamoto M, Nagasawa M, Mogi K, Kikusui T, Ohtani N, Ohta M. Urinary oxytocin as a noninvasive biomarker of positive emotion in dogs. Horm Behav. 2011;60:239–43.
Miyakawa M, Matsui T, Kishikawa H, Murayama R, Uchiyama I, Itoh T, Yoshida T. Salivary chromogranin A as a measure of stress response to noise. Noise Health. 2006;8:108.
Mizoguchi K, Ishige A, Takeda S, Aburada M, Tabira T. Endogenous glucocorticoids are essential for maintaining prefrontal cortical cognitive function. J Neurosci. 2004;24:5492–9.
Moberg GP. How behavioral stress disrupts the endocrine control of reproduction in domestic animals. J Dairy Sci. 1991;74:304–11.
Moberg G. Biological responses to stress: implications for animal welfare. In: Moberg GP, Mench JA, editors. The biology of animal stress basic principles and implications for animal welfare. Oxon: CABI publishing; 2000. p. 1–21.
Moberg GP, Mench JA, editors. The biology of animal stress. Basic principles and implications for animal welfare. Oxon: CABI Publishing; 2000.
Moe RO, Stubsjøen SM, Bohlin J, Flø A, Bakken M. Peripheral temperature drop in response to anticipation and consumption of a signaled palatable reward in laying hens (*Gallus domesticus*). Physiol Behav. 2012;106:527–33.
Moll RJ, Millspaugh JJ, Beringer J, Sartwell J, Woods RJ, Vercauteren KC. Physiological stress response of captive white-tailed deer to video collars. J Wildl Manage. 2009;73:609–14.
Mommersteeg PM, Heijnen CJ, Kavelaars A, van Doornen LJ. Immune and endocrine function in burnout syndrome. Psychosom Med. 2006;68:879–86.
Monclús L, Ballesteros-Cano R, De La Puente J, Lacorte S, Lopez-Bejar M. Influence of persistent organic pollutants on the endocrine stress response in free-living and captive red kites (*Milvus milvus*). Environ Pollut. 2018;242:329–37.

Montali RJ, Mikota SK, Cheng LI. *Mycobacterium tuberculosis* in zoo and wildlife species. Rev Sci Tech Off int Epiz. 2001;20:291–303.

Monti F, Duriez O, Dominici JM, Sforzi A, Robert A, Fusani L, Grémillet D. The price of success: integrative long-term study reveals ecotourism impacts on a flagship species at a UNESCO site. Anim Conserv. 2018;21:448–58.

Moodie EM, Chamove AS. Brief threatening events beneficial for captive tamarins? Zoo Biol. 1990;9:275–86.

Moore IT, Jessop TS. Stress, reproduction, and adrenocortical modulation in amphibians and reptiles. Horm Behav. 2003;43:39–47.

Moreira N, Brown JL, Moraes W, Swanson WF, Monteiro ELA. Effect of housing and environmental enrichment on adrenocortical activity, behavior and reproductive cyclicity in the female tigrina *(Leopardus tigrinus)* and margay *(Leopardus wiedii)*. Zoo Biol. 2007;26:441–60.

Moreno J, Merino S, Martínez J, Sanz J, Arriero E. Heterophil/lymphocyte ratios and heat-shock protein levels are related to growth in nestling birds. Ecosci. 2002;9:434–9.

Morgan KN, Tromborg CT. Sources of stress in captivity. Appl Anim Behav Sci. 2007;102:262–302.

Morgan CA, Southwick S, Hazlett G, Rasmusson A, Hoyt G, Zimolo Z, Charney D. Relationships among plasma dehydroepiandrosterone sulfate and cortisol levels, symptoms of dissociation, and objective performance in humans exposed to acute stress. Arch Gen Psychiatry. 2004;61:819–25.

Morgan D, Mundry R, Sanz C, Ayina CE, Strindberg S, Lonsdorf E, Kühl HS. African apes coexisting with logging: comparing chimpanzee (*Pan troglodytes troglodytes*) and gorilla (*Gorilla gorilla gorilla*) resource needs and responses to forestry activities. Biol Conserv. 2018;218:277–86.

Morici LA, Elsey RM, Lance VA. Effects of long-term corticosterone implants on growth and immune function in juvenile alligators, *Alligator mississippiensis*. J Exp Zool. 1997;279:156–62.

Mormède P, Andanson S, Aupérin B, Beerda B, Guémené D, Malmkvist J, Manteca X, Manteuffel G, Prunet P, van Reenen CG, Richard S, Veissier I. Exploration of the hypothalamic–pituitary–adrenal function as a tool to evaluate animal welfare. Physiol Behav. 2007;92:317–39.

Morrison S, Shenassa ED, Mendola P, Wu T, Schoendorf K. Allostatic load may not be associated with chronic stress in pregnant women, NHANES 1999-2006. Ann Epidemiol. 2013;23:294–7.

Morrow-Tesch J, McGlone J, Norman R. Consequences of restraint stress on natural killer cell activity, behavior, and hormone levels in rhesus macaques (*Macaca mulatta*). Psychoneuroendocrinology. 1993;18:383–95.

Mostl E, Palme R. Hormones as indicators of stress. Domest Anim Endocrinol. 2002;23:67–74.

Muehlenbein MP, Watts DP. The costs of dominance: testosterone, cortisol and intestinal parasites in wild male chimpanzees. Biopsychosoc Med. 2010;4:1–12.

Müllner A, Linsenmair KE, Wikelski M. Exposure to ecotourism reduces survival and affects stress response in hoatzin chicks (*Opisthocomus hoazin*). Biol Conserv. 2004;118:549–58.

Mumby HS, Mar KU, Hayward AD, Htut W, Htut-Aung Y, Lummaa V. Elephants born in the high stress season have faster reproductive ageing. Sci Rep. 2015;5:1–11.

Munck A, Koritz SB. Studies on the mode of action of glucocorticoids in rats I. Early effects of cortisol on blood glucose and on glucose entry into muscle, liver and adipose tissue. Biochim Biophys Acta. 1962;57:310–7.

Muneta Y, Yoshikawa T, Minagawa Y, Shibahara T, Maeda R, Omata Y. Salivary IgA as a useful non-invasive marker for restraint stress in pigs. J Vet Med Sci. 2010;72:1295–300.

Muneta Y, Minagawa Y, Nakane T, Shibahara T, Yoshikawa T, Omata Y. Interleukin-18 expression in pig salivary glands and salivary content changes during acute immobilization stress. Stress. 2011;14:549–56.

Munson L, Nesbit JW, Meltzer DGA, Colly LP, Bolton L, Kriek NPJ. Diseases of captive cheetahs (*Acinonyx jubatus jubatus*) in South Africa: a 20-year retrospective survey. J Zoo Wildl Med. 1999;30:342–7.

Murphy HW, Danforth MD, Clyde VL. The Great Ape Heart Project. Int Zoo Yearb. 2018;52:1–10.

Murray CM, Stanton MA, Wellens KR, Santymire RM, Heintz MR, Lonsdorf EV. Maternal effects on offspring stress physiology in wild chimpanzees. Am J Primatol. 2018;80:1–12.

Nakane H, Asami O, Yamada Y, Harada T, Matsui N, Kanno T, Yanaihara N. Salivary chromogranin A as an index of psychosomatic stress response. Biomed Res. 1998;19:401–6.
Nakayama K, Goto S, Kuraoka K, Nakamura K. Decrease in nasal temperature of rhesus monkeys (*Macaca mulatta*) in negative emotional state. Physiol Behav. 2005;84:783–90.
Narayan E, Molinia F, Christi K, Morley C, Cockrem J. Urinary corticosterone metabolite responses to capture, and annual patterns of urinary corticosterone in wild and captive endangered Fijian ground frogs (*Platymantis vitiana*). Aust J Zool. 2010;58:189–97.
Narayan EJ, Cockrem JF, Hero J-M. Urinary corticosterone metabolite responses to capture and handling in two closely related species of free-living Fijian frogs. Gen Comp Endocrinol. 2012;177:55–61.
Nater UM, Rohleder N. Salivary alpha-amylase as a non-invasive biomarker for the sympathetic nervous system: current state of research. Psychoneuroendocrinology. 2009;34:486–96.
Nater UM, La Marca R, Florin L, Moses A, Langhans W, Koller MM, Ehlert U. Stress-induced changes in human salivary alpha-amylase activity—associations with adrenergic activity. Psychoneuroendocrinology. 2006;31:49–58.
Nater UM, Rohleder N, Schlotz W, Ehlert U, Kirschbaum C. Determinants of the diurnal course of salivary alpha-amylase. Psychoneuroendocrinology. 2007;32:392–401.
Navarro-Castilla Á, Mata C, Ruiz-Capillas P, Palme R, Malo JE, Barja I. Are motorways potential stressors of roadside wood mice (*Apodemus sylvaticus*) populations? PLoS One. 2014;9:e91942.
Nelson RJ. An introduction to behavioral endocrinology. 4th ed. Sunderland: Sinauer Associates; 2011.
Nelson KM, Reiber G, Kohler T, Boyko EJ. Peripheral arterial disease in a multiethnic national sample: the role of conventional risk factors and allostatic load. Ethn Dis. 2007;17:669–75.
Neumann I, Johnstone H, Hatzinger M, Liebsch G, Shipston M, Russell J, Landgraf R, Douglas A. Attenuated neuroendocrine responses to emotional and physical stressors in pregnant rats involve adenohypophysial changes. J Physiol. 1998;508:289–300.
Newman AE, Pradhan DS, Soma KK. Dehydroepiandrosterone and corticosterone are regulated by season and acute stress in a wild songbird: jugular versus brachial plasma. Endocrinology. 2008;149:2537–45.
Newton R, Kuitert L, Slater D, Adcock I, Barnes P. Cytokine induction of cytosolic phospholipase A2 and cyclooxygenase-2 mRNA is suppressed by glucocorticoids in human epithelial cells. Life Sci. 1996;60:67–78.
Ng V, Koh D, Chan G, Ong H, Chia S, Ong C. Are salivary immunoglobulin A and lysozyme biomarkers of stress among nurses? J Occup Environ Med. 1999;41:920–7.
Nganvongpanit K, Soponteerakul R, Kaewkumpai P, Punyapornwithawa V, Buddhachat K, Nomsiri R, Kaewmong P, Kittiwatanawong K, Chawangwongsanukun R, Angkawanish T, Thitaram C, Mahakkanukrauh P. Osteoarthritis in two marine mammals and 22 land mammals: learning from skeletal remains. J Anat. 2017;231:140–55.
Niebylski A, Boccolini A, Bensi N, Binotti S, Hansen C, Yaciuk R, Gauna H. Neuroendocrine changes and natriuresis in response to social stress in rats. Stress Health. 2012;28:179–85.
Nisenbaum LK, Zigmond MJ, Sved AF, Abercrombie ED. Prior exposure to chronic stress results in enhanced synthesis and release of hippocampal norepinephrine in response to a novel stressor. J Neurosci. 1991;11:1478–84.
Noda K, Akiyoshi H, Aoki M, Shimada T, Ohashi F. Relationship between transportation stress and polymorphonuclear cell functions of bottlenose dolphins, *Tursiops truncatus*. J Vet Med Sci. 2007;69:379–83.
Nussey DH, Kruuk LEB, Morris A, Clutton-Brock TH. Environmental conditions in early life influence ageing rates in a wild population of red deer. Curr Biol. 2007;17:R1000–R1.
O'Connor JC, André C, Wang Y, Lawson MA, Szegedi SS, Lestage J, Castanon N, Kelley KW, Dantzer R. Interferon-γ and tumor necrosis factor-α mediate the upregulation of indoleamine 2, 3-dioxygenase and the induction of depressive-like behavior in mice in response to bacillus Calmette-Guérin. J Neurosci. 2009;29:4200–9.
O'Toole D, Grossman A, Gross D, Delle Fave G, Barkmanova J, O'Connor J, Pape U-F, Plöckinger U. ENETS consensus guidelines for the standards of care in neuroendocrine tumors: biochemical markers. Neuroendocrinology. 2009;90:194–202.

Okeson DM, Garner MM, Taus NS, Li H, Coke RL. Ibex-associated malignant catarrhal fever in a bongo antelope (*Tragelaphus euryceros*). J Zoo Wildl Med. 2007;38:460–4.
Olefsky JM, Kimmerling G. Effects of glucocorticoids on carbohydrate metabolism. Am J Med Sci. 1976;271:202–10.
Olster DH, Ferin M. Corticotropin-releasing hormone inhibits gonadotropin-secretion in the ovariectomized rhesus monkey. J Clin Endocrinol Metab. 1987;65:262–7.
Onderka DK, Finlayson MC. Salmonellae and salmonellosis in captive reptiles. Can J Comp Med. 1985;49:268–70.
Ordóñez-Gómez JD, Cristóbal-Azkarate J, Arroyo-Rodríguez V, Santillán-Doherty AM, Valdez RA, Romano MC. Proximal and distal predictors of the spider monkey's stress levels in fragmented landscapes. PLoS One. 2016;11:e0149671.
Orr TE, Taylor MF, Bhattacharyya AK, Collins DC, Mann DR. Acute immobilization stress disrupts testicular steroidogenesis in adult male rats by inhibiting the activities of 17-alpha-hydroxylase and 17,20-lyase without affecting the binding of LH/HCG receptors. J Androl. 1994;15:302–8.
Ott S, Soler L, Moons C, Kashiha M, Bahr C, Vandermeulen J, Janssens S, Gutiérrez A, Escribano D, Cerón J. Different stressors elicit different responses in the salivary biomarkers cortisol, haptoglobin, and chromogranin A in pigs. Res Vet Sci. 2014;97:124–8.
Otten W, Puppe B, Kanitz E, Schön PC, Stabenow B. Physiological and behavioral effects of different success during social confrontation in pigs with prior dominance experience. Physiol Behav. 2002;75:127–33.
Palagi E, Norscia I. Bonobos protect and console friends and kin. PLoS One. 2013;8:e79290.
Palagi E, Chiarugi E, Cordoni G. Peaceful post-conflict interactions between agressors and bystanders in captive lowland gorillas (*Gorilla gorilla gorilla*). Am J Primatol. 2008;70:949–55.
Palme R. Measuring fecal steroids: guidelines for practical application. Ann N Y Acad Sci. 2005;1046:75–80.
Palme R. Monitoring stress hormone metabolites as a useful, non-invasive tool for welfare assessment in farm animals. Anim Welf. 2012;21:331.
Palme R. Non-invasive measurement of stress hormones: advances and problems. Physiol Behav. 2019;199:229–43.
Papargiris MM, Rivalland ETA, Hemsworth PH, Morrissey AD, Tilbrook AJ. Acute and chronic stress-like levels of cortisol inhibit the oestradiol stimulus to induce sexual receptivity but have no effect on sexual attractivity or proceptivity in female sheep. Horm Behav. 2011;60:336–45.
Paramastri Y, Royo F, Eberova J, Carlsson HE, Sajuthi D, Fernstrom AL, Pamungkas J, Hau J. Urinary and fecal immunoglobulin A, cortisol and 11-17 dioxoandrostanes, and serum cortisol in metabolic cage housed female cynomolgus monkeys (*Macaca fascicularis*). J Med Primatol. 2007;36:355–64.
Parga ML, Pendl H, Forbes NA. The effect of transport on hematologic parameters in trained and untrained Harris's hawks (*Parabuteo unicinctus*) and peregrine falcons (*Falco peregrinus*). J Avian Med Surg. 2001;15:162–9.
Park SE, Park D, Song K-I, Seong J-K, Chung S, Youn I. Differential heart rate variability and physiological responses associated with accumulated short-and long-term stress in rodents. Physiol Behav. 2017;171:21–31.
Parker VJ, Douglas AJ. Stress in early pregnancy: maternal neuro-endocrine-immune responses and effects. J Reprod Immunol. 2010;85:86–92.
Parrott R, Hall S, Lloyd D. Heart rate and stress hormone responses of sheep to road transport following two different loading procedures. Anim Welf. 1998;7:257–67.
Pearce G, Paterson A, Hughes P. Effect of short-term elevations in plasma cortisol concentration on LH secretion in prepubertal gilts. J Reprod Fertil. 1988;83:413–8.
Penfold LM, Powell D, Traylor-Holzer K, Asa CS. "Use it or lose it": characterization, implications, and mitigation of female infertility in captive wildlife. Zoo Biol. 2014;33:20–8.
Peppler R, Jacobs JJ. The effect of adrenalectomy on ovulation and follicular development in the rat. Biol Reprod. 1976;15:173–8.
Pereia RJG, Granzinolli MAM, De Barros FM, Duarte JMB. Influence of radiotransmitters on fecal glucocorticoid levels of free-ranging male American kestrels. J Wildl Manage. 2009;73:772–8.

Peric T, Corazzin M, Romanzin A, Bovolenta S, Prandi A, Montillo M, Comin A. Cortisol and DHEA concentrations in the hair of dairy cows managed indoor or on pasture. Livest Sci. 2017;202:39–43.

Peters A, McEwen BS, Friston K. Uncertainty and stress: why it causes diseases and how it is mastered by the brain. Prog Neurobiol. 2017;156:164–88.

Petrovic D, Pivin E, Ponte B, Dhayat N, Pruijm M, Ehret G, Ackermann D, Guessous I, Younes SE, Pechére-Bertschi A, Vogt B, Mohaupt M, Martin P-Y, Paccaud F, Burnier M, Bochud M, Stringhini S. Sociodemographic, behavioral and genetic determinants of allostatic load in a Swiss population-based study. Psychoneuroendocrinology. 2016;67:76–85.

Phillips DM, Lakshmi V, Monder C. Corticosteroid 11-beta-dehydrogenase in rat testis. Endocrinology. 1989;125:209–16.

Piccione G, Casella S, Giannetto C, Giudice E, Fazio F. Utility of acute phase proteins as biomarkers of transport stress in ewes. Small Rumin Res. 2012;107:167–71.

Pihl L, Hau J. Faecal corticosterone and immunoglobulin A in young adult rats. Lab Anim. 2003;37:166–71.

Pineiro M, Pineiro C, Carpintero R, Morales J, Campbell FM, Eckersall PD, Toussaint MJ, Lampreave F. Characterisation of the pig acute phase protein response to road transport. Vet J. 2007;173:669–74.

Piñeiro A, Barja I, Silván G, Illera JC. Effects of tourist pressure and reproduction on physiological stress response in wildcats: management implications for species conservation. Wildl Res. 2012;39:532–9.

Poessel SA, Biggins DE, Santymire RM, Livieri TM, Crooks KR, Angeloni L. Environmental enrichment affects adrenocortical stress responses in the endangered black-footed ferret. Gen Comp Endocrinol. 2011;172:526–33.

Pohlin F, Brabender K, Fluch G, Stalder G, Petit T, Walzer C. Seasonal variations in heart rate variability as an indicator of stress in free-ranging pregnant Przewalski's horses (*E. ferus przewalskii*) within Hortobágy National Park in Hungary. Front Physiol. 2017;8:1–11.

Pomerantz O, Terkel J. Effects of positive reinforcement training techniques on the psychological welfare of zoo-housed chimpanzees (*Pan troglodytes*). Am J Primatol. 2009;71:687–95.

Poole TB. Meeting a mammal's psychological needs: basic principles. In: Shepherdson DJ, Mellen JD, Hutchins M, editors. Second nature: environmental enrichment for captive animals. Washington: Smithsonian Institution Press; 1998. p. 83–94.

Powell DM. Collection planning for the next 100 years: what will we commit to save in zoos and aquariums? Zoo Biol. 2019;38:139–48.

Prado NA, Keady M, Oestmann A, Steinbeiser CM, Brown JL. Hyperprolactinemic African elephant (*Loxodonta africana*) females exhibit elevated dopamine, oxytocin and serotonin concentrations compared to normal cycling and non-cycling, low prolactin elephants. Biol Reprod. 2019; https://doi.org/10.1093/biolre/ioz036.

Pressman SD, Cohen S. Does positive affect influence health? Psychol Bull. 2005;131:925–71.

Price EE, Stoinski TS. Group size: determinants in the wild and implications for the captive housing of wild mammals in zoos. Appl Anim Behav Sci. 2007;103:255–64.

Price K, Kittridge C, Damby Z, Hayes SG, Addis EA. Relaxing life of the city? Allostatic load in yellow-bellied marmots along a rural-urban continuum. Conserv Physiol. 2018;6:coy070.

Prior L, Manley D, Jones K. Stressed out? An investigation of whether allostatic load mediates associations between neighbourhood deprivation and health. Health Place. 2018;52:25–33.

Proctor HS, Carder G. Nasal temperatures in dairy cows are influenced by positive emotional state. Physiol Behav. 2015;138:340–4.

Prunier A, Leterrier C. How should pain in farm animals be assessed? Adv Anim Biosci. 2014;5:310–8.

Pryce CR, Dettling AC, Spengler M, Schnell CR, Feldon J. Deprivation of parenting disrupts development of homeostatic and reward systems in marmoset monkey offspring. Biol Psychiatry. 2004;56:72–9.

Qiu X, Mainka SA. Review of mortality of the giant panda (*Ailuropoda melanoleuca*). J Zoo Wildl Med. 1993;24:425–9.

Quenby S, Farquharson R, Young M, Vince G. Successful pregnancy outcome following 19 consecutive miscarriages: case report. Hum Reprod. 2003;18:2562–4.

Quenby S, Kalumbi C, Bates M, Farquharson R, Vince G. Prednisolone reduces preconceptual endometrial natural killer cells in women with recurrent miscarriage. Fertil Steril. 2005;84:980–4.

Rabin DS, Johnson EO, Brandon DD, Liapi C, Chrousos GP. Glucocorticoids inhibit estradiol-mediated uterine growth—possible role of the uterine estradiol-receptor. Biol Reprod. 1990;42:74–80.

Radwan J, Kuduk K, Levy E, LeBas N, Babik W. Parasite load and MHC diversity in undisturbed and agriculturally modified habitats of the ornate dragon lizard. Mol Ecol. 2014;23:5966–78.

Raharivololona BM, Ganzhorn JU. Gastrointestinal parasite infection of the gray mouse lemur (*Microcebus murinus*) in the littoral forest of Mandena, Madagascar: effects of forest fragmentation and degradation. Madag Conserv Dev. 2009;4:103–12.

Rainville P, Bechara A, Naqvi N, Damasio AR. Basic emotions are associated with distinct patterns of cardiorespiratory activity. Int J Psychophysiol. 2006;61:5–18.

Rakotoniaina JH, Kappeler PM, Ravoniarimbinina P, Pechouskova E, Hämäläinen AM, Grass J, Kirschbaum C, Kraus C. Does habitat disturbance affect stress, body condition and parasitism in two sympatric lemurs? Conserv Physiol. 2016;4:cow034.

Ralph C, Tilbrook A. The usefulness of measuring glucocorticoids for assessing animal welfare. J Anim Sci. 2016;94:457–70.

Ramaley JA. Adrenal-gonadal interactions at puberty. Life Sci. 1974;14:1623–33.

Raouf SA, Smith LC, Brown MB, Wingfield JC, Brown CR. Glucocorticoid hormone levels increase with group size and parasite load in cliff swallows. Anim Behav. 2006;71:39–48.

Read S, Grundy E. Allostatic load—a challenge to measure multisystem physiological dysregulation. London: National Centre for Research Methods; 2012. p. 1–9.

Reddacliff LA, Kirkland PD, Hartley WJ, Reece RL. Encephalomyocarditis virus infections in an Australian Zoo. J Zoo Wildl Med. 1997;28:153–7.

Rehbinder C, Hau J. Quantification of cortisol, cortisol immunoreactive metabolites, and immunoglobulin A in serum, saliva, urine, and feces for noninvasive assessment of stress in reindeer. Can J Vet Res. 2006;70:151–4.

Reidarson TH, McBain JF. Hematologic, biochemical, and endocrine effects of dexamethasone on bottlenose dolphins (*Tursiops truncatus*). J Zoo Wildl Med. 1999;30:310–2.

Richman LK, Montali RJ, Cambre RC, Schmitt D, Hardy D, Hildbrandt T, Bengis RG, Hamzeh FM, Shahkolahi A, Hayward GS. Clinical and pathological findings of a newly recognized disease of elephants caused by endotheliotropic herpesviruses. J Wildl Dis. 2000a;36:1–12.

Richman LK, Montali RJ, Hayward GS. Review of a newly recognized disease of elephants caused by endotheliotropic herpesviruses. Zoo Biol. 2000b;19:383–92.

Riemer S, Assis L, Pike TW, Mills DS. Dynamic changes in ear temperature in relation to separation distress in dogs. Physiol Behav. 2016;167:86–91.

Rimbach R, Link A, Heistermann M, Gómez-Posada C, Galvis N, Heymann EW. Effects of logging, hunting, and forest fragment size on physiological stress levels of two sympatric ateline primates in Colombia. Conserv Physiol. 2013;1:cot031.

Rivier C, Rivest S. Effect of stress on the activity of the hypothalamic-pituitary-gonadal axis—peripheral and central mechanisms. Biol Reprod. 1991;45:523–32.

Robbins MM, Gray M, Fawcett KA, Nutter FB, Uwingeli P, Mburanumwe I, Kagoda E, Basabose A, Stoinski TS, Cranfield MR, Byamukama J, Spelman LH, Robbins AM. Extreme conservation leads to recovery of the Virunga mountain gorillas. PLoS One. 2011;6:e19788.

Robert KA, Schwanz LE. Monitoring the health status of free-ranging tammar wallabies using hematology, serum biochemistry, and parasite loads. J Wildl Manage. 2013;77:1232–43.

Robertson T, Beveridge G, Bromley C. Allostatic load as a predictor of all-cause and cause-specific mortality in the general population: evidence from the Scottish Health Survey. PLoS One. 2017;12:1–14.

Robinson LM, Altschul DM, Wallace EK, Úbeda Y, Llorente M, Machanda Z, Slocombe KE, Leach MC, Waran NK, Weiss A. Chimpanzees with positive welfare are happier, extraverted, and emotionally stable. Appl Anim Behav Sci. 2017;191:90–7.

Robinson LM, Coleman K, Capitanio JP, Gottlieb DH, Handel IG, Adams MJ, Leach MC, Waran NK, Weiss A. Rhesus macaque personality, dominance, behavior, and health. Am J Primatol. 2018;80:e22739.

Rodriguez EJ, Livaudais-Toman J, Gregorich SE, Jackson JS, Nápoles AM, Pérez-Stable EJ. Relationships between allostatic load, unhealthy behaviors, and depressive disorder in U.S. adults, 2005-2012 NHANES. Prev Med. 2018;110:9–15.

Romero LM, Wikelski M. Corticosterone levels predict survival probabilities of Galapagos marine iguanas during El Nino events. Proc Natl Acad Sci U S A. 2001;98:7366–70.

Romero LM, Wikelski M. Severe effects of low-level oil contamination on wildlife predicted by the corticosterone-stress response: preliminary data and a research agenda. Spill Sci Technol Bull. 2002;7:309–13.

Rosen JB, Schulkin J. Adaptive fear, allostasis, and the pathology of anxiety and depression. In: Schulkin J, editor. Allostasis, homeostasis, and the costs of physiological adaptation. Cambridge: Cambridge University Press; 2004. p. 164–227.

Rosenberg N, Park CG, Eldeirawi K. Relationship of serum carotenoid concentrations with allostatic load as a measure of chronic stress among middle-aged adults in the USA. Public Health Nutr. 2014;18:313–21.

Ross SR. Issues of choice and control in the behaviour of a pair of captive polar bears (*Ursus maritimus*). Behav Processes. 2006;73:117–20.

Rossdale P, Burguez P, Cash R. Changes in blood neutrophil/lymphocyte ratio related to adrenocortical function in the horse. Equine Vet J. 1982;14:293–8.

Rosvall M, Engström G, Janzon L, Berglund G, Hedblad B. The role of low grade inflammation as measured by C-reactive protein levels in the explanation of socioeconomic differences in carotid atherosclerosis. Eur J Public Health. 2006;17:340–7.

Rowden LJ, Rose PE. A global survey of banteng (*Bos javanicus*) housing and husbandry. Zoo Biol. 2016;35:546–55.

Royo F, Bjork N, Carlsson HE, Mayo S, Hau J. Impact of chronic catheterization and automated blood sampling (Accusampler) on serum corticosterone and fecal immunoreactive corticosterone metabolites and immunoglobulin A in male rats. J Endocrinol. 2004;180:145–53.

Royo F, Lyberg K, Abelson KS, Carlsson HE, Hau J. Effect of repeated confined single housing of young pigs on faecal excretion of cortisol and IgA. Scand J Lab Anim Sci. 2005;32:33–7.

Rubenstein DR, Shen S-F. Reproductive conflict and the costs of social status in cooperatively breeding vertebrates. Am Nat. 2009;173:650–61.

Ruby S, Buchanan-Smith HM. The effects of individual cubicle research on the social interactions of individual behavior of brown capuchin monkeys (*Sapajus apella*). Am J Primatol. 2015;77:1097–108.

Sabbah W, Watt RG, Sheiham A, Tsakos G. Effects of allostatic load on the social gradient in ischaemic heart disease and periodontal disease: evidence from the Third National Health and Nutrition Examination Survey. J Epidemiol Community Health. 2008;62:415–20.

Salas M, Temple D, Abáigar T, Cuadrado M, Delclaux M, Enseñat C, Almagro V, Martínez-Nevado E, Quevedo MÁ, Carbajal A, Tallo-Parra O, Sabés-Alsina M, Amat M, Lopez-Bejar M, Fernández-Bellon H, Manteca X. Aggressive behavior and hair cortisol levels in captive Dorcas gazelles (*Gazella dorcas*) as animal-based welfare indicators. Zoo Biol. 2016;35:467–73.

Samaras A, Santo CE, Papandroulakis N, Mitrizakis N, Pavlidis M, Höglund E, Pelgrim TNM, Zethof J, Spanings FAT, Vindas MA, Ebbesson LOE, Flik G, Gorissen M. Allostatic load and stress physiology in European seabass (*Dicentrarchus labrax L.*) and gilthead seabream (*Sparus aurata L.*). Front Endocrinol. 2018;9:451.

Sannen A, Van Elsacker L, Heistermann M, Eens M. Urinary testosterone metabolite levels and aggressive behaviors in male and female bonobos (*Pan paniscus*). Aggress Behav. 2004;30:425–34.

Santymire R, Manjerovic M, Sacerdote-Velat A. A novel method for the measurement of glucocorticoids in dermal secretions of amphibians. Conserv Physiol. 2018;6:coy008.

Sapolsky RM. Stress-induced suppression of testicular function in the wild baboon—role of glucocorticoids. Endocrinology. 1985;116:2273–8.

Sapolsky RM. Individual differences and the stress response. Semin Neurosci. 1994;6:261–9.
Saruta J, Tsukinoki K, Sasaguri K, Ishii H, Yasuda M, Osamura YR, Watanabe Y, Sato S. Expression and localization of chromogranin A gene and protein in human submandibular gland. Cells Tissues Organs. 2005;180:237–44.
Scannapieco FA, Torres G, Levine MJ. Salivary α-amylase: role in dental plaque and caries formation. Crit Rev Oral Biol Med. 1993;4:301–7.
Scarlata CD, Elias BA, Godwin JR, Powell RA, Shepherdson D, Shipley LA, Brown JL. Relationship between fecal hormone concentrations and reproductive success in captive pygmy rabbits (*Brachylagus idahoensis*). J Mammal. 2012;93:759–70.
Scarlata CD, Elias BA, Godwin JR, Powell RA, Shepherdson D, Shipley LA, Brown JL. Influence of environmental conditions and facility on faecal glucocorticoid concentrations in captive pygmy rabbits (*Brachylagus idahoensis*). Anim Welf. 2013;22:357–68.
Schaefer SA, Steklis HD. Personality and subjective well-being in captive male western lowland gorillas living in bachelor groups. Am J Primatol. 2014;76:879–89.
Schaftenaar W, Reid C, Martina B, Fickel J, Osterhaus ADME. Nonfatal clinical presentation of elephant endotheliotropic herpes virus discovered in a group of captive Asian elephants (*Elephas maximus*). J Zoo Wildl Med. 2010;41:626–32.
Scheier MF, Bridges MW. Person variables and health: personality predispositions and acute psychological states as shared determinants for disease. Psychosom Med. 1995;57:255–68.
Schenk HM, Jeronimus BF, van der Krieke L, Bos EH, de Jonge P, Rosmalen JGM. Associations of positive affect and negative affect with allostatic load: a lifelines cohort study. Psychosom Med. 2018;80:160–6.
Schenker JG, Meirow D, Schenker E. Stress and human reproduction. Eur J Obstet Gynecol Reprod Biol. 1992;45:1–8.
Scherer K, Dan E, Flykt A. What determines a feeling's position in affective space? A case for appraisal. Cognit Emot. 2006;20:92–113.
Schiml PA, Rissman EF. Cortisol facilitates induction of sexual behavior in the female musk shrew (*Suncus murinus*). Behav Neurosci. 1999;113:166.
Schluessel V. Who would have thought that 'Jaws' also has brains? Cognitive functions in elasmobranchs. Anim Cogn. 2015;18:19–37.
Schmid PG, Eckstein JW, Abboud FM. Comparison of effects of deoxycorticosterone and dexamethasone on cardiovascular responses to norepinephrine. J Clin Invest. 1967;46:590–8.
Schoof VAM, Jack KM, Ziegler TE. Male response to female ovulation in white-faced capuchins (*Cebus capucinus*): variation in fecal testosterone, dihydrotestosterone, and glucocorticoids. Int J Primatol. 2014;35:643–60.
Schreck CB. Stress and fish reproduction: the roles of allostasis and hormesis. Gen Comp Endocrinol. 2010;165:549–56.
Schreiber JR, Nakamura K, Erickson GF. Rat ovary glucocorticoid receptor: identification and characterization. Steroids. 1982;39:569–84.
Schultner J, Kitaysky AS, Welcker J, Hatch S. Fat or lean: adjustment of endogenous energy stores to predictable and unpredictable changes in allostatic load. Funct Ecol. 2013;27:45–55.
Schultz R, Isola J, Parvinen M, Honkaniemi J, Wikström A-C, Gustafsson J-Å, Pelto-Huikko M. Localization of the glucocorticoid receptor in testis and accessory sexual organs of male rat. Mol Cell Endocrinol. 1993;95:115–20.
Scott K, Heistermann M, Cant MA, Vitikainen EI. Group size and visitor numbers predict faecal glucocorticoid concentrations in zoo meerkats. R Soc Open Sci. 2017;4:161017.
Seaward BL. Managing stress: principles and strategies for health and well being. 5th ed. Boston: Jones and Bartlett; 2006.
Seeman TE. Social ties and health: the benefits of social integration. Ann Epidemiol. 1996;6:442–51.
Seeman TE, McEwen BS. Impact of social environment characteristics on neuroendocrine regulation. Psychosom Med. 1996;58:459–71.
Seeman TE, Berkman LF, Blazer D, Rowe JW. Social ties and support and neuroendocrine function: the MacArthur studies of successful aging. Ann Behav Med. 1994;16:95–106.

Seeman TE, Singer BH, Rowe JW, Horwitz RI, McEwen BS. Price of adaptation—allostatic load and its health consequences. Arch Intern Med. 1997;157:2259–68.
Seeman TE, McEwen BS, Rowe JW, Singer BH. Allostatic load as a marker of cumulative biological risk: MacArthur studies of successful aging. Proc Natl Acad Sci U S A. 2001;98:4770–5.
Seeman TE, Singer BH, Ryff CD, Love GD, Levy-Storms L. Social relationships, gender, and allostatic load across two age cohorts. Psychosom Med. 2002;64:395–406.
Seeman TE, Crimmins EM, Huang M-H, Singer BH, Bucur A, Gruenewald TL, Berkman LF, Reuben DB. Cumulative biological risk and socio-economic differences in mortality: MacArthur studies of successful aging. Soc Sci Med. 2004a;58:1985–97.
Seeman TE, Glei DA, Goldman N, Weinstein M, Singer BH, Lin Y-H. Social relationships and allostatic load in Taiwanese elderly and near elderly. Soc Sci Med. 2004b;59:2245–57.
Seeman TE, Epel E, Gruenewald TL, Karlamangla AS, McEwen BS. Socio-economic differentials in peripheral biology: cumulative allostatic load. Ann N Y Acad Sci. 2010;1186:223–39.
Seeman M, Merkin SS, Karlamangla A, Koretz B, Teresa S. Social status and biological dysregulation: the "status syndrome" and allostatic load. Soc Sci Med. 2014;118:143–51.
Seplaki CL, Goldman N, Weinstein M, Lin Y-H. How are biomarkers related to physical and mental well-being? J Gerontol A Biol Sci Med Sci. 2004;59A:201–17.
Setchell JM, Smith T, Wickings EJ, Knapp LA. Social correlates of testosterone and ornamentation in male mandrills. Horm Behav. 2008;54:365–72.
Shepherdson DJ. Tracing the path of environmental enrichment in zoos. In: Shepherdson DJ, Mellen JD, Hutchins M, editors. Second nature: environmental enrichment for captive animals. Washington: Smithsonian Institution Press; 1998. p. 1–12.
Shepherdson DJ, Carlstead KC, Wielebnowski N. Cross-institutional assessment of stress responses in zoo animals using longitudinal monitoring of faecal corticoids and behaviour. Anim Welf. 2004;13:S105–S13.
Sheriff MJ, Dantzer B, Delehanty B, Palme R, Boonstra R. Measuring stress in wildlife: techniques for quantifying glucocorticoids. Oecologia. 2011;166:869–87.
Shields GS, Lam JCW, Trainor BC, Yonelinas AP. Exposure to acute stress enhances decision-making competence: evidence for the role of DHEA. Psychoneuroendocrinology. 2016;67:51–60.
Shini S, Kaiser P. Effects of stress, mimicked by administration of corticosterone in drinking water, on the expression of chicken cytokine and chemokine genes in lymphocytes. Stress. 2009;12:388–99.
Shini S, Huff GR, Shini A, Kaiser P. Understanding stress-induced immunosuppression: exploration of cytokine and chemokine gene profiles in chicken peripheral leukocytes. Poult Sci. 2010;89:841–51.
Shirotsuki K, Izawa S, Sugaya N, Yamada KC, Ogawa N, Ouchi Y, Nagano Y, Nomura S. Salivary cortisol and DHEA reactivity to psychosocial stress in socially anxious males. Int J Psychophysiol. 2009;72:198–203.
Shively CA, Day SM. Social inequalities in health in nonhuman primates. Neurobiol Stress. 2015;1:156–63.
Shutt K, Heistermann M, Kasim A, Todd A, Kalousova B, Profosouva I, Petrzelkova K, Fuh T, Dicky J-F, Bopalanzognako J-B. Effects of habituation, research and ecotourism on faecal glucocorticoid metabolites in wild western lowland gorillas: implications for conservation management. Biol Conserv. 2014;172:72–9.
Siegel HS. Effects of behavioral and physical stressors on immune responses. In: Wiepkema PR, Van Adrichem PWM, editors. Biology of stress in farm animals: an integrative approach. Dordrecht: Springer; 1987. p. 39–54.
Silk JB, Alberts SC, Altmann J. Social bonds of female baboons enhance infant survival. Science. 2003;302:1231–4.
Silva de Souza Matos L, Palme R, Silva Vasconcellos AÉ. Behavioural and hormonal effects of member replacement in captive groups of blue-fronted amazon parrots (*Amazona aestiva*). Behav Processes. 2017;138:160–9.

Simpson GM, Fuller G, Lukas KE, Kuhar CW, Fitch-Snyder H, Taylor J, Dennis PM. Sources of morbidity in lorises and pottos in North American zoos: a retrospective review, 1980-2010. Zoo Biol. 2018;37:245–57.

Skandakumar S, Stodulski G, Hau J. Salivary IgA—a possible stress marker in dogs. Anim Welf. 1995;4:339–50.

Smiley T, Spelman L, Lukasik-Braum M, Mukherjee J, Kaufman G, Akiyoshi DE, Cranfield M. Noninvasive saliva collection techniques for free-ranging mountain gorillas and captive eastern gorillas. J Zoo Wildl Med. 2010;41:201–9.

Smith KP, Christakis NA. Social networks and health. Annu Rev Sociol. 2008;34:405–29.

Smith TE, French JA. Psychosocial stress and urinary cortisol excretion in marmoset monkeys (*Callithrix kuhli*). Physiol Behav. 1997;62:225–32.

Smith AS, Birnie AK, French JA. Social isolation affects partner-directed social behavior and cortisol during pair formation in marmosets, *Callithrix geoffroyi*. Physiol Behav. 2011;104:955–61.

Smith BK, Remis MJ, Dierenfeld ES. Nutrition of the captive western lowland gorilla (*Gorilla gorilla gorilla*): a dietary survey. Zoo Biol. 2014;33:419–25.

Snowdon CT. The criteria for successful captive propagation of endangered primates. Zoo Biol. 1989;1:149–61.

Snyder RJ, Perdue BM, Powell DM, Forthman DL, Bloomsmith MA, Maple TL. Behavioral and hormonal consequences of transporting giant pandas from China to the United States. J Appl Anim Welf Sci. 2012;15:1–20.

Snyder-Mackler N, Sanz JÍ, Kohn JN, Brinkworth JF, Morrow S, Shaver AO, Grenier J-C, Pique-Regi R, Johnson ZP, Wilson ME, Barreiro LB, Tung J. Social status alters immune regulation and response to infection in macaques. Science. 2016;354:1041–5.

Sosa C, Forcada F, Meikle A, Abecia J. Increase in ovine plasma cortisol at oestrus and its relation with the metabolic status during the sexual cycle in sheep. Biol Rhythm Res. 2013;44:445–9.

Speeg K Jr, Harrison RW. The ontogony of the human placental glucocorticoid receptor and inducibility of heat-stable alkaline phosphatase. Endocrinology. 1979;104:1364–8.

Spencer KA, Evans NP, Monaghan P. Postnatal stress in birds: a novel model of glucocorticoid programming of the hypothalamic-pituitary-adrenal axis. Endocrinology. 2009;150:1931–4.

Sperry J, Butler L, Romero L, Weatherhead P. Effects of parasitic infection and radio-transmitters on condition, hematological characteristics and corticosterone concentrations in Texas ratsnakes. J Zool. 2009;278:100–7.

Sterling P. Principles of allostasis. In: Schulkin J, editor. Allostasis, homeostasis, and the costs of physiological adaptation. Cambridge: Cambridge University Press; 2004. p. 17–64.

Sterling P. Allostasis: a model of predictive regulation. Physiol Behav. 2012;106:5–15.

Sterling P, Eyer J. Allostasis: a new paradigm to explain arousal pathology. In: Fisher S, Reason J, editors. Handbook of life stress, cognition and health. Chichester: Wiley; 1988. p. 629–49.

Stewart M, Schaefer A, Haley D, Colyn J, Cook N, Stafford K, Webster J. Infrared thermography as a non-invasive method for detecting fear-related responses of cattle to handling procedures. Anim Welf. 2008a;17:387–93.

Stewart M, Stafford K, Dowling S, Schaefer A, Webster J. Eye temperature and heart rate variability of calves disbudded with or without local anaesthetic. Physiol Behav. 2008b;93:789–97.

Støen O-G, Ordiz A, Evans AL, Laske TG, Kindberg J, Fröbert O, Swenson JE, Arnemo JM. Physiological evidence for a human-induced landscape of fear in brown bears (*Ursus arctos*). Physiol Behav. 2015;152:244–8.

Stoinski TS, Lukas KE, Kuhar CW, Maple TL. Factors influencing the formation and maintenance of all-male gorilla groups in captivity. Zoo Biol. 2004;23:189–203.

Strasser EH, Heath JA. Reproductive failure of a human-tolerant species, the American kestrel, is associated with stress and human disturbance. J Appl Ecol. 2013;50:912–9.

Strong VJ, Grindlay D, Redrobe S, Cobb M, White K. A systematic review of the literature relating to captive great ape morbidity and mortality. J Zoo Wildl Med. 2016;47:697–710.

Strong VJ, Martin M, Redrobe S, White K, Baiker K. A retrospective review of great ape cardiovascular disease epidemiology and pathology. Int Zoo Yearb. 2018;52:113–25.

Sullivan K, van Heugten E, Ange-van Heugten K, Poore MH, Dierenfeld ES, Wolfe B. Analysis of nutrient concentrations in the diet, serum, and urine of giraffe from surveyed North American zoological institutions. Zoo Biol. 2010;29:457–69.

Svec F, Lopez A. Antiglucocorticoid actions of dehydroepiandrosterone and low concentrations in Alzheimer's disease. Lancet. 1989;334:1335–6.

Svobodova I, Chaloupkova H, Koncel R, Bartos L, Hradecka L, Jebavy L. Cortisol and secretory immunoglobulin A response to stress in German shepherd dogs. PLoS One. 2014;9:e90820.

Takai N, Yamaguchi M, Aragaki T, Eto K, Uchihashi K, Nishikawa Y. Effect of psychological stress on the salivary cortisol and amylase levels in healthy young adults. Arch Oral Biol. 2004;49:963–8.

Takai N, Yamaguchi M, Aragaki T, Eto K, Uchihashi K, Nishikawa Y. Gender-specific differences in salivary biomarker responses to acute psychological stress. Ann N Y Acad Sci. 2007;1098:510–5.

Takatsuji K, Sugimoto Y, Ishizaki S, Ozaki Y, Matsuyama E, Yamaguchi Y. The effects of examination stress on salivary cortisol, immunoglobulin A, and chromogranin A in nursing students. Biomed Res. 2008;29:221–4.

Takeshita RS, Bercovitch FB, Huffman MA, Kinoshita K. Development and validation of an enzyme immunoassay for fecal dehydroepiandrosterone sulfate in Japanese macaques (*Macaca fuscata*). Int J Primatol. 2018;39:208–21.

Tamashiro KLK, Nguyen MMN, Sakai RR. Social stress: from rodents to primates. Front Neuroendocrinol. 2005;26:27–40.

Tartu S, Gabrielsen GW, Blévin P, Ellis H, Bustnes JO, Herzke D, Chastel O. Endocrine and fitness correlates of long-chain perfluorinated carboxylates exposure in Arctic breeding black-legged kittiwakes. Environ Sci Technol. 2014;48:13504–10.

Taylor SE. Tend and befriend: biobehavioral bases of affiliation under stress. Curr Dir Psychol Sci. 2006;15:273–7.

Taylor J, McFarland MJ, Carr DC. Age, perceptions of mattering, and allostatic load. J Aging Health. 2018:1–20.

Tébar M, Bellido C, Sánchez-Criado JE. Luteinizing hormone (LH) and corticosterone in proestrous afternoon restore the follicle-stimulating hormone secretion at early estrus in adrenalectomized LH-releasing hormone antagonist-treated rats. Biol Reprod. 1995;52:63–7.

Tennessen JB, Parks SE, Langkilde T. Traffic noise causes physiological stress and impairs breeding migration behaviour in frogs. Conserv Physiol. 2014;2:cou032.

Tetsuka M. Actions of glucocorticoid and their regulatory mechanisms in the ovary. Anim Sci J. 2007;78:112–20.

Tetsuka M, Milne M, Simpson GE, Hillier SG. Expression of 11β-hydroxysteroid dehydrogenase, glucocorticoid receptor, and mineralocorticoid receptor genes in rat ovary. Biol Reprod. 1999;60:330–5.

Thatcher HR, Downs CT, Koyama NF. Using parasitic load to measure the effect of anthropogenic disturbance on vervet monkeys. Ecohealth. 2018;15:676–81.

Thayer Z, Barbosa-Leiker C, McDonell M, Nelson L, Buchwald D, Manson S. Early life trauma, post-traumatic stress disorder, and allostatic load in a sample of American Indian adults. Am J Hum Biol. 2017;29:1–10.

Thibier M, Rolland O. The effect of dexamethasone (DXM) on circulating testosterone (T) and luteinizing hormone (LH) in young postpubertal bulls. Theriogenology. 1976;5:53–60.

Thiel D, Jenni-Eiermann S, Braunisch V, Palme R, Jenni L. Ski tourism affects habitat use and evokes a physiological stress response in capercaillie *Tetrao urogallus*: a new methodological approach. J Appl Ecol. 2008;45:845–53.

Thoma MV, Kirschbaum C, Wolf JM, Rohleder N. Acute stress responses in salivary alpha-amylase predict increases of plasma norepinephrine. Biol Psychol. 2012;91:342–8.

Thompson LA, Spoon TR, Goertz CE, Hobbs RC, Romano TA. Blow collection as a non-invasive method for measuring cortisol in the beluga (*Delphinapterus leucas*). PLoS One. 2014;9:e114062.

Thompson NA, Higham JP, Heistermann M, Vogel E, Cords M. Energy balance but not competitive environment corresponds with allostatic load during development in an Old World monkey. Horm Behav. 2019;108:30–41.

Tilbrook AJ, Turner AI, Clarke IJ. Effects of stress on reproduction in non-rodent mammals: the role of glucocorticoids and sex differences. Rev Reprod. 2000;5:105–13.
Tilbrook AJ, Turner AI, Clarke IJ. Stress and reproduction: central mechanisms and sex differences in non-rodent species. Stress. 2002;5:83–100.
Tiller WA, McCraty R, Atkinson M. Cardiac coherence: a new, noninvasive measure of autonomic nervous system order. Altern Ther Health Med. 1996;2:52–65.
Toda M, Kusakabe S, Nagasawa S, Kitamura K, Morimoto K. Effect of laughter on salivary endocrinological stress marker chromogranin A. Biomed Res. 2007;28:115–8.
Tokarz RR, Summers CH. Stress and reproduction in reptiles. In: David ON, Kristin HL, editors. Hormones and reproduction of vertebrates. London: Academic Press; 2011. p. 169–213.
Toker S, Shirom A, Shapira I, Berliner S, Melamed S. The association between burnout, depression, anxiety, and inflammation biomarkers: C-reactive protein and fibrinogen in men and women. J Occup Health Psychol. 2005;10:344.
Torner L. Actions of prolactin in the brain: from physiological adaptations to stress and neurogenesis to psychopathology. Front Endocrinol. 2016;7:25.
Touma C, Sachser N, Mostl E, Palme R. Effects of sex and time of day on metabolism and excretion of corticosterone in urine and feces of mice. Gen Comp Endocrinol. 2003;130:267–78.
Travain T, Colombo ES, Grandi LC, Heinzl E, Pelosi A, Previde EP, Valsecchi P. How good is this food? A study on dogs' emotional responses to a potentially pleasant event using infrared thermography. Physiol Behav. 2016;159:80–7.
Tress U, Suchodolski JS, Williams DA, Steiner JM. Development of a fecal sample collection strategy for extraction and quantification of fecal immunoglobulin A in dogs. Am J Vet Res. 2006;67:1756–9.
Troïanowski M, Mondy N, Dumet A, Arcanjo C, Lengagne T. Effects of traffic noise on tree frog stress levels, immunity, and color signaling. Conserv Biol. 2017;31:1132–40.
Tsujita S, Morimoto K. Secretory IgA in saliva can be a useful stress marker. Environ Health Prev Med. 1999;4:1–8.
Tung J, Archie EA, Altmann J, Alberts SC. Cumulative early life adversity predicts longevity in wild baboons. Nat Commun. 2016;7:1–7.
Uchino BN. Social support and health: a review of physiological processes potentially underlying links to disease outcomes. Behav Med. 2006;29:377–87.
Uchino BN, Cacioppo JT, Kiecolt-Glaser JK. The relationship between social support and physiological processes: a review with emphasis on underlying mechanisms and implications for health. Psychol Bull. 1996;119:488–531.
Ulrich-Lai YM, Herman JP. Neural regulation of endocrine and autonomic stress responses. Nat Rev Neurosci. 2009;10:397–409.
Upchurch DM, Rainisch BW, Chyu L. Greater leisure time physical activity is associated with lower allostatic load in white, black, and Mexican American midlife women: findings from the national health and nutrition examination survey, 1999 through 2004. Womens Health Issues. 2015a;25:680–7.
Upchurch DM, Stein J, Greendale GA, Chyu L, Tseng C-H, Huang M-H, Lewis TT, Kravitz HM, Seeman TE. A longitudinal investigation of race, socioeconomic status, and psychosocial mediators of allostatic load in midlife women: findings from the study of women's health across the nation. Psychosom Med. 2015b;77:402–12.
Uphoff Meteyer C, Gonzales BJ, Heuschele WP, Howard EB. Epidemiologic and pathologic aspects of an epizootic of malignant catarrhal fever in exotic hoofstock. J Wildl Dis. 1989;25:280–6.
Uphouse L. Stress and reproduction in mammals. In: David ON, Kristin HL, editors. Hormones and reproduction of vertebrates. London: Academic Press; 2011. p. 117–38.
Valera M, Bartolomé E, Sánchez MJ, Molina A, Cook N, Schaefer A. Changes in eye temperature and stress assessment in horses during show jumping competitions. J Equine Vet Sci. 2012;32:827–30.
Valkanova V, Ebmeier KP, Allan CL. CRP, IL-6 and depression: a systematic review and meta-analysis of longitudinal studies. J Affect Disord. 2013;150:736–44.

van Ree JM, Gerrits MA, Vanderschuren LJ. Opioids, reward and addiction: an encounter of biology, psychology, and medicine. Pharmacol Rev. 1999;51:341–96.

Vidya TNC, Sukumar R. The effect of some ecological factors on the intestinal parasite loads of the Asian elephant (*Elephas maximus*) in southern India. J Biosci. 2002;27:521–8.

Vigil JM, Geary DC, Granger DA, Flinn MV. Sex differences in salivary cortisol, alpha-amylase, and psychological functioning following Hurricane Katrina. Child Dev. 2010;81:1228–40.

Viijoen JJ, Ganswindt A, du Toit JT, Langbauer WR. Translocation stress and faecal glucocorticoid metabolite levels in free-ranging African savanna elephants. S Afr J Wildl Res. 2008;38:146–52.

Vitazkova SK, Wade SE. Effects of ecology on the gastrointestinal parasites of *Alouatta pigra*. Int J Primatol. 2007;28:1327–43.

Von Borell E, Langbein J, Després G, Hansen S, Leterrier C, Marchant-Forde J, Marchant-Forde R, Minero M, Mohr E, Prunier A. Heart rate variability as a measure of autonomic regulation of cardiac activity for assessing stress and welfare in farm animals—a review. Physiol Behav. 2007;92:293–316.

von Känel R, Bellingrath S, Kudielka BM. Association between burnout and circulating levels of pro-and anti-inflammatory cytokines in schoolteachers. J Psychosom Res. 2008;65:51–9.

Wade GN, Schneider JE. Metabolic fuels and reproduction in female mammals. Neurosci Biobehav Rev. 1992;16:235–72.

Walker BG, Boersma PD, Wingfield JC. Field endocrinology and conservation biology. Integr Comp Biol. 2005;45:12–8.

Wallace ME, Harville EW. Allostatic load and birth outcomes among white and black women in New Orleans. Matern Child Health J. 2013;17:1025–9.

Wallace M, Harville E, Theall K, Webber L, Chen W, Berenson G. Neighborhood poverty, allostatic load, and birth outcomes in African American and white women: findings from the Bogalusa Heart Study. Health Place. 2013a;24:260–6.

Wallace M, Harville E, Theall K, Webber L, Chen W, Berenson G. Preconception biomarkers of allostatic load and racial disparities in adverse birth outcomes: the Bogalusa heart study. Paediatr Perinat Epidemiol. 2013b;27:587–97.

Wang F, Riley JCM, Behrman HR. Immunosuppressive levels of glucocorticoid block extrauterine luteolysins in the rat. Biol Reprod. 1993;49:66–73.

Wang X, Byars SG, Stearns SC. Genetic links between post-reproductive lifespan and family size in Framingham. Evol Med Public Health. 2013;2013:241–53.

Waples KA, Gales NJ. Evaluating and minimising social stress in the care of captive bottlenose dolphins (*Tursiops aduncus*). Zoo Biol. 2002;21:5–26.

Wasserman MD, Chapman CA, Milton K, Goldberg TL, Ziegler TE. Physiological and behavioral effects of capture darting on red colobus monkeys (*Procolobus rufomitratus*) with a comparison to chimpanzee (*Pan troglodytes*) predation. Int J Primatol. 2013;34:1020–31.

Watanuki S, Kim Y-K. Physiological responses induced by pleasant stimuli. J Physiol Anthropol Appl Human Sci. 2005;24:135–8.

Watters JV, Powell DM. Measuring animal personality for use in population management in zoos: suggested methods and rationale. Zoo Biol. 2012;31:1–12.

Waylen KA, McGowan PJK, Milner-Gulland EJ, Pawi Study G. Ecotourism positively affects awareness and attitudes but not conservation behaviours: a case study at Grande Riviere, Trinidad. Oryx. 2009;43:343–51.

Webster Marketon JI, Glaser R. Stress hormones and immune function. Cell Immunol. 2008;252:16–26.

Webster JI, Tonelli L, Sternberg EM. Neuroendocrine regulation of immunity. Annu Rev Immunol. 2002;20:125–63.

Webster K, Narayan E, De Vos N. Fecal glucocorticoid metabolite response of captive koalas (*Phascolarctos cinereus*) to visitor encounters. Gen Comp Endocrinol. 2017;244:157–63.

Weinstein M, Goldman D, Hedley A, Yu-Hsuan L, Seeman TE. Social linkages to biological markers of health among the elderly. J Biosoc Sci. 2003;35:433–53.

Weissenbock NM, Weiss CM, Schwammer HM, Kratochvil H. Thermal windows on the body surface of African elephants (*Loxodonta africana*) studied by infrared thermography. J Therm Biol. 2010;35:182–8.
Wells SK, Gutter AE, Soike KF, Baskin GB. Encephalomyocarditis virus: epizootic in a zoological collection. J Zoo Wildl Med. 1989;20:291–6.
Wendelaar Bonga SE. The stress response in fish. Physiol Rev. 1997;77:591–625.
West G. Tuberculosis in captive exotic animals. In: Thoen CO, Steele JH, Gilsdorf MJ, editors. *Mycobacterium bovis* infection in animals and humans. 2nd ed. Ames: Blackwell; 2006. p. 248–57.
Whirledge S, Cidlowski JA. A role for glucocorticoids in stress-impaired reproduction: beyond the hypothalamus and pituitary. Endocrinology. 2013;154:4450–68.
Whitehouse J, Micheletta J, Powell LE, Bordier C, Waller BM. The impact of cognitive testing on the welfare of group housed primates. PLoS One. 2013;8:e78308.
Whitham JC, Wielebnowski N. New directions for zoo animal welfare science. Appl Anim Behav Sci. 2013;147:247–60.
Wielebnowski NC, Fletchall N, Carlstead K, Busso JM, Brown JL. Noninvasive assessment of adrenal activity associated with husbandry and behavioral factors in the North American clouded leopard population. Zoo Biol. 2002a;21:77–98.
Wielebnowski NC, Ziegler K, Wildt DE, Lukas J, Brown JL. Impact of social management on reproductive, adrenal and behavioral activity in the cheetah (*Acinonyx jubatus*). Anim Conserv. 2002b;5:291–301.
Wikelski M, Cooke SJ. Conservation physiology. Trends Ecol Evol. 2006;21:38–46.
Wildt DE. Effect of transportation on sexual behavior of cats. Lab Anim Sci. 1980;30:910–2.
Willis SL, Tennstedt SL, Marsiske M, Ball K, Elias J, Koepke KM, Morris JN, Rebok GW, Unverzagt FW, Stoddard AM, Wright E. Long-term effects of cognitive training on everyday functional outcomes in older adults. JAMA. 2006;296:2805–14.
Wilson S, Fell L, Colditz I, Collins D. An examination of some physiological variables for assessing the welfare of beef cattle in feedlots. Anim Welf. 2002;11:305–16.
Wingfield JC, Sapolsky RM. Reproduction and resistance to stress: when and how. J Neuroendocrinol. 2003;15:711–24.
Witeska M. Stress in fish-hematological and immunological effects of heavy metals. Electron J Ichthyol. 2005;1:35–41.
Wittig RM, Crockford C, Weltring A, Deschner T, Zuberbuhler K. Single aggressive interactions increase urinary glucocorticoid levels in wild male chimpanzees. PLoS One. 2015;10:e0118695.
Wojtaszek J, Dziewulska-Szwajkowska D, Łozińska-Gabska M, Adamowicz A, Dżugaj A. Hematological effects of high dose of cortisol on the carp (*Cyprinus carpio L.*): cortisol effect on the carp blood. Gen Comp Endocrinol. 2002;125:176–83.
Wolf OT. HPA axis and memory. Best Pract Res Clin Endocrinol Metab. 2003;17:287–99.
Wooddell LJ, Hamel AF, Murphy AM, Byers KL, Kaburu SSK, Meyer JS, Suomi SJ, Dettmer AM. Relationships between affiliative social behavior and hair cortisol concentrations in semi-free ranging rhesus monkeys. Psychoneuroendocrinology. 2017;84:109–15.
Woof JM, Kerr MA. IgA function—variations on a theme. Immunology. 2004;113:175–7.
Woolley LA, Millspaugh JJ, Woods RJ, Van Rensburg SJ, Page BR, Slotow R. Intraspecific strategic responses of African elephants to temporal variation in forage quality. J Wildl Manage. 2009;73:827–35.
Yamauchi A, Takei I, Kasuga A, Kitamura Y, Ohashi N, Nakano S, Takayama S, Nakamoto S, Katsukawa F, Saruta T. Depression of dehydroepiandrosterone in Japanese diabetic men—comparison between non-insulin-dependent diabetes mellitus and impaired glucose tolerance. Eur J Endocrinol. 1996;135:101–4.
Yeates JW, Main DC. Assessment of positive welfare: a review. Vet J. 2008;175:293–300.
Yin YJ, Nie CY, Liu WS, Zou Q, Zhai JC, Han HS, Li HP. Non-invasive determination of the immune physiological state of reindeer (*Rangifer tarandus*) in the Greater Khingan Mountains, China. Genet Mol Res. 2015;14:6664–73.

Yon L, Kanchanapangka S, Chaiyabutr N, Meepan S, Stanczyk FZ, Dahl N, Lasley B. A longitudinal study of LH, gonadal and adrenal steroids in four intact Asian bull elephants (*Elephas maximus*) and one castrate African bull (*Loxodonta africana*) during musth and non-musth periods. Gen Comp Endocrinol. 2007a;151:241–5.

Yon L, Kanchanapangka S, Chaiyabutr N, Stanczyk F, Meepan S, Lasley B. ACTH stimulation in four Asian bull elephants (*Elephas maximus*): an investigation of androgen sources in bull elephants. Gen Comp Endocrinol. 2007b;151:246–51.

Yoshida C, Ishikawa T, Michiue T, Quan L, Maeda H. Postmortem biochemistry and immunohistochemistry of chromogranin A as a stress marker with special regard to fatal hypothermia and hyperthermia. Int J Leg Med. 2011;125:11–20.

Young KM, Walker SL, Lanthier C, Waddell WT, Monfort SL, Brown JL. Noninvasive monitoring of adrenocortical activity in carnivores by fecal glucocorticoid analyses. Gen Comp Endocrinol. 2004;137:148–65.

Zabka T, Goldstein T, Cross C, Mueller R, Kreuder-Johnson C, Gill S, Gulland F. Characterization of a degenerative cardiomyopathy associated with domoic acid toxicity in California sea lions (*Zalophus californianus*). Vet Pathol. 2009;46:105–19.

Zhang D, Lavaux T, Sapin R, Lavigne T, Castelain V, Aunis D, Metz-Boutigue M-H, Schneider F. Serum concentration of chromogranin A at admission: an early biomarker of severity in critically ill patients. Ann Med. 2009;41:38–44.

Zilioli S, Slatcher RB, Ong AD, Gruenewald TL. Purpose in life predicts allostatic load ten years later. J Psychosom Res. 2015;79:451–7.

Zilioli S, Imami L, Ong AD, Lumley MA, Gruenewald T. Discrimination and anger control as pathways linking socioeconomic disadvantage to allostatic load in midlife. J Psychosom Res. 2017;103:83–90.

Zota AR, Shenassa ED, Morello-Frosch R. Allostatic load amplifies the effect of blood lead levels on elevated blood pressure among middle-aged U.S. adults: a cross-sectional study. Environ Health. 2013;12:1–11.

Zumoff B, Walter L, Rosenfeld RS, Strain JJ, Degen K, Strain GW, Levin J, Fukushima D. Subnormal plasma adrenal androgen levels in men with uremia. J Clin Endocrinol Metab. 1980;51:801–5.

Zwijacz-Kozica T, Selva N, Barja I, Silván G, Martínez-Fernández L, Illera JC, Jodłowski M. Concentration of fecal cortisol metabolites in chamois in relation to tourist pressure in Tatra National Park (South Poland). Acta Theriol. 2013;58:215–22.

Chapter 6
Physiological Thresholds in the Context of Marine Mammal Conservation

Karina Acevedo-Whitehouse

Abstract There is growing evidence of the adverse effects of global environmental change on marine mammals, particularly in terms of changes in abundance, distribution, habitat use, migratory phenology, feeding habits, risk of infectious diseases, bioaccumulation of contaminants, declines in reproductive success, and reductions in genetic diversity. These anthropogenic stressors have led to an evident conservation crisis: a quarter of the extant marine mammal species that have been assessed and now considered at risk of extinction. However, we still know very little about the subtle, subclinical effects of environmental stressors on various aspects of physiology that could impacts their survival and long-term fertility, or that could impact the performance of future generations. This chapter argues for the need of considering physiological thresholds when examining how drivers of global environmental change can impact marine mammal populations. It also poses questions regarding our understanding of individual phenotypic plasticity and resilience of species in the face of environmental stressors, particularly under abrupt and unpredictable environmental changes. Specifically, I present evidence of how drivers of environmental change can exert effects at different levels of biological organization to influence the viability of individuals and populations, and discuss which aspects of environmental change could be the most likely to shift the physiological thresholds of different marine mammal species. Finally, the chapter identifies areas of further research in marine mammal conservation physiology in the current context of global environmental change.

K. Acevedo-Whitehouse (✉)
Unit for Basic and Applied Microbiology, School of Natural Sciences, Autonomous University of Querétaro, Santiago de Querétaro, Mexico
e-mail: karina.acevedo.whitehouse@uaq.mx

P. Comizzoli et al. (eds.), *Reproductive Sciences in Animal Conservation*, Advances in Experimental Medicine and Biology 1200,
https://doi.org/10.1007/978-3-030-23633-5_6

1 Introduction

The current rate of global environmental change, defined here as any biotic or abiotic factor related to human activities that affects the atmosphere, oceans and land systems with widespread effects (Matesanz et al. 2010), is unprecedented. One of the most notorious drivers of global environmental change is climate change: during the twentieth century, the average surface temperature of the planet increased by approximately 0.6 °C, with the majority of this increase occurring in the final quarter of the century (NOAA 2017). This trend worsened since the turn of the century, with a further increase of around 0.5 °C (Sánchez-Lugo et al. 2018). The effects of climate change are exacerbated, as synergies among this driver of environmental change and other ecosystem stressors can further accelerate ecosystem transformation, altering species composition and disrupting habitats and, ultimately, impacting wildlife conservation (Côté et al. 2016). Whether in the form of climate change, habitat loss or degradation, accumulation of contaminants or pathogen spillover, drivers of global change create novel environmental scenarios and different selective pressures that affect species in various synergistic, often unpredictable ways (Thomas and Klaper 2004; Hoffmann and Willi 2008; Acevedo-Whitehouse and Duffus 2009). Individuals respond to these stressors and new scenarios with changes in their physiology (Hofmann and Todgham 2010), the result of which has consequences of different magnitudes for their populations, biotic interactions and community composition.

As a consequence of direct and indirect effects of human pressures, we are presently experiencing a mass extinction event (Barnosky et al. 2011). More than 300 mammal species have already gone extinct since the start of this event in the late Pleistocene (Redding and Mooers 2015). The rate of extinction has been particularly severe in the past 50 years, with a number of unique lineages vanishing completely (Sandom et al. 2014) and various species becoming seriously depleted (Davis et al. 2018). The drivers of loss of mammalian diversity include overexploitation, habitat loss and degradation, invasion by exotic species (Bellard et al. 2016; McDonald et al. 2017) and climate change (Luo et al. 2015; Morueta-Holme et al. 2010). More concerningly, approximately 10% of mammal species not currently considered to be 'at risk', inhabit environments that are likely to predispose them towards extinction, or have biological traits that, under pressure by environmental changes, could increase their extinction risk (Davidson et al. 2017). Marine mammals are not an exception to the conservation crisis. Less than a decade ago, according to The International Union for the Conservation of Nature (IUCN), 25% of the 132 extant marine mammal species for which there is sufficient data available, were at risk of extinction (Davidson et al. 2012). The oceanic ecosystems inhabited by marine mammals are currently experiencing great instability due to various human activities, with warming sea surface temperature (SST), melting sea ice, and acidification occurring at rates never before reported (Worm et al. 2006; Hoegh-Guldberg and Bruno 2010). In addition, high levels of persistent pollutants are being accumulated and biomagnified in the tissues of marine mammals, while concurrently their populations are being overharvested and their prey are being depleted by the alteration of entire food webs (Halpern et al. 2008).

There is growing evidence of the adverse effects of different drivers of global environmental change on marine mammals, particularly in terms of changes in population abundance, distribution and habitat use (Laidre et al. 2015; Rode et al. 2015; Younger et al. 2016; Hamilton et al. 2019), migration (Moore 2008; Block et al. 2011; Chambault et al. 2018), feeding habits (Elorriaga-Verplancken et al. 2016; Szpak et al. 2018), susceptibility to infectious diseases (Van Bressem et al. 2009; Di Guardo et al. 2018; Denkinger et al. 2015; Burek et al. 2008; Alava et al. 2014), bioaccumulation of persistent contaminants (Desforges et al. 2018; Schnitzler et al. 2018), alterations due to increasing sound stressors (i.e. military sonar signals; Wensveen et al. 2019), or seismic surveys (Parente et al. 2007), declines in reproductive success (Leaper et al. 2006; Seyboth et al. 2016; Clausius et al. 2017), and reductions in genetic diversity (Stoffel et al. 2018; O'Corry-Crowe 2008). However, we know very little about the subtle, subclinical effects of environmental stressors on various aspects of physiology that can have long-term impacts for survival and fertility, or that may even impact the performance of future generations with practically no observable effects in their own lifetime (Donelson et al. 2018; Cavieres et al. 2019). To date, there are relatively few studies on conservation physiology of marine mammals, and even fewer of these are long-term studies, let alone transgenerational studies. Considering the rapid rate at which marine environments are changing, and the striking rate of biodiversity loss, there is an urgent need to increase our understanding of how individuals and species can respond to an increasingly variable and unpredictable environment (Botero et al. 2014). Such information could help us identify which of the extant species will be able to cope with such rapid rates of change (Evans and Hofmann 2012). This is particularly important given that some species, such as the vaquita, *Phocoena sinus*, are already on the brink of extinction (Morell 2008; Jaramillo-Legorreta et al. 2016), and others, including Pacific populations of blue, *Balaenoptera musculus*, fin, *B. physalus*, and southern right whales, *Eubalaena australis*, are unlikely to recover under current environmental scenarios (Tulloch et al. 2019).

2 State of the Art

2.1 Physiological Thresholds in a Rapidly-Changing Environment

For more than a century, researchers have been interested in understanding how organisms respond to their natural habitat. Following initial studies on plant and animal physiology, a new discipline, termed 'ecological physiology', also known as *ecophysiology* and *field physiology*, emerged during the first half of the twentieth century. This new discipline combined comparative physiology, behavioral ecology, evolutionary ecology and evolutionary biology, allowing for studies that intended to understand how physiological mechanisms permit an organism to respond to fluctuations in its environment (Costa and Sinervo 2004). Our understanding of feeding biomechanics, osmoregulation, thermoregulation, and diving physiology of marine mammals greatly expanded under this new approach (Castellini et al. 2002; Castellini and Castellini 2004). As available tools and methodological approaches modernized and non-invasive sampling techniques were developed (Altimiras and Anderson 2016), the field began to include molecular and cell biology, enhancing our understanding of physiology across diverse environmental conditions and different spatial and temporal scales. Then, in the midst of the current conservation and environmental crisis, 'conservation physiology' was proposed formally as a new field in 2006, with the main goal of gathering and interpreting physiological data that will provide solid arguments to aid conservation decisions (Wikelski and Cooke 2006).

Conservation physiology operates under four central premises: (1) anthropogenic activities are affecting the abiotic environment on a global scale; (2) the abiotic environment influences various physiological mechanisms, including molecular, biochemical and cellular pathways, of an organism; (3) the alteration of an organism's physiology by various environmental stressors can affect individual fitness, and (4) these intricate associations can have profound implications for processes at the population- and ecosystem level (Metcalfe et al. 2012) (see Fig. 6.1). To properly address questions under these central premises, conservation physiology encompasses comparative physiology, environmental physiology, evolutionary physiology, neurophysiology, endocrinology, bioenergetics, reproductive physiology, immunology, physiological genomics, epidemiology, among other disciplines (Cooke et al. 2013).

Interestingly, conservation physiology is still not fully considered in the majority of studies about the current and future impacts of global change on our planet's ecosystems. For instance, although the concept of environmental tipping points thresholds (Dakos et al. 2018) or has been developed as a means to help inform ecosystem management programs (Somero 2011; Monaco and Helmuth 2011), and it has proposed that the evolutionary response to global environmental change could be predicted to some extent (Botero et al. 2014), most available models disregard effects at the physiological and biological hierarchical scales. Such an omission can potentially lead to erroneous conclusions (Monaco and Helmuth 2011; Seebacher and Franklin 2012). This is because physiology encompasses all cellular, biochemical and molecular mechanisms used by an organism to cope with the internal and external environment that they live in, and alterations in physiology might impact the performance of the individual, with consequences to populations, communities, and ecosystems. Based on studies published to date it is evident that there exists a discrete range of environmental conditions under which the performance of an individual is optimal, and thresholds beyond which various physiological mechanisms are compromised (Hofmann and Todgham 2010), leading to disease or reproductive impairment, both of which can have serious consequences for a given population, particularly those that are already reduced (Holt et al. 2014). An unambiguous

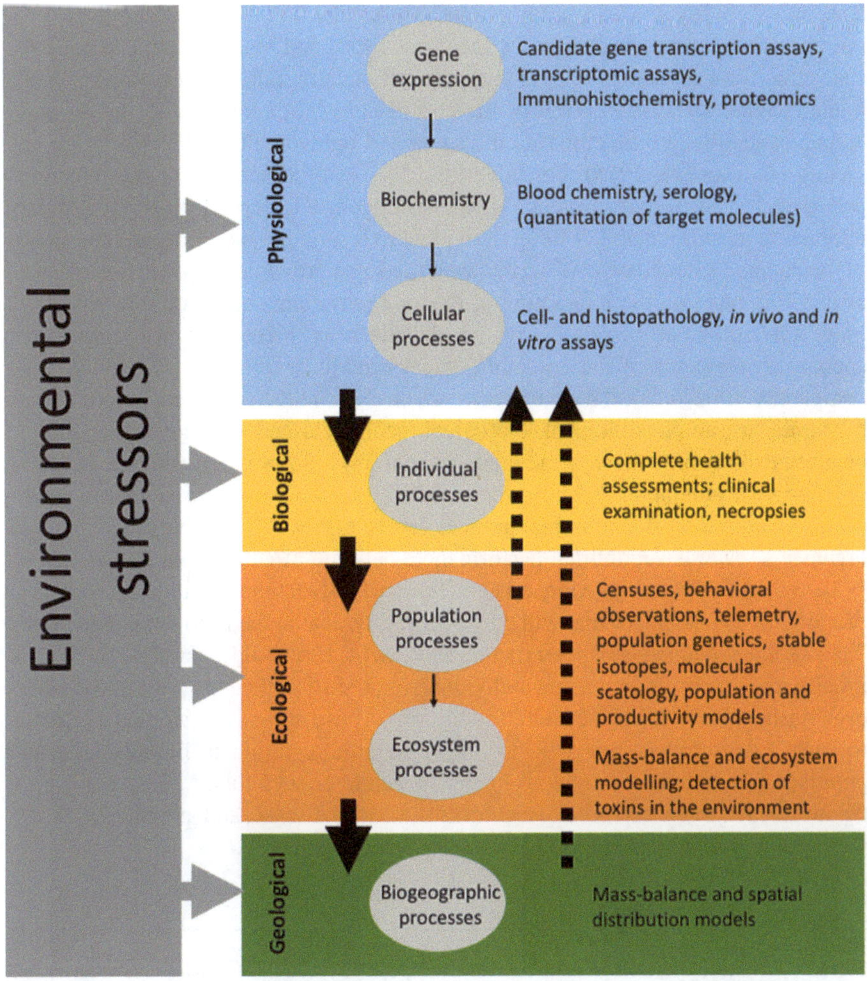

Fig. 6.1 The hierarchical levels of biological complexity. Biological processes are identified as grey circles. The impact of environmental stressors on each level of complexity are indicated by grey arrows. The cascade of effects throughout the levels of complexity is indicated by black arrows. Methods and techniques that can be used to assess suboptimal physiological performance are indicated for each level. Adapted from Metcalfe et al. (2012) and Le Quesne and Pinnegar (2012)

identification of these thresholds for different taxa would allow a more informed assessment of the relative vulnerability to environmental change of a given species (Monaco and Helmuth 2011), and thus help inform conservation assessments and management plans.

Individual- and population level physiological thresholds of a species in the context of environmental change could be better understood by merging the traditional concept of the epidemiological triad (also known as 'disease triangle') with the concept of environmental tipping points. The epidemiological triad, initially proposed to understand plant diseases (McNew 1960) and since adapted to animals, states that for a disease to appear in a population there must be an overlap in traits related to the hosts' susceptibility, the etiological agent's infectivity, and environmental states that are favorable to the agent and unfavorable to the host's immune system (James et al. 2015). For infectious diseases, the duration of the infectious period, prevalence, and virulence of the pathogen will depend on this three-way interaction as will the ability of an individual to respond to the threat and the extent of the damage (Scholthof 2007). However, the same principles apply to any disease, as it will depend on the interaction of external (environmental) and internal (host) traits. Physiological thresholds can be visualized as three concentric circles that encompass elements of the epidemiological triad at different levels of biological complexity (Fig. 6.2a). The most external circle would be defined by conditions pertaining to the physiological, biological, ecological and geological hierarchical levels that allow organisms to maintain optimal physiological performance and their populations to maintain viability; the middle circle would comprise altered conditions pertaining to the four hierarchical levels that will shift the physiology of organisms towards suboptimal performance, but with effects that might not yet be evident in the population (Monaco and Helmuth 2011), which could be reversed if the conditions change, or that will only become evident in the following generations due to epigenetic mechanisms (Herman et al. 2014; Eirin-Lopez and Putnam 2018). Finally, the inner circle depicts individuals that shifted their physiological status even further as a result of the altered conditions. These individuals would be affected to such a degree that they will no longer be able to return to homeostasis, and population-level effects would no longer be subtle or sublethal; rather, they would be observed as lethal reproductive outcomes, serious disease, and, potentially, death, all of which can have marked consequences for a population.

Fig. 6.2 (continued) physiological performance and populations to excel or at least maintain stability; the middle (yellow) circle comprises altered conditions that shift organisms towards suboptimal physiological performance, with subtle or reversible effects; the inner (orange) circle describes serious organismal consequences stemming from the previously altered conditions. Individuals that cross this physiological threshold would be affected to such a degree that they will no longer be able to return to homeostasis, and effects would be observed as lethal reproductive outcomes, serious disease, or death. (**b**) Representation of the physiological thresholds for a species with no conservation concern, where the majority of individuals will have an optimal physiological performance, a small fraction of individuals will have suboptimal physiological performance, and an even smaller fraction will succumb to disease or not be able to breed due to physiological limitations or stochasticity. (**c**) Representation of the physiological thresholds for a species challenged by environmental and intrinsic stressors; here, most individuals will have suboptimal performance and the consequences, in the form of disease, dramatically-reduced fertility or mortality, will be more evident for a larger proportion of individuals. (**d**) Representation of the physiological thresholds for a species with high resilience to environmental stressors. Here, most individuals have suboptimal performance, but very few show signs of disease or reduced reproduction. (**e**) Representation of the physiological thresholds for a species with poor resilience to environmental stressors. Here, most individuals are weak, diseased or have serious reproductive impairment

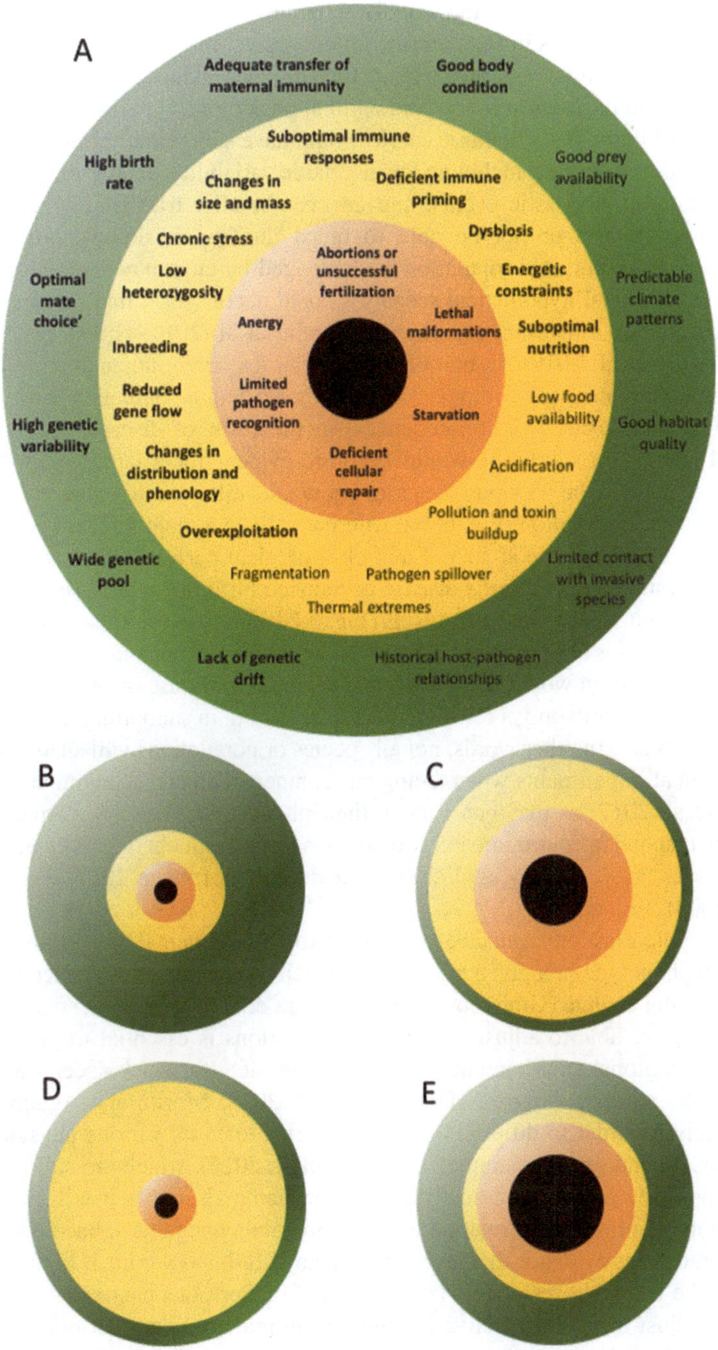

Fig. 6.2 Proposed representation of physiological thresholds under global environmental change. (**a**) The external (green) circle encompasses conditions that allow organisms to maintain optimal

In an ideal scenario, most individuals belonging to a species with no conservation concern that resides in an environment not greatly impacted by anthropogenic-related stressors, would have an adequate physiological status, and selection should favor those with the highest fitness and optimal physiological performance. A small proportion of individuals in this scenario would have suboptimal physiological status due to various factors, including lower genetic variability, expression of deleterious alleles, and stochastic events, and an even smaller fraction of these would succumb to disease or not be able to breed due to physiological limitations (Fig. 6.2b). However, if a population is challenged by environmental and intrinsic stressors, additional hypothetical scenarios could occur, given that species vary in their tolerance to environmental change (Santini et al. 2019; Gibbin et al. 2017; Muletz-Wolz et al. 2019). In one of these scenarios, environmental stressors can push organisms across physiological thresholds, leading a population towards suboptimal performance. The consequences, in the form of disease, dramatically-reduced fertility or mortality, would be more evident for a larger proportion of individuals than in the previous scenario, but would still not reflect the status of the majority of the population (Fig. 6.2c). Alternatively, in a more resilient population, individuals could operate at suboptimal physiological performance having shifted the first threshold, although the disease or reduced birth rate would be observed in only a small number of these organisms (Fig. 6.2d). The inverse situation could also occur if the species has limited resilience to environmental change. Here, the majority of the population would be driven directly towards disease or serious reproductive impairment, with only a few individuals remaining in suboptimal physiological status (Fig. 6.2e). In other words, not all species or populations will share the same physiological impairments when facing environmental change (Santini et al. 2019; Gibbin et al. 2017) as they can vary in their phenotypic plasticity, defined as the ability to respond to environmental stressors via the adjustment of genotypic expression (Relyea 2017; Nussey et al. 2007; Chevin and Hoffmann 2017; Piersma and Drent 2003) (Fig. 6.3).

Phenotypic plasticity enhances responses to environmental change by altering the mean phenotype in a given population, which may increase resilience to global environmental change (Gibbin et al. 2017; Oostra et al. 2018). Understanding how populations are able to adjust to changing conditions is essential to predict their resilience to global environmental change and to identify which species are more likely to be severely affected (Donelson et al. 2018; Moritz and Agudo 2013). Regrettably, in order to do this, we need empirical data on various physiological, phenological and genetic parameters (Urban et al. 2016), which are still relatively scarce for most species. Specifically, for marine mammals, only a handful of studies on phenotypic plasticity in relation to environmental change have been conducted. However, there is evidence that investment in key life-history traits is highly dependent on the available resources. For instance, Southern elephant seals, *Mirounga leonina*, adjust their reproductive expenditure in maternal care not only based on their mass, but also on the prevailing environmental conditions (McMahon et al. 2017), and there is evidence of plasticity in recovery from winter food deprivation

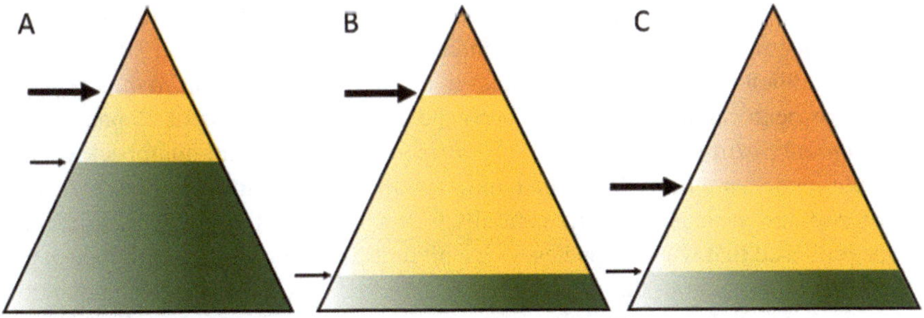

Fig. 6.3 The 'iceberg' effect when studying the effect of environmental change on disease or reproductive failure of a population. The bottom (green) level represents healthy, non-compromised individuals; the middle (yellow) level represents individuals with subtle, subclinical manifestations of compromised physiological performance; the top (orange) level represents individuals with evident manifestations of deficient physiological performance. The thin black arrow shows the threshold to suboptimal physiological performance; the thick black arrow shows the threshold to seriously compromised physiological performance. (**a**) In this scenario, roughly 10% of the population is unhealthy, and traditional methods to assess disease or reproductive failure would identify only a few of the sampled individuals (supposing an unbiased sampling of individuals). (**b**) In this scenario, although the same percentage of the population is unhealthy, there is a larger proportion of individuals with (subtle or undetectable) suboptimal physiological status. Unbiased sampling and traditional diagnostic methods would identify roughly the same number of unhealthy individuals, although there is a potentially larger problem in this population. (**c**) In this scenario, individuals are more susceptible to the environmental stressors and a large proportion of them have crossed the threshold to a poor physiological status. Here, a higher number of unhealthy individuals would be detected by the same sampling strategy

in polar bears, *Ursus maritimus*, depending on whether they stay on land or follow the ice (Whiteman et al. 2018).

One way to help inform the current uncertainty regarding resilience to environmental change, could be to consider the life-history of a given species and make predictions based on the life-history theory (LHT). Under this theory, organisms will optimize their reproductive success within their specific evolutionary constraints (Roff 2001; Stearns 1992), a strategy that explains the dramatic variations in developmental times, age of sexual maturity, number of offspring, parental investment, and longevity of different species. Life-history theory strives to understand how organisms will optimize their survival and reproduction under constraints that have been shaped by natural selection and other evolutionary forces (Roff 2001). Thus, LHT focuses on an optimization problem, where biotic and abiotic factors of an organism's habitat (including anthropogenic stressors) will affect an individual's probability of reproduction and survival due to trade-offs that arise because of restricted resources available to allocate among different life-history traits (Zera and Harshman 2001). In the context of global environmental change, variation in life-history traits can impact the ecological and evolutionary responses of populations to such rapid and unpredictable changes (Snell-Rood et al. 2015).

A study that examined drivers of extinction risk for marine mammals found that the primary predictors of risk were life-history traits, such as body mass at weaning and rate of biomass production (i.e. the size of their offspring at weaning multiplied by the frequency of breeding, (Davidson et al. 2012; Hamilton et al. 2011). These variables influence the capacity of species to recover after the threats to their populations have been removed. This has been the case for the northern elephant seal, *Mirounga angustirostris*, gray whale, *Eschrichtius robustus*, and humpback whale, *Megaptera novaeangliae*, all of which have fast life-histories for their body size, and whose populations increased exponentially following protection (Lotze et al. 2011). It was proposed that when a species with a high rate of biomass production does not recover after protection, the environmental stressors that affected their population have not been mitigated, or other stressors are now affecting their populations (Davidson et al. 2012). Based on LHT, one could question which are the optimal values and combinations of traits that will maximize reproductive success and survival for a given species based on its past (evolutionary) and present (ecological) constraints. Then, models could be developed to answer specific questions pertaining to reproductive success and survival, such as *how many times should a given species reproduce in a warmer environment to maximize its fitness?* or *given climate-induced alterations in prey availability or repertoire and consequential energetic restrictions, what species would be able to maintain maximum fitness if they have to divert resources for immune responses?* Considering the aforementioned environmental stressors known to be of conservation concern for marine mammals, it is reasonable to predict that species with a slow life pace, a low rate of biomass production, restricted distributions, narrow foraging ranges, or those with specialist feeding habits, would be more likely to have limited resilience to environmental change, and thus, be more likely to shift physiological thresholds, particularly when the changes are abrupt and unexpected (Ratajczak et al. 2018). The 2013–2015 anomalous climatic event, known as 'The Blob' that occurred in the northeastern Pacific ocean, and that was exacerbated by the 2015 El Niño event, gave some indication that this prediction could certainly be true. A study about California sea lions, *Zalophus californianus*, in the region affected by The Blob had pre-clinical physiological alterations. The species is a coastal otariid pinniped with annual breeding, and high investment of resources to their pups during at least the first 5–6 months of age. As the feeding range of adult female California sea lions is limited to around 100 km from their breeding colony (Kuhn and Costa 2014), the climatic anomaly led to changes in their feeding ecology, reflected by more distant trips from shore (Elorriaga-Verplancken et al. 2016) which had an impact on their pups' body condition, affected basic processes associated with developmental hematopoiesis (Flores-Morán et al. 2017), and limited blood glucose levels, which are an essential energetic source to implement immune responses (Banuet-Martínez et al. 2017). As expected based on the observed physiological responses, there was a marked reduction in pup survival (Elorriaga-Verplancken et al. 2016; Banuet-Martínez et al. 2017). These studies suggest that this particular species has a limited phenotypic plasticity for key physiological processes when under duress by climate change. An independent study of another otariid species, the South American fur seal, *Arctocephalus australis*, found similar evidence regarding the negative effect of years with high SST on maternal attendance and immune capacity (Seguel et al. 2018). If the cost of exerting immune responses during periods of resource limitation is similarly shaped in other otariid species, this could have implications for vulnerable species. For instance, it has been argued that, after being at the brink of extinction 100 years ago due to exploitation, the population recovery of the Guadalupe fur seal, *Arctocephalus philippii townsendii*, will be closely related to extrinsic factors such as resource availability (Weber et al. 2004) driven by environmental variations (Juárez-Ruiz et al. 2018). Interestingly, the northern elephant seal, *Mirounga angustirostris*, a phocid pinniped with extremely high sexual dimorphism, short lactation periods, notorious lack of genetic diversity and a wider feeding range that also inhabits the areas affected by that climatic anomaly, did not reveal any clinical observable effect similar to that registered for the California sea lion (unpublished data). While still early to establish unequivocally, it would appear that both pinniped species, characterized by entirely different evolutionary histories and life-history traits (Pomeroy 2011; Berta et al. 2015), have distinct physiological thresholds when faced with a similar climatic anomaly. More studies will be needed to understand these physiological thresholds associated with anomalous climatic events for pinnipeds and other marine mammals.

2.2 *Host Genetics in the Context of Global Environmental Change*

One of the main drivers of environmental change and mammalian extinction is habitat fragmentation (Bellard et al. 2016; McDonald et al. 2017). Habitat fragmentation can isolate populations and reduce gene flow, leading to increased population structuring, reduced genetic diversity and, often, increased risk of consanguineous mating (Keyghobadi 2007). Genetic erosion is of concern, given that genetic variability is one of the intrinsic factors that shape various aspects of an individual's physiology, and thus impact performance (Leroy et al. 2017). Populations with high genetic diversity are more likely to contain advantageous alleles of genes under selection by a given environmental stressor (Pauls et al. 2013; Bijlsma and Loeschcke 2012); thus, maintaining genetic variation is critical for most populations to persist.

Inbreeding, or mating between closely-related individuals, increases the frequency of homozygosity in the offspring, who tend to have reduced fitness (i.e. inbreeding depression) because harmful recessive alleles are more likely to be expressed (Charlesworth and Charlesworth 1999). Well known examples of inbreeding depression in humans are increased fetal death, various physical deformities, and high cancer rates (Fareed and Afzal 2017; Baris et al. 2016; Lurie et al. 1984). Furthermore, studies on inbred mice have shown that higher homozygosity is even associated with suboptimal muscular performance and endurance (Kilikevicius et al. 2013), traits that are essential for the survival of free-living individuals. Incestuous mating can also lead to suboptimal immune responses to pathogens.

This is because inbred offspring that are homozygous at genes under balancing selection, such as the major histocompatibility genes (O'Brien and Evermann 1988; Sutton et al. 2011), Toll-like receptor genes (Kloch et al. 2018), and B-defensin genes (Chapman et al. 2016), involved in antigen or pathogen recognition, would be less likely to recognize different pathogen-derived antigens or molecular patterns associated to pathogens (Acevedo-Whitehouse and Cunningham 2006).

Levels of heterozygosity at several genetic markers (typically 10–15) can be used as an indirect measure of inbreeding in individuals with unknown pedigrees (Coltman and Slate 2003; Chapman et al. 2009). Based on these measures of multi-locus heterozygosity, it has been shown that consanguineous mating can impact survival, disease resistance, and reproductive success of marine mammals. Examples include lower juvenile survival of more inbred harbor seals, *Phoca vitulina* (Coltman et al. 1998), harp seals, *P. groenlandica* (Kretzmann et al. 2006), and grey seals, *Halichoerus grypus* (Bean et al. 2004), and decreased male reproductive success in Antarctic fur seals, *Arctocephalus gazella* (Hoffman et al. 2004). Selection against homozygosity appears to be strong in Antarctic fur seals, particularly in light of the current environmental crisis, as low levels of multi-locus heterozygosity are also associated with low birth weight and reduced survival as prey availability becomes reduced in the Antarctic due to climate change (Forcada and Hoffman 2014). Low heterozygosity due to consanguineous matings has also been related to higher parasite loads and parasite-related lesions in California sea lion, *Zalophus californianus*, pups (Acevedo-Whitehouse et al. 2006) and juvenile harbor seals (Rijks et al. 2008), reduced natural antibody levels in Galapagos sea lion, *Zalophus wollebaeki*, pups (Brock et al. 2015), increased risk of severe infections and urogenital carcinoma in adult California sea lions (Acevedo-Whitehouse et al. 2003). The effects of inbreeding on disease susceptibility can be particularly relevant for a population during a disease outbreak. For instance, a morbillivirus outbreak in the Mediterranean Sea that decimated the striped dolphin, *Stenella coeruleoalba*, population, affected more inbred dolphins first (Valsecchi et al. 2004), suggesting that these individuals can act as entry points of a given pathogen into a population. Given that (1) fragmentation and over exploitation lead to reduced gene flow (Monteiro et al. 2019) and higher chances of consanguineous mating (Casas-Marce et al. 2013), (2) environmental stressors increase the risk of infectious disease outbreaks (Wood 2014; Machalaba and Karesh 2018), (3) inbreeding reduces immune detection and elimination of pathogens, and (4) a number of pathogens, such as morbillivirus, can have a devastating impact for marine mammal populations (Di Guardo et al. 2018; Van Bressem et al. 2014), inbreeding depression could be of relevance to conservation of marine mammals in the context of environmental changes.

It appears that the relevance of inbreeding depression to a population might be dependent on the environment. For example, the influence of multi-locus heterozygosity on white blood cell counts in juvenile Galapagos sea lions was observed in a colony that was impacted by urban development, pollution and introduced species, but not in a colony that lacked these environmental stressors (Brock et al. 2015). Considering that different environmental stressors are currently affecting marine mammal populations, it would be important to investigate their role in shaping the mechanistic effect of reduced genetic variability, and in defining physiological thresholds for a population. This means that we must avoid making widespread generalizations about observable detrimental effects of inbreeding on individual physiological performance and population viability across different marine mammal species, without first bearing in mind (1) their life-history traits, (2) the drivers of environmental change they are facing, and (3) the fact that effects could be subtle and not yet observable. For example, it has recently been predicted that killer whale, *Orcinus orca*, populations, especially those near industrialized areas or feeding at high trophic levels, are at high risk of collapsing due to the extremely high levels of lipophilic organochlorine contaminants in their blubber, which exert a strong effect on the species' immune competence (Desforges et al. 2018). The species already has low worldwide genetic diversity (Hoelzel et al. 2002), and has cultural traits (e.g. matrifocal social structuring, differential habitat use and strong resource specialization) that have promoted sufficient isolation of their populations (Moura et al. 2014). If selection and drift were to reduce the genetic pool even further and if the species were to experience some level of consanguineous mating due to isolation, then it is plausible to expect detrimental effects of inbreeding on their physiological performance to be detectable and, possibly, severe, given that at least some aspects of their fitness are already compromised by contaminants. A similar situation could plausibly occur for some polar bear subpopulations (Rode et al. 2014; Wiig et al. 2008). Polar bears are a K-selected species, that has late sexual maturity, extremely reduced litter size and high maternal investment. Higher temperatures and erratic weather patterns are increasing across their range, and as polar bears depend on the Arctic sea ice for access to their prey, they are vulnerable to a habitat that exists as a function of atmospheric temperature and SST (Regehr et al. 2016). In addition, despite living in regions that are relatively untouched by human activities, various industrial contaminants and by-products enter their ecosystem due to convective currents and the Arctic Ocean Boundary Current (AOBC). Being top predators, polar bears are therefore exposed to high levels of persistent contaminants, which bio-magnify and reach high concentrations in their tissues (Desforges et al. 2017; Letcher et al. 2010).

As natural populations are depleted, captive breeding programs are depended upon to reintroduce individuals (Kleiman 1996). Considering that reduced genetic variability can have negative impacts for an individual, regardless of whether the detrimental effects are subtle or evident, such programs need to plan their breeding colonies very carefully. Captive breeding programs of marine mammals must not ignore the detrimental effects of inbreeding on disease susceptibility, particularly given that the numbers of captive-bred California sea lions, harbor seals, and bottlenose dolphins, *Tursiops truncatus*, have greatly increased in zoos and aquaria in North America (Cornell and Asper 2007), and for of these species, breeding groups already have 3rd generation offspring. If relatively more inbred animals will tend to have suboptimal physiological performance under harsher environments, care should be taken to avoid incestuous mating as much as possible. For this, genetic profiling of individuals is crucial prior to deciding which males to use as fathers when setting up breeding colonies. On the one hand this would help decrease potential health problems in their collections, but it would also avoid the reintroduction of inbred organisms to the wild (Venesky et al. 2012).

2.3 Tools and Strategies to Detect Physiological Thresholds

We are currently at a challenging moment for conservation sciences. The rates of environmental change, population declines and extinction are overwhelming, and there is a pressing need to have adequate information about the effects of environmental stressors on different hierarchal levels of biological complexity in order to make better decisions about which populations or species are more likely to be impacted. However, the tools and strategies available to gather this information, particularly in terms of identifying physiological thresholds have increased markedly in the past decades (Madliger et al. 2018). There is now a myriad of options available to explore even subtle changes in the physiology of organisms (Hunt et al. 2013, 2015). Physiological traits of marine mammals can now be measured in a wide array of biological tissues, including traditional sources, such as blood, fur, urine, feces, saliva, baleen, mucosal swabs, and skin and blubber biopsies (Espinosa-de Aquino et al. 2017; Kershaw et al. 2018), but also in samples that can be collected in a less invasive way. For instance, exhaled breath condensate has proven to be extremely useful as a source of physiologically-relevant information of free-ranging cetaceans, from the analysis of respiratory pathogens (Acevedo-Whitehouse et al. 2010) and microbiome (Apprill et al. 2017) to the quantitation of reproductive (Richard et al. 2017) and stress hormones (Burgess et al. 2018; Thompson et al. 2014). As expected, the type of physiological information that can be determined will depend on the biological material being used. Some, like blood, will provide us with a 'photograph' of the status (hormonal, immune, health, etc.) of the individual near to the time of sampling (Robeck and Nollens 2013; Gilbert et al. 2013), while others, such as baleen, will reflect decades of the life of that organism (Hunt et al. 2014). Blood samples and their liquid components (plasma and serum) will still be considered the 'gold standard' when assessing health via traditional methods, such as hematology and serology (Gilbert et al. 2013; Johnstone et al. 2017). However, clinically-relevant changes in hematological profiles will often be seen only when the physiological threshold has been crossed towards a disease status, not allowing for the detection of subclinical, subtle, changes in physiology. Measuring levels of gene transcription can be more informative in this regard. For instance, a recent study of wild adult California sea lions detected spatial patterns in the expression of genes related to Th1 (pro-inflammatory) responses and immune modulation, which suggest differential exposure to pathogens and contaminants across the species' breeding distribution, even though all of the sea lions sampled were clinically healthy (Peñín et al. 2019). Similar evidence of pre-clinical physiological alterations based on gene expression assays has been reported for species such as killer whales (Sitt et al. 2016), bottlenose dolphins (Van Dolah et al. 2015; Morey et al. 2016; Neely et al. 2018), Indo-Pacific humpback dolphins, *Sousa chinensis* (Gui et al. 2013), and sea otters, *Enhydra lutris* (Bowen et al. 2012; Miles et al. 2012).

Beyond candidate-marker designed real time quantitative PCR (RT-qPCR) assays, functional genomics approaches allow the simultaneous quantitation of the expression of all gene transcripts in a given tissue, which can be very useful to study different physiological processes on a wide ecological scale (Gracey and Cossins 2003). Different methods, which vary in terms of storage requirements, cost, reproducibility and amount of work demanded, are available for this purpose. These methods, which include RNAseq, cDNA subtraction, serial analysis of gene expression, and microarrays have been reviewed extensively elsewhere (Thomas and Klaper 2004; Acevedo-Whitehouse and Bowen 2018). Thankfully, what was initially fraught with difficulties due to the lack of assembled reference genomes is changing, and currently, various transcriptome assemblies are available for different tissues of a number of marine mammal species, including the Northern elephant seal (Khudyakov et al. 2015), the ringed seal, *Pusa hispida* (Brown et al. 2017), and the bottlenose dolphin (Neely et al. 2018). These resources are ideal to complement other techniques, such as protein expression studies, hormonal assays, stable isotope analysis, molecular scatolology, pathogen detection, and in vivo or in vitro immune challenges, etc. to enable more integrative studies on the physiological status and tipping points of a population or species challenged by environmental stressors (see Fig. 6.1).

3 Priorities for the Future

We are only just beginning to grasp the subtle ways in which even apparently 'minor' changes can impact the physiology of an organism and push individuals towards suboptimal performance. Studies conducted in other taxonomic groups have shown that even slight shifts in the gut microbiota (i.e. dysbiosis) during critical periods of development can have long-term effects on various traits related to fitness (Warne et al. 2019), and exposure to different environmental stressors shapes the allocation of resources between development and immunity, altering the susceptibility of individuals to the stressors, impacting mass, lipid reserves and oxygen consumption (Kirschman et al. 2017a) and increasing their risk of disease (Kirschman et al. 2017b). We need to expand the range and scope of studies in marine mammals to attain a better understanding of subtle changes in physiology that could be relevant for determining the ability of a population or species to persist in the face of environmental change, and to determine which is the physiological threshold at which harmful effects will become evident. When conducting such studies, we must not ignore the importance of assessing genetic factors, given the known importance of reduced genetic variation of an individual on its physiological performance under different environmental scenarios. One of the major challenges to marine mammal populations is adapting to quickly-changing environments within an appropriate time frame.

To conclude, the rate of environmental changes across our planet's ecosystems, as well as the current extinction rate, is staggering. If the biological consequences of such changes are not fully understood and the knowledge used to identify vulnerable populations or species (Coria-Galindo et al. 2009), it is likely that the number of endangered, threatened and extinct species will continue to increase in our lifetime. Given the current status of a large proportion of marine mammal populations, conservation programs and studies that aim to understand the biological impact of global environmental change at the organism- and population-level would benefit greatly from including concepts, tools, and applications of Conservation Physiology. The examples, references and discussions presented here are meant to interest the readers in conducting research in conservation physiology, and understanding of the importance of physiological thresholds for assessing the impact of global environmental change. The culmination of these efforts would be a more complete understanding of the effects that different drivers of environmental change can have on marine mammal health, reproductive success, and ultimately, the viability and persistence of their populations, which is dependent on the ability of animals to adapt to altered environments as well as to sudden and unexpected environmental changes. Detecting subtle deviations from health that occur when individuals shift to a suboptimal physiological status could provide an early warning system for populations and species at risk from changes in the ocean ecosystem. Evidently, a baseline will need to be established, but exciting opportunities await those interested in understanding physiological impact of global change.

Acknowledgments Fernando Elorriaga-Verplancken and Lizabeth Bowen kindly reviewed this chapter and made insightful comments, all of which were incorporated in the final version. The author is funded by *CONACYT Ciencia Básica* Program CB 2017–2018 (Research Grant A1-S-16417). The funding organization had no role in the topic, decision to publish, or preparation of the manuscript.

References

Acevedo-Whitehouse K, Bowen L. Genetics. In: Dierauf L, Gulland FMD, editors. RC handbook of marine mammal medicine: health, disease and rehabilitation. 3rd ed. Boca Raton: CRC Press; 2018. p. 231–48.

Acevedo-Whitehouse K, Cunningham AA. Is MHC enough for understanding wildlife immunogenetics? Trends Ecol Evol. 2006;21:433–8.

Acevedo-Whitehouse K, Duffus ALJ. Effects of environmental change on wildlife health. Philos Trans R Soc Lond B Biol Sci. 2009;364(1534):3429–38.

Acevedo-Whitehouse K, Gulland F, Greig D, Amos W. Inbreeding: disease susceptibility in California sea lions. Nature. 2003;422(6927):35.

Acevedo-Whitehouse K, Spraker TR, Lyons E, Melin SR, Gulland F, Delong RL, et al. Contrasting effects of heterozygosity on survival and hookworm resistance in California sea lion pups. Mol Ecol. 2006;15(7):1973–82.

Acevedo-Whitehouse K, Rocha-Gosselin A, Gendron D. A novel non-invasive tool for disease surveillance of free-ranging whales and its relevance to conservation programs. Anim Conserv. 2010;13(2):217–25.

Alava JJ, Palomera C, Bendell L, Ross PS. Pollution as a threat for the conservation of the Galapagos marine reserve: environmental impacts and management perspectives. In: Denkinger J, Vinueza L, editors. The Galapagos marine reserve: a dynamic socio-ecological system. New York: Springer; 2014. p. 247–83.

Altimiras J, Anderson WG. Ecophysiology methods: refining the old, validating the new and developing for the future. Comp Biochem Physiol A Mol Integr Physiol. 2016;202:1–2.

Apprill A, Miller CA, Moore MJ, Durban JW, Fearnbach H, Barrett-Lennard LG. Extensive core microbiome in drone-captured whale blow supports a framework for health monitoring. mSystems. 2017;2(5):e00119–7.

Banuet-Martínez M, Espinosa-de-Aquino W, Elorriaga-Verplancken FR, Flores-Morán A, García OP, Camacho M, et al. Climatic anomaly affects the immune competence of California sea lions. PLoS One. 2017;12(6):0179359.

Baris HN, Barnes-Kedar I, Toledano H, Halpern M, Hershkovitz D, Lossos A, et al. Constitutional mismatch repair deficiency in Israel: high proportion of founder mutations in MMR genes and consanguinity. Pediatr Blood Cancer. 2016;6(3):418–27.

Barnosky AD, Matzke N, Tomiya S, Wogan GO, Swartz B, Quental TB, et al. Has the earth's sixth mass extinction already arrived? Nature. 2011;471(7336):51–7.

Bean K, Amos W, Pomeroy PP, Twiss SD, Coulson TN, Boyd IL. Patterns of parental relatedness and pup survival in the grey seal (*Halichoerus grypus*). Mol Ecol. 2004;13(8):2365–70.

Bellard C, Cassey P, Blackburn TM. Alien species as a driver of recent extinctions. Biol Lett. 2016;12:20150623.

Berta A, Sumich JL, Kovacs KM. Marine mammals: evolutionary biology. 3rd ed. Amsterdam: Academic Press; 2015.

Bijlsma R, Loeschcke V. Genetic erosion impedes adaptive responses to stressful environments. Evol Appl. 2012;5(2):117–29.

Block BA, Jonsen ID, Jorgensen SJ, Winship AJ, Shaffer SA, Bograd SJ, et al. Tracking apex marine predator movements in a dynamic ocean. Nature. 2011;475(7354):86–90.

Botero CA, Weissing FJ, Wright J, Rubenstein DR. Evolutionary tipping points in the capacity to adapt to environmental change. Proc Natl Acad Sci. 2014;112(1):184–9.

Bowen L, Miles AK, Murray M, Haulena M, Tuttle J, van Bonn W, et al. Gene transcription in sea otters (*Enhydra lutris*); development of a diagnostic tool for sea otter and ecosystem health. Mol Ecol Resour. 2012;12:67–74.

Brock PM, Goodman SJ, Hall AJ, Cruz M, Acevedo-Whitehouse K. Context-dependent associations between heterozygosity and immune variation in a wild carnivore. BMC Evol Biol. 2015;15(242):1–10.

Brown TM, Hammond SA, Behsaz B, Veldhoen N, Birol I, Helbing CC. De novo assembly of the ringed seal (*Pusa hispida*) blubber transcriptome: a tool that enables identification of molecular health indicators associated with PCB exposure. Aquat Toxicol. 2017;185:48–57.

Burek KA, Gulland FMD, O'Hara TM. Effects of climate change on arctic ecosystems. Ecol Appl. 2008;18(2):S126–34.

Burgess EA, Hunt KE, Kraus SD, Rolland RM. Quantifying hormones in exhaled breath for physiological assessment of large whales at sea. Sci Rep. 2018;8(1):10031.

Casas-Marce M, Soriano L, López-Bao JV, Godoy JA. Genetics at the verge of extinction: insights from the Iberian lynx. Mol Ecol. 2013;22(22):5503–15.

Castellini MA, Castellini JM. Defining the limits of diving biochemistry in marine mammals. Comp Biochem Physiol B Biochem Mol Biol. 2004;139(3):509–18.

Castellini MA, Rivera PM, Castellini JM. Biochemical aspects of pressure tolerance in marine mammals. Comp Biochem Physiol A Mol Integr Physiol. 2002;133(3):893–9.

Cavieres G, Alruiz JM, Medina NR, Bogdanovich JM, Bozinovic F. Transgenerational and within-generation plasticity shape thermal performance curves. Ecol Evol. 2019;9(4):2072–82.

Chambault P, Albertsen CM, Patterson TA, Hansen RG, Tervo O, Laidre KL, et al. Sea surface temperature predicts the movements of an Arctic cetacean: the bowhead whale. Sci Rep. 2018;8(1):9658.

Chapman JR, Nakagawa S, Coltman DW, Slate J, Sheldon BC. A quantitative review of heterozygosity-fitness correlations in animal populations. Mol Ecol. 2009;18(13):2746–65.

Chapman JR, Hellgren O, Helin AS, Kraus RH, Cromie RL, Waldenström J. The evolution of innate immune genes: purifying and balancing selection on β-defensins in waterfowl. Mol Biol Evol. 2016;33(12):3075–87.

Charlesworth B, Charlesworth D. The genetic basis of inbreeding depression. Genet Res. 1999;74(3):329–40.

Chevin LM, Hoffmann AA. Evolution of phenotypic plasticity in extreme environments. Philos Trans R Soc Lond B Biol Sci. 2017;372(1723):20160138.

Clausius E, Mcmahon CR, Hindell MA. Five decades on: use of historical weaning size data reveals that a decrease in maternal foraging success underpins the long-term decline in population of southern elephant seals (*Mirounga leonina*). PLoS One. 2017;12(3):e0173427.

Coltman DW, Slate J. Microsatellite measures of inbreeding: a meta-analysis. Evolution. 2003;57(5):971–83.

Coltman DW, Bowen WD, Wright JM. Birth weight and neonatal survival of harbour seal pups are positively correlated with genetic variation measured by microsatellites. Proc R Soc B Biol Sci. 1998;265(1398):803–9.

Cooke SJ, Sack L, Franklin CE, Farrell AP, Beardall J, Wikelski M, et al. What is conservation physiology? Perspectives on an increasingly integrated and essential science. Conserv Physiol. 2013;1(1) https://doi.org/10.1093/conphys/cot001.

Coria-Galindo E, Rangel-Huerta E, Verdugo-Rodríguez A, Brousset D, Salazar S, Padilla-Noriega L. Rotavirus infections in Galapagos sea lions. J Wildl Dis. 2009;45(3):722–8.

Cornell LH, Asper ED. Census up-date: captive marine mammals in North America. Int Zoo Yearb. 2007;18:220–4.

Costa DP, Sinervo B. Field physiology: physiological insights from animals in nature. Annu Rev Physiol. 2004;66(1):209–38.

Côté IM, Darling ES, Brown CJ. Interactions among ecosystem stressors and their importance in conservation. Proc R Soc B Biol Sci. 2016;283(1824):1–9.

Dakos V, Matthews B, Hendry A, Levine J, Loeuille N, Norberg J, et al. Ecosystem tipping points in an evolving world. Nat Ecol Evol. 2018;3(3):355–62.

Davidson AD, Boyer AG, Kim H, Pompa-Mansilla S, Hamilton MJ, Costa DP, et al. Drivers and hotspots of extinction risk in marine mammals. Proc Natl Acad Sci. 2012;109(9):3395–400.

Davidson AD, Shoemaker KT, Weinstein B, Costa GC, Brooks TM, Ceballos G, et al. Geography of current and future global mammal extinction risk. PLoS One. 2017;12(11):e0186934.

Davis M, Faurby S, Svenning J-C. Mammal diversity will take millions of years to recover from the current biodiversity crisis. Proc Natl Acad Sci. 2018;115(44):11262–7.

Denkinger J, Gordillo L, Montero-Serra L, Murillo JC, Guevara N, Hirschfeld M, et al. Urban life of Galapagos sea lions (*Zalophus wollebaeki*) on San Cristobal Island, Ecuador: colony trends and threats. J Sea Res. 2015;105:10–4.

Desforges JP, Levin M, Jasperse L, De Guise S, Eulaers I, Letcher RJ, et al. Effects of polar bear and killer whale derived contaminant cocktails on marine mammal immunity. Environ Sci Technol. 2017;51(19):11431–9.

Desforges JP, Hall A, McConnell B, Rosing-Asvid A, Barber JL, Brownlow A, et al. Predicting global killer whale population collapse from PCB pollution. Science. 2018;361(6409):1373–6.

Di Guardo G, Centelleghe C, Mazzariol S. Cetacean host-pathogen interaction(s): critical knowledge gaps. Front Immunol. 2018;9:2815.

Donelson JM, Salinas S, Munday PL, Shama LNS. Transgenerational plasticity and climate change experiments: where do we go from here? Glob Chang Biol. 2018;24(1):13–34.

Eirin-Lopez JM, Putnam HM. Marine environmental epigenetics. Ann Rev Mar Sci. 2018;11(1):335–68.

Elorriaga-Verplancken FR, Sierra-Rodríguez GE, Rosales-Nanduca H, Acevedo-Whitehouse K, Sandoval-Sierra J. Impact of the 2015 El Niño-southern oscillation on the abundance and foraging habits of Guadalupe fur seals and California sea lions from the San Benito Archipelago, Mexico. PLoS One. 2016;11(5):e0155034.

Espinosa-de Aquino W, Olvera-Ramírez A, Arellano-Carbajal F, Lanz-Mendoza H, Villagrán-Herrera E, Acevedo-Whitehouse K. Protein and RNA extraction from mucosal swabs: a minimally invasive source of ecological data for studies of natural populations. Methods Ecol Evol. 2017;8(3):370–8.

Evans TG, Hofmann GE. Defining the limits of physiological plasticity: how gene expression can assess and predict the consequences of ocean change. Philos Trans R Soc Lond B Biol Sci. 2012;367(1596):1733–45.

Fareed M, Afzal M. Genetics of consanguinity and inbreeding in health and disease. Ann Hum Biol. 2017;44(2):99–107.

Flores-Morán A, Banuet-Martínez M, Elorriaga-Verplancken FR, García-Ortuño LE, Sandoval-Sierra J, Acevedo-Whitehouse K. Atypical red blood cells are prevalent in California sea lion pups born during anomalous sea surface temperature events. Physiol Biochem Zool. 2017;90(5):564–74.

Forcada J, Hoffman JI. Climate change selects for heterozygosity in a declining fur seal population. Nature. 2014;511(7510):462–5.

Gibbin EM, Massamba N'Siala G, Chakravarti LJ, Jarrold MD, Calosi P. The evolution of phenotypic plasticity under global change. Sci Rep. 2017;7:17253.

Gilbert AT, Fooks AR, Hayman DT, Horton DL, Müller T, Plowright R, et al. Deciphering serology to understand the ecology of infectious diseases in wildlife. Ecohealth. 2013;10(3):298–313.

Gracey AY, Cossins AR. Application of microarray technology in environmental and comparative physiology. Annu Rev Physiol. 2003;65:231–59.

Gui D, Jia K, Xia J, Yang L, Chen J, Wu Y, et al. *De novo* assembly of the Indo-Pacific humpback dolphin leucocyte transcriptome to identify putative genes involved in the aquatic adaptation and immune response. PLoS One. 2013;8(8):e72417.

Halpern BS, Walbridge S, Selkoe KA, Kappel CV, Micheli F, D'Agrosa C, et al. A global map of human impact on marine ecosystems. Science. 2008;319:948–52.

Hamilton MJ, Davidson AD, Sibly RM, Brown JH. Universal scaling of production rates across mammalian lineages. Proc R Soc B Biol Sci. 2011;278(1705):560–6.

Hamilton CD, Vacquié-Garcia J, Kovacs KM, Ims RA, Kohler J, Lydersen C. Contrasting changes in space use induced by climate change in two Arctic marine mammal species. Biol Lett. 2019;15(3):20180834.

Herman JJ, Spencer HG, Donohue K, Sultan SE. How stable "should" epigenetic modifications be? Insights from adaptive plasticity and bet hedging. Evolution. 2014;68(3):632–43.

Hoegh-Guldberg O, Bruno JF. The impact of climate change on the world's marine ecosystems. Science. 2010;328:1523–9.

Hoelzel AR, Natoli A, Dahlheim ME, Olavarria C, Baird RW, Black NA. Low worldwide genetic diversity in the killer whale (*Orcinus orca*): implications for demographic history. Proc R Soc B Biol Sci. 2002;269(1499):1467–73.

Hoffman JI, Boyd JANL, Amosl W. Exploring the relationship between parental relatedness and male reproductive success in the Antarctic fur seal *Arctocephalus gazelle*. Evolution. 2004;58(9):2087–99.

Hoffmann AA, Willi Y. Detecting genetic responses to environmental change. Nat Rev Genet. 2008;9(6):421–32.

Hofmann GE, Todgham AE. Living in the now: physiological mechanisms to tolerate a rapidly changing environment. Annu Rev Physiol. 2010;72(1):127–45.

Holt WV, Brown JL, Comizzoli P. Reproductive sciences in animal conservation. Adv Exp Med Biol. 2014;753:3–14.

Hunt KE, Moore MJ, Rolland RM, Kellar NM, Hall AJ, Kershaw J, et al. Overcoming the challenges of studying conservation physiology in large whales: a review of available methods. Conserv Physiol. 2013;1(1):cot006.

Hunt KE, Stimmelmayr R, George C, Hanns C, Suydam R, Brower H, et al. Baleen hormones: a novel tool for retrospective assessment of stress and reproduction in bowhead whales (*Balaena mysticetus*). Conserv Physiol. 2014;2(1):cou030.

Hunt KE, Rolland RM, Kraus SD. Conservation physiology of an uncatchable animal: the North Atlantic right whale (*Eubalaena glacialis*). Integr Comp Biol. 2015;55(4):577–86.

James TY, Toledo LF, Rödder D, da Silva Leite D, Belasen AM, Betancourt-Román CM, et al. Disentangling host, pathogen, and environmental determinants of a recently emerged wildlife disease: lessons from the first 15 years of amphibian chytridiomycosis research. Ecol Evol. 2015;5(18):4079–97.

Jaramillo-Legorreta A, Cardenas-Hinojosa G, Nieto-Garcia E, Rojas-Bracho L, Ver Hoef J, Moore J, et al. Passive acoustic monitoring of the decline of Mexico's critically endangered vaquita. Conserv Biol. 2016;31(1):183–91.

Johnstone CP, Lill A, Reina RD. Use of erythrocyte indicators of health and condition in vertebrate ecophysiology: a review and appraisal. Biol Rev Camb Philos Soc. 2017;92(1):150–68.

Juárez-Ruiz A, Elorriaga-Verplancken FR, Moreno-Sánchez XG, Aguíniga-García S, Amador-Capitanachi MJ, Gálvez C. Diversification of foraging habits among Guadalupe fur seals from their only well-established breeding colony, Guadalupe Island, Mexico. Mar Biol. 2018;165:86.

Kershaw JL, Botting CH, Brownlow A, Hall AJ. Not just fat: investigating the proteome of cetacean blubber tissue. Conserv Physiol. 2018;6(1):coy003.

Keyghobadi N. The genetic implications of habitat fragmentation for animals. Can J Zool. 2007;85(10):1049–64.

Khudyakov JI, Preeyanon L, Champagne CD, Ortiz RM, Crocker DE. Transcriptome analysis of northern elephant seal (*Mirounga angustirostris*) muscle tissue provides a novel molecular resource and physiological insights. BMC Genomics. 2015;16:64.

Kilikevicius A, Venckunas T, Zelniene R, Carroll AM, Lionikaite S, Ratkevicius A, et al. Divergent physiological characteristics and responses to endurance training among inbred mouse strains. Scand J Med Sci Sports. 2013;23(5):657–68.

Kirschman LJ, McCue MD, Boyles JG, Warne RW. Exogenous stress hormones alter energetic and nutrient costs of development and metamorphosis. J Exp Biol. 2017a;220(18):3391–7.

Kirschman LJ, Crespi EJ, Warne RW. Critical disease windows shaped by stress exposure alter allocation trade-offs between development and immunity. J Anim Ecol. 2017b;87(1):235–46.

Kleiman DG. Reintroduction programs. In: Wild mammals in captivity. Chicago: University of Chicago Press; 1996. p. 297–305.

Kloch A, Wenzel MA, Laetsch DR, Michalski O, Welc-Falęciak R, Piertney SB. Signatures of balancing selection in toll-like receptor (TLRs) genes—novel insights from a free-living rodent. Sci Rep. 2018;8(1):8361.

Kretzmann M, Mentzer L, DiGiovanni R, Leslie MS, Amato G. Microsatellite diversity and fitness in stranded juvenile harp seals (*Phoca groenlandica*). J Hered. 2006;97(6):555–60.

Kuhn CE, Costa DP. Interannual variation in the at-sea behavior of California sea lions (*Zalophus californianus*). Mar Mamm Sci. 2014;30:1297–319.

Laidre KL, Stern H, Kovacs KM, Lowry L, Moore SE, Regehr EV, et al. Arctic marine mammal population status, sea ice habitat loss, and conservation recommendations for the 21st century. Conserv Biol. 2015;29(3):724–37.

Le Quesne WJF, Pinnegar JK. The potential impacts of ocean acidification: scaling from physiology to fisheries. Fish Fish. 2012;12:333–44.

Leaper R, Cooke J, Trathan P, Reid K, Rowntree V, Payne R. Global climate drives southern right whale (*Eubalaena australis*) population dynamics. Biol Lett. 2006;2(2):289–92.

Leroy G, Carroll EL, Bruford MW, DeWoody JA, Strand A, Waits L, et al. Next-generation metrics for monitoring genetic erosion within populations of conservation concern. Evol Appl. 2017;11(7):1066–83.

Letcher RJ, Bustnes JO, Dietz R, Jenssen BM, Jørgensen EH, Sonne C, et al. Exposure and effects assessment of persistent organohalogen contaminants in arctic wildlife and fish. Sci Total Environ. 2010;408(15):2995–3043.

Lotze HK, Coll M, Magera AM, Ward-Paige C, Airoldi L. Recovery of marine animal populations and ecosystems. Trends Ecol Evol. 2011;26(11):595–605.
Luo Z, Jiang Z, Tang S. Impacts of climate change on distributions and diversity of ungulates on the Tibetan Plateau. Ecol Appl. 2015;25(1):24–38.
Lurie IW, Prytkov A, Meldere L. Meckel syndrome in different populations. Am J Med Genet. 1984;18(4):661–9.
Machalaba CM, Karesh WB. Emerging infectious disease risk: shared drivers with environmental change. Rev Sci Tech. 2018;36(2):435–44.
Madliger CL, Love OP, Hultine KR, Cooke SJ. The conservation physiology toolbox: status and opportunities. Conserv Physiol. 2018;6(1):coy029.
Matesanz S, Gianoli E, Valladares F. Global change and the evolution of phenotypic plasticity in plants. Ann N Y Acad Sci. 2010;1206:35–55.
McDonald PJ, Nano CEM, Ward SJ, Stewart A, Pavey CR, Luck GW, et al. Habitat as a mediator of mesopredator-driven mammal extinction. Conserv Biol. 2017;31(5):1183–91.
McMahon CR, Harcourt RG, Burton HR, Daniel O, Hindell MA. Seal mothers expend more on offspring under favourable conditions and less when resources are limited. J Anim Ecol. 2017;86(2):359–70.
McNew L. The nature, origin, and evolution of parasitism. In: Horsfall JG, Dimond AE, editors. Plant pathology—an advanced treatise. New York: Academic Press; 1960. p. 19–69.
Metcalfe JD, le Quesne WJF, Cheung WWL, Righton DA. Conservation physiology for applied management of marine fish: an overview with perspectives on the role and value of telemetry. Philos Trans R Soc Lond B Biol Sci. 2012;367(1596):1746–56.
Miles AK, Bowen L, Ballachey B, Bodkin JL, Murray M, Estes JL, et al. Variations of transcript profiles between sea otters *Enhydra lutris* from Prince William Sound, Alaska, and clinically normal reference otters. Mar Ecol Prog Ser. 2012;451:201–12.
Monaco CJ, Helmuth B. Tipping points, thresholds and the keystone role of physiology in marine climate change research. Adv Mar Biol. 2011;60:123–60.
Monteiro WP, Veiga JC, Silva AR, Carvalho C d S, Lanes ÉCM, Rico Y, et al. Everything you always wanted to know about gene flow in tropical landscapes (but were afraid to ask). PeerJ. 2019;7:e6446.
Moore SE. Marine mammals as ecosystem sentinels. J Mammal. 2008;89(3):534–40.
Morell V. Conservation biology: can the vaquita be saved? Science. 2008;321(5890):767.
Morey JS, Neely MG, Lunardi D, Anderson PE, Schwacke LH, Campbell M, et al. RNA-Seq analysis of seasonal and individual variation in blood transcriptomes of healthy managed bottlenose dolphins. BMC Genomics. 2016;17:720.
Moritz C, Agudo R. The future of species under climate. Science. 2013;341(6154):504–8.
Morueta-Holme N, Fløjgaard C, Svenning JC. Climate change risks and conservation implications for a threatened small-range mammal species. PLoS One. 2010;5(4):e10360.
Moura AE, Kenny JG, Chaudhuri R, Hughes MA, Welch A J, Reisinger RR, et al. Population genomics of the killer whale indicates ecotype evolution in sympatry involving both selection and drift. Mol Ecol. 2014;23(21):5179–92.
Muletz-Wolz CR, Barnett SE, DiRenzo GV, Zamudio KR, Toledo LF, James TY, et al. Diverse genotypes of the amphibian-killing fungus produce distinct phenotypes through plastic responses to temperature. J Evol Biol. 2019; https://doi.org/10.1111/jeb.13413.
Neely MG, Morey JS, Anderson P, Balmer BC, Ylitalo GM, Zolman ES, et al. Skin transcriptomes of common bottlenose dolphins (*Tursiops truncatus*) from the northern Gulf of Mexico and southeastern U.S. Atlantic coasts. Mar Genomics. 2018;38:45–58.
NOAA. Centers for Environmental Information, State of the Climate: global climate report for annual 2017; 2017. Published online Jan 2018; Accessed 19 Dec 2018.
Nussey DH, Wilson AJ, Brommer JE. The evolutionary ecology of individual phenotypic plasticity in wild populations. J Evol Biol. 2007;20(3):831–44.
O'Brien SJ, Evermann JF. Interactive influence of infectious disease and genetic diversity in natural populations. Trends Ecol Evol. 1988;3(10):254–9.

O'Corry-Crowe G. Climate change and the molecular ecology of Arctic marine mammals. Ecol Appl. 2008;18(2):S56–76.

Oostra V, Saastamoinen M, Zwaan BJ, Wheat CW. Strong phenotypic plasticity limits potential for evolutionary responses to climate change. Nat Commun. 2018;9(1):1005.

Parente CL, Araújo JP, Araújo ME. Diversity of cetaceans as tool in monitoring environmental impacts of seismic surveys. Biota Neotrop. 2007;7(1):49–56.

Pauls SU, Nowak C, Bálint M, Pfenninger M. The impact of global climate change on genetic diversity within populations and species. Mol Ecol. 2013;22(4):925–46.

Peñín I, Figueroa-Cabañas ME, Guerrero-de la Rosa F, Soto-García LA, Álvarez-Martínez R, Flores-Morán A, et al. Transcriptional profiles of California sea lion peripheral NK and CD+8 T cells reflect ecological regionalization and infection by oncogenic viruses. Front Immunol. 2019;10:413.

Piersma T, Drent J. Phenotypic flexibility and the evolution of organismal design. Trends Ecol Evol. 2003;18(5):228–33.

Pomeroy P. Reproductive cycles of marine mammals. Anim Reprod Sci. 2011;124(3–4):184–93.

Ratajczak Z, Carpenter SR, Ives AR, Kucharik CJ, Ramiadantsoa T, Stegner MA, et al. Abrupt change in ecological systems: inference and diagnosis. Trends Ecol Evol. 2018;33(7):513–26.

Redding DW, Mooers AO. Ranking mammal species for conservation and the loss of both phylogenetic and trait diversity. PLoS One. 2015;10(12):e0141435.

Regehr EV, Laidre KL, Akcakaya HR, Amstrup SC, Atwood TC, Lunn NJ, et al. Conservation status of polar bears (*Ursus maritimus*) in relation to projected sea-ice declines. Biol Lett. 2016;12(12):20160556.

Relyea RA. Costs of phenotypic plasticity. Am Nat. 2017;159(3):272.

Richard JT, Robeck TR, Osborn SD, Naples L, McDermott A, LaForge R, et al. Testosterone and progesterone concentrations in blow samples are biologically relevant in belugas (*Delphinapterus leucas*). Gen Comp Endocrinol. 2017;246:183–93.

Rijks JM, Hoffman JI, Kuiken T, Osterhaus ADME, Amos W. Heterozygosity and lungworm burden in harbour seals (*Phoca vitulina*). Heredity. 2008;100(6):587–93.

Robeck TR, Nollens HH. Hematological and serum biochemical analytes reflect physiological challenges during gestation and lactation in killer whales (Orcinus orca). Zoo Biol. 2013;32(5):497–509.

Rode KD, Regehr EV, Douglas DC, Durner G, Derocher AE, Thiemann GW, et al. Variation in the response of an Arctic top predator experiencing habitat loss: feeding and reproductive ecology of two polar bear populations. Glob Chang Biol. 2014;20(1):76–88.

Rode KD, Wilson RR, Regehr EV, St Martin M, Douglas DC, Olson J. Increased land use by Chukchi Sea polar bears in relation to changing sea ice conditions. PLoS One. 2015;10(11):e0142213.

Roff DA. Life history evolution. Oxford: Oxford University Press; 2001.

Sánchez-Lugo A, Morice C, Berrisford P, Argüez A. Temperature [in state of the climate in 2017]. Bull Am Meteorol Soc. 2018;99(8):S11–3.

Sandom C, Faurby S, Sandel B, Svenning J-C. Global late quaternary megafauna extinctions linked to humans, not climate change. Proc R Soc B Biol Sci. 2014;281(1787):20133254.

Santini L, González-Suárez M, Russo D, Gonzalez-Voyer A, von Hardenberg A, Ancillotto L. One strategy does not fit all: determinants of urban adaptation in mammals. Ecol Lett. 2019;22(2):365–76.

Schnitzler JG, Reckendorf A, Pinzone M, Autenrieth M, Tiedemann R, Covaci A, et al. Supporting evidence for PCB pollution threatening global killer whale population. Aquat Toxicol. 2018;206:102–4.

Scholthof KB. The disease triangle: pathogens, the environment and society. Nat Rev Microbiol. 2007;5(2):152–6.

Seebacher F, Franklin CE. Determining environmental causes of biological effects: the need for a mechanistic physiological dimension in conservation biology. Philos Trans R Soc Lond B Biol Sci. 2012;367(1596):1607–14.

Seguel M, Montalva F, Perez-Venegas D, Gutiérrez J, Paves HJ, Müller A, et al. Immune-mediated hookworm clearance and survival of a marine mammal decrease with warmer ocean temperatures. Elife. 2018;7:e38432.

Seyboth E, Groch KR, Dalla Rosa L, Reid K, Flores PA, Secchi ER. Southern right whale (*Eubalaena australis*) reproductive success is influenced by krill (*Euphausia superba*) density and climate. Sci Rep. 2016;6:28205.

Sitt T, Bowen L, Lee CS, Blanchard MT, McBain J, Dold C, Stott JL. Longitudinal evaluation of leukocyte transcripts in killer whales (Orcinus Orca). Vet Immunol Immunopathol. 2016;175:7–15. https://doi.org/10.1016/j.vetimm.2016.04.011.

Snell-Rood E, Cothran R, Espeset A, Jeyasingh P, Hobbie S, Morehouse NI. Life-history evolution in the anthropocene: effects of increasing nutrients on traits and trade-offs. Evol Appl. 2015;8(7):635–49.

Somero GN. Comparative physiology: a "crystal ball" for predicting consequences of global change. Am J Physiol Regul Integr Comp Physiol. 2011;301:R1–14.

Stearns SC. The evolution of life histories. Oxford: Oxford University Press; 1992.

Stoffel MA, Humble E, Paijmans AJ, Acevedo-Whitehouse K, Chilvers BL, Dickerson B, et al. Demographic histories and genetic diversity across pinnipeds are shaped by human exploitation, ecology and life-history. Nat Commun. 2018;9(1):4836.

Sutton JT, Nakagawa S, Robertson BC, Jamieson IG. Disentangling the roles of natural selection and genetic drift in shaping variation at MHC immunity genes. Mol Ecol. 2011;20(21):4408–20.

Szpak P, Buckley M, Darwent CM, Richards MP. Long-term ecological changes in marine mammals driven by recent warming in northwestern Alaska. Glob Chang Biol. 2018;24(1):490–503.

Thomas MA, Klaper R. Genomics for the ecological toolbox. Trends Ecol Evol. 2004;19(8):439–45.

Thompson LA, Spoon TR, Goertz CEC, Hobbs RC, Romano TA. Blow collection as a non-invasive method for measuring cortisol in the beluga (*Delphinapterus leucas*). PLoS One. 2014;9(12):e114062.

Tulloch VJD, Plagányi ÉE, Brown C, Richardson AJ, Matear R. Future recovery of baleen whales is imperiled by climate change. Glob Chang Biol. 2019;25(4):1263–81.

Urban MC, Bocedi G, Hendry AP, Mihoub JB, Pe'er G, Singer A, et al. Improving the forecast for biodiversity under climate change. Science. 2016;353(6304):aad8466.

Valsecchi E, Amos W, Raga JA, Podesta M, Sherwin W. The effects of inbreeding on mortality during a morbillivirus outbreak in the Mediterranean striped dolphin (*Stenella coeruleoalba*). Anim Conserv. 2004;7(2004):139–46.

Van Bressem MF, Raga JA, Di Guardo G, Jepson PD, Duignan PJ, Siebert U, et al. Emerging infectious diseases in cetaceans worldwide and the possible role of environmental stressors. Dis Aquat Organ. 2009;86(2):143–57.

Van Bressem M, Duignan P, Banyard A, Barbieri M, Colegrove K, De Guise S, et al. Cetacean morbillivirus: current knowledge and future directions. Viruses. 2014;6(12):5145–81.

Van Dolah FM, Neely MG, McGeorge LE, Balmer BC, Ylitalo GM, Zolman ES, et al. Seasonal variation in the skin transcriptome of common bottlenose dolphins (*Tursiops truncatus*) from the northern Gulf of Mexico. PLoS One. 2015;10(6):e0130934.

Venesky MD, Mendelson JR, Sears BF, Stiling P, Rohr JR. Selecting for tolerance against pathogens and herbivores to enhance success of reintroduction and translocation. Conserv Biol. 2012;26(4):586–92.

Warne RW, Kirschman L, Zeglin L. Manipulation of gut microbiota during critical developmental windows affect host physiological performance and disease susceptibility across ontogeny. J Anim Ecol. 2019; https://doi.org/10.1111/1365-2656.12973.

Weber DS, Stewart BS, Lehman N. Genetic consequences of a severe population bottleneck in the Guadalupe fur seal (*Arctocephalus townsendi*). J Hered. 2004;95(2):144–53.

Wensveen PJ, Isojunno S, Hansen RR, Von Benda-Beckmann AM, Kleivane L, Van Ijsselmuide S, et al. Northern bottlenose whales in a pristine environment respond strongly to close and distant navy sonar signals. Proc R Soc B Biol Sci. 2019;286:20182592.

Whiteman JP, Harlow HJ, Durner GM, Regehr EV, Amstrup SC, Ben-David M. Heightened immune system function in polar bears using terrestrial habitats. Physiol Biochem Zool. 2018;92(1):1–11.

Wiig Ø, Aars J, Born EW. Effects of climate change on polar bears. Sci Prog. 2008;91(2):151–73.

Wikelski M, Cooke SJ. Conservation physiology. Trends Ecol Evol. 2006;21:38–46.

Wood CL. Environmental change and the ecology of infectious disease. Science. 2014;346(6214):1192.

Worm B, Barbier EB, Beaumont N, Duffy JE, Folke C, Halpern BS, et al. Impacts of biodiversity loss on ocean ecosystem services. Science. 2006;314(5800):787–90.

Younger JL, Emmerson LM, Miller KJ. The influence of historical climate changes on Southern Ocean marine predator populations: a comparative analysis. Glob Chang Biol. 2016;22(2):474–93.

Zera AJ, Harshman LG. The physiology of life history trade-offs in animals. Annu Rev Ecol Syst. 2001;32:95–126.

Chapter 7
The Role of Reproductive Sciences in the Preservation and Breeding of Commercial and Threatened Teleost Fishes

Ian Mayer

Abstract The teleost fishes are the largest and most diverse vertebrate group, accounting for nearly half of all known vertebrate species. Teleost fish exhibit greater species diversity than any other group of vertebrates and this is reflected in the unique variety of different reproductive strategies displayed by fish. Fish have always been an important resource for humans worldwide, especially as food. While wild capture fisheries have historically been the main source of fish, the farming of fish (aquaculture) is increasingly becoming the more dominant source of food fish, and is predicted to account for 60% of total global fish production by 2030.

Fishes are increasingly threatened by a wide range of anthropogenic impacts, including loss of habitat, pollution, invasive species and over-exploitation. In addition, climate change, especially the consequences of global warming, can impact fish at all levels of biological organization from the individual to the population level, influencing both physiological and ecological processes in a variety of direct and indirect ways. As such, there is an urgent need to protect and conserve the huge genetic diversity offered by this diverse vertebrate group, not just as a source of genes for contemporary breeding and for protection against the consequences of climate change and disease, but also as part of our national heritage. While the cryopreservation of reproductive cells is a means of achieving these objectives, currently only fish sperm can be successfully frozen. Due to their large size, large yolk compartment, low membrane permeability and high chilling sensitivity, successful and reproducible protocols for the cryopreservation of fish oocytes and embryos still remains elusive. However, significant advances have been made in the cryopreservation of primordial germ cells as an alternative means of conserving both paternal and maternal genomes. Although more research needs to be carried out on how these cells can be optimally applied to emerging reproductive technologies, including transplantation techniques and surrogate broodstock technologies, the successful

I. Mayer (✉)
Norwegian University of Life Sciences, Faculty of Veterinary Medicine, Oslo, Norway
e-mail: ian.mayer@nmbu.no

P. Comizzoli et al. (eds.), *Reproductive Sciences in Animal Conservation*, Advances in Experimental Medicine and Biology 1200,
https://doi.org/10.1007/978-3-030-23633-5_7

cryopreservation of fish germ cells, and the establishment of genetic resource banks, offers the possibility of both conserving and restoring threatened species. Further, current and future conservation efforts need to consider the impact of climate change in both *in situ* conservation and reintroduction efforts.

In conclusion, it is anticipated that the successful cryopreservation of fish germplasm will result in a range of economic, ecological and societal benefits. In partnership with emerging assisted reproductive technologies, the successful cryopreservation of fish germplasm will lead to more efficient reproduction in aquaculture, assist selective breeding programmes, and be of crucial importance to future species conservation actions.

1 Introduction

Teleost fishes or Teleostei, the largest infraclass in the class Actinopterygii (ray-finned fishes), are the largest and most diverse vertebrate group with nearly 30,000 species, covering about 96% of all extant fish species. The teleost fishes, or more simply fish, are ubiquitous throughout the fresh water and marine environments, and occupy almost every aquatic environment from the highest mountain streams to the ocean depths. The diverse range of habitats occupied by fish is characterised by an amazing range of chemical and physical water properties, including temperature, salinity, oxygen and pressure.

Fish have always been an important resource for humans worldwide, especially as food. While wild capture fisheries have historically been the main source of fish, the farming of fish (aquaculture) is increasingly becoming the more dominant source of food fish. Intensive aquaculture only started in the late 1970s, but since then has been the fastest growing sector of animal production with an annual growth rate of nearly 8%, and now represents not only a vital component of global food security but also helps alleviate poverty in many of the world's poorest regions. By 2016, aquaculture production had risen to 80.0 million tonnes of food fish, representing 47% of total global fish production (FAO 2018). As wild capture fisheries have little capacity for future expansion, the continued growth in fish production is expected to come from aquaculture, with production predicted to reach 109 million tonnes by 2030 (FAO 2018).

Teleost fish exhibit greater species diversity than any other group of vertebrates and this is reflected in the unique variety of different reproductive strategies displayed by fish (Smith and Wootton 2016). While the vast majority of fishes are gonochoristic and display external fertilisation following release of the gametes into the water, a small number of species are viviparous. Compared to other vertebrate groups a significant number of species display hermaphrodism, both protoandry and protogyny. Females can release either pelagic or demersal eggs, and fecundity can range from a few to tens of millions. Elaborate reproductive behaviour is common, and some species even display parental care, both maternal and paternal care. It has been argued that this remarkable reproductive diversity simply reflects the extremely wide range of aquatic habitats that fishes occupy, although other reasons may include the inherent plasticity of fish reproductive systems and the occurrence of whole genome duplication (Smith and Wootton 2016). This remarkable reproductive diversity of fishes has created a number challenges for the aquaculture industry as production regimes, including broodstock management and spawning protocols, have to be developed on a species-to-species basis. In 2016, a total of 369 fish species (including 5 hybrids) were reported to be farmed commercially worldwide, although the 20 most common species accounted for 84.2% of total fish production (FAO 2018).

There is growing concern over the global loss of species biodiversity, and according to the International Union for Conservation of Nature (IUCN; www.iucn.org) more than 27% of all assessed species, covering all animal groups, are now threatened with extinction. Current rates of extinction of vertebrate species have been estimated to be about 1000 times the likely background rate of extinction (Pimm et al. 2014), and as such many argue that we have entered the Earth's sixth mass extinction episode (Ceballos et al. 2017). An increasing number of studies are now reporting the severity of this accelerated extinction process, both in terms of declining population levels (Ceballos and Ehrlich 2002) and global biodiversity (Rockstrom et al. 2009). While most attention has focused on terrestrial species, especially iconic mammalian species, the continued loss in species biodiversity is particularly prevalent amongst aquatic species including the teleost fishes (Clausen and York 2008). Fishes are increasingly threatened by a wide range of anthropogenic impacts, including loss of habitat, pollution and over-exploitation, as well as the consequences of climate change. Undoubtedly, the main factor that threatens fish biodiversity globally, particularly marine species, is over fishing. For example, worldwide over 40 local populations of fish species have gone extinct as a result of over-exploitation. These local losses represent a significant erosion of global biodiversity. Moreover, stresses due to other factors such as climate change, habitat loss, invasive species, eutrophication and pollution can accentuate fishing-induced declines and inhibit or prevent recoveries. In general, freshwater species appear to be more vulnerable than marine species. For example, it was recently reported that mean fish extinction rates of freshwater fishes in Western Europe and the USA, two of the most heavily impacted regions on Earth, during the last 110 years is ~112 times higher than calculated natural extinction rates (Dias et al. 2017).

As of July 2017, the IUCN lists 455 fish species as being critically endangered, including 87 species identified as *possibly extinct*. Additionally 3191 fish species (21% of those evaluated) are listed as *data deficient*, meaning there is insufficient information for a full assessment of conservation status. As these species typically have small distributions and/or populations, they are intrinsically likely to be threatened, according to the IUCN. A good example illustrating the plight of fish populations are the sturgeons (family Acipenseridae), a family of ancient fishes. According to the IUCN, 23 of the 27 species of sturgeon are now classified as at risk of extinction, making them more critically endangered than any other group of species.

Declining fish biodiversity and the threatened status of many of the world's fish stocks highlights the need for the development of new reproductive technologies, both for the advancement of aquaculture, and as a means to preserve future fish biodiversity. Central to these objectives are advancements in germplasm cryopreservation techniques. Germplasm cryobanking will be of critical importance for future biodiversity preservation. Preserving genetic material from endangered fish species will secure the possibility of reconstructing the original strain or population of these species. On a broader note, advancements in germplasm cryopreservation techniques will be also of crucial importance to the continued expansion of aquaculture production, both as a key component of increasingly important assisted reproductive technologies, and through its role in genetic selection programs. While the diverse range of reproductive strategies displayed by fish has important implications for aquaculture it has also created a number challenges in the development of optimal cryopreservation protocols. For example, the extreme reproductive diversity evident in fish species has resulted in significant species differences in gamete (both eggs and sperm) morphology and biology, which generally requires the design of species-specific cryopreservation procedures and protocols.

1.1 Reproductive Dysfunction in Captive Fish

As mentioned above, fish display a unique range of reproductive strategies, and as such effective broodstock management and successful reproduction of any captive stock has to be developed very much on a species-to-species basis. For most fish species, reproduction is a seasonal event, with spawning occurring when environmental conditions are most favourable for the growth and survival of offspring. While sexual maturation is under the control of the endocrine reproductive axis, essentially the hormones of the brain-pituitary-gonad (BPG) axis, the timing of seasonal reproduction is mediated by environmental cues. While photoperiod is undoubtedly the most important environmental cue involved in the timing of seasonal events such as reproduction, other environmental factors can be important including temperature, lunar phase, salinity and water velocity. The majority of fishes occupying temperate regions spawn during the spring-early summer period, when the photoperiod is long (long-day spawners), although some species including most salmonids, spawn during the autumn (short-day spawners).

The sensory and endocrine systems of fishes have evolved to recognize when both external and internal conditions are optimal for the successful completion of reproductive development and spawning. The most important internal signal is nutritional status, that is, has the adult fish acquired sufficient energy reserves for the successful completion of gonadal development. Maturation is energetically very costly, and peripheral hormone signals effectively relay information on the energy status to the central brain regions. If an individual adult fish has sufficient energy reserves to enable the completion of gonadal development, these peripheral signals will activate the BPG-axis leading to the onset of gonadal development (Taranger et al. 2010; Copeland et al. 2011; Trombley and Schmitz 2013). Conversely, under sub-optimal conditions when an individual is likely to have depleted energy reserves, the endocrine system can serve to arrest gonadal development, in essence postponing maturation until favourable conditions return. As such, the sensory and endocrine systems in fishes have evolved to ensure the optimal survival (fitness) of both adults and their offspring.

As in other production animals, an essential prerequisite for the successful propagation of any aquaculture species is having full control of its life cycle, that is, the harvesting of eggs and juveniles following successful reproduction in captive broodstocks. The inability to close the life cycle of any aquaculture species is considered the main constraint for the development of mass production of that species. While the majority of fish species maintained in captivity exhibit some form of reproductive dysfunction, the development and application of a number of reproductive technologies, in particular hormone therapies, has facilitated successful breeding in captive broodstocks, ensuring production of seed. The failure to close the life-cycle only persists in a small number of economically important aquaculture species, such as the European and Japanese eels (Family Anguillidae), and for this reason their production still relies, for the time being, on the capture of wild seed.

1.2 Reasons for Reproductive Dysfunction

The degree of reproductive dysfunction differs greatly between species, and can vary from the total absence of gonadal development to impaired spawning and reduced gamete quality (both eggs and sperm). Under captive conditions, females generally exhibit more prevalent reproductive dysfunction compared to males, with the most common dysfunction being the failure to undergo final oocyte maturation and ovulation. While the reason why fish held in captivity consistently display reproductive dysfunction is undoubtedly due to a multitude of reasons, a number of key causal factors can be identified, including (1) inappropriate environmental conditions, (2) poor or inappropriate nutrition, and (3) lack of behavioural, physiological or pheromonal cues. In captivity, fish are maintained under conditions far different than that experienced in nature, including environmental parameters such as photoperiod, temperature and water quality, as well as confinement and high stocking densities. Fish reared under these unnatural conditions often experience chronic stress, which in turn can lead to impaired growth and reproduction (Schreck et al. 2001). As mentioned above, good nutrition is a critical prerequisite for effective broodstock management. Captive broodstocks have to be fed a diet that is both quantitatively and qualitatively appropriate to enable successful gonadal development (Izquierdo et al. 2001). Finally, fish maintained in captivity are generally unable to express their full range of behaviours, be it migratory behaviour, or more importantly, reproductive behaviour. It is now well established that successful reproduction in many fish species is dependent upon the expression of appropriate behavioural cues. The failure of final maturation and successful spawning (gamete release) in fish broodstocks can be attributed to the lack of male-female interactions (courtship, spawning behaviour, pheromonal cues) under captive conditions

(Fleming and Huntingford 2012). Finally, it should be mentioned that the process of domestication may modify reproductive traits such as behaviour, either inadvertently or as a consequence of selective breeding for desirable production traits.

1.3 *Strategies to Mitigate Reproductive Dysfunction in Captivity*

1.3.1 Photoperiod Manipulation

It is now well established that seasonal changes in photoperiod are primarily responsible for synchronising the timing of reproduction in most temperate species. From an early date, it was demonstrated that photoperiod manipulation, especially in combination with temperature, could be effectively used to advance or delay the spawning period in captive broodstocks, including salmonids (Bromage et al. 2001). Initial strategies relied on either compressing or expanding the seasonal photoperiod cycle, which resulted in either advancing or delaying the spawning season respectively. This development had a major economic impact on aquaculture as it allowed the out-of-season production of eggs. A more recent development has been to replace the use of seasonally changing daylength with periods of constant long and short daylength. This method has now been successfully employed to manipulate the spawning season in a wide range of important aquaculture species, including Atlantic salmon (*Salmo salar*), rainbow trout (*Oncorhynchus mykiss*), sea bass (*Dicentrarchus labrax*), seabream (*Sparus aurata*), Atlantic cod (*Gadus morhua*) and Senegal sole (*Solea senegalensis*). However, a limitation of photoperiod manipulation is that this technique can result in reduced gamete quality, including reduced fecundity and egg size, as well as an increase in the incidence of egg deformities (Carrillo et al. 1989; Penney et al. 2006; Bonnet et al. 2007).

1.3.2 Hormonal Treatment

In many aquaculture species, reproductive constraints persist even after improvements to nutrition and environmental conditions. For these species, successful reproduction in captivity has relied on the application of hormone therapies. The application of appropriate hormone therapies not only allows for successful reproduction, but also facilitates improvements in the reproductive performance of broodstocks. This ability to induce spawning in captive stocks has been a major factor in the successful aquaculture production of several economically important species (Zohar and Mylonas 2001; Mylonas et al. 2010). The benefits of hormonal treatment are twofold. Firstly, hormone therapies can stimulate reproduction in those species that display reproductive dysfunction in captivity, allowing final oocyte maturation (FOM) and successful spawning. Secondly, hormone therapies allow for the synchronisation of spawning, a major advantage in broodstock management.

The benefits conveyed in being able to synchronise broodstock spawning are illustrated by the fact that hormone treatments are even used for some species that do spawn in captivity, such as seabream and sea bass (Barbaro et al. 1997; Fornies et al. 2001). The inhibition of final oocyte maturation and spawning has primarily been related to the inhibition of luteinizing hormone (LH) release from the pituitary (Zohar and Mylonas 2001). To mitigate this endocrine block in the BPG-axis, early hormone therapies relied on the use of gonadotropin (GTH) preparations, initially relying on crude pituitary extracts before advancing on to the use of purified or recombinant GTHs (Zohar and Mylonas 2001). This was followed by a second generation of more effective hormone therapies based on the use of the gonadotropin-releasing hormone (GnRH), the hypothalamic hormone responsible for the regulation of pituitary GTH release. Currently, the most widely used hormone therapy used to stimulate final maturation in aquaculture species is the use of synthetic GnRH analogues (GnRHa), often in combination with a dopamine blocker (Mylonas et al. 2010; Mylonas and Zohar 2001). The main advantage of GnRHa is that, compared to native GnRH, it has increased GTH releasing potency, and is more resistant to enzymatic degradation. Controlled GnRHa administration, either via injection or the implantation of microcapsules, is now routinely used to effectively stimulate FOM and spawning in female broodstocks. While hormone therapies are primarily directed at female broodstocks, these treatments can also be beneficial in the management of male broodstocks, primarily to enhance sperm quality (Mylonas et al. 2017). For example, GnRHa treatment has been shown to increase sperm volume (milt), and also to decrease milt viscosity, which can be a major constraint in collecting semen samples (Vermeirsen et al. 1998; Moon et al. 2003).

1.4 Reproductive Technologies Designed to Enhance Production Performance in Aquaculture Species

A number of innovative technologies have also been developed with the specific aim of improving performance traits in aquaculture species. These include technologies to induce sterilization, to manipulate gender (monosex production), and for the production of transgenic fish. Most of these reproductive technologies are now well established, and will only be briefly covered.

1.4.1 Sterilization

The occurrence of early sexual (precocious) maturation is a major constraint in today's aquaculture industry. Many fish species display alternative male reproductive tactics, characterised by differing phenotypic traits, either maturing at a large size together with pronounced secondary sexual characters, or alternatively maturing precociously at a small size (Taborsky 2008). While the occurrence of alternative male reproductive tactics occurs naturally in a number of fish families, the incidence of precocious maturation, where both size and age at puberty is reduced, is often greatly enhanced under intensive aquaculture conditions. Early sexual maturation has undesired impacts on the health, growth, and product quality of farmed fish (Copeland et al. 2011; Aksnes et al. 1986). In order to avoid these adverse production traits, maturation is commonly delayed by exposing fish to constant light regimes, which have the effect of disrupting natural circannual rhythms, including the timing of puberty (Copeland et al. 2011; Wang et al. 2010). As such, the application of constant light regimes at critical times of the year has effectively been employed to delay sexual maturation in a number of important aquaculture species, including Atlantic salmon, Atlantic cod and sea bass (Taranger et al. 2006; Felip et al. 2008; Leclercq et al. 2011).

A number of alternative strategies have been developed to mitigate early sexual maturation in aquaculture, including induced polyploidy, interspecific hybridization and the application emerging-omics technologies (Liu et al. 2007; Piferrer et al. 2009; Iversen et al. 2016). The term polyploidy refers to the condition where an individual has more than two sets of chromosomes. While triploidy (three sets of chromosomes) can be induced through interspecific hybridization (Liu et al. 2007; Gomelsky 2003), or by crossing diploid with tetraploid fish (Francescon et al. 2004; Nam and Kim 2004), the most cost effective means of inducing triploidy is through physical induction, usually by the application of pressure shock. The application of a short pressure shock to newly fertilized eggs results in the retention of the second polar body, leading to the triploid condition. Although triploids show some gonadal growth they are functionally sterile. The induction of triploidy by pressure shock is routinely 100% effective, and has now been successfully employed to a number of important aquaculture species, including Atlantic salmon, rainbow trout and sea bass (Piferrer et al. 2009; Rasmussen and Morrissey 2007; Felip et al. 1997). However, even though triploidy is an effective way of achieving sterile stocks, there is still some reluctance by the industry to use triploids in preference to diploids. This is because in some species, notably Atlantic salmon, the incidence of some adverse health issues, including skeletal deformities and heart abnormalities are more pronounced in triploid compared to diploid fish (Fraser et al. 2012, 2014; Fjelldal and Hansen 2010). Further, at the time of harvest, triploid salmon display inferior growth compared to diploids (Fraser et al. 2013), although this seems to be related to the dietary phosphate content (Fjelldal et al. 2016).

A further incentive for using triploid fish in aquaculture is to mitigate the growing concerns over the environmental impact of escaped farmed fish. Of particular concern is the threat of escaped farmed Atlantic salmon on wild salmon through genetic introgression (Karlsson et al. 2016). Widespread genetic introgression between farmed and wild Atlantic salmon has recently been documented in most Norwegian salmon rivers, resulting in reduced fitness of these already threatened wild populations (Reed et al. 2015). As such, the widespread adoption of triploid salmon by the aquaculture industry would help mitigate the environmental concerns of genetic introgression between farmed and wild fish (Benfey 2019).

Recently, knockdown/out technologies have been used to target the *dead end* gene (*dnd*) which is a vertebrate-specific gene encoding an RNA-binding protein which is crucial for migration and survival of primordial germ cells (PGCs). In vertebrates, PGCs migrate from their site of specification towards the developing gonads, where they differentiate into the gametes. In fish, targeting *dnd* causes removal of PGCs and ultimately affects sex differentiation and fertility (Baloch et al. 2019). Sterility in various fish species, including Atlantic salmon and sturgeon, can be achieved by knockdown or knockout of *dnd* (Linhartova et al. 2015; Wargelius et al. 2016). However, the gene knockdown and knockout techniques, involving microinjection, will likely prove an impractical method for the mass production of sterile fish. Recently, a novel method has been proposed for the mass production of sterile fish using surrogate broodstock technology (Nagasawa et al. 2018). Using medaka, it was demonstrated that it was possible to mass produce sterile individuals with disrupted follicle-stimulating hormone (FSH) receptor genes by transplanting germ cells from a mutant (sterile) medaka donor into a surrogate parent that carried the wild-type FSH receptor. These transplanted germ cells are nursed by the supporting cells in the gonad of the recipient and eventually produce gametes carrying the mutation. Fertilization of recipient eggs with sperm of a pseudo-male of a FSH-receptor-mutant medaka (masculinized XX genotype medaka), results in XX offspring (and all-sterile females).

1.4.2 Monosex Production

Many aquaculture species display pronounced gender differences in a number of important production traits such as growth rate and age/size of sexual maturation. For this reason, for some aquaculture species it is favourable for production to be based on the cultivation of the more desirable sex, that is, monosex production. For example, most tilapia aquaculture is now based on all-male production because males grow faster than females (Beardmore et al. 2001), while rainbow trout aquaculture favours all-female production because females grow slightly faster and are less prone to early sexual maturation (Sheehan et al. 1999). Several approaches can be used to control gender, depending on the species. The plastic nature of sex determination in fishes, together with the sensitivity of early life stages to sex steroid administration is the basis of most methods employed to control gender (Devlin and Nagahama 2002). Direct sex reversal is well established in aquaculture, where the application of exogenous sex steroids is used to alter the phenotypic sex: larval exposure to either androgens or estrogens results in hormone-induced masculinization or feminization respectively (Piferrer 2001; Beardmore et al. 2001). However, given consumer concerns over hormone use, a more popular procedure involves an indirect method of hormone-induced sex reversal, often in combination with triploidy. For example, all-female triploid rainbow trout are produced by first treating female fry (XX) with androgens to produce neomales (XX), which when crossed with normal diploid females produce all-female progeny, whose eggs can be used to produce all-female triploids following pressure shock treatment of the newly fertilised eggs (Piferrer et al. 2009; Lincoln and Scott 1983).

1.4.3 Transgenic Fish

A transgenic fish is **one that contains genes from another species**. Transgenic fish have been developed using a variety of transgenes, methods, and species since the first transgenic fish was produced in China in 1985 (Zhu et al. 1985). Typically, genes of interest from one or more donor-species are isolated, and spliced into artificially constructed infectious agents, which act as vectors to carry the genes into the cells of recipient species. Transgenic technologies have resulted in the production of novel strains of fish with a number of commercially beneficial traits, typically growth enhancement, but also fish quality and disease resistance (Beardmore and Porter 2003). The development of transgenic fish has mostly focused on the major aquaculture species including salmon, trout, carp, tilapia and catfish (Devlin et al. 1994, 1995; Dunham et al. 1999; Hinits and Moav 1999; Martínez et al. 2000). Improvements in growth performance can be dramatic, for example, transgenic coho salmon given a growth hormone (GH) construct derived from sockeye salmon grew on average 11 times faster than their non-transgenic siblings (Devlin et al. 1994). However, the interest in fast growing transgenic fish for commercial aquaculture is limited due in part to the dramatic improvement in growth performance displayed by many highly domesticated stocks. For example, Devlin et al. (2001) found that transgenic rainbow trout modified to overexpress GH did not grow any faster or larger than a commercial strain of rainbow trout that had undergone traditional selection for fast growth.

To date, the commercialization of transgenic fish is very limited due to public unease over potential ecological, food safety and welfare concerns (Muir and Howard 2002; Aerni 2004; Hallerman et al. 2007). In particular, major concerns over the ecological impact of escaped fish has resulted in the implementation of strict containment procedures, and growing demands for the production of only sterile transgenic fish (Maclean and Laight 2000).

2 State of the Art

While the primary incentive in the development of reproductive technologies, including hormone therapies, has been to optimize aquaculture production through improved breeding efficiency, a new generation of reproductive technologies are now being developed, which are expected to be of critical importance to conservation initiatives aimed at protecting fish biodiversity. Central to these reproductive technologies is germplasm cryobanking. Cryopreservation or cryoconservation is a process where cells or whole tissues are preserved by cooling to sub-zero temperatures (−196 °C) using liquid nitrogen. The cryopreservation of different cellular types, including sperm, oocytes, embryos or germ cells, has now become an indispensable tool for the preservation of genetic resources of a wide range of species. With the help of rapidly evolving reproductive technologies, germplasm cryopreservation has become a key component of genetic selection programs, assisted reproductive technologies (ARTs), and in the preservation of fish biodiversity. The cryobanking of genetic material from endangered fish species will secure the possibility of reconstructing the original strain or population of these species. However, technological advancements in the cryopreservation of fish germplasm, including the validation of effective freeze-thaw protocols, are limited for many cell types, and certainly not as advanced as for mammalian vertebrates. This chapter will review the current state of the art in the cryopreservation of fish germplasm, and detail how these advancements, with the help of rapidly evolving reproductive biotechnologies, can play a key role in the preservation of fish biodiversity.

2.1 Sperm

While research on fish germplasm cryopreservation has been carried out on different cell types, to date the greatest success has been achieved with sperm. Successful cryopreservation of fish sperm has now been reported in over 200 species, most of which are of commercial importance including salmonids, cyprinids and sturgeons. A number of reviews have been published in recent years detailing freezing protocols for both commercial species and small model species such as zebrafish (Suquet et al. 2000; Kopeika et al. 2007; Tiersch et al. 2007, 2011; Asturiano et al. 2017; Martínez-Páramo et al. 2017). The significant advances in the cryopreservation of fish sperm, as compared to other types of germplasm including oocytes and embryos, can be attributed to a number of advantages offered by sperm. Sperm are small in size and have a large surface-to-volume ratio, which facilities the movement of both cryoprotectants and water into and out of the cells. Sperm also possess a highly condensed nucleus and little cytoplasm, making them relatively easy to freeze. In addition, compared to other types of germplasm, sperm from most fish species can be collected in very large numbers and are relatively easy to obtain. Further, for many fish species, motile sperm can be obtained over a much longer time period compared to mature oocytes. For example, in the Senegalese sole (*Solea senegalensis*), which has two defined spawning seasons (spring and autumn), motile sperm can be obtained from captive males over the entire year (Beirão et al. 2011).

Due to the favourable characteristics of sperm, controlled-rate slow freezing has become the standard method of cryopreserving fish sperm. However, the last decade has witnessed increasing interest in preserving fish sperm by vitrification, also known as ice-free cryopreservation (reviewed by Xin et al. 2017; Magnotti et al. 2018). The process of vitrification involves liquid being rapidly transformed into an amorphous, glass-like solid, free of any crystalline structure. To prevent intracellular ice crystal formation, the vitrification process involves very rapid cooling and the use of high concentrations of permeable cryoprotectants. The major advantage of vitrification over slow freezing is that it is low cost and relatively low tech and simple to use, making it a good alternative for use in the field. The main disadvantage of vitrification is that the rapid freezing rates can be difficult to achieve with larger volumes, therefore restricting the process to small volumes (2–4 μl). To date, limited success has been achieved in using vitrification to preserve sperm from a number of important aquaculture species, including Atlantic salmon (Figueroa et al. 2015), rainbow trout (Figueroa et al. 2013), Eurasian perch, *Perca fluviatilis* and European eel, *Anguilla Anguilla* (Kasa et al. 2017), Southern flounder *Paralichthys lethostigma* (Cuevas-Uribe et al. 2017), red snapper, *Lutjanus campechanus*, spotted seatrout, *Cynoscion nebulosus*, and red drum, *Sciaenops ocellatus* (Cuevas-Uribe et al. 2015). However, in view of the fact that vitrification is limited to small volumes its most practical future application may be in the cryopreservation of sperm of small model species (Cuevas-Uribe et al. 2011).

Although the cryopreservation of fish sperm is now a routine procedure for many commercial species an ongoing constraint to the wider application of frozen sperm in the aquaculture industry is the lack of standardization in cryopreservation protocols (Asturiano et al. 2017; Torres et al. 2016). The most likely explanation for this is that the extreme reproductive diversity shown by fish has also resulted in significant species differences in sperm morphology and biology, leading to the necessity of designing species-specific cryopreservation procedures and protocols. Future priorities should include the standardization of cryopreservation procedures, including use of extenders, cryoprotectants and their concentrations, and sperm concentration. Further standardization is also needed for freeze-thaw procedures, such as controlled-rate freezing versus vitrification, thawing system, and use of straws or vials. Finally, greater standardization is needed in the evaluation of sperm quality. While the standard method of evaluating sperm quality is by computer assisted sperm analysis (CASA), there are currently numerous different CASA systems commercially available, leading to difficulties in comparing results obtained by the different systems (Boryshpolets et al. 2013; Amann and Waberski 2014). Overall, the lack of standardization leads to difficulties, both in the replication of cryopreservation procedures and the interpretation of results.

Although the lack of standardization is an ongoing concern, procedures for the evaluation of sperm quality are rapidly evolving, allowing for a better understanding of the underlying processes resulting in reduced sperm quality and viability following cryopreservation procedures. In addition to advances in the capabilities of new CASA systems, advances in a range of new methods are now being developed to assess sperm quality (reviewed by Cabrita et al. 2014), primarily markers of oxidative stress including quantification of ROS, lipid peroxidation, mitochondria dysfunction and DNA damage. As the increase in oxidative stress during cryopreservation procedures is considered the primary cause of cryodamage, the addition of antioxidants to the extender solution has become a common strategy to reduce oxidative damage in frozen sperm (Lahnsteiner et al. 2011; Kutluyer et al. 2014; Liu et al. 2015). However, results have been inconsistent, depending on species, concentration of antioxidant used, and whether they were added in combination. For example, while an earlier study reported that a mixture of 10 antioxidants did not improve post-thaw quality of sperm from brook trout and rainbow trout (Lahnsteiner et al. 2011), improved sperm quality, in terms on sperm motility, was reported in a later study on rainbow trout and red seabream following addition of antioxidants (Kutluyer et al. 2014; Liu et al. 2015).

Advances in proteome analysis has increased our understanding in sperm physiology, including new insights into the causes of cryodamage (Nynca et al. 2015; Zilli and Vilella 2012). Proteome analysis has been demonstrated to be a good indicator of cryopreservation success, and identified proteins represent potential markers of cryodamage. For example, proteome analysis of rainbow trout semen has provided information about specific proteins released from sperm and their relation to sperm structure and function, as a result of cryopreservation (Nynca et al. 2015). Proteome markers have also been found in the seminal plasma of the European perch and sterlet, and changes in the proteome over the spawning season could be correlated to sperm quality traits, including sperm motility (Shaliutina et al. 2012, 2013).

While sperm banking under liquid nitrogen is undoubtedly the most widely used technique for the long-term preservation of fish sperm, as with any germplasm, this confers a number of logistical problems. Storage of sperm samples under liquid nitrogen is very demanding in terms of storage space, equipment, trained personnel, and most importantly the constant need of large quantities of liquid nitrogen. An alternative preservation method, which would alleviate most of these logistical demands, is dry storage following freeze-drying. Freeze-drying is achieved by sublimation of the ice after freezing the sample to subzero temperatures. To date, reports of successful freeze-drying of fish sperm are very limited (Poleo et al. 2005), as this procedure is damaging to the cell membrane, and rehydrated cells are generally devoid of biological activity, motility and viability. Further, fertilization can only be achieved following intracytoplasmic sperm injection (ICSI) of freeze-dried sperm.

2.2 *Oocytes*

The ability to establish cryobanks for the full range of fish germplasm, including sperm, oocytes and embryos, would be a powerful tool for use in aquaculture, reproductive technologies and species conservation. Unfortunately, while the cryopreservation of sperm is now well established for many fish species, the freezing of both oocytes and embryos still remains a major challenge in fish (reviewed by Zhang and Lubzens 2009; Martínez-Páramo et al. 2017). Certainly in comparison with mammals, where significant progress has been made in the cryopreservation of both oocytes and embryos, similar advances in non-mammalian vertebrates (fishes, birds, reptiles and amphibians) are noticeably lacking (Saragusty 2012; Saragusty and Arav 2011). The ability to preserve the maternal genome is especially important for species conservation as several genetic factors are inherited maternally in the oocyte cytoplasm such as mitochondrial DNA, and mRNA that determine the early stages of embryonic development. The reasons why successful cryopreservation of fish oocytes has proven to be so difficult can be attributed to a number of factors.

Firstly, oocytes are considerably larger in fish than in mammals, resulting in a lower surface-to-volume ratio, and thus poorer water and CP movement across the cellular membrane during chilling, freezing and thawing. For example, while the diameter of the mouse and human oocyte is 80 μm and 120 μm respectively, the diameter of the zebrafish oocyte is 750 μm, and that of most salmonids is >5000 μm. The large size of fish oocytes also means that a considerably larger volume of water has to be vitrified, thus increasing the risk for intracellular ice formation and cell damage. Finally, later stage (post-vitellogenic) oocytes contain large quantities of yolk, and the different behaviour of yolk during freezing makes cryopreservation very complex.

Even given the above mentioned difficulties, more attention has focused on attempting to cryopreserve fish oocytes and ovarian follicles, as they are considered better candidates than the larger fish embryos for cryopreservation. While earlier studies focused primarily on slow controlled-rate freezing, in most cases using zebrafish oocytes (Isayeva et al. 2004; Plachinta et al. 2004; Lubzens et al. 2005; Tsai et al. 2009), more recent studies have reported on the vitrification of early stage oocytes (Guan et al. 2010; Godoy et al. 2013). In general, studies on the cryopreservation of fish oocytes indicate that better results have been obtained with early developmental stages. For example, it was demonstrated in zebrafish that early stage ovarian follicles (stages I and II) were less sensitive to chilling injury than late stage ovarian follicles (Tsai et al. 2009). In summary, while some limited success has been reported, the development of successful protocols fish oocyte cryopreservation still remains elusive.

Recently, an alternative approach for preserving the maternal genome has been described involving the air-drying of nuclei (also known as germinal vesicles). In this case, germinal vesicles (GV) were isolated from immature cat oocytes and preserved by air-drying in the presence of the disaccharide trehalose (Graves-Herring et al. 2013). Following storage at 4 °C (up to 32 weeks) the GVs could be successfully revived, including the resumption of meiosis, following microinjection into a fresh conspecific cytoplast (enucleated oocytes). Given the ongoing impasse in the successful cryopreservation of fish oocytes and embryos, this technique could be an alternative method for the long-term preservation of the female genome of threatened fish species (Holt 2013). In practical terms, this methodology would also negate the need for costly freezers or liquid nitrogen containers.

2.3 Embryos

As with oocytes, the successful cryopreservation of fish embryos is a major challenge, and has yet to be achieved (reviewed by Martínez-Páramo et al. 2017; Robles et al. 2009). The ability to cryopreserve fish embryos would be a major benefit to species conservation as frozen embryos would ensure the preservation of both the paternal and maternal genome. Many of the ongoing challenges facing the successful cryopreservation of fish embryos are largely the same as with oocytes. Compared to mammalian embryos, those of fish are considerably larger with a low surface-to-volume ratio, large yolk compartment, and low membrane permeability (Martínez-Páramo et al. 2017; Saragusty and Arav 2011). In addition, fish embryos have at least three membrane structures – the yolk syncytial layer (YSL), plasma membrane of the developing embryo, and the chorion membrane which surrounds the perivitelline space (Saragusty 2012). Each of these membranes has a different permeability coefficient for both water and CPs, resulting in water permeability in the range of one order of magnitude lower in fish embryos compared to mammals – 0.022–0.1 μm × min per atm in zebrafish compared to 0.43 in mice (Hagedorn et al. 1997a). In addition, the different compartments in fish embryos have a different water content and different osmotically inactive water content (Hagedorn et al. 1997b). As a result, fish embryos have a high chilling sensitivity, making cryopreservation extremely complicated and difficult. The high sensitivity to chilling is further compounded by the high lipid content of fish embryos (Robles et al. 2009). The high yolk content of fish embryos increases the likelihood of membrane disruption by intracellular ice formation upon cooling. Studies on yolk removal by micro-aspiration has demonstrated that it is possible to decrease chilling sensitivity in zebrafish embryos by reducing the yolk content (Liu et al. 2001).

The cryopreservation of fish embryos has been studied in over 20 fish species (reviewed by Martínez-Páramo et al. 2017). Although both slow controlled-rate freezing and vitrification have been used for fish embryo cryopreservation, vitrification is now generally considered to be the freezing technique offering the highest probability of success (de Carvalho et al. 2014). To date, embryo survival following vitrification has been reported in zebrafish and two marine species (Chen and Tian 2005; Robles et al. 2005; Martínez-Páramo et al. 2009), although in all cases embryo survival was very limited. A recent review on fish embryo vitrification protocols established that the major aspects to be considered in the development of vitrification protocols are cryoprotectant toxicity, embryo developmental stage, permeability of the cryoprotectant, and vitrification devices (de Carvalho et al. 2014).

As the permeability of the plasma membrane plays a crucial role in the successful cryopreservation of both fish oocytes and embryos, several efforts have been made to improve the permeability of both water and CPs across the plasma membrane, including the application of hydrostatic pressure (Routray et al. 2002), ultrasound (Rahman et al. 2017), electroporation (Rahman et al. 2013), by direct microinjection of cryoprotectants (Janik et al. 2000), and by dechlorination (Hagedorn et al. 1997c). Although most of these procedures resulted in enhanced CP uptake, there was little improvement in cryopreservation success. A more novel approach to overcome the low membrane permeability in both fish oocytes and embryos has been to microinject the aquaporin-3 water channel protein. The microinjection of aquaporin-3 RNA resulted in increased membrane permeability to both water and CPs in both medaka and zebrafish oocytes (Seki et al. 2007; Valdez et al. 2006, 2013), as well as zebrafish embryos (Hagedorn et al. 2002), although again no significant improvement in cryopreservation success was reported.

High chilling sensitivity is one on the main obstacles to the successful cryopreservation of fish embryos. While earlier studies have shown that chilling sensitivity can be addressed by rapid cooling (Hagedorn et al. 1997c), the large size of the yolk compartment of fish embryos impedes rapid cooling and warming, leading to damaging ice crystal formation during cryopreservation. Recently, a novel technique involving laser-assisted gold nanoparticle (GNR) warming has been shown to increase the viability of cryopreserved zebrafish embryos (Khosla et al. 2017). The technique involves microinjecting the embryos with a combination of cryoprotectant and GNRs prior to being rapidly cooled at a rate of 90,000 °C/min in liquid nitrogen. The irradiation of the frozen sample with a laser pulse causes excitation of the GNRs, allowing for rapid warming (1.4×10^7 °C/min), which facilitates the outrunning of ice formation, resulting in reduced embryo damage. Using this technique, viable embryos with consistent structure at 1 h (31%) and continuing development at 3 h (17%) and movement at 24 h (10%) postwarming were reported (Khosla et al. 2017). This technology has broad implications for improving the cryopreservation of germplasm including larger samples of gonadal tissue, not just for teleost fishes but for other vertebrate systems.

Although significant advances have been made, especially in the application of vitrification techniques, successful cryopreservation of fish embryos still remains extremely complicated and often practically impossible. As such, the development of successful and reproducible protocols for the cryopreservation of fish embryos still seems far off.

2.4 *Ovarian Tissue*

From the perspective of aquaculture and species conservation, the cryopreservation of ovarian tissue would be an alternative strategy for the preservation of female genetic material. The continuing lack of success in the cryopreservation of both fish oocytes and embryos has resulted in increased interest in the potential of cryopreserving fish ovarian tissue. However, while the cryopreservation of ovarian tissue has several advantages over oocyte or embryo cryopreservation, it also comes with its own unique challenges. Successful vitrification of ovarian tissue has already been reported in avian and mammalian species (Wang et al. 2008; Liu et al. 2010), opening the possibility of attempting ovarian tissue cryopreservation in fish. Ovarian tissue from fish is available over a long time span, and affords a source of both oocytes at different developmental stages, as well as germ cells. For example, Godoy et al. (2013) studied the vitrification of stage III zebrafish follicles in ovarian fragments, and reported that follicles located in the middle of the fragments were more protected from cryodamage, and some showed good morphological appearance 2 h post-warming. However, the mitochondria integrity of the granulosa cells was clearly damaged by the vitrification process. Following vitrification of zebrafish ovarian fragments, improved survival rates have been reported for immature stage I follicles (Marques et al. 2015). Recently, viable germ cells for use in transplantation studies (see Sect. 2.5 below) have been successfully harvested from cryopreserved whole ovaries of rainbow trout and Siberian sturgeon, *Acipenser baerii*

(Lee et al. 2016, Pšenička et al. 2016). Recently, the successful vitrification of ovarian tissue from a salmonid species has been reported (Lujić et al. 2017). In this study, viable oogonia (up to 40%) were obtained following the vitrification of ovarian tissue from juvenile brown trout (*Salmo trutta*).

Given the ongoing difficulties in cryopreserving both oocytes and embryos in fish, most recent studies have focused on the cryopreservation of ovarian tissue. However, progress has also been made on the cryopreservation of testicular tissue. For example, Lee et al. (2013) previously described the production of functional rainbow trout sperm and eggs following transplantation of type A spermatogonia obtained from cryopreserved whole testes.

2.5 Germ Cells

Germ cells are considered immortal, and have the potential to reproduce indefinitely (Cinalli et al. 2008). During embryonic development in fish, germ cell precursors known as primordial germ cells (PGCs) form at the fringe of the embryo proper and migrate to the genital ridges, where they will later develop into mature germ cells—sperm in males and oocytes in females (Yoshizaki et al. 2002). Transplantation studies suggest that molecular mechanisms that control PCG migration are conserved among most fish species. From an applied science perspective, PGCs are a practical alternative to other reproductive cells for cryobanking as they conserve both the paternal and maternal genome, and due to their small size (c. 15 μm in diameter) and low lipid content, are relatively easy to cryopreserve. As PGCs are the precursors of gametes and have the ability to establish or recover active gametogenesis they are increasingly being used for the cryopreservation of genetic resources and surrogate reproduction via germ-line chimera.

Although germ cell transplantation techniques are relatively well established in mammals, its application to teleost fishes has only advanced during the last 20 years. Several different approaches have now been developed for germ cell transplantation in fish, using recipients at different life history stages, from embryos to sexually mature adults (reviews Lacerda et al. 2013; Robles et al. 2017; de Siqueira-Silva et al. 2018; Yoshizaki and Lee 2018; Yoshizaki and Yazawa 2019). Surrogate broodstock technology consists of producing donor-derived gametes in a surrogate host (recipient individual) by transplanting germ cells of a donor into a recipient of a different strain or species, aiming for the production of heterologous gametes. Given the difficulties of isolating individual PGCs, this technology has benefited by being able to transplant cell suspensions from the testis or ovary containing PGCs, which will become sperm or eggs, respectively, into larvae immediately after hatching. Even though allogeneic or xenogeneic donor cells are transplanted into recipients, rejection is avoided because newly hatched recipient larvae do not have a mature immune system (Okutsu et al. 2007). Following microinjection of the cell suspension into the intraperitoneal cavity only germline stem cells migrate to the recipient's genital ridges for incorporation, while the remaining cells eventually die in the abdominal cavity. In most cases the success of germ cell transplantation procedures depends on the availability of a sterile host because donor cells require a niche within the germinal epithelium to colonize and develop without the competition from host germ cells. Sterilization can be achieved by a number of different procedures, including hybridization and triploidization, germline gene knockdown, hyperthermia, and chemical treatment (de Siqueira-Silva et al. 2018).

The first reported instance of intraperitoneal transplantation of PGCs used Green Fluorescent Protein-transgenic trout as the donor and wild-type trout as the recipient (Takeuchi et al. 2003). The host-derived PGCs were microinjected into the peritoneal cavity of recipient embryos (Fig. 7.1), and following migration to the genital ridge began proliferation and gametogenesis. At maturation, the recipient wild-type rainbow trout produced donor-derived eggs and sperm, depending on the sex of the recipient fish (Takeuchi et al. 2003). The feasibility of transferring PGCs between different salmonid species (xenotransplantation) was soon confirmed; PGCs from rainbow trout were transplanted into newly hatched larvae of masu salmon (*Oncorhynchus masou*), which on reaching maturity produced donor-derived offspring (Takeuchi et al. 2004). Since then, successful xenotransplantation of PGCs has been reported for a range fish species (Robles et al. 2017; Kobayashi et al. 2007).

In light of the success in transplanting PGCs, researchers soon started investigating the feasibility of using other types of reproductive cells in transplantation studies, including spermatogonial cells. In one of the first studies, type A spermatogonia (ASG) isolated from adult male *vasa-Gfp* rainbow trout were microinjected into the peritoneal cavity of newly hatched wild-type rainbow trout larvae. After the recipients reached maturity, donor spermatogonia differentiated into spermatozoa and fully functional eggs in male and female fish respectively (Okutsu et al. 2006). This study indicated that the testicular germ cells possessed a high degree of sexual plasticity, and that the resulting sex of the germ cells is determined primarily by the somatic environment rather than in a cell-autonomous manner (Okutsu et al. 2006).

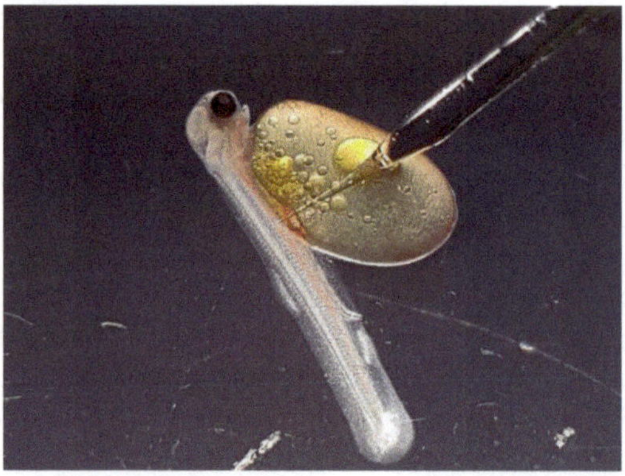

Fig. 7.1 Microinjection of donor-derived germline stem cells into peritoneal cavity of rainbow trout hatchlings; from Yoshizaki and Yazawa (2019)

Successful ASG transplantation between species was soon confirmed (Okutsu et al. 2007). In this study, ASGs from *vasa-Gfp* rainbow trout donors were xenotransplanted into the peritoneal cavity of newly hatched sterile triploid masu salmon; this resulted in the transplanted trout spermatogonia colonizing the gonads of the recipients, which eventually produced donor-derived eggs or sperm, depending on the sex of the recipients (Fig. 7.2). Successful spermatogonial germ cell xenotransplantation has now been achieved in range of fish species, even between different families (Yoshizaki et al. 2011; Yoshizaki and Yazawa 2019).

One of the limitations of germ cell transplantation methodologies involving fish embryos and larvae is that it takes a relatively long time to produce viable gametes. Germ cell transplantation into adult mature fish would allow a faster means of obtaining eggs and sperm. Lacerda et al. (2006) were the first to report successful germ cell transplantation in an adult fish, the Nile tilapia (*Oreochromis niloticus*). In this first study it was initially reported that an injection of an enriched ASG cell suspension, labelled with a fluorescent dye, into the testes of the recipient male did not evoke immune-rejection. A subsequent study demonstrated that following transplantation, cryopreserved tilapia ASGs were able to proliferate and differentiate into spermatozoa in the testes of the recipient male (Lacerda et al. 2010). Successful xenogeneic transplantation of ASGs was reported by Majhi et al. (2009): transplanted ASGs isolated from juvenile pejerrey (*Odontesthes bonariensis*) were able to colonize the gonads of adult Patagonian pejerrey (*Odontesthes hatchery*) and generate donor-derived sperm within 6 months post transplantation.

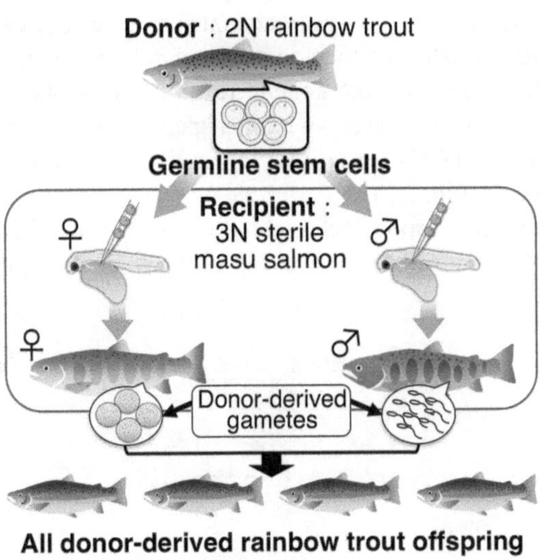

Fig. 7.2 Production of masu salmon surrogate broodstock and next generation of rainbow trout using spermatogonial stem cells (ASGs). ASGs from male rainbow trout were transplanted into the peritoneal cavity of newly hatched sterile triploid masu salmon; this resulted in the transplanted trout spermatogonia colonizing the gonads of the recipients, which eventually produced donor-derived eggs or sperm, depending on the sex of the recipients: from Yoshizaki and Yazawa (2019)

As the ovaries of adult female fish retain large numbers of mitotic germ cells, and some can behave as oogonial stem cells (Nakamura et al. 2010), the protocol developed for rainbow trout ASG transplantation was soon applied to oogonial transplantation (Yoshizaki et al. 2010). An ovarian cell suspension prepared from *vasa-Gfp* rainbow trout was transplanted into the peritoneal cavity of sterile triploid rainbow trout. The donor oogonia successfully migrated to the gonads of the recipient, where they were eventually incorporated. The donor-derived oogonia initiated spermatogenesis and oogenesis in male and female recipients, respectively, and after 2 years produced functional gametes. Thus, oogonia are capable of differentiating into functional sperm, indicating that oogonia exhibit the same sexual plasticity as PGSs and ASGs. The validation of oogonial transplantation expanded the versatility of fish germ cell transplantation technologies in fish (Yoshizaki et al. 2010). The primary advantage of oogonial transplantation is the transference of mitochondrial DNA and germplasm constituents, since maternally inherited genetic information is lost in spermatogonial cell transplantation. As such, the cryopreservation of oogonial stem cells would appear to be a valuable contribution to the preservation of genetic resources in fish.

Germ cell transplantation techniques, especially involving surrogate broodstock technology, are expected to have major implications to both aquaculture and fish conservation. In aquaculture, these reproductive technologies will facilitate the efficient production of offspring carrying superior genetic traits by transplanting donor germ cells from a single selected fish with superior traits into many recipient fish. Also, this technology will reduce the time required to produce gametes of a species with a long generation time by using a recipient surrogate species with a short generation time. For future fish conservation, this technology might represent a "silver bullet" for preserving the valuable genetic resources of endangered fish species. Restoration of a threatened, or even an extinct species, could be achieved by transplanting cryopreserved PGCs (or ASGs) into recipient surrogate hatchlings of a closely related species (Fig. 7.3).

Advances in surrogate broodstock technology are particularly advantageous in the conservation of threatened species having a long generation time, and/or species

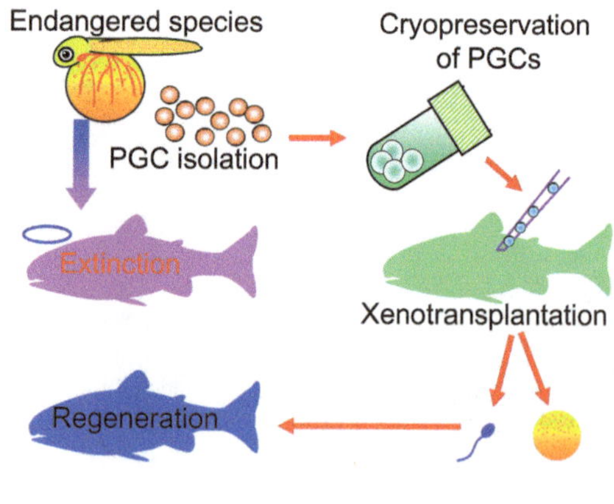

Fig. 7.3 Importance of cryopreserved PGCs to genetic resource banking. Restoration of a threatened, or even an extinct fish species, could be achieved by transplanting cryopreserved PGCs (or ASGs) into recipient surrogate hatchling of a closely related species: from Yoshizaki and Lee (2018)

which mature at a large body size, life history traits which dramatically increases the complexity and cost of maintaining broodstocks. The tunas and sturgeons are two examples where surrogate broodstock technology will be particularly beneficial to conservation initiatives, as well as aquaculture production. By employing surrogate broodstock technologies, seed production of these threatened large-bodied, commercially important species, is expected to become possible using small-bodied surrogate parents with short generation times. The bluefin tuna (*Thunnus orientalis*), arguably the world's most valuable fish, is now listed as critically endangered because of extensive overfishing. Although the life cycle of the bluefin tuna has been closed (Sawada et al. 2005), tuna farming still relies on the capture of large numbers of wild juveniles (capture-based aquaculture), which only serves to accelerate the depletion of wild tuna stocks. The successful aquaculture of bluefin tuna is considered a prerequisite to alleviate pressure on wild tuna stocks. However, it takes 3–5 years for bluefin tuna to reach sexual maturity, by which time the body weight has reached several hundred kilograms. Therefore, expansive rearing spaces are required for bluefin tuna broodstocks, leading to high labour and production costs. Surrogate broodstock technology now offers the possibility of producing eggs and sperm of bluefin tuna using a smaller species from the same Scombridae family as the surrogate recipient, such as the chub mackerel (*Scomber japonicas*) which weighs only 300 g and matures in 1 year (Yazawa et al. 2010, 2013).

Surrogate broodstock technology is also being employed to help in the conservation of sturgeons, the most threatened of all fish groups. For example, the Chinese sturgeon (*Acipenser sinensis*), one of three sturgeon species endemic to the Yangtze River, is considered critically endangered or possibly extinct (IUCN 2018). Unfortunately, due to its late maturity (8–26 years) and large size (200–400 kg), captive breeding programmes are economically prohibitive. Recently, Ye et al. (2017) validated a germ cell transplantation technique for the endangered Chinese sturgeon using the closely related but significantly smaller Dabry's surgeon (*A. dabryanus*) as the recipient, which matures in captivity at 3–5 years (c. 16 kg). Similarly, Pšenička et al. (Pšenička et al. 2015) demonstrated that donor germ cells isolated from the Siberian sturgeon (*Acipenser baerii*) ovary and testes could colonize and proliferate in the gonads of a smaller recipient species, the sterlet (*Acipenser ruthenus*). The donor Siberian sturgeon is again a large late maturing species (18–28 years) and listed as endangered, while the recipient sterlet is one of the most common and smallest sturgeon species, and has a short reproductive cycle (about 5 years).

An alternative to primordial germ cells is to use blastomeres, which are defined as any type of cell produced by cleavage (cell division) of the zygote after fertilization, and importantly lack yolk deposits. Blastomeres obtained from fish embryos retain pluripotency (Ho and Kimmel 1993) and cryopreserved blastomeres maintain the genetic diversity of both the nuclear genome and mitochondrial DNA. The successful creation of germ-line chimeras following transplantation of blastomeres has been reported in a number of fish species, including goldfish (Kusuda et al. 2004), zebra fish (Lin et al. 1992), medaka (Wakamatsu et al. 2001) and rainbow trout (Takeuchi et al. 2001) embryos. The transplantation of frozen-thawed blastomeres into goldfish embryos led to the differentiation of blastomeres into primordial germ cells (Kusuda et al. 2004). These early studies demonstrate that blastomeres that survive after cryopreservation retain their pluripotency and are able to differentiate into both somatic and germ cell lines.

In conclusion, until significant advances are made in the cryopreservation of fish oocytes and embryos, the cryopreservation of PCGs, ASGs or blastomeres represents the most practical way of preserving both the male and female genomes, and as such, will remain a vital component of genetic resource banking aimed at the conservation of threatened species. Further, advances in surrogate broodstock technology now allows the possibility of restoring threatened, or even an extinct fish species whose reproductive cells have been preserved.

2.6 Fish Generation by Nuclear Transfer

An alternative to the use of germ cells for the preservation of fish genetic resources is to use somatic cells. Somatic cells are diploid, so have the advantage of transmitting both the maternal and paternal genome, and can be collected from any individual at all life history stages. The skin and fin are good candidates for tissue collection because of their regenerative capacity, thus allowing for the sampling of endangered or unique individuals. Following cryobanking of somatic cells, fish can be reconstructed by microinjecting the somatic cell nucleus into an oocyte of a closely related species, a procedure called somatic cell nuclear transfer (SCNT), also called cloning. Although SCNT has been used to produce a small number of viable offspring (Le Bail et al. 2010; Tanaka et al. 2012), difficulties still persist with reconstruction technologies (Chenais et al. 2014). To date, most groups have reported moderate success in the development of reconstructed fish embryos up to the mid-blastula stage, with later stages characterized by high mortality (Le Bail et al. 2010; Luo et al. 2011). As transcription of the embryonic genome is activated at around the mid-blastula stage, developmental failures after this stage is a strong indication that the reprogramming of the somatic chromatin into an embryonic pattern was not optimal.

One of the major advantages of regeneration by nuclear transfer is that the recipient can belong to a species which is different from that of the donor. This could have major implications for future conservation initiatives as it allows the possibility of reconstructing individuals from rare or threatened species using oocytes from easily farmed species. However, in the context of genome resource preservation it is important to ensure that the genome of the valuable individual is correctly represented in the reconstructed offspring. Therefore, it is important to characterize the genetic background of the clones after nuclear transfer. Although limitations in regeneration technologies still persist, nuclear transfer appears to be a promising technology for the future preservation of fish genetic resources. As such, the cryobanking of fin or skin somatic cells should be a widely adopted strategy in fish biodiversity conservation initiatives.

3 Consequences of Climate Change

While continuing declines in both global fish populations and fish biodiversity can be attributed to a variety of anthropogenic factors including habitat loss and increased habitat fragmentation, pollution, invasive species and over-exploitation, the impact of climate change is increasingly being recognized as a contributing factor (Brander 2007; Graham and Harrod 2009; Rijnsdorp et al. 2009; Pörtner and Peck 2010). There are an increasing number of published reports of species declines and even extinctions correlated with climate change, with rare or endangered fish species being particularly vulnerable (Dias et al. 2017; Pörtner and Knust 2007). Climate change, especially global warming, will continue to affect fish at all levels of biological organization, from the individual to the population level, influencing physiological and ecological processes in a number of direct and indirect ways. Overall, continued climate change, including rising seawater temperatures, declining oxygen concentrations and ocean acidification is predicted to be a major threat to fish biodiversity (Brander 2007).

3.1 Global Warming

Global warming has resulted in mean sea surface temperatures rising by 0.13 °C per decade since 1979. This increase in temperature can affect fish at multiple levels. Temperature directly controls metabolic processes, and is a key determinant of a range of physiological processes, including growth, reproduction and immune response. Increasing temperatures will have profound effects on fish reproduction, including alterations in the reproductive endocrine system, such as the inhibition of ovarian estrogen production (reviewed by Pankhurst and Munday 2011). Juvenile stages are generally more sensitive than adults to changes in environmental conditions, especially temperature. Global warming will influence the timing and duration of embryonic development, and impact both egg and larval survival rates. As larval growth and survival are the primary determinants of recruitment success, any changes brought about by climate change can potentially have a profound impact on fish population dynamics and long-term biodiversity.

As all fish have an optimal thermal range, global warming is now resulting in pronounced shifts in distribution patterns, with many marine species showing a rapid polewards shift (Perry et al. 2005). Generally, global warming impacts have favoured fish populations occupying the colder temperate regions. For example, it has been reported that species richness in the North Sea has increased since 1985, and that this increase could be related to an increase in water temperature (Hiddink and Hofstede 2008). Specifically, it was found that eight times more fish species displayed increased distribution ranges in the North Sea (mainly small species of southerly origin such as anchovy, *Engraulis encrasicolus* and red mullet, *Mullus surmuletus*) compared with those whose range decreased (mainly large and northerly species such as wolfish, *Anarhichas lupus* and ling, *Molva molva*). At the same time average winter bottom temperatures in the North Sea increased by 0.7 °C decade^{-1}.

Ocean warming reduces the solubility of oxygen, as well as raising metabolic rates, which accelerates the rate of oxygen consumption. As a consequence, oxygen concentrations in both the open oceans and coastal waters have been declining since the 1950s. As oxygen is fundamental to biological processes its decline can cause major changes in ocean productivity and biodiversity. A reduction in aerobic scope, as a consequence of elevated temperatures and reduced oxygen concentrations, is having a detrimental impact on fish populations (Pörtner and Knust 2007), especially diadromous species such as salmonids and sturgeons, which undergo energetically demanding migrations (Friedland et al. 2003; Reynolds et al. 2005). For example, the collapse of the 2004 Fraser River sockeye salmon run was associated with unusually high temperatures during the summer spawning migration (Farrell et al. 2008). Although evidence suggests that the warming river temperature is exerting strong selective pressure for physiological adaptation in the Fraser river salmon (Eliasen et al. 2011), it is questionable whether the rate of physiological adaptation will allow them to adapt quickly enough to cope with the current rate of temperature increase, increasing the likelihood of extinction of local populations.

Climate change can impact the abundance and biodiversity of fish populations indirectly through its effect on aquatic food chains and ocean circulation patterns. Unfavourable shifts in plankton communities, driven by climate change can have a dramatic impact on fish recruitment (Graham and Harrod 2009), while recent changes in productivity and the distribution of a number of fish species can be accredited to changes to major ocean circulatory phenomena such as the North Atlantic Oscillation and the El Nino-Southern Oscillation (Brander 2007). Climate change, especially global warming, may also make conditions more suitable for invasive species. Invasive species can threaten native species through competition for resources, predation, or disease. Lakes and inland waters are particularly vulnerable to extinction threats from introduced species. For example, the intentional introduction of the Nile perch (*Lates niloticus*) to Lake Victoria in East Africa in the 1950s was largely responsible for the extinction of about 200 species of cichlids (Witte et al. 1992).

3.2 Ocean Acidification and Deoxygenation (Hypoxia)

While most attention has focused on global warming, climate change is also driving other environmental changes, including deoxygenation and ocean acidification. The three processes are mechanistically linked and thus it is often their combined effects that shape marine ecosystems. Since 1950, more than 500 coastal areas have reported oxygen concentrations $\leq$2 mg l^{-1}, a threshold often used to delineate hypoxia (Breitburg et al. 2018). While the primary cause of these so-called 'dead zones' is nutrient pollution (**eutrophication), global** warming is predicted to exacerbate oxygen depletion in many nutrient-enriched coastal systems through mechanisms similar to those of the open ocean: increased intensity and duration of stratification, decreased oxygen solubility, and accelerated respiration. Low oxygen levels can have a number of deleterious effects on fish, including reduced growth and survival, altered behaviour, and impaired gamete production (Diaz and Rosenberg 2008; Thomas et al. 2015). As a consequence, oxygen deficient coastal areas are characterized by severely reduced fish abundance and biodiversity.

Ocean acidification is the ongoing decrease in the pH of the Earth's oceans, caused by the uptake of carbon dioxide (CO_2) from the atmosphere. Surface ocean pH is estimated to have decreased by 0.1 units to 8.14 in the last 200 years, and is predicted to drop a further 0.3–0.5 units by 2100 (The Royal Society 2005). Ocean acidification is predicted to affect marine ecosystems in a number of complex ways, including modification of fish behaviour, such as impaired auditory behaviour and foraging behaviour (Simpson et al. 2011; Jiahuan et al. 2018).

Interactions between climate change, ozone depletion, and ultraviolet (UV) radiation are altering exposure to UV radiation in aquatic ecosystems. Aquatic organisms are particularly vulnerable to increased levels of UV radiation, especially in the high latitudes (Arctic and Antarctic) where levels are at their highest. High egg and larval mortality rates have been reported in a number of fish species which spawn in shallow clear lakes and streams, which are characterized by high UV penetration rates (Häder et al. 2007).

On a regional scale, changes in climate are highly variable in both space and time, leading to widespread floods in wetter regions, more severe droughts in drier regions, and increases in extreme storm events. Freshwater species are particularly sensitive to changes in the frequency, duration, and timing of extreme precipitation events, such as floods or droughts. Again, anadromous species such as salmonids are particularly vulnerable to extremes in water flow rates, impacting migrating adults and the survival of eggs and larvae (Jonsson and Jonsson 2009). Climate change is likely to further stress sensitive freshwater species, which are already adversely affected by a variety of other human impacts, such as altered flow regimes, habitat fragmentation and deterioration of water quality.

In conclusion, evidence now strongly supports the view that climate change, in particular the consequences of global warming, is a threat to fish biodiversity. Climate change will continue to impact fish at all levels of biological organization from the individual to the population and community levels, influencing both physiological and ecological processes in a variety of direct and indirect ways. As such, current and future conservation efforts need to consider the impact of climate change in both *in situ* conservation and reintroduction efforts. There needs to be an awareness of the threats posed by climate change to fish biodiversity, especially as the consequences of climate change can be exacerbated by other stressors such as habitat fragmentation, pollution and the impact of invasive species. As a consequence, since 2008 the IUCN has begun incorporating projections of future risk from climate change into its red-list rankings. While protection of fish biodiversity will still rely heavily on *in situ* conservation approaches, the likelihood of future extinctions highlights the importance of establishing genetic resource banks for fish species.

4 Priorities for the Future

4.1 Cryopreservation of Germplasm for Conservation of Fish Biodiversity

Even though the conservation status of only 5% of the nearly 30,000 fish species have been assessed (IUCN 2017), the scale and threat to the survival of fish species, both in terms of population size and declining biodiversity, is arguably greater than any other vertebrate group. As of July 2017, the IUCN lists 455 fish species as being critically endangered, including 87 species identified as *possibly extinct*. The state of California alone has recently listed 36 species or subspecies of fish as either threatened or endangered, two of which are now believed to be extinct, the thicktail chub and tecopa pupfish (www.wildlife.ca.gov/Conservation/Fishes). Freshwater species generally appear to be at greater risk than marine species, for example an earlier study reported that 47% of Europe's 287 native freshwater fish species are classified as threatened with extinction (Reynolds et al. 2005). With the growing concerns over the continued loss in fish biodiversity the last three decades has witnessed the establishment of several European cryobanks with the specific aim of preserving the genetic diversity of Europe's fish species. This cryopreserved germplasm can be used, without limitations of time and space, employing a variety of assisted reproduction techniques, in future breeding programmes aimed at conserving threatened species.

Fish Cryobanks

One of the first fish cryobanks was established in 1981 at the National Academy of Science of Ukraine, in the former USSR. The genetic resources preserved in this cryobank are now recognized as being of unique importance. The bank contains frozen sperm from an impressive range of fishes, primarily from carp, sturgeon and salmonid species collected from the different republics of the former USSR. Many of the samples are from species listed as critically endangered or close to extinction, including the ship or spiny sturgeon (*Acipenser nudiventris*), formerly abundant in the Black, Aral and Caspian seas, but now listed as critically endangered (CR), the Russian sturgeon (*Acipenser gueldenstaedtii*) also listed as CR, and the green sturgeon (*Acipenser medirostris*), which is native to the Pacific Ocean. It is widely acknowledged that this cryobank, which is maintained on a very limited budget, represents a unique resource for future breeding programmes, or restoration programmes for endangered species.

One of the more successful biobanking initiatives working to halt the loss of animal biodiversity is the Frozen Ark (www.frozenark.org), which has its headquarters at the University of Nottingham (UK). This charitable organization was establish in 2004, and now has 27 partners in 12 countries and currently holds tissue samples collected from c. 5000 species. In partnership with the LIRANS Institute of Research in Applied Natural Sciences (University of Bedford), the Frozen Ark has now cryobanked tissue samples from well over 100 fish species endemic to the UK, including a number of critically endangered species (Rawson et al. 2011). In the USA, a new program was initiated in 2015 called the Aquatic Germplasm and Genetic Resources Center (AGGRC). The AGGRC (www.aquaticgermplasm.com) was initiated to provide global leadership in repository development and commercialization of genetic resources of aquatic species, although primarily for use in research rather than for fish conservation. Finally, in Brazil, which holds 20% of the world's fish species, several cryobanks have recently been established in an attempt to conserve native fish species, especially the multitude of species native to the Amazon basin (Godinho and Viveiros 2011; Viveiros et al. 2014).

Salmonid species are particularly vulnerable to over exploitation, habitat destruction and the effects of climate change, factors generally compounded by their life cycle characteristics of long-range migrations between freshwater and marine habitats. In addition, wild populations of some salmonids, such as Atlantic salmon, are increasingly threatened by the rapid growth of aquaculture. Recent studies have confirmed that widespread genetic introgression between wild salmon and escaped farmed conspecifics results in reduced offspring fitness (McGinnity et al. 2003; Fraser et al. 2010), potentially accelerating the extinction of unique local populations. In Norway, which has over 400 unique salmon populations, the biotechnology company Cryogenetics (www.cryogenetics.com) in partnership with the Norwegian Environment Agency, has committed to a comprehensive biodiversity strategy aimed at conserving as much as possible the genetic diversity of Norway's wild Atlantic salmon populations. To date, this cryobank comprises of sperm samples from more than 6200 individual males from 172 populations. In addition to its cryobank, Cryogenetics also offers professional cryopreservation services for number of commercially important species, including Atlantic salmon, rainbow trout, sockeye salmon (*Oncorhynchus nerka*), Arctic charr (*Salvelinus alpinus*) and sturgeon (*Acipenser sp.*)

Unfortunately most cryobanks of fish germplasm, notably the earlier ones, were established as national initiatives, and as such are characterized by a lack of continuity and standardization in cryopreservation procedures. It has long been advocated that genetic resource banks whose primary aim is to conserve wildlife biodiversity should be organized on a global basis (Wildt 1992; Holt and Pickard 1999). As such, a future priority should be to strengthen international collaboration between the major cryobanks in order to catalogue a comprehensive inventory of all available cells and tissues, as well as to standardize protocols both in the collection and storage of germplasm.

4.2 Repositories for Small Model Fish

Advances in the cryopreservation of fish germplasm is important to international repositories responsible for maintaining important strains of laboratory model fish species, such as zebrafish (*Danio rerio*), medaka (*Oryzias latipes*) and swordtails (*Xiphophorus sp*). With their unique range of advantages, these small fish species are increasingly becoming an indispensable vertebrate model in most areas of biomedical research. The dramatic increase in the use of model fish species has led to the production of thousands of specific strains and lines, and the number is increasing every day. In order to preserve this unique genetic resource it soon became apparent that high-throughput international repositories needed to be established (Hagedorn et al. 2019). These include the Zebrafish International Resource Center (ZIRC) at the University of Oregon (https://zfin.org) which provides a central repository for wild-type and mutant strains of zebrafish. The ZIRC, which opened in 2000, holds the largest biomedical collection in the US with over 60,000 samples from zebrafish representing some 9000 lines. Other model fish repositories include the *Xiphophorus* Genetic Stock Center housed at Texas State University (www.xiphophorus.txstate.edu) and the European Zebrafish Resource Center (EZRC), situated in Karlsruhe in Germany. The EZRC houses the largest experimental fish facility in Europe and provides an extensive repository of zebrafish resources (http://www.ezrc.kit.edu). These zebrafish resource centers have benefited from the development of a standardized protocol for high-throughput sperm cryopreservation procedures (Yang et al. 2007).

4.3 Epigenetic Changes

Information on whether the cryopreservation of germplasm can induce epigenetic changes is still largely lacking. It has been proposed that cryopreservation procedures can produce molecular alterations in key genes and transcripts undetectable by traditional assays, and these modifications could interfere with normal embryo development (Urrego et al. 2014; Labbé et al. 2017). One of the potential limitations identified in nuclear transfer technologies is the likelihood of induced epigenetic changes in the clones (Chenais et al. 2014). Also, the application of surrogate production techniques, involving the cryopreservation and inter-species transplantation of PGCs or spermatogonia (see Sect. 2.5) could induce epigenetic modifications that could have a detrimental impact on the progeny (Robles et al. 2017). In the first published study, Riesco and Robles (2013) demonstrated that cryopreservation caused genetic and epigenetic changes in the genital ridges of zebrafish embryos. It is recognized that potential epigenetic changes brought about by the cryoprotective agents and techniques used cannot be studied effectively until enough offspring are produced from these procedures to provide a representative population. With their ability to produce hundreds of embryos from a single egg batch and sperm sample, it has been proposed that fish would be a good model system to investigate how cryopreservation procedures and assisted reproductive technologies can impact offspring performance through epigenetic changes (Labbé et al. 2017).

4.4 Biosecurity Considerations

One of the major limitations in the trade of germplasm, both nationally and internationally, is the lack of effective biosecurity measures (Tiersch and Jenkins 2003). New biosecurity measures need to be implemented to address the growing societal concerns surrounding pathogens and disease in the aquaculture industry. Germplasm, notably semen, traded and used in assisted reproductive technologies (ARTs), both in aquaculture and conservation efforts, is a potential vector for pathogen transfer. Currently, there are two main areas of biosecurity concern: (1) potential pathogens carried with the germplasm, and (2) potential hazards transmitted via the cryoextender. As such, future technologies and procedures in the cryopreservation and storage of fish germplasm for use in assisted reproductive technologies (ARTs) need to incorporate improved biosecurity measures.

4.5 Concluding Remarks

Preservation of animal genetic resources, including teleost fishes, is now an international priority. The teleost fishes are the largest and most diverse vertebrate group, accounting for nearly half of all known vertebrate species. As such, there is an urgent need to protect and conserve the huge genetic diversity offered by this diverse vertebrate group, not just as a source of genes for contemporary breeding and for protection against the consequences of climate change and disease, but also as part of our national heritage. While the cryopreservation of reproductive cells is a means of achieving these objectives, currently only fish sperm, but not eggs, can be successfully frozen. Due to their large size, large yolk compartment, low membrane permeability and high chilling sensitivity, successful and reproducible protocols for the cryopreservation of fish oocytes and embryos still remains elusive. However, significant advances have been made in the cryopreservation of primordial germ cells as an alternative means of conserving both paternal and maternal genomes. Although more research needs to be carried out on how these cells can be optimally applied to emerging reproductive technologies, including transplantation techniques, the successful cryopreservation of fish germ cells offers the possibility of both conserving and restoring threatened species. In conclusion, it is anticipated that the successful cryopreservation of fish germplasm will result in a range of economic, ecological and societal benefits. In partnership with emerging assisted reproductive technologies, the successful cryopreservation of fish germplasm will lead to more efficient reproduction in aquaculture, assist selective breeding programmes, and be of crucial importance to future species conservation actions.

References

Aerni P. Risk, regulation and innovation: the case of aquaculture and transgenic fish. Aquat Sci. 2004;66:327–41.

Aksnes A, Gjerde B, Roald SO. Biological, chemical and organoleptic changes during maturation of farmed Atlantic salmon, *Salmo salar*. Aquaculture. 1986;53:7–20.

Amann RP, Waberski D. Computer-assisted sperm analysis (CASA): capabilities and potential developments. Theriogenology. 2014;81:5–17.

Asturiano JF, Cabrita E, Horvath A. Progress, challenges and perspectives on fish gamete cryopreservation: a mini review. Gen Comp Endocrinol. 2017;245:69–76.

Baloch AR, Franěk R, Saito T, Pšenička M. Dead-end (dnd) protein in fish—a review. Fish Physiol Biochem. 2019; https://doi.org/10.1007/s10695-018-0606-x.

Barbaro A, Francescon A, Bozzato G, Merlin A, Belvedere P, Colombo L. Induction of spawning in gilthead seabream, Sparus aurata L., by long-acting GnRH agonist and its effect on egg quality and daily timing of spawning. Aquaculture. 1997;154:349–59.

Beardmore JA, Porter JS. Genetically modified organisms and aquaculture. Fisheries circular no. 989. Rome: FAO; 2003.

Beardmore JA, Mair GC, Lewis RI. Monosex male production in finfish as exemplified by tilapia: applications, problems, and prospects. Aquaculture. 2001;197(1/4):283–301.

Beirão J, Soares F, Herráez MP, Dinis MT, Cabrita E. Changes in *Solea senegalensis* sperm quality throughout the year. Anim Reprod Sci. 2011;126:122–9.

Benfey TJ. Effectiveness of triploidy as a management tool for reproductive containment of farmed fish: Atlantic salmon (*Salmo salar*) as a case study. Rev Aquac. 2019;8:264–82.

Bonnet E, Montfort J, Esqurre J, Hugot K, Fostier A, Bobe J. Effect of photoperiod manipulation on rainbow trout (*Oncorhynchus mykiss*) egg quality: a genomic study. Aquaculture. 2007;268:13–22.

Boryshpolets S, Kowalski RK, Dietrich GJ, Dzyuba B, Ciereszko A. Different computer-assisted sperm analysis (CASA) systems highly influence sperm motility parameters. Theriogenology. 2013;80:758–65.

Brander KM. Global fish production and climate change. PNAS. 2007;104:19709–14.

Breitburg D, Levin LA, Oschlies A, et al. Declining oxygen in the global ocean and coastal waters. Science. 2018;359.

Bromage N, Porter M, Randall C. The environmental regulation of maturation in farmed finfish with special reference to the role of photoperiod and melatonin. Aquaculture. 2001;197:63–98.

Cabrita E, Martínez-Páramo S, Gavaia PJ, Riesco MF, Valcarce DG, Sarasquete C, Herráez MP, Robles V. Factors enhancing fish sperm quality and emerging tools for sperm analysis. Aquaculture. 2014;432:389–401.

Carrillo M, Bromage NR, Zanuy S, Serrano R. The effect of modifications in photoperiod on spawning time, ovarian development and egg quality in the sea bass (*Dicentrarchus labrax* L.). Aquaculture. 1989;81:351–65.

Ceballos G, Ehrlich PR. Mammal population losses and the extinction crisis. Science. 2002;296(5569):904–7.

Ceballos G, Ehrlich PR, Dirzo R. Biological annihilation via the ongoing sixth mass extinction signaled by vertebrate population losses and declines. PNAS. 2017;114:E6089–96.

Chen SL, Tian YS. Cryopreservation of flounder (*Paralichthys olivaceus*) embryos by vitrification. Theriogenology. 2005;63:1207–19.

Chenais N, Depince A, Le Bail PY, Labbe C. Fin cell cryopreservation and fish reconstruction by nuclear transfer stand as promising technologies for the preservation of finfish genetic resources. Aquac Int. 2014;22:63–76.

Cinalli RM, Rangan P, Lehmann R. Germ cells are forever. Cell. 2008;132:559–62.

Clausen R, York R. Global biodiversity decline of marine and freshwater fish: a cross-national analysis of economic, demographic and ecological influences. Soc Sci Res. 2008;37:1310–20.

Copeland DL, Duff RJ, Liu Q, Prokop J, Londraville RL. Leptin in teleost fishes: an argument for comparative study. Front Physiol. 2011;2:26. https://doi.org/10.3389/fphys.2011.00026.

Cuevas-Uribe R, Yang H, Daly J, Savage MG, Walter RB, Tiersch TR. Production of F1 offspring with vitrified sperm from a live-bearing fish, the green swordtail *Xiphophorus hellerii*. Zebrafish. 2011;8:167–79.

Cuevas-Uribe R, Chesney EJ, Daly J, Tiersch TR. Vitrification of sperm from marine fish: effect on motility and membrane integrity. Aquacult Res. 2015;46:1770–84.

Cuevas-Uribe R, Hu E, Daniels H, Gill AO, Tiersch TR. Vitrification as an alternative approach for sperm cryopreservation in marine fishes. N Am J Aquac. 2017;79:187–96.

de Carvalho AFS, Ramos SE, de Carvalho TSG, de Souza YCP, Zangeronimo MG, Pereira LJ, Murgas LDS. Efficacy of fish embryo vitrification protocols in terms of embryo morphology—a systematic review. CryoLetters. 2014;35:361–70.

de Siqueira-Silva DH, Saito T, Dos Santos-Silva AP, da Silva Costa R, Psenicka M, Yasui GS. Biotechnology applied to fish reproduction: tools for conservation. Fish Physiol Biochem. 2018;44:1469–85.

Devlin RH, Nagahama Y. Sex determination and sex differentiation in fish: an overview of genetic, physiological, and environmental influences. Aquaculture. 2002;208:191–364.

Devlin RH, Yesaki TY, Biagi CA, Donaldson EM, Swanson P, Chan W-K. Extraordinary salmon growth. Nature. 1994;371:209–10.

Devlin RH, Yesaki TY, Donaldson EM, Du SJ, Hew CL. Production of germline transgenic Pacific salmonids with dramatically increased growth performance. Can J Fish Aquat Sci. 1995;52:1376–84.

Devlin RH, Biagi CA, Yesaki TY, Smailus DE, Byatt JC. Growth of domesticated transgenic fish—a growth-hormone transgene boosts the size of wild but not domesticated trout. Nature. 2001;409:781–2.

Dias MS, Tedesco PA, Hugueny B, Jézéquel C, Beauchard O, Brosse S, Oberdorff T. Anthropogenic stressors and riverine fish extinctions. Ecol Indic. 2017;79:37–46.

Diaz RJ, Rosenberg R. Spreading dead zones and consequences for marine ecosystems. Science. 2008;321:926–9.

Dunham RA, Chitmanat C, Nichols A, Argue B, Powers DA, Chen TT. Predator avoidance of transgenic channel catfish containing salmonid growth hormone genes. Marine Biotechnol. 1999;1:545–51.

Eliasen EJ, Clark TD, Hague MJ, Hanson LM, Gallagher ZS, Jeffries KM, Gale MK, Patterson DA, Hinch SC, Farrell AP. Differences in thermal tolerance among sockeye salmon populations. Science. 2011;332:109–12.

FAO. The state of world fisheries and aquaculture. Rome: FAO; 2018.

Farrell AP, Hinch SG, Cooke SJ, Patterson DA, Crossin GT, Lapointe M, Mathes MT. Pacific salmon in hot water: applying aerobic scope models and biotelemetry to predict the success of spawning migration. Physiol Biochem Zool. 2008;81:697–708.

Felip A, Zanuy S, Carrillo M, Martinez G, Ramos J, Piferrer F. Optimal conditions for the induction of triploidy in the sea bass (*Dicentrarchus labrax* L.). Aquaculture. 1997;152:287–98.

Felip A, Zanuy S, Muriach B, Cerdá-Reverter JM, Carrillo M. Reduction of sexual maturation in male *Dicentrarchus labrax* by continuous light both before and during gametogenesis. Aquaculture. 2008;275(1–4):347–55.

Figueroa E, Merno O, Risopatron J, Isachenko V, Sánchez R, Effer B, Isachenko E, Farias JG, Val debenito I. Effect of seminal plasma on Atlantic salmon (*Salmo salar*) sperm vitrification. Theriogenology. 2015;83:238–45.

Figueroa E, Risopatron J, Sanchez R, Isachenko E, Merino O, Isachenko V, Valdebenito I. Spermatozoa vitrification of sex-reversed rainbow trout (*Oncorhynchus mykiss*): effect of seminal plasma on physiological parameters. Aquaculture. 2013;372–375:119–26.

Fjelldal PG, Hansen T. Vertebral deformities in triploid Atlantic salmon (*Salmo salar* L.) underyearling smolts. Aquaculture. 2010;309:131–6.

Fjelldal PG, Hansen T, Lock E-J, Wargelius A, Fraser TWK, Sambraus F, El-Mowafi A, Albrektsen S, Waagbø R, Ørnsrud R. Increased dietary phosphorous prevents vertebral deformities in triploid Atlantic salmon (*Salmo salar* L.). Aquacult Nutr. 2016;22:72–90.

Fleming IA, Huntingford F. Reproductive behaviour. In: Huntingford F, Jobling M, Kadri S, editors. Aquaculture and behaviour. Oxford: Blackwell; 2012. p. 286–321.

Fornies MA, Mananos E, Carrillo M, Rocha A, Laureau S, et al. Spawning induction of individual European sea bass females (*Dicentrarchus labrax*) using different GnRHa-delivery systems. Aquaculture. 2001;202:221–34.

Francescon A, Barbaro A, Bertotto D, Libertini A, Cepollaro F, Richard J, Belvedere P, Colombo L. Assessment of homozygosity and fertility in meiotic gynogens of the European sea bass (*Dicentrarchus labrax* L.). Aquaculture. 2004;243:93–102.

Fraser DJ, Minto C, Calvert AM, Eddington JD, Hutchings JA. Potential for domesticated-wild interbreeding to induce maladaptive phenology across multiple populations of wild Atlantic salmon (*Salmo salar*). Can J Fish Aquat Sci. 2010;67:1768–75.

Fraser T, Fjelldal PG, Hansen T, Mayer I. Welfare considerations of triploid fish. Rev Fish Sci. 2012;20:192–211.

Fraser TWK, Skjæraasen JE, Mayer I, Sambraus F, Hansen T, Fjelldal PG. The effect of triploidy on the culture performance, deformity prevalence, and heart morphology in Atlantic salmon. Aquaculture. 2013;416–417:255–64.

Fraser TWK, Hansen T, Mayer I, Skjæraasen JE, Glover KA, Sambraus F, Fjelldal PG. The effect of triploidy on vaccine side-effects in Atlantic salmon. Aquaculture. 2014;433:481–90.

Friedland KD, Reddin DG, McMenemy JR, Drinkwater KF. Multidecadal trends in North American Atlantic salmon (*Salmo salar*) stocks and climate trends relevant to juvenile survival. Can J Fish Aquat Sci. 2003;60:563–83.

Godinho HP, Viveiros ATM. Current status of sperm cryopreservation of Brazilian characiform fishes. In: Tiersch TR, Green CC, editors. Cryopreservation in aquatic species. 2nd ed. Baton Rouge: World Aquaculture Society; 2011. p. 875–84.

Godoy LC, Streit DP Jr, Zampolla T, Bos-Mikich A, Zhang T. A study on the vitrification of stage III zebrafish (*Danio rerio*) ovarian follicles. Cryobiology. 2013;67:347–54.

Gomelsky B. Chromosome set manipulation and sex control in common carp: a review. Aquat Living Resour. 2003;16:408–15.

Graham CT, Harrod C. Implications of climate change for the fishes of the British Isles. J Fish Biol. 2009;74:1143–205.

Graves-Herring JE, Wildt DE, Comizzoli P. Retention of structure and function of the cat germinal vesicle after air-drying and storage at suprazero temperature. Biol Reprod. 2013;88:1–7.

Guan G, Rawson DM, Zhang T. Cryopreservation of zebrafish (*Danio rerio*) oocytes by vitrification. Cryo Letters. 2010;31:230–8.

Häder DP, Kumar HD, Smith RC, Worrest R. Effects of solar UV radiation on aquatic ecosystems and interactions with climate change. Photochem Photobiol Sci. 2007;6:267–85.

Hagedorn M, Hsu E, Kleinhans FW, Wildt DE. New approaches for studying the permeability of fish embryos: toward successful cryopreservation. Cryobiology. 1997a;34:335–47.

Hagedorn M, Kleinhans FW, Freitas R, Liu J, Hsu EW, Wildt DE, Rall WF. Water distribution and permeability of zebrafish embryos, *Brachydanio rerio*. J Exp Zool. 1997b;278:356–71.

Hagedorn M, Kleinhans FW, Wildt DE, Rall WF. Chill sensitivity and cryoprotectant permeability of dechlorinated zebrafish embryos, *Brachydanio rerio*. Cryobiology. 1997c;34:251–63.

Hagedorn M, Lance SL, Fonseca DM, Kleinhans FW, Artimov D, Fleischer R, Hoque AT, Hamilton MB, Pukazhenthi BS. Altering fish embryos with aquaporin-3: an essential step toward successful cryopreservation. Biol Reprod. 2002;67:961–6.

Hagedorn M, Varga Z, Walter RB, Tiersch TR. Workshop report: cryopreservation of aquatic biomedical models. Cryobiology. 2019;86:120–9.

Hallerman EM, McLean E, Fleming IA. Effects of growth hormone transgenes on the behavior and welfare of aquacultured fishes: a review identifying research needs. Appl Anim Behav Sci. 2007;104:265–94.

Hiddink JG, Hofstede RT. Climate induced increases in species richness of marine species. Glob Chang Biol. 2008;14:453–60.

Hinits Y, Moav B. Growth performance studies in transgenic *Cyprinus carpio*. Aquaculture. 1999;173:285–96.

Ho RK, Kimmel CB. Commitment of cell fate in the early zebrafish embryo. Science. 1993;261:109–11.
Holt WV. Who needs cytoplasm? Genomic preservation for the 21st century. Biol Reprod. 2013;88:1–2.
Holt WV, Pickard AR. Role of reproductive technologies and genetic resource banks in animal conservation. Rev Reprod. 1999;4:143–50.
Isayeva A, Zhang T, Rawson DM. Studies on chilling sensitivity of zebrafish (*Danio rerio*) oocytes. Cryobiology. 2004;49:114–22.
Iversen M, Myhr AI, Wargelius A. Approaches for delaying sexual maturation in salmon and their possible ecological and ethical implications. J Appl Aquac. 2016;28:330–69.
Izquierdo MS, Fernandez-Palacios H, Tacon A. Effect of brood stock nutrition on reproductive performance of fish. Aquaculture. 2001;197:25–42.
Janik M, Kleinhans FW, Hagedorn M. Overcoming a permeability barrier by microinjecting cryoprotectants into zebrafish embryos (*Brachydanio rerio*). Cryobiology. 2000;41:25–34.
Jiahuan R, Wenhao S, Xiaofan G, Wei S, Shanjie Z, Maolong H, Haifeng W, Guangxu L. Ocean acidification impairs foraging behavior by interfering with olfactory neural signal transduction in black sea bream, *Acanthopagrus schlegelii*. Front Physiol. 2018;9:1592.
Jonsson B, Jonsson N. A review of the likely effects of climate change on anadromous Atlantic salmon *Salmo salar* and brown trout *Salmo trutta*, with particular reference to water temperature and flow. J Fish Biol. 2009;75:2381–447.
Karlsson S, Diserud OH, Fiske P, Hindar K. Widespread genetic introgression of escaped farmed Atlantic salmon in wild salmon populations. ICES J Mar Sci. 2016;73:2488–98.
Kasa E, Bernath G, Kollar T, Zarski D, Lujic J, Marinovic Z, Bokor Z, Hegyi A, Urbanyi B, Vilchez MC, Morini M, Penaranda S, Perez L, Asturiano JF, Horvath A. Development of sperm vitrification protocols for freshwater fish (Eurasian perch, *Perca fluviatilis*) and marine fish (European eel, *Anguilla anguilla*). Gen Comp Endocrinol. 2017;245:102–7.
Khosla K, Wang Y, Hagedorm M, Qin Z, Bischof J. Gold nanorod induced warming of embryos from the cryogenic state enhances viability. ASC Nano. 2017;11:7869–78.
Kobayashi T, Takeuchi Y, Takeuchi T, et al. Generation of viable fish from cryopreserved primordial germ cells. Mol Reprod Dev. 2007;74:207–13.
Kopeika E, Kopeika J, Zhang T. Cryopreservation of fish sperm. Methods Mol Biol. 2007;368:203–17.
Kusuda S, Teranishi T, Koide N, Nagai T, Arai K, Yamaha E. Pluripotency of cryopreserved blastomeres of the goldfish. J Exp Zool. 2004;301A:131–8.
Kutluyer F, Kayim M, Öğretmen F, Büyükleblebici S, Tuncer PB. Cryopreservation of rainbow trout *Oncorhynchus mykiss* spermatozoa: effects of extender supplemented with different antioxidants on sperm motility, velocity and fertility. Cryobiology. 2014;69:462–6.
Labbé C, Robles V, Herraez MP. Epigenetics in fish gametes and early embryo. Aquaculture. 2017;472:93–106.
Lacerda SMSN, Batlouni SR, Silva SBG, et al. Germ cells transplantation in fish: the Nile-tilapia model. Anim Reprod. 2006;3:146–59.
Lacerda SMSN, Batlouni SR, Costa GMJ, et al. A new and fast technique to generate offspring after germ cells transplantation in adult fish: the Nile tilapia (*Oreochromis niloticus*) model. PLoS One. 2010;5:e10740.
Lacerda SMSN, Costa GMJ, Campos-Junior PHA, Segatelli TM, Yazawa R, Takeuchi Y, Morita T, Yoshizaki G, Franca LR. Germ cell transplantation as a potential biotechnological approach to fish reproduction. Fish Physiol Biochem. 2013;39:3–11.
Lahnsteiner F, Mansour N, Kunz FA. The effect of antioxidants on the quality of cryopreserved semen in two salmonid fish, the brook trout (*Salvelinus fontinalis*) and the rainbow trout (*Oncorhynchus mykiss*). Theriogenology. 2011;76:882–90.
Le Bail PY, Depince A, Chenais N, Mahe S, Maisse G, Labbe C. Optimization of somatic cell injection in the perspective of nuclear transfer in goldfish. BMC Dev Biol. 2010;10:64. https://doi.org/10.1186/1471-213X-10-64.

Leclercq E, Taylor JF, Sprague M, Migaud H. The potential of alternative lighting-systems to suppress pre-harvest sexual maturation of 1+ Atlantic salmon (*Salmo salar*) post-smolts reared in commercial sea-cages. Aquacult Eng. 2011;44:35–47.
Lee S, Iwasaki Y, Shikina S, Yoshizaki G. Generation of functional eggs and sperm from cryopreserved whole testes. Proc Natl Acad Sci U S A. 2013;110:1640–5.
Lee S, Katayama N, Yoshizaki G. Generation of juvenile rainbow trout derived from cryopreserved whole ovaries by intraperitoneal transplantation of ovarian germ cells. Biochem Biophys Res Commun. 2016;478:1478–83.
Lin S, Long W, Chen J, Hopkins N. Production of germ-line chimeras in zebrafish by cell transplants from genetically pigmented to albino embryos. Proc Natl Acad Sci U S A. 1992;89:4519–23.
Lincoln RF, Scott AP. Production of all-female triploid rainbow trout. Aquaculture. 1983;30:375–80.
Linhartova Z, Saito T, Kaspar V, Rodina M, Praskova E, Hagihara S, Psenicka M. Sterilization of sterlet *Acipenser ruthenus* by using knockdown agent, antisense morpholino oligonucleotide, against dead end gene. Theriogenology. 2015;84:1246–55.
Liu XH, Zhang T, Rawson DM. Effect of cooling rate and partial removal of yolk on the chilling injury in zebrafish (*Danio rerio*) embryos. Theriogenology. 2001;55:1719–31.
Liu SJ, Qin QB, Xiao J, Lu W, Shen J, Li W, et al. The formation of the polyploid hybrids from different subfamily fish crossing and its evolutionary significance. Genetics. 2007;176:1023–34.
Liu J, Song Y, Cheng KM, Silversides FG. Production of donor-derived offspring from cryopreserved ovarian tissue in Japanese quail (*Corurnix japonica*). Biol Reprod. 2010;83:15–9.
Liu Q, Wang X, Wang W, Zhang X, Xu S, Ma D, Zhizhong X, Xiao Y. Effect of the addition of six antioxidants on sperm motility, membrane integrity and mitochondrial function in red seabream (*Pagrus major*) sperm cryopreservation. Fish Physiol Biochem. 2015;41:413–22.
Lubzens E, Pekarsky I, Blais I, Cionna C, Carnevali O. Cryopreservation of oocytes from a marine fish: achievements and obstacles. Cryobiology. 2005;51:385.
Lujić J, Marinović Z, Bajec SS, Djurdjevič I, Kása E, Urbányi B, Horváth A. First successful vitrification of salmonid ovarian tissue. Cryobiology. 2017;76:154–7.
Luo DJ, Hu W, Chen SP, Zhu ZY. Critical developmental stages for the efficiency of somatic cell nuclear transfer in zebrafish. Int J Biol Sci. 2011;7:476–86.
Maclean N, Laight RJ. Transgenic fish: an evaluation of benefits and risks. Fish Fish. 2000;1:146–72.
Magnotti C, Cerqueira V, Lee-Estevez M, Farias JG, Valdebenito I, Figueroa E. Cryopreservation and vitrification of fish semen: a review with special emphasis on marine species. Rev Aquac. 2018;10:15–25.
Majhi SK, Hattori RS, Yokota M, et al. Germ cell transplantation using sexually competent fish: an approach for rapid propagation of endangered and valuable germlines. PLoS One. 2009;4:e6132.
Marques LS, Bos-Mikich A, Godoy LC, Silva LA, Maschio D, Zhang T, Streit DP Jr. Viability of zebrafish (*Danio rerio*) ovarian follicles after vitrification in a metal container. Cryobiology. 2015;71:367–73.
Martínez R, Juncal J, Zaldívar C, Arenal A, Guillén I, Morera V, et al. Growth efficiency in transgenic tilapia (*Oreochromis sp.*) carrying a single copy of an homologous cDNA growth hormone. Biochem Biophys Res Commun. 2000;267:466–72.
Martínez-Páramo S, Barbosa V, Pérez-Cerezales S, Robles V, Herráez MP. Cryoprotective effects of antifreeze proteins delivered into zebrafish embryos. Cryobiology. 2009;58:128–33.
Martínez-Páramo S, Horvath A, Labbe C, et al. Cryobanking of aquatic species. Aquaculture. 2017;472:156–77.
McGinnity P, Prodöhl P, Ferguson A, Hynes R, Maoiléidigh NÓ, Rogan G, Taggart J, et al. Fitness reduction and potential extinction of wild populations of Atlantic salmon, *Salmo salar*, as a result of interaction with escaped farm salmon. Proc R Soc Lond B. 2003;270:2443–520.
Moon SH, Lim HK, Kwon JY, Lee JK, Chang YJ. Increased plasma 17-hydroxyprogesterone and milt production in response to gonadotropin-releasing hormone agonist in captive male starry flounder, *Platichthys stellatus*. Aquaculture. 2003;218:703–16.
Muir WM, Howard D. Assessment of possible ecological risks and hazards of transgenic fish with implications for other sexually reproducing organisms. Transgenic Res. 2002;11:101–14.

Mylonas CC, Zohar Y. Use of GnRHa-delivery systems for the control of reproduction in fish. Rev Fish Biol Fish. 2001;10:463–91.

Mylonas CC, Fostier A, Zanuy S. Broodstock management and hormonal manipulations of fish reproduction. Gen Comp Endocrinol. 2010;165:516–34.

Mylonas CC, Duncan NJ, Asturiano JF. Hormonal manipulations for the enhancement of sperm production in cultured fish and evaluation of sperm quality. Aquaculture. 2017;472:21–44.

Nagasawa K, Ishida M, Octavera A, Kusano K, Kezuka F, Kitano T, Yoshiura Y, Yoshizaki G. Novel method for mass producing genetically sterile fish from surrogate broodstock via spermatogonial transplantation. Biol Reprod. 2018;100:535–46.

Nakamura S, Kobayashi K, Nishimura T, Higashijima S, Tanaka M. Identification of germline stem cells in the ovary of the teleost medaka. Science. 2010;328:1561–3.

Nam YK, Kim DS. Ploidy status of progeny from the crosses between tetraploid males and diploid females in mud loach (*Misgurnus mizolepis*). Aquaculture. 2004;236:575–82.

Nynca J, Arnold G, Frölich J, Ciereszko A. Cryopreservation-induced alterations in protein composition of rainbow trout semen. Proteomics. 2015;15:2643–54.

Okutsu T, Suzuki K, Takeuchi Y. Testicular germ cells can colonize sexually undifferentiated embryonic gonad and produce functional eggs in fish. Proc Natl Acad Sci U S A. 2006;103:2725–272.

Okutsu T, Shikina S, Kanno M, et al. Production of trout offspring from triploid salmon parents. Science. 2007;317:1517.

Pankhurst NW, Munday PL. Effects of climate change on fish reproduction and early life history stages. Mar Freshw Res. 2011;62:1015–26.

Penney W, Lush PL, Wade AJ, Brown JA, Burton MPM. Effect of Photoperiod manipulation on broodstock spawning, fertilization success, and egg developmental abnormalities in Atlantic cod, *Gadus morhua*. J World Aquacult Soc. 2006;37:273–81.

Perry AL, Low PJ, Ellis JR, Reynolds JD. Climate change and distribution shifts in marine fishes. Science. 2005;308:1912–4.

Piferrer F. Endocrine sex control strategies for the feminization of teleost fish. In: Lee C-S, Donaldson EM, editors. Reproductive biotechnology in finfish aquaculture. New York: Elsevier; 2001. p. 229–81.

Piferrer F, Beaumont A, Falguiere JC, Flajshans M, Haffray P, Colombo L. Polyploid fish and shellfish: production, biology and applications to aquaculture for performance improvement and genetic containment. Aquaculture. 2009;293:125–56.

Pimm SL, Jenkins CN, Abell R, Brooks TM, et al. The biodiversity of species and their rates of extinction, distribution, and protection. Science. 2014;344 https://doi.org/10.1126/science.1246752.

Plachinta M, Zhang T, Rawson DM. Preliminary studies on cryopreservation of zebrafish (*Danio rerio*) vitellogenic oocytes using controlled slow cooling. Cryobiology. 2004;49:347.

Poleo GA, Godke RR, Tiersch TR. Intracytoplasmic sperm injection using cryopreserved, fixed, and freeze-dried sperm in eggs of Nile tilapia. Marine Biotechnol. 2005;7:104–11.

Pörtner O, Knust R. Climate change affects marine fishes through the oxygen limitation of thermal tolerance. Science. 2007;315:95–7.

Pörtner O, Peck MA. Climate change effects on fishes and fisheries: towards a cause-and-effect understanding. J Fish Biol. 2010;77:1745–79.

Pšenička M, Saito T, Linhartová Z, Gazo I. Isolation and transplantation of sturgeon early-stage germ cells. Theriogenology. 2015;83:1085–92.

Pšenička M, Saito T, Rodina M, Dzyuba B. Cryopreservation of early stage Siberian sturgeon *Acipenser baerii* germ cells, comparison of whole tissue and dissociated cells. Cryobiology. 2016;72:119–22.

Rahman SM, Strüssmann CA, Suzuki T, Watanabe M. Electroporation enhances permeation of cryoprotectant (dimethyl sulfoxide) into Japanese whiting (*Sillago japonica)* embryos. Theriogenology. 2013;79:853–8.

Rahman SM, Strüssmann CA, Suzuki T, Majhi SK, Hattori RS, Alam MA. Effects of ultrasound on permeation of cryoprotectants into Japanese whiting *Sillago japonica* embryos. Cryobiology. 2017;77:19–24.

Rasmussen RS, Morrissey MT. Biotechnology in aquaculture: transgenics and polyploidy. Compr Rev Food Sci Food Saf. 2007;6:2–16.

Rawson DM, Reid GM, Lloyd RE. Conservation rationale, research applications and techniques in the cryopreservation of lower vertebrate biodiversity from marine and freshwater environments. Int Zoo Yearb. 2011;45:108–23.

Reed TE, Prodöhl P, Hynes R, Cross T, Ferguson A, McGinnity P. Quantifying heritable variation in fitness-related traits of wild, farmed and hybrid Atlantic salmon families in a wild river environment. Heredity. 2015;115:173–84.

Reynolds JD, Webb TJ, Hawkins LA. Life history and ecological correlates of extinction risk in European freshwater fishes. Can J Fish Aquat Sci. 2005;62:854–62.

Riesco MF, Robles V. Cryopreservation causes genetic and epigenetic changes in zebrafish genital ridges. PLoS One. 2013;8(6):e67614.

Rijnsdorp AD, Peck MA, Engelhard GH, Möllmann C, Pinnegar JK. Resolving the effect of climate change on fish populations. ICES J Mar Sci. 2009;66:1570–83.

Robles V, Cabrita E, Fletcher GL, Shears MA, King MJ, Herráez MP. Vitrification assays with embryos from a cold tolerant sub-Arctic fish species. Theriogenology. 2005;64:1633–46.

Robles V, Cabrita E, Acker JP, Herráez P. Embryo cryopreservation: what we know until now. In: Cabrite E, Robles V, Herráez P, editors. Methods in reproductive aquaculture: marine and freshwater species. Boca Raton: CRC Press; 2009. p. 265–94.

Robles V, Riesco MF, Psenicka M, Saito T, Valcarce DG, Cabrita E, Herraez P. Biology of teleost primordial germ cells (PGCs) and spermatogonia: biotechnological applications. Aquaculture. 2017;472:4–20.

Rockstrom J, Steffen W, Noone K, Persson A, et al. A safe operating space for humanity. Nature. 2009;461:472–5.

Routray P, Suzuki T, Strüssmann CA, Takai R. Factors affecting the uptake of DMSO by the eggs and embryos of medaka, *Oryzias latipes*. Theriogenology. 2002;58:1483–96.

The Royal Society. Ocean acidification due to increasing atmospheric carbon dioxide. London: The Royal Society; 2005. Policy Document 12/05.

Saragusty J. Genome banking for vertebrates wildlife conservation. In: Katkov I, editor. Current frontiers in cryobiology. Rijeka: InTech; 2012. ISBN: 978-953-51-0191-8.

Saragusty J, Arav A. Current progress in oocyte and embryo cryopreservation by slow freezing and vitrification. Reproduction. 2011;141:1–19.

Sawada Y, Okada T, Miyashita S, Murata O, Kumai H. Completion of the Pacific bluefin tuna, *Thunnus orientalis*, life cycle under aquaculture conditions. Aquacult Res. 2005;36:413–21.

Schreck CB, Contreras-Sanchez Martin W, Fitzpatrick S. Effects of stress on fish reproduction, gamete quality and progeny. In: Lee C-S, Donaldson EM, editors. Reproductive biotechnology in finfish aquaculture. New York: Elsevier; 2001. p. 3–24.

Seki S, Kouya T, Hara T, Valdez DM Jr, Jin B, Kasai M, Edashige K. Exogenous expression of rat aquaporin-3 enhances permeability to water and cryoprotectants of immature oocytes in the zebrafish (*Danio rerio*). J Reprod Dev. 2007;53:597–604.

Shaliutina A, Hulak M, Dzyuba B, Linhart O. Spermatozoa motility and variation in the seminal plasma proteome of Eurasian perch (*Perca fluviatilis*) during the reproductive season. Mol Reprod Dev. 2012;79:879–87.

Shaliutina A, Hulak M, Li P, Sulc M, Dzyuba B, Linhart O. Comparison of protein fractions in seminal plasma from multiple sperm collections in sterlet (*Acipenser ruthenus*). Reprod Domest Anim. 2013;48:156–9.

Sheehan RJ, Shasteen SP, Suresh AV, Kapuscinski AR, Seeb JE. Better growth in all-female diploid and triploid rainbow trout. Trans Am Fish Soc. 1999;128:491–8.

Simpson SD, Munday PL, Wittenrich ML, Manassa R, Dixson DL, Gagliano M, Yan HY. Ocean acidification erodes crucial auditory behaviour in a marine fish. Biol Lett. 2011;7:917–20.

Smith C, Wootton RJ. The remarkable reproductive diversity of teleost fishes. Fish Fish. 2016;17:1208–15.

Suquet M, Dreanno C, Fauvel C, Cosson J, Billard R. Cryopreservation of sperm in marine fish. Aquacult Res. 2000;31:231–43.

Taborsky M. Alternative reproductive tactics in fish. In: Oliveira RF, Taborsky M, Brockmann HJ, editors. Alternative reproductive tactics: an integrative approach. Cambridge: Cambridge University Press; 2008. p. 251–99.

Takeuchi Y, Yoshizaki G, Takeuchi T. Production of germ-line chimeras in rainbow trout by blastomere transplantation. Mol Reprod Dev. 2001;59:380–9.

Takeuchi Y, Yoshizaki G, Takeuchi T. Generation of live fry from intraperitoneally transplanted primordial germ cells in rainbow trout. Biol Reprod. 2003;69:1142–9.

Takeuchi Y, Yoshizaki G, Takeuchi T. Surrogate broodstock produces salmonids. Nature. 2004;430:629–30.

Tanaka D, Takahashi A, Takai A, Ohta H, Ueno K. Attempt at cloning high-quality goldfish breed 'Ranchu' by fin-cultured cell nuclear transplantation. Zygote. 2012;20:79–85.

Taranger GL, Aardal L, Hansen T, Kjesbu OS. Continuous light delays sexual maturation and increases growth of Atlantic cod (*Gadus morhua* L.) in sea cages. ICES J Mar Sci. 2006;63:365–75.

Taranger GL, Carrillo M, Schulz RW, Fontaine P, Zanuy S, et al. Control of puberty in farmed fish. Gen Comp Endocrinol. 2010;165:483–515.

Thomas P, Rahman MS, Picha ME, Tan W. Impaired gamete production and viability in Atlantic croaker collected throughout the 20,000 km^2 hypoxic region in the northern Gulf of Mexico. Mar Pollut Bull. 2015;101:182–92.

Tiersch TR, Jenkins JA. Biosecurity considerations for cryopreserved gametes and early life stages of aquatic species. In: Lee CS, O'Bryen PJ, editors. Biosecurity in aquaculture production systems: exclusion of pathogens and other undesirables. Baton Rouge: World Aquaculture Society; 2003.

Tiersch TR, Yang H, Jenkins JA, Dong Q. Sperm cryopreservation in fish and shellfish. Soc Reprod Fertil Suppl. 2007;65:493–508.

Tiersch TR, Yang H, Hu E. Outlook for development of high-throughput cryopreservation for small-bodied biomedical model species. Comp Biochem Physiol C. 2011;154:76–81.

Torres L, Hu E, Tiersch TR. Cryopreservation in fish: current status and pathways to quality assurance and quality control in repository development. Reprod Fertil Dev. 2016;28:1105–15.

Trombley S, Schmitz M. Leptin in fish: possible role in sexual maturation in male Atlantic salmon. Fish Physiol Biochem. 2013;39:103–6.

Tsai T, Rawson DM, Zhang T. Development of cryopreservation protocols for early stage zebrafish (*Danio rerio*) ovarian follicles using controlled slow cooling. Theriogenology. 2009;71:1226–33.

Urrego R, Rodriguez-Osorio N, Niemann H. Epigenetic disorders and altered gene expression after use of assisted reproductive technologies in domestic cattle. Epigenetics. 2014;9:803–15.

Valdez DM Jr, Hara T, Miyamoto A, Seki S, Jin B, Kasai M, Edashige K. Expression of aquaporin-3 improves the permeability to water and cryoprotectants of immature oocytes in the medaka (*Oryzias latipes*). Cryobiology. 2006;53:160–8.

Valdez DM Jr, Tsuchiya R, Seki S, Saida N, Niimi S, Koshimoto C, Matsukawa K, Kasai M, Edashige K. A trial to cryopreserve immature medaka (*Oryzias latipes*) oocytes after enhancing their permeability by exogenous expression of aquaporin 3. J Reprod Dev. 2013;59:205. https://doi.org/10.1262/jrd.2012-179.

Vermeirsen ELM, Scott AP, Mylonas CC, Zohat Y. Gonadotrophin-releasing hormone agonist stimulates milt fluidity and plasma concentrations of 17,20β-dihydroxylated and 5β-reduced, 3α-hydroxylated C_{21} steroids in male plaice (*Pleuronectes platessa*). Gen Comp Endocrinol. 1998;112:163–77.

Viveiros ATM, Orfão LH, Leal MC. Biologia e conservação de espermatozoides. In: Baldisserotto B, Cyrino JEP, Urbinati EC, editors. Biologia e Fisiologia de Peixes Neotropicais de Água Doce. Jaboticabal: FUNEP, UNESP; 2014. p. 307–27.

Wakamatsu Y, Pristyaznhyuk I, Ju B, Niwa K, Ladygina T, Kinoshit M, et al. Fertile and diploid nuclear transplants derived from embryonic cells of a small laboratory fish, medaka (*Oryzias Latipes*). Proc Natl Acad Sci U S A. 2001;98:1071–6.

Wang Y, Xiao Z, Li L, Fan W, Li SW. Novel needle immersed vitrification: a practical and convenient method with potential advantages in mouse and human ovarian tissue cryopreservation. Hum Reprod. 2008;23:2256–65.

Wang N, Teletchea F, Kestemont P, Milla S, Fontaine P. Photothermal control of the reproductive cycle in temperate fishes. Rev Aquac. 2010;2:209–22.

Wargelius A, Leininger S, Skaftnesmo KO, Kleppe L, Andersson E, Taranger GL, Schulz RW, Edvardsen RB. *Dnd* knockout ablates germ cells and demonstrates germ cell independent sex differentiation in Atlantic salmon. Sci Rep. 2016;6:21284. https://doi.org/10.1038/srep21284.

Wildt DE. Genetic resource banks for conserving wildlife species: justification, examples and becoming organized on a global basis. Anim Reprod Sci. 1992;28:247–57.

Witte F, Goldschmidt T, Wanink J, van Oijen M, Goudswaard K, Witte-Maas E, Bouton N. The destruction of an endemic species flock: quantitative data on the decline of the haplochromine cichlids of Lake Victoria. Environ Biol Fishes. 1992;34:1–28.

Xin M, Siddique MAM, Dzyuba B, Cuevas-Uribe R, Shaliutina-Kolesova A, Linhart O. Progress and challenges of fish sperm vitrification: a mini review. Theriogenology. 2017;98:16–22.

Yang H, Carmichael C, Varga ZM, Tiersch TR. Development of a simplified and standardized protocol with the potential for high-throughput for sperm cryopreservation in zebrafish, *Danio rerio*. Theriogenology. 2007;68:128–36.

Yazawa R, Takeuchi Y, Higuchi K, Yatabe T, Kabeya N, Yoshizaki G. Chub mackerel gonads support colonization, survival, and proliferation of intraperitoneally transplanted xenogenic germ cells. Biol Reprod. 2010;82:896–904.

Yazawa R, Takeuchi Y, Morita T, Ishida M, Yoshizaki G. The Pacific bluefin tuna (*Thunnus orientalis*) dead end gene is suitable as a specific molecular marker of type A spermatogonia. Mol Reprod Dev. 2013;80:871–80.

Ye H, Li C-J, Yue H-M, Hao D, Yang XG, Yoshino T, Hayashida T, Takeuchi Y, Wei Q-W. Establishment of intraperitoneal germ cell transplantation for critically endangered Chinese sturgeon *Acipenser sinensis*. Theriogenology. 2017;94:37–47.

Yoshizaki G, Lee S. Production of live fish derived from frozen germ cells via germ cell transplantation. Stem Cell Res. 2018;29:103–10.

Yoshizaki G, Yazawa R. Application of surrogate brookstock technology in aquaculture. Fish Sci. 2019; https://doi.org/10.1007/s12562.

Yoshizaki G, Takeuchi Y, Kobayashi T, et al. Primordial germ cells: the blueprint for a piscine life. Fish Physiol Biochem. 2002;26:3–12.

Yoshizaki G, Ichikawa M, Hayashi M, et al. Sexual plasticity of ovarian germ cells in rainbow trout. Development. 2010;137:1227–30.

Yoshizaki G, Fujinuma K, Iwasaki Y, Okutsu T, Shikina S, Yazawa R, Takeuchi Y. Spermatogonial transplantation in fish: a novel method for the preservation of genetic resources. Comp Biochem Physiol Part D Genomics Proteomics. 2011;6:55–6.

Zhang T, Lubzens E. Cryopreservation of fish oocytes. In: Cabrite E, Robles V, Herráez P, editors. Methods in reproductive aquaculture: marine and freshwater species. Boca Raton: CRC Press; 2009. p. 251–63.

Zhu Z, Li G, He L, Chen S. Novel gene transfer into fertilized eggs of goldfish (*Carassius auratus* L 1758). J Appl Ichthyol. 1985;1:31–4.

Zilli L, Vilella S. Effect of cryopreservation on bio-chemical parameters, DNA integrity, protein profile and phosphorylation state of proteins of seawater fish spermatozoa. In: Katkov I, editor. Current frontiers in cryobiology. Rijeka: InTech; 2012. p. 391–414.

Zohar Y, Mylonas CC. Endocrine manipulations of spawning in cultured fish: from hormones to genes. Aquaculture. 2001;197:99–136.

Chapter 8
Reproductive Microbiomes in Wild Animal Species: A New Dimension in Conservation Biology

Pierre Comizzoli and M. Power

Abstract Communities of microbes have coevolved in animal organisms and are found in almost every part of the body. Compositions of those communities (microbiota) as well as their genomes and genes (microbiomes) are critical for functional regulations of the body organ systems—the digestive or 'gut' microbiome being the most described so far. Based on extensive research in humans, microbiomes in the reproductive tract may play a role in reproductive functions and pregnancy. However, in wild animal species, those microbiomes have been poorly studied, and as a result, little is known about their involvement in fertility or parental/offspring health. This emerging research area is highly relevant to conservation biology from captive breeding management to successful reintroduction or maintenance of wild populations. The objective of this chapter is to review current knowledge about reproductive microbiomes in healthy wild animal species. While recognizing the current technical limits of microbial identification in all animal species, we also explore the link between microbial communities (within female or male reproductive systems) and fertility, from conception to birth outcome. In addition, it is critical to understanding how reproductive microbiomes are affected by environmental factors (including captivity, contact with other individuals, or changes in the ecosystem) to optimize conservation efforts. Thus, reproductive microbiomes represent a novel dimension in conservation biology that will likely gain importance in the future.

1 Introduction: Definition and Role of Microbiomes

A fundamental paradigm shift in organismal biology is that few, if any, organs or tissues are truly sterile. As demonstrated in the human body, just about every part investigated so far harbors a specific microbial community. Although microbes only

P. Comizzoli (✉) · M. Power
Smithsonian Conservation Biology Institute, National Zoological Park, Washington, DC, USA
e-mail: Comizzolip@si.edu

P. Comizzoli et al. (eds.), *Reproductive Sciences in Animal Conservation*, Advances in Experimental Medicine and Biology 1200,
https://doi.org/10.1007/978-3-030-23633-5_8

account for about 3% of human body mass, we have ten times more microbial cells living on and within us than our own cells. Most importantly, the number of expressed microbial genes that make up our microbiome far exceed the number of our genome (Ursell et al. 2012). Many molecules important to health and well-being are produced by microorganisms that have coevolved with human and animal species (McFall-Ngai et al. 2013). Thus, the microbiome has been called the second genome and as such has been shown to have significant effects on the immune system, metabolism, and disease risks (Grice and Segre 2012). The role of some microbial gene products has long been understood. For example, vertebrates do not produce enzymes that can break down cellulose. Most vertebrate herbivores and omnivores host populations of coevolved fungi, bacteria and/or protozoa symbionts in their guts, which produce the cellulases needed to digest fiber. The herbivorous life styles of cows and gorillas could not exist without their coevolved gut microbiomes. The entire vertebrate food web actually rests upon the digestive enzymes produced by gut microbes (Hanning and Diaz-Sanchez 2015). The microbiome affects biological functions in many ways beyond nutrition. Our understanding of biology is evolving through our growing understanding of the extent to which animal physiology, metabolism, and immune function are affected by microbial partners.

The taxonomic structure of a microbial community defines the 'microbiota' while the term microbiome technically refers to the totality of the genomes and genes of the microbiota (Ursell et al. 2012). For instance, the vaginal microbiome properly refers to the suite of genomes and genes from the vaginal microbiota, the microbial species that make up the communities living in the vagina. This distinction is not often made and the term vaginal microbiome is instead used to refer to both. In addition, the totality of microbial partners are often referred to as 'the microbiome', but more realistically, there are multiple microbiomes on and in animal bodies (the oral cavity, the intestinal tract, the skin and the respiratory tract among others). Extensive studies are now conducted to catalogue those microbes and annotate their functional roles. So far, the gut microbiome is the best characterized in terms of function and composition. Some organs (e.g. brain) are better protected from microbial colonization than others, and even though still controversial, low abundance of microbial DNA has been detected in the uterus and the placenta (Power et al. 2017). There is controversy over whether these microbial assemblages are truly self-sustaining microbial communities (microbiota) as opposed to invaders or opportunistic 'tourists' that manage to reach these organs but do not actually colonize them (Baker et al. 2018). There also is no clear evidence of the presence of live bacteria in most cases since just DNA is detected (Theis et al. 2019).

The various microbiomes of an animal are spatially separate, but often interconnected, with the species composition of one community being associated with the species compositions of other communities within the same organism (and even with the microbial communities of other individuals sharing the same space). There also are many symbiotic interactions between microbes, their products, and the host immune system. A healthy microbial community system is essential for normal development of the host. For example, microbiome-free rodent models display many anatomical, physiological, metabolic and immunological deficiencies

(Gensollen et al. 2016). Similarly, experiments demonstrate that germ-free fish larvae fail to develop properly in the absence of microbiomes (Merrifield and Rodiles 2015). Furthermore, the development of the neonatal immune system appears highly influenced by the presence and composition of the microbiome (Gensollen et al. 2016). For example, the maturation of the intestinal epithelium, especially its associated immune system, appears to be strongly influenced by commensal colonizing microbes (Spor et al. 2011; Geuking et al. 2014). Importantly, mammalian offspring are first exposed to the female reproductive microbiomes, and this interaction is thought to be critical, as offspring born by c-section do not fare as well as those born vaginally (Goltsman et al. 2018). There appear to be critical time windows in which the developing immune system learns to be tolerant of its commensal microbes (Macpherson et al. 2017). The lack of exposure to a normal microbiome during these windows can have long lasting negative consequences (Macpherson et al. 2017).

Providing resistance to pathogenic microbial colonization is another prime benefit of a healthy microbiome as it resists colonization by non-commensal microbes. For example, infection by the pathogen *Clostridium difficile* often is associated with recent antibiotic treatments, while a normal fecal microbiota has been shown to be associated with enhanced resistance to *Clostridium difficile* colonization. Interestingly, fecal transplantation from healthy donors has been shown to be effective in treating *C. difficile* infection (van Nood et al. 2013).

The core of animal conservation is the maintenance of genetic diversity within populations that are large enough to be sustainable and thriving (Holt et al. 2014). In addition to protecting viable populations in their natural habitat (*in situ* conservation), it is critical to maintain populations in captivity (*ex situ* conservation) for eventual reinforcements or reintroductions. Reproductive sciences therefore play a fundamental role in biodiversity conservation efforts. Understanding and sustaining healthy reproductive microbiomes in wild species is an emerging discipline that could optimize wildlife preservation. The objective of this chapter is to review current knowledge about the link between microbiomes and reproduction in healthy wild animals and its relevance to conservation biology. While recognizing the current technical limits of studying microbiomes, we specifically explore how microbial communities within male and female reproductive systems may impact fertility, from conception to implantation to birth outcome. We then focus on relations between physiological status and reproductive microbiomes. Lastly, we explore the effects of external factors (including stress, captivity, and environmental changes) on reproductive microbiomes and ultimately on conservation efforts.

2 Relevance of Microbiomes to Conservation Biology

We know little about microbiomes in animal species compared to humans. Most reports on the microbiomes of non-human species are on domesticated and laboratory animals, mostly mammals, but an increasing number of studies have been conducted on wildlife over the last decade (Redford et al. 2012; Bahrndorff et al. 2016; Colston and Jackson 2016). For instance, recent research has demonstrated how social interactions are an important determinant of gut microbiome composition in wild populations of baboons. This connection has important ramifications for understanding how social networks influence health, as well as the evolution of primate groups (Tung et al. 2015). Integrating new understanding of the patterns of microbial diversity and early signs of impending microbial disruption offers valuable tools for informing conservation strategies and ultimately monitoring primate health (Stumpf et al. 2016). Interestingly, the best illustrations of the link between microbiome and conservation are in non-mammalian species. We briefly review here what is known about the microbiomes of various non-mammalian vertebrate lineages, starting with avian species.

Birds represent over 30% of known tetrapod diversity and possess their own suite of adaptations relevant to the microbiome. In vultures, a strong association between gut bacteria to food source has been shown, exemplifying a specialized host–microbial alliance (Roggenbuck et al. 2014). Transmission of maternal, cloacal microbiome to the offspring has been reported in avian species (van Veelen et al. 2017). Fifty nine species of birds display immense variations in their microbiomes but bird taxonomy and ecology are most correlated with the gut microbiome (Hird 2017). Thus, collecting and analyzing wild microbiomes contributes to the main goals of evolutionary biology and to understanding the specific ways that avian species are unique microbial hosts. This exemplifies that to truly understand the evolution of a host, we really need to understand the millions of microorganisms that inhabit it (Hird 2017).

Amphibian skin differs from that of reptiles, birds and mammals. The amphibian skin microbiome plays an important role in physiological homeostasis and could even defend the host against pathogens (including the deadly chytrid fungus). It is therefore vital to understand what drives variation among individuals in their skin microbiome (Jiménez and Sommer 2017; Campbell et al. 2018).

In fishes, microbial communities inhabit the skin (mucosa), gills, and gastrointestinal tract as well. Those microbiomes are affected by a range of factors including temperature, seasonality, host genetics, and diet (Merrifield and Rodiles 2015). Interestingly, the gut microbiome contains core components that are commonly found in individuals of the same fish species regardless of location or living conditions (wild or captive) (Merrifield and Rodiles 2015). Fish therefore are good models for studying animal social behavior in relation to microbial transmission strategies, and can provide insights into the roles of microbiome in shaping social behaviors (Soares et al. 2018).

Some attention also has been paid to the microbiome of insects including important vectors of pathogens of human and veterinary importance, agricultural pests, and model species (bees or wasps). Collective results suggest that the microbiome of insects is highly dependent on environment, species, or population and may affect the reproductive fitness of host species (McFall-Ngai et al. 2013). For example, in drosophilae, male and female gut bacteria interact to modulate their daughters', but not sons' body masses, revealing a new trans-generational effect of parental gut microbiota during mating and reproduction (Morimoto et al. 2017). These results also are important for our understanding of invasion of non-native species, responses to pathogens, and responses to chemicals and global changes in the present and future (Bahrndorff et al. 2016).

3 Limits in Microbial Detections and Identifications

Many microbial symbionts and commensals cannot be effectively cultured. Indeed, it is estimated that only approximately 1% of bacteria are culturable (Cicinelli et al. 2009).

Our enhanced understanding of interconnected microbial communities stems from recent advances in DNA sequencing methods, the production of 16S ribosomal RNA (rRNA) gene amplicon libraries, and statistical methods that have been developed to analyze large sequence datasets (Davidson and Epperson 2018). Microbiomes are characterized by their gene-expression potential, usually estimated by whole genome shotgun sequencing, or even by sequencing the transcriptome (messenger RNAs). High-throughput sequencing of 16S rRNA gene amplicons and shotgun metagenomics enable extensive culture-independent characterizations of microbial diversity and functional potential. However, an alternative way to study community structures of naturally occurring microbial populations, along with the added benefit of recovering information about their physiological states, could rely on high throughput sequencing of transfer RNA (tRNA) transcripts (Schwartz et al. 2018). Transfer RNA-seq represents another direction in microbiome research as it relies on an actively expressed gene product that is core to protein synthesis machinery, universally abundant, and phylogenetically informative. It offers a complementary approach to the sequencing of 16S rRNA gene amplicons and genome-resolved metagenomics for broader characterization of environmental microbiomes.

As mentioned above, the taxonomic structure of a microbial community defines the microbiota. However, taxonomy is poorly known and identification remains impossible for a large proportion of microbes (up to 90% in some cases) (Menke et al. 2015). While the term microbiome technically refers to the totality of the genomes and genes of the microbiota (Ursell et al. 2012), the distinction of the microbiota can be important because DNA sequencing techniques cannot distinguish between alive and dead microbes. If microbes invade a body space but cannot colonize, they potentially will be detected as part of the microbiota based on the presence of their DNA. Also, individuals have been shown to share suites of microbial genes that produced functionally similar metabolic pathways despite differing in the taxonomic units of microbes in their gut microbiota (Turnbaugh et al. 2009). Thus, different microbiota can be functionally similar (Ursell et al. 2012). Conversely, strains of the same bacterial species often cannot be distinguished by their 16S rRNA genes, but can have different gene encoded on their chromosome, or the same strain may express genes differently depending on environmental conditions (Vital et al. 2015), thus be functionally different. This complexity is partly due to the horizontal gene transfer that can occur between bacteria (genetic material transferred between microbes of different species). In essence, microbiota are distinguished by their microbial composition based on the 16S rRNA gene, which is transmitted vertically through reproduction. Horizontal gene transfer will affect the function of the microbiome, not the microbiota.

4 Reproductive Microbiomes

4.1 Female Microbiomes

Microbiomes in the female reproductive tract are less characterized than those of the gut (Moreno et al. 2016). In women, there appear to be distinct microbial communities in the cervix and the vagina. This reflects a microbiota continuum along the human female reproductive tract (Chen et al. 2017), which we assume is representative of other animals.

The human vaginal microbiome has been investigated for over 125 years (beginning with Albert Döderlein in 1892) and it has been shown that *Lactobacillus* species often dominate the human vaginal microbiota and produce copious amount of lactic acid. The resulting acidic pH < 4 in the vaginal environment has long been considered critical to vaginal health (Fettweis et al. 2012). However, a *Lactobacillus* dominated vaginal microbiota appears to be unique to humans and not even ubiquitous to all populations (Ravel et al. 2011). Interestingly, vaginal microbiomes vary widely in primate species (Yildirim et al. 2014). Non-human primate species, such as baboons, macaques and even our closest relative the chimpanzee, have a higher bacterial species diversity in their vaginal microbiota with few if any *Lactobacillus* (Yildirim et al. 2014; Spear et al. 2010, 2012; Stumpf et al. 2013). Somewhat surprising, unlike findings regarding the fecal microbiome, microbial species patterns do not align with host phylogeny. Closely related host species do not necessarily have similar vaginal microbial communities. However, species less related to humans are more likely to harbor microbiomes that cannot be identified to the genus level, suggesting that phylogeny might be related to harboring unique taxa. In addition, in the only captive-wild comparison, no difference was observed between the vaginal microbiota of captive and wild vervet monkeys, suggesting that the vaginal microbiota is under strong regulatory control in this species (Yildirim et al. 2014). Another wild species whose vaginal microbiota has been studied is the giant panda, but results are very preliminary (see below) (Yang et al. 2017). Note that inter-individual diversity of the vaginal microbial community, coupled with its intra-individual variation based on phase of the ovarian cycle, exposure to semen from different males, and other factors, provides a significant barrier to characterizing a "normal" vaginal microbiome for any species, except in broad generalization.

Existing data suggest the presence of a commensal microbial community within the uterine cavity as well. However, even in women, there is still no consensus on whether a physiological microbiota exists in the upper reproductive tract outside of infection (Garcia-Grau et al. 2018). The assessment of the uterine microbiota has suffered from many limitations, including the need for larger and better-defined cohorts and standardized procedures to determine the uterine microbiome under physiological conditions (Garcia-Grau et al. 2018). Sampling the uterine environment without contamination from the lower genital tract is challenging and has confounded these studies. Nevertheless, resident microbiomes have been explored in uteruses from mares and sheep (Heil et al. 2018; Smith and Garza 2016). The impact of the uterine microbiome on infertility in cows has been examined as well (Carneiro et al. 2016). Authors suggest that to develop strategies to limit the impact of uterine disease on fertility, future work should focus on determining which bacteria and virulence factors cause endometritis, and understanding how the host response to infection is regulated in the endometrium (Carneiro et al. 2016). Preliminary data also have been reported in the giant panda uterus to explore negative pregnancy outcomes (Yang et al. 2017).

Less is known about the placental microbiome, although the association between the placenta's bacterial makeup and preterm labor or other pregnancy complications is being investigated in women (Pelzer et al. 2017). Interestingly, the human placental microbiome appears to be most similar to the maternal oral microbiome (Pelzer et al. 2017). But, as mentioned above, these results still need to be confirmed (Theis et al. 2019).

The milk/mammary microbiome shares clear similarities with the maternal gut microbiome (Power et al. 2017). The mammary/milk microbiome appears to be influenced by maternal characteristics and may play a role in inoculating the infant but may also be affected by the infant's oral microbiome (Gomez-Gallego et al. 2016). However, little is known about the milk microbiome and its role in offspring physiology, immune function and health outcome.

Lastly, there is an interconnectivity between female reproductive microbiomes, other maternal microbiomes, and developing infant microbiomes with potential roles of each in maternal, fetal, and neonatal health or disease (Chen et al. 2017).

4.2 Male Microbiomes

The male reproductive tract microbiota are still poorly studied. Microbial communities exist in the human male lower genital tract, mostly in urethra and coronal sulcus, with high inter-subject variability (Mändar 2013). Semen and vaginal samples from couples showed that semen had a greater bacterial diversity, but lower bacterial count (Mändar et al. 2015). Most semen samples from sperm donors and infertile men contain diverse microbial communities, with many of the taxa present associated with bacterial vaginosis, a known dysbiosis of the vaginal microbiota in women (Hou et al. 2013). There is only one study of human testicular microbiome. It shows that normozoospermic men have small amounts of bacteria in the testis, with Actinobacteria, Bacteroidetes, Firmicutes, and Proteobacteria as the dominant phyla (Alfano et al. 2018). Bacteria harbored in the male reproductive system may influence reproductive function and health of the male and result in developmental origins of adult health and disease effects in offspring (Javurek et al. 2016). One rare example in animal species is the penile microbiota studies in Dorper rams showing that it is associated with pathology rather than fertility (Marie and Courchay 2017).

The microbiota of the male reproductive tract also could serve as a source of microbes for the female genital microbiota. Semen is enriched in carbohydrate with a pH of between 7.1 and 8.0, providing an excellent environment for microorganisms (Altmäe 2018).

5 Relations Between Normal Physiology and Reproductive Microbiomes

There are some consistencies between how various parameters affect both human and baboon vaginal microbiota, such as how both species exhibit profound changes in microbial composition in response to puberty, reproductive state, ovarian cycle phase, and sexual behavior. For example, both baboon and human vaginal microbial communities become less diverse after puberty (Uchihashi et al. 2015). Major shifts in composition during ovulation may have implications for disease risk and conception success. There is a need for future studies to account for fine-scale differences in reproductive state, particularly microbiota differences that occur during the various phases of the ovarian cycle (Miller et al. 2017). A holistic approach is important since there is an interconnectivity between the female reproductive microbiomes and other maternal microbiomes, such that disruptions and changes to the microbial community on one part of the body has the possibility of affecting multiple microbiomes (Chen et al. 2017).

Human endometrial microbiome exists and is independent of hormonal regulation. Endometrial microbiota appear to be highly stable during the acquisition of endometrial receptivity. The existence of non-*Lactobacillus* bacteria in the endometrium could be correlated with negative impacts on reproductive function in women and should be considered as an emerging cause of implantation failure and pregnancy loss (Moreno et al. 2016). More studies are required though, as women who have low or even no vaginal and endometrial *Lactobacillus* can become pregnant without any problems. Anyway, most other species have low levels of *Lactobacillus* in the female reproductive tract, indicating humans may be a poor model for other mammals. For example, *Lactobacillus* represented less than 0.05% of the giant panda vaginal microbiota (Yang et al. 2017).

Microbiota remain relatively stable at each body site during pregnancy. However, disruptions in maternal microbiomes may be linked to pregnancy complications and maternal, fetal, and neonatal health. Thus, pregnancy outcomes might be predicted by features of the microbiota early in gestation (DiGiulio et al. 2015).

Interestingly in humans, the vaginal microbiome changes regardless of normal delivery or caesarian section (Goltsman et al. 2018). However, not surprisingly, the gut microbiome of an infant born from caesarian section is different in infants born vaginally (Bäckhed et al. 2015).

Reproductive microbiomes can also be affected by mating. As mentioned earlier, semen and vaginal samples from human couples show a strong concordance in microbial taxa (Mändar et al. 2015). The development of biofilms in mare endometritis (leading to subfertility) is directly related to the male microbiome (Leblanc and Causey 2009). Another relevant example is the cloacal microbiome in birds that is affected by mating (Hird 2017).

Besides the influence of reproductive cycles and natural mating, other physiological functions may interact with reproductive microbiomes. There is a clear interplay between microbiota and the endocrine system including the production of steroid hormones that affect reproduction (Neuman et al. 2015). Although the precise pathways of microbiota-hormonal signaling have not yet been deciphered, the microbiota produces and secretes hormones, responds to host hormones and regulates expression levels of host hormones. The gut microbiome is a virtual endocrine organ with implications on reproductive physiology as shown in farm and domestic animals (O'Callaghan et al. 2016). A recent study in Southern White Rhinos also suggest that, in addition to species differences in estrogen receptor sensitivity to phytoestrogens, reproductive outcomes may be driven by the gut microbiota's transformation of dietary phytoestrogens in captive females (Williams et al. 2019). Microbiota dysbiosis may be an additional input into the concept of allostatic load, in which cumulative life-time stresses result in physiological dysregulation that leads to poor health.

6 Environmental Factors Directly and Indirectly Influencing Reproductive Microbiomes

Like humans, animals acquire their microbiota from their parents and from the immediate environment in which they reside, which likely includes conspecifics, food, nesting materials, and other species-specific contacts. Considering the difficulty of breeding endangered species in captivity, altered microbiomes might be an important factor contributing to the sub-fertility in animals housed in zoos. In most cases, captivity likely alters many if not most of the inputs from which the microbiomes derive. For instance, captive Tasmanian devils have significantly lower levels of gut bacterial diversity compared to wild individuals (Cheng et al. 2015). Anthropogenic influence can dramatically reshape equid gastrointestinal microbiomes, which has broader implications for the conservation management of endangered mammals (Metcalf et al. 2017). In contrast, gut microbiomes of free-ranging and captive Namibian cheetahs are comparable (Wasimuddin et al. 2017). The gut microbiome is the most variable and adaptable because it is constantly influenced by both internal and external food sources. In contrast, reproductive microbiomes are expected to be more stable since they do not receive direct influence from food or the environment. In women, the evidence of variation in the vaginal microbiota over the menstrual cycle is mixed, with evidence for microbiota stability (Chaban et al. 2014) and for rapid change during menstruations in some women (Gajer et al. 2012). As mentioned earlier, one study about reproductive microbiomes reports that there is no difference between the vaginal microbiota of captive and wild vervet monkeys (Yildirim et al. 2014). This also is aligned with the fact that the core component of the normal reproductive microbiome (vagina or mammary glands) is the same for women living in completely different geographical areas. However, similar comparative studies should be performed in other species as they could address breeding difficulties or pregnancy loss.

Protecting, relocating, reinforcing and reintroducing rare species in their natural habitat also has to take into account the composition and potential roles of environmental microbiomes (soil, water, ecosystems, rural vs. urban areas (Redford et al. 2012)) and the presence of disrupting factors (e.g. antibiotics) that can potentially influence reproductive microbiomes. In that sense, marine animal microbiome research is profiting from the elevation in microbiome research funding and attention. There could also be growing interest in using marine animals as models for examining resilience, promoted by the fact that alterations in the ocean conditions are often outpacing those in terrestrial environments. Given the phylogenetic breadth of animals in the ocean, coupled with the many diverse ocean environments, there is certainly a wealth of research opportunities available to study their host–microbiome interactions (Apprill 2017).

Lastly, assisted reproductive techniques (administration of exogenous hormones, handling of gametes/embryos in vitro, artificial insemination, or embryo transfer) that are developed to better manage populations of endangered species may also modify the reproductive microbiomes through introduction of contaminating microbes. This has been explored minimally in humans (Sirota et al. 2014; Franasiak et al. 2016) but should become a huge area of interest in the coming years since it is expected to be closely related to success of implantation, healthy pregnancy, and healthy offspring.

7 Conclusions

The main emphasis of this chapter was to provide a summary of information on reproductive microbiomes in healthy wild animal species, which is unfortunately limited. As itemized in Table 8.1, there is still a lot to do in terms of basic microbiome studies but we also have to keep in mind priorities and new horizons. In the coming years, it is expected that more reports will be published in wild animal species but we need to make sure that understanding, monitoring and even manipulating reproductive microbiomes is properly and efficiently integrated into conservation efforts. There are also undeniable lessons to learn from human studies to better explore and understand reproductive microbiomes in wild animal species. A better

Table 8.1 Reproductive microbiomes in wild animal species—priorities and new horizons

Priorities
• Comprehensive microbe taxonomy for better identification
• Study of reproductive microbiomes in more, targeted taxa (including domestic species used as models) For instance, marsupials, amphibian (cloacal and spermic urine microbiome), invertebrates (oysters, corals)
• Characterize the male microbiomes in all animal species
• Investigate seasonality and microbiomes
• Monitor microbiomes during estrus cycles and pregnancies (future non-invasive markers)
• Understanding the role of microbiomes during embryo development (oviduct), implantation, delayed implantation, pregnancy (uterus, placenta), and delivery
• Influence of captive breeding on reproductive microbiomes—'domestication' of the microbiomes

New horizons
• Use systems biology to understand the connection within and between microbiomes as well as with other biological networks
• Microbiomes and assisted reproductive techniques For example, effects of semen microbiota on in vitro fertilization, impact of artificial insemination, or embryo transfer on microbiomes during gestation
• Use microbiomes as new biomarkers For example, early signs of impending microbial disruption for informing conservation strategies and ensuring health, or allostatic load
• Manipulating or bioengineering microbiomes to enhance or suppress fertility
• Conservation policy or legislation to include microbiome assessments (for relocation and reintroduction in natural habitats)

knowledge of reproductive microbiomes and their sources of variability and dysbiosis will enhance conservation efforts from management of captive or wild populations to reintroductions or relocations.

What we know is that the reproductive tract harbors different microbiomes that are related to the regulation of physiological status (Fig. 8.1). Interactions with other microbiomes (especially the gut microbiome) are also critical in homeostasis. Importantly, microbiomes are transmitted from the mother to her offspring and then can be exchanged between individuals through social interactions or natural mating (Fig. 8.1). Environmental factors (as well as other species sharing the same habitat) can also influence the composition and function of reproductive microbiomes (Fig. 8.1).

It is important to briefly mention the vast body of research about the influence of diseases, pathogens, or medical conditions on microbiomes. In those cases, disruption of microbiomes may influence reproduction and vice versa. Again, this has been clearly demonstrated in humans where microbiomes can be used as biomarkers of pathological conditions (e.g. Poly-cystic ovarian syndrome, ovarian cancer) (Neuman et al. 2015). Another example is the association between human vaginal microbiomes and the risk for bacterial vaginosis (Power et al. 2017). Researchers are currently in the early discovery phase for developing probiotic treatments for

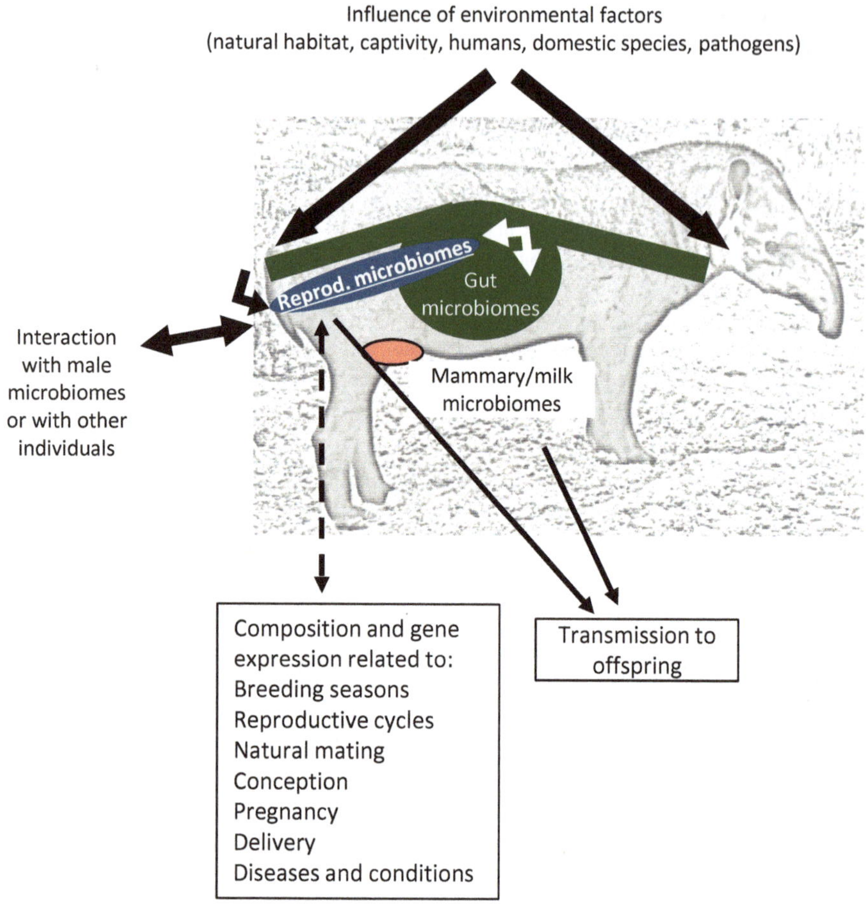

Fig. 8.1 Relations of reproductive microbiomes with other microbiomes, physiological status, and environmental factors (double white arrow indicates interactions between microbiomes)

protection against infectious diseases, especially for the new antibiotic-resistant strains that are becoming common (Pamer 2016). Exploring reproductive microbiomes in detail and pairing this information with an individual's detailed medical history will provide a more complete picture of the status and importance of the microbial network to health (Power et al. 2017). Microbiomes should also be considered as relevant biomarkers in the measurement of allostatic load on animal welfare. Mitigating imbalances in microbiomes has already been experimented in human and veterinary medicine using fecal microbiome from healthy individuals to treat inflammatory bowel disease (Browne and Kelly 2017). Probiotic therapies such as "vaginal seeding", in which the neonate is swabbed with vaginal fluid after a cesarean birth, offer potential health benefits, but require more rigorous testing, as there are also potential dangers.

References

Alfano M, Ferrarese R, Locatelli I, et al. Testicular microbiome in azoospermic men-first evidence of the impact of an altered microenvironment. Hum Reprod. 2018;33:1212–7. https://doi.org/10.1093/humrep/dey116.

Altmäe S. Commentary: uterine microbiota: residents, tourists, or invaders? Front Immunol. 2018;9:1874.

Apprill A. Marine animal microbiomes: toward understanding host–microbiome interactions in a changing ocean. Front Mar Sci. 2017;4 https://doi.org/10.3389/fmars.2017.00222.

Bäckhed F, Roswall J, Peng Y, et al. Dynamics and stabilization of the human gut microbiome during the first year of life. Cell Host Microbe. 2015;17:690–703.

Bahrndorff S, Alemu T, Alemneh T, Lund Nielsen J. The microbiome of animals: implications for conservation biology. Int J Genomics. 2016;2016:1–7. https://doi.org/10.1155/2016/5304028.

Baker JM, Chase DM, Herbst-Kralovetz MM. Uterine microbiota: residents, tourists, or invaders? Front Immunol. 2018;9 https://doi.org/10.3389/fimmu.2018.00208.

Browne AS, Kelly CR. Fecal transplant in inflammatory bowel disease. Gastroenterol Clin N Am. 2017;46:825–37.

Campbell LJ, Hammond SA, Price SJ, Sharma MD, Garner TWJ, Birol I, Helbing CC, Wilfert L, Griffiths AGF. A novel approach to wildlife transcriptomics provides evidence of disease-mediated differential expression and changes to the microbiome of amphibian populations. Mol Ecol. 2018;27:1413–27.

Carneiro LC, Cronin JG, Sheldon IM. Mechanisms linking bacterial infections of the bovine endometrium to disease and infertility. Reprod Biol. 2016;16:1–7.

Chaban B, Links MG, Jayaprakash T, et al. Characterization of the vaginal microbiota of healthy Canadian women through the menstrual cycle. Microbiome. 2014;2:23.

Chen C, Song X, Wei W, et al. The microbiota continuum along the female reproductive tract and its relation to uterine-related diseases. Nat Commun. 2017;8:875. https://doi.org/10.1038/s41467-017-00901-0.

Cheng Y, Fox S, Pemberton D, Hogg C, Papenfuss AT, Belov K. The Tasmanian devil microbiome-implications for conservation and management. Microbiome. 2015;3:76.

Cicinelli E, De Ziegler D, Nicoletti R, Tinelli R, Saliani N, Resta L, Bellavia M, De Vito D. Poor reliability of vaginal and endocervical cultures for evaluating microbiology of endometrial cavity in women with chronic endometritis. Gynecol Obstet Investig. 2009;68:108–15.

Colston TJ, Jackson CR. Microbiome evolution along divergent branches of the vertebrate tree of life: what is known and unknown. Mol Ecol. 2016;25:3776–800.

Davidson RM, Epperson LE. Microbiome sequencing methods for studying human diseases. Methods Mol Biol. 2018;1706:77–90.

DiGiulio DB, Callahan BJ, McMurdie PJ, et al. Temporal and spatial variation of the human microbiota during pregnancy. Proc Natl Acad Sci. 2015;112:11060–5.

Fettweis JM, Serrano MG, Sheth NU, Mayer CM, Glascock AL, Brooks JP, Jefferson KK, Buck GA. Species-level classification of the vaginal microbiome. BMC Genomics. 2012;13(Suppl 8):S17.

Franasiak JM, Werner MD, Juneau CR, Tao X, Landis J, Zhan Y, Treff NR, Scott RT. Endometrial microbiome at the time of embryo transfer: next-generation sequencing of the 16S ribosomal subunit. J Assist Reprod Genet. 2016;33:129–36.

Gajer P, Brotman RM, Bai G, et al. Temporal dynamics of the human vaginal microbiota. Sci Transl Med. 2012;4:132ra52.

Garcia-Grau I, Simon C, Moreno I. Uterine microbiome—low biomass and high expectations. Biol Reprod. 2018; https://doi.org/10.1093/biolre/ioy257.

Gensollen T, Iyer SS, Kasper DL, Blumberg RS. How colonization by microbiota in early life shapes the immune system. Science. 2016;352:539–44.

Geuking MB, Köller Y, Rupp S, McCoy KD. The interplay between the gut microbiota and the immune system. Gut Microbes. 2014;5:411–8.

Goltsman DSA, Sun CL, Proctor DM, et al. Metagenomic analysis with strain-level resolution reveals fine-scale variation in the human pregnancy microbiome. bioRxiv. 2018; https://doi.org/10.1101/266700.

Gomez-Gallego C, Garcia-Mantrana I, Salminen S, Collado MC. The human milk microbiome and factors influencing its composition and activity. Semin Fetal Neonatal Med. 2016;21:400–5.

Grice EA, Segre JA. The human microbiome: our second genome. Annu Rev Genomics Hum Genet. 2012;13:151–70.

Hanning I, Diaz-Sanchez S. The functionality of the gastrointestinal microbiome in non-human animals. Microbiome. 2015;3:51.

Heil BA, Thompson SK, Kearns TA, Davolli GM, King G, Sones JL. Metagenetic characterization of the resident equine uterine microbiome using multiple techniques. J Equine Vet Sci. 2018;66:111.

Hird SM. Evolutionary biology needs wild microbiomes. Front Microbiol. 2017;8 https://doi.org/10.3389/fmicb.2017.00725.

Holt WVWV, Brown JLJL, Comizzoli P. Reproductive science as an essential component of conservation biology. Adv Exp Med Biol. 2014;753:3–14.

Hou D, Zhou X, Zhong X, Settles ML, Herring J, Wang L, Abdo Z, Forney LJ, Xu C. Microbiota of the seminal fluid from healthy and infertile men. Fertil Steril. 2013;100:1261–9.

Javurek AB, Spollen WG, Ali AMM, Johnson SA, Lubahn DB, Bivens NJ, Bromert KH, Ellersieck MR, Givan SA, Rosenfeld CS. Discovery of a novel seminal fluid microbiome and influence of estrogen receptor alpha genetic status. Sci Rep. 2016;6 https://doi.org/10.1038/srep23027.

Jiménez RR, Sommer S. The amphibian microbiome: natural range of variation, pathogenic dysbiosis, and role in conservation. Biodivers Conserv. 2017;26:763–86.

Leblanc M, Causey R. Clinical and subclinical endometritis in the mare: both threats to fertility. Reprod Domest Anim. 2009;44:10–22.

Macpherson AJ, de Agüero MG, Ganal-Vonarburg SC. How nutrition and the maternal microbiota shape the neonatal immune system. Nat Rev Immunol. 2017;17:508–17.

Mändar R. Microbiota of male genital tract: impact on the health of man and his partner. Pharmacol Res. 2013;69:32–41.

Mändar R, Punab M, Borovkova N, et al. Complementary seminovaginal microbiome in couples. Res Microbiol. 2015;166:440–7.

Marie A, Courchay M. Molecular characterization of the penile microbiome of Dorper rams (Ovis aries). Thesis presented in fulfilment of the requirements for the degree of masters in the Faculty of Science at Stellenbosch University; 2017.

McFall-Ngai M, Hadfield MG, Bosch TCG, et al. Animals in a bacterial world, a new imperative for the life sciences. Proc Natl Acad Sci U S A. 2013;110:3229–36.

Menke S, Meier M, Sommer S. Shifts in the gut microbiome observed in wildlife faecal samples exposed to natural weather conditions: lessons from time-series analyses using next-generation sequencing for application in field studies. Methods Ecol Evol. 2015;6:1080–7.

Merrifield DL, Rodiles A. The fish microbiome and its interactions with mucosal tissues. In: Peatman E, editor. Mucosal health in aquaculture. San Diego: Academic Press; 2015. p. 273–95.

Metcalf JL, Song SJ, Morton JT, et al. Evaluating the impact of domestication and captivity on the horse gut microbiome. Sci Rep. 2017;7:15497.

Miller EA, Livermore JA, Alberts SC, Tung J, Archie EA. Ovarian cycling and reproductive state shape the vaginal microbiota in wild baboons. Microbiome. 2017;5:8. https://doi.org/10.1186/s40168-017-0228-z.

Moreno I, Codoñer FM, Vilella F, et al. Evidence that the endometrial microbiota has an effect on implantation success or failure. Am J Obstet Gynecol. 2016;215:684–703. https://doi.org/10.1016/j.ajog.2016.09.075.

Morimoto J, Simpson SJ, Ponton F. Direct and trans-generational effects of male and female gut microbiota in Drosophila melanogaster. Biol Lett. 2017;13:20160966. https://doi.org/10.1098/rsbl.2016.0966.

Neuman H, Debelius JW, Knight R, Koren O. Microbial endocrinology: the interplay between the microbiota and the endocrine system. FEMS Microbiol Rev. 2015;39:509–21.

O'Callaghan TF, Ross RP, Stanton C, Clarke G. The gut microbiome as a virtual endocrine organ with implications for farm and domestic animal endocrinology. Domest Anim Endocrinol. 2016;56:S44–55.

Pamer EG. Resurrecting the intestinal microbiota to combat antibiotic-resistant pathogens. Science. 2016;352:535–8.

Pelzer E, Gomez-Arango LF, Barrett HL, Nitert MD. Review: maternal health and the placental microbiome. Placenta. 2017;54:30–7.

Power ML, Quaglieri C, Schulkin J. Reproductive microbiomes: a new thread in the microbial network. Reprod Sci. 2017;24:1482–92.

Ravel J, Gajer P, Abdo Z, et al. Vaginal microbiome of reproductive-age women. Proc Natl Acad Sci. 2011;108:4680–7.

Redford KH, Segre JA, Salafsky N, Del Rio CM, Mcaloose D. Conservation and the microbiome. Conserv Biol. 2012;26:195–7.

Roggenbuck M, Schnell IB, Blom N, Bælum J, Bertelsen MF, Pontén TS, Sørensen SJ, Gilbert MTP, Graves GR, Hansen LH. The microbiome of new world vultures. Nat Commun. 2014; https://doi.org/10.1038/ncomms6498.

Schwartz MH, Wang H, Pan JN, Clark WC, Cui S. Microbiome characterization by high-throughput transfer RNA sequencing and modification analysis. Nat Commun. 2018;9:5353. https://doi.org/10.1038/s41467-018-07675-z.

Sirota I, Zarek SM, Segars JH. Potential influence of the microbiome on infertility and assisted reproductive technology. Semin Reprod Med. 2014;32:35–42.

Smith KE, Garza AL, Robinson C, Ashley RL, Ivey SL.WS Influence of sampling location and pregnancy on composition of the microbiome associated with the reproductive tract of the ewe. Journal of Animal Science, suppl. 2016 Supplement 5; 94: 498. https://doi.org/10.2527/jam2016-1039

Soares MC, Cable J, Lima-Maximino M, Maximino C, Xavier R. Using fish models for assessing the role of sociality on the microbiome: the next step for translational microbiome research? 2018; https://doi.org/10.20944/PREPRINTS201809.0387.V1.

Spear GT, Gilbert D, Sikaroodi M, Doyle L, Green L, Gillevet PM, Landay AL, Veazey RS. Identification of rhesus macaque genital microbiota by 16S pyrosequencing shows similarities to human bacterial vaginosis: implications for use as an animal model for HIV vaginal infection. AIDS Res Hum Retrovir. 2010;26:193–200.

Spear G, Rothaeulser K, Fritts L, Gillevet PM, Miller CJ. In captive rhesus macaques, cervicovaginal inflammation is common but not associated with the stable polymicrobial microbiome. PLoS One. 2012;7:e52992.

Spor A, Koren O, Ley R. Unravelling the effects of the environment and host genotype on the gut microbiome. Nat Rev Microbiol. 2011;9:279–90.

Stumpf RM, Wilson BA, Rivera A, Yildirim S, Yeoman CJ, Polk JD, White BA, Leigh SR. The primate vaginal microbiome: comparative context and implications for human health and disease. Am J Phys Anthropol. 2013;152:119–34.

Stumpf RM, Gomez A, Amato KR, Yeoman CJ, Polk JD, Wilson BA, Nelson K, White BA, Leigh SR. Microbiomes, metagenomics, and primate conservation: new strategies, tools, and applications. Biol Conserv. 2016;199:56–66.

Theis KR, Romero R, Winters AD, et al. Does the human placenta delivered at term have a microbiota? Results of cultivation, quantitative real-time PCR, 16S rRNA gene sequencing, and metagenomics. Am J Obstet Gynecol. 2019;220:267.e1–267.e39.

Tung J, Barreiro LB, Burns MB, Grenier JC, Lynch J, Grieneisen LE, Altmann J, Alberts SC, Blekhman R, Archie EA. Social networks predict gut microbiome composition in wild baboons. elife. 2015;2015:1–18.

Turnbaugh PJ, Ridaura VK, Faith JJ, Rey FE, Knight R, Gordon JI. The effect of diet on the human gut microbiome: a metagenomic analysis in humanized Gnotobiotic mice. Sci Transl Med. 2009;1:6ra14.

Uchihashi M, Bergin IL, Bassis CM, Hashway SA, Chai D, Bell JD. Influence of age, reproductive cycling status, and menstruation on the vaginal microbiome in baboons (*Papio anubis*). Am J Primatol. 2015;77:563–78.

Ursell LK, Metcalf JL, Parfrey LW, Knight R. Defining the human microbiome. Nutr Rev. 2012;70:S38–44.

van Nood E, Dijkgraaf MGW, Keller JJ. Duodenal infusion of feces for recurrent *Clostridium difficile*. N Engl J Med. 2013;368:2143–5.

van Veelen HPJ, Falcao Salles J, Tieleman BI. Multi-level comparisons of cloacal, skin, feather and nest-associated microbiota suggest considerable influence of horizontal acquisition on the microbiota assembly of sympatric woodlarks and skylarks. Microbiome. 2017;5:156.

Vital M, Harkema JR, Rizzo M, Tiedje J, Brandenberger C. Alterations of the murine gut microbiome with age and allergic airway disease. J Immunol Res. 2015;2015:1–8.

Wasimuddin MS, Melzheimer J, Thalwitzer S, Heinrich S, Wachter B, Sommer S. Gut microbiomes of free-ranging and captive Namibian cheetahs: diversity, putative functions and occurrence of potential pathogens. Mol Ecol. 2017;26:5515–27.

Williams CL, Ybarra AR, Meredith AN, Durrant BS, Tubbs CW. Gut microbiota and phytoestrogen-associated infertility in southern white rhinoceros. MBio. 2019;10 https://doi.org/10.1128/mBio.00311-19.

Yang X, Cheng G, Li C, et al. The normal vaginal and uterine bacterial microbiome in giant pandas (Ailuropoda melanoleuca). Microbiol Res. 2017;199:1–9.

Yildirim S, Yeoman CJ, Janga SC, Thomas SM, Ho M, Leigh SR, Consortium PM, White BA, Wilson BA, Stumpf RM. Primate vaginal microbiomes exhibit species specificity without universal lactobacillus dominance. ISME J. 2014;8:2431–44.

Part II
Progress in Understanding, Assisting, or Suppressing Reproduction in Wild Species

Chapter 9
Update on Comparative Biology of Elephants: Factors Affecting Reproduction, Health and Welfare

Janine L. Brown

Abstract Asian (*Elephas maximus*) and African (*Loxodonta africana*) elephants serve as important keystone, umbrella and flagship species. Despite that, population numbers are declining, due mainly to poaching and habitat destruction. Understanding reproductive mechanisms is vital to effective management, particularly insurance populations in captivity, and to that end, long-term biological databases are key to understanding how intrinsic and extrinsic factors affect reproductive function at individual and population levels. Through decades of hormonal and ultrasonographic monitoring, many unique aspects of zoo elephant reproduction have been identified, including differences in luteal steroidogenic activity, follicular maturation, pituitary gonadotropin secretion, fetal development and reproductive tract anatomy. Reproductive problems also hamper captive propagation efforts, particularly those related to abnormal or lack of ovarian cyclicity. Recent large-scale, multi-institutional studies and use of epidemiological approaches have identified factors important for good welfare and reproduction, which include enrichment, feeding diversity, good elephant-keeper relations, social compatibility, exercise, and not being obese. There are notable differences in reproductive mechanisms between Asian and African elephants, as well as the factors that influence reproduction and welfare, suggesting species-targeted management approaches are needed to maximize fitness. In the first edition, we discussed reproductive function in male and female elephants. Since then, a number of significant advances have been made primarily in female elephants, which will be the focus of this updated review.

Keywords African and Asian elephant · Reproduction · Endocrinology · Ultrasonography · Ovarian cycle · Pregnancy · Welfare · Health · Obesity · Metabolic function · Fertility

J. L. Brown (✉)
Smithsonian Conservation Biology Institute, National Zoological Park, Front Royal, VA, USA
e-mail: brownjan@si.edu

P. Comizzoli et al. (eds.), *Reproductive Sciences in Animal Conservation*, Advances in Experimental Medicine and Biology 1200,
https://doi.org/10.1007/978-3-030-23633-5_9

1 Introduction

Elephants serve as keystone species and modify habitat by converting forests to grassland, creating water holes in times of drought, and spreading the seeds of plants. They also act as umbrella species, as the conservation of elephants preserves not only habitat, but the species therein. As iconic flagship species, elephants raise awareness for action and funding of broader conservation efforts. However, despite their importance, wild elephants are threatened with extinction throughout most of their natural range, due primarily to poaching (Wittemyer et al. 2014). The Great Elephant Census, conducted between 2007 and 2014 was the first continent-wide, standardized survey of African savannah elephants, and estimated the extant population to be about 350,000; in areas with historical data, numbers had decreased by an estimated 144,000 (Chase et al. 2016). Furthermore, African elephant populations are currently shrinking by 8% per year (Wittemyer et al. 2014), despite the fact that roughly 84% live in protected areas (Chase et al. 2016). The Asian elephant (*Elephas maximus*) is endangered, with less than 55,000 remaining among the 13 range states; those population too are declining (IUCN 2018). Though not a panacea for human exploitation, captive populations are increasingly viewed as a component of species conservation (Conde et al. 2011; Olive and Jansen 2017; Che-Castaldo et al. 2018). Moreover, captive elephants contribute to important research, and can serve as a source of animals for reintroduction (e.g., Angkawanish and Thitaram 2012) if wild populations decrease to crisis levels.

Globally, there are about 1000 African elephants in captivity, mostly in zoos, and upwards of 16,000 Asian elephants in zoos, circuses, sanctuaries, logging and tourist camps. Unfortunately, many captive populations are not self-sustaining due to high mortality and low birth rates, including in range countries (Thitaram 2012). In the U.S., only 3.5 births to five deaths occur annually, with demographic modeling suggesting increasing reproduction is more important than decreasing mortality for sustainability, although some importation will be necessary in the short-term, at least for Asian elephants (Faust and Marti 2011a, b). By contrast, Asian elephants in Europe are now self-sustaining, while the African population is close (A.-K. Oerke, pers. comm., European Elephant Group, data archive). Some captive populations in Asia also are breeding well (Pushpakumara et al. 2016; C. Thitaram, pers. comm.), although for others, supplementation by illegal wild capture is needed and still ongoing. There is little doubt that increasing reproduction is key to population sustainability, especially when coupled with efforts to maximize welfare, as the latter has a direct impact on reproductive fitness.

Efforts to stabilize captive elephant populations are hampered by numerous challenges. Some are logistical, like not housing fertile males and females together, while others are physiological, and include ovarian cycle abnormalities, uterine pathologies, gestational difficulties and bull subfertility. There also are behavioral issues, such as mate incompatibility and poor libido that thwart breeding success. Through advances in endocrine and health monitoring, and ultrasonographic imaging, we are beginning to understand the complex mechanisms involved in controlling reproductive function in female elephants. As a result, through nearly three decades of study, we have an expansive database for both Asian and African elephants. Several reproductive traits appear unique to Elephantidae, which have at times helped and at other times impeded breeding efforts. Adapting domestic, laboratory and other wildlife techniques to elephants has not always been straightforward; however, a successful artificial insemination approach has been developed that continues to aid captive propagation efforts.

This chapter updates current knowledge of female Asian and African elephant reproductive biology, expanding on the previous review (Brown 2014), with attention to species differences in reproductive function, fertility problems, welfare requirements, and how use of assisted reproductive techniques are enhancing reproductive efficiency and genetic management.

2 General Reproductive Biology

2.1 *Life History Traits*

In the wild, elephant females reach puberty around 10–14 years of age, but in captivity can mature much earlier, perhaps due to better nutrition and reaching a body weight capable of supporting reproductive activity at a younger age (Kaplowitz 2008; Mendis et al. 2017). In U.S. zoos, the youngest Asian and African elephants to give birth were 5 and 10 years old, respectively (Glaeser et al. 2012; Keele et al. 2014; Olson 2014). Captive elephants in range countries mature earlier as well, like those at the Pinnawala Elephant Orphanage in Sri Lanka, where females have begun cycling as young as 4.5 and conceived by 8 years of age (Mendis et al. 2017; Pushpakumara et al. 2016), while records on >8000 working elephants in the Myanma Timber Enterprise (Mar 2007) document a birth in a 5.3-year-old (Hayward et al. 2014). In Europe, the earliest age at first reproduction was 5 years, with a mean age of first reproduction being 11–15 years (range: 5–30 years), and inter-birth interval of 5.5 years (range: 1.0–11.7 years). In the Udawalawe National Park, Sri Lanka the average age at first calving was about 13 years (de Silva et al. 2013), in line with other wild populations of both Asian and African elephants (Sukumar 2003; Allen 2006; Lee et al. 2016).

African elephants that commenced reproduction early had higher overall reproductive rates, compared to those that commenced later (Lee et al. 2016). The early onset of reproduction did not affect survival, nor did their calves experience reduced longevity. This contrasts with data on timber elephants in Myanmar, where an association between reproduction and adult survival was observed to be positive in early, but negative in later life (Robinson et al. 2012). Reproduction and survival trade-offs were greater after peak reproduction was achieved, and investing in offspring after the age of 30 years decreased survival probabilities. Thus, long-lived Asian females in this population produced fewer offspring over their lifetimes. Whether these life trait differences are species-related is not known because extensive databases of longevity, reproduction, and/or survival are only available for a few populations.

Age at last reproduction varies between species, global regions, and wild versus captive populations (e.g., Clubb et al. 2009), with rare instances of females giving birth into their 50s (Pinnawala Elephant Orphanage, Sri Lanka, Pushpakumara et al. 2016; Udawalawe National Park, Sri Lanka, de Silva et al. 2013; Amboseli, Africa, Lee et al. 2016) and even 60s (Myanmar timber elephants, Hayward et al. 2014; Lahdenperä et al. 2014). In Western zoos, age at last reproduction for Asian and African elephants is 41 and 32 years for the U.S. (Keele et al. 2014; Olson 2014) and 48 and 45 years in Europe (A-K Oerke, pers. comm., European Elephant Group, data archive). Evidence that elephants experience a true menopause as in humans; i.e., the termination of ovarian cyclicity in midlife, is mixed. In Myanmar, records on >1000 individuals indicate reductions in fertility parallel declines in survivorship and health (Lahdenperä et al. 2014). By contrast, Lee et al (2016) reported a mean period of post-fertile viability of 16 years in Amboseli African females, based on an average fertility decline at 49 years and cohort longevity of 65 years; about a third of that population survived past the period of fertility decline. Thus, post-reproductive survival, and the potential for the expression of some type of menopause may differ between Asian and African elephants.

The presence of grandmothers appears to be important to reproductive fitness for both Asian and African elephants. In the Myanmar timber elephant population, grandmothers increase inclusive fitness by enhancing daughter's reproductive success and survival regardless of their own reproductive history (Lahdenperä et al. 2016). The authors theorize that fitness enhancing grandmaternal effects are widespread, which is an important factor to consider given high calf mortality in captivity—25–50% in some populations (Mar 2007; Clubb et al. 2009; Faust and Marti 2011a; Mar et al. 2012). Similarly, in wild African elephants, higher rates of reproduction were observed in daughters with simultaneously reproducing mothers (Lee et al. 2016). Thus, there are reproduction and survival advantages for maternal and grand maternal co-existence for both species, whether mothers are still reproducing or not (Lahdenperä et al. 2016; Lee et al. 2016).

2.2 Ovarian Cycle

Hormonal patterns throughout the elephant estrous cycle are now well established (see Brown 2014) and summarized in Fig. 9.1. Females are polyestrous and exhibit the longest spontaneous cycle of any mammal; 13–17 weeks in duration, with an 8–10-week luteal phase and 4–7-week follicular phase. An unusual characteristic is the production of 5α-reduced pregnanes by the corpus luteum rather than progesterone (e.g., 5α-pregnane-3,20-dione, 5α-pregnane-3-ol-20 one, 17α-hydroxyprogesterone), which will be referred to herein as 'progestagens'. Elephants are easily trained for non-stressful blood collection; however, ovarian

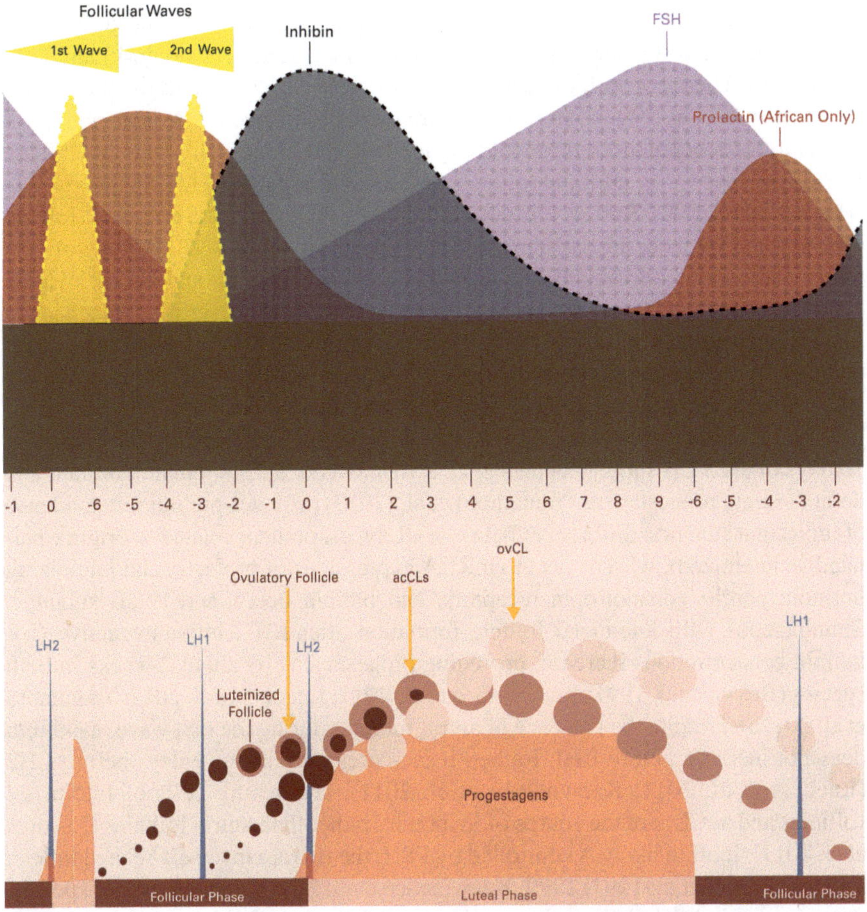

Fig. 9.1 Model of the elephant ovarian cycle, showing relationships among the secretion of estradiol, progestagens, LH, FSH, prolactin and inhibin during two follicular waves that culminate in an LH surge (first surge, LH1; second surge, LH2), and through development of follicles and corpora lutea (CL), both ovulatory (ovCL) and accessory (acCL). Prolactin is presented for African elephants only, as it is not influenced by stage of the cycle in Asian elephants. Adapted from Brown (2000), Hildebrandt et al. (2006) and Lueders et al. (2010, 2011). Reprinted from Brown (2014)

activity also can be monitored non-invasively using urine, feces, or saliva (Wasser et al. 1996; Czekala et al. 2003; Illera et al. 2014). Generally, the first pubertal rise in progestagens is quantitatively and qualitatively similar to subsequent cycles (Glaeser et al. 2012; Mendis et al. 2017). Though not a common reproductive strategy, occasional periods of estrous synchronization have been observed among zoo elephants (Asian: Glaeser et al. 2012; Edwards et al. 2016; African: Bechert et al. 1999; Weissenbock et al. 2009).

A distinguishing feature of the elephant ovarian cycle is the "double LH surge"; two precisely timed LH peaks associated with follicular waves that occur 3 weeks apart during the follicular phase, referred to herein as LH1 (anovulatory) and LH2 (ovulatory). These two surges occur in both species (African: Kapustin et al. 1996; Asian: Brown et al. 1999), but generally are more pronounced in Asians, with peak concentrations being up to three-fold higher than in Africans (Brown et al. 2004a). During the first wave, multiple follicles develop and trigger LH1, but none ovulate (Hermes et al. 2000). Rather, it is not until the second wave that a dominant follicle emerges and ovulates in response to LH2. Based on a series of studies utilizing transrectal ovarian ultrasound and hormonal measurements throughout the ovarian cycle in Asian elephants, Lueders et al. (2010, 2011, 2012) describes a mechanism for dominant follicle selection. After the first wave of the estrous cycle, rather than completely regress, some follicles luteinize and form accessory CL (acCL). During the second wave, multiple follicles develop, but only one becomes dominant, ovulates and forms the ovulatory CL (ovCL). Follicle luteinization is apparent within 10 days of each LH surge (Lueders et al. 2010), but the acCLs remain dormant and do not secrete progestagens. Yamamoto et al. (2017) provides preliminary evidence of molecular function differences between CL types, perhaps related to origin (ovulated vs luteinized), with respect to mRNA expressions of prolactin and luteinizing hormone/chorio gonadotropin receptors, and inhibin betaA and betaB subunits. Simultaneous with luteinized follicle formation after LH1, immunoreactive (ir)-inhibin concentrations increase, preceding progestagens by about 2 weeks in both species (Brown et al. 1991; Kaewmanee et al. 2011; Lueders et al. 2011; Yamamoto et al. 2012a). Despite the presence of antral follicles during the first wave, ir-inhibin does not increase before LHI. Rather, it increases only about 9 days before LH2 (Lueders et al. 2011; Kaewmanee et al. 2011), suggesting cells of luteinized follicles and acCLs are the source of ir-inhibin, rather than antral follicles (Lueders et al. 2011). Inhibin likely is controlled by FSH; the two are inversely related in both species (Brown et al. 1991, 1999; Kaewmanee et al. 2011), and as in other species, the reduction in FSH likely facilitates dominant follicle deviation with the transition of follicles from an FSH-dependent to an independent state. FSH concentrations decline throughout the follicular phase, reaching nadir concentrations at the time of LH2 (Brown et al. 1991; Kaewmanee et al. 2011). Thus, dominant follicle selection begins about 5–6 days before ovulation when inhibin concentrations are high and FSH is low.

Prolactin is folliculogenic in many species (Freeman et al. 2000; Frasor and Gibori 2003), and in African elephants is elevated during the non-luteal phase of the cycle, inversely related to progestagens, and phase-shifted by about 4 weeks (Yamamoto et al. 2010; Dow and Brown 2012) (Fig. 9.1). By contrast, in Asian elephants, prolactin concentrations are unvaried throughout the cycle and remain at baseline concentrations, representing a major species difference (Brown et al. 2004a). As discussed below, it may be significant that ovarian cycle problems associated with abnormal prolactin secretion are common in African, but not Asian elephants (Brown et al. 2004a; Dow and Brown 2012). More recently, cortisol has been shown to exhibit a distinct pattern in Asian elephants (Fanson et al. 2014), increasing throughout the second half of the follicular phase, with low concentrations during the luteal phase; the significance of which is unknown.

Elephants are not obligatory seasonal breeders, but conceptions can be influenced by rainfall and resource availability, and possibly latitude, more in line with a long-day breeder (Sukumar 2003; Hufenus et al. 2018). A group of African elephants housed indoors because of extreme cold weather showed prolonged non-luteal phases in winter before re-initiating normal ovarian cycles in the spring (Schulte et al. 2000). Seasonal effects on Asian elephant reproduction tend to be more subtle, but globally more conceptions occur after February, with a peak in June and limited activity from October to January (Hufenus et al. 2018). However, in a semi-captive population at the Pinnawala Elephant Orphanage in Sri Lanka, conception peaks occurred in the wettest months from September to October, which is the short-day period (Pushpakumara et al. 2016), so seasonality may not be consistent across regions. Studying the largest semi-captive population of Asian elephants in Myanmar, Mumby et al. (2015) found seasonal effects related to higher fecal glucocorticoids and lower birth rates at the start of the monsoon season (June–August) when logging activities are highest. Furthermore, females born in high stress months exhibited faster reproductive senescence and significantly reduced lifetime reproductive success than cohorts born at other times of year. A detailed study of captive elephants in Thailand showed females cycled year-round, but exhibited slightly longer follicular phases during the rainy season (June–October) (Thitaram et al. 2008). Specifically, it was the duration from the progestagen drop to LH1 that varied seasonally, averaging 33 days during the rainy season, compared to 22 days in the winter and 19 days during the summer. The interval between LH1 and LH2 was consistent at 19 days, so seasonal variation in estrous cycle length is mediated by events during the first follicular wave leading up to LH1 (Thitaram et al. 2008). By contrast, Glaeser et al. (2012) found no seasonality in ovarian cycle lengths in zoo females over a 20-year period in the Pacific Northwest. Variability was higher for the follicular than the luteal phase, at least in Asian elephants (Thitaram et al. 2008; Glaeser et al. 2012), and the two were negatively correlated, suggesting a regulatory role of the follicular phase in maintaining relatively consistent cycle duration within individuals. In the study of Glaeser et al. (2012), Asian elephants were highly resilient to numerous life events (births, deaths, transfers in and out, changes in herd structure), which had minimal effects on cycle dynamics, a finding that agrees with 25 years of unpublished data on over 60 Asian females evaluated at SCBI (J. L. Brown). By contrast, our data on over 80 African females provide numerous examples of events related to temporary or permanent suppression of estrous cyclicity (e.g., translocations, changing blood collection frequency, altered herd dynamics, keeper changes), suggesting that Asian and African elephants may differ in biological responses to changes in the captive environment, something that deserves further investigation in the context of optimizing conditions based on species-specific needs.

2.3 Pregnancy and Parturition

Elephants have the longest gestation period, lasting 20–23 months on average (see Brown 2014). The placenta is chorioallantoic, zonary endothelialchorial; implantation is central and superficial, with a mesometrical orientation of the yolk sac; and the embryo is antimesometrial in location (Allen 2006; Hildebrandt et al. 2006). Although the placenta itself is endocrinologically inert, fetal gonads enlarge during the second half of gestation and synthesize 5α-dihydroprogesterone and other 5α-pregnane derivatives from cholesterol and pregnenolone (Allen et al. 2005; Allen 2006; Stansfield and Allen 2012). Based on longitudinal transrectal ultrasound monitoring, both species exhibit an initial period of comparatively slow embryonic development that has been compared to delayed implantation (Hildebrandt et al. 2006; Drews et al. 2008). The embryonic vesicle is visible at ~8 weeks post-conception, which is much smaller (~10 mm) than that of other species at that stage (cattle, 40 mm; sheep, 70 mm; horse, 40 mm). Likewise, time of implantation is estimated to be ≤20 days in humans, dogs and sheep, but ≥50 days in the elephant. Placentation occurs during the second to third month of gestation (Drews et al. 2008). In other delayed implanters (e.g., mustelids, bears, roe deer), a significant rise in serum progestagens occurs at implantation, which is observed in elephants at 6–8 weeks post-conception (Meyer et al. 2004). The embryo then doubles in size between the fourth and fifth month, increasing from 60 to 120 mm (Drews et al. 2008). Organogenesis is completed at about 110–120 days, when the end of the embryonic period is reached. It is possible to sex the fetus after about a year of gestation with near 100% accuracy by measuring circulating maternal testosterone concentrations, at least in Asian elephants (Duer et al. 2002; Brown et al. 2004b). Presumably elevated testosterone is of fetal testicular origin, although the CL could also be a source (Castracane et al. 1998). Interestingly, this technique has proven less accurate for African elephants (J. L. Brown, unpubl).

A model for the endocrinology of pregnancy is depicted in Fig. 9.2. Diagnosis and monitoring of pregnancy is easily done by longitudinal analysis of circulating progestagens or the relevant metabolites in urine, feces or saliva (see Brown 2014). Elephants and horses share some gestational traits, such as the presence of multiple large CL in the maternal ovaries (reviewed by Stansfield and Allen 2012). However, there are notable differences. The mare produces one CL at ovulation, with additional accessory CLs formed as the result of post-conception ovulations (Squires and Ginther 1975) due to the LH-like activity of equine chorionic gonadotropin (eCG) that causes ovulation/luteinization of mature follicles (Urwin and Allen 1982). By contrast, in elephants, additional follicles are not formed during gestation (see Lueders et al. 2012), nor is there evidence of gestational gonadotropin-like activity in serum or placental extracts (Meyer et al. 2004; Allen 2006). Rather, acCLs in the elephant are the product of luteinized follicles (Lueders et al. 2010, 2011) and/or ovulations (Stansfield and Allen 2012) from preceding cycles. In both Asian and African elephants, acCLs and the ovCL begin to regress about 5–6 weeks after conception (the normal luteal phase length), but then rebound and grow

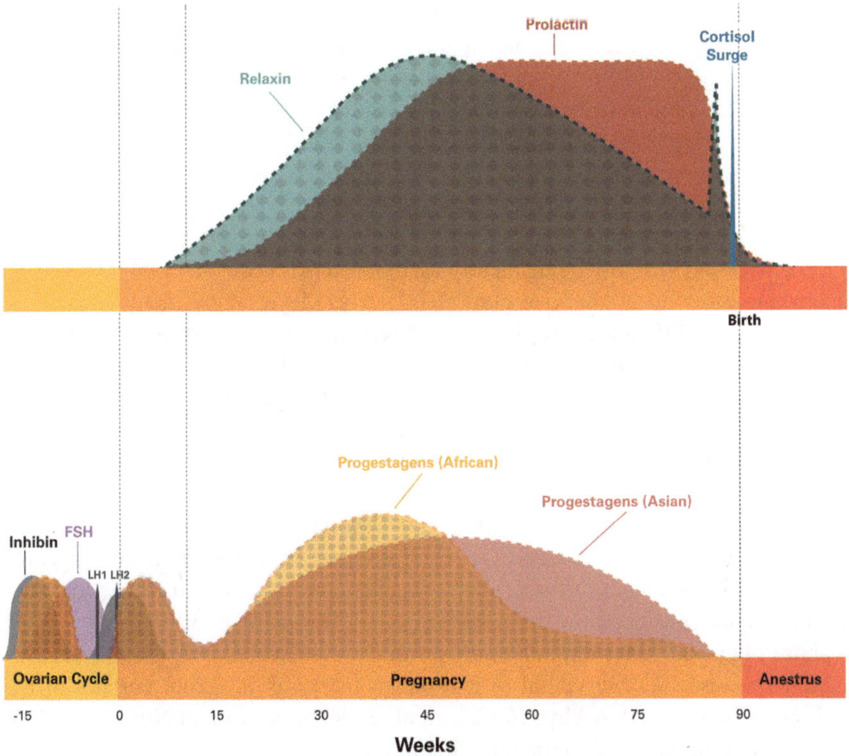

Fig. 9.2 Model of progestagen, LH, FSH, prolactin, relaxin, cortisol, and inhibin secretion during the estrous cycle and gestation in the elephant. Adapted from Meyer et al. (2004), Yamamoto et al. (2012a) and Lueders et al. (2012). Preprinted from Brown (2014)

significantly larger than those in a non-conceptive luteal phase (Lueders et al. 2012), commensurate with the marked secondary rise in progestagen concentrations in the maternal circulation after the second gestational month (see Meyer et al. 2004; Lueders et al. 2012).

After 4–7 months of gestation, prolactin immunoactivity (ir-prolactin) increases up to 100-fold, peaks at 11–14 months and remains high until birth in both species (Brown and Lehnhardt 1995; Meyer et al. 2004; Yamamoto et al. 2011, 2012b). Prolactin and placental lactogens are luteotropic in other species, and enhance CL progestagen production (Freeman et al. 2000; see Takahashi 2006), which would be important for elephants because the placenta is steroidogenically inactive (African: Allen et al. 2002). The source of high ir-prolactin during gestation is primarily placental (Yamamoto et al. 2011), similar to lactogenic hormones in other species (Forsyth and Wallis 2002). Measurement of serum ir-prolactin past 7 months of gestation is a reliable pregnancy test, even with a single sample, unlike progestagens that require longitudinal sampling. Serum relaxin also can be used diagnostically in both species (Meyer et al. 2004; Niemuller et al. 1998), as concentrations are significantly elevated after 5 months of gestation. A recent paper by Lueders et al. (2018) provides evidence of prolonged luteal phases in three unmated elephants that lasted 5–16 months, with one female displaying increased prolactin secretion after 3 months. Each female had varying degrees of uterine pathology, so damage to the endometrium may have temporarily blocked the luteolytic signal from the uterus. Thus, though rare, elevated progestagens and prolactin may not necessarily indicate pregnancy, requiring an ultrasound exam to confirm.

Gestation lengths range from 620–690 days, with differences by species, sex of calf, dam age or parity not being clearly established (Meyer et al. 2004; Dale 2010; Pushpakumara et al. 2016; Kiso et al. 2017). The Meyer study identified several notable species differences in gestational hormone patterns. While overall mean progestagen concentrations were similar, temporal profiles differed. Concentrations were higher in African elephants during the first half of gestation, but then declined to levels below those observed in Asian elephants (Fig. 9.2). There was a fetal gender effect in Asian, but not African elephants, with progestagen concentrations being higher in Asian cows carrying male calves as compared to those carrying females (Meyer et al. 2004). It is curious that significant fetal gender differences in maternal steroids (androgens and progestagens) are observed in Asian elephants only, suggesting a species difference in gonadal and/or placental function. Both species have zonary placentation; however, far more information is available on African elephant placental function, and comparable studies in Asians are not available. In both species, the observation of significant increases in serum cortisol in the week or so before, and at birth (Meyer et al. 2004; Kajaysri and Nokkaew 2013) suggests an important role in the initiation of parturition.

In many species, particularly primates, inhibin produced by follicles, CLs and/or the placenta (Knight 1996) is involved in the establishment and maintenance of pregnancy (Florio et al. 2010). In the elephant, whereas CLs are a major source of inhibin during the estrous cycle (Kaewmanee et al. 2011; Yamamoto et al. 2012a), they do not appear to produce inhibin during gestation (Yamamoto et al. 2012a) (Fig. 9.2). On closer inspection, inhibin in fact is increased for the first 8 weeks post-conception, mimicking the normal luteal phase pattern. After that, although progestagens rebound to even higher concentrations on average, inhibin concentrations decline. This pattern suggests a shift in luteal cell function during early gestation, and a deviation in the secretory ability of CLs between cycling and pregnant elephants. Taken together, a role for inhibin in elephant pregnancy seems unlikely.

3 Pituitary-Ovarian Problems

Factors limiting the number of breeding female elephants vary by species, but include problems related to high rates of ovarian acyclicity, behavioral incompatibility, reduced libido and reproductive tract pathologies (see Brown 2014). Now more than ever, we need to identify factors related to reproductive performance of captive elephants, and make changes in management to ensure population survival.

Some are the result of failure to breed elephants at a young age and keep them reproductively active, while others are related to social management and husbandry. Several large-scale studies involving surveys of reproduction in North America (Freeman et al. 2009; Brown et al. 2016) and Europe (Hartley and Stanley 2016) identify similarities and differences in reproductive challenges regionally. In particular, nearly all Asian populations are impacted by advancing age and reproductive pathologies, but for Africans, age, obesity and ovarian acyclicity are risk factors for reproductive failure in the U.S. (Morfeld and Brown 2014; Morfeld et al. 2016), but not in Europe (Hartley and Stanley 2016).

3.1 Reproductive Tract Pathologies

Reproductive track pathologies are serious problems impacting captive elephants globally, especially older females (reviewed by Hartley and Stanley 2016; Hildebrandt et al. 2006; Brown 2014). Vestibular cysts occur in both species, whereas vestibular polyps are observed only in Africans, with an incidence of about 70% in females >30 years of age. In both species, vaginal cysts and neoplastic formations may be extensive and fill the vaginal lumen, blocking semen flow after mating and causing discomfort during estrus and mating. Periodic vaginal discharge containing mucus and clotted blood is a symptom of this condition. Asian and African elephants both develop endometrial hyperplasia, whereas Asian elephants develop multiple benign uterine leiomyomas. In general, reproductive tract pathologies are more prevalent in older (>30 years of age) nulliparous cows, and those where reproduction has not occurred within 10–15 years. In a recent survey, more than half of zoo females with documented tract pathologies had no previous breeding history (Dow et al. 2011a). The occurrence of urogenital pathologies in older females is termed 'asymmetric reproductive aging' (Hermes et al. 2004), and believed to be the result of continuous ovarian cyclicity of non-bred females. Repetitive remodeling and exposure of the endometrium to ovarian steroids likely has a negative and cumulative effect on reproductive health (Hermes et al. 2008). In the wild, most females are either pregnant or lactating and thus experience comparatively few reproductive cycles in their lifetime. Though data are limited, these pathologies do not appear to be common in wild elephants (Hildebrandt et al. 2006; Freeman et al. 2008).

3.2 Ovarian Acyclicity

In the U.S., the problem of abnormal or a lack of ovarian cycles has been recognized for nearly three decades, and in the most recent survey involving 90% of elephants in American Zoo and Aquarium Association (AZA) zoos, rates of normal cycling, acyclicity and irregular cycling were 73.2, 22.5 and 4.2% for Asian, and 48.4, 37.9

and 13.7% for African elephants, respectively, which differed between species in overall percentages, and in the proportion of females affected among age groups (Brown et al. 2016) (Fig. 9.3). For both species, ovarian cycle problems were more prevalent in older age females, but for African elephants, high rates were observed in reproductive age females (13–35 years) as well. By contrast, in Europe, acyclicity rates are only 13.5 and 4.7% for Asian and African elephants, respectively (Hartley and Stanley 2016). However, of the entire survey population (122 Asian, 67 African), progestagen cyclicity data were only available for 37 and 43 females, respectively. Thus, it is possible ovarian cycle problems are underestimated in Europe if tested females are primarily those involved in breeding programs or at zoos interested in participating in hormone monitoring. Often there is less incentive to continue progestagen testing once a female has been diagnosed as acyclic, which is unfortunate.

Several conditions known to be associated with infertility in other species are not related to acyclicity in elephants, such as 'stress' (e.g., increased cortisol; Brown et al. 2004a), hyperandrogenism (Mouttham et al. 2011), hyperestrogenism (Prado-Oviedo et al. 2013) and thyroid dysfunction (Brown et al. 2004a). Concentrations of anti-Müllerian hormone (AMH), a measure of ovarian follicle reserves and diagnosis of menopause in women (Rico et al. 2009; Grynnerup et al. 2012) also are similar between cycling and non-cycling African elephants, suggesting primary ovarian insufficiency due to constant cycling is not a problem, and that ovaries may be capable of responding to ovulation induction therapies (Dow et al. 2011b).

One commonality among these studies is that they generally relied on an analysis of a single factor, so an important advance has been the use of multi-factorial, holistic approaches to understand what is most important for good reproduction and welfare. These studies have relied on well-accepted, animal-based welfare indicators (outcomes) of physical, behavioral and physiological condition, such as body condition,

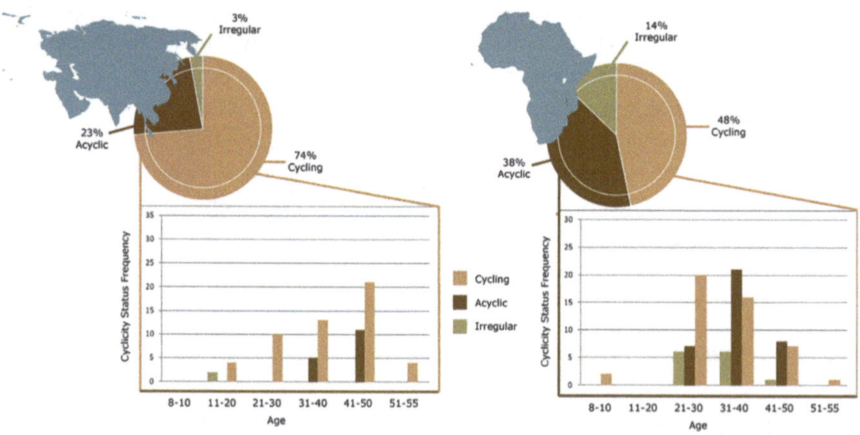

Fig. 9.3 Pie charts showing proportions of normal, irregular and non-cycling elephants for 71 Asian (left) and 95 African (right) elephants in AZA zoos surveyed in 2013, with a breakdown of cyclicity status by age categories for each species. Reprinted from Brown et al. (2016)

foot and joint health, walking rates and recumbence, stereotypy, ovarian cycling, prolactin levels, and glucocorticoids that are analyzed using univariate and multivariable models in relation to a broad range of facility- and husbandry-based measures (inputs). Though not all studies measure the same indicators, a number of relationships have been identified that can be used to make improvements in management and husbandry to increase fitness of captive elephant populations.

The first study to use this approach was conducted by Freeman et al. (2009) and involved 91 African elephants with known hormonal data at 46 U.S. zoos (~90% of the AZA population). Using Principal Components and Multiple Logistic Regression Analyses, one significant factor associated with ovarian acyclicity was a high social dominance rank, which confirmed earlier (Freeman et al. 2004) and subsequent (Freeman et al. 2010) studies. In the wild, the largest, oldest female in a herd is the matriarch and is crucial to their survival. In captivity, dominance is still important for maintaining social harmony, but it is possible the energy that goes into peace keeping within a captive herd of unrelated, and sometimes incompatible females may be compromising ovarian function (Freeman et al. 2009). Reproductive suppression is a natural strategy for many species in the wild, including elephants under some situations where resources are limited (Wasser and Barash 1983; Creel and MacDonald 1995), but when it occurs in captivity it could be indicative of suboptimal conditions (Wielebnowski 1998). The study of Freeman et al. (2009) also was the first to find a relationship between a high body mass index indicative of obesity with ovarian acyclicity, which will be discussed in detail later.

A second epidemiological study with 75 Asian and 95 African females was conducted in the U.S. to further identify management factors related to reproduction function. For African elephants, univariate modeling found the following management and housing factors were associated with ovarian acyclicity ($P < 0.15$): Age (positive), Percent Time with Restricted Physical Access (positive), and Enrichment Diversity (negative), Outdoor Space Experience (negative), and Percent Time Housed Separately (positive). Multi-variable logistic predictive modeling found that the combination of Age and Enrichment Diversity to be most predictive of whether a female elephant would be acyclic (Brown et al. 2016). Modeling for Asian elephants was not included in that study, but as shown in Table 9.1, a number of univariate factors were significant risk factors for acyclicity: Age (negative), Birth Origin (captive), Animal Contact (negative), Space Experience (negative), In/Out Choice (negative), and Time Spent in Mixed Sex Herds (negative). None of these factors remained in the multivariable model; however, data reinforce the relationship between age and ovarian cycle problems, and agree with a previous study that bull presence is associated with higher cyclicity rates (Dow et al. 2011a), although that could be related to those zoos focusing more on breeding and having reproductively viable females.

A multi-variable study conducted in Europe did not find an effect of age, uterine pathologies, social ranking, or recent changes in herd composition on ovarian acyclicity in 43 African elephants (Hartley and Stanley 2016). Rather, herd stability and compatibility, allomothering, and maternal experience were main factors related to reproductive success. In Asian elephants (n = 37), reproductive failure was much

Table 9.1 Univariate analysis of ovarian acyclicity using logistic regression in Asian elephants (n = 68)

Variables	Beta	OR	P-value
Age	0.06	1.06	0.037*
Birth origin (ref = captive)	−1.71	0.18	0.113*
Herd size	−0.20	0.82	0.219
Animal contact	−0.30	0.74	0.109*
Enrichment diversity	0.24	1.28	0.919
Enrichment program	0.31	1.36	0.294
Exercise diversity	0.77	2.16	0.342
Exercise hours per week	−0.13	0.88	0.489
% time indoors	0.00	1.00	0.735
% time outdoors	−0.00	0.99	0.782
Space experience	−0.03	0.97	0.079*
Space experience in/out choice	−0.03	0.97	0.038*
% time in mixed sex herd	−0.03	0.97	0.049*
% time with restricted physical access	0.00	1.00	0.838
% time alone	0.00	1.00	0.858
Social group contact	0.06	1.06	0.711
Social bond score	0.15	1.16	0.365
Walk/week	−0.09	0.92	0.685

Confounder analysis: no confounders were identified
OR odds ratio
*P < 0.15

less common, but when it did occur, it was associated with age and reproductive pathologies (Hartley and Stanley 2016), as in other studies. Another cause of reproductive failure was lack of physical access to a compatible bull (97.6% of African, 89.1% of Asian). So, it would appear that a very small percentage of females are responsible for reproduction in that population. In a separate study of the European population (326 Asian, 192 African), Schiffmann et al. (2018) found breeding status was correlated to elephant group size, enclosure size, and a diet with fewer vegetables.

A growing concern today is how many captive elephants are overweight or obese, especially in Western zoos (56–74%) (Harris et al. 2008; Morfeld and Brown 2014; Morfeld et al. 2016; Schiffmann et al. 2018) when compared to captive elephants in Thailand (38%; Norkaew et al. 2018) and those in the wild (Ange et al. 2001; Morfeld and Brown 2014; Schiffmann et al. 2018). For some time, researchers have speculated that reproductive problems may be caused in part by metabolic derangements associated with excessive body fat or weight (Clubb et al. 2009; Mason and Veasey 2010). Indeed, one significant finding with African (though not Asian) elephants in the U.S. was a positive relationship between ovarian acyclicity and a high body condition score (BCS) (Morfeld and Brown 2014). In captive Asian (Norkaew et al. 2018) and African (Morfeld and Brown 2014, 2017) elephants, altered lipid and/or sugar metabolism also is observed in overweight elephants, including higher insulin, lower glucose to insulin ratio, and higher cholesterol and lipoproteins. Leptin RNA and protein expression has been demonstrated in African elephant adipose tissue (Nilsson et al. 2014), with circulating concentrations positively associated with BCS (Morfeld and Brown 2014) and higher rates of ovarian acyclicity (Morfeld and Brown 2017). These findings agree with studies in horses and humans showing obesity can lead to metabolic changes that impair fertility (Vick et al. 2006; Miller et al. 2008; Jungheim and Moley 2010). For example, obese mares experience an extended interval between successive ovulations, and amenorrhea is common in obese women, not unlike the irregular cycles observed in elephants (Brown 2000). The European population may be different, however, because a clear relationship between body condition and ovarian cycle activity based on hormone monitoring is less evident, although breeding status was reportedly correlated to lower BCSs (Schiffmann et al. 2018). Stillbirths and dystocias also are common in obese women and horses, and are major causes of calf mortality in elephants (Clubb et al. 2009). Thus, evidence suggests that elephants may be experiencing metabolic and possible fertility problems associated with obesity, which should be a concern for elephant managers given how many captive elephants are overweight or obese.

In studies of female elephants in the U.S., Europe, and Thailand, management factors affecting body condition include negative associations (i.e., better body condition) with exercise and activity (African, Morfeld et al. 2016; Morfeld and Brown 2017; Asian, Norkaew et al. 2018), unpredictable feeding schedules (African, Morfeld et al. 2016; Morfeld and Brown 2017), reduced amount of bread in the diet (African, Schiffmann et al. 2018), feeding enrichment (African, Schiffmann et al. 2018), and larger enclosures (Asian, Schiffmann et al. 2018). Positive associations were found between BCS and increased feeding diversity and frequency (African, Morfeld et al. 2016; Asian, Schiffmann et al. 2018), and amount of vegetables in the diet (Asian, Schiffmann et al. 2018). In the U.S., elephants that experienced 14 or more hours of staff-directed walking per week were more likely to have a normal BCS as compared to elephants that exercised less. Similarly, in Thailand, tourist elephants involved in riding activities (with or without saddles) had lower, more ideal BCSs (Norkaew et al. 2018). Feeding diversity may increase the risk of obesity due to higher caloric intake, while implementing an "unpredictable or random" feeding schedule may decrease the risk by encouraging more exploration and locomotion (Holdgate et al. 2016). Thus, instead of simply spreading food on the ground, use of alternative feeding methods that include presenting food by hanging it above the enclosure, hiding it within the enclosure, or putting it in an enrichment device such as a puzzle feeder might improve elephant health.

In the U.S. population, a variety of demographic, management, housing, and social zoo variables were investigated as predictors of leptin and G:I levels. Significant relationships with leptin were found for: Walk Week (negative), Feed Diversity (positive), Space Experience per Elephant (negative), Animal Contact (negative), and Social Group Contact (negative). For the G:I, significant relationships were found for: Age (negative), Sex (negative), Species (negative), Exercise Week (negative), Walk Week (negative), Feeding Predictability (positive), Feed Diversity (negative),

Percent Time Indoor (negative), Space Experience per Elephant (positive), Space Experience (positive), and Relative Positive Reinforcement and Negative Punishment (positive) (Morfeld and Brown 2017). Taken together, good diets, proper exercise and feeding regimens, and enriched environments are important factors in reducing the risk of obesity and metabolic derangements, and in some populations, may also benefit reproduction.

Clearly, we need to understand more about why some elephant females do not cycle normally and why there appears to be a difference in acyclicity rates between the U.S. and Europe. Prolonged acyclicity does not appear to occur in wild African elephants based on physiological studies that show females can cycle into their 50s, although capacity does decline with age (Freeman et al. 2008, 2011, 2013). It is unlikely that any one management factor is responsible, as 52% of zoos house both cycling and non-cycling females (Freeman et al. 2009). Some elephants even alternate between cyclic and non-cyclic periods (Brown 2006). Rather, there likely are multiple etiologies, so we need to identify the best approach to ameliorate them, on both a population and individual level (e.g., exercise programs for overweight elephants, compatible social groups, creation of multi-generational herds, enriched space, etc.). While more work is needed to develop targeted treatments, given the complexity of controlling endocrine function, it is more important to identify underlying causes so that mitigating steps can be taken to prevent problems from occurring in the first place.

3.3 Hyperprolactinemia

Of particular interest is an association between ovarian cycle problems and a hormonal imbalance—hyperprolactinemia, which occurs in African, but not Asian elephants (Brown et al. 2004a; Yamamoto et al. 2010; Dow and Brown 2012). In the most recent survey, out of 101 African elephant females, 57% (58/101) were acyclic, and of those, 48% (28/58) were hyperprolactinemic (Brown et al. 2016). In other species, hyperprolactinemia is associated with infertility (Aron et al. 1985; Yuen 1992; Zacur 1998), and in women it is the most common disorder of the hypothalamic-pituitary axis; up to 40% presenting with secondary amenorrhea are hyperprolactinemic (Serri et al. 2003; Wang et al. 2012). In over 20 years of monitoring the North American captive elephant population, only non-cycling African elephants have been diagnosed with hyperprolactinemia, suggesting these two conditions are linked (Brown and Lehnhardt 1995; Brown et al. 2004a; Dow and Brown 2012; Prado-Oviedo et al. 2013; Brown et al. 2016).

Prolactin is produced by lactotroph cells in the anterior pituitary and is under inhibitory control by dopamine (Schuff et al. 2002; Melmed 2003; Bachelot and Binart 2007). Dopamine concentrations are reduced in patients with idiopathic hyperprolactinemia (Sarkar et al. 1984; Grattan 2015), currently making dopamine agonists the most common treatment for this condition. Any lesion interfering with dopamine synthesis or release can increase prolactin secretion significantly, quickly and for prolonged periods of time (Zacur 1998). A common cause of hyperprolactinemia in women is a prolactinoma, which is the most common type of pituitary tumor, and associated with amenorrhea, galactorrhea, and hypogonadism (Melmed 2003). Prolactinomas retain intact trophic control, so they may develop in an environment with reduced dopamine concentrations, reduced dopamine sensitivity, or as a result of vasculature isolation that prevents dopamine from reaching the lactotrophs (Schuff et al. 2002; Melmed 2003). However, we lack data on the incidence of prolactin-secreting tumors in African elephants, as pituitary histopathology is rarely performed at necropsy.

There is convincing evidence in elephants that prolactin is controlled at least in part through dopamine negative feedback based on findings that a dopamine agonist (cabergoline) decreases (Morfeld et al. 2014) and an antagonist (domperidone) increases (J. Brown and T. Dow, unpubl) prolactin secretion. Treatments that modulate prolactin secretion have been tested in African elephants as a means to correct ovarian dysfunction. Cabergoline is an effective treatment for hyperprolactinemia-induced infertility in women (Verhelst et al. 1999), so a clinical trial was conducted in eight hyperprolactinemic elephants (1–2 mg twice weekly oral cabergoline for 4–12 months), which resulted in a significant and immediate reduction in prolactin during the treatment period; however, no females resumed cycling and prolactin increased to levels as high or higher after treatment withdrawal (Morfeld et al. 2014). Perhaps increasing the dose or extending the treatment period would be more effective, but until such trials are conducted, it is not clear if ovarian acyclicity can be corrected by merely reducing prolactin.

In addition to hypothalamic inhibition by dopamine, stimulatory factors can be involved in fine tuning prolactin secretion in response to physiological (i.e., pregnancy, suckling), environmental (i.e., light-dark cycles) or physiological (stress) changes (Lamberts and Macleod 1990; Freeman et al. 2000). Estradiol is one of these factors, but it has been shown to have no relation to elevated prolactin (Prado-Oviedo et al. 2013). Evidence in other species suggests that oxytocin, one of the oldest neuromodulators in the animal world, may be a major prolactin releasing factor (Richard et al. 1991; Tabak et al. 2010) to aid in stimulating prolactin surges during proestrus or in response to suckling (Richard et al. 1991; Tabak et al. 2010). Furthermore, human and animal studies have shown there is a positive feedback mechanism between prolactin and oxytocin, and concentrations are known to increase with hyperprolactinemia (Sarkar et al. 1992; Emiliano and Fudge 2004). Serotonin also stimulates prolactin release from hypothalamic neurons, and its stimulatory role is compounded by increasing the release of other prolactin releasing factors, such as thyrotropin-releasing hormone and vasoactive intestinal peptide (Lamberts and Macleod 1990; Jorgensen 2007). Correspondingly, serotonin concentrations are increased in hyperprolactinemic patients and animal models (Emiliano and Fudge 2004; Ferrari et al. 1978; Cocchi et al. 1978). In fact, both oxytocin and serotonin are elevated in hyperprolactinemic elephants, so it appears stimulatory factors may be involved (Prado et al. 2019a). By contrast, rather than being reduced as hypothesized, urinary dopamine was elevated in hyperprolactinemic elephants compared to cycling and baseline prolactin groups. This finding suggests that dopamine synthesis or secretion may not be impaired in these elephants, but in fact may be augmented (Prado-Oviedo 2015; Prado et al. 2019a).

Because of the link between the two conditions, the assumption has always been that hyperprolactinemia precedes acyclicity in elephants. However, a recent longitudinal analysis of prolactin and progestagen found the majority of African elephants had been acyclic for at least 5 years before prolactin began to rise, and in a number of cases ($n = 9$) for over 10 years (Prado et al. 2019b) (see Fig. 9.4). By contrast, non-cycling elephants with baseline prolactin had been acyclic for less than 5 years. Finally, there was a positive association between the duration of acyclicity and mean prolactin concentrations. Thus, the chance an elephant will develop hyperprolactinemia increases the longer an individual exhibits baseline progestagen concentrations, suggesting the etiology differs from that in women. Rather, long periods of low prolactin that lack a normal cyclical pattern may be insufficient to support normal follicular function, and could represent an important transitory state in the etiology of infertility in some elephants. It may be that lack of ovarian or other negative feedback signals from the ovaries over time results in the eventual rise in observed prolactin concentrations, possibly through an increase in neural stimulatory factors (Prado et al. 2019b).

Stress may also play a role in excess prolactin secretion and ovarian inactivity under some conditions. Human studies have shown that stressors such as social conflict, a new job, death of a loved one, divorce, separation from a parent during childhood, and academic pressures increase the secretion of prolactin and may predispose individuals to hyperprolactinemia and infertility (Assies et al. 1992; Sobrinho et al. 1984; Sobrinho 2003; Sonino et al. 2004). A consequence of the

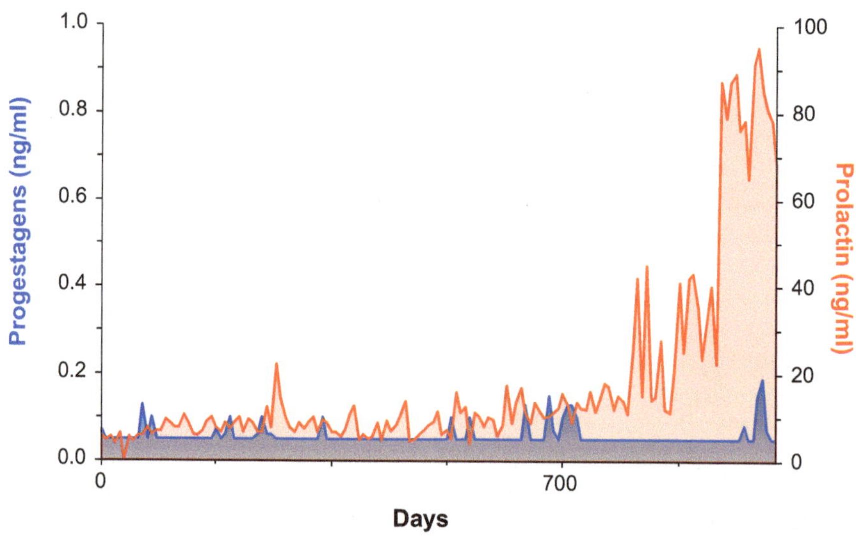

Fig. 9.4 Representative progestagen and prolactin profiles of an acyclic elephant with low prolactin that developed hyperprolactinemia. Adapted from Prado et al. (2019b)

stress response can be reduced dopamine secretion and its tonic inhibition of prolactin, or an up regulation of prolactin stimulating factors (Calogero et al. 1998). Cortisol has not been found to be related to hyperprolactinemia in elephants (Brown et al. 2004a), so it might be more beneficial to examine how individual elephant temperaments, social relationships and life events relate to prolactin secretion and reproductive status, and if management or husbandry changes could help an elephant better cope with its captive environment, similar to that described for other species (Wielebnowski 1998; Wielebnowski et al. 2002a, b; Mellen 2005; Carlstead et al. 2009). Indeed, in the U.S. welfare study, multi-variable modeling determined Age (positive), Enrichment Diversity (negative), Alternate Feeding Methods (negative) and Social Group Contact (positive) were predictors of hyperprolactinemia (Brown et al. 2016). Thus, providing elephants with a more enriched environment may support normal pituitary-ovarian function. By contrast, although elephants are highly social species, those that spent time in a greater number of different social groups had a greater risk of hyperprolactinemia. This strategy may be needed to mitigate problems with socially incompatible elephants, so dividing a herd into multiple groups and housing elephants in a variety of social configurations may yield a more demanding social environment to which they respond with elevated prolactin production. Finally, in terms of temperament, consistently baseline prolactin (i.e., low) was positively associated with a more fearful and apprehensive temperament, while elevated prolactin was positively associated with a more popular and caring temperament (Prado-Oviedo 2015). This agrees with the finding that cyclic elephants are more likely to be rated as herd peacekeepers (Freeman et al. 2004). So, while increased levels of prolactin may help elephants cope with increased social demands, it could occur at the cost of normal reproductive function.

4 Control of Reproduction

4.1 Estrous Synchronization

Ovarian function is under hormonal control; therefore, methods to alter HPG activity might provide a means of regulating reproduction (Thitaram and Brown 2018). Estrous synchronization by hormonal manipulation has been attempted in a few elephants. Prostaglandin $F_{2\alpha}$ at doses of 80–125 mg on day 25, 30, 45 and 60 post-estrus reduced progestagen secretion temporarily, but had no luteolytic effect (Hildebrandt et al. 2011). When administered during the follicular phase, the GnRH-agonist buserelin-acetate (80 μg) stimulated LH release (i.e., LH1), and if given 13–42 days after progestagens reached baseline, actually induced a spontaneous secondary surge ~20 days later (i.e., LH2) resulting in ovulation (Thitaram et al. 2009). By contrast, no LH2 occurred when GnRH was administered before day 12. One conception from natural breeding resulted from this double LH surge induction, so it could be useful for scheduling multiple elephants for AI.

4.2 GnRH Vaccines

GnRH vaccines have been used to control gonadal function for reproductive management and contraception. These act by stimulating the production of anti-GnRH antibodies that block the binding of endogenous GnRH to receptors in the pituitary gland, which leads to a reduction in LH release from the anterior pituitary and, in turn, reduced stimulation of gonadal steroidogenic activity. The first GnRH vaccine study in elephants involved administration of Repro-BLOC® (Amplicon Vaccine, Washington, USA) at doses of 3–30 mg over five boosters, which resulted in high titers of GnRH antibodies within 7 months that successfully down-regulated the HPG axis and resolved life-threatening reproductive tract pathologies (Boedeker et al. 2012). Today, 11 years later, at the age of 73, this female is healthy, although she never resumed cycling. By contrast, a similar GnRH vaccine (Improvac®; Pfizer Animal Health, South Africa) at a dose of 600 μg, with one booster 7 weeks later, failed to induce anestrus in free-ranging African females (Valades et al. 2012). That could have been due to ineffective dosing because another study reported successful suppression of estrus, breeding and conception at a dose of 1000 μg in captive African elephants with ongoing boosters every 5–6 months (Bertschinger and Lueders 2018). GnRH vaccination of Asian and African elephant bulls (Improvac® or Equity®; Pfizer Animal Health, Australia) at doses of 400 and 1200 μg per elephant with boosters over months to years also has successfully postponed or suppressed the HPA axis and symptoms associated with musth (Somgird et al. 2016; Lueders et al. 2014, 2017; Bertschinger and Lueders 2018). Success of immunocontraception depends on the time of vaccination and age of the treated animal, with younger animals potentially having greater responses. Reversibility of GnRH vaccines needs to be investigated further, but permanent effects have been observed in bulls, particularly young ones, after prolonged treatment (Bertschinger and Sills 2013; Lueders et al. 2014; Bertschinger and Lueders 2018).

4.3 Porcine Zona Pellucida Vaccines

The most widely used contraceptive agent in elephants is an immunogenic preparation of porcine zona pellucida (pZP) harvested from pig ovaries, which acts by blocking oocyte sperm receptors; it does not shut down ovarian cyclicity (Delsink et al. 2002, 2006; Fayrer-Hosken et al. 1999). In one field study in Africa, cows vaccinated with an initial dose of 400 μg i.m., followed by two 200 μg boosters at 5–7 week intervals, and then subsequent 200 μg annual boosters decreased population growth from 10 to 0% in just a few years (Ahlers et al. 2012). The PZP vaccine has many advantages: (1) high contraceptive efficacy (at least 90%); (2) safe and cost effective; and (3) does not pass into to the food chain (Delsink et al. 2007, 2013;

Kirkpatrick et al. 1996). Over the course of 11 years, PZP immunocontraception has had no detectable effect on behavior or social contexts, providing a convincing argument for the use of sustained immunocontraception for elephant population control (Delsink and Kirkpatrick 2012). There are limited data on reversibility of long-term PZP vaccination in elephants, but a few contracepted for 4–6 years were shown to reverse and conceived within 1.5–4 years after the last treatment (Bertschinger et al. 2018).

4.4 Artificial Insemination

One the most significant advances in elephant reproductive management was the development of an artificial insemination (AI) technique (Brown et al. 2004b) pioneered by Dr. Thomas Hildebrandt (Institute for Zoo and Wildlife Research, Berlin, Germany) in the 1990s. The non-surgical procedure involves a custom-made balloon catheter (2.5-cm diameter 140 cm length) lubricated with non-spermicidal sterile gel, inserted into the vestibule to slightly distend the reproductive tract for optimal visualization and allow passage of a flexible 3.0-m endoscope containing a customized disposable insemination catheter (3-mm diameter, 300 cm length). Both endoscopic and transrectal ultrasonographic visualizations are used to guide the insemination catheter to the distal vagina in front of the cervical opening or intracervically, where semen deposition can be visualized ultrasonographically to verify placement. The timing of AI is based on identifying LH1 in daily blood samples during the follicular phase and then conducting 2–3 inseminations ~18–22 days later to coincide with LH2 (Brown et al. 2004b). The first elephant AI birth was in 1999 (Saragusty et al. 2009); since then over 50 births have resulted worldwide, with the majority (~70%) being in African elephants. As a management tool, AI has been in use for over 20 years; however, the utility of this technique is hampered by: (1) the inability to consistently obtain good quality semen samples (Kiso et al. 2013; Hermes et al. 2013); (2) poor spermatozoa survival after liquid storage and transport (Kiso et al. 2011); and (3) poor post-thaw recovery of cryopreserved sperm (Kiso et al. 2012). There have been three reported pregnancies from frozen-thawed sperm to date: a stillborn calf after 17 months in Thailand (Thongtip et al. 2009); an African pregnancy that terminated after 5 months in the U.S. (Dennis L. Schmitt, pers. comm.); and an African elephant inseminated with frozen semen imported from South Africa that successfully gave birth in Austria (Hildebrandt et al. 2012). There is little doubt that AI is a significant management tool for captive elephants, especially if consistently good quality samples can be reliably obtained and cryopreserved for use with AI (Hermes et al. 2013), and would be even more effective if sex-sorted sperm were available to select for females (Hermes et al. 2009), which are easier to manage in captivity and more important for population sustainability.

5 Conclusions and Future Priorities

Since the last review, new revelations about female elephant biology and how individuals cope with challenges in their environment are being applied to improve reproductive success. Although fertility problems still hamper propagation efforts, like urogenital tract pathologies, ovarian acyclicity, and hyperprolactinemia, new approaches to management that aim to meet physical and physiological species-specific needs have real potential to better conserve these species. Particular attention to creating compatible mixed-sex social groups, providing proper nutrition and exercise, and ensuring environments are complex and enriching could mitigate many of the problems associated with early asymmetric reproductive aging, gonadal and pituitary dysfunction, and poor pregnancy outcomes. Knowing that continuous ovarian cyclicity of non-bred females is having negative and cumulative effects on reproductive health, emphasis should be on breeding females soon after puberty and continuing throughout the reproductive years. It is quite possible that in the near future, most captive populations will be self-sustaining, and in fact within the past decade many already are.

Needed are more advanced methods to assess fertility, pregnancy, health, stress and welfare status, preferably using non-invasive samples, like urine and/or feces. Measures of biomarkers across multiple organs, such as those involved in cardiac, inflammatory, immune, neural and metabolic systems are informative individually, but could offer a more holistic view of animal health and welfare when combined to create allostatic load indexes (i.e., a tool to measure the wear and tear of the body caused by chronic stress; see Edwards et al. this volume).

Rapid or dipstick tests to monitor reproductive, welfare and nutritional status would be particularly useful for assessing environmental and/or anthropogenic effects on elephant physiology and would permit monitoring of elephants in the field or where laboratory capabilities are limited, as is the case in most range countries. Finally, development of automatic collection devices, perhaps in collars or wearable packs, would allow for remote sampling, and would be even more valuable if multiple analytes could be quantified concurrently.

Elephants are not always easy to study. Long reproductive cycle and gestation lengths hinder the speed at which information can be generated, which is one reason it has taken over three decades to understand the complex dynamics of the HPG axis. On the other hand, the ability to collect serial blood samples combined with ultrasonographic examinations has produced an unprecedented amount of reproductive data. These advances have been possible only through the cooperation of elephant-holding facilities, and because of how easily elephants can be trained for non-stressful reproductive assessments. Among wildlife, there is no match to the biological knowledge we have on elephants, which will only grow in the years to come.

Acknowledgments We thank all of the technical, veterinary, zoo and research staff who contributed to the reproductive work described herein, and all participating elephants. This work was funded or supplied with in-kind donations by the Abbott Fund, Anela Kolohe Foundation, AZA Conservation Endowment Fund, Friends of the National Zoo, Institute of Museum and Library

Services, International Elephant Foundation, Morris Animal Foundation, Pfizer Corporation, Siemens Medical Solutions Diagnostics, Shared Earth Foundation, Smithsonian Competitive Grants Program, Smithsonian Scholar Studies Program, Smithsonian Women's Committee, Susan Fulton and FBB Capital Partners, and the Wolcott Fund.

References

Ahlers MJ, Ganswindt A, Münscher S, Bertschinger HJ. Fecal 20-oxo-pregnane concentrations in free-ranging African elephants (*Loxodonta africana*) treated with porcine zona pellucida vaccine. Theriogenology. 2012;78:77–85.

Allen WR. Ovulation, pregnancy, placentation and husbandry in the African elephant (*Loxodonta africana*). Philos Trans R Soc B. 2006;361:821–34.

Allen WR, Mathias SS, Wooding FBP, Skidmore JA, van Aarde RJ. Placentation in the African elephant, *Loxodonta africana. I.* Endocrinological aspects. Reprod Suppl. 2002;60:105–16.

Allen WR, Mathias S, Ford M. Placentation in the African elephant, *Loxodonta africana.* IV. Growth and function of the fetal gonads. Reproduction. 2005;130:713–20.

Ange K, Crissey SD, Doyle C, Lance K, Hintz H. A survey of African (*Loxodonta afriana)* and Asian *(Elephas maximus)* elephant diets and measured body dimensions compared to their estimated nutrient requirements. Proc Nutr Advisory Group Conf. 2001;4:5–14.

Angkawanish T, Thitaram C. Behavioral study and monitoring of Asian elephant (*Elephas maximus*) reintroduction under the Queen's initiative. In: Aranovich M, Dufresne O, editors. Elephants: ecology, behavior and conservation. New York: Nova Science; 2012. p. 133–44.

Aron DC, Tyrrell JB, Wilson CB. Pituitary tumors: current concepts in diagnosis and management. West J Med. 1985;162:340–52.

Assies J, Vingerhoets JJM, Poppelaars K. Psychosocial aspects of hyperprolactinemia. Psychoneuroendocrinology. 1992;17:673–9.

Bachelot A, Binart N. Reproductive role of prolactin. Reproduction. 2007;113:361–9.

Bechert US, Swanson L, Wasser SK, Hess DL, Stormshak F. Serum prolactin concentrations in the captive female African elephant *(Loxodonta africana)*: potential effects of season and steroid hormone interactions. Gen Comp Endocrinol. 1999;114:269–78.

Bertschinger H, Sills ES. Contraceptive applications of GnRH-analogs and vaccines for wildlife mammals of southern Africa: current experience and future challenges. In: Sills EC, editor. Gonadotropin-releasing hormone (GnRH): production, structure and function. 1st ed. New York: Nova Science; 2013. p. 85–108.

Bertschinger HJ, Lueders I. Use of anti-gonadotropin-releasing hormone vaccines in African elephants (*Loxodonta africana*): a review. Bothalia. 2018;48:a2320.

Bertschinger HJ, Delsink A, van Altena JJ, Kirkpatrick JF. Porcine zona pellucida vaccine immunocontraception of African elephant (*Loxodonta africana*). A review of 22 years of research. Bothalia Afr Biodivers Conserv. 2018;48(2):a2324.

Boedeker NC, Hayek LC, Murray S, de Avila DM, Brown JL. Effects of a gonadotropin-releasing hormone vaccine on ovarian cyclicity and uterine morphology of an Asian elephant (*Elephas maximus*). J Zoo Wildl Med. 2012;43:603–14.

Brown JL. Reproductive endocrine monitoring of elephants: an essential tool for assisting captive management. Zoo Biol. 2000;19:347–67.

Brown JL. Reproductive endocrinology. In: Fowler ME, Mikota SK, editors. Biology, medicine and surgery of elephants. Ames: Blackwell; 2006. p. 377–88.

Brown JL. Comparative reproductive biology of elephants. In: Holt WV, Brown JL, Comizzoli P, editors. Reproductive sciences in animal conservation—progress and prospects, Advances in experimental medicine and biology. New York: Springer; 2014. p. 135–69.

Brown JL, Lehnhardt J. Serum and urinary hormones during pregnancy and the peri- and postpartum period in an Asian elephant *(Elephas maximus)*. Zoo Biol. 1995;14:555–64.

Brown JL, Citino SB, Bush M, Lehnhardt J, Phillips LG. Cyclic patterns of luteinizing hormone, follicle stimulating hormone, inhibin, and progesterone in the Asian elephant *(Elephas maximus)*. J Zoo Wildl Med. 1991;22:49–57.

Brown JL, Schmitt DL, Bellem A, Graham LH, Lehnhardt J. Hormone secretion in the Asian elephant (*Elephas maximus*): characterization of ovulatory and anovulatory luteinizing hormone surges. Biol Reprod. 1999;61:1294–9.

Brown JL, Walker SL, Moeller T. Comparative endocrinology of cycling and noncycling Asian *(Elephas maximus)* and African (*Loxodonta africana*) elephants. Gen Comp Endocrinol. 2004a;136:360–70.

Brown JL, Goritz F, Pratt-Hawkes N, Hermes R, Galloway M, Graham LH, Gray C, Walker SL, Gomez A, Moreland R, Murray S, Schmitt DL, Howard J, Lehnhardt J, Beck B, Bellem A, Montali R, Hildebrandt TB. Successful artificial insemination of an Asian elephant at the National Zoological Park. Zoo Biol. 2004b;23:45–63.

Brown JL, Paris S, Prado-Oviedo NA, Meehan C, Hogan J, Carlstead K. Reproductive health assessment of female elephants in North American zoos and association of husbandry practices with reproductive dysfunction in African elephants (*Loxodonta africana*). PLoS One. 2016;11(7):e0145673.

Calogero AE, Badgy G, D'Agata R. Mechanisms of stress on reproduction—evidence for a complex intra-hypothalamic circuit. Ann N Y Acad Sci. 1998;1:364–70.

Carlstead K. A comparative approach to the study of keeper-animal relationships in the zoo. Zoo Biol. 2009;8:589–608.

Castracane VD, Stewart DR, Gimpel T, Overstreet JW, Lasley BL. Maternal serum androgens in human pregnancy: early increases within the cycle of conception. Hum Reprod. 1998;13:460–4.

Chase MJ, Schlossberg S, Griffin CR, Bouche PJC, Djene SW, Elkan PW, Ferreira S, Grossman F, Kohi EM, Landen K, Omondi P, Peltier A, Selier SAJ, Sutcliffe R. Continent-wide survey reveals massive decline in African savannah elephants. PeerJ. 2016;4:e2354.

Che-Castaldo JP, Grow SA, Faust LJ. Evaluating the contribution of North American zoos and aquariums to endangered species recovery. Sci Rep. 2018;8:9789.

Clubb R, Rowcliffe M, Lee P, Mar KU, Moss C, Mason GF. Fecundity and population viability in female zoo elephants: problems and possible solutions. Anim Welfare. 2009;18:237–47.

Cocchi D, Locatelli V, Carminati R, Muller EE. Mechanisms underlying the prolactin-lowering effect of metergoline in the rat. Life Sci. 1978;23:927–36.

Conde DA, Flesness N, Colchero F, Jones OR, Scheuerlein A. An emerging role of zoos to conservation biodiversity. Science. 2011;331:1390–1.

Creel S, MacDonald D. Sociality, group size, and reproductive suppression among carnivores. Adv Study Behav. 1995;24:203–57.

Czekala NM, MacDonald EA, Steinman K, Walker S, Garrigues NW II, Olson D, Brown JL. Estrogen and LH dynamics during the follicular phase of the estrous cycle in the Asian elephant. Zoo Biol. 2003;22:443–54.

Dale RH. Birth statistics for African (*Loxodonta africana*) and Asian (*Elephas maximus*) elephants in human care: history and implications for elephant welfare. Zoo Biol. 2010;29:87–103.

de Silva S, Webber CE, Weerathunga US, Pushpakumara TV, Weerakoon DK, Wittemyer G. Demographic variables for wild Asian elephants using longitudinal observations. PLoS One. 2013;8(12):e82788.

Delsink AK, Kirkpatrick JF. Free-ranging African elephant immunocontraception. A new paradigm for elephant management. Cape Town: Trident Press; 2012. p. 36.

Delsink AK, van Altena JJ, Kirkpatrick J, Grobler D, Fayrer-Hosken RA. Field applications of immunocontraception in African elephants (*Loxodonta africana*). Reprod Suppl. 2002;60:117–24.

Delsink AK, van Altena JJ, Kirkpatrick J, Grobler D, Bertschinger H, Kirkpatrick J, Slotow R. Regulation of a small, discrete African elephant population through immunocontraception in the Makalali conservancy, Limpopo, South Africa. South Afr J Sci. 2006;102:403–5.

Delsink AK, van Altena JJ, Grobler D, Bertschinger HJ, Kirkpatrick JF, Slotow R. Implementing immunocontraception in free-ranging African elephants at Makalali conservancy. J S Afr Vet Assoc. 2007;78:25–30.

Delsink AK, Kirkpatrick J, van Altena JJ, Bertschinger HJ, Ferreira SM, Slotow R. Lack of spatial and behavioral responses to immunocontraception application in African elephants (*Loxodonta africana*). J Zoo Wildl Med. 2013;44:52–74.

Dow TL, Brown JL. Evidence that hyperprolactinemia is associated with ovarian acyclicity in female zoo African elephants (*Loxondonta africana*). Reprod Fertil Dev. 2012;24:1019–27.

Dow TL, Holaskova I, Brown JL. Results of the third reproductive assessment survey of North American Asian (*Elephas maximus*) and African (*Loxodonta africana*) female elephants. Zoo Biol. 2011a;30:699–711.

Dow TL, Roudebush W, Parker FN, Brown JL. Influence of age and gender on secretion of anti-Müllerian hormone in Asian (*Elephas maximus*) and African (*Loxodonta africana*) elephants. Theriogenology. 2011b;75:620–7.

Drews B, Hermes R, Göritz F, Gray C, Kurz J, Lueders I, Hildebrandt TB. Early embryo development in the elephant assessed by serial ultrasound examinations. Theriogenology. 2008;69:1120–8.

Duer C, Carden M, Schmitt D, Tomasi T. Utility of maternal serum testosterone analysis for fetal gender determination in Asian elephants (*Elephas maximus*). Anim Reprod Sci. 2002;69:47–52.

Edwards KL, Trotter J, Jones M, Brown JL, Steinmetz HW, Walker SL. Investigating temporary acyclicity in a captive group of Asian elephants (*Elephas maximus*): relationship between management, adrenal activity and social factors. Gen Comp Endocrinol. 2016;225:104–16.

Edwards KL, Edes A, Brown JL. Stress, well-being and reproductive success. In: Holt WV, Brown JL, Comizzoli P, editors. Reproductive sciences in animal conservation—progress and prospects, Advances in experimental medicine and biology. 2nd ed. New York: Springer; this volume.

Emiliano ABF, Fudge JL. From galactorrhea to osteopenia: rethinking serotonin-prolactin interactions. Neuropsychopharmacology. 2004;29:833–46.

Fanson KV, Keeley T, Fanson BG. Cyclic changes in cortisol across the estrous cycle in parous and nulliparous Asian elephants. Endocr Connect. 2014;3:57–66.

Faust L, Marti K. Technical report on zoo risk modeling of the North American Asian elephant population, AZA TAG/SSP. Chicago: Lincoln Park Zoo; 2011a.

Faust L, Marti K. Technical report on zoo risk modeling of the North American African elephant population, AZA TAG/SSP. Chicago: Lincoln Park Zoo; 2011b.

Fayrer-Hosken RA, Bertschinger HJ, Kirkpatrick JF, Grobler D, Lamberski N, Honneyman G, Ulrich T. Contraceptive potential of the porcine zona pellucidavaccine in the African elephant (*Loxodonta africana*). Theriogenology. 1999;52:835–46.

Ferrari C, Caldara R, Rampini P, Telloli P, Romussi M, Bertazzoni A, Polloni G, Mattei A, Crosignani PG. Inhibition of prolactin release by serotonin antagonists in hyperprolactinemic subjects. Metabolism. 1978;127:1499–504.

Florio P, Gabbanini M, Borges LE, Bonaccorsi L, Pinzauti S, Reis FM, Boy Torres P, Rago G, Litta P, Petraglia F. Activins and related proteins in the establishment of pregnancy. Reprod Sci. 2010;17:320–30.

Forsyth IA, Wallis M. Growth hormone and prolactin—molecular and functional evolution. J Mammary Gland Biol Neoplasia. 2002;7:291–312.

Frasor J, Gibori G. Prolactin regulation of estrogen receptor expression. Trends Endocrinol Metab. 2003;14:118–23.

Freeman ME, Kanyicska B, Lerant A, Nagy G. Prolactin: structure, function and regulation of secretion. Physiol Rev. 2000;80:1524–631.

Freeman EW, Weiss E, Brown JL. Examination of the interrelationships of behavior, dominance status, and ovarian activity in captive Asian and African elephants. Zoo Biol. 2004;23:431–48.

Freeman EW, Whyte I, Brown JL. Reproductive evaluation of elephants culled in Kruger National Park, South Africa between 1976 and 1995. Afr J Ecol. 2008;47:192–201.

Freeman EW, Guagnano G, Olson D, Keele M, Brown JL. Social factors influence ovarian acyclicity in captive African elephants (*Loxodonta africana*). Zoo Biol. 2009;28:1–15.

Freeman EW, Schulte BA, Brown JL. Investigating the impact of rank and ovarian activity on the social behavior of captive female African elephants. Zoo Biol. 2010;29:154–67.

Freeman EW, Meyer JM, Putman S, Schulte BA, Brown JL. Using a simplified field progestagen method to assess ovarian activity in female African elephants. Biol Conserv. 2011;44:2105–11.

Freeman EW, Meyer JM, Putman S, Schulte BA, Brown JL. Ovarian cycle status varies with respect to age and social status in free-ranging elephants in Addo Elephant National Park, South Africa. Cons Physiol. 2013;1:cot025.

Glaeser SS, Hunt KE, Martin MS, Finnegan M, Brown JL. Investigation of individual and group variability in estrous cycle characteristics in female Asian elephants (*Elephas maximus*) at the Oregon Zoo. Theriogenology. 2012;78:285–96.

Grattan DR, Kokay IC. Prolactin: a pleiotropic neuroendocrine hormone. J Neuroendocrinol. 2008;20:752–63.

Grynnerup AG, Lindhard A, Sorensen S. The role of anti-Müllerian hormone in female fertility and infertility—an overview. Acta Obstet Gynecol Scand. 2012;91:1252–60.

Harris M, Sherwin C, Harris S. The welfare, housing and husbandry of elephants in UK zoos. Final report. Bristol: University of Bristol; 2008.

Hartley M, Stanley CR. Survey of reproduction and calf rearing in Asian and African elephants in European zoos. J Zoo Aquar Res. 2016;4:139–46.

Hayward AD, Mar KU, Lahdenpera M, Lummaa V. Early reproductive investment, senescence and lifetime reproductive success in female Asian elephants. J Evol Biol. 2014;27:7772–83.

Hermes R, Olson D, Göritz F, Brown JL, Schmitt DL, Hagan D, Peterson JS, Fritsch G, Hildebrandt TB. Ultrasonography of the estrous cycle of female African elephants (*Loxodonta africana*). Zoo Biol. 2000;19:369–82.

Hermes R, Hildebrandt TB, Göritz F. Reproductive problems directly attributable to long-term captivity: asymmetric reproductive aging. Anim Reprod Sci. 2004;82–83:49–60.

Hermes R, Sargusty J, Schaftenaar W, Goeritz F, Schmitt DL, Hildebrandt TB. Obstetrics in elephants. Theriogenology. 2008;70:131–44.

Hermes R, Behr B, Hildebrandt TB, Blottner S, Sieg B, Frenzel A, Knieriem A, Saragusty J, Rath D. Sperm sex-sorting in the Asian elephant (*Elephas maximus*). Anim Reprod Sci. 2009;112:390–6.

Hermes R, Saragusty J, Göritz F, Bartels P, Potier R, Baker B, Streich JW, Hildebrandt TB. Freezing African elephant semen as a new population management tool. PLoS One. 2013;8:e57616.

Hildebrandt TB, Göritz F, Hermes R, Reid C, Dehnhard M, Brown JL. Aspects of the reproductive biology and breeding management of Asian and African elephants. Int Zoo Yearb. 2006;40:20–40.

Hildebrandt TB, Lueders I, Hermes R, Göritz F, Sargusty J. Reproductive cycle of the elephant. Anim Reprod Sci. 2011;124:176–83.

Hildebrandt TB, Hermes R, Saragusty J, Potier R, Schwammer HM, Balfanz F, Vielgrader H, Baker B, Bartels P, Göritz F. Enriching the captive elephant population genetic pool through artificial insemination with frozen-thawed semen collected in the wild. Theriogenology. 2012;78:1398–404.

Holdgate MR, Meehan CL, Hogan JN, Miller LJ, Soltis J, Andrews J, Shepherdson DJ. Walking behavior of zoo elephants: associations between GPS-measured daily walking distances and environmental factors, social factors, and welfare indicators. PLoS One. 2016;11(7):e0150331.

Hufenus R, Schiffmann C, Hatt JM, Muller DWH, Lackey LB, Clauss M, Zerbe P. Seasonality of reproduction in Asian elephants *Elephas maximus* and African elephants *Loxodonta africana*: underlying photoperiodic cueing? Mammal Rev. 2018;48:261–76.

Illera JC, Silván G, Cáceres S, Carbonell MD, Gerique C, Martinez-Fernández L, Munro C, Cesares M. Assessment of ovarian cycles in the African elephant (*Loxodonta africana*) by measurement of salivary progesterone metabolites. Zoo Biol. 2014;33:245–9.

IUCN Red List of Threatened Species. 2018. Version 2018-2. www.iucnredlist.org.

Jorgensen HS. Studies on the neuroendocrine role of serotonin. Dan Med J. 2007;54:266–88.

Jungheim ES, Moley KH. Current knowledge of obesity's effects in the pre- and periconceptional periods and avenues for future research. Am J Obstet Gynecol. 2010;203:525–30.

Kaewmanee S, Watanabe G, Kishimoto M, Zhu Jin W, Yamamoto Y, Yamamoto T, Nagaoka K, Narushima E, Komiya T, Taya K. Secretion of inhibin during the estrous cycle in the female Asian elephant (*Elephas maximus*). J Vet Med Sci. 2011;73:77–82.

Kajaysri J, Nokkaew W. Assessment of pregnancy status of Asian elephants (Elephas maximus) by measurement of progestagen and glucocorticoid and their metabolite concentrations in serum and feces, using enzyme immunoassay (EIA). J Vet Med Sci. 2013;76:363–8.

Kaplowitz PB. Link between body fat and the timing of puberty. Pediatrics. 2008;121:S208–17.

Kapustin N, Critser JK, Olson D, Malven PV. Nonluteal estrous cycles of 3-week duration are initiated by anovulatory luteinizing hormone peaks in African elephants. Biol Reprod. 1996;55:1147–54.

Keele M, Lewis K, Owens T. Asian elephant (*Elephas maximus*) North American Regional Studbook. Portland: Oregon Zoo; 2014.

Kirkpatrick JF, Turner JW Jr, Liu IK, Fayrer-Hosken R. Applications of pig zona pellucida immunocontraception to wildlife fertility control. J Reprod Fertil Suppl. 1996;50:183–9.

Kiso W, Brown JL, Siewerdt SDL, Olson D, Crichton EG, Pukazhenthi BS. Liquid semen storage in elephants (*Elephas maximus* and *Loxodonta africana*): species differences and storage optimization. J Androl. 2011;21:420–31.

Kiso WK, Asano A, Travis AJ, Schmitt DL, Brown JL, Pukazhenthi BS. Pretreatment of Asian elephant (*Elephas maximus*) spermatozoa with cholesterol-loaded cyclodextrins and glycerol addition at 4°C improves cryosurvival. Reprod Fertil Dev. 2012;24:1134–42.

Kiso WK, Selvaraj V, Nagashima J, Asano A, Brown JL, Schmitt DL, Leszyk J, Travis AJ, Pukazhenthi BS. Lactotransferrin in Asian elephant (*Elephas maximus*) seminal plasma correlates with semen quality. PLoS One. 2013;8:e71033.

Kiso W, Wiedner E, Isaza R, Lindsay W, Aria J, Jacobson G, Jacobson K, Schmitt S. Reproductive parameters and birth statistics for a herd of Asian elephants (*Elephas maximus*) in North America over a 20-year period. J Zoo Wildl Med. 2017;48:987–97.

Knight PG. Roles of inhibins, activins, and follistatin in the female reproductive system. Front Neuroendocrinol. 1996;17:476–509.

Lahdenperä M, Mar KU, Lummaa V. Reproductive cessation and post-reproductive lifespan in Asian elephants and pre-industrial humans. Front Zool. 2014;11:54–68.

Lahdenperä M, Mar KU, Lummaa V. Nearby grandmother enhances calf survival and reproduction in Asian elephants. Sci Rep. 2016;6:27213.

Lamberts SWJ, Macleod RM. Regulation of prolactin secretion at the level of the lactotroph. Physiol Rev. 1990;70:279–98.

Lee PC, Fishlock V, Webber CE, Moss CJ. The reproductive advantages of a long life: longevity and senescence in wild female African elephants. Behav Ecol Sociobiol. 2016;70:337–45.

Lueders I, Niemuller C, Gray C, Rich P, Hildebrandt TB. Luteogenesis during the estrous cycle in Asian elephants (*Elephas maximus*). Reproduction. 2010;140:777–86.

Lueders I, Taya K, Watanabe G, Yamamoto Y, Yamamoto T, Kaewmanee S, Niemuller C, Gray C, Streich W, Hildebrandt TB. Role of the double luteinizing hormone peak, luteinizing follicles, and the secretion of inhibin for dominant follicle selection in Asian elephants (*Elephas maximus*). Biol Reprod. 2011;85:714–20.

Lueders I, Niemuller C, Rich P, Gray C, Hermes R, Göritz F, Hildebrandt TB. Gestation for 22 months: luteal development and pregnancy maintenance in elephants. Proc R Soc B. 2012;279:3687–96.

Lueders I, Hildebrandt T, Gray C, Botha S, Rich P, Niemuller C. Suppression of testicular function in a male Asian elephant (*Elephas maximus*) treated with gonadotropin-releasing hormone vaccines. J Zoo Wildl Med. 2014;45:611–9.

Lueders I, Young D, Maree L, van der Horst G, Luther I, Botha S, Tindall B, Fosgate G, Ganswindt A, Bertschinger HJ. Effects of GnRH vaccination in wild and captive African elephant bulls (*Loxodonta africana*) on reproductive organs and semen quality. PLoS One. 2017;12:e0178270.

Lueders I, Niemuller C, Steinmetz HW, Bouts T, Gray C, Knauf-Witzens T, Taya K, Watanabe G, Yamamoto Y, Oerke AK. Prolonged luteal lifespan and pseudopregnancy in Asian elephants (*Elephas maximus*). Anim Reprod Sci. 2018;197:58–66.

Mar KU. The demography and life-history strategies of timber elephants in Myanmar. PhD thesis. London: University College; 2007.

Mar KU, Lahdenpera M, Lummaa V. Causes and correlates of calf mortality in captive Asian elephants (*Elephas maximus*). PLoS One. 2012;7(3):e32335.

Mason GJ, Veasey JS. What do population-level welfare indices suggest about the well-being of zoo elephants? Zoo Biol. 2010;29:256–73.

Melmed S. Mechanisms for pituitary tumorigenesis: the plastic pituitary. J Clin Invest. 2003;112:1603–18.

Mellen JD. Factors influencing reproductive success in small captive exotic felids (Felis spp.): a multiple regression analysis. Zoo Biol. 2005;10:95–110.

Mendis S, Jayasekera NK, Rajapakse RC, Brown JL. Endocrine correlates of puberty in female Asian elephants (*Elephas maximus*) at the Pinnawala Elephant Orphanage, Sri Lanka. BMC Zool. 2017;2:1. https://doi.org/10.1186/s40850-016-0012-8.

Meyer JM, Walker SL, Freeman EW, Steinetz BG, Brown JL. Species and fetal gender effects on the endocrinology of pregnancy in elephants. Gen Comp Endocrinol. 2004;138:263–70.

Miller PB, Forstein DA, Styles S. Effect of short-term diet and exercise on hormone levels and menses in obese, infertile women. J Reprod Med. 2008;53:315–9.

Morfeld KA, Brown JL. High body condition scores and elevated insulin and leptin are associated with ovarian acyclicity in zoo African elephants (*Loxodonta africana*). Reprod Fertil Dev. 2014;28:640–7.

Morfeld K, Brown JL. Metabolic health assessment of zoo elephants: management factors predicting leptin levels and the glucose-to-insulin ratio and their associations with health parameters. PLoS One. 2017;12(11):e0188701.

Morfeld KA, Ball RL, Brown JL. Recurrence of hyperprolactinemia and continuation of ovarian acyclicity in captive African elephants (*Loxodonta africana*) treated with cabergoline. J Zoo Wildl Med. 2014;45:569–76.

Morfeld KA, Meehan C, Hogan J, Brown JL. Assessment of body condition in African (*Loxodonta africana*) and Asian (*Elephas maximus*) elephants in North American zoos and management practices associated with high body condition scores. PLoS One. 2016;11(7):e0155146.

Mouttham L, Buhr M, Brown JL. Ovarian acyclicity and dominance status of African elephants in zoos is not associated with increased serum testosterone production. Zoo Biol. 2011;126:115–21.

Mumby HS, Mar KU, Thitaram C, Courtiol A, Towiboon P, Min-Oo Z, Htut-Aung Y, Brown JL, Lummaa V. Stress and body condition are associated with climate and demography in Asian elephants. Conserv Physiol. 2015; https://doi.org/10.1093/conphys/cov030.

Niemuller CA, Gray C, Cummings E, Liptrap RM. Plasma concentrations of immunoreactive relaxin activity and progesterone in the pregnant Asian elephant (*Elephas maximus*). Anim Reprod Sci. 1998;53:119–31.

Nilsson EM, Fainberg HP, Choong SS, Giles TC, Sells J, May S, Stansfield FJ, Allen WR, Emes RD, Mostyn A, Mongan NP, Yon L. Molecular characterization of adipose tissue in the African elephant (*Loxodonta africana*). PLoS One. 2014;9(3):e91717.

Norkaew T, Brown JL, Bansiddhi P, Somgird C, Thitaram C, Punyapornwithaya V, Punturee KK, Vongchan P, Nopphamas S, Khonmee J. Body condition and adrenal glucocorticoid activity affects metabolic marker and lipid profiles in working female tourist elephants in Thailand. PLoS One. 2018;13(10):e0204965.

Olive A, Jansen K. The contribution of zoos and aquaria to Aichi Biodiversity Target 12: a case study of Canadian Zoos. Glob Ecol Conserv. 2017;10:103–13.

Olson D. North American Region Studbook for the African elephant (*Loxodonta africana*). Azle: International Elephant Foundation; 2014. p. 1–71.

Prado NA, Keady M, Oestmann A, Steinbeiser C, Brown JL. Hyperprolactinemic African elephant (*Loxodonta africana*) females exhibit elevated dopamine, oxytocin and serotonin concentrations compared to normal cycling and non-cycling, low prolactin elephants. Biol Reprod. 2019a;100:1549–60.

Prado NA, Paris SW, Brown JL. Prolonged ovarian acyclicity is associated with a higher likelihood of developing hyperprolactinemia in captive female African elephants (*Loxodonta africana*). Zoo Biol. 2019b; https://doi.org/10.1002/zoo.21464.

Prado-Oviedo NA. Hyperprolactinemia and ovarian acyclicity in captive African elephant females (*Loxodonta africana*). George Mason University, College of Science Archival Resources; 2015.

Prado-Oviedo NA, Mallow EJ, Deng X, Brown JL. Hyperprolactinemia is not associated with hyperestrogenism in noncycling African elephants (*Loxodonta africana*). Gen Comp Endocrinol. 2013;89C:7–14.

Pushpakumara PGA, Rajapakse RC, Perera BMAO, Brown JL. Reproductive performance of the largest captive Asian elephant (*Elephas maximus*) population in Sri Lanka. Anim Reprod Sci. 2016;174:93–9.

Richard P, Moos F, Freund-Mercier MJ. Central effects of oxytocin. Physiol Rev. 1991;71:331–55.

Rico C, Fabre S, Medigue C, di Clemente N, Clement F, Bontoux M, Touze JL, Dupont M, Briant E, Remy B, Beckers JF, Monniaux D. Anti-Müllerian hormone is an endocrine marker of ovarian gonadotropin-responsive follicles and can help to predict superovulatory responses in the cow. Biol Reprod. 2009;80:50–9.

Robinson MR, Mar KU, Lummaa V. Senescence and age-specific trade-offs between reproduction and survival in female Asian elephants. Ecol Lett. 2012;15:260–6.

Saragusty J, Hermes R, Göritz F, Schmitt DL, Hildebrandt TB. Skewed birth sex ratio and premature mortality in elephants. Anim Reprod Sci. 2009;115:247–54.

Sarkar DK, Gottschall PE, Meites J. Decline of tuberoinfundibular dopaminergic function resulting from chronic hyperprolactinemia in rats. Endocrinology. 1984;115:1269–74.

Sarkar DK, Frautschy SA, Mitsugi N. Pituitary portal plasma levels of oxytocin during the estrous cycle, lactation and hyperprolactinemia. Ann N Y Acad Sci. 1992;652:397–410.

Schiffmann C, Clauss M, Fernando P, Pastorini J, Wendler P, Ertl N, Hoby S, Hatt JM. Body condition scores of European zoo elephants (*Elephas maximus* and *Loxodonta africana*): status quo and influencing factors. J Zoo Aquar Res. 2018;6:91–103.

Schuff KG, Hentges ST, Kelly MA, Binart N, Kelly PA, Iuvone PM, Asa SL, Low MJ. Lack of prolactin receptor signaling in mice results in lactotroph proliferation and prolactinomas by dopamine dependent and independent mechanisms. J Clin Invest. 2002;110:973–81.

Schulte BA, Feldman E, Lambert R, Oliver R, Hess DL. Temporary ovarian inactivity in elephants: relationship to status and time outside. Physiol Behav. 2000;71:123–31.

Serri O, Chik CL, Ur E, Ezzat S. Diagnosis and management of hyperprolactinemia. Can Med Assoc J. 2003;169:575–81.

Sobrinho LG. Prolactin, psychological stress and environment in humans: adaptation and maladaptation. Pituitary. 2003;6:35–9.

Sobrinho LG, Nunes MCP, Calhaz-Jorge C, Afonso AM, Pereira MC, Santos MA. Hyperprolactinemia in women with paternal deprivation during childhood. Obstet Gynecol. 1984;64:465–8.

Somgird C, Homkong P, Sripiboon S, Brown JL, Stout TA, Colenbrander B, Mahasawangkul S, Thitaram C. Potential of a gonadotropin-releasing hormone vaccine to suppress musth in captive male Asian elephants (*Elephas maximus*). Anim Reprod Sci. 2016;164:111–20.

Sonino N, Navarrini C, Ruini C, Fallo F, Boscaro M, Fava GA. Life events in the pathogenesis of hyperprolactinemia. Eur J Endocrinol. 2004;151:61–5.

Squires EL, Ginther OJ. Follicular and luteal development in pregnant mares. J Reprod Fertil Suppl. 1975;23:429–33.

Stansfield FJ, Allen WR. Luteal maintenance of pregnancy in the African elephant (*Loxodonta africana*). Reproduction. 2012;143:845–54.

Sukumar R, editor. The living elephants: evolution, ecology, behavior and conservation. Oxford: Oxford University Press; 2003. p. 478.

Tabak J, Gonzalez-Iglesias AE, Toporikova N, Bertram R, Freeman ME. Variations in the response of pituitary lactotrophs to oxytocin during the rat estrous cycle. Neuroendocrinology. 2010;151:1806–13.

Takahashi T. Biology of the prolactin family in bovine placenta. I. Bovine placental lactogen: expression, structure and proposed roles. Anim Sci J. 2006;77:10–7.

Thitaram C. Breeding management of captive Asian elephant (*Elephas maximus*) in range countries and zoos. Jpn J Zoo Wildl Med. 2012;17:91–6.

Thitaram C, Brown JL. Control and monitoring of ovarian function in elephants. Theriogenology. 2018;109:42–7.

Thitaram C, Brown JL, Pongsopawijit P, Chansitthiwet S, Wongkalasin W, Daram P, Roongsri R, Kulmapijit A, Mahasawangkul S, Rojanasathien S, Colenbrander B, van der Weijden GC, van Eerdenburg FJCM. Seasonal effects on the endocrine pattern of the semi-captive female Asian elephant (*Elephas maximus*): timing of the luteinizing hormone surge determines the length of the estrous cycle. Theriogenology. 2008;69:237–44.

Thitaram C, Pongsopawijit P, Chansitthiwet S, Brown JL, Nimtragul N, Boonprasert K, Homkong P, Mahasawangkul S, Rojanasthien S, Colenbrander B, van der Weijden GC, van Eerdenburg FJ. Induction of the ovulatory LH surge in Asian elephants (*Elephas maximus*): a novel aid in captive breeding management of an endangered species. Reprod Fertil Dev. 2009;21:672–8.

Thongtip N, Mahasawangkul S, Thitaram C, Pongsopavijitr P, Kornkaewrat K, Pinyopummin A, Angkawanish T, Jansittiwate S, Rungsri R, Boonprasert K, Songkalasin W, Homkong P, Dejchaisri S, Wajjwalku W, Saikhun K. Successful artificial insemination in the Asian elephant (*Elephas maximus*) using chilled and frozen-thawed semen. Reprod Biol Endocrinol. 2009;7:75–81.

Urwin VE, Allen WR. Pituitary and chorionic gonadotrophin control of ovarian function during early pregnancy in equids. J Reprod Fertil Suppl. 1982;32:371–82.

Valades G, Ganswindt A, Annandale H, Schulman M, Bertschinger H. Non-invasive assessment of the reproductive cycle in free-ranging female African elephants (*Loxodonta africana*) treated with a gonadotropin-releasing hormone (GnRH) vaccine for inducing anoestrus. Reprod Biol Endocrinol. 2012;10:63.

Verhelst J, Abs R, Maiter D, van den Bruel A, Vandeweghe M, Velkeniers B, Mockel J, Lamberigts G, Petrossians P, Coremans P, Mahler C, Stevenaert A, Verlooy J, Raftopoulos C, Beckers A. Cabergoline in the treatment of hyperprolactinemia: a study in 455 patients. J Clin Endocrinol Metab. 1999;84:2518–22.

Vick MM, Sessions DR, Murphy BA, Kennedy EL, Reedy SE, Fitzgerald BP. Obesity is associated with altered metabolic and reproductive activity in the mare: effects of metformin on insulin sensitivity and reproductive cyclicity. Reprod Fertil Dev. 2006;18:609–17.

Wang AT, Mullan RJ, Lane MA, Hazem A, Prasad C, Gathaiya NW, Fernandez-Balsells MM, Bagatto A, Coto-Yglesias F, Carey J, Elraiyah TA, Erwin PJ, Gandhi GY, montori VM, Murad MH. Treatment of hyperprolactinemia: a systematic review and meta-analysis. Syst Rev. 2012; https://doi.org/10.1186/2046-4053-1-33.

Wasser SL, Barash DP. Reproductive suppression among female mammals: implications for biomedicine and sexual selection theory. Q Rev Biol. 1983;58:513–38.

Wasser SL, Papageorge S, Foley C, Brown JL. Excretory fate of estradiol and progesterone in the African elephant (*Loxodonta africana*) and pattern of fecal steroid concentrations throughout the estrous cycle. Gen Comp Endocrinol. 1996;102:255–62.

Weissenbock NM, Schwammer HM, Ruf T. Estrous synchrony in a group of African elephants (*Loxodonta africana*) under human care. Anim Reprod Sci. 2009;113:322–7.

Wielebnowski N. Contribution of behavioral studies to captive management and breeding of rare and endangered animals. In: Caro T, editor. Behavioral ecology and conservation biology. New York: Oxford University Press; 1998. p. 130–62.

Wielebnowski NC, Fletchall N, Carlstead K, Busso JM, Brown JL. Non-invasive assessment of adrenal activity associated with husbandry and behavioral factors in the North American clouded leopard population. Zoo Biol. 2002a;21:77–98.

Wielebnowski NC, Ziegler K, Wildt DE, Lukas J, Brown JL. Impact of social management on reproductive, adrenal and behavioural activity in the cheetah (*Acinonyx jubatus*). Anim Conserv. 2002b;5:291–301.

Wittemyer G, Northrup JM, Blanc J, Douglas-Hamilton I, Omondi P, Burnham KP. Illegal killing for ivory drives global decline in African elephants. Proc Natl Acad Sci. 2014;111(36):13117–21.

Yamamoto Y, Yamamoto T, Watanabe G, Yuto N, Keio M, Narushima E, Katayanagi M, Nakao R, Morikubo S, Sakurai Y, Kaneko M, Kaewmanee S, Taya K. Prolactin secretion and ovarian function in cycling and non-cycling African female elephants (*Loxodonta africana*). J Vet Med Sci. 2010;72:845–52.

Yamamoto Y, Yamamoto T, Taya K, Watanabe G, Stansfield FJ, Allen WR. Placentation in the African elephant (*Loxodonta africana*). V. The trophoblast secretes placental lactogen. Placenta. 2011;32:506–10.

Yamamoto Y, Yuto N, Yamamoto T, Kaewmanee S, Shiina O, Mouri Y, Narushima E, Katayanagi M, Sugimura K, Nagaoka K, Watanabe G, Taya K. Secretory pattern of inhibin during the estrous cycle and pregnancy in African (*Loxodonta africana*) and Asian (*Elephas maximus*) elephants. Zoo Biol. 2012a;31:511–22.

Yamamoto Y, Yamamoto T, Yuto N, Hildebrandt TB, Lueders I, Wibbelt G, Shiina O, Mouri Y, Sugimura K, Sakamoto S, Kaewmanee S, Nagoaka K, Watanabe G, Taya K. The secretory pattern and source of immunoreactive prolactin in pregnant African (*Loxodonta africana*) and Asian (*Elephas maximus*) elephants. J Reprod Dev. 2012b;58:105–11.

Yamamoto Y, Nagaoka K, Kamite Y, Watanabe G, Allen T, Stansfield F, Taya K. Different origins of two corpora lutea recovered from a pregnant African elephant (*Loxodonta africana*). Reprod Domest Anim. 2017;52(6):1138–41.

Yuen BH. Hyperprolactinemia in women of reproductive age: etiology, diagnosis and management. Can Fam Physician. 1992;38:367–79.

Zacur HA. Hyperprolactinemia. In: Knobil E, Neill J, editors. Encyclopedia of reproduction, vol. 2. New York: Academic Press; 1998. p. 725–35.

Chapter 10
Comprehensive Breeding Techniques for the Giant Panda

Meghan S. Martin-Wintle, David C. Kersey, Nathan J. P. Wintle, Copper Aitken-Palmer, Megan A. Owen, and Ronald R. Swaisgood

Abstract The dramatic growth of the captive giant panda (*Ailuropoda melanoleuca*) population exemplifies how the application of scientific findings to animal care and reproductive management can improve conservation breeding outcomes. Detailed behavioral studies of giant panda estrus, pregnancy and cub rearing have demonstrated the importance of husbandry management that supports natural reproductive behavior to enhance breeding success. Natural breeding has been valuably augmented by the development of assisted reproductive techniques founded through detailed studies of the reproductive physiology of the giant panda and outlining fundamental information about reproductive seasonality, male fertility and characterization of the estrous cycle. The resultant holistic understanding of giant panda reproduction has improved reproductive success in the captive population to such an extent that it is now self-sustaining and provides surplus animals for reintroduction. Despite these significant advances, there are knowledge gaps and remaining challenges to be addressed. Pregnancy detection remains the single biggest challenge when determining if natural mating or assisted breeding have been successful. Because pregnancy can only be determined in the few weeks prior to parturition, there are gaps in understanding and detecting delayed implantation and early embryonic loss. Additionally, dynamic management practices and standard of care

M. S. Martin-Wintle (✉)
PDXWildlife, Portland, OR, USA

Institute for Conservation Research, San Diego Zoo Global, Escondido, CA, USA
e-mail: meg@pdxwildlife.com

D. C. Kersey
College of Veterinary Medicine, Western University of Health Sciences, Pomona, CA, USA

N. J. P. Wintle
PDXWildlife, Portland, OR, USA

C. Aitken-Palmer
Chicago Zoological Society, Brookfield, IL, USA

M. A. Owen · R. R. Swaisgood
Institute for Conservation Research, San Diego Zoo Global, Escondido, CA, USA

P. Comizzoli et al. (eds.), *Reproductive Sciences in Animal Conservation*, Advances in Experimental Medicine and Biology 1200,
https://doi.org/10.1007/978-3-030-23633-5_10

for reproductive assistance needs to be developed. Only large breeding centers in China have the ability to promote normal reproductive behaviors and allow mate choice for the giant panda. These challenges need to be addressed in the near future in order to maintain a self-sustaining, genetically diverse and behaviorally competent captive population. This chapter documents the development of successful giant panda managed breeding programs by focusing on three key areas, (1) the development of science-driven reproductive techniques to improve fecundity in a species where the mating system was poorly understood, (2) how targeted research and adaptive management of social settings surrounding estrus and breeding improved reproductive success, and (3) insights and solutions to challenges faced across the program's history with future directions for research.

Keywords Giant panda · *Ailuropoda melanoleuca* · Assisted breeding technology · Reproductive biology · Semen collection · Semen cryopreservation · Estrus detection · Artificial insemination · Natural mating · Mate choice · Behavioral management

1 Introduction

Conservation breeding is an important species recovery tool, but faces many challenges associated with bringing at-risk species into an artificially managed environment that may not allow for normal expression of reproduction for that species. A scientific understanding of normal reproduction is the best tool to allow for natural reproduction and to identify when to use assisted reproduction techniques for giant pandas in managed care. Facilities that are entrusted with these special species must commit to evidenced-based conservation breeding programs to ensure successful reproduction and contribute to species conservation goals. This commitment is not easy and requires constant flexibility to adapt to new scientific findings. The use of science and technology to augment fertility is essential to maximize success, especially for species in which reproductive knowledge is limited under managed care. The unique reproductive biology of the giant panda posed challenges that initially compromised reproductive success within the population. These challenges were overcome by decades of research to illuminate the normal behavior and physiology of reproduction in this species. Due to focused research detailing reproductive behavior and physiology, the giant panda is one of the best understood species in managed care and has supported the expansive growth of the *ex situ* population. These studies also emphasized the importance of health and well-being to maximize reproduction. Specifically, studies have indicated the integration of bamboo-based nutrition, veterinary medicine, endocrinology, expression of normal reproductive and communication behaviors, environmental enrichment, mate choice, cub rearing, gamete analysis, gamete preservation and artificial insemination were necessary to support the dramatic growth of the *ex situ* giant panda population over the past 25 years (Aitken-Palmer et al. 2012; Charlton et al. 2009a; Edwards et al. 2006; Jin et al. 2011; Kersey et al. 2016; Lindburg et al. 2001; Loeffler et al. 2006;

Martin-Wintle et al. 2015; Snyder et al. 2004; Swaisgood et al. 2003). Despite this population growth and an overall improvement in giant panda health, well-being and fertility, there is still room for improvement. Although the giant panda is an interesting species, being the oldest extant bear species, a monogastric animal subsisting almost exclusively on a plant-based diet, with a striking black-white pelage, the reproductive biology is the most fascinating aspect of this species. This chapter documents the development of current-day knowledge on giant panda reproductive biology and application of reproductive techniques in the giant panda over multiple decades and suggests areas of research to continue improvements into the future.

2 Conservation Status of the Giant Panda

First listed as Rare on the IUCN Red List in 1986 and Endangered in 1990, the giant panda now occupies a rare position as one of the few species downlisted from endangered (Hoffmann et al. 2010), and is currently designated as *Vulnerable* (Swaisgood et al. 2016). Subsequently, following similar criteria, the Chinese government downlisted the panda to vulnerable on its Biodiversity Red List (Zhang et al. 2017b). However, the giant panda should not be seen as "recovered," but rather viewed as recovering (Swaisgood et al. 2018). Continued conservation efforts are vital to keep the giant panda on the path to recovery.

The giant panda's status assessment is fortunate to be much less data deficient than most other at-risk species. China's State Forestry Administration conducts a large-scale survey about every 10 years to monitor giant panda populations and habitat. There have now been four surveys (survey periods ending in 1977, 1988, 2001 and 2014) and each survey covers all suitable habitat and involves tens of thousands of worker days, making it an intensive effort to ascertain a species' status. The surveys showed an initial sharp decline in giant panda population, followed by a period of stabilization and gradual increase. With an estimated 1,864 giant pandas remaining in the wild during the most recent survey, combined with a stable or increasing population, pandas do not qualify for Endangered status under IUCN criteria. Although there have been some methodological changes from the first survey, the last two surveys were almost identical in methodology and therefore the observed population increase between the third and fourth surveys is considered robust. The population increase is also consistent with findings of an expanded range of giant pandas, increased habitat extent, and, in some cases, improved habitat conditions (Swaisgood et al. 2018; Wei et al. 2018b; Xu et al. 2017).

This population increase for the giant panda, iconic for its Endangered status, should not come as a surprise given the large-scale efforts to conserve this species (Wei et al. 2018a, 2015). The Chinese government invested in a growing protected area system, increasing the number of giant panda reserves by more than fivefold in the past two decades from ~10 to 67; this reserve system covers 58% of giant panda habitat. Further, two of the world's largest ecocompensation programs provide protections to giant panda habitat outside of reserves. The National Forest Conservation

Program bans logging throughout the giant panda's range and the Grain to Green Program incentivizes local communities to reforest farmlands (Yang et al. 2015). These polices have been associated with increasing forest cover and improving giant panda habitat (Li et al. 2013; Wei et al. 2018b). Efforts to relocate human populations outside of reserves, alongside economic trends motivating people to leave the countryside for employment in urban areas, have supported a trend for decreasing human populations in much of the giant panda habitat (Xu et al. 2017).

Yet, giant pandas face many continuing and emerging threats that place its recovery at risk (Kang and Li 2016; Swaisgood et al. 2016, 2018; Wei et al. 2018b; Xu et al. 2017). Habitat fragmentation is a legacy of the rapid development of China's countryside over the past several hundred years of human population expansion, and today giant panda habitat exists in a network of relatively isolated fragments. Without intervention, the fragmented giant panda populations, especially the smaller ones, are vulnerable to demographic and stochastic events leading to extirpation and loss of genetic diversity through increased inbreeding (Hu et al. 2017). Interventions are underway to address this threat, including corridor establishment to connect fragmented populations (mostly through the Giant Panda National Park), and translocations and release of captive-reared giant pandas to some of the smallest, most vulnerable populations (Yang et al. 2018). More worrisome are recent increases in a number of threats, including road construction, tourism, farming, and livestock encroachment, that while less impactful on giant pandas than logging, still have the potential to steadily degrade giant panda habitat (Li et al. 2017a; Wei et al. 2018b; Xu et al. 2017; Zhang et al. 2017a).

A combination of continued ecocompensation and improved enforcement of existing regulations is required to address these anthropogenic activities that threaten giant panda recovery. More hope is also found in the proposed expansion of the protected area system, through the creation of a Giant Panda National Park, which is a network of parks that will limit human activities in most of the remaining giant panda habitat and establish corridors between isolated populations. Where sufficient corridors cannot be established, continued genetic rescue efforts involving release of giant pandas bred in conservation breeding centers or attempts at artificial insemination or semen collection for gamete cryopreservation will ensure that these populations persist. If these measures are sufficient, the giant panda should remain vulnerable, or perhaps someday be removed from the IUCN list of threatened species altogether.

3 Giant Panda Reproductive Biology

Before addressing conservation breeding challenges and solutions it is necessary to understand the basic reproductive biology of the species. Giant panda females have reproductive tracts that resemble other Ursids in general form (see Fig. 17.5 in Hildebrandt et al. 2006). Females have a hairless anogenital opening containing glands that omit short-chain fatty acid and aromatic compounds used in olfactory communication (Swaisgood et al. 2004). Urine is also used for olfactory communication, typically conveying more short-term motivational information regarding reproductive readiness (Nie et al. 2012a). The vaginal opening is narrow and measures 85 mm deep with strong circular mucosal walls characterized by a pink mucosa with enlarged pink colored labia around the time of estrus (Raven et al. 1936). The urethral opening is visible within the vaginal vault near the fornix and is ventral to the cervical opening. The cervix measures about 4 cm in length and is slightly convoluted with the cone shaped portio vaginalis appearing swollen and containing an opaque viscous mucus during estrus (Hildebrandt et al. 2006). The uterus is bicornuate with a corpus uteri that measures half that of the uterine horns or cornu uteri (Davis 1964), and is lined with two endometrial layers (Hildebrandt et al. 2006). There are scant descriptions of giant panda ovaries and follicular development. One study indicates ovaries measure 30 mm long by 23 mm wide (Raven et al. 1936), but there is likely some individual variation to size and methodology used to measure (i.e., ultrasound versus necropsy measurements). There is one report describing giant panda follicular development where during the follicular phase the ovary contained anywhere from 3–10 Graafian follicles and luteal ovary had an equal number of corpora lutea that measured 5–7 mm in diameter (Hildebrandt et al. 2006). Similar to the ovarian measurements, there is likely variation in follicular size among females. Broader studies of ovarian measurements and normal follicular development combined with endocrine monitoring are needed to understand normal versus abnormal ovarian size and follicular development. Ovaries are inactive during the nonbreeding season (Hildebrandt et al. 2006).

Male giant panda body morphometrics can be utilized to evaluate for normal secondary sex characteristics. Male body mass, chest girth, and abdominal girth are consistent among adult giant pandas and do not differ across reproductive seasons (Aitken-Palmer et al. 2012). Male giant pandas have a penis that is entirely withdrawn within a puckered prepuce, surrounded by anogenital glands similar to females (see Fig. 17.8 in Hildebrandt et al. 2006). This prepuce forms the covering of the glans of the penis, which only covers the body of the penis during an erection leaving the glans (head of penis) extending 50–70 mm beyond the prepuce (Raven et al. 1936). The glans penis is relatively small in relation to body size and measures about 70 mm long by 25 mm wide in a full-grown male with slight variation in size among males (Raven et al. 1936). The penis is supported by multiple muscles (M. retractor penis, sphincter, externus, bulbocavernosus, and sphincter urethrae membranaceae) and suspensory ligaments that attach to the pelvis (Raven et al. 1936). The body of the penis is made up of three cavernous compartments: two corpora fibrosa and the corpus spongiosum (Raven et al. 1936). These are all encased in connective tissue that make up the exterior of the penis beneath the skin (Raven et al. 1936). Further support for the penis is provided by the baculum or "os penis". The baculum is about 24 mm long and comprised of a shaft-like body with flattened, downward angled, distal extensions, which form an inverted slot where the urethra sits (Raven et al. 1936).

The testes sit about 70 mm caudal to the prepuce and are ovate in structure. Dramatic changes in testicular size occurs within the reproductive and non-reproductive season; total testicular volume can increase nearly twofold from the nonbreeding to prebreeding season (Aitken-Palmer et al. 2012). A further increase in testicular size has been documented to occur during the peak breeding period that was sustained through the late breeding interval (Aitken-Palmer et al. 2012). Testicular size should be symmetrical in the normal adult giant panda with no differences between the size of the right and left testis (Aitken-Palmer et al. 2012). The epididymis is large and is divided into three separate regions: caput, corpus, and cauda. The ductus deferens continues off the tail of the epididymis and loops back toward the dorsum of the bladder. Both ducts are enlarged as they approach the neck of the bladder and unite 5 mm before entering the urethra. At the terminus is the 100 mm long urethra that begins as a thick, muscular walled vessel and ends longitudinally ridged. Although Davis (1964) reported that male giant pandas do not have a prostate gland, later advancements in monitoring using ultrasound (Hildebrandt et al. 2006) showed a distinctive prostate comprised of two lateral lobes next to the urethra. The prostate can also be palpated rectally and is bulbous in shape similar to other carnivore species (Rong et al. 2006).

4 Reproductive Success: A Holistic Approach

There are recognized fundamental elements for successful breeding and cub survival in the giant panda captive breeding program: (1) Understanding how to identify fertility (or subfertility) in the seasonal male and female, (2) understanding how to optimize breeding introductions for success and reduced aggression, (3) understanding when and how to use semen collection and timed artificial insemination instead of natural breeding, (4) identifying pregnancy and evaluating why an animal may not become pregnant, and (5) successful rearing of the cub by the dam, by hand, or a combination of the two approaches when necessary (i.e., 'twin swapping'). For artificial insemination one also needs additional steps; (1) knowledge of the male reproductive biology to collect viable semen for evaluation, preservation and insemination; (2) identifying timing of ovulation by interpretation of urinary estrogen and progestogen results to determine when the cervix will be open; and (3) the technique necessary to deposit viable sperm in the receptive uterus via transcervical intrauterine insemination (IUI). An understanding and comprehensive view of successful natural and captive reproductive behavior, structural and functional anatomy, endocrine and gamete physiology, and application of assisted reproductive techniques is needed at each facility that cares for giant pandas to form a successful, sustainable managed breeding program. For the giant panda, this reproductive knowledge has accumulated over four decades. The following overview of giant panda reproductive biology focuses on captive animals.

A species' ability to successfully reproduce is the culmination of environment, health, evolution/biology, and nutrition. Li et al. (2017b) have noted successful reproductive rates of the giant panda under human care to be below what is considered optimal. In a species reliant on reproductive strategy with an extremely short fertile female window (72 h every 2 years, devoting only 1% of their lifetime to mating activity; Kersey et al. 2016) this species should, in theory, have a high rate of fertility and reproductive success to maximize its limited opportunities for reproduction. Indeed, reproductive rates in wild giant pandas are sufficient to sustain the population, and compromised breeding in captivity is a byproduct of captive management (i.e. inappropriate social management, abnormal health, suboptimal nutrition and husbandry, compromised welfare) not inherent reproductive deficiencies in giant pandas (Lü et al. 2000; Snyder et al. 2003). Poor ability to detect early pregnancy also limits the ability to determine if breeding or artificial insemination has been successful. Instead, we rely on only the birth of a cub, or lack of birth. However, conception and delayed implantation with early embryonic loss likely contributes to the inaccurate impression that giant pandas reproduce poorly in captivity. Without the ability to definitively diagnose pregnancy true reproductive success cannot be determined.

When sub-optimal fertility is suspected, several factors should be methodically evaluated to determine if a cause can be identified. This process, used broadly for all species, should include a meticulous evaluation of health, husbandry, anatomy, nutrition, behavior, welfare, and reproductive biology of the individual giant panda. Routine (i.e. annual, or every few years) health exams of both sexes are the first step to identify problems that may affect fertility and should be conducted prior to the breeding season so that if problems are noted, medical intervention may be taken to resolve the problem. A physical exam in which vitals (i.e., weight, blood pressure, heart rate, respiratory rate, temperature) and each body system are evaluated should be part of the exam; additionally, specialized dental and breeding soundness exams should be standard for this species. A thorough dental exam is necessary as giant pandas are prone to dental disease (Loeffler et al. 2007). Breeding soundness evaluations are necessary to confirm gender and evaluate for normal or abnormal anatomy. Each wellness exam and breeding soundness evaluation should include venipuncture with complete blood counts (hematology), serum biochemical analysis and diagnostic imaging (radiographs, ultrasound, and CT where available) as part of a minimum database of health information (Miller and Fowler 2014). For males, there should be additional rectal exams, semen collection, and comprehensive semen analysis included to assess fertility. Cryptorchidism and anatomic abnormalities can impact fertility (Howard et al. 2006a) and are known to occur in the giant panda. Normal seasonal variation in male fertility is more protracted than the female breeding season, males produce sperm from November through May in the Northern Hemisphere with maximal ejaculate quality shown to coincide with female peak estrus March/April (Aitken-Palmer et al. 2012). Season of semen collection should be considered when ejaculate quality is evaluated in order to avoid misdiagnoses of suboptimal sperm production or infertility (Aitken-Palmer et al. 2012). The ideal time for male fertility assessment is early February through the end of May in the Northern hemisphere (Aitken-Palmer et al. 2012). Additionally, sexual maturity, normal age of maturity, and senescence of fertility should be considered when attempting to determine why an individual may have sub-optimal fertility. Male pandas reach sexual maturity at approximately 8 years of age with expression of adult behaviors (i.e., complete hand-stand scent marking) developing in the last years of maturation (Aitken-Palmer 2010). Normal behavior is essential for natural mating in both sexes and should be monitored and considered as part of the breeding soundness evaluation. Additionally, serial longitudinal monitoring of urine (i.e. estrogen, progesterone) or fecal (i.e. thyroid, androgens) hormones should be evaluated for normal versus abnormal seasonal patterns and metabolic function.

Artificial insemination or other assisted reproductive techniques should not be used in place of natural mating and should not occur prior to a routine wellness exam or a breeding soundness evaluation. This is because artificial insemination cannot overcome deficits in impaired health, nutrition, infectious diseases, or anatomical abnormalities. Artificial insemination has the most utility when geographical barriers occur (i.e., solo housed individual), incompatibility between planned pairs, or other management challenge prevents natural mating. For example, AI should be considered when male aggression or behavioral incompetence is observed in attempted mating pairings, there exists no ability to allow mate choice among individuals, or genetically related individuals are housed within the same facility.

5 Behavioral Management for Natural Mating

We start with the fundamental premise that natural mating should be prioritized over assisted reproduction techniques, where feasible, and that assisted reproduction is an important backup for specific cases where natural mating fails or other specific objectives necessitate assisted reproduction. Although natural mating in giant pandas is often associated with higher pregnancy rates (Li et al. 2017b), there are additional reasons to prioritize natural mating. These reasons include the greater cost efficiency of natural mating, the opportunity to study and gain knowledge about reproductive behavior with insights for wild pandas, and the welfare benefits of allowing animals to perform their natural behavioral repertoire. The benefits of assisted reproduction techniques are discussed later in this chapter, but here we address the history, challenges, and solutions for natural mating, detailing lessons learned primarily at the China Research and Conservation Center for Giant Pandas (CCRCGP) comprised of the historical Wolong Breeding Center and its subsidiary facilities.

Retrospective analysis of more than 300 insemination events over 21 years demonstrated that natural mating had the highest likelihood of producing cubs (60.7%) compared to 50.6% for natural mating followed directly by AI, and 18.5% for AI only. Within the combined AI/natural insemination technique group, genetic paternity analysis revealed that 81.8% of births were attributed to natural mating. The low success rates for AI may be attributed to physiological and health factors such as the use of frozen sperm and inaccurate timing of AI. Nonetheless, it appears that under most circumstances and because natural mating requires significantly less financial and human labor resources, natural mating should be the method of choice for breeding of this species. AI is still an important tool to breed pairs that are housed in separate facilities, to incorporate genetic diversity from poorly represented founders and from pairs whose offspring are desirable for genetic management, and to maintain heterozygosity above 97% (Li et al. 2017b).

Historically, the giant panda was notoriously difficult to breed in captive conditions and until the late 1990s overall reproductive rates were insufficient to sustain the captive population (Swaisgood et al. 2006). To get past these obstacles, early research addressed several behavioral factors that might limit reproductive success, including enrichment and well-being (Swaisgood et al. 2001), communication in reproductive contexts (Charlton et al. 2009a; Swaisgood et al. 2000, 2002), and maternal care (Snyder et al. 2003; Zhang et al. 2000). A series of investigations into the use of chemical signals demonstrated that pandas possess a sophisticated chemical communication system and that exposure to certain signals increased indices of sexual motivation and decreased aggressive motivation in both males and females (Swaisgood et al. 2004). Incorporation of these findings into management included timed exposure to scent at the beginning of the female estrus via human-assisted exchange of scent "postcards" and pen swapping between the sexes. In the latter, male giant pandas were placed in female pens with the female removed to avoid direct contact and aggression and vice-versa. A retrospective time series analysis of cub production success rates from natural matings revealed a significant increase after these management modifications were implemented indicating that they considerably impacted reproductive success of the program (Li et al. 2017b). Years later similar experimentation demonstrated the importance of acoustic signals for identifying individuals, sexual status, and sexual receptivity (Charlton et al. 2009a, 2010, 2018). These management interventions to enhance communication—along with improved welfare, better hormonal and behavioral monitoring to determine the fertile period, and promotion of maternal care, as well as non-behavioral improvements such as health, nutrition, and hand-rearing of twins—contributed to large increases in the rate of successful natural mating and a rapid population increase (Swaisgood et al. 2006).

A basic understanding of giant panda reproductive ecology, behavior and physiology underpinned the development of behavioral management for reproduction at the CCRCGP. Sexual maturity has a range of 3–6.5 years for female giant pandas (Kersey et al. 2010a; Kleiman et al. 1979; Schaller et al. 1985; Steinman et al. 2006) and 5.5 with consistent sperm production for males (Aitken-Palmer 2010). Pandas are sexually mature and capable of reproduction all of the way into their early 20s with the oldest, successful captive panda birth at 23 years old (SB# 544). Females experience a single annual estrus occurring in the spring, and males show seasonal behavioral and physiological changes during this time (Aitken-Palmer et al. 2012; Kersey et al. 2010a; Kleiman 1983; Nie et al. 2012b, c). Males occupy large home ranges that overlap extensively with several home ranges of neighboring conspecifics of both sexes (Hull et al. 2015; Schaller et al. 1985; Wenshi 2014). Female giant pandas begin to advertise their reproductive status several weeks before ovulation through scent deposition (Swaisgood et al. 2000, 2004; White et al. 2002) and vocal signals (Charlton et al. 2009a, b, 2010, 2011, 2018). Approximately 5 days prior to ovulation, two to three males congregate near the female and compete for reproductive access to the female while the female stays sequestered in a tree near the competition site (Nie et al. 2012b). Nie et al. (2012b) reported that a dominant male is always established by day 2 of the competition and will engage in mate guarding behavior near the base of the tree containing the female. This mate guarding includes deterring subordinate males who remained near the breeding site and chasing the female back up the tree if she attempts to leave (Nie et al. 2012b). The dominant male wins sole breeding rights to the female and all animals disperse from the mating site after copulation occurrs (Nie et al. 2012b).

In captivity, adult giant pandas are managed most of the year based on the solitary nature of the species. They are typically housed individually and only allowed occasional contact with neighboring conspecifics, requiring that fertility and sexual readiness be monitored closely using behavioral and physiological variables to determine the correct time to conduct mating introductions (Fig. 10.1; Swaisgood et al. 2006). Behavioral indicators include increased locomotion, water play, proximity to the opposite sex, several vocalizatons (especially chirps, bleats, moans), scent marking and investigation of conspecific scent, and several behaviors associated with courtship (female tail up, anogenital presentation, and lordosis, and male genital inspection and erections; McGeehan et al. 2002; Owen et al. 2016; Swaisgood et al. 2006). When the onset of estrus is detected (see Sect. 6 below), the female is often relocated to provide access to potential mates. Four to 5 days prior to estrus, neighboring animals of the opposite sex are routinely swapped between enclosures to provide them with olfactory communication opportunties (e.g. deposition of scent marks, urine, and feces) that have been shown to provide important information about individual identification (Swaisgood et al. 1999), sex and reproductive status (Swaisgood et al. 2000), reproductive maturity (White et al. 2003), and competitive status (White et al. 2002). The timing to attempt pairing of potential mates

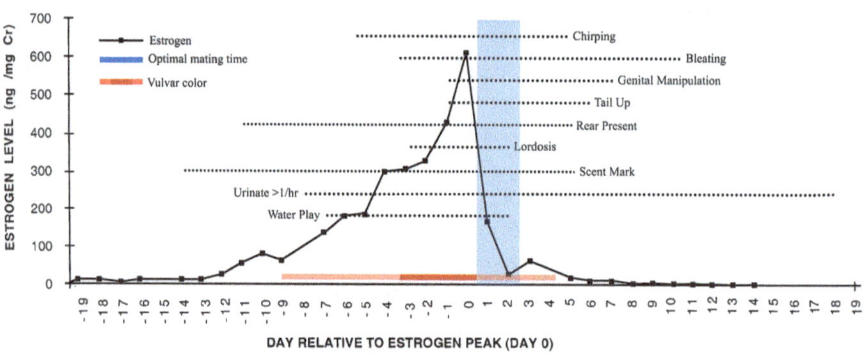

Fig. 10.1 Composite graph indicating estrogen profile (solid line and squares), vulva coloring (pink and red horizontal bar), and temporal patterns of behavior (dotted lines) during mating season in a giant panda female. The light blue vertical bar indicates ideal mating window for successful cub production. Graph is modified from Lindburg et al. (2001), Swaisgood et al. (2003) and Zhang et al. (2004) and behavioral time frames are averaged from Fig. 1 of all three publications after centering data at estrogen peak (assuming day of mating was 2 days after estrogen peak). Vulvar coloration time periods are averaged from Durrant et al. (2003) and Zhang et al. (2004)

is based on a combination of hormonal (Kersey et al. 2010b), behavioral (Owen et al. 2013, 2016; Swaisgood et al. 2006), and morphological (Durrant et al. 2003) monitoring, which together predict both the timing of ovulation and behavioral motivation to mate. If male-female interactions look promising, the pair is always given the chance to mate naturally, although both successful and unsuccessful natural mating attempts are commonly followed with artificial insemination.

Once the decision has been made to move forward with a mating introduction (see Fig. 10.2 for decision tree on monitoring ovulation, breeding introductions, and artificial insemination), the mating attempt is monitored closely for sexual motivation, excessive aggression, and copulation success. Males will approach the female, manipulate her into position through nape biting and pushing with the shoulders and forelegs, while at the same time establishing penile erection. Receptive females will walk backwards, raise the tail and adopt the lordotic posture encouraging the male to mount the female, and thrust to achieve intromission. Giant pandas produce a distinctive "orgasm bleat" when intromission is achieved (Charlton et al. 2018) and the male will attempt to hold the female in a mounted position even after the female attempts to break away. The male achieves this in part by adopting a "roll back" position—pulling the female towards him while sitting on his hindquarters— typically when copulation lasts longer than 10–30 s (Fig. 10.3). Once the female dislodges (the female almost invariably breaks contact) the pair is separated by staff. Another pairing attempt may be made within a few hours to increase probability of fertilization. If excessive aggression is observed or the pair is unsuccessful at achieving copulation, breeding managers will swap the unsuccessful male for another and attempt a mating introduction with the new male as long as male-female behavioral interactions suggest compatibility during protected contact in adjacent enclosures. Through this process females will usually be given mating introduction opportunities with 1–6 males during estrus.

Evidence suggests that deficits in sexual motivation and copulatory performance are not the result of poor timing of mating introductions (Zhang et al. 2004) but rather failures are due to behavioral or physical abnormality in the male or the lack of mate compatibility between the pair (Martin-Wintle et al. 2015). Indeed, most females mate successfully when given access to a different male (Martin-Wintle et al. 2015; Zhang et al. 2004) and giant pandas have been shown to assortatively mate based on personality profiles (Martin-Wintle et al. 2017; Powell et al. 2008). Housing male and female pandas adjacent to one another provides important opportunities for mate familiarization and for evaluation of pair compatibility. An example of a facility designed to managed mate pairings is at the CCRCGP's breeding facility, where a specialized complex comprised of multiple, small enclosures with shift doors adjoining neighboring pens allow animal managers to initally familiarize mates without physical contact (Fig. 10.4). Pens are arranged in a circular fashion so that animals can be moved from pen to pen for quick access to potential mates. Each female entering estrus is placed in an enclosure adjacent to one or two males who are selected based on a Mate Suitability Index (MSI), which combines mean kinship, inbreeding, and difference between mean kinship value of the male and female (Li et al. 2017b). Pairs with an MSI less than 4 are genetically valuable pairs

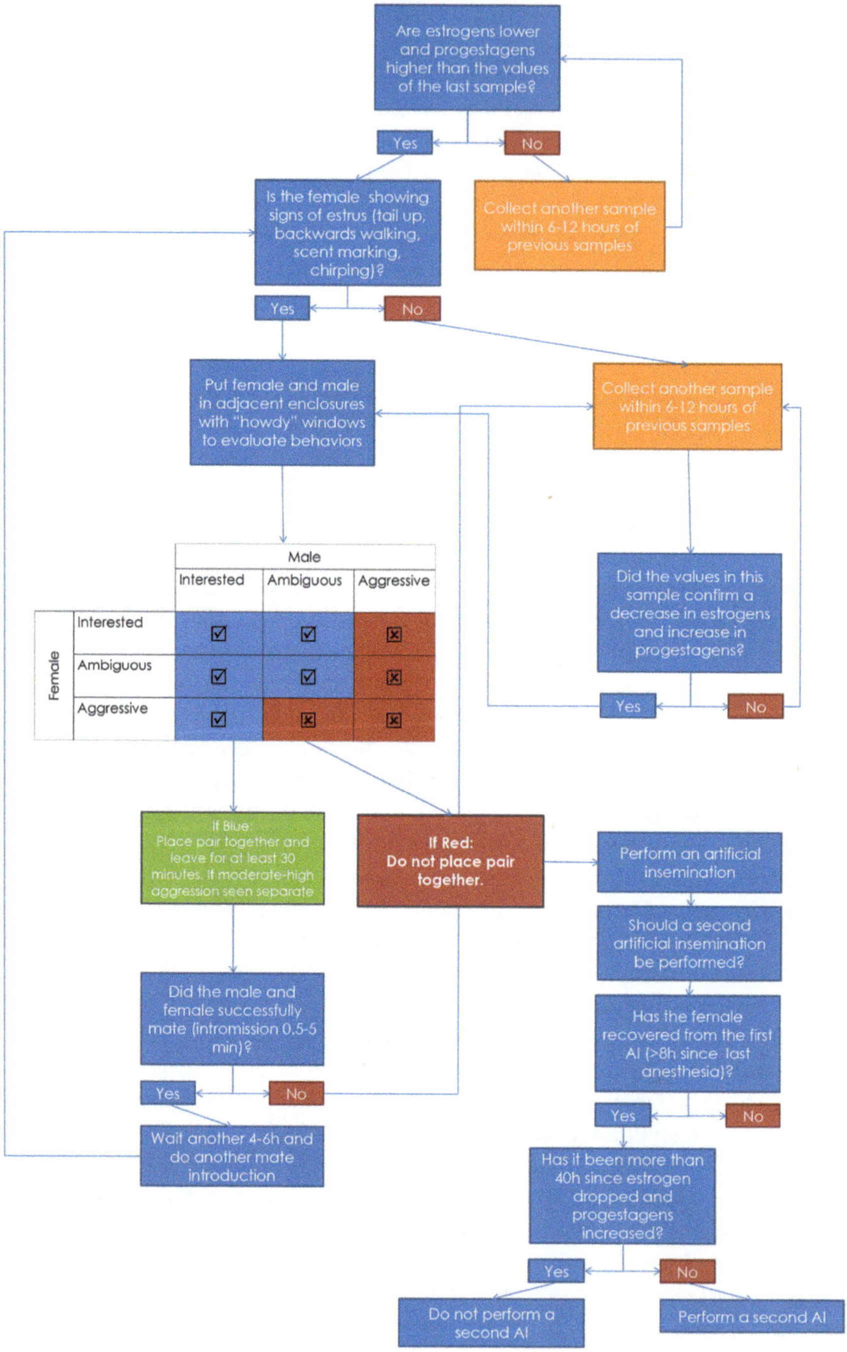

Fig. 10.2 Mating introduction and artificial insemination decision tree

Fig. 10.3 Giant pandas breeding at the Bifengxia Chinese Conservation and Research Center for the Giant Panda (CCRCGP) facility in 2017. (**a**) Successful mounting with the female correctly positioned in a slight lordosis and tail up. (**b**) "Rollback" achieved approximately 10–30 s after the beginning of a successful intromission

Fig. 10.4 Giant panda breeding center at the Chinese Conservation and Research Center for the Giant Panda (CCRCGP) Shenshuping facility. (**a**) An aerial view of the breeding center demostrating the circular layout. Groups of five enclosures are connected via shift doors to aid in movement of animals. (**b**) View of enclosure showing one howdy window, one shift door, and elevated walkways for observing animals. The opposite side of the enclosure has an identical layout to determine female mate preference. (**c**) Male and female giant panda at the shift door directly prior to mating introduction

and are a priority to breed while breeding managers try to avoid pairs with MSIs higher than four (Traylor-Holzer and Ballou 2017). Breeding enclosures have no trees or man-made enrichment to allow quick separation if the mate introduction becomes aggressive. However, enclosures provide opportunties for familirization of females and males through "howdy" openings in enclosure walls that provide visual, olfactory, acoustic, and limited tactile contact (Martin-Wintle et al. 2015; Zhang et al. 2004).

Familiarization is achieved through moving females from their home enclosures between the two highest MSI ranked males approximately 1–2 weeks prior to estrus and performing pen swaps on a regular basis to gauge both male and female interest. Females are normally located at the same facility as males but sometimes are brought in from other facilities (it is unknown whether excessive movement during estrus disrupts ovulation so movement should be kept to a minimum). In addition to providing opportunities for communication and mate familiarization, this set up also allows the role of behavioral preferences and mate choice to be evaluated (Martin-Wintle et al. 2015). Recent findings indicate that when both giant panda partners prefer each other (preferred$_{♀}$-preferred$_{♂}$) there is an 83.3% chance of the pairing having intromission success as well as a 90% chance of successful intromissions producing a cub. Success rates of intromission and cub production are both reduced to approximately 50% if only one individual in the pairing is preferred (nonpreferred$_{♀}$-preferred$_{♂}$ or preferred$_{♀}$-nonpreferred$_{♂}$) and to 0% intromission success if mate introductions were attempted where both partners were non-preferred (Martin-Wintle et al. 2015). Giant pandas have also been shown to assortatively mate based on behavioral personality types. In combination, these studies suggest that behavioral mechanisms of mate choice may serve to optimize genetic compatibility (Martin-Wintle et al. 2015, 2017).

6 Ovulation Detection

Detection of ovulation is a useful component for determining the optimal time for mate-pairing for natural breeding; however, knowing the time of ovulation is essential to time AI.

During the single, annual ovulatory event the female giant panda will release one to three oocytes that are viable for approximately 3 days (Czekala et al. 2003). Identification of this critical, albeit brief, period in the reproductive cycle is important for investigation of the species' unique reproductive biology and maximizing breeding and/or artificial insemination (AI) efforts. To this end, various approaches have been employed to determine the timing of ovulation in the giant panda.

Identifying ovulation starts with detecting the behavioral and morphological changes associated with the receptivity phase of estrus. Estrus is the defining behavioral period of the female reproductive cycle, and the earliest descriptions of giant panda reproduction concerned female behavioral changes around this time (Kleiman et al. 1979). All female giant pandas that enter the follicular phase engage in a compliment of olfactory, vocal, visual and tactile communications that convey her attractivity, proceptivity, or receptivity to mating (Fig. 10.1; Lindburg et al. 2001; McGeehan et al. 2002; Owen et al. 2016). The expression of the compliment of these behaviors lasts 1–2 weeks (Steinman et al. 2006), is also termed proestrus, which corresponds with proceptivity, and estrus, which corresponds with receptivity. However, the number, duration and intensity of the behaviors, and their relationship with hormones are variable within and among females (Lindburg et al. 2001; McGeehan et al. 2002; Owen et al. 2016; Snyder et al. 2004). For example, the receptive behavior "tail-up" may be of 1–6 days duration around the time of estrogen peak, and not all females express this behavior (McGeehan et al. 2002). Vocal behavior of females change dramatically across the estrous period, which influences the frequency (Owen et al. 2016) and structure of vocalizations (Charlton et al. 2010). For example, the duration and harshness of chirps differ before and during the fertile period (Charlton et al. 2010), but not all females will chirp during peak fertility (McGeehan et al. 2002). Casual observations during routine care of the female are sufficient to detect the onset of the reproductive cycle; however more rigorous assessments are warranted thereafter to note the subtle changes associated with the proceptive phase of estrus. Additionally, supplementing the behavioral/morphological observations with other ovulation detection methods strengthens the ability to identify ovulation correctly.

In the giant panda, assessment of the ovarian hormones, estrogens and progestagens can be achieved by immunoassay measure of the serum (Bonney et al. 1982), urine (Monfort et al. 1989) and feces (Kersey et al. 2010b). Characteristically, the onset of the follicular phase is denoted by an increase in estrogen values above baseline 7–14 days prior to peak values, after which there is a 1–4 days decline to baseline (Fig. 10.5; Kersey et al. 2010b, 2016; Steinman et al. 2006). Coincident

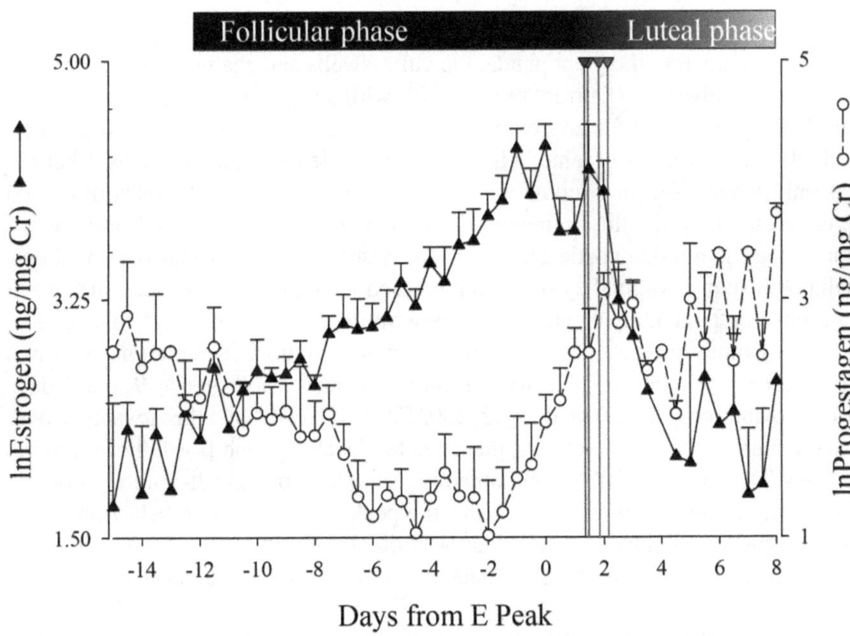

Fig. 10.5 Female giant panda natural log (ln)-transformed urinary estrogen and progestagen follicular and early luteal phase composite. Five years of hormone data (2006, 2008, 2010, 2013, 2016) from one female (SB452) were arranged by day from estrogen peak (Day 0) and averaged (+SEM) every 12 h from Day 0. The follicular phase begins on Day −12 and transitions to the luteal phase between days 0 and 4. The vertical dashed lines topped with solid upside-down triangles denote the time of the first AI in each year with a range of 35.0–51.6 h after estrogen peak. In each year, AI resulted in pregnancy as noted by birth

with the initial increase in estrogen at the follicular phase onset, progestagens decline below baseline, and remain so until the periovulatory estrogen drop, at which time progestagens rise above baseline (Kersey et al. 2016). Because estrogen values fluctuate during the follicular phase as a result of changing physiology and/or artifact of analysis, the measurement of both estrogen and progestagen is necessary to accurately assess the changes in physiology. For detection of these trends, daily urine collection is preferred for on-site and rapid data analysis to time breeding and/or AI (Czekala et al. 1998; Kersey et al. 2016), whereas, fecal measures are advantageous for long-term studies (Kersey et al. 2010b). The advantage of steroid assessment is that their abundance is a direct reflection of follicle growth and ovulation, and urinary hormone data can be obtained within a few hours of sample collection. The disadvantage is the technical complexity and expense of hormone assessment, and that the hormone trends are not accurate predictors, but more retrospective assessors, of ovulation. Alternative to the complex steroid immunoassays is a rapid, semiquantitative snap test to assess the ovulatory luteinizing hormone (LH) surge. In the giant panda, LH has been assessed in the urine and serum (Durrant et al. 2006b); however, because the diagnostic is only semi-quantitative and can be affected by urine concentration, this approach is best to support, but not supplant, estrogen and progestagen analysis.

Estrogen produced during follicular growth affects the morphology of the female reproductive tract. In the giant panda, the vulva swells and changes color as the follicular phase advances (Durrant et al. 2003), with a peak in vulvar score (size and color) being sustained for approximately 3 days periovulation (Fig. 10.6; Durrant et al. 2003). This easily obtained visual cue of female reproductive status is beneficial only when used in conjunction with other assessment techniques discussed here, because of variability in intensity, duration, and proximity to ovulation, vulva morphology is not diagnostic for ovulation. Assessment of exfoliated vaginal epithelia with stained vaginal cytology can be used to augment other modalities when assessing for impending ovulation. Among approaches discussed in this section, vaginal cytology uniquely offers predictive determination of ovulation, with two characteristic cellular stain uptake (chromic) changes occurring at 9 and 2 days prior to ovulation (Durrant et al. 2002, 2003, 2006a). Yet despite this advantage over other ovulation detection methods, there can be sizeable variability in these chromic changes (Durrant et al. 2006a) and obtaining samples from a well-trained, cooperative female decreases as she approaches the peak in her receptive behaviors. Unto itself, vaginal cytology can determine the follicular phase status of a female but is of limited utility to determine ovulation if employed without urinary hormone monitoring.

The approaches described above are the most common means of assessing follicular status in the giant panda, and their prevalence is largely driven by utility during breeding/AI timing; however, application of established and emerging technologies for giant panda ovulation detection warrant consideration. Transrectal ultrasonography can be used to visualize the ovaries of anesthetized females (Dehnhard et al. 2006; Hildebrandt et al. 2003, 2006), but because of chemical restraint, this approach is reserved to confirm ovarian status and inseminating

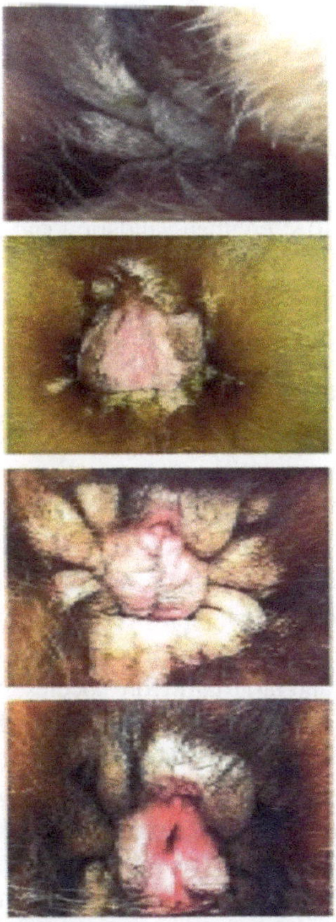

Fig. 10.6 Presentation of the female giant panda vulva in (**a**) anestrus, (**b**) early proceptive phase, (**c**) proceptive phase, and (**d**) receptive phase

catheter placement during AI (Hildebrandt et al. 2006). With advancements in technology for ultrasound machines and improved diagnostic capabilities of trained veterinary radiologists and ultrasonographers, further research is needed to explore the future utility of this diagnostic imaging modality. Ovarian function and follicular development may be visible by transabdominal imaging with the right expertise and animal positioning, however the large gastrointestinal tract of the giant panda can impede diagnostic imaging attempts. Additionally, training female giant pandas to allow ultrasonographic imaging without anesthesia can be challenging.

Urinary scent marking provides reliable cues that males use to discriminate female estrus (Swaisgood et al. 2002), and chemical analysis revealed a marked increase in expression of chemo-signals such as decanoic acid coincident with the onset of the follicular phase (Dehnhard et al. 2006). Although this study did not reveal any volatile chemicals associated with ovulation, the notion deserves further exploration to consider a broader array of chemicals in the urine periovulation (Wilson et al. 2018).

Finally, recent investigations have demonstrated a corollary between traditional urinary estrogen measures and near infrared spectroscopy (NIRS) of the urine (Kinoshita et al. 2010, 2012). Although NIRS offers greater mobility and speed of assessment over hormone analytics, there are several factors regarding NIRS that need to be resolved, including the cost of the NIRS units, the accuracy of detecting peak estrogen concentrations (Kinoshita et al. 2012), and incorporation of other hormonal measures (e.g. progestagens, LH) to strengthen validation and identification of ovulation.

7 Semen Collection, Evaluation, and Cryopreservation

Semen collection is part of the male breeding soundness exam. Typically, semen collection occurs under general anesthesia. It is important to carefully plan for and monitor the vitals of the giant panda under general anesthesia. There has been at least one giant panda that died potentially due to anesthetic causes (Foster 2010). Despite the potential risks of general anesthesia, there have been hundreds of male giant pandas safely anesthetized for semen collection. Similar considerations should be made for the giant panda as other ursidae and large carnivore species. Anesthetic monitoring using pulse oximetry, blood gas monitoring, ECG monitoring and capnography are considered standard of care for giant panda. There are few studies that evaluate the impact of anesthetics on semen collection; however, protocols with ketamine used as a solo agent for induction and maintenance (10–12 mg/kg ketamine hydrochloride, I.M. injection), or ketamine used as the induction agent and isoflurane gas delivered via a face mask or endotracheal tube for maintenance of general anesthesia successfully result in high quality ejaculate with electroejaculation (Aitken-Palmer et al. 2012). Telazol (5 mg/kg) has also been used successfully to anesthetize giant pandas for semen collection (Kersey et al. 2016). Once under anesthesia, testicular and body morphometry can be evaluated as part of the breeding soundness exam. Specifically, the length and width of each testicle can be measured using a digital caliper and volume of each testis calculated using the formula volume = $4/3\ \pi\ ab^2$ (a is 1/2 length; b is 1/2 width; Howard et al. 2006a). Total testicular volume per male is determined by combining the right and left testis volumes (average testicular for peak breeding season from March 22–April 15 is $370 \pm 31\ cm^3$; Aitken-Palmer et al. 2012).

Electroejaculation conducted during general anesthesia is well-described (Aitken-Palmer et al. 2012; Howard et al. 2006a). In brief, this involved use of a 2.6- to 4.0-cm-diameter rectal probe with three longitudinal electrodes and a 60-Hz sine-wave stimulator (P.T. Electronic) or similar equipment. Standardized sets of low voltage stimulations (2–5 V) over three series of 30 stimuli each delivered over 20 min to elicit an erection and ejaculation into a temperature-controlled (23 °C) container (Aitken-Palmer et al. 2012; Howard et al. 2006a). Expected seminal volume of ejaculate from an average sized male giant panda during the breeding season is approximately 1 ml to >3 ml. Ejaculate pH should be ≥7.5 from normal giant panda (Aitken-Palmer et al. 2012). Ejaculate pH <7.5 indicates urine contamination and should be discarded (Aitken Palmer 2010). Sperm quality within the ejaculate should be high during the breeding season with 70–90% motile sperm within a sample and 3.5–4.5 forward progressive motility (scale, 0–5; 5 = fast forward cellular trajectory) when viewed under phase-contrast microscopy (200–400×; Aitken-Palmer et al. 2012; Howard et al. 2006a). Sperm concentration calculated using a standard hemocytometer method should indicate highly concentrated sperm ($1–2 \times 10^9$ sperm/ejaculate).

Sperm morphology (200 sperm per sample) assessed using phase contrast microscopy (1000×) (Howard et al. 2006a) from normal giant panda ejaculates obtained during the breeding season should have predominately structurally normal spermatozoa, with most malformations occurring in the midpiece or flagellum (Aitken-Palmer et al. 2012). Flagellar deformities were documented to be most prevalent during the prebreeding and may give an indication of male readiness during the seasonal period (Aitken-Palmer et al. 2012). Advanced age has no impact the giant panda's ability to produce high quality sperm/ejaculate (Aitken-Palmer 2010), however age associated health abnormalities can. The giant panda is known to get osteoarthritis with advancing age, this can affect the male's ability to perform normal behaviors or even copulation. Additional testicular tumors (seminoma) have been reported in male giant panda with advancing age.

When electroejaculation was attempted during the non-breeding season (i.e., August in the N. Hemisphere) no seminal fluid or spermic ejaculate was produced and individuals tended not to develop an erection during the electroejaculation process. In contrast, all males produced spermic ejaculate during the early, peak, and late breeding seasons. The sperm concentration of ejaculates varied throughout the year and was consistent with increases in total testes volume and mean weekly androgen excretion (Aitken-Palmer et al. 2012).

Sperm cryopreservation, is a successful management tool for the giant panda to preserve male genetic material for future use. Giant panda sperm is cryotolerant when diluted in a TEST egg-yolk diluent containing 3.3% glycerol, cooled slowly to 5 °C over 4 h, loaded in 0.25 ml straws and cryopreserved at various rates (−35 °C/min; −100 °C/min; −40 °C/min, then −100 °C/min; Aitken-Palmer et al. 2007). Giant panda sperm had good post-thaw quality when frozen using a field-ready method where samples were placed directly into a dry shipper for 10 min (−35 °C/min) and then thawed in a waterbath at 50 °C for 10 s (Aitken-Palmer et al. 2007). Giant panda sperm cryopreserved using the pellet method and thawed at 37 °C had excellent post thaw motility and acrosomal integrity (Howard et al. 2006a).

8 Artificial Insemination

Maintaining a species of conservation concern *ex situ* requires a commitment to maximize the genetics of the population through a captive breeding program. Artificial insemination (AI) is an assisted reproductive technique that supplants coitus as the means of inducing semen to the female reproductive tract, and is useful in captive breeding programs when males are no longer capable of breeding (e.g. age, health or death), are not present at the same facility of the female, or the male and female are behaviorally incompatible for mating (Wildt 2003). Implementing AI requires basic knowledge about the reproductive biology of the species, most importantly the reproductive cycle and ovulatory patterns of the female, when and how to collect semen from the male, and the most efficacious place to deposit the semen. An early report (Liu et al. 1979) suggested the potential use of AI in breeding of giant pandas, however it wasn't until more knowledge of the species' biology became available that AI became a standard tool in the captive breeding program (Howard et al. 2006b).

The utility of artificial insemination, the most successful assisted reproductive technology used for the giant panda, has been debated in the literature with recently noted low rates of success identified at the China Conservation and Research Center for the Giant Panda (Li et al. 2017b). This low rate of success suggests that AI may not be as efficient as natural mating for achieving pregnancies, but other approaches to AI have met with better success and it is important to maintain and improve this tool to enhance reproduction when natural mating fails (Li et al. 2017b; see also Sect. 4). Broadly differing methods used for artificial insemination in giant pandas may also account for the varied success across facilities or the inability to diagnose early pregnancy may affect reported success rates. AI success can be greatly improved with the use of fresh ejaculate, standardization of the quality and quantity of sperm sample used, and consistency of AI technique performed (Howard et al. 2008). The single biggest challenge with determining giant panda AI success rates is the ability to determine early conception and pregnancy. Ultrasound has proven effective and is considered the gold standard to determine pregnancy but is challenged by delayed implantation and the fact that pregnancy can only be diagnosed in the 2 weeks prior to parturition (Cai et al. 2017; Kersey et al. 2016). As a result, ultrasound cannot be used to determine early pregnancy. New diagnostics using 13, 14, dihydro-15-keto-prostaglandin F2α (PGFM) concentrations have shown promise for early pregnancy detection (Roberts et al. 2018), but these diagnostics need further study. If successful, this method may be able to identify conception and used when monitoring for when early embryonic loss occurs (Sutherland-Smith et al. 2004). Female giant pandas, in which AI may be successful, but early embryonic loss occurs, should be further examined for infectious or environmental causes of pregnancy loss. Factors contributing to successful conception, early embryonic loss, and successful parturition need further study. Although the success rate of AI varies widely (18.5–100%) based on the means of ovulation detection, the quality of semen used, and strategic use as a follow up to natural breeding (Howard et al. 2006b; Huang et al. 2001; Li et al. 2017b), the approach has shown to be valuable for the giant panda captive population in complementing natural breeding to fulfill breeding recommendations (Li et al. 2017b).

The section above (see Sect. 7) describes the best practices for obtaining and handling of fresh semen and the techniques necessary to maintain the seminal characteristics through the cryopreservation and thawing procedures. It is evident that fresh semen is of better quality in all aspects (live, motile, progression and longevity) even when semen quality is well-maintained post-thaw from cryopreservation. The benefit of fresh semen is that it will have the best odds of fertilization; however, the drawbacks are that the male must be present to undergo the anesthetic collection procedure, and this can also delay the timing of AI. The benefits of frozen-thawed semen are that the male does not need to be present and the semen can be prepared for insemination while the female is anesthetically induced for the AI; the drawback is that quality and longevity of the semen is reduced, potentially reducing fertilization rate.

As mentioned above (see Sect. 6), urinary steroid hormone (estrogen and progestagen) assessments provide the greatest accuracy in pinpointing the time of ovulation (Czekala et al. 2003; Howard et al. 2006b, 2008; Huang et al. 2012a, b; Kersey et al. 2016). For hormonal measures, multiple urine samples must be collected daily to monitor ovarian follicle changes. Urinary estrogen values reflect physiology but can also be affected by urine concentration and collection artifacts. However, when estrogen values are paired with progestagen trends and behavioral indices, accuracy of ovulation detection is increased. Based on these factors, the greatest success of AI is achieved when the procedure is conducted within 40 h of the estrogen decline (Fig. 10.5; Howard et al. 2006b, 2008; Huang et al. 2012b; Kersey et al. 2016).

The AI procedure begins with anesthetizing the female (Howard et al. 2006b; Huang et al. 2012a, b; Kersey et al. 2016; Masui et al. 1989; Moore et al. 1984), and positioning her in dorsal recumbency with the hind quarters elevated (Huang et al. 2012b; Moore et al. 1984). Although there are reports where semen was deposited in the vagina and cervix (Knight et al. 1985; Masui et al. 1989; Moore et al. 1984), these structures pose significant physical and biological barriers for the sperm to traverse before entering the uterus. Bypassing the vagina and cervix via transcervical intrauterine insemination (IUI) increases the likelihood of the semen reaching the site of fertilization, requires less semen for deposition, and fewer procedures (Howard et al. 2008). For this procedure the cervical os may be approached via endoscope (Howard et al. 2008) or vaginal speculum (Howard et al. 2006b; Huang et al. 2012a; Masui et al. 1989), and a catheter advanced through the cervical canal into the uterine body. During peak estrus immediately following ovulation, the cervical os is easily catheterized, however if attempted outside the fertile window it is difficult to place a transcervical catheter for IUI. Ultrasound during the procedure is beneficial to confirm follicular stage and affirm the terminal placement of the catheter in the uterine body (Hildebrandt et al. 2003; Kersey et al. 2016). For insemination, 2–2.5 ml of semen is prepared in a syringe, attached to the insemination catheter, injected, and the length of the catheter flushed with media Maintaining the hindquarters elevated post insemination limits backflow of semen. Although a single, well-timed AI can be successful (Howard et al. 2008), more commonly multiple (2–5) AIs (Czekala et al. 2003; Kersey et al. 2016; Lindburg et al. 2001) are conducted as a theoretical practice to increase fertilization rate (Huang et al. 2012b; Li et al. 2017b).

9 Maternal Behavior and Cub Survivorship

Even though multiple studies have shed light on the physiological changes taking place during the reproductive cycle of giant pandas, many questions remain in regards to gestation and fetal development (Zhang et al. 2009). Cubbing primarily occurs from June to October with a gestation period of 85–185 days in length (Zhu et al. 2001). However, since females can enter embryonic diapause or delayed implantation and can also experience pseudopregnancy or false pregnancy, the precise timing of implantation and details of embryo development are difficult to pinpoint.

Neonates in the family *Ursidae* have the smallest litter weight relative to maternal size of all eutherian mammals, and among bears giant pandas are the most extreme in this regard (Ramsay and Dunbrack 1986), with cub weight at birth ranging from 90–130 g (Zhang et al. 2009). In regards to altriciality (i.e., relatively undeveloped at birth and requiring parental care to survive) giant pandas are also the extreme among bears (Spady et al. 2007). With ears and eyes closed and very sparse pelage, giant panda neonates need maternal support for their thermal, nutritional and excretory needs (Schaller et al. 1985). To meet these needs, and in contrast to bears of the genus *Ursus*, giant panda mothers demonstrate what can be described as an 'active' maternal care style. Responsive to the plaintive vocalizations cubs emit from birth (Baotic et al. 2014), giant panda mothers actively hold and manipulate their cubs, positioning them frequently for suckling, warmth and comfort (Snyder et al. 2003). While the denning period for giant panda mothers and cubs typically lasts about 3 months, an especially critical period of maternal care has been documented in studies of captive giant pandas, with the highest frequency maternal-attentive behaviors seen during the first 2–3 weeks post-partum (Martin-Wintle et al. in prep). A key aspect of maternal care in the first weeks of life is keeping the cub on the body and covered.

While maternal failure to rear cubs has been less problematic than mating failure, inadequate maternal care has been an occasional obstacle to successful reproduction (Swaisgood et al. 2006). Some females, particularly those with behavioral problems, fail to care for their cubs, but can be successfully trained to accept and rear their cubs using a combination of olfactory and acoustic cues for the cub to desensitize the female to cub stimuli, and provision of a surrogate cub (stuffed panda toy) to allow the mother to practice maternal care behavior (Zhang et al. 2000).

There are limited data describing maternal care behavior in wild giant pandas (Zhu et al. 2001), however studies of captive giant pandas have afforded a very detailed view of maternal care and cub behavior, the repertoire and variability of maternal behaviors, and factors associated with cub survival. Giant panda mothers will position their cubs to facilitate nursing or to ensure warmth and lick the anogenital region to stimulate excretion of feces and urines. Mothers will also lick other parts of the cub pelage to keep neonates clean. Giant panda cubs frequently vocalize when not nursing or sleeping, exhibiting a repertoire of acoustically and behaviorally distinct calls that include squawks, squalls, and croaks (Baotic et al. 2014; Stoeger et al. 2012). Giant panda mothers are attentive to cub vocalizations, with intense squawks and squalls engendering a more urgent response than 'croaks.' Thus, each call type conveys a different degree of plaintiveness that provides information regarding an individual's arousal state (Charlton et al. 2018), which is critical for eliciting the maternal care behavior to support cub survival (Stoeger et al. 2012).

While pandas give birth to twins approximately 50% of the time (average 1.45 cubs per year; Li et al. 2017b), most evidence suggests that they will only attempt to raise one cub (Schaller et al. 1985; Zhu et al. 2001). Indeed, the demanding and highly manipulative style of maternal care that giant pandas exhibit may limit the capacity of mothers to raise two cubs simultaneously from birth, though there are limited observations of giant panda mothers in the wild attempting to raise twins (R. R. Swaisgood, personal communication) and in 2016 one captive female successfully raised twins without human intervention (Tan 2016). Historically, when twins were born in the captive setting one neonate from the litter would be removed soon after birth, and attempts would be made to hand rear this cub exclusively in the nursery. These efforts were met with near complete failure (Edwards et al. 2006). However, since the late 1990s the technique of 'twin swapping' has been developed and refined resulting in a dramatic increase in the rate of survival for both cubs born in a twin litter. Twin swapping entails rotating each cub in a twin litter at regular intervals between maternal-rearing and nursery-rearing (Edwards et al. 2006). The timing of cub swapping varies by individual need but in the early weeks happens multiple times a day and is gradually increased as the cubs get older until they are both with their mother full time at approximately 4 months of age. This technique has been highly successful and has a profound impact on the growth of the captive panda population in recent decades.

The degree and intensity of plaintive cub vocalizations and maternal care is highest during the first 3 weeks post-partum, and drop off markedly after cubs are 14 weeks of age, or the end of the post-partum denning phase (Snyder et al. 2003). For wild giant panda cubs, physical growth and behavioral development during the denning phase are critical for ensuring that cubs can meet the demands of life outside the den during winter in the mountains of Sichuan (Schaller et al. 1985; Zhu et al. 2001). While cubs gain increasing independence and capacity once outside the den, they remain primarily dependent on mother's milk for nutrition through the first year of life (Snyder et al. 2003), slowly integrating bamboo into their diets from 6 months on. The best information available suggests that giant panda cubs are weaned and dissociate from their mothers in the wild approximately between 18 and 30 months of age (Lu et al. 1994; Schaller et al. 1985). However, traditional captive rearing practices have meant that cubs have been housed in peer-group nursery settings and removed from maternal care at 4–6 months of age which allows female giant pandas to be bred annually instead of every 2 years. It is unknown whether these practices compromise cub survival or if peer-reared cubs have lesser reproductive success as adults, but research has shown cubs that are left with their mothers for rearing longer (i.e., 12–18 months) demonstrate enhanced behavioral skill acquisition and more active behaviors, especially playfighting and other forms of play (Snyder et al. 2003) and social behavior.

10 Looking to the Future: Potential Impacts of Climate Change

Worldwide, climate change is driving both increasing variability and directional shifts in ecoregion specific temperature and weather patterns, including precipitation and snowmelt (Brown et al. 2016), constituting a major threat to global biodiversity (Pecl et al. 2017). These changes have been in turn associated with changes in the phenology and life history traits of an array of species, including distinct changes in the timing or success of reproductive events (Levy et al. 2016). Presumably, the influence of climate change on reproductive phenology will be proportionate to how responsive the species reproductive biology is to changes in environmental cues (e.g. photoperiod, rainfall, temperature; Lany et al. 2016). While species may benefit, at least in the short term, from plasticity, they may also suffer declines in individual fitness or population size if there is a loss of synchrony with environmental resources that are critical for maintaining body condition and energy levels to successfully reproduce or for offspring survival.

The influence of climate change on giant panda will be multifactorial, but early models show that the species success will be closely linked to availability to bamboo (Tuanmu et al. 2013). A primary concern related to changes in the distribution of bamboo is that reserve boundaries may no longer hold adequate resources to sustain growing panda populations, with models predicting declines of habitat ranging from 37 to 100% (Fan et al. 2014; Li et al. 2015; Shen et al. 2015). However, recent more biologically complex models suggest the loss of bamboo habitat will be much less than the initial projections (Wang et al. 2018). Because reserves offer greater protection from human activities (Wei et al. 2018b) and contain other habitat features that are critical for giant pandas (e.g., old-growth trees and tree cavities used for maternal dens; Swaisgood and Schulte 2010; Wei et al. 2019), climate change may indeed be a challenge for recovering giant panda populations if habitat management, policy and planning do not address this emerging threat.

Also linked to the availability of bamboo is the potential for a temporal mismatch between the seasonal peak in breeding and births and the peak of foraging conditions. Both breeding and raising offspring are metabolically demanding activities for an energy-limited species specializing on nutrient-poor bamboo (Schaller et al. 1985). Thus, both activities should take place at the time of optimal conditions and should be closely linked to availability of food supplies. Increased global temperature will influence timing of food availability, particularly growth of bamboo leaves, culms, and shoots. This, in turn, will limit the ability of food resources available during the breeding season, embryo implantation and maternal care limiting a female's ability to bear and raise offspring. Decoupling of reproductive windows and primary food sources has recently been shown to be problematic for offspring production and development in other species (Visser and Both 2005). These changes may be brought about not only by mean temperature changes but also temperature patterns across the season (Peñuelas and Filella 2001). While the estrous cycles of female giant pandas seem to be regulated by photoperiod (Kersey et al. 2010b;

Snyder et al. 2004; Steinman et al. 2006) these studies have never decoupled photoperiod from concurrent increases in temperature. Thus, it is unknown how other factors that have the potential to regulate seasonality (e.g. temperature and rainfall) may affect the timing of estrus and cubbing in the giant panda. Recent research suggests that Andean bears have slight plasticity in seasonal reproduction implying that they have some capacity to adjust to climate-related shifts in resources (Appleton et al. 2018) making it probable that giant pandas may also possess this ability. Defining the environmental cues that influence the breeding and cubbing seasons will provide information on the impact climate change will have on these two key reproductive events.

Beyond habitat change and access to necessary nutrition, infectious diseases may comprise a significant threat associated with global climate change. Climate change can exacerbate disease outbreaks by increasing the range in which pathogens are viable or altering disease-carrying human and domestic animal distributions on the landscape. The giant panda is extremely susceptible to infectious diseases carried by human populations (Li et al. 2014) and domestic animals (Zhang et al. 2017c; Zhao et al. 2017). Canine distemper (CDV) is a significant threat to wild giant panda populations as areas bordering giant panda habitat change via increased human impact. Exposure to CDV results in death for a giant panda (Feng et al. 2016). CDV outbreak during the mating season, a period when giant pandas are less solitary and have more social interaction, could result in catastrophic loss of fragmented wild populations. Giant panda habitats in Shaanxi, Sichuan, and Gansu provinces are adjacent to three regions with high CDV prevalence in canines (Jin et al. 2017) and a CDV outbreak within these regions would threaten giant panda populations. It is important to note, CDV is preventable. Vaccination strategies of domestic dogs within at-risk areas are probably needed to protect wild giant panda populations similar to canine vaccination buffer zones in wild parts of Africa. These buffer zones have been used to combat infection for regional at-risk wild species such as rabies in Ethiopian wolves and CDV in lions and African wild dogs (Prager 2011; Sterner and Smith 2006; Woodroffe et al. 2012). The global threat posed by CDV to a number of endangered species has prompted mandatory vaccination of dogs in Europe (Di et al. 2015). Another prevention strategy is to vaccinate other known CDV reservoir species with recombinant vectored vaccines developed for those wild species. Regardless of the strategies employed, there is a need for long-term monitoring and further studies of how infectious disease will impact wild giant panda populations.

11 Conclusion and Future Directions

The reproductive biology of the giant panda is without a doubt the most fascinating aspect of this charismatic species. Although at first blush, the reproductive strategy is bizarre, with a female that has only one estrus per breeding season of approximately 1–2 days of fertility, a conceptus that does not grow for most of gestation, and a cub that is 700–1000 times smaller than the dam; yet the giant panda is reproductively successful. Being the oldest extant bear species, the giant panda reproductive strategy, at closer look, is well crafted and has afforded the species to persist for millions of years. Indeed, once the initial hurdles surrounding the giant panda's unique biology and behavior had been overcome, the panda has proven a successful breeder even in captivity where it was once thought to be notoriously difficult to breed. Today 548 pandas reside in 93 zoological institutions worldwide with a reported pregnancy rate of 75% and cub survival rate of 97% at the CCRCGP (Xie 2018). This chapter provides the most comprehensive description of giant panda reproductive knowledge to date and provides a road map on how to successfully manage captive breeding for the species. That being said, there are significant portions of giant panda reproductive biology and behavior that are still unknown and deserve further investigation.

As mentioned above, one of the last unknown biological time periods of giant panda reproduction is fertilization of the oocyte and delayed implantation of the fetus. To this end, there is a great need for development of early, non-invasive pregnancy detection diagnostics preferably using either urine or feces. The ability to distinguish pregnancy from pseudopregnancy and to predict the timing of parturition will improve caretaker management and help them prepare for the arrival of cubs potentially decreasing maternal stress due to early denning of pregnant females or unnecessary denning of non-pregnant females, improve maternal and fetal care during gestation, as well as increasing neonatal survival. Accurate prediction of parturition is critical because it facilitates staffing of animal care support which is key to ensuring cub survival as the highly altricial giant panda infant has a high probability of mortality within hours of birth due to rejection, neglect, or insufficient lactation. Furthermore, accurately detecting and visualizing pregnancy will enable managers and scientists to research delayed implantation and the determinates of implantation in the giant panda giving further insights into predictors of pregnancy success and loss.

Another area of captive management that needs further investigation is the reproductive ramifications of the social housing environment. As giant pandas are housed in artificially high population densities in captivity, more research should be conducted on inhibition of reproductive physiology or behavior in response to social housing conditions. Conspecifics can influence reproductive suppression through mechanisms where individuals inhibit their own reproduction and induce reproductive suppression in others. It is unknown how housing environments during development, breeding, and cubbing influence eventual reproductive success, but these avenues should be investigated more thoroughly. For example, it is common practice for cubs to be peer-raised in communal nurseries from 6-months to 2-years of age at which point they are raised in pairs until adulthood. Whether this cohabitation negatively impacts future reproduction between cohoused individuals is unknown. Very little research has been conducted on the effects of kin-selection mechanisms or inbreeding avoidance in the captive population. Furthermore, when conducting this work there is a dire need for accurate, genetically-based paternity assignments. It is also unknown whether housing gestational females in close proximity affects cubbing success probabilities and whether housing mothers next to each other affects maternal care. Lastly, it is possible that social suppression occurs during breeding if males or females are housed next to a dominant conspecific. All of these factors have the potential to impact not only reproductive success in the captive population but also normal behavioral development of developing cubs.

There is still much to be learned from wild giant panda populations so that natural breeding systems can be applied to managed breeding programs. For example, very little is known about how male-male dominance is achieved or how it influences reproductive success or how weaning occurs in nature. Applying the technology of sperm collection to wild populations to establish a frozen genome bank of giant panda spermatozoa from wild males would give the managed breeding program more opportunities to both incorporate the genetics into the current breeding practices and also serve as a genetic reservoir. With emerging disease threats to the population, researchers would also benefit from a better understanding of disease ecology which would provide insight to potential disease transfer. It would be helpful to know what species giant pandas encounter in the wild and/or share food sources with and with what type of frequency. Concurrently and with every wild and captive death it would be beneficial to develop and broadly implement gross and histological necropsy practices so that we can cultivate a better understanding of disease susceptibility within the species.

The lack of successful reproduction during the early years of the giant panda conservation breeding program, coupled with the urgent need to establish a managed insurance population, spurred the large scale and concerted efforts to thoroughly understand the reproductive biology of giant pandas. Once frequently described as enigmatic and poorly understood, the giant panda is now considered something of a model species, having been the focus of targeted research efforts designed to bolster the *ex situ* population. As the giant panda program transitions to a maintenance phase for the captive population and focuses on producing individuals that will contribute to reintroduction efforts, it will be important to increase the proportion of animals that express natural behaviors, adaptability, and high cognitive abilities. Key to this effort will be increasing the proportion of individuals that naturally mate as well as maternally rear cubs and being able to reliably transfer these traits to potential reintroduction candidates.

Acknowledgements The success of the giant panda managed breeding program has been achieved through the dedication and support of numerous scientific collaborators, undergraduate and post-graduate students and zoological institutions. We are extremely grateful for the contribution and support of all the staff at the Chinese Conservation and Research Center for the Giant Panda, the Chengdu Giant Panda Base, San Diego Zoo, and the Smithsonian National Zoo.

References

Aitken-Palmer C. Assessment of male giant panda seasonal reproduction, sexual maturity and comparative sperm cryotolerance. College Park: University of Maryland; 2010.

Aitken-Palmer C, Hou R, Wildt D, Ottinger MA, Spindler R, Howard J. Giant panda sperm tolerate cryopreservation at rapid freezing and thawing rates. Biol Reprod. 2007;77(Suppl. 1):107–8.

Aitken-Palmer C, Hou R, Burrell C, Zhang Z, Wang C, Spindler R, Wildt DE, Ottinger MA, Howard J. Protracted reproductive seasonality in the male giant panda (*Ailuropoda melanoleuca*) reflected by patterns in androgen profiles, ejaculate characteristics, and selected behaviors. Biol Reprod. 2012;86:195.

Appleton R, Van Horn R, Noyce K, Spady T, Swaisgood R, Arcese P. Phenotypic plasticity in the timing of reproduction in Andean bears. J Zool. 2018;305:196–202.

Baotic A, Stoeger AS, Li D, Tang C, Charlton BD. The vocal repertoire of infant giant pandas (*Ailuropoda melanoleuca*). Bioacoustics. 2014;23:15–28.

Bonney RC, Wood DJ, Kleiman DG. Endocrine correlates of behavioural oestrus in the female giant panda (*Ailuropoda melanoleuca*) and associated hormonal changes in the male. J Reprod Fertil. 1982;64:209–15.

Brown CJ, O'connor MI, Poloczanska ES, Schoeman DS, Buckley LB, Burrows MT, Duarte CM, Halpern BS, Pandolfi JM, Parmesan C. Ecological and methodological drivers of species' distribution and phenology responses to climate change. Glob Chang Biol. 2016;22:1548–60.

Cai K, Yie S, Zhang Z, Wang J, Cai Z, Luo L, Liu Y, Wang H, Huang H, Wang C. Urinary profiles of luteinizing hormone, estrogen and progestagen during the estrous and gestational periods in giant pandas (*Ailuropda melanoleuca*). Sci Rep. 2017;7:40749.

Charlton B, Huang Y, Swaisgood RR. Vocal discrimination of potential mates by female giant pandas (*Ailuropoda melanoleuca*). Biol Lett. 2009a;5:597–9.

Charlton BD, Zhihe Z, Snyder RJ. The information content of giant panda, *Ailuropoda melanoleuca*, bleats: acoustic cues to sex, age and size. Anim Behav. 2009b;78:893–8.

Charlton BD, Keating JL, Rengui L, Huang Y, Swaisgood RR. Female giant panda (*Ailuropoda melanoleuca*) chirps advertise the caller's fertile phase. Proc R Soc B Biol Sci. 2010;277:1101–6.

Charlton BD, Keating JL, Kersey D, Rengui L, Huang Y, Swaisgood RR. Vocal cues to male androgen levels in giant pandas. Biol Lett. 2011;7:71–4.

Charlton BD, Martin-Wintle MS, Owen MA, Zhang H, Swaisgood RR. Vocal behaviour predicts mating success in giant pandas. R Soc Open Sci. 2018;5:181323.

Czekala N, Lindburg D, Durrant B, Swaisgood R, He T, Tang C. The estrogen profile, vaginal cytology, and behavior of a giant panda female during estrus. In: Zhang A, He G, editors. Proceedings of the international symposium on the protection of the giant panda (*Ailuropoda melanoleuca*). Chengdu: Sichuan Publishing House of Science and Technology; 1998. p. 111–3.

Czekala N, McGeehan L, Steinman K, Xuebing L, Gual-Sil F. Endocrine monitoring and its application to the management of the giant panda. Zoo Biol. 2003;22:389–400.

Davis DD. The giant panda: a morphological study of evolutionary mechanisms. Fieldiana Zoology Memoirs. 1964;3:1.

Dehnhard M, Hildebrandt T, Knauf T, Jewgenow K, Kolter L, Göritz F. Comparative endocrine investigations in three bear species based on urinary steroid metabolites and volatiles. Theriogenology. 2006;66:1755–61.

Di DS, Savini G, Lorusso A. Canine distemper and endangered wildlife: is it time for mandatory vaccination of dogs? Vaccine. 2015;33:6519.

Durrant B, Czekala N, Olson M, Anderson A, Amodeo D, Campos-Morales R, Gual-Sill F, Ramos-Garza J. Papanicolaou staining of exfoliated vaginal epithelial cells facilitates the prediction of ovulation in the giant panda. Theriogenology. 2002;57:1855–64.

Durrant BS, Olson MA, Amodeo D, Anderson A, Russ KD, Campos-Morales R, Gual-Sil F, Garza JR. Vaginal cytology and vulvar swelling as indicators of impending estrus and ovulation in the giant panda (*Ailuropoda melanoleuca*). Zoo Biol. 2003;22:313–21.

Durrant BS, Olson MA, Anderson A, Gual-Sil F, Li D, Huang Y. The value and significance of vaginal cytology. In: Wildt DE, Zhang A, Zhang Z, Janssen DL, Ellis S, editors. Giant pandas: biology, veterinary medicine and management. Cambridge: Cambridge University Press; 2006a. p. 231–44.

Durrant BS, Ravida N, Spady T, Cheng A. New technologies for the study of carnivore reproduction. Theriogenology. 2006b;66:1729–36.

Edwards MS, Wei R, Hawes J, Sutherland-Smith M, Tang C, Li D, Hu D, Zhang G. The neonatal giant panda: hand-rearing and medical management. In: Wildt DE, Zhang A, Zhang H, Jassen DL, Ellis S, editors. Giant pandas biology, veterinary medicine and management. Cambridge: Cambridge University Press; 2006. p. 315.

Fan J, Li J, Xia R, Hu L, Wu X, Li G. Assessing the impact of climate change on the habitat distribution of the giant panda in the Qinling Mountains of China. Ecol Model. 2014;274:12–20.

Feng N, Yu Y, Wang T, Wilker P, Wang J, Li Y, Sun Z, Gao Y, Xia X. Fatal canine distemper virus infection of giant pandas in China. Sci Rep. 2016;6:27518.

Foster P. China poised to sue japan for compensation over dead panda. The Telegraph; 2010. https://www.telegraph.co.uk/news/worldnews/asia/china/7999715/China-poised-to-sue-Japan-for-compensation-over-dead-panda.html. Accessed 17 Nov 2018.

Hildebrandt TB, Brown JL, Hermes R, Goritz F. Ultrasound for analysis of reproductive function in wildlife species. In: Holt WV, Pickard AR, Rodger JC, Wildt DE, editors. Reproductive science and integrated conservation. Cambridge: Cambridge University Press; 2003. p. 166–82.

Hildebrandt TB, Brown JL, Goritz F, Ochs A, Morris P, Sutherland-Smith M. Ultrasonography to assess and enhance health and reproduction in the giant panda. In: Wildt DE, Zhang A, Zhang H, Janssen DL, Ellis S, editors. Giant pandas: biology, veterinary medicine and management. Cambridge: Cambridge University Press; 2006. p. 410–39.

Hoffmann M, Hilton-Taylor C, Angulo A, Böhm M, Brooks TM, Butchart SHM, Carpenter KE, Chanson J, Collen B, Cox NA, Darwall WRT, Dulvy NK, Harrison LR, Katariya V, Pollock CM, Quader S, Richman NI, Rodrigues ASL, Tognelli MF, Vié J-C, Aguiar JM, Allen DJ, Allen GR, Amori G, Ananjeva NB, Andreone F, Andrew P, Ortiz ALA, Baillie JEM, Baldi R, Bell BD, Biju SD, Bird JP, Black-Decima P, Blanc JJ, Bolaños F, Bolivar GW, Burfield IJ, Burton JA, Capper DR, Castro F, Catullo G, Cavanagh RD, Channing A, Chao NL, Chenery AM, Chiozza F, Clausnitzer V, Collar NJ, Collett LC, Collette BB, Fernandez CFC, Craig MT, Crosby MJ, Cumberlidge N, Cuttelod A, Derocher AE, Diesmos AC, Donaldson JS, Duckworth JW, Dutson G, Dutta SK, Emslie RH, Farjon A, Fowler S, Freyhof JR, Garshelis DL, Gerlach J, Gower DJ, Grant TD, Hammerson GA, Harris RB, Heaney LR, Hedges SB, Hero J-M, Hughes B, Hussain SA, Icochea MJ, Inger RF, Ishii N, Iskandar DT, Jenkins RKB, Kaneko Y, Kottelat M, Kovacs KM, Kuzmin SL, La Marca E, Lamoreux JF, Lau MWN, Lavilla EO, Leus K, Lewison RL, Lichtenstein G, Livingstone SR, Lukoschek V, Mallon DP, McGowan PJK, McIvor A, Moehlman PD, Molur S, Alonso AM, Musick JA, Nowell K, Nussbaum RA, Olech W, Orlov NL, Papenfuss TJ, Parra-Olea G, Perrin WF, Polidoro BA, Pourkazemi M, Racey PA, Ragle JS, Ram M, Rathbun G, Reynolds RP, Rhodin AGJ, Richards SJ, Rodríguez LO, Ron SR, Rondinini C, Rylands AB, Mitcheson YSd, Sanciangco JC, Sanders KL, Santos-Barrera G, Schipper J, Self-Sullivan C, Shi Y, Shoemaker A, Short FT, Sillero-Zubiri C, Silvano DL, Smith KG, Smith AT, Snoeks J, Stattersfield AJ, Symes AJ, Taber AB, Talukdar BK, Temple HJ, Timmins R, Tobias JA, Tsytsulina K, Tweddle D, Ubeda C, Valenti SV, van Dijk PP, Veiga LM, Veloso A, Wege DC, Wilkinson M, Williamson EA, Xie F, Young BE, Akçakaya HR, Bennun L, Blackburn TM, Boitani L, Dublin HT, Fonseca GABd, Gascon C, Lacher Jr TE, Mace GM, Mainka SA, McNeely JA, Mittermeier RA, Reid GM, Rodriguez JP, Rosenberg AA, Samways MJ, Smart J, Stein BA, Stuart SN. The impact of conservation on the status of the world's vertebrates. Science. 2010;330:1503–1509.

Howard J, Zhang Z, Li D, Huang Y, Hou R, Li G, Zhang M, Ye Z, Zhang J, Huang S. Male reproductive biology in giant pandas in breeding programmes in China. In: Giant pandas: biology, veterinary medicine and management. Cambridge: Cambridge University Press; 2006a. p. 159.

Howard JG, Huang Y, Wang P, Li D, Zhang G, Hou R, Zhang Z, Durrant BS, Spindler RE, Zhang H, Zhang A, Wildt DE. Role and efficiency of artificial insemination and genome resource banking. In: Wildt DE, Zhang A, Zhang H, Janssen DL, Ellis S, editors. Giant pandas: biology, veterinary medicine and management. Cambridge: Cambridge University Press; 2006b. p. 469–94.

Howard JG, Kersey DC, Aitken-Palmer C, Monfort SL, Wildt DE. Capacity of the giant panda to give birth after a single intrauterine insemination using precise ovulation detection. Biol Reprod. 2008;78:203.

Hu Y, Nie Y, Wei W, Ma T, Horn RV, Zheng X, Swaisgood RR, Zhou Z, Zhou W, Yan L. Inbreeding and inbreeding avoidance in wild giant pandas. Mol Ecol. 2017;26:5793–806.

Huang Y, Wang PY, Zhang HM, Zhang GQ, Li DS, Tang CX, Wei RP, Hu DM, Howard JG. Efficiency of artificial insemination in giant pandas at the Wolong Breeding Center. In: Proceedings of the American Society of Andrology. Seattle: American Society of Andrology; 2001. p. 118.

Huang Y, Li D, Zhou Y, Zhou Q, Li R, Wang C, Huang Z, Hull V, Zhang H. Factors affecting the outcome of artificial insemination using cryopreserved spermatozoa in the giant panda (*Ailuropoda melanoleuca*). Zoo Biol. 2012a;31:561–73.

Huang Y, Zhang H, Li D, Zhang G, Wei R, Huang Z, Zhou Y, Zhou Q, Liu Y, Wildt DE. Relationship of the estrogen surge and multiple mates to cub paternity in the giant panda (*Ailuropoda melanoleuca*): implications for optimal timing of copulation or artificial insemination. Biol Reprod. 2012b;87:112.

Hull V, Zhang J, Zhou S, Huang J, Li R, Liu D, Xu W, Huang Y, Ouyang Z, Zhang H. Space use by endangered giant pandas. J Mammal. 2015;96:230–6.

Jin K, Xue C, Wu X, Qian J, Zhu Y, Yang Z, Yonezawa T, Crabbe MJC, Cao Y, Hasegawa M. Why does the giant panda eat bamboo? A comparative analysis of appetite-reward-related genes among mammals. PLoS One. 2011;6:e22602.

Jin Y, Zhang X, Ma Y, Qiao Y, Liu X, Zhao K, Zhang C, Lin D, Fu X, Xu X. Canine distemper viral infection threatens the giant panda population in China. Oncotarget. 2017;8:113910.

Kang D, Li J. Premature downgrade of panda's status. Science. 2016;354:295.

Kersey DC, Wildt DE, Brown JL, Huang Y, Snyder RJ, Monfort SL. Parallel and seasonal changes in gonadal and adrenal hormones in male giant pandas (*Ailuropoda melanoleuca*). J Mammal. 2010a;91:1496–507.

Kersey DC, Wildt DE, Brown JL, Snyder RJ, Huang Y, Monfort SL. Endocrine milieu of periestrus in the giant panda (*Ailuropoda melanoleuca*) as determined by noninvasive hormone measures. Reprod Fertil Dev. 2010b;22:901.

Kersey DC, Aitken-Palmer C, Rivera S, Willis EL, Liang LY, Snyder RJ. The birth of a giant panda: tracking the biological factors that successfully contribute to conception through to postnatal development. Theriogenology. 2016;85:671–7.

Kinoshita K, Morita H, Miyazaki M, Hama N, Kanemitsu H, Kawakami H, Wang P, Ishikawa O, Kusunoki H, Tsenkova R. Near infrared spectroscopy of urine proves useful for estimating ovulation in giant panda (*Ailuropoda melanoleuca*). Anal Methods. 2010;2:1671.

Kinoshita K, Miyazaki M, Morita H, Vassileva M, Tang C, Li D, Ishikawa O, Kusunoki H, Tsenkova R. Spectral pattern of urinary water as a biomarker of estrus in the giant panda. Sci Rep. 2012;2:856.

Kleiman DG. Ethology and reproduction of captive giant pandas (*Ailuropoda melanoleuca*). Z Tierpsychol. 1983;62:1.

Kleiman DG, Karesh WB, Chu PR. Behavioural changes associated with oestrus in the giant panda *Ailuropoda melanoleuca* with comments on female proceptive behaviour. Int Zoo Yearb. 1979;19:217–24.

Knight JA, Bush M, Celma M, Garcia del Campo AL, Goltenboth R, Hearn JP, Hodges J, Jones DM, Klos HG, Monsalve L, Montali R, Moore HDM. Veterinary aspects of reproduction in the giant panda (*Ailuropoda melanoleuca*). Bongo. 1985;10:93.

Lany NK, Ayres MP, Stange EE, Sillett TS, Rodenhouse NL, Holmes RT. Breeding timed to maximize reproductive success for a migratory songbird: the importance of phenological asynchrony. Oikos. 2016;125:656–66.

Levy O, Buckley LB, Keitt TH, Angilletta MJ Jr. Ontogeny constrains phenology: opportunities for activity and reproduction interact to dictate potential phenologies in a changing climate. Ecol Lett. 2016;19:620–8.

Li Y, Viña A, Yang W, Chen X, Zhang J, Ouyang Z, Liang Z, Liu J. Effects of conservation policies on forest cover change in giant panda habitat regions, China. Land Use Policy. 2013;33:42–53.

Li D, Zhu L, Cui H, Ling S, Fan S, Yu Z, Zhou Y, Wang T, Qian J, Xia X. Influenza a (h1n1) pdm09 virus infection in giant pandas, China. Emerg Infect Dis. 2014;20:480–3.

Li R, Xu M, Wong MHG, Qiu S, Sheng Q, Li X, Song Z. Climate change-induced decline in bamboo habitats and species diversity: implications for giant panda conservation. Divers Distrib. 2015;21:379–91.

Li BV, Pimm SL, Li S, Zhao L, Luo C. Free-ranging livestock threaten the long-term survival of giant pandas. Biol Conserv. 2017a;216:18–25.

Li D, Wintle NJ, Zhang G, Wang C, Luo B, Martin-Wintle MS, Owen MA, Swaisgood RR. Analyzing the past to understand the future: natural mating yields better reproductive rates than artificial insemination in the giant panda. Biol Conserv. 2017b;216:10–7.

Lindburg DG, Czekala NM, Swaisgood RR. Hormonal and behavioral relationships during estrus in the giant panda. Zoo Biol. 2001;20:537–43.

Liu W, Ye J, Li C, Liao G. Artificial insemination experiment on the giant panda. Chin Zoo Annu. 1979;2:20.

Loeffler IK, Montali RJ, Rideout BA. Diseases and pathology of giant pandas. In: Wildt DE, Zhang A, Zhang H, Janssen DL, Ellis S, editors. Giant pandas: biology, veterinary medicine and management. Cambridge: Cambridge University Press; 2006.

Loeffler IK, Howard J, Montali RJ, Hayek L-A, Dubovi E, Zhang Z, Yan Q, Guo W, Wildt DE. Serosurvey of ex situ giant pandas (*Ailuropoda melanoleuca*) and red pandas (Ailurus fulgens) in China with implications for species conservation. J Zoo Wildl Med. 2007;38:559–66.

Lu Z, Pan W, Harkness J. Mother-cub relationships in giant pandas in the Qinling Mountains, China, with comment on rescuing abandoned cubs. Zoo Biol. 1994;13:567–8.

Lü Z, Pan W, Zhu X, Wang D, Wang H. What has the panda taught us? In: Entwistle A, Dunstone N, editors. Priorities for the conservation of mammalian diversity: has the panda had its day? Cambridge: Cambridge University Press; 2000. p. 325–34.

Martin-Wintle MS, Shepherdson D, Zhang G, Zhang H, Li D, Zhou X, Li R, Swaisgood RR. Free mate choice enhances conservation breeding in the endangered giant panda. Nat Commun. 2015;6:10125.

Martin-Wintle MS, Shepherdson D, Zhang G, Huang Y, Luo B, Swaisgood RR. Do opposites attract? Effects of personality matching in breeding pairs of captive giant pandas on reproductive success. Biol Conserv. 2017;207:27–37.

Martin-Wintle MS, Owen MA, Zhou X, Zhang G, Wintle NJP, Hall SH, Swaisgood RR. Working smarter not harder: Experienced mothers strategically allocate maternal care effort. in prep.

Masui M, Hiramatsu H, Nose N, Nakazato R, Sagawa Y, Tajima H, Saito K. Successful artificial insemination in the giant panda (*Ailuropoda melanoleuca*) at ueno zoo. Zoo Biol. 1989;8:17–26.

McGeehan L, Li X, Jackintell L, Huang S, Wang A, Czekala NM. Hormonal and behavioral correlates of estrus in captive giant pandas. Zoo Biol. 2002;21:449–66.

Miller RE, Fowler ME. Fowler's zoo and wild animal medicine, vol. 8. St. Louis: Elsevier Health Sciences; 2014.. e-book.

Monfort SL, Dahl KD, Czekala NM, Stevens L, Bush M, Wildt DE. Monitoring ovarian function and pregnancy in the giant panda (*Ailuropoda melanoleuca*) by evaluating urinary bioactive FSH and steroid metabolites. J Reprod Fertil. 1989;85:203–12.

Moore HD, Bush M, Celma M, Garcia A-L, Hartman TD, Hearn JP, Hodges JK, Jones DM, Knight JA, Monsalve L, Wildt DE. Artificial insemination in the giant panda (*Ailuropoda melanoleuca*). J Zool. 1984;203:269.

Nie Y, Swaisgood RR, Zhang Z, Hu Y, Ma Y, Wei F. Giant panda scent-marking strategies in the wild: role of season, sex and marking surface. Anim Behav. 2012a;84:39–44.

Nie Y, Swaisgood RR, Zhang Z, Liu X, Wei F. Reproductive competition and fecal testosterone in wild male giant pandas (*Ailuropoda melanoleuca*). Behav Ecol Sociobiol. 2012b;66:721–30.

Nie Y-G, Zhang Z-J, Swaisgood RR, Wei F-W. Effects of season and social interaction on fecal testosterone metabolites in wild male giant pandas: implications for energetics and mating strategies. Eur J Wildl Res. 2012c;58:235–41.

Owen MA, Swaisgood RR, McGeehan L, Zhou X, Lindburg DG. Dynamics of male–female multimodal signaling behavior across the estrous cycle in giant pandas (*Ailuropoda melanoleuca*). Ethology. 2013;119:869–80.

Owen MA, Swaisgood RR, Zhou X, Blumstein DT. Signalling behaviour is influenced by transient social context in a spontaneously ovulating mammal. Anim Behav. 2016;111:157–65.

Pecl GT, Araújo MB, Bell JD, Blanchard J, Bonebrake TC, Chen I-C, Clark TD, Colwell RK, Danielsen F, Evengård B. Biodiversity redistribution under climate change: impacts on ecosystems and human well-being. Science. 2017;355:eaai9214.

Peñuelas J, Filella I. Responses to a warming world. Science. 2001;294:793–5.

Powell D, Hong L, Carlstead K, Kleiman D, Zhang H, Zhang G, Zhang Z, Yu J, Zhang J, Lu Y. Relationships between temperament, husbandry, management, and socio-sexual behavior in captive male and female giant pandas (*Ailuropoda melanoleuca*). Acta Zool Sin. 2008;54:169.

Prager KC. An investigation of infectious disease dynamics in African carnivores: identifying reservoirs and risk factors and investigating control strategies. Davis: University of California; 2011.

Ramsay MA, Dunbrack RL. Physiological constraints on life history phenomena: the example of small bear cubs at birth. Am Nat. 1986;127:735–43.

Raven HC, Carter TD, Sage D. Notes on the anatomy of the viscera of the giant panda (Ailuropoda melanoleuca). New York: American Museum of Natural History; 1936.

Roberts BM, Brown JL, Kersey DC, Snyder RJ, Durrant BS, Kouba AJ. Use of urinary 13, 14, dihydro-15-keto-prostaglandin f2α (PGFM) concentrations to diagnose pregnancy and predict parturition in the giant panda (*Ailuropoda melanolecua*). PLoS One. 2018;13:e0195599.

Rong H, Zhang Z, Liu Y, Wang C, Lan J, Huang X, Yang Z, Li M, Aitken-Palmer C. Combined electroejaculation with transrectal massage to facilitate semen collection in giant pandas. J Guangxi Agric Biol Sci. 2006;25:78.

Schaller GB, Jinchu H, Wenshi P, Jing Z. The giant pandas of Wolong. Chicago: University Of Chicago Press; 1985.

Shen G, Pimm SL, Feng C, Ren G, Liu Y, Xu W, Li J, Si X, Xie Z. Climate change challenges the current conservation strategy for the giant panda. Biol Conserv. 2015;190:43–50.

Snyder RJ, Zhang AJ, Zhang ZH, Li GH, Tian YZ, Huang XM, Luo L, Bloomsmith MA, Forthman DL, Maple TL. Behavioral and developmental consequences of early rearing experience for captive giant pandas (*Ailuropoda melanoleuca*). J Comp Psychol. 2003;117:235–45.

Snyder RJ, Lawson DP, Zhang A, Zhang Z, Luo L, Zhong W, Huang X, Czekala NM, Bloomsmith MA, Forthman DL, Maple TL. Reproduction in giant pandas: hormone and behavior. In: Lindburg DG, Baragona K, editors. Giant pandas: biology and conservation. Berkeley: University of California Press; 2004. p. 125–32.

Spady TJ, Lindburg DG, Durrant BS. Evolution of reproductive seasonality in bears. Mammal Rev. 2007;37:21–53.

Steinman K, Monfort SL, McGeehan L, Kersey D, Gual-Sil F, Snyder R, Wang P, Nakao T, Czekala N. Endocrinology of the giant panda and application of hormone technology to species management. In: Wildt DE, Zhang A, Zhang H, Janssen DL, Ellis S, editors. Giant pandas: biology, veterinary medicine and management. Cambridge: Cambridge University Press; 2006. p. 198–230.

Sterner RT, Smith GC. Modelling wildlife rabies: transmission, economics, and conservation. Biol Conserv. 2006;131:163–79.

Stoeger AS, Baotic A, Li D, Charlton BD. Acoustic features indicate arousal in infant giant panda vocalisations. Ethology. 2012;118:896–905.

Sutherland-Smith M, Morris PJ, Silverman S. Pregnancy detection and fetal monitoring via ultrasound in a giant panda (*Ailuropoda melanoleuca*). Zoo Biol. 2004;23:449–61.. Published in affiliation with the American Zoo and Aquarium Association.

Swaisgood RR, Schulte BA. Applying knowledge of mammalian social organization, mating systems, and communication to management. In: Kleiman DG, Thompson KV, Baer CK, editors. Wild mammals in captivity: principles and techniques for zoo management. Chicago: The University of Chicago Press; 2010.
Swaisgood RR, Lindburg DG, Zhou X. Giant pandas discriminate individual differences in conspecific scent. Anim Behav. 1999;57:1045–53.
Swaisgood RR, Lindburg DG, Zhou X, Owen MA. The effects of sex, reproductive condition and context on discrimination of conspecific odours by giant pandas. Anim Behav. 2000;60:227–37.
Swaisgood RR, White AM, Zhou X, Zhang H, Zhang G, Wei R, Hare VJ, Tepper EM, Lindburg DG. A quantitative assessment of the efficacy of an environmental enrichment programme for giant pandas. Anim Behav. 2001;61:447–57.
Swaisgood RR, Lindburg DG, Zhang H. Discrimination of oestrous status in giant pandas (*Ailuropoda melanoleuca*) via chemical cues in urine. J Zool. 2002;257:381–6.
Swaisgood RR, Zhou X, Zhang G, Lindburg DG, Zhang H. Application of behavioral knowledge to conservation in the giant panda. Int J Comp Psychol. 2003;16:65–84.
Swaisgood RR, Lindburg D, White AM, Zhang H, Zhou X. Chemical communication in giant pandas: experimentation and application. In: Lindburg D, Baragona K, editors. Giant pandas: biology and conservation. Berkeley: University of California Press; 2004. p. 106–20.
Swaisgood RR, Zhang G, Zhou X, Zhang H. The science of behavioural management: creating biologically relevant living environments in captivity. In: Wildt DE, Zhang A, Zhang H, Janssen DL, Ellis S, editors. Giant pandas: biology, veterinary medicine and management. Cambridge: Cambridge University Press; 2006. p. 274–98.
Swaisgood RR, Wang D, Wei F. *Ailuropoda melanoleuca*. In IUCN red list of threatened species; 2016. http://www.Iucnredlist.Org/details/712/0.
Swaisgood RR, Wang D, Wei F. Panda downlisted but not out of the woods. Conserv Lett. 2018;11:e12355.
Tan A. Giant panda surprises vienna zoo with twins after keepers find extra cub. ABC News: ABC News Network; 2016. abcnews.go.com/International/giant-panda-surprises-vienna-zoo-twins-keepers-find/story?id=41424190. Accessed 17 Nov 2018.
Traylor-Holzer K, Ballou J. 2017 breeding and management recommendations and summary of the status of the giant panda ex situ population: Conservation Specialist Planning Group; 2017.
Tuanmu M-N, Viña A, Winkler JA, Li Y, Xu W, Ouyang Z, Liu J. Climate-change impacts on understorey bamboo species and giant pandas in China's Qinling mountains. Nat Clim Chang. 2013;3:249–53.
Visser ME, Both C. Shifts in phenology due to global climate change: the need for a yardstick. Proc R Soc Lond B Biol Sci. 2005;272:2561–9.
Wang F, Zhao Q, McShea WJ, Songer M, Huang Q, Zhang X, Zhou L. Incorporating biotic interactions reveals potential climate tolerance of giant pandas. Conserv Lett. 2018;11:e12592.
Wei F, Swaisgood R, Hu Y, Nie Y, Yan L, Zhang Z, Qi D, Zhu L. Progress in the ecology and conservation of giant pandas. Conserv Biol. 2015;29:1497–507.
Wei F, Costanza R, Dai Q, Stoeckl N, Gu X, Farber S, Nie Y, Kubiszewski I, Hu Y, Swaisgood R, Yang X, Bruford M, Chen Y, Voinov A, Qi D, Owen M, Yan L, Kenny DC, Zhang Z, Hou R, Jiang S, Liu H, Zhan X, Zhang L, Yang B, Zhao L, Zheng X, Zhou W, Wen Y, Gao H, Zhang W. The value of ecosystem services from giant panda reserves. Curr Biol. 2018a;28:2174–2180.e7.
Wei W, Swaisgood RR, Dai Q, Yang Z, Yuan S, Owen MA, Pilfold NW, Yang X, Gu X, Zhou H, Han H, Zhang J, Hong M, Zhang Z. Giant panda distributional and habitat-use shifts in a changing landscape. Conserv Lett. 2018b;11(6):e12575.
Wei W, Swaisgood RR, Owen MA, Pilfold NW, Han H, Hong M, ... Zhang Z. The role of den quality in giant panda conservation. Biol Conserv. 2019;231:189–196.
Wenshi P. A chance for lasting survival: ecology and behavior of wild giant pandas. Washington: Smithsonian Institution Scholarly Press; 2014.

White A, Swaisgood R, Zhang H. The highs and lows of chemical communication in giant pandas: effect of scent deposition height on signal discrimination. Behav Ecol Sociobiol. 2002;51:519–29.

White AM, Swaisgood RR, Zhang H. Chemical communication in the giant panda (*Ailuropoda melanoleuca*): the role of age in the signaller and assessor. J Zool. 2003;259:171–8.

Wildt DE. The role of reproductive technologies in zoos: past, present and future. Int Zoo Yearb. 2003;38:111–8.

Wilson AE, Sparks DL, Knott KK, Kouba AJ, Willard S, Brown A. Behavioral, semiochemical and androgen responses by male giant pandas to the olfactory sexual receptivity cues of females. Theriogenology. 2018;114:330–7.

Woodroffe R, Prager KC, Munson L, Conrad PA, Dubovi EJ, Mazet JA. Contact with domestic dogs increases pathogen exposure in endangered African wild dogs (*Lycaon pictus*). PLoS One. 2012;7:e30099.

Xie Z. The 2018 international studbook for giant panda (*Ailuropoda melanoleuca*). Beijing: Chinese Association of Zoological Gardens. 2018.

Xu W, Viña A, Kong L, Pimm SL, Zhang J, Yang W, Xiao Y, Zhang L, Chen X, Liu J, Ouyang Z. Reassessing the conservation status of the giant panda using remote sensing. Nat Ecol Evol. 2017;1:1635–8.

Yang B, Busch J, Zhang L, Ran J, Gu X, Zhang W, Du B, Xu Y, Mittermeier RA. China's collective forest tenure reform and the future of the giant panda. Conserv Lett. 2015;8:251–61.

Yang Z, Gu X, Nie Y, Huang F, Huang Y, Dai Q, Hu Y, Yang Y, Zhou X, Zhang H, Yang X, Wei F. Reintroduction of the giant panda into the wild: a good start suggests a bright future. Biol Conserv. 2018;217:181–6.

Zhang GQ, Swaisgood RR, Wei RP, Zhang HM, Han HY, Li DS, Wu LF, White AM, Lindburg DG. A method for encouraging maternal care in the giant panda. Zoo Biol. 2000;19:53–63.

Zhang G, Swaisgood RR, Zhang H. Evaluation of behavioral factors influencing reproductive success and failure in captive giant pandas. Zoo Biol. 2004;23(1):15–31.

Zhang H, Li D, Wang C, Hull V. Delayed implantation in giant pandas: the first comprehensive empirical evidence. Reproduction. 2009;138:979–86.

Zhang J, Hull V, Ouyang Z, Li R, Connor T, Yang H, Zhang Z, Silet B, Zhang H, Liu J. Divergent responses of sympatric species to livestock encroachment at fine spatiotemporal scales. Biol Conserv. 2017a;209:119–29.

Zhang L, Luo Z, Mallon D, Li C, Jiang Z. Biodiversity conservation status in China's growing protected areas. Biol Conserv. 2017b;210:89–100.

Zhang W, Yang S, Shan T, Hou R, Liu Z, Li W, Guo L, Wang Y, Chen P, Wang X. Virome comparisons in wild-diseased and healthy captive giant pandas. Microbiome. 2017c;5:90.

Zhao N, Li M, Luo J, Wang S, Liu S, Wang S, Lyu W, Chen L, Su W, Ding H. Impacts of canine distemper virus infection on the giant panda population from the perspective of gut microbiota. Sci Rep. 2017;7:39954.

Zhu X, Lindburg DG, Pan W, Forney KA, Wang D. The reproductive strategy of giant pandas (*Ailuropoda melanoleuca*): infant growth and development and mother–infant relationships. J Zool. 2001;253:141–55.

Chapter 11
Marsupials: Progress and Prospects

John C. Rodger

Abstract The chapter provides a review of the application of reproductive science to technologies for marsupial conservation and population management and discusses prospects for the future. This includes the status of technologies such as sperm freezing, artificial insemination, and exogenous hormone treatments to stimulate ovarian activity and cycling in the female. Fertility-based population management for introduced pest species and over-abundant native marsupials remain an elusive goal. Immune-contraceptive approaches, despite demonstration of basic effectiveness, have not progressed to field deliverable agents. Emerging genetic technologies such as gene drives offer great promise, but gene modifications face major challenges to be broadly accepted both socially and politically. A main theme is the potential advantages, both genetic and economic, of integrating frozen stored genomic material, such as sperm, into the captive breeding component of threatened species strategies. However, the sperm of many marsupial species display no or very poor recovery of motility on thawing. For this reason, it is proposed that the traditional assisted breeding paradigm for conservation—cervical artificial insemination with thawed frozen sperm, based on cattle breeding—is not a viable default strategy. Rather, techniques such as sperm injection and emerging stem cell technologies that utilize stored frozen cells, and in the case of sperm, immotile cells, are better candidates for the development of a more generic approach. In addition, this change in focus encourages wide scale proactive genome storage when genetic diversity is greatest, without the need to demonstrate success in traditional sperm cryopreservation and thawing. However, the promise of the potential of reproductive science to conservation and non-lethal population management is problematic without far greater recognition of, and investment in, the needs of wildlife by society.

J. C. Rodger (✉)
FAUNA Research Alliance, School of Environmental and Life Sciences,
The University of Newcastle, Callaghan, NSW, Australia
e-mail: john.rodger@newcastle.edu.au

P. Comizzoli et al. (eds.), *Reproductive Sciences in Animal Conservation*,
Advances in Experimental Medicine and Biology 1200,
https://doi.org/10.1007/978-3-030-23633-5_11

1 Introduction

1.1 Initial Steps

Progress and prospects is a very appropriate subtitle for this book and chapter because it is now just over 40 years since Paul Watson organised the first international meeting to focus on assisted breeding for non-domestic animals held by the Zoological Society of London in 1977 (Watson 1978). Our contribution on marsupials (Rodger and White 1978) reviewed the limited knowledge relevant to application in conservation in the late 1970s. This included: The removal of suckling pouch young (RPY) as a reliable, but imprecise, tool to induce female cycling and ovulation; early observations of ovarian responses to exogenous hormone treatment and; some quite basic attempts at electroejaculation of field-shot animals to assess male fertility in ecological studies of seasonal breeding. Our paper's main focus and subsequent book chapter was on our establishment of electroejaculation as a reliable method of semen collection in anaesthetised animals, semen handling and biochemistry. Ironically, a passing comment based on very limited observations in one species and no actual data, suggesting that marsupial sperm were not susceptible to cold-shock, and thus promising candidates for frozen storage, became enshrined as 'gospel' till the mid-2000s (see later discussion).

The 1970s and 1980s were active times for research on more fundamental aspects of marsupial reproduction, gamete biology and fertilization. However, this was arguably of little direct relevance to conservation applications (Reviewed in Rodger 1991). This was also true of marsupial reproductive endocrinology which, although arguably the largest area of marsupial reproductive research at the time, was primarily focused on macropods and the phenomenon of embryonic diapause, corpus luteum reactivation, embryonic development and birth (Tyndale-Biscoe and Renfree 1987; Hinds 1990; Tyndale-Biscoe 2005). The main focus of assisted breeding, endocrine regulation of ovarian follicular development, ovulation, mating and conception events, were not major areas of study. The late 1980s saw a renewed interest in developing assisted breeding for marsupials.

1.2 Fertility-Based Population Control

In the 1990s a second area of wildlife reproductive research opened up focused on managing populations of pest or overabundant species by targeting fertility. In Australia the target species were introduced invasive species (e.g. fox), that are major predators of native species including marsupials. It also included environmental degrading competitors (e.g. European rabbit) with impacts on both marsupial habitat and agriculture (Rodger 2003). Overabundance of wildlife also extended to some marsupials (e.g. kangaroos) that are now very abundant, due to the changed land management practices. In New Zealand the introduced marsupial the Australian brushtail possum had become a major pest with environmental and agricultural impacts (Montague 2000). This led to the establishment of two collaborative research centres funded through the Australian Government's Cooperative Research Centres program with the goal of fertility-based non-lethal population management. Most of this contraceptive vaccine research targeted the zona pellucida or gonadotrophin releasing hormone production (Rodger 2003; Mate and Hinds 2003).

1.3 The Current Context for Marsupial Reproductive Research

Underlying both the theme of assisted breeding for conservation of threatened species and fertility-based population management is the centrality of reproductive science. Assisted reproductive technologies (ART) have been employed as a means to overcome breeding problems in wildlife, just as they are in treating human infertility (Holt et al. 2003). However, for the bulk of threatened marsupials assisting fertility and breeding *per se* is not the issue. The majority of threatened marsupial species in conservation captive breeding programs breed well naturally, often excessively so, leading to inbreeding. Arguably, the strongest justification for assisted breeding for marsupial conservation is genome storage and use of this resource for the management of genetic diversity. The goals being: to manage inbreeding in small and fragmented populations of threatened marsupials, conserve desirable alleles in shrinking populations; and to serve as, arguably, the major long-term defence against future emerging threats such as disease and climate change.

In the case of fertility-based marsupial population management the context is the declining availability, and social acceptability of lethal (in New Zealand mainly toxin-based) population control (Rodger 2003). It is quite reasonable to expect that fertility based, non-lethal, control options for wildlife generally will be the only socially and ultimately politically acceptable methods of managing even introduced pest species in the foreseeable future. This constraint is likely to be even greater for overabundant native species such as kangaroos.

This chapter reviews the current state of play for applied reproductive technologies for marsupials. Much has been achieved, but in key aspects progress has been very limited, and initial promise has not been realized in technical proof of concept, let alone practical outcomes. Arguably our reach has often exceeded our grasp. In the area of assisted breeding the question is posed: Is the paradigm that has driven most of research science for conservation through assisted breeding the right one as we move into the third decade of the 2000s?

2 State of the Art: A Strategic Overview for Marsupials

There have been several detailed reviews of assisted breeding for marsupials over recent years (Rodger et al. 2009; Witt and Rodger 2018; Johnston and Holt 2014). In this chapter the focus is on a strategic overview of achievements and limitations in the current state of play as a prelude to discussion of future roles for ART in the conservation and management of marsupials. No marsupial has been domesticated so that there are no obvious model species through which to develop reproductive technologies as in eutherian mammals. However, two Australian species; the tammar wallaby (formerly *Macropus eugenii* now *Notamacropus eugenii* (Eldridge et al. 2017)), a 6–10 kg macropod, and the small shrew-like fat-tailed dunnart [*Sminthopsis carassicaudata*], a 15–20 g dasyurid, and the South American gray short-tailed opossum [*Monodelphis domestica*], a 60–100 g didelphid, have been established in research colonies. These three have allowed research to cover a wide range of marsupial diversity with the goal of establishing generic principles.

2.1 Cryo-Preservation of Sperm

As in eutherian ('placental') mammals the sperm of marsupials can be cryopreserved using a variety of cryoprotectant and freezing and thawing protocols [comprehensively reviewed in Charlton et al. (2018)]. This is presumably also the case for American didelphid and caenolestid marsupials, but they have not been the subjects of sperm cryopreservation, or any ART directed research. Thus far however no live births have followed insemination with frozen-thawed sperm of any marsupial, although this has been achieved with chilled sperm in the koala (Johnston and Holt 2014; Allen et al. 2008). As with eutherians the sperm of marsupials show a range of recovery of vitality and motility following freezing and thawing from a high percentage motility (70–80% motile in brushtail possum, and common wombat (Molinia and Rodger 1996; Taggart et al. 1996; MacCallum and Johnston 2005)), around 40–50% in the Koala (Johnston et al. 2006) and bandicoots (Taggart et al. 1996) through to zero motility post thaw in some dasyurids (three species) (Czarny et al. 2009; Czarny and Rodger 2010). Of the other species examined sperm motility post thaw is low to moderate around 5–20% and a significant component of this motility is weak or non-progressive (Macropods (Molinia and Rodger 1996; McClean et al. 2008), Potoroids, Pseudocherids (Taggart et al. 1996), dasyurid Tasmanian devil (Keeley et al. 2012)).

Although cryoprotectant toxicity is undoubtedly an issue for some marsupial species, and more complex mixtures of penetrating and non-penetrating protective agents have yielded improvements in post-thaw recovery, the gains are small and it remains that for species like the brushtail possum very simple diluents and high glycerol were very effective (Rodger et al. 1991). As mentioned earlier, although in most species sperm appear to be resistant to 'cold-shock' (very rapid cooling to around 0 °C (Mann 1964)), sperm of at least one dasyurid, the fat-tailed dunnart, are highly susceptible (Czarny and Rodger 2010) and this issue needs to be more critically examined for other marsupials.

In the context of the very mixed results for marsupial sperm cryo-preservation it is worthwhile looking at two eutherian species. It was susceptibility to cold-shock, and poor motility post-thaw, which held back the use of cryopreserved sperm in sheep breeding and led eventually, after major research investment, to the use of intrauterine insemination rather than conventional vaginal/cervical AI (Salamon and Maxwell 2000). Similarly, despite the importance of the mouse for biomedical research and considerable research investment, live births with frozen sperm have only been achieved reliably using intra-cytoplasmic sperm injection (ICSI) (reviewed in Charlton et al. 2018).

2.2 *Artificial Insemination (AI) in Marsupials*

A reasonably reliable method to achieve live birth following AI has only been established for one marsupial, the koala; and this was in conscious, tractable fauna park animals (Allen et al. 2008). AI is likely to remain a major limiting step in applying ART to marsupials. Most species will require some form of sedation, yet the effects of sedation on the peri-ovulatory and conception processes are unknown (Rodger et al. 2009). Further, the complex anatomy of the marsupial vaginal complex restricts access to the cervices (Rodger et al. 2009). This is further complicated in that the cervices and uteri are completely separate, and many species are monovulatory alternate ovulators (Rodger et al. 2009; Tyndale-Biscoe and Rodger 1978). In the latter cases not only must the AI sperm be delivered effectively to the site of fertilization via the cervix/uterus/oviduct but to the side of the tract with the ovulating ovary. Even in a well-studied species, the tammar wallaby, with quite precise natural means of timing ovulation, AI with fresh sperm was not a reliable means of achieving live births (Paris et al. 2005). This is of particular concern and points to major, as yet unidentified, issues in applying AI for marsupials more generally.

2.3 *Marsupial Oocytes and Embryos*

In a range of marsupials, preovulatory oocytes can be induced to develop, mature and ovulate by RPY or by exogenous hormone treatment (reviewed in Witt and Rodger 2018). Repeated superovulation does not impair subsequent fertility in the tammar wallaby (Magarey et al. 2003). Marsupial oocytes from exogenous hormone-primed females will mature in vitro (Mate and Rodger 1993). Thus, the options for obtaining marsupial oocytes for IVF or ICSI are similar to that of eutherian mammals. Marsupial oocytes do, however, have unique preovulatory features not seen in eutherians (Mate 1996); late oocyte growth (Rodger et al. 1992; Falconnier and Kress 1992) and, in some species, late formation of the cortical granules (Falconnier and Kress 1992; Mate et al. 1992). These are factors that need to be taken into consideration. IVF, cleavage and development to late embryo stages has been achieved in the didelphid gray short-tailed opossum with quite simple culture conditions (Moore and Taggart 1993). The only other marsupial where any progress on IVF has been achieved is the common brushtail possum and this required a complex 'capacitation' process of culture of sperm in oviduct epithelial cell-conditioned medium containing approximately 1 mM ionic calcium to achieve sperm-zona binding and then spermic penetration and fusion with the oocyte (Mate et al. 2000; Sidhu et al. 2003).

Intracytoplasmic sperm injection has been successfully attempted in three species of marsupial; common wombat (West et al. 2007), tammar wallaby (Magarey and Mate 2003; Richings et al. 2004) and fat-tailed dunnart (N. Czarny and J. Rodger, unpublished observations). The latter is significant because of the relatively large size of the dasyurid oocyte and sperm compared with those of other marsupials and eutherians. However, although these posed physical challenges for ICSI, they were not insurmountable. No live births have been achieved in any marsupial after IVF or ICSI but in most of the research reported here the end point and goal of the experiments was early conception and embryo development events.

Embryo transfer (ET) has been used in basic studies of marsupial development using early stage cleaving embryos obtained after natural mating in two macropod marsupials, the quokka (Tyndale-Biscoe 1963) and tammar wallaby (Tyndale-Biscoe 1970) and a dasyurid, the fat-tailed dunnart (Breed and Leigh 1996). In all cases a simple surgical technique was used, and the embryos were placed through the uterine wall using a finely drawn hand-held pipette. Where the timing of the recipient's endocrine status was appropriate, development to the expanded blastocyst stage was supported but live births were not achieved. Embryo transfer to surrogate mothers has not been attempted as part of an ART strategy either within a species or between related species.

2.4 *Managing the Female Cycle in Marsupials*

Control of the female cycle as a source of oocytes, embryos or ET recipients has been achieved using a variety of techniques. Removal of suckling pouch young is a very reliable, and quite precise, means of inducing follicular cycles and mating in some species, in particular the tammar wallaby (Hinds et al. 1990). However, in most marsupials the subsequent ovulation and mating can occur over a range of several days (reviewed in Rodger et al. 2009; Witt and Rodger 2018). Although long established as a research tool, RPY generally involves the removal and death of a suckling young which raises ethical concerns, and is counterproductive for use in conservation breeding.

In some marsupials, exogenous gonadotrophin hormone treatment results in quite predictable follicular development, ovulation and mating (dunnart (Rodger et al. 1992), tammar wallaby (Hinds et al. 1996)). However, in the monovular species this results in 'superovulation' so that a large number of follicles develop, but not all ovulate, and there are multiple corpora lutea produced (ovarian stimulation by exogenous hormones in marsupials was recently reviewed in Witt and Rodger (2018)). Thus, there are concerns over the endocrine status of such females especially as ET recipients. Recently there have been attempts to develop a potentially generic approach to female synchronisation based on downregulation of hypothalamic-pituitary gonadotrophin secretion using GnRH analogues to reset female cycling (Ballantyne et al. 2016; Witt et al. 2016). The rationale for this approach is that ovarian-targeted approaches, widely used in eutherian mammals to synchronise female activity, such as progesterone or prostaglandins, are not luteolytic in marsupials (Witt et al. 2016; Crawford et al. 2011). These approaches show initial promise with conception and births following a single treatment with 'Lucrin Depot' a microsphere preparation of Leuprorelin acetate in fat-tailed dunnarts (Witt et al. 2018a) and tammar wallabies (Witt et al. 2018b). The unusual biology of the marsupial corpus luteum (CL), which persists for essentially the same period in pregnant and non-pregnant cycles, has been known for many decades (Tyndale-Biscoe 2005).

3 Case Studies in Reproductive Technologies

3.1 Genome Storage Must Be Proactive

The Tasmanian devil (*Sarcophilus harrisii*) is an example of a lost opportunity in the application of reproductive science to conservation. The devil, the last surviving large marsupial carnivore, had been the subject of limited reproductive research and no research directed at genome storage or ART for conservation until the late 1990s when it was recognised that the species faced an imminent extinction threat from a highly unusual contagious facial tumour disease. This, in part, was due to the restriction of the species to Tasmania and the challenges of working with the species compared with more tractable, and easily mass-reared shrew-like dunnarts, which have been the main dasyurid experimental model. The case of the Black-footed Ferret, although demonstrating sub-optimal very late action, is an example of the value of preserving genomic resources (in that case frozen sperm) decades before the necessary ART was developed to utilise these sperm to successfully supplement the genetic health of the captive population (Howard et al. 2016).

Both these cases point to the major limitation in the realisation of the potential of genome storage in marsupial conservation, or conservation generally. To preserve optimal genetic diversity proactive strategies, need to be implemented before crises are recognised or populations crash, not as *ad hoc* 'band-aid' attempts at intervention when species are under threat of extinction and inevitably have reduced or depauperate genetic diversity.

In essentially all cases, this will mean that genomes would be stored before the ART to use them exists. This has arguably been a major impediment to the uptake of genome storage into conservation policy. However, implementation of such a proactive approach would mean that a genetic resource of highest possible value could be preserved and, as threat of extinction loomed, these resources could be directed to its use as part of recovery programs that managed the declining living population, and much larger frozen populations, as an integrated whole.

3.2 Genome Storage and Captive Breeding of Threatened Species

There are strong practical and economic arguments for genome banking for conservation. Captive breeding is an accepted part of threatened species recovery plans in Australia, including those for marsupials. However, they are expensive, and in the case of some species, such as the termite eating numbat, extremely challenging. In addition, captive populations are usually relatively small (reflecting maintenance costs and housing requirements) and thus do not preserve the optimal number of individuals that genetic theory recommends (Frankham et al. 2010). The cost of captive breeding programs for a range of vertebrates, including threatened marsupials, is currently the subject of a PhD research project at the University of Newcastle. Annualised estimates (set-up, depreciation of facilities and running cost) for two threatened marsupial species at Perth Zoo (colony sizes ranging from 30–70) are as follows: the Numbat (A$150,000–$200,000), and the Dibbler (A$150,000–$175,000). These are the only captive breeding colony for both species (Raw unpublished data curtesy of Dr. Peter Mawson, Perth Zoo, analysed by Lachlan Howell). Upscaling of these facilities to even approach the recommended size to manage inbreeding (minimum 500), or indeed the optimal to conserve genetic diversity (thousands) is unlikely to ever be a realistic option for most species (Traill et al. 2007; Brook et al. 2011). However, a strategy that integrated cryo-stored genomes with captive population management would make thousand-individual populations feasible because the threatened species effective population would be the total number of living and frozen individuals. Set-up costs for a multi-species wildlife cryo-storage facility have been estimated at A$20,000–$30,000 (Zoos Victoria 2014) and, operating costs are modest based on published annual cryo-storage charges for biomedically-valuable genomic material (A$5000) (Australian Phenomics Facility 2018).

The captive breeding costs cited here for threatened marsupials are being integrated into a modelling exercise with genetic population viability theory, in what will be the first study, that I am aware of, that seeks to quantify the benefits and costs of integrating genome storage into threatened species management. The model, based on published Greater Vancouver Zoo captive breeding costs and effective population size data for the Oregon Spotted frog (Kissel et al. 2017; Phillipsen et al. 2009), set as its conservation management goal meeting the IUCN's 100 years, 90% heterozygosity criteria (no more than 10% inbreeding). Not surprisingly total costs of captive breeding to achieve such high genetic diversity were high (C$490 million after 100 years). However, integration of stored genome banked sperm into the breeding program significantly reduced this cost. If frozen sperm were used each generation (3 years) from day 1, the total cost over the 100 years was reduced to C$17 million and if each 5th generation (15 years) to C$70 million (Unpublished analysis: L. G. Howell, R. Frankham, J. C. Rodger, S. Clulow, R. M. O. Upton, and J. Clulow). This analysis does not yet include the costs of genome storage and use but their quantum (A$10,000s) are not expected to alter the conclusions. These data show that utilizing biobanking in captive breeding colonies allows genetic diversity to be retained, meeting high IUCN criteria, at substantial cost savings that allow limited economic resources to be diverted into additional captive breeding programs for endangered species. The knowhow and technology may not currently exist to use this stored genomic resource in most cases, but the modelling shows significant cost advantages even if the stored sperm are not used until after 5 generations (15 years for the Oregon Spotted frog). This is a very realistic time frame in which time to develop the needed knowhow and technology and consistent with the experience in the Black-footed Ferret (Howard et al. 2016).

3.3 Population Control for Marsupials and Threatening Introduced Species

Reproduction-based population control for free-ranging wildlife has not been successful except for the proof of concept experiments using ZP-based immuno-contraception in the mouse (Redwood et al. 2007). In this case, although promising, it has not been trialled in the field largely due to concerns about the safety of GMO virus vectors. Viral-vectored ZP-based immuno-contraception has also been researched but with less promising results in rabbits and foxes (Hardy et al. 2006). In marsupials (Brushtail Possum (Duckworth et al. 2007), eastern grey kangaroo, tammar wallaby (Kitchener et al. 2009a; Snape et al. 2013) and koala (Kitchener et al. 2009b)) fertility has been compromised with injectable vaccines against zona pellucida or GnRH antigens. There has also been considerable research on hormonal suppression of marsupial fertility using GnRH analogues and long acting gestagens, either as injections or implants (e.g. eastern grey kangaroo (Coulson et al. 2008), tammar wallaby (Herbert et al. 2005), brush-tail possum (Eymann et al. 2007)). Despite these successes, field delivery, preferably of a 'one-shot' treatment, and longevity of fertility suppression—critical criteria for an effective population management agent—remain enormous challenges (Cowan et al. 2003).

Immuno-contraception for practical wildlife management, including for marsupials, remains beyond our grasp largely because oral delivery is likely to be the only practical method for free-ranging animals and oral immunisation is, despite early success with viral vaccines, a problematic area generally. For marsupials this is exacerbated by our rudimentary knowledge of their immunology and in particular its practical manipulation. It is more likely that fertility control candidates will arise via ever emerging gene technologies, such as Crispr and Gene Drives but, like all gene modification technologies, these face challenges of social and political acceptability. However, the fundamental field delivery challenges will remain the same; acceptable specificity, safety broadly, sufficient reach to impact on the population of target free-ranging animals, and species appropriate longevity of effects. Arguably, as genetic approaches favour sterility this may increase the possibility of use for pest species but is a problem for overabundant, and ultimately desirable, native species where the goal is more humane control not eradication.

4 Priorities for the Future

4.1 *Rethinking the Conservation ART Paradigm?*

Cattle breeding using frozen semen has, like many areas where applied science has been a great practical success, had profound impacts on how we, as researchers, think about how we would approach similar problems in other species. I would argue, and have argued (Rodger et al. 2009) but now with more force, that this is a flawed paradigm when it comes to wildlife. It has been long recognised, and eloquently argued by David Wildt and his colleagues (Wildt et al. 2003), that shortage of basic information on reproductive biology is a major limiting factor in applying ART for conservation purposes. But I would argue the problem is more intractable. The cattle paradigm is based on AI using frozen-thawed sperm. There are variants, but this remains the major focus for the majority of wildlife, including marsupial ART, research.

As discussed earlier, for marsupials, and I would argue most target species, that strategies that ultimately depend on the motility of thawed sperm (% motile, quality of motility and longevity of motility) are problematic. AI with frozen-thawed sperm may be a strategy for the common brushtail possum and wombat, but is high risk for most other species where all three aspects of post thaw motility are poor. In addition, for marsupials there are the anatomical restrictions on reaching the cervices, due to the complex vaginal anatomy and the uncertain effects of the necessary sedation (Rodger et al. 2009). In sheep these challenges were met by intra-uterine insemination, and this AI approach has been used effectively to achieve conception in the common brushtail possum (Molinia et al. 1998), but has not yet been used to produce live young. In the mouse, attempts to use AI were eventually abandoned and ICSI is the standard form of ART used to conserve valuable mouse models and strains (Charlton et al. 2018). The move to ICSI has not only achieved the required breeding outcome, but facilitated the use of alternative forms of sperm preservation (sperm freeze drying, freezing without cryoprotectants, freezing of testes and epididymis at −80 °C and whole mice 15 years at −20 °C (Ogonuki et al. 2006)). Given the cost and complexity of conventional sperm storage in liquid nitrogen, such alternatives are very attractive for wildlife genome storage.

4.2 *Learning from a Case Study of One Form of Human Infertility*

My thesis is that our focus should change such that conception using immotile sperm should become the standard goal of conservation ART. It is worth reflecting on how ICSI for human IVF changed the paradigm of male-based infertility treatment. This is an area that I was very familiar with because of the research of colleagues in my Department on infertility treatments, who focused on men with anti-sperm antibodies. Not unlike some sperm cryo-preservation work, this had been going on for some years with small incremental progress, but generally poor outcomes, attempts to remove attached antibodies by sperm 'washing' techniques were pursued for many years until in the early 1990s ICSI made anti-sperm antibodies, and even sperm motility, irrelevant to IVF-based conception. My argument is that conventional AI using motile sperm should still form part of the conservation ART tool-box, but it should not be seen as the default approach. This removes the long and laboured, and often heroic efforts, to recover acceptable motility for frozen-thawed sperm from the equation. Sperm and other potential genome sources can be frozen using a range of technologies and then used for ICSI to produce embryos for transfer to recipient females as argued above. With, I expect, a trend towards freeze drying as the uptake of that approach grows. Presumably there would remain a need to conventionally frozen-store supporting tissues and serum to be used for genetic and biosecurity/disease purposes. Arguably, and perhaps most importantly, removing the focus on post-thaw motility of sperm as a key criterion of success, to high quality freezing and storage *per se* of reproductive and reference tissues. This change in paradigm means that proactive global genome banking becomes a far more justifiable conservation policy strategy, and one that is more consistent with the continuing and anticipated, if as yet unidentified, emerging extinction threats.

4.3 ICSI a Core Technology

It seems, based on the reasonably wide range of species in which ICSI has been applied (including marsupials as discussed above), that this technology can be seen as essentially generic and that the expertise and equipment is now widely available. This means that the focus of intensive research for Advanced ART should in future be on what has always been arguably the most critical component of ART for mammals, effective manipulation and control of the female cycle of both oocyte donors and embryo recipients. For marsupials, the reproductive biology knowledge base will always be very limited compared to eutherian mammals. At least for eutherians there are the very well-studied domestic species, biomedical models like rodents and companion animals to use as experimental models for wildlife. Surrogacy, particularly as oocyte donors (as discussed later for white rhino research), and the use of embryo recipients, is also an area deserving active examination. This is particularly important for threatened species where, if a suitable surrogate mother were available, the productivity of the target threatened species could potentially be increased significantly. This concept has been at the base of the marsupial pouch young cross fostering work. Which has used the pouch of a common but related species as the recipient of the transferred young of the threatened marsupial. In the best case this resulted in an eightfold increase in production of offspring of the threatened species, which normally produce only single young in a breeding season (Taggart et al. 2005, 2010).

5 Conclusion

Recently sperm injection (ICSI), in vitro oocyte maturation and embryonic stem cell techniques have been used to successfully produce embryos in the rhinoceros (Hildebrandt et al. 2018). Given such progress it seems likely that eventually live births will be achieved by this international collaboration. This high-tech approach was taken because of the repeated failure of conventional ART in the rhino. But its success argues for a much more technically ambitious approach to assisted reproduction for conservation that draws on the rapidly developing tools of developmental biology including stem cell technologies. Hildebrandt has coined some new terms which may help to signal this changing paradigm 'Advanced ART' and SCAT (Stem Cell Associated Technologies) (Thomas Hildebrandt, personal communication). SCAT seems a little awkward, given the word's normal meaning, and perhaps more simply Advanced ART contains all, or Advanced ARDT to signal the likely increasing importance of developmental biology are better alternatives.

Such technologies are not without risk. However, despite the now decadal experience in the use of Advanced ART in treating human infertility, and the very close attention given to the children produced, there has been no evidence for major causes of concern. Indeed, problems that have been recorded cannot be separated from the underlying infertility that brought the parents into treatment (Farhi and Fisch 2007). Furthermore, the most significant problems seemed to be associated with multiple embryo transfer (Nardelli et al. 2014). This experience suggests in my view, that although unexpected outcomes must be expected and monitoring policies developed, the risks of applying Advanced ART/ARDT in conservation are acceptable in the face of the scale of the global extinction threat.

Marsupials, like most wildlife, face a very uncertain future due to multiple and growing threats; from globalisation, the spread of invasive competitor and predator species, disease agents and the ever-growing demand of the human population for land for urban development, resources and agriculture. These unfortunately very familiar problems are only amplified by ever more alarming predictions of extreme climate change driven habitat change and loss. All these pressures mean there is now, more than ever, a need to better understand and learn to practically manipulate the reproduction of animals if humanity seriously hopes that even a reasonable sample of current wild species are to continue to co-exist with us. As discussed earlier, the limited knowledge base of marsupial biology generally, and the absence of well-studied domestic or lab species so important in ART development for eutherian mammals, exacerbate the challenge. This scenario presents a challenge for policy makes world-wide, but particularly for Australia, which is home to two thirds of the world's extant marsupial species. To develop the evidence-based strategies and tools for innovation in conservation needed to halt current declines, meet emerging challenges and where possible return wildlife to depauperate ecosystems, will require significant increases in investment in research (Howell and Rodger 2018).

Ultimately, this poses an ethical or values dilemma. We seem very able to find the very significant resources via government, non-government organisations and philanthropy to support research and innovation to find solutions and tools to address our own species' social and health problems. Often driven by our personal and family experiences. Can we find more meaningful empathy, engagement and investment for our evolutionary relatives; iconic, attractive and unattractive, as they face this perfect storm of challenges to their survival due to humanities overt actions and ignorance?

Acknowledgements My research has had the input of many collaborators; in particular Karen Mate and Frank Molinia for the earlier work on superovulation, gamete biology, AI and sperm freezing, and Anne Kitchener, Carmen McCartney and David Kay for the work on marsupial fertility control vaccines. More recently Ryan Witt and Lyn Hinds have made major contributions to the research on managing the female cycle and Lachlan Howell and John Clulow to research on captive breeding costs and practice and genome storage policy options.

My research on marsupial reproduction has been supported over the years by the Australian governments CRC and ARC programs, New Zealand government funded possum control programs, The Morris Animal Foundation and the Holsworth Wildlife Research Endowment.

My colleagues Carmen McCartney and Ryan Witt read the manuscript and gave good advice as did two referees. Ryan also introduced me to Endnote and generously made his data set available. The opinions expressed are my own.

References

Allen CD, Burridge M, Mulhall S, Chafer ML, Nicolson VN, Pyne M, et al. Successful artificial insemination in the Koala (*Phascolarctos cinereus*) using extended and extended-chilled semen collected by electroejaculation. Biol Reprod. 2008;78:661–6.

Australian Phenomics Facility. Costs. 2018. http://pbstatic.apf.edu.au/phenbank/Costs.htm. Accessed 14 Dec 2018.

Ballantyne K, Anderson ST, Pyne M, Nicolson V, Mucci A, Lisle AT, et al. Use of the gonadotrophin-releasing hormone antagonist azaline B to control the oestrous cycle in the koala (*Phascolarctos cinereus*). Reprod Fertil Dev. 2016;28:1686–94.

Breed WG, Leigh CM. Successful embryo transfer in a small dasyurid marsupial, *Sminthopsis crassicaudata*. Theriogenology. 1996;45:1075–80.

Brook BW, Bradshaw CJA, Traill LW, Frankham R. Minimum viable population size: not magic, but necessary. Trends Ecol Evol. 2011;26:619–20.

Charlton JC, Nielsen MB, Pedersen CR, Thomsen L, Kristjansen MP, Sorensen TB, et al. Strong heterogeneity in advances in cryopreservation techniques in the mammalian orders. Zool Sci. 2018;35:1–22.

Coulson G, Nave CD, Shaw G, Renfree MB. Long-term efficacy of levonorgestrel implants for fertility control of eastern grey kangaroos (*Macropus giganteus*). Wildl Res. 2008;35:520–4.

Cowan PE, Pech R, Curtis P. Field applications of fertility control for wildlife management. In: Holt WV, Pickard AR, Rodger JC, Wildt DE, editors. Reproductive science and integrated conservation. Cambridge: Cambridge University Press; 2003. p. 305–18.

Crawford JL, McLeod BJ, Eckery DC. The hypothalamic–pituitary–ovarian axis and manipulations of the oestrous cycle in the brushtail possum. Gen Comp Endocrinol. 2011;170:424–48.

Czarny NA, Rodger JC. The spermatozoa of the dasyurid marsupial, *Sminthopsis crassicaudata*, are highly susceptible to cold shock. Reprod Fertil Dev. 2010;22:580–5.

Czarny NA, Harris MS, De Luliss GN, Rodger JC. Acrosomal integrity, viability, and DNA damage of sperm from dasyurid marsupials after freezing or freeze drying. Theriogenology. 2009;72:817–25.

Duckworth JA, Wilson K, Cui X, Molinia FC, Cowan PE. Immunogenicity and contraceptive potential of three infertility relevant zona pellucida 2 epitopes in the marsupial brushtail possum (*Trichosurus vulpecula*). Reproduction. 2007;133:177–86.

Eldridge MDB, Miller EJ, Neaves LE, Zenger KR, Herbert CA. Extensive genetic differentiation detected within a model marsupial, the tammar wallaby (*Notamacropus eugenii*). PLoS One. 2017;12:1–23.

Eymann J, Herbert CA, Thomson BP, Trigg TE, Cooper DW, Eckery DC. Effects of deslorelin implants on reproduction in the common brushtail possum (*Trichosurus vulpecula*). Reprod Fertil Dev. 2007;19:899–909.

Falconnier C, Kress A. Ultrastructural aspects of oocyte growth in the marsupial *Monodelphis domestica* (gray short-tailed opossum). J Anat. 1992;181:481–98.

Farhi J, Fisch B. Risk of major congenital malformations associated with infertility and its treatment by extent of iatrogenic intervention. Pediatr Endocrinol Rev. 2007;2007:352–7.

Frankham R, Ballou J, Briscoe D. Introduction to conservation genetics. Cambridge: Cambridge University Press; 2010.

Hardy CM, Hinds LA, Kerr PJ, Lloyd ML, Redwood AJ, Shellam GR, et al. Biological control of vertebrate pests using virally vectored immunocontraception. J Reprod Immunol. 2006;71:102–11.

Herbert CA, Trigg TE, Renfree MB, Shaw G, Eckery DC, Cooper DW. Long-term effects of deslorelin implants on reproduction in the female tammar wallaby (*Macropus eugenii*). Reproduction. 2005;129:361–9.

Hildebrandt TB, Hermes R, Colleoni S, Diecke S, Holtze S, Renfree MB, et al. Embryos and embryonic stem cells from the white rhinoceros. Nat Commun. 2018;9:2589.

Hinds LA. Control of pregnancy, parturition and luteolysis in marsupials. Reprod Fertil Dev. 1990;2:535–52.

Hinds LA, Poole WE, Tyndale-Biscoe CH, van Oorschot RAH, Cooper DW. Reproductive biology and the potential for genetic studies in the tammar wallaby, *Macropus Eugenii*. Aust J Zool. 1990;37:223–34.

Hinds LA, Fletcher TP, Rodger JC. Hormones of oestrus and ovulation and their manipulation in marsupials. Reprod Fertil Dev. 1996;8:661–72.

Holt WV, Pickard AR, Rodger JC, Wildt DE, editors. Reproductive science and integrated conservation. Cambridge: Cambridge University Press; 2003.

Howard JG, Lynch C, Santymire RM, Marinari PE, Wildt DE. Recovery of gene diversity using long-term cryopreserved spermatozoa and artificial insemination in the endangered black-footed ferret. Anim Conserv. 2016;19:102–11.

Howell LG, Rodger JC. An examination of terrestrial vertebrate fauna research funding from Australian federal government sources. Pac Conserv Biol. 2018;24:142–7.

Johnston SD, Holt WV. The Koala (*Phascolarctos cinereus*): a case study in the development of reproductive technology in a marsupial. In: Holt WV, Brown JL, Comizzoli P, editors. Reproductive sciences in animal conservation: progress and prospects. London: Springer; 2014. p. 171–204.

Johnston SD, MacCallum C, Blyde D, McClean R, Lisle A, Holt WV. An investigation into the similarities and differences governing the cryopreservation success of koala (*Phascolarctos cinereus*: goldfuss) and common wombat (*Vombatus ursinus*: shaw) spermatozoa. Cryobiology. 2006;53:218–28.

Keeley T, McGreevy PD, O'Brien JK. Cryopreservation of epididymal sperm collected postmortem in the Tasmanian devil (*Sarcophilus harrisii*). Theriogenology. 2012;78:315–25.

Kissel AM, Palen WJ, Govindarajulu P. A decision-theory approach to cost-effective population supplementation for imperiled species. Ecol Econ. 2017;142:194–202.

Kitchener AL, Harman A, Kay DJ, McCartney CA, Mate KE, Rodger JC. Immunocontraception of Eastern Grey kangaroos (*Macropus giganteus*) with recombinant brushtail possum (*Trichosurus vulpecula*) ZP3 protein. J Reprod Immunol. 2009a;79:156–62.

Kitchener AL, Kay DJ, Walters B, Menkhorst P, McCartney CA, Buist JM, et al. The immune response and fertility of koalas (Phascolarctos cinereus) immunised with porcine zonae pellucidae or recombinant possum ZP3 protein. J Reprod Immunol. 2009b;82:40–7.

MacCallum C, Johnston SD. Studies on the cryopreservation of common wombat (*Vombatus ursinus*) spermatozoa. Reprod Fertil Dev. 2005;17:727–32.

Magarey GM, Mate KE. Timing and ultrastructural events following intracytoplasmic sperm injection in a marsupial, the tammar wallaby, (*Macropus eugenii*). Reprod Fertil Dev. 2003;15:397–406.

Magarey GM, Rodger JC, Buist JM, Mate KE. Effects of repeated superovulation and surgical oocyte collection on ovarian response and natural breeding ability of the tammar wallaby (*Macropus eugenii*). Reproduction. 2003;125:701–7.

Mann T. The biochemistry of semen and the male reproductive tract. London: Methuen; 1964.

Mate KE. Cytoplasmic maturation of the marsupial oocyte during the periovulatory period. Reprod Fertil Dev. 1996;8:509–19.

Mate KE, Hinds LA. Contraceptive vaccine development. In: Holt WV, Pickard AR, Rodger JC, Wildt DE, editors. Reproductive science and integrated conservation. Conservation biology. Cambridge: Cambridge University Press; 2003. p. 291–304.

Mate KE, Rodger JC. In vitro maturation of oocytes from a marsupial, the tammar wallaby (*Macropus eugenii*). Mol Reprod Dev. 1993;34:329–36.

Mate KE, Giles I, Rodger JC. Evidence that cortical granule formation is a periovulatory event in marsupials. J Reprod Fertil. 1992;95:719–28.

Mate KE, Sidhu KS, Molinia FC, Glazier AM, Rodger JC. Sperm binding and penetration of the zona pellucida *in vitro* but not sperm-egg fusion in an Australian marsupial, the brushtail possum (*Trichosurus vulpecula*). Zygote. 2000;8:189–96.

McClean R, Zee YP, Holt WV, Johnston SD. Cryopreservation of kangaroo spermatozoa using alternative approaches that reduce cytotoxic exposure to glycerol. Cryobiology. 2008;57:304–7.

Molinia FC, Rodger JC. Pellet-freezing spermatozoa of two marsupials: the tammar wallaby, *Macropus eugenii*, and the brushtail possum, *Trichosurus vulpecula*. Reprod Fertil Dev. 1996;8:681–4.

Molinia FC, Gibson RJ, Brown AM, Glazier AM, Rodger JC. Successful fertilisation after superovulation and laparoscopic intrauterine insemination of the brushtail possum, *Trichosurus vulpecula*, and tammar wallaby, *Macropus eugenii*. J Reprod Fertil. 1998;112:9–17.

Montague TL. The Brushtail Possum—biology, impact and management of an introduced marsupial. Lincoln: Manaaki Whenua Press; 2000. p. 292.

Moore HDM, Taggart DA. In vitro fertilization and embryo culture in the grey short-tailed opossum, Mondelphis domestica. J Reprod Fertil. 1993;98:267–74.

Nardelli AA, Stafinski T, Motan T, Klein K, Menon D. Assisted reproductive technologies (ARTs): evaluation of evidence to support public policy development. Reprod Health. 2014;11:76.

Ogonuki N, Mochida K, Miki H, Inoue K, Fray M, Iwaki T, et al. Spermatozoa and spermatids retrieved from frozen reproductive organs or frozen whole bodies of male mice can produce normal offspring. Proc Natl Acad Sci U S A. 2006;103:13098–103.

Paris DBBP, Taggart DA, Shaw G, Temple-Smith PD, Renfree MB. Birth of pouch young after artificial insemination in the tammar wallaby (*Macropus eugenii*). Biol Reprod. 2005;72:451–9.

Phillipsen IC, Bowerman J, Blouin M. Effective number of breeding adults in Oregon spotted frogs (*Rana pretiosa*): genetic estimates at two life stages. Conserv Genet. 2009;11:737–45.

Redwood AJ, Smith LM, Lloyd ML, Hinds LA, Hardy CM, Shellam GR. Prospects for virally vectored immunocontraception in the control of wild house mice (*Mus domesticus*). Wildl Res. 2007;34:530–9.

Richings NM, Shaw G, Temple-Smith PD, Renfree MB. Intracytoplasmic sperm injection in a marsupial. Reproduction. 2004;128:595–605.

Rodger JC. Fertilization of marsupials. In: Dunbar BS, O'Rand MG, editors. A comparative overview of mammalian fertilization. New York: Plenum Press; 1991. p. 117–35.

Rodger JC. Fertility control for wildlife. In: Holt WV, Pickard AR, Rodger JC, Wildt DE, editors. Reproductive science and integrated conservation. Cambridge: Cambridge University Press; 2003. p. 281–90.

Rodger JC, White IG. The collection, handling and some properties of marsupial semen. In: Watson PF, editor. Artificial breeding of non-domestic species. Symposium of the Zoological Society of London, vol. 43. London: Academic; 1978. p. 289–301.

Rodger JC, Cousins SJ, Mate KE. A simple glycerol-based freezing protocol for the semen of a marsupial, *Trichosurus vulpecula*, the common brushtail possum. Reprod Fertil Dev. 1991;3:119–25.

Rodger JC, Giles I, Mate KE. Unexpected oocyte growth after follicular antrum formation in four marsupial species. J Reprod Fertil. 1992;96:755–63.

Rodger JC, Paris DBBP, Czarny NA, Harris MS, Molinia FC, Taggart DA, et al. Artificial insemination in marsupials. Theriogenology. 2009;71:176–89.

Salamon S, Maxwell WMC. Storage of ram semen. Anim Reprod Sci. 2000;62:77–111.

Sidhu KS, Mate KE, Molinia FC, Berg DK, Rodger JC. Ionic calcium levels in oviduct explant-conditioned media from an Australian marsupial, the brushtail possum (*Trichosurus vulpecula*) and its relevance to *in vitro* fertilization. Zygote. 2003;11:285–91.

Snape MA, Hinds LA, Davey CC, Henry SJ, Fletcher C, Wimpenny C, et al. Effects of GonaCon™ vaccination on reproduction, growth and behavior in two Australian macropodid marsupials. J Zoo Wildl Med. 2013;44. Abstract.

Taggart DA, Leigh CM, Steele VR, Breed WG, Temple-Smith PD, Phelan J. Effect of cooling and cryopreservation on sperm motility and morphology of several species of marsupial. Reprod Fertil Dev. 1996;8:673–9.

Taggart DA, Schultz D, White C, Whitehead P, Underwood G, Phillips K. Cross-fostering, growth and reproductive studies in the brush-tailed rock-wallaby, *Petrogale penicillata* (Marsupialia: Macropodidae): efforts to accelerate breeding in a threatened marsupial species. Aust J Zool. 2005;53:313–23.

Taggart DA, Schultz DJ, Fletcher TP, Friend JA, Smith IG, Breed WG, et al. Cross-fostering and short-term pouch young isolation in macropodoid marsupials: implications for conservation and species management. In: Coulson G, Eldridge MDB, editors. Macropods: the biology of Kangaroos, wallabies and rat-kangaroos. Collingwood: CSIRO Publishing; 2010.

Traill LW, Bradshaw CJA, Brook BW. Minimum viable population size: a meta-analysis of 30 years of published estimates. Biol Conserv. 2007;139:159–66.

Tyndale-Biscoe CH. Blastocyst transfer in the marsupial *Setonix brachyurus*. J Reprod Fertil. 1963;6:41–8.

Tyndale-Biscoe CH. Resumption of development of quiescent blastocysts transferred to primed ovariectomised recipients in the marsupial *Macropus eugenii*. J Reprod Fertil. 1970;23:25–32.

Tyndale-Biscoe CH. Life of marsupials. Collingwood: CSIRO Publishing; 2005.

Tyndale-Biscoe CH, Renfree MB. Reproductive physiology of marsupials. Melbourne: Cambridge University Press; 1987.

Tyndale-Biscoe CH, Rodger JC. Differential transport of spermatozoa into the two sides of the genital tract of a monovular marsupial, the tammar wallaby (*Macropus eugenii*). J Reprod Fertil. 1978;52:37–43.

Watson PF. Artificial breeding of non-domestic animals. Symposium of the Zoological Society of London, vol. 43. London: Academic; 1978.

West M, Lachlam-Kaplan O, Cleary M, Galloway D, Shaw J, Trounson AO. *In vitro* maturation and cytoplasmic sperm injection of oocytes collected from hormonally stimulated common wombats, *Vombatus ursinus*. Anim Reprod Sci. 2007;98:311–21.

Wildt DE, Ellis S, Janssen D, Buff J. Toward more effective reproductive science for conservation. In: Holt WV, Pickard AR, Rodger JC, Wildt DE, editors. Reproductive science and integrated conservation. Cambridge: Cambridge University Press; 2003.

Witt RR, Rodger JC. Recent advances in tools and technologies for monitoring and controlling ovarian activity in marsupials. Theriogenology. 2018;109:58–69.

Witt RR, Forbes IR, McBain J, Rodger JC. Ovarian suppression in a marsupial following single treatment with a gonadotrophin-releasing hormone agonist in microspheres. Reprod Fertil Dev. 2016;28:1964–73.

Witt RR, Rodger JJ, Rodger JC. Breeding in the fat tailed dunnart following ovarian suppression with the gonadotrophin-releasing hormone agonist Lucrin® Depot. Reprod Fertil Dev. 2018a;30:507–18.

Witt RR, Hinds LA, Rodger JC. Delayed return to oestrus following treatment with the gonadotrophin-releasing hormone agonist, Lucrin® Depot, in the tammar wallaby. Theriogenology. 2018b;115:108–16.

Zoos Victoria. 2014–19 wildlife conservation master plan. 2014. https://www.zoo.org.au/sites/default/files/ZVConsPlan14-19_FA_1.pdf. Accessed 12 Dec 2018.

Chapter 12
Using the Koala (*Phascolarctos cinereus*) as a Case Study to Illustrate the Development of Artificial Breeding Technology in Marsupials: an Update

Stephen D. Johnston and William V. Holt

Abstract The successful development and application of an assisted breeding program in any animal relies primarily on a thorough understanding of the fundamental reproductive biology (anatomy, physiology and behaviour) of the species in question. Surely, the ultimate goal and greatest hallmark of such a program is the efficacious establishment of a series of reliable techniques that facilitate the reproductive and genetic management of fragmented populations, both in captivity and in the wild. Such an achievement is all that more challenging when knowledge of the reproductive biology of that species is essentially rudimentary and without adequate models to compare to. Using the koala (*Phascolarctos cinereus*) as a case study, this chapter provides insights into the development of a concept that began as small undergraduate student project but that subsequently evolved into the first-ever successful artificial insemination of a marsupial. Apart from this historical perspective, we shall also provide a brief review of the current reproductive biology of the koala, discuss technical elements of current assisted breeding technology of this species, its application to the closely related wombat, and the potential role it might play in helping to conserve wild koala populations in the form of a live koala genome bank. There is little doubt that the unique reproductive biology and tractability of the koala has been a benefit rather than a hindrance to the success of artificial breeding in this species.

S. D. Johnston (✉)
Wildlife Biology Unit, School of Agriculture and Food Sciences, The University of Queensland, Gatton, Australia
e-mail: s.johnston1@uq.edu.au

W. V. Holt
Academic Unit of Reproductive and Developmental Medicine, The University of Sheffield, Sheffield, UK

P. Comizzoli et al. (eds.), *Reproductive Sciences in Animal Conservation*, Advances in Experimental Medicine and Biology 1200,
https://doi.org/10.1007/978-3-030-23633-5_12

Keywords Koala · *Phascolarctos cinereus* · Assisted breeding technology · Reproductive biology · Semen collection · Semen cryopreservation · Oestrous detection · Induction of ovulation · Artificial insemination

1 Introduction

Reproductive technology offers significant advantages for the genetic management and propagation of captive wildlife but only in a small proportion of species is it currently making a significant contribution to conservation (Holt 2008). The application of reproductive technology to marsupial species is even less well developed and has the added complication of being restricted by a lack of fundamental knowledge with respect to their unusual mode of reproduction (Johnston and Holt 2001). Only in the koala is the use of artificial insemination sufficiently advanced, that it is being used regularly for the production of pouch young (Johnston et al. 2003; Rodger et al. 2009). Nonetheless, and despite significant advances in the establishment of koala reproductive technologies, there are still some significant technical problems that need to be solved; this is especially true regarding the use of semen cryopreservation and the general implementation of the technology into koala conservation. The koala represents an excellent case study of what it ultimately takes to develop an assisted breeding program in a species where only limited reproductive information is available. This chapter provides an opportunity to document the development of reproductive technology in a novel wildlife species over a 25-year period; a project that began in 1992 as a modest undergraduate investigation but which has expanded to the most successful artificial insemination program in any marsupial species.

2 Conservation Status of the Koala

Woinarski and Burbidge (2016) have summarised the numerous threats to the survival of the koala in Eastern Australia; these range from temperature extremes [heat waves], increase fire frequency/intensity, disease and dog attacks as medium impact threats and habitat fragmentation [urban development, forestry], drought and car strike as low impact threats. Although the latest (2014) IUCN assessment listed koala conservation status as vulnerable (Woinarski and Burbidge 2016), koala conservation is a vexed issue because it depends largely upon on which Australian populations are being considered and the degree of human intervention involved in their management. Nevertheless, Australian state-based estimates of koala population densities (Threatened Species Scientific Committee 2012; McAlpine et al. 2012, 2015) agree on an overall decline of koala numbers over three generations of nearly 30%; a koala generation being 5–8 years. Expert knowledge was used to estimate Queensland koala population trends for past three generations and the next three generations by Adams-Hosking et al. (2016), in which they report an average decline of 53%. The plight of wild koalas in South-east Queensland is particularly dire, as a modelling study of koala populations, incorporating data collected from 1996 to 2014 confirmed strong evidence for a rapid decline of population densities in the order of 80% (Rhodes et al. 2015). In New South Wales (NSW), trends in koala populations vary across the state (NSW Chief Scientist and Engineer 2016) but overall there has been a 26% average decline estimated over the past three generations and next three generations (Adams-Hosking et al. 2016); while populations on the NSW coast have been recorded as stable, declining gradually or rapidly, other populations west of the Great Dividing Range have actually expanded (Adam-Hosking 2017). Although koalas in Victoria occur over much of the southern and eastern lowlands, and while population densities are artificially high in some areas (Raymond Island, Snake Island, French Island, parts of the Otway Ranges, Framlingham, Mount Eccles National Park, Tower Hill Game Reserve) (Natural Resource Management Ministerial Council 2009), McAlpine et al. (2012) have still predicted a 14.3% overall decline in Victorian populations over the next three generations. The densities of koalas are lower in the dry forests and woodlands in northern parts of Victoria where the habitat is of lower quality and much of the habitat remaining in the state is fragmented so that many populations are isolated (Natural Resource Management Ministerial Council 2009). Koalas in South Australia can now be found in a greater range and abundance than at the time of European settlement, because of introductions both within its natural range and areas where it did not occur naturally (Woinarski and Burbidge 2016; Natural Resource Management Ministerial Council 2009); nevertheless, deliberate management of over-abundant koala populations on Kangaroo Island has resulted in population density estimates declining in the order of 39% (Threatened Species Scientific Committee 2012) and 2.9% (McAlpine et al. 2012).

3 Why Develop Assisted Reproductive Technology in the Koala?

The management and conservation benefits of assisted reproductive technology (ART) to wildlife species have been promoted heavily (e.g. Pukazhenthi et al. 2006) and in theory at least, are worthy of pursuit, but are in reality, overstated (Monfort 2014). Poor correlation between the potential of ART and its practical application to species conservation is no better illustrated than its role in the management of marsupial populations (Johnston et al. 2003; Rodger et al. 2009), whereby a lack of knowledge of the somewhat unique reproductive biology of this taxon and an associated lack of funding of its investigation has limited studies of its application to only a few species. Consequently, many questions still remain as to the specific uses of such tools in marsupials and how they may be implemented both in captive and wild populations. As the application of ART requires a detailed understanding of reproductive biology, it is likely that a significant, but probably most overlooked contribution that ART can make to conservation, is the acquisition of novel fundamental biology (Holt 2008; Johnston and Holt 2001; Wildt et al. 2003); this includes basic information such as reproductive seasonality, oestrous cycle characterisation, mating behaviour and mate choice—all of which impact on behavioural ecology (Johnston et al. 2013). Given the current and future crisis faced by the koala, and those that manage these populations, the development of reliable ART for small population genetic management and the associated reproductive knowledge required to establish these techniques will become increasingly important for koala conservation in South-east Queensland (Rhodes et al. 2017).

A major concern for captive koala management is the maintenance of appropriate genetic diversity within the closed population; consequently, techniques such as artificial insemination [AI] are likely to play an important role. Semen cannot only be transported across borders and physical boundaries between zoos, but also, with the use of cryopreservation technology, it can be stored through time, prolonging the generation interval of genetically important sires. Discussions on the use of this technology for the genetic management of koalas is still poorly developed; we have recently advocated the use of what we refer to as "live" and "frozen" genome banks for the koala populations in South-east Queensland most likely to need urgent human intervention in order to survive (Johnston et al. 2013, 2018); these specific concepts will be discussed later in the chapter. There are also significant animal welfare benefits to the koala associated with the use of ART; while some of the procedures such electro-ejaculation and artificial insemination are mildly invasive, the transportation of semen, rather than whole animals between institutions [national and international] is likely to be substantially less stressful in terms of animal welfare. Associated with the transport of semen is the lower comparative cost of this approach, and the fact that many semen samples can be sent and stored in the one shipment container. Thus far, we have only used chilled koala semen transported between zoological institutions on the Australian Gold Coast to successfully produce a joey by artificial insemination (Allen et al. 2008a).

ART techniques are also very useful for identifying and overcoming reproductive problems or specific anatomical, physiological and behavioural blocks to reproduction (Johnston et al. 1999). Any knowledge that contributes to the fundamental reproductive biology of the koala, adds to a better understanding of what is "normal" and what is "abnormal". This has always been a major issue for those of us working in the area of wildlife or exotic species reproduction, with this type of database information difficult to find or accumulate, especially when compared to domestic animals or humans. The assessment of what represents a normal semen sample in the koala is a classic example of this phenomenon; normative seminal characteristics in the koala, boar, bull, stallion or human are likely to be completely different in those animals with very divergent reproductive strategies and selective pressure. For example, to date, there have been no studies on marsupials that have established a relationship between semen quality and fertility.

Artificial insemination is an example of an ART that could be used to overcome mating problems in koalas with physical injuries or behavioural incompatibilities. Pathological investigations (Dief 2011; Palmieri et al. 2018) have shown that male koalas positive for chlamydial related prostatitis or urethritis can still possess epididymidal tissue free of the organism—these animals would be ideal for gamete recovery and AI. Similarly, non-infected koalas that are euthanised by koala hospitals following dog attacks or car accidents are suitable candidates for gamete recovery. ART thus has the potential to recover important alleles from individuals that would otherwise not make a contribution.

Another important reason for developing ART in the koala was because at the time of commencing the initial project, there were still significant numbers of animals in the wild and in captivity that could be utilised to develop the technology. Too often the development of ART is implemented as an afterthought, or at a stage in conservation management where there are insufficient animals to conduct the fundamental science. Instead, a proactive approach to the development of ART is required—in the koala, we have largely been able to achieve this. Now that AI is a relatively routine procedure in the koala, it will be necessary to see how the technology can be further refined and improved so that it has an important role in the conservation of wild populations.

4 The Koala as a Research Animal

Irrespective of its iconic status, the successful and rapid development of assisted breeding in the koala has been very much dependent on the unique biology of the species. Although perhaps regarded as a difficult animal to maintain in captivity, in our hands, we have found the koala to be an ideal research animal for ART. Its relatively passive nature, size, tractability and adaptability to captivity, mean that handling the animal for blood collection and reproductive procedures can occur without need for anaesthesia or sedation. While the koala is no doubt a "fussy" and expensive animal to feed, their ability to habituate to human presence while in captivity has allowed us to develop and conduct procedures that would not normally be possible in other marsupials—for example, semen collection by means of an artificial vagina (Johnston et al. 1997) and AI (Johnston et al. 2003; Allen et al. 2008a) can both be accomplished successfully without the need for anaesthesia. In addition, and probably most importantly, the koala possesses an extremely overt and prolonged [Average—10 days] behavioural oestrus (Johnston et al. 2000a; Feige et al. 2007) and an induced ovulatory response (Johnston et al. 2004), so that this makes predicting the timing of insemination a relatively straightforward process.

5 How to Make a Koala Pouch Young

There are three fundamental components to the development of any AI program (Johnston and Holt 2001). These include (a) an understanding of male reproductive biology in order to collect viable semen for evaluation, preservation and insemination, (b) characterisation of the oestrous cycle, with particular focus on expression of oestrus and the type and timing of ovulation and (c) identification of the most appropriate site and mechanism of semen deposition into the female's reproductive tract. Each of these components are of course based on building a framework of koala reproductive biology composed of structural and functional anatomy, endocrinology and gamete physiology and natural and captive reproductive behavioural observations, all of which are continually refined over many years. The following summary of koala reproductive biology will be used as the basis on which to discuss the development of ART in this species and will primarily focus on captive animals.

6 The Reproductive Biology of the Female Koala

The female reproductive tract of the koala (Fig. 12.1) is, like that of all marsupials, small in size and possessing a urogenital sinus that opens into a cloaca alongside a vascular clitoris and the rectum. Johnston (1999) also noted the presence of vestibular or Bartholin's glands opening into the urogenital sinus. The urogenital sinus gives rise to the urethra ventrally and dorsally into two small poorly defined ostia that open into the muscular lateral vaginae. The marsupial possesses two vaginae because the Mullerian ducts are prevented from fusing on the mid-line due to the medial migration of the ureters during organogenesis (Tyndale-Biscoe 2005). The lateral vaginae open into a vaginal cul-de-sac that receives the simple but muscular cervices. The vaginal cul-de-sac in the koala is partitioned by a medial septum that completely separates the left and right sides, which therefore means that for semen to enter both uteri it must travel up through both lateral vaginae separately; this has obvious implications for the koala AI procedure. The duplex uteri each open into a

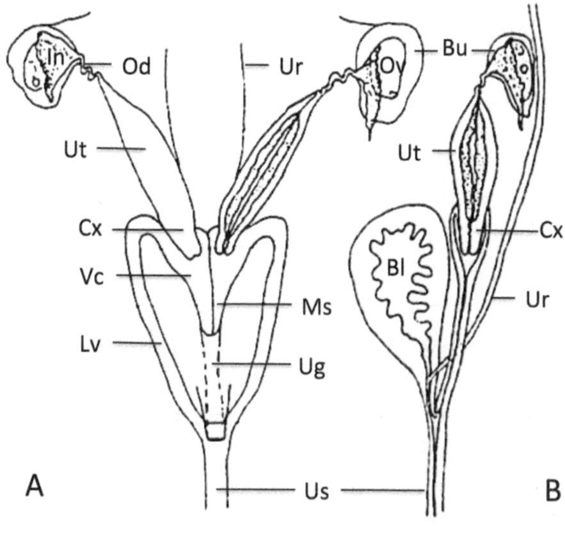

Fig. 12.1 Schematic drawing of female koala reproductive tract; (**a**) Ventral aspect and (**b**) Lateral aspect. *Bl* bladder, *Bu* bursa, *Cx* cervix, *Fi* fimbria, *In* infundibulum, *Lv* lateral vagina, *Ms* medial septum, *Od* oviduct, *Ug* urogenital strand, *Ur* ureter, *Us* urogenital sinus, *Vc* vaginal cul-de-sac (Modified from Obendorf 1988)

convoluted oviduct that terminates in the infundibulum. Depending on the stage of the oestrous cycle the ovary is ovoid and typically 10–12 mm long, 7–9 mm wide and 3–5 mm deep. The morphology of folliculogenesis in the koala ovary is unremarkable apart from the large size [up to 7 mm] of the mature Graafian follicles, which represent some of the largest recorded follicles in any marsupial; the active corpora luteum are of similar size or larger (Johnston 1999). Of particular interest is the thickness of the koala zona pellucida [14.2 μm], which is over twice that recorded for any marsupial, being more similar in thickness to that of the zonae pellucidae of eutherian species (Johnston 1999; Chapman et al. 2006). Johnston (1999) has commented that a lack of coitus during oestrus in the koala presumably leads to an "over-ripening" of the pre-ovulatory follicle, subsequent atresia and formation of haemorrhagic follicles, similar to what has been observed in the rabbit ovary and described large numbers of haemorrhagic and/or atretic follicles in varying stages of formation in the koala ovary at all phases of the reproductive cycle. Interestingly, part of the remaining theca interna and granulosal tissue also undergoes hyperplasia, with cells of the theca interna ultimately being transformed into what appear to be ovarian "interstitial-like" tissue; the functional significance of this tissue and its endocrinology is presently unknown. The koala ovary is surrounded by a translucent ovarian bursa; the oviduct and bursa are often sites of inflammation associated with *Chlamydia* infection, leading to infertility possibly via fluid pressure atrophy on the ovary and/or stenosis of the oviduct (Blanshard and Bodley 2008). Although no direct observations have been documented, histological observations (Johnston 1999) have suggested that birth of the koala foetus most likely occurs through the terminal portion of the vaginal cul-de-sac, into the pre-softened oedemic tissue of the urogenital strand, with the neonate exiting to the exterior through the urogenital sinus and cloaca. The pouch young then climbs unaided into the pouch where it firmly attaches to the teat, suckles and continues to develop. Presumably, the reproductive tract heals over after each parturition event.

The reproductive cycle of the koala is reasonably well understood (Johnston et al. 2000a, b, 2004); endocrinology of the pregnant and non-pregnant koala cycle is illustrated in Fig. 12.2. While ovulation appears to be induced by coitus, the exact mechanism of induction is still unresolved as there is some evidence that semen may possess some type of ovulation-inducing biochemical, similar to what has been described in camelids (Silva et al. 2010). Like the felidae, there are three possible types of koala oestrous cycle. The first involves the follicular phase without induction of ovulation and a luteal phase and is referred to as inter-oestrus (Fig. 12.2a; Johnston et al. 2000a). The female displays oestrus on average for 10 days and if mating does not occur, oestradiol secretion subsides, and the female takes approximately 33 days to come back into oestrus again if measured from the first day of her previous oestrus. If the female mates with a male, an LH surge occurs approximately 28 h after mating (Johnston et al. 2004) and plasma progesterone levels commence to rise and peak about 28 days after mating (Fig. 12.2b). If pregnancy is not successful, the female will come back into oestrus approximately 50 days from the previous mating and ovulation (Fig. 12.2b). If a pregnancy does result, gestation will occur over a period of 35 days (Fig. 12.2b) and, following parturition, the

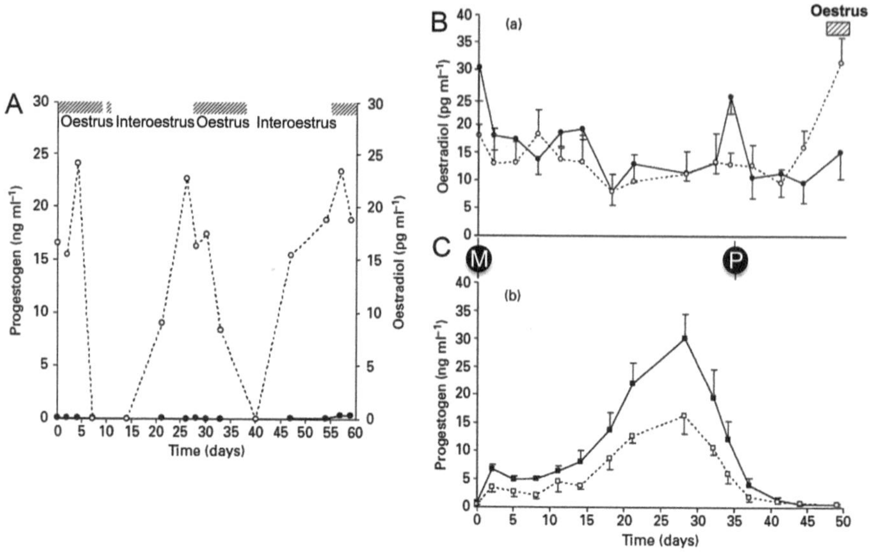

Fig. 12.2 Reproductive endocrinology of the koala. (**a**) Oestradiol 17β (open circles) and progestogen (closed circles) secretion during an anovulatory oestrous cycle; (**b**) Oestradiol secretion post-mating during pregnancy (closed circle) and a non-sterile oestrous cycle (including luteal phase; open circle); (**c**) Progestogen secretion post-mating during pregnancy and a non-sterile oestrous cycle (including luteal phase). *Cross-hatching* behavioural oestrus, *M* mating, *P* parturition (From Johnston et al. 2000a)

female will fail to return to oestrus, probably because the suckling stimulus of the pouch young prevents further ovarian activity. This detailed understanding of the physiology of the koala oestrous cycle and ovulatory pattern has been fundamental to the success of the AI program, in terms of the most appropriate timing for insemination, but also in developing procedures to induce ovulation. Recent molecular analysis of koala reproductive protein hormones (GnRH, FSH and LH) and their receptors (Busby et al. 2014) and publication of the koala genome (Johnson et al. 2018) will likely lead to further characterisation of koala reproductive physiology, especially for the purposes of developing specific antisera so that these protein hormones can be monitored throughout the cycle, pregnancy and early lactation; this information will also help us to design or select the most appropriate GnRH analogues for work on oestrous synchronisation. An ability to monitor reproductive status in the koala using faecal hormone metabolites also allows non-invasive assessment in both wild and captive populations. To date these procedures have been used to monitor both progesterone and testosterone metabolites (Kusuda et al. 2009, 2013) but a reliable faecal oestrogen metabolite assay has yet to be validated.

Oestrus in the captive koala is one of the most visually and auditory overt behaviours reported for any marsupial (Feige et al. 2007; Johnston et al. 2000c). In a procedure known as "teasing", an adult male is brought into an all-female enclosure. The male is initially presented to each female by the zookeeper by holding the male at eye-level with each female so that all females are aware of the male's presence. Those females that are not in oestrus will typically reject the male with an aggressive vocalisation and or by physically striking out at the male with their forelimb. Females in oestrus will show interest in the male and some will respond by immediately initiating oestrus-related behaviours. The male is then placed on the enclosure floor where he undergoes prescriptive pre-copulatory behaviours including urination, scent marking a tree pole within the enclosure with his sternal gland, and most characteristically, vocalising with deep guttural bellows (Charlton et al. 2013; Ellis et al. 2015). The bellow of the male is typically the most significant trigger that initiates the expression of oestrus in the female and on hearing the bellow the oestrus female's excitement level rapidly increases. Interestingly, the female will not necessarily approach the bellowing male but more typically, locates other females in the enclosure with whom she attempts to mount and copulate. The oestrous female will normally take on the male role during pseudo-copulation, establishing a neck bite to the nap of the neck to stabilise the female that is being mounted and orientating herself as if she were the male performing the copulatory act (Feige et al. 2007) (Fig. 12.3). The mounting female will display the full range of male copulatory behaviours, including pelvic thrusting, a period of stillness, which is similar to the period when the male would be ejaculating, and finally, a neck bite to the shoulder as a signal of disengagement; some females will even bellow at the end of the pseudo-mating attempt, similar to what occurs in the male post-copulation. Remarkably, the female that is being mounted will take on the normal female role

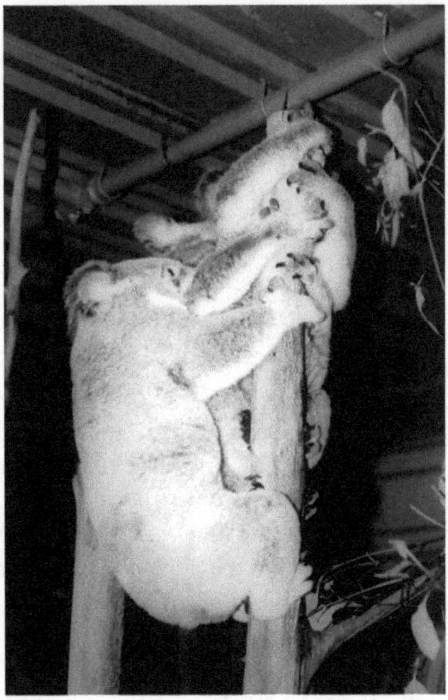

Fig. 12.3 Pseudomale copulatory behavior in three female koalas in oestrus. Note the homosexual behaviour of the middle and lower females (From Johnston 1999)

during this process and demonstrate lordosis [arching of the spine and backward tilt of the head], a period of stillness while the female is mimicking male thrusting behaviour and then a period where she shows convulsive jerking of her body associated what is presumably a mimic of ejaculatory behaviour (Feige et al. 2007). Typically, the oestrous female seeks out other females that are either in oestrus or coming into or out of oestrus, as females that have been mounted will then proceed to serve the female who has just finished mounting. Given that oestrus in the koala has an average duration of 10 days [range 0–19 days], within a 33-day (range 25–48) cycle (Johnston et al. 2000a), it is not surprising that there will be more than one or two oestrous females in the enclosure at the same time. Other behaviours that are associated with koala oestrus include, bellowing behaviour by the female, convulsive jerking of the body in what has been likened to uncontrollable "hiccoughing", urination and increased agitation or restlessness (Johnston et al. 2000a). Koalas are normally very sedentary in captivity but there is a notable change in movements and activity during oestrus. As in a range of domestic species, koalas also appear to demonstrate a standing oestrus in which the female will stand still [receptivity] while the male ensues a copulatory position. In some cases, the female may even back down "rump first" into the face of the male on the same pole in what appears to be a form of "cloacal presentation". All of these behaviours are unmistakable indicators of oestrus for the zookeeper, who along with help from the teaser male koala, can determine the daily reproductive status of the female from behavioural observations. Some zookeepers have even learnt to mimic the bellow of the male koala or play recordings of the male bellow to the female and this has been used for the effective detection of oestrus. All these oestrous behaviours have been linked to an elevated secretion of oestradiol (Johnston et al. 2000a). The significance of koala homosexual behaviour is difficult to comprehend but is probably an artefact of captivity, as the behaviour has never been observed in the wild. Feige et al. (2007) have suggested that it may be associated with sexual excitement linked to high levels of oestradiol in the systemic circulation and that this may be the physio-behavioural mechanism that causes wild koalas to seek out males outside of their normal home range (Martin and Handasyde 1999).

7 The Reproductive Biology of the Male Koala

The reproductive anatomy of the male koala has been well described by Temple-Smith and Taggart (1990) so that only a brief overview will be presented here. As for most marsupials, the male koala has a pre-penile scrotum; the scrotal; skin being non-pigmented and covered with short hairs. The importance of the scrotum for thermoregulation of the testes has not yet been investigated. The flaccid penis is maintained within the prepuce in an s-shaped configuration. The glans penis is covered proximally with prominent keratinised spines but the precise role of these spines has yet to be clearly determined—we have speculated that they are likely to have role in stimulating reflex ovulation along with ovulation factors in the semen (Johnston et al. 2004). In the centre of the chest of the male is a sternal gland which produces an oily sebaceous and odorous secretion; the scent gland appears to be androgen dependent and its activity increases leading up to and during the breeding season (Allen et al. 2010).

The testis of the koala is supplied by a rete mirabile of over 100 blood vessels and 5 large lymphatics. The testis is ellipsoidal and small in comparison to that of other marsupials, weighing only approximately 0.07% of total body weight (Johnston 1994); the significance of small testicular size on the reproductive strategy and behavioural ecology of the koala requires further investigation. Oishi et al. (2013) have described the quantitative testicular histology and the dynamics of the seminiferous cycle in the koala and wombat in which they report both species possessing eight stages of cellular associations. Interestingly, the koala (≈33%) differs from the Southern Hairy-nosed Wombat (*Lasiorhinus latifrons*) in typically having a greater proportion of interstitial testicular parenchyma. Temple-Smith and Taggart (1990) refer to the koala as possessing a Type 3 pattern of organisation with large tracts of Leydig cells completely isolating adjacent seminiferous tubules. The Sertoli cells of the koala contain nuclei that are 5× larger than that those in eutherian species and which possess unusual crystalloid inclusions, the precise function of which is still unknown. Testicular volume of koalas in SE Queensland changes throughout the year in both wild and captive populations with an increase over spring and summer and a decrease in autumn and winter (Allen et al. 2010).

Spermiogenesis in the koala has a number of unusual features including; the formation of a proacrosomal granule within the acrosomal vacuole, an uneven condensation of chromatin and a unique flattening of the sperm nucleus which results in the wide variation observed in sperm head morphology (Temple-Smith and Taggart 1990; Johnston et al. 1994; Wildt et al. 1991). The acrosome is located in a dorsal nuclear cleft and there is an unusual ventral neck insertion of the mid-piece into the sperm head (Harding and Aplin 1990; Harding et al. 1979, 1987). Harding and Aplin (1990) have used these unusual features of the spermatozoon to review the phylogenetic position of the koala confirming a close relationship with the wombat.

The gross anatomy of the koala epididymis and associated vasculature is similar to that described for other diprotodontid marsupials (Temple-Smith and Taggart 1990). The *caput epididymidis* is expanded in shape and connected via a *corpus* segment to a bulbous *cauda epididymidis*. The epithelium of koala epididymis consists of four basic cell types; principal, basal, mitochondrial rich and electron-translucent, but they have no specialised regions of phagocytic principal cells. As for most mammals, koala spermatozoa appear to gain the capacity to become motile as they move into the *cauda epididymidis*. Fluid absorption in the cauda epididymidis results in a higher sperm concentration than that found in the caput epididymidis (Temple-Smith and Taggart 1990). While there is a marked increase in the curvature of the sperm head and a folding of the acrosomal surface and condensation of the accessory cytoplasmic droplet to within the hook of the sperm head (Hughes 1977), there appear to be no other morphological changes to the koala spermatozoa during epididymidal transit; this is in sharp contrast to changes in sperm morphology that occur during epididymidal transit in other marsupials (Harding et al. 1983; Rodger and Mate 1993).

The ductus deferentia run from the neck of the scrotum through the inguinal canal and into the abdominal cavity, emptying finally into the cranial portion of the prostatic urethra. The ductus deferens is non-specialised, non-glandular, with no seminal vesicles or ampullae [sperm storage organs in eutherian mammals]. The koala prostate is pyriform and can be divided into three histologically distinct regions (Rodger and Hughes 1973), the functional significance of which have yet to determined. A short membranous urethra connects the prostate to the penile urethra. At the distal extremity of the membranous urethra and in close association with the penile crura and bulbs are three pairs of bulbourethral glands. Glands BII and BIII are filled with a thick, clear, mucin-like secretion that is strongly eosinophilic (Temple-Smith and Taggart 1990). Although Mitchell (1990) noted the presence of paracloacal glands in koalas, he gave no anatomical details or references. It is likely that the prostate provides the bulk of seminal plasma in koala semen, while the bulbo-urethral gland secretions have an important role in producing a seminal plug. In SE Queensland, a study of post-mortem specimens entering koala hospitals revealed no seasonal change in the size of the prostate but an increase in bulbo-urethral gland volume over spring, a decrease over summer and autumn and an increase towards the end of winter (Allen et al. 2010).

Like most mammals, plasma testosterone secretion in the koala is highly episodic (Johnston 1999; McFarlane 1990). The highest concentration of testosterone appears to occur in the koala during periods of male dispersal just prior to the breeding season, and not, as may have been expected, during the breeding season (Allen et al. 2010; McFarlane 1990; Handasyde et al. 1990). A rise in testosterone concentration is coincident with the onset of bellowing, prior to the commencement of breeding activity (Handasyde et al. 1990). In order to obtain a more reliable measure of the testosterone secretory capacity of the koala testis, a GnRH or hCG stimulation test can be employed (Allen et al. 2006). This approach was used to demonstrate a seasonal change in testosterone secretion of both captive and wild koalas in southeast Queensland with a peak in spring and nadir in autumn (Allen et al. 2010). Kusuda et al. (2010) have also used measurements of faecal androgens metabolite to assess age-related and seasonal changes in male reproductive endocrinology.

8 Semen Collection

Our first major step towards the development of a successful AI program in the koala was the ability to reliably collect semen in sufficient volumes to be used for insemination. Semen collection in the koala using electro-ejaculation was first described by Wildt et al. (1991) and later by Johnston (1994, 1999); the procedure has been extremely effective with greater than 90% of semen collection attempts resulting in spermatozoa. The procedure is field applicable (Fig. 12.4a) and a study

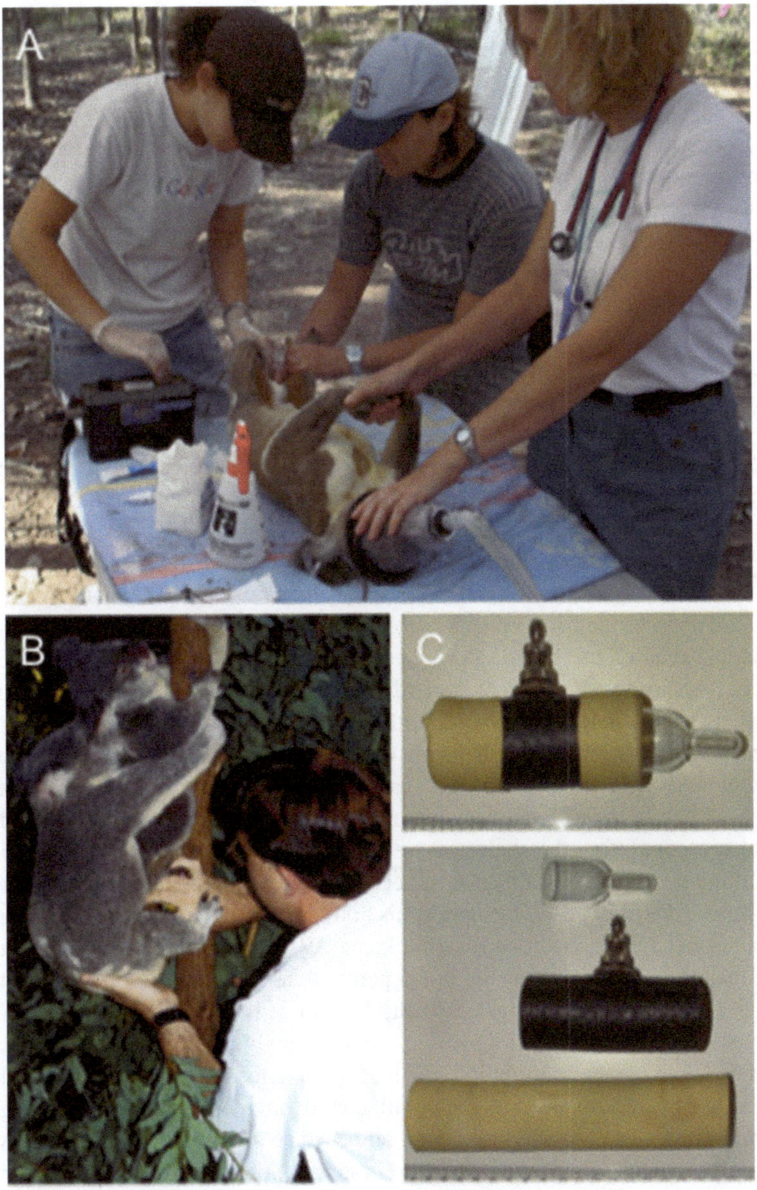

Fig. 12.4 Semen collection in the Koala. (**a**) Electro-ejaculation can be used in the field; (**b** and **c**) Use of a modified sheep artificial vagina for captive koalas (From Johnston et al. 1997; Johnston 1999)

by Allen et al. (2010) showed that it was possible to collect semen repeatedly from both captive [monthly] and wild koalas [every 6 weeks] from the same individuals. However, there are also disadvantages with respect to the use of the procedure that need to be acknowledged, including the requirement for the koala to be anaesthetised [which can be problematic], production of semen with variable compositions of seminal plasma, a lower sperm concentration, urine contamination and in very rare cases, rectal trauma through improper placement of or overstimulation from the rectal probe. We have used the electro-ejaculation procedure to determine semen quality and to provide semen for studies of sperm physiology, preservation and artificial insemination.

A major breakthrough in determining the appropriate parameters of koala semen for AI was the ability to collect semen using an artificial vagina (AV) (Johnston et al. 1997) (Fig. 12.4b, c). Before attempting the procedure, we carefully observed the natural mating behaviour of the koala to inform us of how we might engage the animal to serve the artificial vagina. A major problem to overcome was the fact that the male koala thrusts his penis into the female's urogenital sinus in a vertical direction, and that semen was presumably ejaculated in a similar trajectory against "gravity". Contributing to the success of the technique was the tractability of both male and female koalas to tolerate human presence during mating activity. In fact, it was not uncommon for zookeepers to assist the male to direct his penis into the urogenital sinus of the female during natural mating. Prior to attempting semen collection with the AV, zookeepers had reported to us that, during movement of koalas from one enclosure to another, the males held against the abdomen of the zookeeper would often ejaculate during transit. Captive koalas are handled from an early age and there is no doubt this early exposure to the zookeeper resulted in animals with a high tolerance to intervention and manipulation. We had observed that during copulation the male was very focussed on securing the female in an appropriate position by establishing a neck bite and using his forearms, but that his hind limbs were rarely used to support his hold. Based on our observations of natural mating behaviour, we constructed a koala AV from a ram artificial vagina that was adjusted to the length of the koala penis. After preparing the AV with warm water [42–45 °C] and inflating it with air, the female is placed on a single tree-pole so that she is secure for mating, and at approximately eye height of the collector. The male koala is then brought into the female's enclosure, placed on the floor and allowed to conduct his normal pre-copulatory behaviour, including sternal gland scent marking, urinating and bellowing, all of which stimulates oestrous behaviour in the female. The male is then placed on the tree pole directly below an oestrous female and allowed to establish his mating position. The male typically manipulates the female into position while at the same time establishes penile erection. Once the penis is fully erect, the collector would direct the male's penis into the warm AV; the male in most cases appeared to be unaware that he was serving the AV rather than the female, and the female continued to display her normal copulatory behaviour. The male then thrusts vigorously into the AV and completes this behaviour with two strong final ejaculatory thrusts, which appear to signal the commencement of ejaculation. At this point, the collector lifts the rump of the male slightly horizontally so as to direct the flow of semen into the collection vial; this manipulation was possible because the hind limb of the male is not typically used for support during copulation. Once ejaculation is completed the male disengages the female by biting her shoulder or arm, and which results in an aggressive response by the female and separation. During disengagement, the collector removes the AV from the male's penis. Semen collected by the AV provides a means of determining a more natural estimate of seminal characteristics, including semen volume [≈1 mL], pH and sperm concentration; all these seminal parameters are crucial for developing appropriate methods of artificial insemination. While this method of semen collection [58% successful] is less reliable than electro-ejaculation, it does offer the additional advantage of observing structural physical defects associated with breeding soundness and the assessment of male sexual drive [libido]. To date, the koala remains the only marsupial for which the AV has been used successfully. A video of the koala AV collection procedure can be found in the documentary series "Born in the Wild" (Episode 2—http://www.windfall-films.com/show/2942/episode-2-kangaroo.aspx).

It is also possible to recover spermatozoa from post-mortem specimens collected from koala hospitals and veterinary clinics and use these gametes in assisted breeding programs (Johnston and Holt 2001; Johnston et al. 2013, 2018). Each year in SE Queensland, hundreds of koalas are euthanised because of disease or trauma; this is an appalling waste of genetic material that could potentially be utilised for genetic exchange programs. While this procedure has been described and proven in the Common Wombat (MacCallum and Johnston 2005) and Southern Hairy-nosed Wombat (Johnston unpublished observations) there are currently no published studies of gamete recovery in the koala. Although the yield of recovered epididymidal spermatozoa is likely to be less than that of the wombat, we nevertheless have preliminary data indicating that this approach is feasible (Schultz and Johnston, unpublished). It has also been noted that while ascending chlamydial infection can cause orchitis and epididymitis in the koala, some animals have prostatic infection only (Dief 2011; Palmieri et al. 2018); spermatozoa collected from these animals can be used for AI or for cryopreservation. We are also currently exploring the use of semen "clean-up" technology to remove or destroy viable chlamydial elementary bodies (EB) (Johnston et al. 2018).

9 Semen Evaluation

The koala ejaculate collected by electro-ejaculation offers some challenges for seminal analysis as the sample is typically viscous and contains a high proportion of prostatic and bulbo-urethral secretions; on rare occasions the semen may even coagulate preventing further manipulation. In order to assess semen quality, the raw semen is diluted in a Tris-citrate glucose or fructose diluent at room temperature. The raw semen sample has little mass activity, the pH is typically neutral and osmolality is in the order of 300–350 $mOsmKg^{-1}$. Sperm concentration of koala semen collected by AV is typically higher [165 × 10^6/mL] than that collected by electro-ejaculation (83 × 10^6/mL).

The assessment of sperm motility is conducted on a diluted sample on a warm stage set to 35 °C [approximate body temperature of the koala] and the percentage of sperm swimming in a progressively forward manner is determined subjectively along with an assessment of the rate of sperm movement [0—no movement to 5—extremely rapid movement]. The progressive motility of spermatozoa collected by electro-ejaculation or by artificial vagina is usually in the order of 70–80%. In early studies, and without the use of fluorescent microscopy, viability [integrity of the plasma membrane] was assessed using nigrosin-eosin stain (Johnston et al. 1994, 1997) but this was later substituted for the use of SYBR-14 [live] and propidium iodide [dead] stains (Zee et al. 2007) as we found nigrosin-eosin somewhat problematic when using diluents containing egg yolk. We have also used the JC-1 stain (5,5′,6,6′-tetrachloro-1,1′,3,3′-tetraethylbezimidazolyl carbocyanine iodide) for the assessment of mitochondrial membrane potential, to provide an additional insight into sperm function (Johnston et al. 2007). The live-dead and JC-1 stains can be used in combination to provide simultaneous assessment of both parameters (Fig. 12.5a). Interestingly, sperm motility of wild and captive koalas was highest in winter, as was the post-thaw survival of cryopreserved spermatozoa from the same animals (Allen et al. 2010).

One of most striking features of the koala spermatozoon is the extent of pleiomorphy in the head morphotype. Temple-Smith and Taggart (1990) identified two extreme and four common intermediate head morphologies among both testicular and epididymal spermatozoa, including a structural abnormality of the neck-midpiece region. Wildt et al. (1991) identified 10 nuclear morphotypes, while Johnston et al. (1994) identified 11 (Fig. 12.5b); mid-piece and principal piece structural abnormalities have also been described (Johnston et al. 1994); but what specifically represents normal sperm head morphology in the koala has yet to be defined, as is the exact mechanism of fertilisation [capacitation and acrosome reaction]. Given the unusual location of the acrosome within the curvature of the sperm head and thickened zona pellucida (Johnston 1999; Chapman et al. 2006) it is still a mystery as to how the sperm nucleus physically passes into the perivitelline space.

An assay for the assessment of sperm DNA fragmentation in the koala (Fig. 12.5c) has been developed (Johnston et al. 2007; Zee et al. 2009a). This was a first for a marsupial and initially developed to explore reasons for post-thaw decondensation of koala sperm DNA. The assay has been appropriately validated with *in situ* nick translation and comet assay and is based on sperm chromatin dispersion in a microgel. Koala spermatozoa are loaded into a microgel on a microscope slide and treated with protein lysing agent to expose loops of DNA. Single stranded and double stranded DNA is then allowed to disperse in the microgel and DNA identified by fluorescence microscopy. Our data thus far suggests that DNA fragmentation occurs at a relatively low incidence (6.7%) (Johnston et al. 2012a) in captive populations of koalas and that the DNA molecule is able tolerate prolonged periods of chilled (5 °C) storage. While our analysis revealed individual koalas with high levels of DNA fragmentation (up to 15.3%) the significance of this phenomenon on fertility

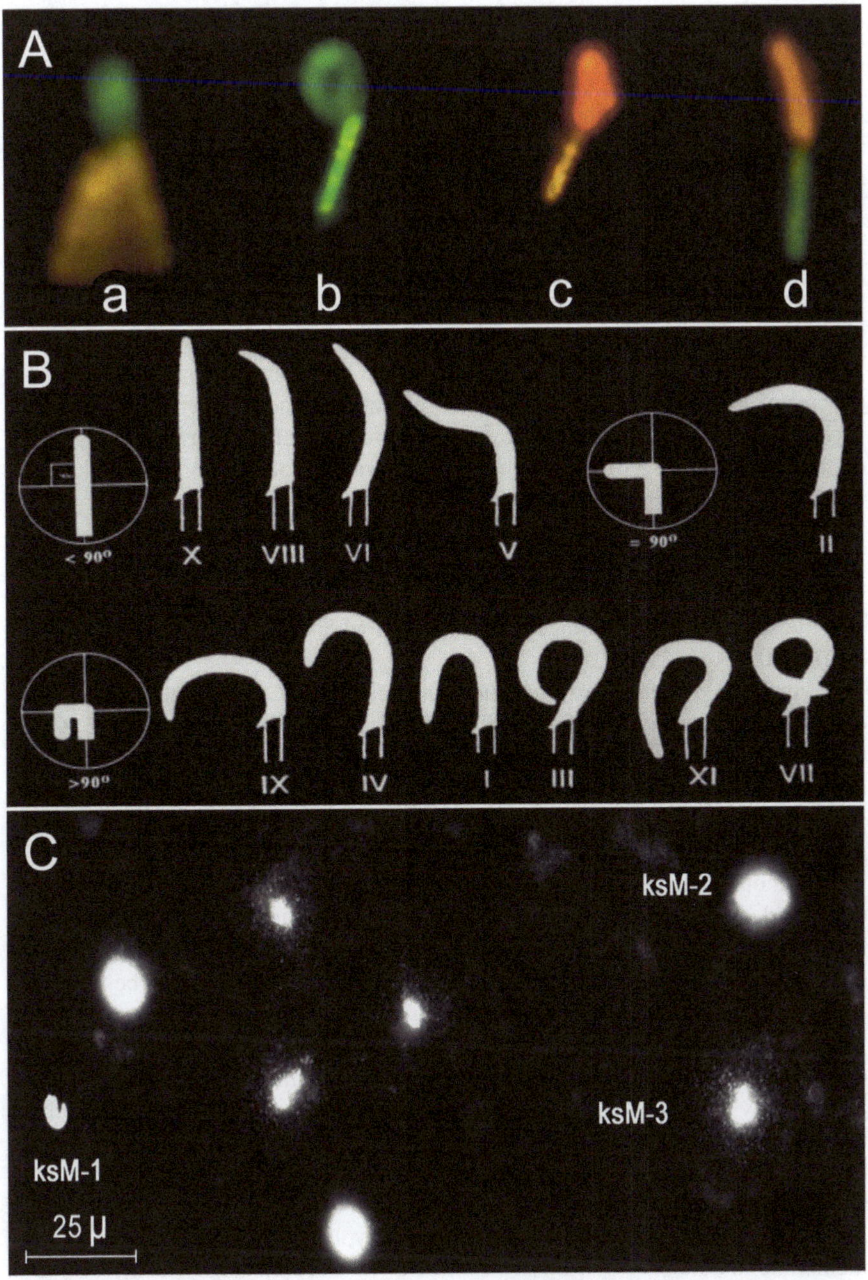

Fig. 12.5 Evaluation of Koala spermatozoa; (**a**) Assessment of live (Sybr14—green) and dead (Propidium Iodide—red) sperm cells in combination with the assessment of mitochondrial membrane potential using the JC-1 stain (orange—high MMP; green—low MMP) (From Zee et al. 2007); (**b**) Koala sperm nuclear morphotypes (I–XI); (**c**) Assessment of koala sperm DNA fragmentation—*KSM-1* rod-shaped nuclei without a halo of chromatin dispersion, *KSM-2* round and rod-shaped nuclei showing a compact halo of chromatin dispersion about a nuclear core, *KSM-3* sperm nuclei with enlarged halo and stellar chromatin corresponding to DNA fragments diffusing from the central core (From Johnston et al. 2007)

has yet to be determined. Given the relationship between chlamydial infection and high levels of DNA fragmentation in human spermatozoa (Gallegos et al. 2008), future studies are planned to investigate this relationship in the koala (Hulse and Johnston unpublished observations).

10 Semen Manipulation and Liquid Preservation

The practicality of an assisted breeding program based on artificial insemination is dependent on the ability to prepare and extend semen for insemination and the period of time that semen can be successfully preserved. Raw spermatozoa typically die rapidly *ex vivo* so that the semen must be rapidly diluted in a medium that is compatible for its evaluation and survival; this requires an analysis of the how spermatozoa tolerate physicochemical conditions such as pH, temperature and osmolality. Early in the development of our artificial insemination program we showed that koala sperm prefer a pH of 7–8, an osmolality of approximately 300 $mOsmKg^{-1}$ and can tolerate a temperature range from 15–35 °C without a loss of motility (Johnston et al. 2000c).

Contrary to evidence in most other marsupials, there is a small but significant decrease in sperm motility after rapid cooling of diluted semen from 35 to 5 °C, but compared to most eutherian species, koala spermatozoa would still be regarded as cold-shock tolerant and can be readily stored and manipulated at room temperature. In an attempt to explain the cold tolerance of marsupial spermatozoa we investigated the sperm membrane fatty acid composition in the koala and compared this to the common wombat and the eastern grey kangaroo (Miller et al. 2004). We discovered that koala sperm membranes have a high ratio of unsaturated/saturated membrane fatty acids compared to wombats, and that sterol levels in marsupial spermatozoa were generally very low. These marsupials were first to be examined for lipid membrane composition and highlight the dearth of information that exists for this taxon. We also employed a cryomicroscope to examine the effect of chilling on the sperm membrane directly (Zee et al. 2007). Using this technique, we were able to view the effect of cooling to 5 °C on the plasma membrane and mitochondrial membrane potential (MMP) simultaneously, but there was no significant effect on the proportion of spermatozoa with high MMP or intact plasma membranes. Given the tolerance of koala sperm to cope with chilled temperatures, we are somewhat ambivalent regards the effectiveness of egg yolk in the protection of the plasma membrane.

Tolerance of spermatozoa to changes in osmotic pressure can provide a guide to the sperm cell's ability to cope with osmotic flux during cryopreservation. We originally determined that koala spermatozoa had a relatively narrow osmotic tolerance and were particularly susceptible to hypo-osmotic media [<250 $mOsmKg^{-1}$]. Later we compared the osmotic tolerance of koala spermatozoa to that of wombat spermatozoa and found that koala spermatozoa were more susceptible to hyperosmotic damage than wombat spermatozoa (Johnston et al. 2006) both in terms of motility and plasma membrane integrity. We have also examined the effect of osmolality on mitochondrial membrane potential, relaxed [swollen or decondensed] chromatin and DNA fragmentation (Johnston et al. 2012a). Plasma membrane integrity, chromatin relaxation and SDF appeared particularly susceptible to extreme hypotonic environments, whereas MMP, while susceptible to extreme hypo- and hypertonic environments, showed an ability to rebound from hypertonic stress when returned to isotonic conditions. The problem of chromatin relaxation, which accompanies the post-thaw swelling of sperm heads after thawing, is a major impediment to the successful cryopreservation of koala spermatozoa.

Another important, but often ignored component in the development of semen preservation technology and artificial insemination, is an understanding of the normal flora and pathogens associated with the external genitalia, prepuce and semen. Prior to determining which antibiotics are required for inclusion in semen diluents, it is first necessary to conduct culture and sensitivity tests upon the organisms that are present. We conducted a screening procedure with koala semen and were surprised by the diversity of microflora, describing a novel *Corynebacterium* spp in the process (Johnston et al. 1997). Once we understood which organisms were present [nine bacteria and two yeasts] and their respective antibiotic sensitivities (Penicillin G 1000 IU/mL and gentamicin 100 ug/mL), we had to ensure that the antibiotics were not spermicidal; this included antibiotic dose response studies on sperm parameters (motility) and validation of antibiotic action to prevent bacterial growth (Johnston et al. 1997).

We then extended this approach to examine the possibility of using antibiotic therapy to successfully kill chlamydial elementary bodies [EBs] following experimental inoculation of cultured EBs in semen samples. These results revealed that penicillin at 25 iu/mL, erythromycin at 1000 ug/mL and tetracycline at 200 ug/mL were highly effective at rendering the chlamydiae non-viable, but streptomycin showed no antichlamydial activity (Bodetti et al. 2003). There was a significant reduction of the motility of spermatozoa extended in diluents containing erythromycin, but spermatozoa incubated with tetracycline up to concentrations of 200 ug/mL were not affected. The use of "clean up" semen using antibiotic therapy or other approaches could prove to be an important mechanism of genetic recovery for animals actively shedding EBs, from post-mortem tissue of animals with clinical signs of the disease or that prove to be PCR positive to chlamydia (Johnston et al. 2018). If there is no adverse effect on sperm viability from these antimicrobials then they should be incorporated as routine therapies prior to AI [Hulse and Johnston, unpublished observations]. We have recently extended our PCR screen test for chlamydia in koala semen (Bodetti et al. 2002) to develop a novel multiplex RT PCR that allows testing for both species of koala *Chlamydia* (*C. pecorum* and *C. pneumoniae*), *Mycoplasma*, *Ureaplasma* and *Bordetella bronchiseptica* and to quantify *C. pecorum* (Hulse et al. 2018).

Short-term preservation of semen offers an alternative for temporary sperm storage in those species where the spermatozoa are difficult to cryopreserve; the koala is certainly one of those species. Other domestic species that fit into this category include the stallion and boar, and in both species, there is a robust and commercially viable industry based on chilled or liquid sperm preservation. We conducted the first short-term preservation studies in 1992 where we naively used phosphate buffered saline to maintain sperm survival for 2 days (Johnston et al. 1992). Later we included antibiotics in a Tris-citrate-glucose diluent to successfully store the spermatozoa at 5 °C for 8 days (Johnston et al. 2000c). Remarkably, the spermatozoa still showed 46% progressive motility after 8 days; but on leaving the spermatozoa in the refrigerator for a total of 22, 35 and 42 days there was still approximately 47%, 31% and 16% motility, respectively. We have also examined the effect of chilled storage for 16 days on sperm motility, membrane integrity, the percentage of spermatozoa with relaxed chromatin and sperm DNA fragmentation. After 16 days storage at 5 °C, the progressive motility, plasma membrane integrity and nuclear relaxed chromatin were 50%, 57% and 25%, respectively (Johnston et al. 2013). We also examined the DNA fragmentation dynamics of the spermatozoa over a 16-day storage period. DNA dynamics is a concept in which DNA of the spermatozoa at defined time periods of storage is examined following incubation of sperm at body temperature in order to mimic *in vivo* conditions within the female reproductive system (Schultz et al. 2018). For the koala, we examined sperm DNA dynamics immediately after ejaculation, 4 h, 1 d, 2 d, 4 d, 8 d and 16 d after chilled storage at 5 °C. At each of these time periods the spermatozoa were incubated at 35 °C and DNA fragmentation assessed at 0, 2, 6, 24 and 48 h. Remarkably, these results revealed that koala sperm DNA fragmentation after 16 d storage at 5 °C, followed by 48 h incubation at 35 °C was still only 15%. These results suggest that koala sperm chromatin is very stable when stored in a chilled state for up to 16 days and, based on this evidence and the motility and viability data, we expect that the spermatozoa remain potentially fertile and could be used for AI. We have already used chilled [5 °C] semen stored for 24 and 72 h to produce two and four koala pouch young, respectively (Allen et al. 2008a). Chilled storage and transport of semen can not only be used for national genetic transfers but also has the potential to be used for international exchange. As the koala has a fertile period of oestrus of approximately 8–10 days (Allen et al. 2008a) it would be possible, given notification of oestrus on day 1 of an animal located in a zoo in the USA or Japan, to collect semen from males in Australia the following day. The semen from those males could then be loaded into a chilled liquid preservation diluent containing antibiotics and antichlamydials; in addition, the semen could be screened for disease using PCR (Hulse et al. 2018). The sample can then be loaded into a commercial shipper that would maintain the sperm at 5 °C for up to 2–4 days during transportation; even after 5 days, there is a high likelihood that the female will still be in heat and ready for insemination. The semen would be examined on arrival and the sample(s) with the best post-chilled parameters used for insemination. Based on our observations (Johnston et al. 1992, 2000c, 2012a, 2013; Schultz et al. 2018), it might be possible that semen stored chilled for up to 40 days could be used for successful fertilization following AI.

11 Sperm Cryopreservation

There is no doubt that sperm cryopreservation for use in successful AI currently remains one of the greatest challenges and limitations to the broad scale use and application of ART in the koala. The ability to freeze-thaw koala spermatozoa would allow genetic management to occur through both time and space and would facilitate the global shipment of gametes and the use of genetic materials recovered from post-mortem animals. We first attempted to freeze koala spermatozoa in 1993 and, at that stage, there had been only one other attempt to cryopreserve marsupial spermatozoa; this was performed by Rodger et al. (1991) and focused on the brush-tail possum (*Trichosurus vulpecula*). Remarkably, this study found that possum spermatozoa had the best post-thaw survival when frozen in 17.5% glycerol. Based on the possum methodology, we conducted a similar protocol for koala spermatozoa using a range of glycerol concentrations (4–18% w/w) in a Tris-citrate 20% egg yolk extender (Johnston et al. 1993). The results of these investigations revealed that glycerol had a detrimental effect on motility when semen was stored for 1 h at room temperature but, more importantly, koala spermatozoa, similar to Brush-tail possum, required a higher concentration of glycerol (14%) in order to produce post-thaw motility of 30%.

Some 13 years later, we returned to investigate sperm cryopreservation in the koala and this time compared the cryopreservation success of koala spermatozoa with that of common wombat spermatozoa (Bodetti et al. 2003). The use of wombat spermatozoa was significant, as both species share a close phylogeny and similar morphology, but wombat spermatozoa showed a remarkable tolerance to cryopreservation (MacCallum and Johnston 2005; Taggart et al. 1998) compared to that of koala spermatozoa. In our comparative study, we confirmed that common wombat spermatozoa showed greater post-thaw survival than koala spermatozoa in terms of motility, plasma membrane integrity and decondensed (relaxed) chromatin. We also showed that both koala and wombat spermatozoa survive freeze-thawing if frozen at slow cooling rates (−6 °C/min) compared to rapid freezing (3 cm above the liquid N_2 interface), and that both species show greatest post-thaw survival when frozen in 14% rather than 8% (w/w) glycerol. Koala spermatozoa showed a large increase in the proportion of spermatozoa with relaxed or swollen chromatin compared to wombats following 2 h incubation post-thaw. The lower tolerance of koala spermatozoa to cope with hyperosmotic media may be contributing to lower post-thaw survival. We also examined both spermatozoa for the presence of F-actin in an attempt to relate poor cryopreservation success with F-actin induced plasma membrane inflexibility—we discounted this hypothesis when we found F-actin in the wombat spermatozoa but not in the koala.

Following some success at using a cryomicroscope for investigating freeze-thaw methodologies in kangaroo spermatozoa (Holt et al. 1999), we subsequently turned our attention to this instrument to investigate the cryopathology of koala spermatozoa (Zee et al. 2007). This work represented a shift away from the traditional empirical approach to investigating cryopreservation towards a more hypotheses-driven perspective. Using a combination of fluorescent probes to directly observe plasma membrane integrity, MMP, lipid raft stability and phosphatidylserine translocation, during the freeze-thaw protocol, we were able to gain a better understanding of koala sperm specific cryopathology. These observations revealed that high MMP declined significantly during rewarming, especially as the temperature increased from 5 to 35 °C. We also concluded that chilling and freezing had no effect on the distribution of ganglioside GM1 in sperm membrane or on plasma membrane lipid asymmetry.

We have also examined the use of cryoprotectants other than glycerol for use with koala spermatozoa, including dimethyl sulfoxide (DMSO), methanol, propylene glycol and dimethylacetamide (DMA) (Zee et al. 2008). These experiments revealed that 10–15% DMA may be a useful alternative to glycerol for cryopreservation. We have also examined individual variability in post-thaw survival with a large captive koala population (Zee et al. 2009b). There were significant differences in post-thaw survival from different animals that were independent of pre-freeze semen quality. We showed that glycerol [14%] was a better cryoprotectant than DMA (12.5%) in terms of maintaining motility, plasma membrane integrity and high MMP, but there was no difference between the two compounds with respect to their ability to prevent chromatin relaxation. We also noted that the efficiency of energy generation by the mitochondria was lowered by cryopreservation. From this study it was postulated that the unpredictability of assessing post-thaw survival from pre-freeze koala semen parameters is likely to be associated with variation in ejaculate composition or inherent genetic differences between animals.

The issue of sperm chromatin relaxation following cryopreservation led us to work in Spain, and a very fruitful collaboration with the laboratory of Prof Jaime Gosalvez at the Autonomous University of Madrid (see Chap. 15 in the first edition of this book (Gosalvez et al. 2014)). We were initially concerned that koala spermatozoa were showing evidence of fragmented DNA after the freeze-thaw procedure, so we developed and validated an assay for the assessment of koala sperm DNA fragmentation based on the sperm chromatin dispersion test (SCDt) (Johnston et al. 2007). In this early work, we reported three koala nuclear morphotypes following the assay (Fig. 12.5c); KSM1 and KSM2 morphotypes showed no or limited sperm DNA fragmentation. We proposed that KSM2 results from DNA that is damaged as part of the normal processing of the spermatozoa in the assay and is primarily a consequence of the lack of cysteine residues, and associated stabilising disulphide bonds, in marsupial nuclear proteins. We also concluded that "True" DNA fragmentation was represented by the KSM3 morphotype that showed massive dispersion of chromatin following the SCDt. This work revealed that the proportion of KSM3 actually increased and the KSM2 decreased following incubation of frozen thawed spermatozoa. The next step in this investigation was to determine whether the sperm DNA fragmentation we were visualising was a result of single stranded or double stranded DNA breaks; our prediction was that KSM2 were being caused as a result of single stranded breaks, whereas KSM3 were associated with the more severe double stranded breaks. This was achieved by developing a double comet assay to assess sperm chromatin damage (Zee et al. 2009a) and comparing these results with

SCDt. Following SCDt, we discovered a continuum of nuclear morphotypes, ranging from no apparent DNA fragmentation to those with highly dispersed chromatin; these morphotypes were mirrored by a similar diversity of comet morphologies that could be further differentiated by double comet assay. Spermatozoa with "true" DNA fragmentation exhibited a continuum of comet morphologies, ranging from a more severe form of alkaline-susceptible DNA with a diffuse single tail to nuclei that exhibited both single and double-stranded breaks with two comet tails. We have also explored the effect of cryopreservation on koala sperm DNA fragmentation. As for our studies on chilled koala spermatozoa, we used a different variation of the SCDt, which examined the DNA fragmentation dynamics of frozen-thawed koala spermatozoa over a 48 h period of incubation at 35 °C (Johnston et al. 2012a) to compare spermatozoa frozen in 14% glycerol and 10% DMA. This study revealed that while the pre-freeze incidence of DNA fragmentation was no different when comparing between spermatozoa frozen in DMA and glycerol, spermatozoa frozen in glycerol showed a higher rate of DNA fragmentation than pre-freeze spermatozoa. This result differed from that of observations on progressive motility, plasma membrane integrity and relaxed chromatin, which were all adversely affected after cryopreservation in both glycerol and DMA. We also found that after thawing, koala sperm chromatin tended to relax, but interestingly, the incidence of sperm DNA fragmentation was not correlated with the incidence of sperm chromatin relaxation after cryopreservation. These results suggested that chromatin relaxation was not necessarily associated with DNA fragmentation (double stranded breaks in sperm DNA).

In attempt to investigate the underlying aetiology of sperm chromatin relaxation associated with koala sperm cryopreservation and post-thaw incubation, we explored an experimental model that mimicked the structural and physiological effects of osmotic flux on the sperm cell (Johnston et al. 2012b). Similar to our previous findings, DNA labelling after *in situ* nick translation of thawed cryopreserved spermatozoa revealed a positive correlation between area of relaxed chromatin in the nucleus and the degree of nuclear labelling, and while the chromatin of some spermatozoa increased more than eight times its normal size, not all sperm nuclei with relaxed chromatin showed evidence of nucleotide incorporation. Preferential staining associated with DNA fragmentation was typically located in the peri-acrosomal and peripheral regions of the sperm head and at the base of the sperm nucleus where it appeared as "hot spots" of damage following cryopreservation. We also compared the effect of exposure to anisotonic media and cryopreservation on the integrity of koala spermatozoa and found that injury induced by exposure to osmotic flux essentially imitated the results found following cryopreservation. Plasma membrane integrity, chromatin relaxation and SDF appeared very susceptible to extreme hypotonic exposure. Although susceptible to both extreme hypo- and hypertonic media, MMP showed an ability to rebound from hypertonic stress when returned to isotonic conditions. Koala spermatozoa exposed to 64 $mOsmKg^{-1}$ media showed an equivalent, or more severe, degree of structural and physiological injury than that of frozen-thawed spermatozoa, supporting the hypothesis that cryoinjury is principally associated with a hypo-osmotic effect. A direct comparison of SDF of thawed cryopreserved spermatozoa and those exposed to a 64 $mOsmKg^{-1}$ excursion showed a significant correlation, but no correlation was found when the percentage of spermatozoa with relaxed chromatin was compared. We concluded that while a cryo-induced osmotic injury model appears to explain post-thaw changes in koala SDF, the mechanisms resulting in relaxed chromatin require further study and it will be essential to understand them before frozen gamete banking becomes a reality for the koala. A lack of correlation between the percentage of spermatozoa with relaxed chromatin and SDF might indicate that the timings of these pathologies are asynchronous; perhaps cryo-induced injury involves a combination of structural damage (rupture of the membrane) and oxidative stress, that first leads to the reduction of MMP and the relaxation of chromatin, which is then ultimately followed by an increase in DNA fragmentation. Recent studies comparing the post-thaw behavior of koala and wombat sperm chromatin may help to solve this mystery (Johnston et al. 2019) as these observations show that despite similarities in sperm morphology, relative differences in their protamine sequence, and therefore their packaging and protection of the DNA molecule, may account for the koala sperm chromatin being more susceptible to chromatin relaxation.

12 Synchronisation of Oestrus and Induction of Ovulation

In most eutherian mammals it is possible to synchronise oestrus by intervening in the endocrine control of the corpus luteum (CL). The inhibitory effects on GnRH secretion imposed by progesterone can be removed by the use of prostaglandin and its analogues to cause regression of the CL or by administration and timed removal of progestogen implants (Allen et al. 2008b). This approach may not be possible in marsupials, as in the species studied thus far, uterine prostaglandin does not appear to result in luteolysis in non-pregnant females (Tyndale-Biscoe and Renfree 1987) and endogenous progesterone [at least in kangaroos and wallabies] does not suppress GnRH secretion (Shaw 2006). This means that other methods of oestrus and ovulation control may be required in marsupials. As LH and FSH are considered to be important hormones in the control of the marsupial reproduction, we have considered that an alternative approach for taking control of the oestrous cycle, without the benefit of being able to manipulate prostaglandin and progesterone, is the use of GnRH antagonists. GnRH antagonists bind to the GnRH receptor on the gonadotroph and typically outcompete natural sequence GnRH but they do not elicit activation of gonadotrophin secretion. This inhibition of FSH and LH secretion should result in the suppression of steroidogenesis, preventing the female from cycling. The female will continue in anoestrus while the GnRH antagonist is administered but once the molecule has been metabolised, the receptors are free to bind to natural GnRH and the cycle commences again. The consistency with which the cycle can be interrupted or recommenced and synchronised depends on the dosage of antagonist administered, its biological half-life and its binding affinity. In a preliminary study using the male koala as our model (Allen et al. 2008b) we used a single injection of the GnRH antagonist acyline [100 ug = 14.3 ug/kg or 500 ug = 71.4 ug/kg] but this had no effect on suppressing anterior pituitary or testicular steroidogenesis. The inability of acyline to suppress the reproductive axis of male koalas may be associated with a low affinity for the GnRH receptor, although it is also possible that acyline is cleared quickly from circulation. Subsequently, Ballantyne et al. (2016a) examined the effectiveness of a gonadotrophin-releasing hormone (GnRH) antagonist, azaline B, to suppress plasma LH and 17β-oestradiol concentrations in koalas and its potential application for oestrous synchronisation. These experiments showed that daily 1 mg sub-cutaneous injections of azaline B for 10 days was able to suppress the anterior pituitary LH response to mGnRH over the 10-day treatment period but unable to suppress basal LH or 17β-oestradiol concentration; while oestrus was delayed in all treatment females by approximately 24 days, the return to oestrus was variable ranging from 9–39 days. Ballantyne et al. (2016b) have also attempted to control the koala oestrous cycle using a 70 mg implant of the synthetic progesterone, levonorgestrel [LNG]. Overall results from this study indicated that LNG implants in koalas can inhibit oestrous behaviour and reduce circulating oestradiol-17β levels before oestrus, most likely by preventing development of the pre-ovulatory follicle but there was no evidence of LH suppression by the LNG implants. Removal of LNG implants resulted in the return to oestrus 13, 14, 17 and 20 days later. Even this modest degree of oestrous synchronisation was unexpected and maybe related to the effect of progesterone on the expression of oestrous behaviours rather than ovarian activity; further studies on a larger koala population are required to validate these findings.

A very significant advantage in the reproductive physiology of the koala that has contributed to the success of AI is that ovulation is induced via coitus (Johnston et al. 2000a). Our initial investigations led us to believe that ovulation was induced by a complete duration of penile thrusting and that the mechanism was essentially a copulo-receptive reflex (Johnston et al. 2000b) similar to which occurs in the felidae. However, further study (Johnston et al. 2004) revealed that mechanical stimulation of the urogenital sinus with a glass rod alone did not appear to induce a luteal phase (0/9 individuals tested), that semen alone with stimulation induced a luteal phase in some koalas (4/9 individuals tested) but that a combination of semen and glass rod simulation was most effective (7/9 individuals tested) at inducing a luteal phase. These results suggested that there might be factors in the semen that promote induction of the luteal phase. We have speculated that the presence of specific ovulation factors in the semen and the mechanical stimulation of the urogenital sinus during coitus, work synergistically to induce ovulation; perhaps the thrusting of the male's penis in the urogenital sinus in some way prepares the urogenital epithelium for exposure to the ovulating factors? We talk about a luteal phase, rather than ovulation because it is difficult to observe the ovary directly in this species. While ultrasonography has been used recently to document normal (Blanshard and Bodley 2008; Larkin et al. 2018) and abnormal reproductive male and female anatomy (Blanshard and Bodley 2008), the ability to effectively document and time ovulation by ultrasound would be an important advance that needs to be perfected in this species and is part of our ongoing research.

While we currently use glass rod stimulation of the urogenital sinus as part of our standard artificial insemination technique, it is also technically possible to use pharmacological induction of the ovulation using either GnRH and human chronic gonadotrophin (hCG) (Johnston et al. 2000b, 2003; Allen et al. 2008a). To date, we have only produced one pouch young following administration of hCG (Johnston et al. 2003). We have also induced what appear to be normal luteal phases in the koala oestrous cycle using the GnRH agonist buserelin (Allen et al. 2008a). If hCG or GnRH agonists are going to be employed to routinely induce ovulation as part of the koala AI protocol, then it might be prudent to inseminate the female approximately 24–32 h before injection of these pharmaceuticals; clearly, there need to be further studies to refine their respective applications. The use of pharmaceutical induction of ovulation is particularly relevant when using cryopreserved or heavily diluted spermatozoa, as dilution with semen extenders or the washing out of cryoprotectants could potentially remove the ovulating factors in the semen. Nevertheless, we regularly use diluted semen 1:1 with Tris-citrate extender as part of AI program and have been successful; it would be interesting to see how far the semen could be extended before the concentration of ovulating factor is insufficient to induce ovulation.

While it is not currently possible to synchronise oestrus precisely in the koala to the same extent as for domestic animals, it is also possible to make a case that this form of oestrus control may not be necessary in order to develop an effective artificial insemination program. For example, given that the non-mated koala has an average oestrous cycle length of 33 days, an oestrus of typically 10 days (Johnston et al. 2000a), and assuming that that the non-lactating females are cycling normally, the majority of koalas should come into oestrus and be artificially inseminated within a 23–33 day period. Females that fail to conceive and do not ovulate will come back into oestrus again around 33 days later, while those who do ovulate but fail to conceive will return in approximately 50 days; as the koala is polyoestrous this represents approximately 5–6 AI opportunities each breeding season.

13 Artificial Insemination

We have developed two different successful techniques for artificial insemination in the koala. Our most commonly used protocol involves the conscious female and insemination using a custom designed Foley insemination catheter (Johnston et al. 2003) (Fig. 12.6a). Typically, we will choose a female for AI who is in day 2–5 of oestrus but could wait up to day 8 of oestrus (Allen et al. 2008a). She is brought into the veterinary surgery and held in a full restraint position by the zookeeper (Fig. 12.6b). The koala's urogenital sinus is positioned slightly upwards and towards the inseminator. A sterile glass rod that mimics the glans penis of the koala is then gently inserted into the urogenital sinus and stroked back and forth with a slight twisting motion to a depth of 40–60 mm; stimulation of the urogenital sinus is based on previous descriptions of natural koala coitus (40 penile thrusts per 20s (Johnston et al. 2000a, b)). The koala AI catheter is then inserted the full depth of the

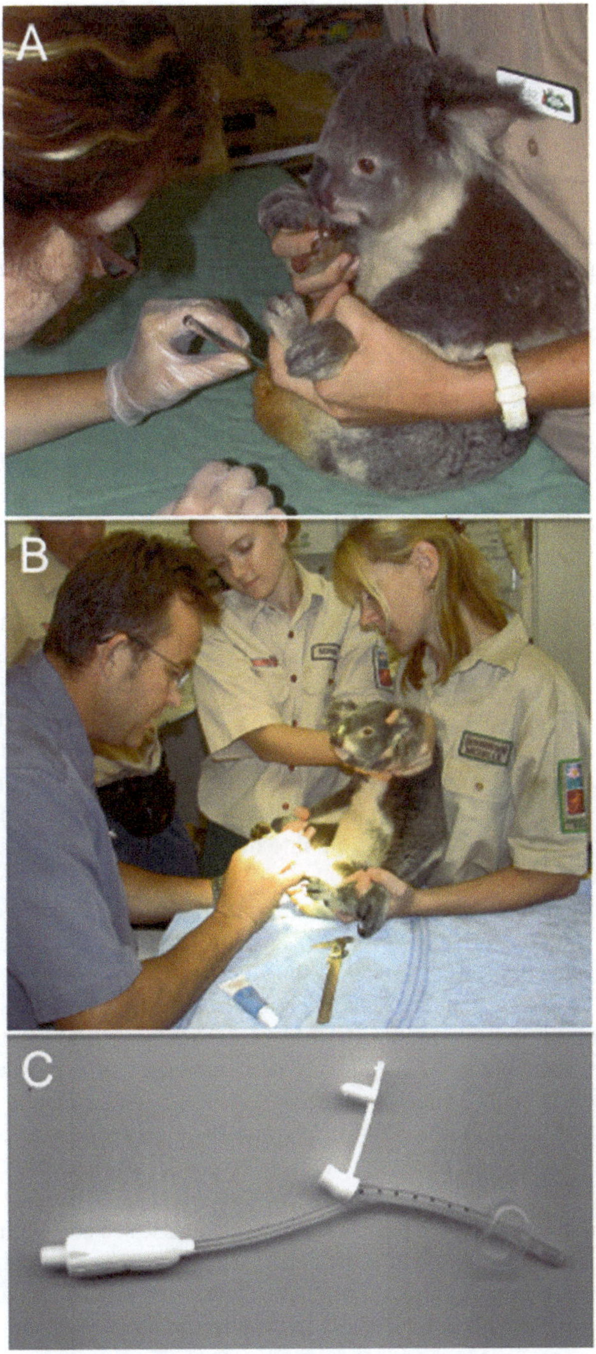

Fig. 12.6 Artificial insemination of the koala. (**a**) Glass rod stimulation of the koala urogenital sinus; (**b**) Artificial insemination of the conscious koala; (**c**) Customised designed koala insemination Foley catheter (From Johnston et al. 2003)

urogenital sinus and the cuff of the catheter inflated so as to produce a seal, thus preventing retrograde flow of semen; the basis of the insemination technique is one of positive displacement of seminal volume into the ostia of the lateral vaginae. We typically use a seminal volume of approximately 1 mL and this can be diluted (1:1) or undiluted semen. We have been successful using semen with a sperm concentration as low as 3×10^6/mL but we would suggest using a sperm concentration of approximately >20×10^6/mL as standard (Allen et al. 2008a). The semen is drawn up into the catheter with a 1 mL syringe and inseminated slowly; following deposition of semen, a second 1 mL syringe of air is used to displace any semen remaining in the bore of the catheter. The bore of the insemination catheter is then capped and the female koala restrained with the catheter in place for another 5 min, after which the catheter cuff is deflated and removed. As the female is conscious, she can then be immediately returned to her enclosure. We also developed another method of insemination, which we refer to as the urogenitoscopic method in which the female is anaesthetised, placed in ventral recumbency on a tilt table, the ostia of the vaginae visualised using an otoscope and speculum and a tom-cat catheter directed into the vaginal ostia (Johnston et al. 2003). In addition to the inconvenience and risk associated with anaesthesia, we found that proliferation of the oestrous urogenital epithelium made visualisation of the ostia virtually impossible. Reliable insemination into both ostia is important as the left and right sides of the koala reproductive tract are separated by a medial septum of the vaginal cul-de-sac, so that semen needs to be deposited into each ostium. As it is extremely difficult to determine which ovary has ovulated without ultrasonography or laparoscopy, we concluded that the use of the catheter had inherently less procedural risk and a greater probability of successful semen delivery into both sides of the tract. Nevertheless, the use of small volumes of semen, or the use of cryopreserved semen, will probably require us to re-think new methods of insemination higher up into the reproductive tract, such as laparoscopic intrauterine insemination.

The most immediate and current use of koala AI has been within and between zoological institutions for the purposes of genetic exchange, animal pairings with behaviour incompatibilities or structural soundness problems that prevent or impede physical mating. However, there is also a greater potential for use of artificial insemination, especially if it is combined with cryopreservation and genome banking principles. For example, we have already demonstrated that it possible to collect semen from wild donors (Allen et al. 2010) so that this semen can be screened for disease, and then used after chilling or cryopreservation for AI into captive reservoirs of female koalas or females in wild isolated genetically restricted habitat fragments. Semen can also be collected, used or stored in genome banks, from animals that require euthanasia or even post-mortem specimens associated with dog and car accident trauma and disease. While any genetic exchange should be carefully monitored with a detailed understanding of genetic history and relatedness of individuals in the population, we are unfortunately getting to the point in certain parts of SE Queensland where intensive genetic management is going to be necessary to prevent localised extinction (Johnston et al. 2013, 2018).

To date we have produced a total of 34 koala pouch by means of artificial insemination, but much of this work has been experimental rather than specifically for genetic management. Consequently, our results have been limited by the constraints of our own experimental approach, so that we have not yet been able to reliably determine the overall efficiency of the procedure, as might be the case for commercial AI in the cattle industry. Nevertheless, the number of offspring produced [i.e. 34] significantly outnumbers what has been produced in other marsupials (Rodger et al. 2009) and rivals that of successful ART programs in other captive wildlife.

It is one thing to develop efficient and reliable protocols for koala AI but another to implement these technologies within commercial zoo practices and the governmental genetic management of wild animals. That is why we were particularly satisfied with an excerpt in the Queensland koala Nature Conservation [Koala] Conservation Plan (2006–2016) (Queensland Department of Environment Heritage and Protection 2006) which mentioned our work in Sect. 4.8 of the policy document regarding the use of artificial insemination "*To reduce the need to export live koalas in the future, the EPA is also working with the University of Queensland, a world leader on koala biological research, on reproductive technologies for koalas. In the future, artificial insemination will provide for the introduction of new genetic diversity to local and overseas zoos and provide more effective, less costly measures for maintaining viable breeding colonies*". It is only with government and public acceptance that we will be able to implement and evaluate all the potential benefits that ART can deliver.

14 The Transfer of Koala Art to the Wombat

From day 1 when the plan for a koala artificial insemination program was first "conceived" it has always been our goal to apply what we learnt from the koala to other species, such as the Northern Hairy-nosed Wombat [*Lasiorhinus kreffeti*]. The establishment of model species is a sound principle for developing an AI technique. The problem that we faced working in the area of marsupial reproduction was that were very few well-studied species for direct comparison. The characterisation of all marsupials into a generalised mode of reproduction has been a major limitation to the development of ART; this point is well illustrated in the koala and wombat. While the close phylogeny of these two species is reflected in their respective reproductive anatomies and gamete ultrastructure, the reproductive physiology and behaviour of these two species, is frustratingly, very different (Johnston 2019). For example, the female reproductive tract of the wombat is similar enough that a slightly larger version of the koala insemination catheter would be ideal for the wombat, but unlike the koala, the wombat ovulates spontaneously and appears to have a relatively short oestrus. These differences, combined with the cryptic, non-tractile and fossorial nature of the wombat, make the application of ART a much more "hands-off" and technically difficult challenge (Hogan et al. 2013). Another interesting difference is that wombat spermatozoa are very tolerant of cryopreservation and yet it is this element of the ART program in the koala that is currently the rate-limiting step for the use of frozen-thawed semen and establishment of a genome resource bank. To date there have been no marsupial young born following AI with frozen-thawed semen.

15 Application of Art to Behavioural Ecology of the Koala

Key physiological and behavioural information accumulated in the development of an ART program can also have direct impact on understanding behavioural ecology of the wild animal; an example of this is the observation that ovulation in the koala is induced by coitus. This information is fundamental to understanding and interpreting female koala movement and home range observations. Although yet to be proven, we would postulate that female koalas in the wild that come into oestrus will show increased levels of physical activity and seek out males for mating, motivated by the high concentration of oestradiol secretion from the pre-ovulatory follicle. Once mated, the follicle would collapse, the female loses sexual interest and returns to her home range. It is only when we understand the physiological parameters of the reproductive biology that we can interpret what we see in the wild. We are currently examining other fundamental aspects of koala biology that are relevant to refining ART but also have a role in understanding ecology [e.g. refractory period of the male].

16 Conclusion and Future Directions

Given that we have produced 34 koala pouch young by means of artificial insemination, there is no doubt that we currently have a reliable means of genetic management for captive koalas based on ART. It would also be safe to say that the koala program has been the most successful for any marsupial (Rodger et al. 2009). Despite this success, there are still a number of missing pieces to the complete ART puzzle; the two most significant being the development of a reliable method of sperm cryopreservation and the direct application of the technology to wild population management. The establishment of a frozen genome bank of koala spermatozoa from representative males from every wild population in Australia is long-term objective of our group. We also recommend developing AI protocols that allow the use of extended chilled semen, as we are confident that combination of chilled liquid preserved semen [2–4 weeks] could be useful for the international exchange of semen between zoos and facilitating the timing of artificial insemination.

In close association with koala ART, we are currently developing genetic exchange programs in the form of a live koala genome bank (LKGB; Johnston et al. 2013; Johnston et al. 2018). This concept is similar to a frozen genome bank except that live female koalas are used as the storage units. The LKGB has two primary aims that include the use of zoo populations; (a) as a reservoir to safe-guard genetic diversity of local wild populations and (b) to serve as breeding centre for the genetic management of local wild populations. In such a program we would use koalas that are genetically selected to be representative of the population under investigation and then manage them closely. These concepts include; (a) *Genetic connectivity* where koalas spend short periods of time in captivity to facilitate genetic exchange between fragmented populations such that animals from isolated fragments could be brought into captivity bred and released back into their respective fragments (Holt et al. 1996); in this we ensure greater genetic diversity and gene flow into both fragments; (b) *Genetic capture* where koalas destined for translocation because of sudden habitat loss or destruction could be brought into captivity into a live genome bank to make a quick genetic deposit; males, for example, could spend a short period of time in a captive facility to breed, or perhaps semen could be collected from wild males and inseminated to dedicated genetically selected females—in this way their genes would be captured and stored in live genome banks; (c) *Genetic rescue*—where gametes from diseased, trauma or post-mortem koalas could be recovered and used for germplasm storage or artificial insemination—this source of genetic material is currently being wasted in SE Queensland hospitals given that between 1997 and 2013, 41,606 koalas were admitted to koala hospitals (Gonzalez-Astudillo et al. 2017), and (d) *Genetic propagation*—where we propose the establishment of a purpose built koala breeding centre that could be used for the generation of genetically compatible koalas for release into either reclaimed or restored habitat. SE Queensland zoos have world's best practice at producing koalas in captivity and a skill set ideally suited for this type or conservation (Rhodes et al. 2017). Underpinning the LKGB concept is a thorough understanding of koala metapopulation genetics, now facilitated by publication of the koala genome (Johnson et al. 2018) and a *Chlamydia* vaccination program to prevent disease to the animals in the LKGB (Johnston et al. 2018).

As Johnston et al. (2018) have recently noted, the final report of the Queensland State Government Koala Expert Panel (Rhodes et al. 2017) has provided recommendations that appear to support the general direction of the LKGB concept and include the integration of zoo-based koala populations more explicitly into the management of wild koala populations in south-east Queensland, particularly in urban areas. This can be accomplished through (a) recognition that some zoo populations may provide a valuable reservoir for future potential re-introductions, (b) supporting the on-going maintenance of studbook records for all koalas held in zoos and wildlife parks in Queensland and (c) engaging zoos, where appropriate, in the management of wild koala populations, drawing on their knowledge and expertise in captive breeding and reintroduction.

Some of these ideas are quite unconventional and might even be regarded as controversial, but there are certain regions in Queensland where intensive management may be the only way that koala populations can ultimately survive. In combination with relevant habitat preservation, predator and social management and policies, we are convinced that ART in its most simple and/or advanced forms has a major role in koala conservation. In our view, the koala is the ideal species for demonstrating the benefits of ART to marsupial conservation biology; its iconic status and public appeal, tractility, unique and amenable reproductive biology and its ability to adapt well to temporary captivity and release, all point towards a high probability for success.

Acknowledgements The success of the koala ART program has largely been achieved through the dedication and support of a wide range of scientific collaborators, post-graduate students and zoological institutions. In alphabetic order, we are extremely grateful for the contribution and support of Dr Camryn Allen, Dr Katie Ballantyne, Ms Michele Barnes, Prof Ken Beagley, Ms Michelle Burridge, Mr Kevin Bradley, Prof Randal Cameron, Prof Frank Carrick, Dr Ron Cox, Dr Jon Curlewis, Mr Robert Douglas, Dr Bill Ellis, Prof Jaime Gosalvez, Ms Lyndal Hulse, Dr Carmen Lopez-Fernandez, Mr Alan Lisle, Dr Tamara Keeley, Prof Michael McGowan, Dr Allan McKinnon, Mr Al Mucci, Mr Paul O'Callaghan, Dr Vere Nicolson, Ms K Nilsson, Dr Nancy Phillips, Dr Michael Pyne, Prof Jennifer Seddon, Mr Peter Theilman and Dr Yeng Peng Zee.

References

Adam-Hosking C. Current status of the koala in Queensland and NSW. World Wildlife Fund-Australia; 2017. https://www.wwf.org.au/ArticleDocuments/353/pub-current-status-of-the-koala-in-queensland-and-new-south-wales-19may17.pdf.aspx. Accessed 25 Feb 2019.

Adams-Hosking C, McBride MF, Baxter G, Burgman M, de Villiers D, Kavanagh R, et al. Use of expert knowledge to elicit population trends for the koala (*Phascolarctos cinereus*). Divers Distrib. 2016;22:249–62.

Allen CD, McKinnon AJ, Lisle AT, D'Occhio MJ, Johnston SD. Use of a GnRH agonist and hCG to obtain an index of testosterone secretory capacity in the koala (*Phascolarctos cinereus*). J Androl. 2006;27:720–4.

Allen CD, Burridge M, Mulhall S, Chafer ML, Nicolson VN, Pyne M, et al. Successful artificial insemination in the koala (*Phascolarctos cinereus*) using extended and extended-chilled semen collected by electroejaculation. Biol Reprod. 2008a;78:661–6.

Allen CD, Burridge M, Chafer ML, Nicolson VN, Jago SC, Booth RJ, et al. Control of the koala (*Phascolarctos cinereus*) anterior pituitary-gonadal axis with analogues of GnRH. Reprod Fertil Dev. 2008b;20:598–605.

Allen CD, de Villiers DL, Manning BD, Dique DS, Burridge M, Chafer ML, et al. Seasonal reproduction in wild and captive male koala (*Phascolarctos cinereus*) populations in south-east Queensland. Reprod Fertil Dev. 2010;22:695–709.

Ballantyne K, Anderson S, Pyne M, Nicolson V, Mucci A, Lisle A, et al. Use of the gonadotrophin-releasing hormone antagonist azaline B to control the oestrous cycle in the koala (*Phascolarctos cinereus*). Reprod Fertil Dev. 2016a;28:1686–94.

Ballantyne K, Anderson S, Pyne M, Nicolson V, Mucci A, Lisle A, et al. The use of a synthetic progesterone, levonorgestrel (LNG), to control the oestrous cycle in the koala. Reprod Fertil Dev. 2016b;28:1526–32.

Blanshard W, Bodley K. Koalas. In: Vogelnest L, Woods R, editors. In medicine of Australian mammals. Collingwood: CSIRO Publishing; 2008. p. 227–328.

Bodetti T, Johnston SD, Pospischil A, Knox C, Timms P. PCR detection of Chlamydia in koala (*Phascolarctos cinereus*) semen. Vet Rec. 2002;151:147–9.

Bodetti TJ, Hengstberger K, Johnston S, Timms P. Evaluation of tetracycline, erythromycin, penicillin and streptomycin for decontaminating koala semen contaminated *in vitro* with chlamydiae. Vet Rec. 2003;153:588–59.

Busby E, Soeta S, Sherwood N, Johnston S. Molecular analysis of the koala reproductive hormones and their receptors: gonadotrophin-releasing hormone (GnRH), follicle-stimulating hormone β and luteinising hormone β with localisation of GnRH. J Neuroendocrinol. 2014;26:870–87.

Chapman JA, Leigh CM, Breed WG. The zona pellucida of the koala (*Phascolarctos cinereus*): its morphogenesis and thickness. J Anat. 2006;209:393–400.
Charlton BD, Frey R, McKinnon AJ, Fritsch G, Fitch WT, Reby D. Koalas use a novel vocal organ to produce unusually low-pitched mating calls. Curr Biol. 2013;23:R1035–6.
Dief HHA. Chlamydial disease of the male koala (*Phascolarctos cinereus*) reproductive tract. Master of philosophy thesis. Gatton: University of Queensland; 2011.
Ellis W, FitzGibbon S, Pye G, Whipple B, Barth B, Johnston S, et al. The role of bioacoustic signals in koala sexual selection: insights from seasonal patterns of associations revealed with GPS-proximity units. PLoS One. 2015;10:e0130657.
Feige S, Nilsson K, Phillips CJC, Johnston SD. Heterosexual and homosexual behaviour and vocalisations in captive female koalas (*Phascolarctos cinereus*). Appl Anim Behav Sci. 2007;103:131–45.
Gallegos G, Ramos B, Santiso R, Goyanes V, Gosálvez J, Fernández JL. Sperm DNA fragmentation in infertile men with genitourinary infection by Chlamydia trachomatis and Mycoplasma. Fertil Steril. 2008;90:328–34.
Gonzalez-Astudillo V, Allavena R, McKinnon A, Larkin R, Henning J. Decline causes of koalas in South East Queensland, Australia: a 17-year retrospective study of mortality and morbidity. Sci Rep. 2017;7:42587. https://doi.org/10.1038/srep42587.
Gosalvez J, Holt WV, Johnston SD. Sperm DNA fragmentation and its role in wildlife conservation. Adv Exp Med Biol. 2014;753:357–84.
Handasyde KA, McDonald IR, Than KA, Michaelides J, Martin RW. Reproductive hormones in the koala. In: Lee AK, Handasyde KA, Sanson GD, editors. Biology of the koala. Chipping North: Surrey Beatty; 1990. p. 203–10.
Harding H, Aplin K. Phylogenetic affinities of the koala (Phascolarctidae; Marsupialia): a reassessment of the spermatozoal evidence. In: Lee AK, Handasyde KA, Sanson GD, editors. Biology of the koala. Chipping North: Surrey Beatty; 1990. p. 1–31.
Harding HR, Carrick FN, Shorey CD. Special features of sperm structure and function in marsupials. Baltimore and Munich: Spermatozoon, Urban & Schwarzenberg; 1979. p. 289–303.
Harding H, Carrick F, Shorey C. Acrosome development during spermiogenesis and epididymal sperm maturation in Australian marsupials. In: The sperm cell. Dordrecht: Springer; 1983. p. 411–4.
Harding H, Carrick F, Shorey C. The affinities of the koala *Phascolarctos cinereus* (Marsupialia: Phascolarctidae) on the basis of sperm ultrastructure and development. In: Archer M, editor. Possums and opossums: studies in evolution. Sydney: Surrey Beatty and Royal Zoological Society of New South Wales; 1987. p. 353–64.
Hogan LA, Janssen T, Johnston SD. Wombat reproduction (Marsupialia; Vombatidae): an update and future directions for the development of artificial breeding technology. Reproduction. 2013;145:R157–73.
Holt WV. Cryobiology, wildlife conservation and reality. CryoLetters. 2008;29:43–52.
Holt WV, Bennett PM, Volobouev V, Watson PF. Genetic resource banks in wildlife conservation. J Zool. 1996;238:531–44.
Holt WV, Penfold LM, Johnston SD, Temple-Smith P, McCallum C, Shaw J, et al. Cryopreservation of macropodid spermatozoa: new insights from the cryomicroscope. Reprod Fertil Dev. 1999;11:345–53.
Hughes R. Light and electron-microscope studies on spermatozoa of koala, *Phascolarctos cinereus* (Marsupialia). J Anat. 1977;124:513.
Hulse LS, Hickey D, Mitchell JM, Beagley KW, Ellis W, Johnston SD. Development and application of two multiplex real-time PCR assays for detection and speciation of bacterial pathogens in the koala. J Vet Diagn Investig. 2018;30:523–9.
Johnson RN, O'Meally D, Chen Z, Etherington GJ, Ho SY, Nash WJ, et al. Adaptation and conservation insights from the koala genome. Nat Genet. 2018;50:1102–11.
Johnston SD. Reproductive biology of the Queensland koala (*Phascolarctos cinereus adustus*: Goldfuss). Unpublished Honours thesis. Brisbane: University of Queensland; 1994.

Johnston SD. Studies towards the development of an artificial insemination protocol in the koala. PhD thesis. Brisbane: University of Queensland; 1999.

Johnston SD. Challenges associated with the development and transfer of assisted breeding technology in marsupials and monotremes: lessons from the koala, wombat and short-beaked echidna. Reprod Fertil Dev. 2019. https://doi.org/10.1071/rd18113

Johnston SD, Holt WV. Germplasm conservation in marsupials. In: Watson PF, Holt WV, editors. Cryobanking the genetic resource wildlife conservation for the future? London: Taylor and Francis; 2001. p. 203–25.

Johnston SD, McGowan MR, Carrick FN, Cameron RDA. Towards the development of an artificial insemination programme for koalas (*Phascolarctos cinereus*). Proc 12th Int Congr Anim Reprod. 1992;4:1954–7.

Johnston SD, McGowan MR, Carrick FN, Tribe A. Preliminary investigations into the feasibility of freezing koala (*Phascolarctos cinereus*) semen. Aust Vet J. 1993;70:424–5.

Johnston SD, McGowan MR, Carrick FN, Cameron RD, Tribe A. Seminal characteristics and spermatozoal morphology of captive Queensland koalas (*Phascolarctos cinereus adustus*). Theriogenology. 1994;42:501–11.

Johnston SD, O'Callaghan P, McGowan MR, Phillips NJ. Characteristics of koala (*Phascolarctos cinereus adustus*) semen collected by artificial vagina. J Reprod Fertil. 1997;109:319–23.

Johnston SD, McGowan MR, O'Callaghan P. Assisted breeding technology for the conservation and propagation of *Phascolarctos cinereus* or how to make a koala pouch young. In: Proceedings of the Postgraduate Committee in veterinary science 1999. Sydney: University of Sydney; 1999. p. 119–216.

Johnston SD, McGowan MR, O'Callaghan P, Cox R, Nicolson V. Studies of the oestrous cycle, oestrus and pregnancy in the koala (*Phascolarctos cinereus*). J Reprod Fertil. 2000a;120:49–57.

Johnston SD, McGowan MR, O'Callaghan P, Cox R, Nicolson V. Natural and artificial methods for inducing the luteal phase in the koala (*Phascolarctos cinereus*). J Reprod Fertil. 2000b;120:59–64.

Johnston SD, McGowan MR, Phillips NJ, O'Callaghan P. Optimal physicochemical conditions for the manipulation and short-term preservation of koala (*Phascolarctos cinereus*) spermatozoa. J Reprod Fertil. 2000c;118:273–81.

Johnston SD, McGowan MR, O'Callaghan P, Cox R, Houlden B, Haig S, et al. Birth of koalas (*Phascolarctos cinereus*) at Lone Pine Koala sanctuary following artificial insemination. Int Zoo Yearb. 2003;38:160–72.

Johnston SD, O'Callaghan P, Nilsson K, Tzipori G, Curlewis JD. Semen-induced luteal phase and identification of a LH surge in the koala (*Phascolarctos cinereus*). Reproduction. 2004;128:629–34.

Johnston SD, MacCallum C, Blyde D, McClean R, Lisle A, Holt WV. An investigation into the similarities and differences governing the cryopreservation success of koala (*Phascolarctos cinereus*: Goldfuss) and common wombat (*Vombatus ursinus*: Shaw) spermatozoa. Cryobiology. 2006;53:218–28.

Johnston SD, Lopez-Fernandez C, Gosalbez A, Zee Y, Holt WV, Allen C, et al. The relationship between sperm morphology and chromatin integrity in the koala (*Phascolarctos cinereus*) as assessed by the sperm chromatin dispersion test (SCDt). J Androl. 2007;28:891–9.

Johnston SD, Zee YP, Lopez-Fernandez C, Gosalvez J. The effect of chilled storage and cryopreservation on the sperm DNA fragmentation dynamics of a captive population of koalas. J Androl. 2012a;33:1007–15.

Johnston SD, Satake N, Zee Y, Lopez-Fernandez C, Holt WV, Gosalvez J. Osmotic stress and cryoinjury of koala sperm: an integrative study of the plasma membrane, chromatin stability and mitochondrial function. Reproduction. 2012b;143:787–97.

Johnston SD, Mucci A, Ellis W. The role of captive koalas in koala conservation. In: Flint N, Melzer A, editors. A conserving Central Queensland's koalas. Rockhampton: Koala Research Centre of Central Queensland, Central Queensland University; 2013. p. 54–9.

Johnston S, Beagley K, Seddon J, Schultz B, Keeley T, et al. A new paradigm for koala conservation in SE Queensland: establishment of a living koala genome bank. In: Proceedings of the Queensland Environmental Law Association 2018 conference, 23–25 May 2018; 2018. p. 9–17.

Kusuda S, Araki S, Makino T, Doi O, Ito H, Hashikawa H, et al. Noninvasive monitoring of reproductive activity based on fecal progestagen profiles and sexual behavior in koalas, *Phascolarctos cinereus*. Biol Reprod. 2009;81:1033–40.

Kusuda S, Hashikawa H, Takeda M, Ito H, Goto A, Oguchi J, et al. Season- and age-related reproductive changes based on fecal androgen concentrations in male koalas, *Phascolarctos cinereus*. J Reprod Dev. 2013;259:308–13.

Larkin R, Palmieri C, Oishi M, Hulse L, Johnston SD. Ultrasonographic assessment of the male koala (*Phascolarctos cinereus*) reproductive tract. Res Vet Sci. 2018;117:219–23.

MacCallum C, Johnston SD. Studies on the cryopreservation of common wombat (*Vombatus ursinus*) spermatozoa. Reprod Fertil Dev. 2005;17:727–32.

Martin R, Handasyde K. The koala: natural history, conservation, management. 2nd ed. Kensington: University of New South Wales Press; 1999.

McAlpine CA, Melzer A, Lunney D, Foley WJ, Adams-Hosking C, Lawler IR, et al. Conserving koalas in the 21st century: synthesising the dynamics of Australia's koala populations. 2012. http://www.uq.edu.au/krn/ACEAS_Workshops_final_report.pdf. Accessed 25 Feb 2019.

McAlpine C, Lunney D, Melzer A, Menkhorst P, Phillips S, Phalen D, et al. Conserving koalas: a review of the contrasting regional trends, outlooks and policy challenges. Biol Conserv. 2015;192:226–36.

McFarlane JR. Endocrine factors influencing reproduction in male non-eutherian mammals. PhD thesis. Brisbane: University of Queensland; 1990.

Miller RR Jr, Sheffer CJ, Cornett CL, McClean R, MacCallum C, Johnston SD. Sperm membrane fatty acid composition in the Eastern grey kangaroo (*Macropus giganteus*), koala (*Phascolarctos cinereus*), and common wombat (*Vombatus ursinus*) and its relationship to cold shock injury and cryopreservation success. Cryobiology. 2004;49:137–48.

Mitchell P. Social behaviour and communication of koalas. In: Lee AK, Handasyde KA, Sanson GD, editors. Biology of the koala. Chipping North: Surrey Beatty; 1990. p. 151–70.

Monfort SL. "Mayday Mayday Mayday". The millennium ark is sinking! Adv Exp Med Biol. 2014;753:15–31.

Natural Resource Management Ministerial Council. National koala conservation and management strategy 2009–2014. Canberra. 2009. https://www.environment.gov.au/system/files/resources/165139fc-3ab5-4c96-8b15-d11a1ad882ab/files/koala-strategy.pdf. Accessed 25 Feb 2019.

NSW Chief Scientist and Engineer. Report of the independent review into the decline of koala populations in key areas of NSW. 2016. http://www.chiefscientist.nsw.gov.au/__data/assets/pdf_file/0010/94519/161202-NSWCSE-koala-report.pdf. Accessed 25 Feb 2019.

Obendorf DL. The pathogensis of urogenital tract disease in the koala. In: Australian wildlife, proceedings number 104. Sydney: University of Sydney Post-Graduate Committee in Veterinary Science; 1988. p. 649–55.

Oishi M, Takahashi M, Amasaki H, Janssen T, Johnston SD. The seminiferous epithelial cycle and microanatomy of the koala (*Phascolarctos cinereus*) and southern hairy-nosed wombat (*Lasiorhinus latifrons*) testis. J Anat. 2013;222:380–9.

Palmieri C, Hulse L, Pagliarani S, Larkin R, Higgins DP, Beagley K, et al. *Chlamydia pecorum* infection in the male reproductive system of koalas (*Phascolarctos cinereus*). Vet Pathol. 2018;56:300–6. https://doi.org/10.1177/0300985818806963.. [Epub ahead of print].

Pukazhenthi B, Comizzoli P, Travis AJ, Wildt DE. Applications of emerging technologies to the study and conservation of threatened and endangered species. Reprod Fertil Dev. 2006;18:77–90.

Queensland Department of Environment Heritage and Protection. Nature conservation (koala) conservation plan 2006 and management program 2006–2016. 2006. http://www.uq.edu.au/krn/QLD_KoalaConservationPlan.pdf. Accessed 25 Feb 2019.

Rhodes JR, Beyer HL, Preece HJ, McAlpine CA. South East Queensland koala population modelling study. Brisbane: UniQuest; 2015. https://environment.des.qld.gov.au/wildlife/koalas/pdf/seq-koala-population-modelling-study.pdf. Accessed 25 Feb 2019.

Rhodes JR, Hood A, Melzer A, Mucci A. Queensland Koala Expert Panel: a new direction for the conservation of koalas in Queensland. A report to the Minister for Environment and Heritage Protection. Brisbane: Queensland Government; 2017. https://environment.des.qld.gov.au/wildlife/koalas/pdf/qld-koala-expert-panel-report-2017.pdf. Accessed 25 Feb 2019.

Rodger JC, Hughes R. Studies of the accessory glands of male marsupials. Aust J Zool. 1973;21:303–20.

Rodger JC, Mate KE. Marsupial gametes and fertilisation. Today's Life Sci. 1993;5:28–33.

Rodger JC, Cousins SJ, Mate KE. A simple glycerol-based freezing protocol for the semen of a marsupial *Trichosurus vulpecula*, the common brushtail possum. Reprod Fertil Dev. 1991;3:119–25.

Rodger JC, Paris DB, Czarny NA, Harris MS, Molinia FC, Taggart DA, et al. Artificial insemination in marsupials. Theriogenology. 2009;71:176–89.

Schultz B, Hulse L, Nicolson V, Larkin R, Bromfield E, Nixon B, et al. Prolonged chilled preservation and preliminary investigations of energy production of koala spermatozoa. In: 13th int symp spermatology, 11–14th May 2018, Stockholm; 2018.

Shaw G. Reproduction. In: Armati PJ, Dickman CR, Hume ID, editors. Marsupials. Cambridge: Cambridge University Press; 2006.

Silva M, Niño MA, Letelier C, Godoy RH, Adams GP, Ratto MH. Is ovulation-inducing factor (OIF) present in the seminal plasma of rabbits? In: Annual meeting of the Society for the Study of Reproduction, Milwaukee; 2010.

Taggart DA, Steele VR, Schultz D, Dibben R, Dibben J, Temple-Smith PD. Semen collection and cryopreservation in the Southern Hairy-nosed Wombat *Lasiorhinus latifrons*: implications for conservation of the Northern Hairy-nosed Wombat *Lasiorhinus krefftii*. In: Wells RT, Pridmore PA, editors. Wombats. Chipping North: Surrey Beatty; 1998. p. 180–91.

Temple-Smith P, Taggart D. On the male generative organs of the koala (*Phascolarctos cinereus*): an update. In: Lee AK, Handasyde KA, Sanson GD, editors. Biology of the koala. Chipping North: Surrey Beatty; 1990. p. 33–54.

Threatened Species Scientific Committee. Listing advice for the koala. 2012. http://www.environment.gov.au/biodiversity/threatened/species/pubs/197-listing-advice.pdf. Accessed 25 Feb 2019.

Tyndale-Biscoe CH. Life of marsupials. Collingwood: CSIRO Publishing; 2005.

Tyndale-Biscoe CH, Renfree M. Reproductive physiology of marsupials. Cambridge: Cambridge University Press; 1987.

Wildt DE, Bush M, O'Brien SJ, Murray ND, Taylor A, Graves JA. Semen characteristics in free-living koalas (*Phascolarctos cinereus*). J Reprod Fertil. 1991;92:99–107.

Wildt DE, Ellis S, Janssen D, Buff J. Toward more effective reproductive science for conservation. In: Holt WV, Pickard AR, Rodger JC, Wildt DE, editors. Reproductive science and integrated conservation. Cambridge: Cambridge University Press; 2003. p. 1–20.

Woinarski J, Burbidge AA. The IUCN Red List of Threatened Species 2016: e.T16892A21960344. 2016. https://doi.org/10.2305/IUCN.UK.2016-1.RLTS.T16892A21960344.en. Accessed 25 Feb 2019.

Zee YP, Holt WV, Allen CD, Nicolson V, Burridge M, Lisle A, et al. Effects of cryopreservation on mitochondrial function and heterogeneity, lipid raft stability and phosphatidylserine translocation in koala (*Phascolarctos cinereus*) spermatozoa. Reprod Fertil Dev. 2007;19:850–60.

Zee YP, Holt WV, Gosalvez J, Allen CD, Nicolson V, Pyne M, et al. Dimethylacetamide can be used as an alternative to glycerol for the successful cryopreservation of koala (*Phascolarctos cinereus*) spermatozoa. Reprod Fertil Dev. 2008;20:724–33.

Zee YP, Lopez-Fernandez C, Arroyo F, Johnston SD, Holt WV, Gosalvez J. Evidence that single-stranded DNA breaks are a normal feature of koala sperm chromatin, while double-stranded DNA breaks are indicative of DNA damage. Reproduction. 2009a;138:267–78.

Zee YP, Holt WV, Nicolson V, Pyne M, Johnston SD. Individual variability in post-thaw sperm survival in a captive koala population. Cryobiology. 2009b;59:69–74.

Chapter 13
Reproductive Science Methods for Wild, Fully-Marine Mammals: Current Approaches and Future Applications

Janet M. Lanyon and Elizabeth A. Burgess

Abstract Determining reproductive rates of marine mammal populations can give insight into their persistence and resilience in changing environments. As our marine environments continue to degrade along developed coastal fringes and as mankind's influences extend across even our widest oceans, there is a concern that the reproductive functioning of marine mammals may be affected adversely. Since many marine mammal species and populations are still in the recovery phase post-commercial harvest, and yet others are endangered or threatened by ongoing pressures, further environmental changes may represent direct or indirect threats to their reproductive potential. In this chapter, we review the current methods employed to investigate various aspects of reproductive science in fully-marine mammals, including direct observation of reproductive behavior and output, endocrinology to determine reproductive state, and assisted reproductive technologies to enhance reproductive outcomes. In particular, we focus on the most recent developments and innovations to reproductively sample marine mammals. Two case studies are presented to illustrate the challenges thrown up to researchers studying free-ranging marine mammals, and to highlight diversity in research approach. The North Atlantic right whale is on the brink of extinction through historic overharvest and present-day entanglement and ship strike. Environmental disruption to their migration routes and declining population health has resulted in reduced reproductive rates. In contrast, the main current threats to the reproductive success and survival of the vulnerable dugong are extreme weather events that affect availability of its seagrass diet. Climate disruption with increasingly severe coastal storms and flooding threaten the health of coastal seagrass beds, and consequently reproductive success and survival of this species. It is anticipated that climate change may have diverse and often serious effects on marine mammal reproduction in populations around the globe.

J. M. Lanyon (✉)
Marine Vertebrate Ecology Research Group, School of Biological Sciences,
The University of Queensland, St Lucia, QLD, Australia
e-mail: j.lanyon@uq.edu.au

E. A. Burgess
Anderson Cabot Center for Ocean Life, New England Aquarium, Boston, MA, USA

P. Comizzoli et al. (eds.), *Reproductive Sciences in Animal Conservation*,
Advances in Experimental Medicine and Biology 1200,
https://doi.org/10.1007/978-3-030-23633-5_13

Keywords Reproduction · Marine mammals · Methods · Endocrinology · Observation · Cetaceans · Sirenians

1 Introduction

As global ocean environments continue to degrade through the activities of human beings (Worm et al. 2006; Halpern et al. 2008; IUCN 2018), the risk to the world's marine mammal fauna worsens (Avila et al. 2018). It has been estimated that approximately 74% of all marine mammals experience high levels of human impact, with the most at-risk species occurring in coastal areas that overlap with human activities (Davidson et al. 2012). However, even open ocean environments—previously considered largely protected from the worst of anthropogenic impact—are becoming industrialized, posing threats to marine mammals (McCauley et al. 2015). The greatest direct threats to marine mammals resulting in mortality and morbidity include incidental catch (including fisheries bycatch), pollution, direct harvest and vessel strike (Avila et al. 2018). Indirect threats such as habitat degradation and loss of food may cause disease and/or include impacts to their reproductive health and behaviour and thus to reproductive rate, and ultimately persistence of populations (Simmonds and Isaac 2007; Schipper et al. 2008). Moreover, the high mobility of many marine mammals may expose them to increased risks and/or cumulative impacts from multiple stressors (Maxwell et al. 2013).

Modelling studies suggest that the most important predictor of risk to marine mammals is their reproductive rate, since the speed at which they can produce offspring determines how fast they can rebound from human impacts (Davidson et al. 2012). Determining reproductive rate is therefore a crucial step required by conservation agencies when calculating extinction risk. Recent estimates of the number of marine mammal species at imminent risk of extinction, based on population size and predicted reproductive rate, range from 32 to 47 species, i.e., 25–37% of the world's marine mammal taxa, respectively (Schipper et al. 2008; Davidson et al. 2012). A further almost 40% of marine mammal species may also be at risk of extinction, but the life history data (including reproductive parameters) that would permit risk modelling are deficient (Davidson et al. 2012; IUCN 2018).

Determining reproductive parameters of wild marine mammals offers particular challenges to the researcher. Species that breed on land (pinnipeds, polar bears and sea otters) tend to be best understood because their terrestrial life stage(s) present opportunities to directly measure reproductive timing, pregnancy rates and recruitment into a population via sampling of individuals of different life stages including pups at haul-out sites (e.g., Jameson and Johnson 1993; Derocher and Stirling 1995; Boyd 2000; McKenzie et al. 2005, Gibbens et al. 2010; Oosthuizen et al. 2012). In contrast, fully aquatic marine mammals, i.e., cetaceans (dolphins, porpoises, whales) and sirenians (manatees and dugongs), offer relatively limited opportunities for direct observation because of their offshore and/or remote habitats, great mobility (including long breeding migrations), patchy distribution and/or cryptic natures that arise from spending large proportions of their lives underwater and out of sight (Costa 1993). For these reasons, the reproductive scientist must employ non-traditional, often indirect, methods to elucidate reproductive parameters (Lanyon and Burgess 2014). These differences in approach are reflected in the type and amount of reproductive information available for the major groups of marine mammals: information for the semi-aquatic pinnipeds is markedly more extensive (Atkinson 1997; Boyd et al. 1999), possibly because of this greater access to individuals at particular life stages. For the entirely-aquatic cetaceans, the bottlenose dolphin's reproductive biology is arguably the best known due to its accessibility to researchers, i.e., it is the species most commonly held in aquaria and is found in coastal waters worldwide, often in close proximity to humans (Connor et al. 2000; Leatherwood and Reeves 2012). In contrast, reproductive parameters of most other cetacean species and of the sirenians are relatively unknown.

Determining the potential of a wild marine mammal population to maintain or increase its size through reproductive growth is contingent on a suite of intrinsic factors including life history parameters (e.g., growth rate, age at maturity), reproductive cycles (e.g., timing of estrous, mating, gestation, birth, lactation, resting periods) and population size and structure (e.g., sex ratio, number of reproductively active adults). Reproductive rate is then tempered by how well individuals reproduce under various levels of resource limitation and external stressors, i.e., their reproductive fitness.

Marine mammals are K-selected species meaning that most are long-lived and slow to mature with protracted reproductive parameters, low birth rates and relatively large investment in their young (Fowler 1981). These life history traits make reproductive studies of marine mammal species challenging to conduct (i.e., requiring long-term datasets) but paramount to conservation practice because most populations are vulnerable to extrinsic mortality. For instance, the past heavy exploitation and incidental by-catch of many cetacean species has caused marked changes to many reproductive parameters of exploited populations (Fowler 1981, 1984). Slow breeding (including increased age at first calving, reduced number of reproductively active females, low calf production, longer inter-birth intervals) is a common response of long-lived iteroparous species such as marine mammals to adverse environmental conditions (DeMaster 1981; Kraus et al. 2001; Hadley et al. 2006). Interestingly though, in some baleen whales, enhanced reproductive capacity (i.e., decreased age at sexual maturation, increased proportion of mature females in ovulation and increased pregnancy rates) appears to have occurred, possibly in response to increased prey availability following reduction in whale numbers after commercial whaling (Lockyer 1984; Lockyer and Smellie 1985: Boyd et al. 1999). Those species with lower age at maturity have consistently been found to have a greater likelihood of recovery (Lotze et al. 2017).

Furthermore, many marine mammal species are distributed over wide latitudinal ranges and across multiple oceans or biogeographic regions (Schipper et al. 2008), so that reproductive parameters of demographically-isolated populations are likely to vary across environmental gradients. Intraspecific populations of marine mammals may therefore exhibit regional differences in asymptotic body size, pre-reproductive period, breeding frequency or season, length of reproductive cycles, and ultimately, reproductive potential and resilience. For example, eastern and northern spinner dolphin (*Stenella longirostris*) populations vary in body length, breeding seasonality and ovulation rates (Perrin and Reilly 1984; Perrin et al. 1985). Populations of pantropical spotted dolphins (*Stenella attenuata*) north and south of the equator show differences in each of timing of breeding seasons, proportion of pregnant females (Barlow 1985) and the average body length/age at puberty (Chivers and Myrick 1993). For some cetacean species, differences in growth and reproductive parameters have been sufficiently distinctive to identify discrete management units (e.g., Botta et al. 2010). Population life history studies have led to a better understanding of the basic biology of marine mammal species, their relationships to local environments (Börjesson and Read 2003; Danil and Chivers 2006), and to their stock structure (Dizon et al. 1994), behavior (Whitehead and Mann 2000), mating systems (Murphy et al. 2005) and resilience to anthropogenic threats (Chivers and Myrick 1993; Wells et al. 2005). When working to conserve marine mammals, it is therefore imperative that life history parameters appropriate to the focal populations are understood and applied to population models. As a fundamental step, reproductive scientists need reliable and useful methodologies to assess reproductive processes of marine mammals.

This chapter reviews the practical approaches that reproductive scientists employ today to examine various aspects of reproductive biology in marine mammals, with a focus on wild (free-ranging) populations of fully-aquatic species i.e., cetaceans and sirenians. Two case studies are presented to illustrate the challenges faced by researchers seeking reproductive information on free-ranging marine mammals. Understanding such reproductive patterns of marine mammals and determining the factors that influence their reproductive potential is critical if we are to identify and manage threats to population survival and growth.

2 How to Study Reproduction in Wild Fully Marine Mammals (Table 13.1)

2.1 Reproduction in Captive Marine Mammals

Marine mammals in captivity provide a unique opportunity to investigate species-specific behavior including vocalizations (Edds et al. 1993; Vergara et al. 2010; Ridgway et al. 2012), cognitive function (Mercado and deLong 2010), post-natal development (Lyamin et al. 2005; Favaro et al. 2013) and physiology (including dive response: Kooyman et al. 1981) that normally would be challenging to study in the wild. Extrapolation of information gained from captive individuals to wild populations can be complicated (see Lambrechts et al. 1999; Jenssen et al. 2001), because extended periods spent in artificial environments may induce variation in physiological traits (e.g., Mellish et al. 2006), including reproduction. Nonetheless, significant insights into reproductive cycles of marine mammals have been made

Table 13.1 Summary of reproductive techniques in both female and male marine mammals with referenced applications in free-ranging populations (where possible)

Technique	Sex	Significance	Example studies
Captive studies	F	Longitudinal monitoring, estrous cycling, gestational changes, growth rate	[a]Biancani et al. (2009) and [a]Steinman et al. (2012)
	M	Longitudinal monitoring, sexual maturity and activity, ejaculate characteristics, growth rate	[a]Robeck and Monfort (2006) and [a]Yuen et al. (2009)
Carcass dissection and histology, immunohistology	F	Ovarian follicular activity, previous ovulations and birth, absolute age	Marsh and Kasuya (1986) and Foote (2008)
	M	Testicular development, stage of spermatogenesis, absolute age	Desportes et al. (1993), Holt et al. (2004), and Murphy et al. (2005)
Direct observation	M/F	Mating behavior, social associations, calf abundance, length of nursing, intercalving period, growth and maturity rates	Würsig and Jefferson (1990), Parks et al. (2007), and de Bruyn et al. (2011)
Sexual dimorphism	M/F	Mating system, indicator of maturity	Whitehead (1994), MacLeod (1998), Burgess et al. (2012b), and Fitzpatrick et al. (2012)
Body scarring	M/F	Aggressive social interactions: sexual coercion, male-male competition, energetics	Scott et al. (2005) and Burgess et al. (2013b)
Endocrinology	F	Estrous cycling, pregnancy diagnosis, lactation, fertility potential	Rolland et al. (2005), Kellar et al. (2006), Burgess et al. (2012a), Wasser et al. (2017), and Pallin et al. (2018)
	M	Sexual maturity and activity, mating behaviour, development of secondary sexual characters, fertility potential	Rolland et al. (2005), Kellar et al. (2009), Wilson et al. (2011), and Burgess et al. (2012b)
Ultrasound	F	Reproductive tract activity, fetal development	[a]Brook (2001), [a]Lacave et al. (2004), and Burgess et al. (2012a)
	M	Gonadal activity	[a]Brook et al. (2000)
Spermatozoa analysis	M	Sperm morphology and morphometrics, ultrastructure, phylogenetic comparisons	Plön and Bernard (2006) and [a]Yuen et al. (2009)
Gamete cryopreservation	F	Oocyte preservation	Fujihira et al. (2006) and Bhuiyan et al. (2009)
	M	Sex sorting, artificial insemination (AI)	[a]O'Brien et al. (2008, 2009) and [a]O'Brien and Robeck (2012b)

[a]Technique primarily applied in captive population of marine mammals

through captive breeding programs, where individuals' reproductive status have been monitored in longitudinal studies of behavior and internal physiological state (e.g., Atkinson 1997; Boyd et al. 1999; Robeck et al. 2001, 2015).

Captive marine mammals for which reproductive parameters have been obtained, including within artificial breeding programs, comprise delphinids (dolphins, killer whales, false killer whales; e.g., Atkinson et al. 1999; Robeck and Monfort 2006; Katsumata 2010; Tizzi et al. 2010; O'Brien and Robeck 2012a), monodontids (beluga; e.g., Katsumata 2010; Steinman et al. 2012), sirenians (manatees, dugongs; e.g., Amaral et al. 2009; Burgess et al. 2013a) and pinnipeds (seals, sea lions, walrus; e.g., Atkinson and Gilmartin 1992; Myers et al. 2010; Kinoshita et al. 2012). Due to limitations of space and provision of adequate nutrition, marine mammals housed in zoological settings are typically commonly found in coastal waters, small (<10 m long) and mostly tractable. Larger marine mammals, such as all mysticete cetaceans (Edds et al. 1993; Sumich et al. 2001), have not been held in captivity past reproductive age. A number of factors may affect the quality and robustness of reproductive data obtained from captive animals. First, numbers of captive individuals are usually small. Second, individuals are often segregated by sex and age so that natural social groupings may be fragmented or nonexistent. Furthermore, the effects of highly artificial physical surroundings on an animal's reproductive behavior and functioning are not well understood.

Despite these limitations, observations of mating and subsequent parturition in captive animals have enabled at least initial estimates of age at maturity, reproductive seasonality, and length of gestation for many smaller species (McBride and Hebb 1948; McBride and Kritzler 1951; Essapian 1962; Cornell et al. 1987; Odell et al. 1995; Blanchet et al. 2006). More recently, training programs using operant conditioning along with advanced monitoring technologies (see Sects. 2.4 and 2.5) have enabled collection of serial biological samples and/or physical examinations under temporal regimes that are adequate to characterize reproductive cycles in live females and males (reviewed by O'Brien and Robeck 2012b).

A great advantage of captive marine mammals is that most can be trained to present voluntarily for collections of body fluids which may include blood (Robeck and Monfort 2006; Bauer et al. 2010), saliva (Pietraszek and Atkinson 1994; Hogg et al. 2005), respiratory exudate (Hogg et al. 2005; Richard et al. 2017), ocular secretions (Amaral et al. 2009), urine (Walker et al. 1988; Robeck et al. 1993, 2005a; Wakai et al. 2002), feces (Biancani et al. 2009), and milk (West et al. 2000), all of which may be useful for longitudinal monitoring of hormone profiles. The relatively noninvasive techniques of fecal sampling (and urinary sampling with training) can be conducted daily, and offer the best prospect for examining endocrine-reproductive relationships. Such investigations of reproductive cycling have been enhanced by monitoring body temperature (Terasawa et al. 1999; Katsumata et al. 2006a, b), examining the reproductive tract using trans-abdominal ultrasound (Brook et al. 2000, 2001, 2004; Lacave et al. 2004), conducting vaginal cytology of females (Pietraszek and Atkinson 1994), collecting semen from males (Schroeder and Keller 1989; Robeck and Monfort 2006; O'Brien et al. 2008; Robeck et al. 2009;

Yuen et al. 2009) and simultaneously documenting reproductive behavior (e.g., Tavolga and Essapian 1957; Shane et al. 1986).

Understanding the reproductive cycles of individuals has led to improvements in captive husbandry practices including management of pregnant females and fetal health (Williamson et al. 1990; Lacave et al. 2004; Katsumata et al. 2006b), establishment of socially cohesive groupings (Shane et al. 1986; Waples and Gales 2002; O'Brien and Robeck 2006), contraception (Atkinson et al. 1993; Briggs 2000; Calle 2005) and captive breeding programs (Robeck et al. 2005a, b). It has enabled the development of assisted reproductive technologies including artificial insemination (Robeck et al. 2004, 2005b, 2009, 2010; O'Brien and Robeck 2006, 2010; O'Brien et al. 2008), estrous synchronization (Robeck et al. 2009), sperm sexing (O'Brien and Robeck 2006; O'Brien et al. 2009), and semen preservation (Robeck and O'Brien 2004; O'Brien et al. 2008; Yuen et al. 2009). Understanding the relationships between reproductive behavior and physiological events for marine mammals in captivity has enabled facilities to determine and/or manipulate readiness for breeding, and to ensure successful conception for several cetacean taxa (Cornell et al. 1987; Robeck et al. 2005b; O'Brien and Robeck 2012a, b). With the increasing need for aquaria to display animals born in captivity rather than wild-caught individuals, the success of captive breeding programs for marine mammals is necessary to sustain *ex situ* populations and manage genetic bottlenecks that can occur in reproductively-isolated facilities (reviewed by O'Brien and Robeck 2012a, b).

2.2 Direct Observations

Direct observations of reproductive activity and behavior at sea of free-ranging, fully marine mammals vary from single or opportunistic sightings (e.g., courtship: Preen 1989; birth: Stacey and Baird 1997; copulation: Adulyanukosol et al. 2007; naso-suckling: Gero and Whitehead 2007) through to longitudinal sightings (Glockner-Ferrari and Ferrari 1990; Scott et al. 1990; Whitehead 1993; Rathbun et al. 1995; Langtimm et al. 1998; Mann et al. 2000a, b; Wells 2000; Rowntree et al. 2001; Burgess et al. 2012a, b). Field studies of all marine mammals must be conducted under research permits and following ethical guidelines, with the stringency of these regulations dictated by conservation status of the species or population, geographic location and or governing agencies. In some cases, these regulatory procedures hamper our ability to approach and study imperiled populations of marine mammals. Unsurprisingly, the quality of behavioral data varies markedly depending on approachability to the species and the observational methods applied. The observation methods should be selected carefully to meet the objectives and scale of the research project and the target species.

Distinguishing sexes of marine mammals through visual means can be challenging for species that exhibit no obvious sexual dimorphism in body size or morphology, e.g., most cetaceans and sirenians. Male marine mammals are testicond (ascrotal) with their testes permanently concealed in the abdomen (excluding otariid seals which have scrotal testes; Atkinson 1997) and the flaccid penis can be retracted into the body wall (Marsh et al. 1984b; Rommel et al. 1992), facilitating streamlining. Thus, for the majority of sexually-monomorphic species, the sex of most observed individuals remains unknown unless genetic sexing is accomplished. For most cetaceans and sirenians, mating has never been confirmed and courtship is difficult to interpret or recognize.

In fact, detailed information on patterns of courtship, mating and parturition in free-ranging animals exists for a small number of fully-marine mammal species only (e.g., Karczmarski et al. 1997; Stacey and Baird 1997; Hückstädt and Antezana 2001; Parks et al. 2007) and is mostly documented for pinnipeds that mate and give birth on land (e.g., Le Boeuf 1991; Cassini 1999; Acevedo et al. 2008; Karamanlidis et al. 2010). A general understanding of reproductive patterns for most entirely-aquatic species has largely been compiled from limited, often opportunistic, data and then analogized according to our knowledge of other mammalian species, e.g., primates (Wells 2003; Connor 2007). For some large odontocetes including beaked and bottlenose whales, their rarity, remote habitats and secretive habits ensure that they are rarely, if ever, observed alive (Heyning 1984; Thompson et al. 2012) so that their reproductive behavior remains a mystery. For many of the mysticete whales, direct observation is often fleeting and opportunistic, and rarely extensive or enlightening in terms of reproductive behavior. Exemplifying this point is the little we know about the reproductive habits of the largest animals that have ever existed, blue whales, *Balaenoptera musculus* (Yochem and Leatherwood 1985) despite having documented extensively various aspects of their ecology. Their elusive nature, deep-water habits and open geographic ranges (i.e., the only clear geographic boundaries to baleen whale movement are large continental land masses) have meant that breeding grounds of southern hemisphere populations remain unverified (Attard et al. 2012, 2016). However, for other large baleen whales (e.g., Kraus et al. 2001; Clapham 2008), some observations regarding reproductive behavior have been conducted at particular locations along predictable migration routes between feeding and breeding grounds (e.g., Everitt and Krogman 1979) or in breeding grounds themselves (e.g., Whitehead 1993; Knowlton et al. 1994; Cole et al. 2013), particularly for species that travel along topographic features such as coastlines or continental shelves. Amongst the best known of the marine mammals are those that spend significant time in coastal habitats (including migratory whales, delphinids, manatees) or at known breeding aggregation points, e.g., around islands, sea mounts, reefs and along mid ocean ridges.

Direct observation methods usually include distant sightings from elevated platforms: vessels, aircraft or land-based vantage points. Even though some fully-marine mammals may be approached closely by boat at times (e.g., manatees: Hartman 1979; dugongs: Preen 1992), most of their reproductive activities probably occur underwater and are unavailable to observers, so that some inference may be required (e.g., sex of observed individuals). In contrast, indirect 'observations' rely on telemetry to track animals, and these methods call for capture and attachment of radio/satellite/GPS/acoustic tags or recoverable data loggers (e.g., Mate et al. 1998;

Noad and Cato 2001; Deutsch et al. 2003). Information on location and movements of tagged individuals has been particularly useful for tracking breeding migrations, determining timing of reproductive cycles, or for examining smaller scale movements, social associations and reproductive behavior within breeding grounds (Mate et al. 1998, 2003; Russell et al. 2013). Most recently, advances in aerial drone (or unmanned aerial vehicles) technology offer new opportunities for remotely studying marine mammals at sea, with a huge potential to improve the safety, cost, and efficiency of data collection (Fiori et al. 2017; Pirotta et al. 2017).

Direct visual observations may also be conducted at the population level and may yield information regarding spatial distribution of animals, density and timing of breeding migrations, adult aggregations, breeding associations, and presence of neonates. In these cases, individual identity of group members is not always necessary. Examples of this approach include aerial surveys of humpback whale adults aggregating within breeding grounds (Andriolo et al. 2006), surveys of right whale nursery areas and adult aggregations (Rowntree et al. 2001; Parks et al. 2007) and calf counts for various taxa as an index of annual recruitment of breeding success (Perryman et al. 2002; Sobtzick et al. 2017). Compilation of these observations over extended temporal scales gives information regarding reproductive timing and breeding patterns.

Behavioral observations are greatly enhanced through recognition of individuals. If individual animals can be distinguished and identified through discriminate marks (permanent pigmentation, dorsal fin shape and/or body scarring, genetic tags) or artificially-applied marks (physical tags), there is the opportunity for longitudinal studies of life history. Recognizable individuals allow for descriptions of basic activity patterns (resting, socializing, travelling, feeding) as well as interpretation of inter-specific behaviors, especially if sex and reproductive condition are known (e.g., Connor et al. 1992; Smolker et al. 1992; Brown et al. 1995). For example, long-term observations of known identified mother-calf pairs of Florida manatees at winter aggregation sites have provided insight into frequency of calving, intercalving period, growth rates of calves, duration of nursing, time to maturity and other reproductive measures critical to population modeling (Langtimm et al. 2004). The value of longitudinal sightings data lies not only in compilation of life history parameters relevant to reproductive biology (including reproductive and total life span if individuals are recognized from an early age), but in elucidation of social behaviors including social associations, mate selection, nurturing of young and other details about a species' breeding behavior (Mann et al. 2000a, b). Individual identification has become a staple of many field studies of cetaceans (reviewed by Würsig and Jefferson 1990).

Despite the technical difficulties in directly observing and identifying mammals in the field, our understanding of reproductive behavior and social interactions is improving. New techniques that allow us to track individuals through time and space, such as the amplification of molecular markers from tissue biopsies for tagging as well as advanced telemetry and biologging (Block 2005; Bograd et al. 2010; McIntyre 2014; Hays et al. 2016; Hückstädt et al. 2018), are providing superior insights into social structure and social associations related to reproduction.

2.3 Dissection of Carcasses

Almost all of the early information regarding basic reproductive patterns of wild cetaceans and sirenians was derived from the examination of reproductive tracts recovered from carcasses obtained from hunted (e.g., Best 1969; Miyazaki 1984; Desportes et al. 1993; Dawbin 1997; Kwan 2002) or stranded animals (e.g., Calzada et al. 1996; Evans and Hindell 2004; Bezerra et al. 2018) or from those incidentally caught and killed in fishing nets (e.g., Read 1990; Van Waerebeek and Read 1994; Hohn et al. 1996), shark nets (e.g., Marsh et al. 1984a, b; Cockcroft and Ross 1990) or from other anthropogenic impacts (Perez 2015). Early studies using post-mortem material provided information on gross morphology and anatomy of the reproductive tract (e.g., Harrison and Ridgway 1971). Functional morphology of reproductive tracts, histology of gonads and mammary glands, and embryology (van der Schoot 1995; Thewissen and Heyning 2007; Chivers 2018) have provided information on the timing of reproductive activity, conception, gestation, birth and lactation, and also information concerning age and/or body size-related life history parameters including first reproduction and numbers of offspring produced over lifetimes (e.g., cetaceans: Laws 1961; Ohsumi and Masaki 1975; Lockyer and Martin 1983; Marsh and Kasuya 1984, 1986; Kato and Sakuramoto 1991; Read and Hohn 1995; sirenians: Marsh et al. 1984a, b; Hernandez et al. 1995; Marmontel 1995; Kwan 2002). For example, if age can be determined (see Read et al. 2018), age at sexual maturity in males can be resolved through assessment of testicular development (including seminiferous tubule diameter, Sertoli and Leydig cell activity), relative weight and length of testes, stage of spermatogenesis and presence/abundance of sperm (e.g., Perrin and Henderson 1984; Perrin and Reilly 1984). Immunocytochemical methods have also been applied to examine maturation of gonads: these include screening for distribution of cytoskeletal proteins (e.g., smooth muscle actin and vimentin) as markers of testis development (Holt et al. 2004). For females, ovarian follicular activity is indicative of maturity, and previous ovulations have been confirmed by presence of *corpus luteum* and/or *corpus albicans*, since in many marine mammals, ovarian scars appear to be persistent (e.g., Marsh and Kasuya 1984; Perrin and Reilly 1984). Pregnancy rate and calving intervals can be determined though counts of placental scars (e.g., dugongs, Marsh et al. 1984a) or ovulatory scars (in cetaceans, most are thought to represent pregnancies, Boyd et al. 1999) regressed against age or body length at death. Timing of sexual activity is suggested by the presence of mature spermatozoa in males and ovarian activity in females (Perrin and Reilly 1984), and reproductive senescence through their absence (Marsh and Kasuya 1986; Ellis et al. 2018).

Carcass analysis, for all of its value, is prone to both temporal and/or spatial biases (Boyd et al. 1999). Carcasses are usually sourced from places where hunting, harvest or other causes of mortality are most likely, or from areas where reportage or recovery of carcasses are most common (e.g., Chittleborough 1954; Marsh et al. 1984a, b; Marsh and Kasuya 1986; Desportes et al. 1993; Kwan 2002). In some cases, certain individuals or life stages are more likely to be sampled, e.g., the targeted fishery for dugongs where hunters preferentially take pregnant rather than resting females (Boyd et al. 1999); or samples obtained at certain times of the year, e.g., from baleen whales during the hunting season (Chittleborough 1954; Laws 1961). Further, the utility of a carcass depends on its degree of decomposition, so that if carcasses from warmer climates decompose faster, less information may be available for these than for animals from higher latitudes. However, the value of carcasses for studying reproduction and other facets of marine mammal biology cannot be understated. Indeed, for some rare marine mammal species, carcass recovery has been the only avenue for direct examination (e.g., Reyes et al. 1991; Thompson et al. 2012). In many cases, however, recoveries are too few to obtain useful information regarding reproductive biology. In contrast, for those species that strand *en masse*, recovery of multiple carcasses has yielded reproductive profiles of social groupings (e.g., Mignucci-Giannoni et al. 2000). When using carcass analysis to obtain information, there is significant value in establishing coordinated responses to marine mammal strandings (e.g., Thompson et al. 2012) at regional or national levels to ensure that, when rescue attempts fail, there is timely collection of tissues (including from reproductive tracts) and other vital data.

2.4 Endocrinology

Since most reproductive processes are mediated through hormones circulated in the bloodstream, and hormonal functions tend to be conserved across mammalian taxa, endocrine analysis is a useful means of assessing reproductive status (e.g., sexual maturity, pregnancy diagnosis) and reproductive activity (e.g., estrous cycles, seasonality), particularly for live individuals.

The most commonly measured hormone in marine mammals is progesterone (Table 13.2) since its diagnostic use indicates the onset of female sexual maturity or pregnancy. Understanding reproductive status of females provides vital information for the management of both wild (e.g., Villegas-Amtmann et al. 2009; Burgess et al. 2012a; Kellar et al. 2013a, 2014, 2017; Wasser et al. 2017; Pallin et al. 2018; Riekkola et al. 2018) and captive populations (e.g., Robeck et al. 2005a; O'Brien and Robeck 2012a, b). The other sex steroid hormones (Table 13.2) that are routinely measured are testosterone (e.g., Atkinson and Gilmartin 1992; Desportes et al. 1994; Kellar et al. 2009; Burgess et al. 2012b) and various forms of estrogen (e.g., Francis-Floyd et al. 1991; Pietraszek and Atkinson 1994; Robeck et al. 2005b; Fragalà et al. 2015). Gonadotropins including luteinizing hormone (LH) and follicle stimulating hormone (FSH), and other protein hormones (Table 13.2) have been measured primarily in cases of assisted reproduction (i.e., artificial insemination or gamete harvesting), often requiring multiple samples to be collected daily (Robeck et al. 2004, 2009; Muraco et al. 2010), although some studies exploiting opportunistic sample collection have been reported (Suzuki et al. 2001; Watanabe et al. 2004; Hao et al. 2007). Recent advances include investigation of oxytocin and its links to maternal behavior in wild marine mammal populations (Robinson et al. 2015) and the use of relaxin for pregnancy diagnosis (Bergfelt et al. 2013; Roebeck et al. 2018).

Table 13.2 Summary of endocrine glands and hormones investigated in marine mammals with referenced applications in free-ranging populations (where possible)

Gland	Hormone	Chemical class	Principal functions	Example studies
Ovary Graafian follicle	Estrogens (e.g. estradiol)	Steroid	Female mating behavior, secondary sex characteristics, maintenance of female duct system, mammary growth	[a]Robeck et al. (2005b), Rolland et al. (2005), and [a]Steinman et al. (2012)
	Inhibin	Protein	Regulates release of FSH from anterior pituitary	Wetzel et al. (2009)
	Anti-Müllerian hormone (AMH)	Protein	Sophisticated biomarker of reproductive potential	Wilson et al. (2011)
Ovary Corpus luteum	Progestins (e.g. progesterone)	Steroid	Maintenance of pregnancy, mammary growth and secretion	Rolland et al. (2005), Burgess et al. (2012a), and [a]O'Brien and Robeck (2012a)
	Relaxin	Polypeptide	Expansion of pelvis, dilation of cervix	Schwabe et al. (1989) and [a]Bergfelt et al. (2011)
Testis Leydig cells	Androgens (e.g. testosterone)	Steroid	Male mating behavior, spermatogenesis, maintenance of male duct system and accessory glands	Rolland et al. (2005), [a]Robeck and Monfort (2006), and Burgess et al. (2012b)
Sertoli cells	Inhibin	Protein	Regulates release of FSH	Miller et al. (2002a) and [a]Katsumata et al. (2012)
	Anti-Müllerian hormone (AMH)	Protein	Sophisticated biomarker of reproductive potential	Wilson et al. (2011)
Adrenal cortex	Glucocorticoids (e.g. cortisol, corticosterone)	Steroid	Stress response, parturition induction, milk synthesis	[a]Hunt et al. (2004, 2006), Myers et al. (2010), and Burgess et al. (2013b)
Pineal gland	Melatonin	Biogenic amine	Control of seasonal reproduction	Barrell and Montgomery (1989), Aarseth et al. (2003), [a]Funasaka et al. (2011), and [a]Panin et al. (2012)

Posterior pituitary	Oxytocin	Octapeptide	Parturition and milk ejection in females	Archer et al. (1964) and Eisert et al. (2013)
Anterior pituitary	Follicle stimulating hormone (FSH)	Glycoprotein	Stimulate follicle growth and estrogen production in females, spermiogenesis in males	[a]Walker et al. (1988), Gardiner et al. (1999), Suzuki et al. (2001), Watanabe et al. (2004), and Hao et al. (2007)
	Luteinizing hormone (LH)	Glycoprotein	Stimulate ovulation, support corpus luteum formation and progesterone secretion, stimulate testosterone synthesis by Leydig cells of testis	Gardiner et al. (1999), Suzuki et al. (2001), Watanabe et al. (2004), [a]Robeck et al. (2005b), and Hao et al. (2007)
	Prolactin	Protein	Development of mammary function, maintain lactation, effects on maternal behavior, post-partum estrous cycling	Boyd (1991) and [a]Steinman et al. (2012)
	Adrenocorticotrophic hormone (ACTH)	Protein	Release of corticosteroids and glucocorticoids, initiate parturition	[a]Schmitt et al. (2010), Tripp et al. (2010), and Schaefer et al. (2011)
Hypothalamus	Gonadotrophic releasing hormone (GnRH)	Decapeptide	Stimulate release of FSH and LH from anterior pituitary	Atkinson et al. (1998), [a]Robeck et al. (2010), and [a]O'Brien and Robeck (2012a)

[a]Technique primarily applied in captive population of marine mammals

In early studies, endocrine analysis relied on collecting blood samples to determine circulating hormone concentrations so that marine mammals held in captivity were sampled more frequently than free-ranging individuals (Yoshioka et al. 1986; Cornell et al. 1987; Duffield et al. 1995), and stranded carcasses were sampled for heart blood post-mortem (Higgins and Noad 2006). The first ground-breaking hormonal study on marine mammals investigated serum testosterone in two captive male bottlenose dolphins (Harrison and Ridgway 1971). Broad-scale blood sampling for endocrine analysis in field studies has been possible in few circumstances such as during haul-out of the semi-aquatic pinnipeds (e.g., Atkinson and Gilmartin 1992; Gardiner et al. 1996; Harcourt et al. 2010; Robinson et al. 2015), or when sampling relatively small numbers of fully aquatic mammals during health assessments (e.g., Wells et al. 2004; Lanyon et al. 2010b; Bonde et al. 2012) or even during lethal harvests (e.g., Suzuki et al. 2001; Kjeld et al. 2004, 2006). The need for physical restraint out-of-water to ensure collection of blood samples uncontaminated by seawater has hampered endocrine evaluations in most species, particularly of large or rare cetaceans (Hunt et al. 2013, 2015). Recently, sampling of diverse media other than blood for hormone profiles has been increasingly recognized as a useful approach for free-ranging aquatic mammals (reviewed by Amaral 2010; Hunt et al. 2013; de Mello and de Oliveira 2016).

Hormones are removed from circulation through metabolic processes and eventually excreted from the body through various pathways (depending on the species), providing alternative opportunities (from blood) for measuring hormones and their byproducts. Hormone concentrations have been measured in a variety of body tissues and excreta from marine mammals, including saliva (Pietraszek and Atkinson 1994; Atkinson et al. 1999; Hogg et al. 2005; Amaral et al. 2009), milk (West et al. 2000), ocular secretion (Atkinson et al. 1999; Amaral et al. 2009), respiratory exudate (Richard et al. 2017; Burgess et al. 2018), adipose tissue (blubber) (Trego et al. 2013; Kellar et al. 2014; Beaulieu-McCoy et al. 2017; Mello et al. 2017), epidermal skin (Bechshoft et al. 2015), muscle (Yoshioka et al. 1994), bone (Charapata et al. 2018), urine (Robeck et al. 1993; Wakai et al. 2002; Constable et al. 2006; Amaral et al. 2009, 2014; Kellar et al. 2013a; Steinman et al. 2012) and feces (Rolland et al. 2005; Biancani et al. 2009; Burgess et al. 2012a, b; Wasser et al. 2017). Stratified or growing tissues have also been sampled successfully; these include earplugs (Trumble et al. 2013), hair/whiskers (Bechshøft et al. 2011; Karpovich et al. 2018) and baleen plates (Hunt et al. 2014b, 2016, 2018). These non-conventional endocrine approaches have been widely applied to monitoring reproductive processes in captive individuals (see Sect. 2.1) and show great promise for live, free-ranging populations. Hormone evaluation for some post-mortem cases can even provide retrospective information (Trumble et al. 2013; Hunt et al. 2016, 2018). In fact, measuring levels of excreted hormones in other media such as feces may offer advantages over blood levels because these represent averaged values pooled over time (e.g., integrated over the gut passage time), rather than a single point-in-time measure. These sample types also present as physiological indicators less influenced by the acute physiological response of sampling blood. Consequently, the development of these less invasive techniques (especially fecal and blubber sampling) has enabled the physiological study of large cetaceans to expand beyond lethal sampling (Fossi and Marsili 1997). However, it is important that for each new species under study, careful biochemical and biological validation of hormones in each sampled medium is conducted (Lasley and Kirkpatrick 1991). Moreover, researchers must carefully consider the processes of hormone integration and preservation in the sample matrix, as well as the biology, ecology and habits of the species, as well as the project's scale, when deciding on which media are most appropriate (see Hunt et al. 2013).

To date, the most commonly used matrices for hormone determination of live, free-ranging cetaceans and sirenians have been feces (e.g., Larkin et al. 2005; Rolland et al. 2005; Amaral et al. 2009; Burgess et al. 2012a, b; Wasser et al. 2017) and blubber (e.g., Kellar et al. 2013a, b, 2014, 2017; Pallin et al. 2018). Obviously, logistical difficulties arise with collection of such samples from elusive and typically uncatchable, large marine mammals (reviewed by Hunt et al. 2013, 2015). For fecal hormone analysis, floating feces voided by large whales have been sampled by dip-net during focal follows of, for example, right whales (Rolland et al. 2005; Wade et al. 2011), blue whales (Valenzuela-Molina et al. 2018), humpback whales (Lefebvre et al. 2002), gray whales (Newell and Cowles 2006), sperm whales (Marcoux et al. 2007; Rolland et al. 2015) and killer whales (Wasser et al. 2017). Sinking feces have also been collected from smaller cetaceans (e.g., bottlenose dolphins and even Blainville's beaked whales), by snorkelers towed by boats travelling in close proximity to mobile groups (Parsons et al. 1999, 2003; Rolland et al. 2015), although a large proportion of fecal material disperses too rapidly for collection. One innovation to facilitate collection has been the use of on-board 'sniffer' dogs trained to detect floating whale feces (Rolland et al. 2006). Another approach has been to collect feces of uncertain origin, either floating or as stools washed ashore, and then conduct molecular analysis for taxon and/or sex identity (Parsons et al. 1999; Green et al. 2007; Gillett et al. 2010). For migratory whales, fecal collection is often restricted to opportunistic sampling in feeding grounds along migration routes, which poses limitations for any study requiring repeated sampling of a targeted individual whale, throughout seasonal movements and/or fasting periods. In live-capture studies of smaller marine mammals, fecal samples have been collected either opportunistically as they are voided during restraint (Lanyon et al. 2005, 2010b; Larkin et al. 2005) or directly by inserting a soft latex tube into the distal rectum (Lanyon et al. 2005; Biancani et al. 2009; Burgess et al. 2012a, b).

For blubber hormone analysis, investigations were initially trialled on dead cetaceans collected from directed hunts, incidental bycatch or strandings (Mansour et al. 2002; Kellar et al. 2006, 2009, 2013a; Trana et al. 2015; Mello et al. 2017). Increasingly, remote biopsy darting has been used to obtain small cores of fatty hypodermis (approximately 4–6 mm in diameter × 5–30 mm in length) from numerous free-swimming cetaceans (Pérez et al. 2011; Kellar et al. 2013b, 2014, 2017; Trego et al. 2013; Vu et al. 2015; Clark et al. 2016; Mello et al. 2017; Hayden et al. 2017; Pallin et al. 2018) and this also has potential application in pinnipeds (e.g., Hoberecht et al. 2006; Beaulieu-McCoy et al. 2017). Biopsy sampling techniques for cetaceans were initially established for non-endocrine studies including genetics, stable isotopes (foraging ecology), contaminant burdens and fatty acid analysis. A variety of approaches has been used to collect biopsies (e.g., crossbow or pneumatic rifle modified with dart tips), with the optimal technique depending on the behavior and morphology of the target species, and the platform from which the sampling is being conducted, e.g., boat or land-based (Noren and Mocklin 2012). Although collection of blubber biopsies is considered less invasive compared to some other approaches, despite its value, it may not be appropriate for vulnerable species, certain life history stages (e.g., mothers and young calves) or populations—especially if conservation concerns deem any risk to an animal's health unwarranted.

Perhaps the most innovative recent approach for measuring hormone levels in marine mammals has been the use of respiratory exudate or exhaled 'blow' (Tizzi et al. 2010; Aksenov et al. 2014; Hunt et al. 2014a; Richard et al. 2017; Burgess et al. 2018), which was first described by Hogg et al. (2005). Cetaceans ventilate a larger percentage of the respiratory tract with greater force than other mammals and their lungs are heavily vascularized, causing expulsion of a substantial volume of lung exudate droplets (along with respiratory gases) with each exhalation (Piscitelli et al. 2013). The appeal of investigating whale blow is based on its entirely non-invasive approach and potential frequency of sample collection in highly mobile aquatic species, since cetaceans must come to the surface of the water to breathe. Respiratory samples can be collected from free-swimming whales using pole-based methods (Hogg et al. 2009; Hunt et al. 2014a; Burgess et al. 2018) (Fig. 13.1) or by

Fig. 13.1 Collecting a blow sample from a North Atlantic right whale, using a polystyrene dish deployed at the end of a long pole and positioned above the exhaling blowholes [Photo was taken under U.S. NOAA Permit #14233 and Canada DFO Permit under Species at Risk Act. Photograph courtesy of the Anderson Cabot Center for Ocean Life at the New England Aquarium]

a small, remote-controlled drone (i.e., unmanned aerial vehicle (Acevedo-Whitehouse et al. 2010; Apprill et al. 2017; Pirotta et al. 2017). Importantly, researchers must carefully consider the collection materials used to catch the aerosol droplets of whale blow, and be aware of the level of hormone assay interference produced by these materials during sample processing, to ensure that signals of biological interest are detectable (Burgess et al. 2016). Technical considerations for blow sampling are especially important for studies on smaller marine mammals, whose collected exhalations may not yield sufficient hormones for robust analysis (e.g., inherent low sample volume is below the limit of analytical detection (JML and EAB unpubl. data)) and/or whose blow hormone measurements may be negatively influenced by exogenous factors (Mingramm et al. 2019). Hormone concentrations of field-collected whale blow samples need to be normalized using a dilution factor (such as urea) (Burgess et al. 2018), since reliable volumetric measurement is not possible due to variable amounts of water vapor and a high potential for seawater contamination of a sample that occurs when the whale breaks the surface of the water for exhalation (Hogg et al. 2009; Hunt et al. 2014a; Burgess et al. 2018). Although sampling whale blow at sea can be challenging, it has proved feasible to collect samples from targeted individuals, as well as to obtain repeated samples over time (Burgess et al. 2018). These major sampling benefits, along with the promise that these may be chemically analyzed to answer a number of biological questions (Hunt et al. 2013), have encouraged the development of this approach for physiological assessment of cetaceans.

Baleen, a taxonomically unique tissue to the mysticete whales, is also proving highly useful for endocrine assessment of the largest and least tractable animals on earth (Hunt et al. 2016, 2017, 2018), and arguably, the most difficult to study from a physiological perspective (Hunt et al. 2015). Baleen comprises a series of keratin plates, suspended from the upper jaw, that comprise the filter-feeding apparatus of mysticetes ('baleen whales'). Similar to mammalian hair, baleen is a cornified tissue that contains circulating endocrine compounds that have been deposited during its growth (Hunt et al. 2017). Baleen grows continuously throughout the life of the whale, and the extensive length of a baleen plate presents an exceptional tissue for retrospective, longitudinal analyses (Hunt et al. 2014b, 2018). A baleen plate is at least 25 cm and up to 300+ cm long, depending on the species—it maintains a constant functional length because the older growth at the distal end steadily erodes due to mechanical wear during feeding. In whale species with shorter baleen (e.g., gray whale, minke whale), the baleen reflects at least a full year of growth, whereas for species with the longest baleen (i.e., bowhead and right whales), a single baleen plate may represent a decade or more of tissue, useful for retrospective analysis (see Hunt et al. 2017; Fleming et al. 2018). These growth durations are sufficiently long to encompass entire pregnancies, estrous cycles, seasonality, inter-calving intervals, and other physiological responses through time (Hunt et al. 2016, 2018). A major caveat to this approach is that baleen must be salvaged from deceased whales, with studies gathering specimens from necropsy investigations and/or subsistence hunts, and even archived specimens. Given that baleen is metabolically inert and preserved dry, the possibility exists to use museum archival collections to gain insights into historical physiological trends of ancient whales.

Another recent endocrine approach that utilizes dead specimens is the analysis of hormones in the concentric layers of cerumen (ear wax) in whale earplugs (Trumble et al. 2013). This method has the advantage of capturing the whale's entire lifetime, but is limited in that earplugs are not produced in all mysticetes, are difficult to excise, decompose post-mortem, and the temporal resolution has proven to be relatively coarse (Trumble et al. 2013).

The discipline of conservation endocrinology is still evolving for free-ranging marine mammals, and will undoubtedly continue to provide new and critical reproductive information to assist management of marine mammal populations into the future. The diversity of marine mammal species, and their habits, may offer researchers unforeseen opportunities to explore novel sample types and develop innovative approaches to achieve physiological sample collection in a challenging environment. For example, the routine collection of fluid samples is logistically difficult for marine mammal taxa, yet one unique study on Weddell seals *Leptonychotes weddellii* analyzed hormones in urine that had serendipitously become preserved in ice (Constable et al. 2006). Perhaps more impressive is that these alternative techniques are enabling endocrine sampling of wild populations in the most remote locations on the planet, such as blubber sampling of free-swimming humpback whales *Megaptera novaeangliae* in Antarctic waters (Pallin et al. 2018). Ultimately, these less-invasive means of hormone monitoring (i.e., using non-blood samples) provide marine mammal scientists with the capability to conduct both basic and applied physiological-endocrine research that can be integrated with other disciplines including genetics, behavior, nutrition, animal health, ecology and evolution. Exciting developments in wildlife endocrine studies are enabling us to move beyond assessing an individual's reproductive status to being able to understand an individual's reproductive potential (i.e., fertility status).

Managers of marine mammal populations need to know not only how many individuals are present and/or how many are of breeding age (i.e., effective population size), but also understand an individual's contribution to population growth. Recently, sophisticated biomarkers (including anti-Müllerian hormone (AMH), inhibin A, inhibin B) have been developed to provide direct measures of an individual's gonadal function as well as fertility potential (ovarian reserve) in both female and male mammals (Table 13.2) (Lee and Donahoe 1993; Tremellen et al. 2006; Kumar et al. 2010). The first reports of detection of AMH and inhibin A and B in marine mammals, including Florida manatees (Wetzel et al. 2009; Wilson et al. 2011), bottlenose dolphins (Schwierzke 2011), beluga whales (Katsumata et al. 2012; Montano et al. 2017) and dugongs (JML and EAB unpubl. data) suggest that these biomarkers offer promise as a means of evaluating the 'reproductive quality' of individuals comprising a population. With careful development and validation, the use of such biomarkers may provide insight into the effects of stressors on critical biological functions such as reproduction, which can then be used to inform effective and focused conservation efforts.

2.5 Diagnostic Imaging

Diagnostic imaging (ultrasonography and endoscopy) has been used to examine the urogenital tracts and implement advanced reproductive technologies (including artificial insemination) of terrestrial wildlife (Hildebrandt et al. 2003). These techniques have been applied to fully marine mammals, but in relatively limited ways due to the need for restraint and immobility during the procedures. Ultrasonography is a useful technique for examination and differentiation of internal soft tissues and compared to other reproductive imaging techniques is arguably the simplest, safest, least invasive and most cost-effective method. It has been useful for obtaining vital information regarding reproductive activity and fetal development in mostly captive cetaceans (e.g., Brook 1997, 1999; Brook et al. 2001, 2004). Recent advances in ultrasound technologies have enabled researchers to examine gonadal activity through development of testes and ovarian follicles, monitor ovulatory cycles, implement artificial insemination and estimate gestational age of fetuses and therefore parturition date for small captive marine mammals (e.g., Brook 1997; Robeck et al. 1998, 2001, 2005b; Brook et al. 2001, 2004; Lacave et al. 2004). Some captive cetaceans have been trained to station for ultrasound examination (e.g., Brook et al. 2001), whilst in a field situation, ultrasound may be applied to individuals removed from the water (e.g., Wells et al. 2004; Goldstein et al. 2006) or even restrained alongside a vessel if a portable device with waterproof transducer is deployed (e.g., Nichols 2005). For larger, but still tractable marine mammals, such as big males or individuals with thick blubber layers, ultrasonography may be relatively challenging and the outcome of examinations doubtful, whilst it is not practical for the largest free-ranging cetaceans.

Perhaps the greatest value of diagnostic sonography with respect to reproduction has been as an adjunct tool to physical examination and endocrinology. The use of ultrasonography for validation of reproductive state has been crucial in developing hormone-based diagnostic tests for pregnancy (e.g., Burgess et al. 2012a) and for confirming pregnancy in captive cetaceans. Pseudo-pregnancy is possible in captive marine mammals (i.e., elevated hormone levels in females that are not carrying a fetus) (Atkinson 1997; West et al. 2000) and it is only through the use of ultrasound examination, that true pregnancy can be discriminated from this condition (Robeck et al. 2001). Furthermore, ultrasound combined with endocrine monitoring offers insight into how hormonal changes directly relate to ovulation (Robeck et al. 2001). Interestingly, the first evidence of facultative-induced ovulation in a cetacean, the beluga, was determined using serial urinary hormone monitoring combined with ovarian ultrasound (Steinman et al. 2012). Reproductive cycles of both males and females of some cetacean taxa have been determined through routine repeat ultrasounds of the gonads (e.g., Indo-Pacific bottlenose dolphins *Tursiops aduncus*: Brook 1997; Indo-Pacific humpback dolphins *Sousa chinensis*: Brook et al. 2004). As ultrasound devices become smaller, more robust and affordable, and their value apparent, their use in marine field situations is only likely to increase.

Endoscopy is a minor surgical procedure and, although more invasive than ultrasonography, allows direct visual internal examination through minimal tissue trauma. Laparoscopy of the abdominal cavity of marine mammals has been conducted for some captive cetaceans and pinnipeds, mostly as a diagnostic tool for disease or for removal of swallowed foreign objects, with limited application in reproductive studies (Dover and van Bonn 2001). Flexible endoscopes have also been used to guide practitioners during artificial insemination operations of captive cetaceans (Robeck et al. 1994; Robeck 2000; Dover and van Bonn 2001).

2.6 Gamete Collection and Cryopreservation

Spermatogenesis is an important indicator of the reproductive status of the male mammal, including the onset of sexual maturity, the start of breeding in seasonally reproducing species and gamete quality. Studies of the morphology of spermatozoa have been extensive for land mammals but remain in their infancy for marine mammals because of limited access to gametes. Detailed studies of the gross morphology of spermatozoa from wild marine mammals have been mostly limited to samples collected post-mortem from the epididymis and/or vas deferens, e.g., in cetaceans (reviewed by Plön and Bernard 2006; Kita et al. 2001), pinnipeds (Cummins and Woodall 1985) and sirenians (Marsh et al. 1984b; Miller et al. 2001). Sperm rescued postmortem from urogenital tracts of variable states of decomposition (dependent on timely collection and preservation, e.g., Mogoe et al. 1998; Miller et al. 2002a, b; Hiwasa et al. 2009), often comprise small samples of variable quality and viability, which makes comparisons between both taxa and ontogenetic stages within a taxon challenging (see Plön and Bernard 2006). Semen ejaculates from live, free-ranging marine mammals can be obtained using electro-ejaculation techniques (e.g., gray seals *Halichoerus grypus*, Lawson et al. 1996) or passively through opportunistic collection of voluntary ejaculate (e.g., dugongs, Burgess et al. 2012b; JML unpubl. data).

Live captive cetaceans have provided the most detailed knowledge of spermatozoa (in terms of morphology, ultrastructure, motility, viability) in fully marine mammals (Miller et al. 2016), with specimens collected via electro-ejaculation (Fleming et al. 1981; Lawson et al. 1996) or conditioned for voluntary semen collection after manual external stimulation (Keller 1986; Miller et al. 2002b; Robeck and O'Brien 2004; O'Brien et al. 2008; Sánchez-Calabuig et al. 2017). The latter method potentially produces higher quality samples more indicative of ejaculates produced during natural matings. Such knowledge has helped elucidate aspects of sperm development and maturation and the fertilization process in marine mammals, and has provided fundamental insights for the development of assisted reproductive technologies.

Assisted reproductive technologies, including semen cryopreservation (Robeck and O'Brien 2004), artificial insemination (Robeck et al. 2005b) and more recently, sex-sorting of spermatozoa (O'Brien and Robeck 2006; Montano et al. 2012), provide a number of benefits to captive breeding programs such as permitting easier genetic exchange between facilities without the need for translocation of animals (which can be risky, expensive and disruptive to the stability of social groups) as well as sex-ratio management. Thus, research into gamete collection and preservation is considered high priority for *in situ* management of captive marine mammals, particularly cetaceans (O'Brien and Robeck 2012b). Due to species-specific ejaculate characteristics and sperm biology, the composition of diluents used to preserve spermatozoa in vitro varies considerably across species (e.g., slightly higher osmolarity of diluents for cetaceans than for terrestrial animals) (reviewed by Fukui et al. 2007) and requires careful experimentation for each new species under study (e.g., Fukui et al. 1996; Robeck and O'Brien 2004; Robeck et al. 2004, 2009; O'Brien and Robeck 2010). Cryopreservation of ejaculated semen and formation of a genome (sperm) resource bank for captive populations has been achieved for bottlenose dolphins (Robeck and O'Brien 2004), killer whales (Robeck et al. 2004), Pacific white-sided dolphins *Lagenorhynchus obliquidens* (Robeck et al. 2009), belugas (O'Brien and Robeck 2010) and dugongs (T. Keeley and R. Bathgate unpubl. data). For wild populations, cryopreservation techniques have also been attempted for rescued spermatozoa from harvested common minke whales *Balaenoptera acutorostrata* (Mogoe et al. 1998) and Bryde's whales *Balaenoptera edeni* (Hiwasa et al. 2009) as well as voluntary ejaculates from live dugongs (JML and T. Keeley unpubl. data).

The potential use of genome banks can be further enhanced by the development of viable methods for oocyte and embryo preservation. To date, all significant research on in vitro oocyte maturation, fertilization and embryo culture has been performed using salvaged post-mortem tissues from freshly harvested whale (mysticete) species (Asada et al. 2001a; Iwayama et al. 2005; Fukui et al. 2007; Watanabe et al. 2007; Bhuiyan et al. 2009). Advanced studies have experimented with frozen-thawed oocytes (Asada et al. 2001a) and vitrified oocytes (Asada et al. 2001b; Fujihira et al. 2006) and have examined the feasibility of using interspecies somatic cell nuclear transfer to produce whale embryos (Lee et al. 2009; Bhuiyan et al. 2010). However, successful embryo preservation has not yet been reported for a marine mammal species.

It is easy to foresee that, as with terrestrial mammals, the collection and cryopreservation of viable gametes, combined with studies to ensure fertility following thawing, will enable the indefinite storage of valuable genetic material from threatened and endangered marine mammals. These approaches have been useful for captive breeding of marine mammals (Robeck et al. 2004, 2009, 2010; O'Brien and Robeck 2006, O'Brien et al. 2008, 2009; Sánchez-Calabuig et al. 2017), and in the future, may also be of value for conservation of wild populations (reviewed by O'Brien and Robeck 2012b).

3 Case Studies

3.1 North Atlantic Right Whale

The North Atlantic right whale (*Eubalaena glacialis*) is one of the rarest species of marine mammal in the world (Thomas et al. 2015) (Fig. 13.2). At the end of commercial whaling in 1935, this species was near extinction. The population size was

Fig. 13.2 North Atlantic right whale skim-feeding at the surface in its New England feeding ground [Photo was taken under U.S. NOAA Permit #14233 and Canada DFO Permit under Species at Risk Act. Photograph courtesy of the Anderson Cabot Center for Ocean Life at the New England Aquarium]

estimated at 270 in 1990, and over the following two decades, population growth occurred at a rate of 2.8% per year, bringing the population to a modern maximum of about 483 individuals in 2010 (Pace et al. 2017). Disappointingly, this initial trajectory of increasing abundance has not continued (Pace et al. 2017); present-day threats caused the population to decline to an estimated 428 whales in 2017 (Pettis et al. 2018). The unprecedented discovery of at least 17 dead right whales in 2017, representing a fivefold increase in annual mortality (Van der Hoop et al. 2013), followed by no documented calves born in the subsequent breeding season (Pettis et al. 2018), have amplified concerns that the North Atlantic right whale is imperiled (Kraus et al. 2016).

The North Atlantic right whale is found along the coastal waters of eastern North America, breeding in the waters off Florida in winter, and migrating to summer feeding grounds off New England and northeastern Canada. Its range overlaps one of the most industrialized stretches of ocean in the world, crowded with direct threats including ships, fishing operations and energy infrastructure. Despite legislative and spatial protections, ship strikes have long been a threat (Laist et al. 2014), and fatal entanglements in commercial fishing gear are taking an increasing toll (Knowlton et al. 2012, 2016). The exposure of whales to such dangers has worsened since 2011, when a portion of the right whale population moved away from their traditional feeding habitats in the Gulf of Maine and into northern waters which were well outside the known and protected areas. This unanticipated shift into higher risk areas (where protective policies were not current for right whales), appears to be a response to a climate-mediated decline in prey availability in the usual foraging grounds (Meyer-Gutbrod et al. 2018a, b). In addition to these threats to survival, population recovery of right whales is further jeopardized by demographic changes and reproductive failure, leaving the right whale vulnerable to extinction (Pace et al. 2017).

The study of population demography and reproductive parameters has revealed critical insights into the compromised recovery of the North Atlantic right whale. The inclusion of demographic data in recent population models has clearly confirmed that the species has entered a modern period of decline (Pace et al. 2017). There are substantially more males than females in the population (Pace et al. 2017). Furthermore, inter-calving periods are protracted, so that females are having calves only every 10 years or more (Pettis et al. 2018) compared with every 3 years in the 1980s (Kraus et al. 2001). Subsequently, calving rates have dropped by nearly 40% (Kraus et al. 2016), with calf production linked to female health (Rolland et al. 2016) and body condition (Miller et al. 2012; Pettis et al. 2017), oceanographic processes (Meyer-Gutbrod and Greene 2018a) and stressors from an urbanized ocean (e.g., repetitive sub-lethal interactions with fishing gear, including the effects of drag from entanglement) (Van der Hoop et al. 2016). This low reproductive rate and recently declining survival probabilities, particularly for breeding females (Pace et al. 2017), has prevented this population from recovering *cf.* southern hemisphere right whales *Eubalaena australis* (Corkeron et al. 2018).

Gaining a detailed understanding of the reproductive biology of large whales is logistically challenging, as discussed earlier in this chapter. Nevertheless, the North Atlantic right whale population has been monitored since 1980 (Kraus and Rolland 2007), and dedicated research efforts have enabled a near complete photographic census of the population over several decades (Pace et al. 2017). In fact, this research program is grounded on the ability to visually identify each individual whale at an early age due to their unique callosity pattern on the head (Hamilton et al. 2007). This research effort is enhanced by a cooperative data sharing commitment through the North Atlantic Right Whale Consortium (http://www.narwc.org) (Pettis et al. 2018), in which data are integrated from various international organizations that conduct substantial boat and aerial-based surveys to photograph whales for identification, collect biological samples, document calving and mortality events, and assess body condition and external health indicators of right whales (Rolland et al. 2016; Pettis et al. 2017). The resulting long-term North Atlantic Right Whale Identification and Sightings Databases hold vast amounts of data on individual whales (i.e., over 75,000 sightings records), including the birth year (for whales sighted as calves), sex, calving history, habitat-use patterns, and the impact of human activities (Hamilton et al. 2007; Pettis et al. 2018). The intensive multi-decadal study of this species has amassed life history data on well-known individual right whales which in turn has provided an invaluable model system for developing and validating a toolbox of innovative techniques for understanding the physiology and reproductive biology of large whales.

The urgent need for new sampling techniques to investigate North Atlantic right whales began in response to concerns about impaired reproduction in the species in the late 1990s (Reeves et al. 2001). The first samples of right whale feces for endocrine study were collected in 1999, kicking off one of the first long-term research programs on conservation physiology of a large whale (Rolland et al. 2007) and resulting in an extensive sample archive (Hunt et al. 2015). The success of fecal hormone analysis (see Sect. 1.1.4 in this chapter) has enabled the accurate determination of sex and reproductive status for most life history stages of free-swimming right whales (Corkeron et al. 2017), including pregnancy diagnosis, male maturity (Rolland et al. 2005, 2007), and intrinsic stress responses associated with reproduction (Hunt et al. 2006; Burgess et al. 2017); as well as other groundbreaking studies on whale stress responses to anthropogenic disturbance (Rolland et al. 2012, 2017). Because fecal sampling of large whales is limited by opportunistic collection, this research program cleverly adapted the use of scent detection dogs at sea to increase sampling rates (Rolland et al. 2006). Locating fecal samples in open water requires a strong partnership between the boat crew and the working dog, involving tactful navigation to orientate the vessel relative to wind and tidal currents, and steer in a direction guided by behavioral cues from the dog—who can smell floating feces as far as one nautical mile away. Despite the significant contribution of fecal sampling to our understanding of reproduction in right whales, there remained a need for a technique that enables repeated sampling of targeted individual whales, throughout seasonal movements (e.g., including non-feeding grounds) and across all life history stages.

Since 2011, researchers have been collecting respiratory vapor samples for endocrine analysis, and this has shown great promise for hormone assessment of free-swimming right whales at a population-level (Hunt et al. 2013; Burgess et al. 2018). Right whale blow samples were collect using a device fastened to the end of a carbon fiber pole (~10 m long), which was mounted to a cantilevered pivot on the foredeck of a research vessel. For sample collection, the vessel slowly approached an individual whale on a gradually converging course at idling speed to minimize disturbance to the whale. On anticipating the surfacing behavior of the whale, the pole was extended and lowered to position the sampling device above the exhaling blowholes (<1 m) to catch a portion of the aerosol droplets (Fig. 13.1). Using this technique, a recent study was able to collect 100 samples for physiological assessment from 46 individual right whales over just 8 days at sea (Burgess et al. 2018). This particular study of right whales has been instrumental in developing this novel and nascent approach for endocrine sampling of wild cetaceans, since robust quality data exist against which to validate hormone levels in respiratory vapor and other biological media. Singularly, the capacity to determine pregnancy from elevated progesterone levels measured in blow samples (Burgess et al. 2018) will be revolutionary in enabling reproductive assessment of free-swimming right whales in near real-time—with temporal detail that until now was unfeasible.

Another innovation in reproductive assessment of mysticete whales, hormone analysis of baleen, has been applied and then validated in the North Atlantic right whale, using the extensive baseline hormone database (see Sect. 1.1.4). Progesterone levels have been measured at regular intervals along the baleen plates collected from two dead-stranded female right whales with well-documented calving histories (Hunt et al. 2016). Increased progesterone levels (two orders of magnitude higher) in regions of the plate corresponded temporally (based on estimated baleen growth rates) with known calving events over the last decade of each whale's life (Hunt et al. 2016). Although baleen hormone analysis is not applicable to living whales, researchers hope that contemporary samples as well as historical archives (baleen plates collected during the era of whaling, which spans nearly 1000 years for right whales) may provide a means to evaluate longer term trends in pregnancy rates, inter-calving intervals, and other aspects of endocrinology in whale populations exposed to a changing marine environment.

North Atlantic right whale research is an interesting case study on the benefits of a well-studied population of known individuals to advance the development of novel cutting-edge approaches to physiologically assess a large uncatchable species. Multi-decadal monitoring involving consistent survey efforts, cooperative research partners, dedicated sample collection and preservation, and the integration of data from various sources and disciplines has greatly facilitated the expansion of knowledge on the reproductive biology of this long-lived and highly mobile species. There are examples of parallel long-term studies using similar multifaceted approaches on other free-swimming cetaceans, include bottlenose dolphins (Wells 2000, 2003; Wells et al. 2004), and killer whales (e.g., Ford et al. 2011; Wasser et al. 2017). Understanding reproductive parameters is critical to any conservation effort, but in the case of North Atlantic right whales, such data have been invaluable for better predicting population abundance trends and demonstrating the pronounced effect of reproductive variability on species viability (Pace et al. 2017). This reproductive study has raised the alarm for urgent management action of this seriously endangered large whale (Kraus et al. 2016).

3.2 Dugongs in Queensland Australia

Understanding reproductive rate and resilience is critically important for the effective management of vulnerable marine mammal species such as the dugong. Dugongs are challenging to study because they are fully aquatic, live in usually turbid coastal waters, spend almost all of their lives underwater foraging on benthic seagrasses (Marsh et al. 2011), are shy and elusive, and have few distinguishing features to discriminate individuals (Lanyon et al. 2002 *cf.* Anderson 1995). Until the past decade or so, all knowledge of the reproductive biology of dugongs (e.g., sexual maturation, gestation, reproductive activity, seasonality) was obtained from analysis of dugong carcasses recovered from tropical north Queensland, Australia (Marsh et al. 2011). Tissue samples obtained from dead dugongs of both sexes, all ages and reproductive states have formed the basis of descriptions of functional anatomy and histology of the reproductive system and timing of breeding cycles (Marsh et al. 1984a, b, c). Direct observations of social associations in wild dugongs including presumed reproductive behaviors (courtship, mating) are few (Preen 1989; Anderson 1997) and remain largely anecdotal because the sexes of individuals involved could not be confirmed. No dugongs have been bred in captivity, and no wild births have been witnessed.

Examination of known-aged reproductive tracts suggests that dugongs may be amongst the slowest reproducers of all the marine mammals, with life history parameters broadly similar to humans. Female dugongs in parts of tropical north Queensland mature at 13–17 years, have one calf at a time and then a long intercalving period of 3–7 years (Marsh et al. 1984a). Male dugongs similarly mature late (>9 years) and have been described as discontinuous breeders due to asynchronous spermatogenesis (Marsh et al. 1984b, c). However, there is evidence of marked inter- and even intra-population variation in reproductive parameters (Marsh et al. 2011) so that dugongs elsewhere in the tropics or at different times, possibly under more favorable nutritional circumstances, appear to grow faster, mature earlier (4–7 years) and reproduce more frequently (Kwan 2002; Marsh and Kwan 2008). The geographically-biased and diverse nature of life history and reproductive parameters suggests that a regionally-specific approach to determining reproductive parameters may be particularly appropriate for this species. Dugongs have a wide geographic distribution between 26° and 27° north and south of the equator, yet until recently, reproductive parameters of dugongs have not been determined in non-tropical regions (i.e., beyond 23° 27′ latitude) where seasons may be more pronounced than in the tropics. Such gaps in our knowledge of dugong populations existed due to a previous lack of non-lethal methodologies to collect quantitative data on reproductive status and activity. Therefore, novel research approaches were warranted to investigate reproductive processes in live wild populations.

At the start of the new century (2001), the first long-term capture-mark-recapture population study of dugongs was initiated, offering an opportunity to examine reproductive biology in live, free-ranging animals (Lanyon et al. 2002, 2006, 2012). During this program, more than 2200 sampling events of dugongs of both sexes and all body sizes, in the Moreton Bay population of ~900 dugongs, have been conducted. Dugongs were captured using an open water technique (Lanyon et al. 2006) and sampled at the water surface for a short period (5–6 min) (Lanyon et al. 2010a, b). Some dugongs were removed from the water each year as part of an annual health assessment program, with clinically valuable samples (blood, urine, faeces) collected for various projects including reproductive endocrinology (Lanyon et al. 2010b) (Fig. 13.3). Life history and reproductive parameters for individual dugongs including growth rate, body size at reproductive maturity and reproductive status have been elucidated through an innovative sampling approach that includes integration of information on molecular identity (Broderick et al. 2007; McHale et al. 2008; Lanyon et al. 2009; Kellogg Hunter et al. 2011), body morphometrics (Lanyon et al. 2010b), body scarring (Athousis 2012) and fecal hormone analysis (Burgess et al. 2012a, b, 2013b).

Fig. 13.3 Abdominal ultrasound for pregnancy of a wild-caught adult female dugong [Photo was taken under The University of Queensland Animal Ethics #ZOO/ENT/737/08/ARC/CRL/SW/AMMC, Australian Scientific Purposes Permit # WISP04937308 and Moreton Bay Marine Park Permit #QS2008/CVL228]

Since most reproductive processes are hormone-dependent, endocrine analysis is an effective and reliable non-lethal method for assessment of reproductive status (Schwarzenberger et al. 1996; Touma and Palme 2005). Concentrations of the steroid hormones testosterone, estrogen and progesterone have been measured in fecal samples collected (via rectal tube) from dugongs of both sexes, all size classes and in all months since inception of the program (almost 20 years), using enzyme-immunoassay (EIA) (Burgess et al. 2012a, b). All EIAs were biologically validated for this new species under study. Fecal hormone concentrations were initially examined against serum levels of a subsample of 50 dugongs sampled out-of-water to ensure that circulating concentrations in the blood were reliably reflected in the feces (Burgess et al. 2012a, b).

A pregnancy test for dugongs based on fecal progesterone in combination with body morphometrics (body length, maximum girth and teat length) was developed, validated through ultrasonography (Fig. 13.3). and then applied across the female population (Burgess et al. 2012a). Body size to first reproduction in females in this wild population could then be determined: compared to dugongs in tropical locations, these subtropical females grew slower, matured later but achieved greater asymptotic body size (Burgess et al. 2012a). For males, increases in fecal testosterone levels and tusk eruption (a secondary sex trait) characterized puberty, with further ontogenetic increases in testosterone indicating late onset of mature sexual activity (Burgess et al. 2012b). Temporal profiles of fecal hormone levels indicated discrete seasonality in reproductive activity, mating and pregnancy (spring-summer months), with synchronous peaks in male testosterone occurring in spring (Burgess et al. 2012b), coincident with heightened stress (fecal glucocorticoids), loss of body condition (Burgess et al. 2013b) and increased body scarring indicative of aggressive conspecific conflict (Athousis 2012; Burgess et al. 2013b): all consistent with competitive scramble mating that has been suggested for this species (Preen 1989). Interestingly, hormone profiles for recaptured individuals has confirmed discontinuous breeding in males but has also indicated the possibility of male senescence with advanced age (Burgess et al. 2012b). Further longitudinal monitoring of individuals in this population will provide information on breeding cycles, interbreeding intervals and lifetime reproductive output in this long-lived species. This is an example of how perseverance and evolving state-of-the-art science can make headway in understanding reproduction and associated processes in difficult-to-study marine mammals. However, whilst this integrated approach for investigating reproductive parameters has been appropriate for a relatively easy to access dugong population, i.e., in a relatively clear water habitat close to a major city (Brisbane), it cannot be applied easily to dugongs living in more remote coastal areas of tropical Australia.

Over the recent past, northern Australia had experienced a number of extreme weather events including severe cyclones (e.g., Cyclone Yasi in early 2011 and Cyclone Debbie in 2017) which have resulted in unprecedented rainfall events with extensive flooding (Perry et al. 2014). Extreme volumes of terrigenous run-off into coastal systems caused catastrophic seagrass dieback due to increased water turbidity (suspended particulate material) and direct smothering by sediments (McKenna et al. 2015; Lanyon 2019). The knock-on effects to seagrass grazers including dugongs and green turtles has been catastrophic with episodic mortality events coinciding with these storms (Meager and Limpus 2014). Recovered dugong carcasses have indicated starvation as likely cause of death, whilst field-based health assessment has found dugongs in poor body condition up to several months after the flood (JML unpubl. data). Routine dedicated aerial surveys conducted in the worst affected regions recorded zero neonate calves in the years directly after the floods, but an increase in calf counts in surveys conducted 5 years later, suggesting some slow recovery in reproductive rate (Sobtzick et al. 2017) (Fig. 13.4). That dugongs are somewhat resilient and able to resume breeding within years of catastrophic seagrass loss is heartening. However, the apparent increasing frequency and intensity of extreme weather events that have directly degraded dugong coastal habitat in Queensland Australia, i.e., three so-called one-in-a hundred-year flood events within a decade, is a real concern. Under such environmental stressors, with their late age to sexual maturity and slow reproductive rate, the risk of extirpation of coastal dugong populations looms larger with time.

Fig. 13.4 Dugong cow with neonate calf in Moreton Bay, Queensland Australia [Photo was taken under The University of Queensland Animal Ethics #ZOO/ENT/737/08/ARC/CRL/SW/AMMC, Australian Scientific Purposes Permit # WISP04937308 and Moreton Bay Marine Park Permit #QS2008/CVL228]

4 The Future

Our knowledge of the reproductive biology of fully marine mammals is very uneven. For those few species that are commonly held captive or for which carcasses have been recovered and analyzed, we have a basic understanding of reproductive processes. However, for the vast majority of fully aquatic marine mammals we know very little. The rarest, largest and/or most elusive of whales will always offer extreme challenges to the reproductive practitioner and we may never discover a great deal about their reproductive behavior or details of reproductive patterns at a population level. However, for many other species, recent advances in reproductive technologies with more 'remote' collection procedures (e.g., molecular and endocrine analysis) present the opportunity to initiate focused and integrated programs. Of particular importance to this field has been the development of wildlife endocrinology. Although initially developed with terrestrial wildlife in mind, the advent of technologies to collect and analyze body tissues other than blood has been nothing short of revolutionary. Now, marine mammal field biologists have opportunities to sample tissues in non-invasive (feces, sloughed skin, exhaled blow) or minimally invasive ways (biopsied skin, blubber) and develop reproductive hormonal profiles for live individuals or across free-ranging populations. With some ingenuity, creativity and a little perseverance, marine mammal biologists are applying these new technologies in innovative and sometimes remarkable ways. Furthermore, the expanding suite of parameters that can be measured (e.g., steroid and non-steroid hormones, and other sophisticated reproductive biomarkers in combination with genetics) permits ever more profound and meaningful questions in marine mammal reproductive science.

With increasing human usage of the world's oceans for commercial, recreational and strategic purposes and the rising incidence of extreme weather events, many populations of marine mammals are in an increasingly parlous state (Kaschner et al. 2011). Their resilience comes down to how fast they can reproduce, and their survival depends on their ability to maintain these reproductive rates in degrading aquatic environments and under climatic stress. Reproductive rate, if it can be determined, provides vital information on natural population fecundity and reproductive capacity, potential population growth rate and minimum viable population size. Having a solid understanding of the way in which marine mammals reproduce will inform us as to what levels of mortality and disturbance, populations or even species, can tolerate and help us to identify those species and populations most at risk as we enter this uncertain future.

References

Aarseth JJ, Van't Hof TJ, Stokkan K-A. Melatonin is rhythmic in newborn seals exposed to continuous light. J Comp Physiol B. 2003;173:37–42.

Acevedo J, Aguayo-Lobo A, Torres D. Fetus presentation and time taken for parturition in Antarctic fur seal, *Arctocephalus gazelle*, at cape Shirreff, Antarctica. Polar Biol. 2008;31:1137–41.

Acevedo-Whitehouse K, Rocha Gosselin A, Gendron D. A novel non-invasive tool for disease surveillance of free-ranging whales and its relevance to conservation programs. Anim Conserv. 2010;13:217–25.

Adulyanukosol K, Thongsukdee S, Hara T, Arai N, Tsuchiya M. Observations of dugong reproductive behavior in Trang Province, Thailand: further evidence of intraspecific variation in dugong behavior. Mar Biol. 2007;151:1887–91.

Aksenov AA, Yeates L, Pasamontes A, Siebe C, Zrodnikov Y, Simmons J, McCartney MM, Deplanque J-P, Wells RS, Davis CE. Metabolite content profiling of bottlenose dolphin exhaled breath. Anal Chem. 2014;86:10616–24.

Amaral RS. Use of alternative matrices to monitor steroid hormones in aquatic mammals: a review. Aquat Mamm. 2010;36:162–71.

Amaral RS, Rosas FCW, Viau P, d'Affonse Neto JA, da Silva VMF, de Oliveira CA. Noninvasive monitoring of androgens in male Amazonian manatee (*Trichechus inunguis*): biologic validation. J Zoo Wildl Med. 2009;40:458–65.

Amaral RS, Rosas FC, Graham LH, da Silva VM, Oliveira CA. First attempt to monitor luteinizing hormone and reproductive steroids in urine samples of the Amazonian manatee (*Trichechus inunguis*). J Zoo Wildl Med. 2014;45:843–51.

Anderson PK. Scarring and photo identification of dugongs in Shark Bay, Western Australia. Aquat Mamm. 1995;21:205–11.

Anderson PK. Shark Bay dugongs in summer. I: Lek mating. Behaviour. 1997;134:433–62.

Andriolo A, Martins CCA, Engel MH, Pizzorno JL, Mas-Rosa S, Freitas AC, Morete ME, Kinas PG. The first aerial survey to estimate abundance of humpback whales (*Megaptera novaeangliae*) in the breeding ground off Brazil (breeding stock a). J Cetacean Res Manag. 2006;8:307–11.

Apprill A, Miller CA, Moore MJ, Durban JW, Fearnbach H, Barrett-Lennard LG. Extensive core microbiome in drone-captured whale blow supports a framework for health monitoring. MSystems. 2017;2:682–15.

Archer R, Chauvet J, Chauvet MT. Isolation of finback whale oxytocin and vasopressin. Nature. 1964;201:191–2.

Asada M, Tetsuka M, Ishikawa H, Ohsumi S, Fukui Y. Improvement on *in vitro* maturation, fertilization and development of minke whale (*Balaenoptera acutorostrata*) oocytes. Theriogenology. 2001a;56:521–33.

Asada M, Wei H, Nagayama R, Tetsuka M, Ishikawa H, Ohsumi S, Fukui Y. An attempt at intracytoplasmic sperm injection of frozen-thawed minke whale (*Balaenoptera bonaerensis*) oocytes. Zygote. 2001b;9:299–307.

Athousis C. Body scarring as an indicator of tusk function in intraspecific social interactions in the dugong (*Dugong dugon*). Unpublished Honours thesis, The University of Queensland, Brisbane; 2012. 69 pp.

Atkinson S. Reproductive biology of seals. Rev Reprod. 1997;2:175–94.

Atkinson S, Gilmartin WG. Seasonal testosterone pattern in Hawaiian monk seals (*Monachus schauinslandi*). J Reprod Fertil. 1992;96:35–9.

Atkinson S, Gilmartin WG, Lasley BL. Testosterone response to a gonadotrophin-releasing hormone agonist in Hawaiian monk seals (*Monachus schauinslandi*). J Reprod Fertil. 1993;97:35–8.

Atkinson S, Ragen TJ, Gilmartin WG, Becker BL, Johanos TC. Use of a GnRH agonist to suppress testosterone in wild male Hawaiian monk seals (*Monachus schauinslandi*). Gen Comp Endocrinol. 1998;112:178–82.

Atkinson S, Combelles C, Vincent D, Nachtogall P, Pawloski J, Breese M. Monitoring of progesterone in captive female false killer whales, *Pseudorca crassidens*. Gen Comp Endocrinol. 1999;115:323–32.

Attard CRM, Beheregaray LB, Jenner KCS, Gill PC, Jenner M-N, Morrice MG, Robertson KM, Möller LM. Hybridization of southern hemisphere blue whale subspecies and a sympatric area off Antarctica: impacts of whaling or climate change? Mol Ecol. 2012;21:5715–27.

Attard CR, Beheregaray LB, Möller LM. Towards population-level conservation in the critically endangered Antarctic blue whale: the number and distribution of their populations. Sci Rep. 2016;6:22291.

Avila IC, Kaschner K, Dormann CF. Current global risks to marine mammals: taking stock of the threats. Biol Conserv. 2018;221:44–58.

Barlow J. Variability, trends, and biases in reproductive rates of spotted dolphins, *Stenella attenuata*. Fish Bull. 1985;83:657–69.

Barrell GK, Montgomery GW. Absence of circadian patterns of secretion of melatonin or cortisol in Weddell seals under continuous natural daylight. J Endocrinol. 1989;122:445–9.

Bauer GB, Colbert DE, Gaspard JC III. Learning about manatees: a collaborative program between new College of Florida and Mote Marine Laboratory to conduct laboratory research for manatee conservation. Int J Comp Psychol. 2010;23:811–25.

Beaulieu-McCoy NE, Sherman KK, Trego ML, Crocker DE, Kellar NM. Initial validation of blubber cortisol and progesterone as indicators of stress response and maturity in an otariid; the California Sea lion (*Zalophus californianus*). Gen Comp Endocrinol. 2017;252:1–11.

Bechshøft TØ, Sonne C, Dietz R, Born EW, Novak MA, Henchey E, Meyer JS. Cortisol levels in hair of East Greenland polar bears. Sci Total Environ. 2011;409:831–4.

Bechshoft T, Wright AJ, Weisser JJ, Teilmann J, Dietz R, Hansen M, Björklund E, Styrishave B. Developing a new research tool for use in free-ranging cetaceans: recovering cortisol from harbour porpoise skin. Conserv Physiol. 2015;3:cov016–9.

Bergfelt DR, Steinetz BG, Lasano S, West KL, Campbell M, Adams GP. Relaxin and progesterone during pregnancy and the post-partum period in association with live and stillborn calves in bottlenose dolphins (*Tursiops truncatus*). Gen Comp Endocrinol. 2011;170:650–6.

Bergfelt DR, Steinetz BG, Reif JS, Schaefer AM, Bossart GD, Mazzoil MS, et al. Evaluation of single-sample analysis of progesterone in combination with relaxin for diagnosis of pregnancy in wild bottlenose dolphins (*Tursiops truncatus*). Aquat Mamm. 2013;39:187.

Best PB. The sperm whale (*Physeter catodon*) off the west coast of South Africa. 3. Reproduction in the male. Invest Rep Div Sea Fish South Africa. 1969;72:1–20.

Bezerra AR, Salmito-Vanderley CS, Bersano PR, Carvalho VL, Meirelles AC, Attademo FL, et al. Histological characterization of reproductive tract and fetal annexes of the West Indian manatee (*Trichechus manatus*) from Brazil. Pesqui Vet Bras. 2018;38:2166–74.

Bhuiyan MMU, Suzuki Y, Watanabe H, Matsuoka K, Fujise Y, Ishikawa H, Ohsumi S, Fukui Y. Attempts at *in vitro* fertilization and culture of *in vitro* matured oocytes in sei (*Balaenoptera borealis*) and Bryde's (*B. edeni*) whales. Zygote. 2009;17:19–28.

Bhuiyan MM, Suzuki Y, Watanabe H, Lee E, Hirayama H, Matsuoka K, Fujise Y, Ishikawa H, Ohsumi S, Fukui Y. Production of sei whale (*Balaenoptera borealis*) cloned embryos by inter- and intra-species somatic cells nuclear transfer. J Reprod Dev. 2010;56:131–9.

Biancani B, Da Dalt L, Lacave G, Romagnoli S, Gabai G. Measuring fecal progestogens as a tool to monitor reproductive activity in captive female bottlenose dolphins (*Tursiops truncatus*). Theriogenology. 2009;72:1282–92.

Blanchet M-A, Desportes G, Nance T, Vanman C. The description of the pregnancies, labours and pre- and post-partum events of two harbour seals in a zoological environment. Aquat Mamm. 2006;32:145–51.

Block BA. Physiological ecology in the 21st century: advancements in biologging science. Integr Comp Biol. 2005;45:305–20.

Bograd SJ, Block BA, Costa DP, Godley BJ. Biologging technologies: new tools for conservation. Introduction. Endanger Species Res. 2010;10:1–7.

Bonde RK, Garrett A, Belanger M, Askin N, Tan L, Wittnich C. Biomedical health assessments of the Florida manatee in Crystal River—providing opportunities for training during the capture, handling, and processing of this endangered aquatic mammal. J Mar Anim Their Ecol. 2012;2:1–12.

Börjesson P, Read AJ. Variation in the timing of conception between populations of harbor porpoise. J Mammal. 2003;84:948–55.

Botta S, Secchi ER, Muelbert MM, Danilewicz D, Negri MF, Cappozzo HL, Hohn AA. Age and growth of franciscana dolphins, Pontoporia blainvillei (Cetacea: Pontoporiidae) incidentally caught off southern Brazil and northern Argentina. J. Mar. Biol. Assoc. UK. 2010;90(8):1493–500.

Boyd IL. Changes in plasma progesterone and prolactin concentrations during the annual cycle, and role of prolactin in the maintenance of lactation and luteal development in the Antarctic fur seal (*Arctocephalus gazella*). J Reprod Fertil. 1991;91:637–47.

Boyd IL. State-dependent fertility in pinnipeds: contrasting capital and income breeders. Funct Ecol. 2000;14:623–30.

Boyd IL, Lockyer C, Marsh HD. Reproduction in marine mammals. In: Reynolds III JE, Rommel SA, editors. Biology of marine mammals. Washington: Smithsonian Institution Press; 1999. p. 218–86.

Briggs M. Contraception in bottlenose dolphins (*Tursiops truncatus*). In: Duffield DA, Robeck TR, editors. The bottlenose dolphin breeding workshop. Silver Springs: American Zoological Association, Marine Mammal Taxon Advisory Group; 2000. p. 201–4.

Broderick D, Ovenden J, Slade R, Lanyon JM. Characterisation of 26 new microsatellite loci in the dugong (*Dugong dugon*). Mol Ecol Notes. 2007;7:1275–7.

Brook FM. The use of diagnostic ultrasound in assessment of the reproductive status of the bottlenose dolphin, *Tursiops aduncus*, in captivity and applications in management of a controlled breeding program. Unpublished PhD thesis, Hong Kong Polytechnic University, Hong Kong; 1997. 339 pp.

Brook FM. Sonographic ovarian and testicular evaluation in bottlenose dolphins, Tursiops truncatus aduncus. In: Duffield DA, Robeck TR, editors. Proceedings bottlenose dolphin reproduction workshop, San Diego; 1999. p. 207–22.

Brook FM. Ultrasonographic imaging of the reproductive organs of the female bottlenose dolphin, *Tursiops truncatus aduncus*. Reproduction. 2001;121:419–28.

Brook FM, Kinoshita RE, Brown B, Metreweli C. Ultrasonographic imaging of the testis and epididymis of the bottlenose dolphin, *Tursiops truncatus aduncus*. J Reprod Fertil. 2000;119:233–40.

Brook FM, van Bonn W, Jensen E. Ultrasonography. In: Dierauf LA, Gulland FMD, editors. CRC handbook of marine mammal medicine. 2nd ed. Boca Raton: CRC Press; 2001. p. 590–620.

Brook FM, Lim EHT, Chua FHC, Mackay B. Assessment of the reproductive cycle of the indo-Pacific humpback dolphin, *Sousa chinensis*, using ultrasonography. Aquat Mamm. 2004;30:137–48.

Brown MR, Corkeron PJ, Hale PT, Schultz KW, Bryden MM. Evidence for a sex-segregated migration in the humpback whale (*Megaptera novaeangliae*). Proc R Soc Lond B. 1995;259:229–34.

Burgess EA, Lanyon JM, Brown JL, Blyde D, Keeley T. Diagnosing pregnancy in free-ranging dugongs using fecal progesterone metabolite concentrations and body morphometrics: a population application. Gen Comp Endocrinol. 2012a;177:82–92.

Burgess EA, Lanyon JM, Keeley T. Testosterone and tusks: maturation and seasonal reproductive patterns of live free-ranging dugongs (*Dugong dugon*) in a subtropical population. Reproduction. 2012b;143:683–97.

Burgess EA, Blanshard WH, Barnes A, Gilchrist S, Chua J, Lanyon JM. Reproductive hormone monitoring of dugongs in captivity: detecting the onset of sexual maturity in a cryptic marine mammal. Anim Reprod Sci. 2013a;140:255–67.

Burgess EA, Brown JL, Lanyon JM. Sex, scarring and stress: understanding seasonal costs in a cryptic marine mammal. Conserv Physiol. 2013b;1:cot014.

Burgess EA, Hunt KE, Kraus SD, Rolland RM. Get the most out of blow hormones: validation of sampling materials, field storage and extraction techniques for whale respiratory vapour samples. Conserv Physiol. 2016;4:cow024.

Burgess EA, Hunt KE, Kraus SD, Rolland RM. Adrenal responses of large whales: integrating fecal aldosterone as a complementary biomarker to glucocorticoids. Gen Comp Endocrinol. 2017;252:103–10.

Burgess EA, Hunt KE, Kraus SD, Rolland RM. Quantifying hormones in exhaled breath for physiological assessment of large whales at sea. Sci Rep. 2018;8:1–14.

Calle PP. Contraception in pinnipeds and cetaceans. In: Wildlife contraception: issues, methods, and applications. Baltimore: John Hopkins University Press; 2005. p. 168–76.

Calzada N, Aguilar A, Sorensen TB, Lockyer C. Reproductive biology of female striped dolphin (*Stenella coeruleoalba*) from the western Mediterranean. J Zool. 1996;240:581–91.

Cassini MH. The evolution of reproductive systems in pinnipeds. Behav Ecol. 1999;10:612–6.

Charapata P, Horstmann L, Jannasch A, Misarti N. A novel method to measure steroid hormone concentrations in walrus bone from archaeological, historical, and modern time periods using LC/MS/MS. Rapid Commun Mass Spectrom. 2018:1–57.

Chittleborough RG. Aspects of reproduction in the male humpback whale *Megaptera nodosa* (Bonnaterre). Aust J Mar Freshwat Res. 1954;6:1–29.

Chivers SJ. Marine mammal life history: guide to biological sample collection and data archive: NOAA Technical Memorandum; 2018.

Chivers SJ, Myrick AC Jr. Comparisons of age at sexual maturity and other reproductive parameters for two stocks of spotted dolphin, *Stenella attenuata*. Fish Bull. 1993;91:611–8.

Clapham PJ. The social and reproductive biology of humpback whales: an ecological perspective. Mammal Rev. 2008;26:27–49.

Clark CT, Fleming AH, Calambokidis J, Kellar NM, Allen CD, Catelani KN, Robbins M, Beaulieu NE, Steel D, Harvey JT. Heavy with child? Pregnancy status and stable isotope ratios as determined from biopsies of humpback whales. Conserv Physiol. 2016;4:1–13.

Cockcroft VG, Ross GJB. Age, growth and reproduction of bottlenose dolphins *Tursiops truncatus* from the east coast of southern Africa. Fish Bull. 1990;88:289–302.

Cole TV, Hamilton P, Henry AG, Duley P, Pace RM III, White BN, Frasier T. Evidence of a North Atlantic right whale *Eubalaena glacialis* mating ground. Endanger Species Res. 2013;21:55–64.

Connor RC. Dolphin social intelligence: complex alliance relationships in bottlenose dolphins and a consideration of selective environments for extreme brain size evolution in mammals. Philos Trans R Soc B. 2007;362:587–602.

Connor RC, Smolker RA, Richards AF. Two levels of alliance formation among male bottlenose dolphins (*Tursiops* sp.). Proc Natl Acad Sci. 1992;89:987–90.

Connor RC, Wells R, Mann J, Read A. The bottlenose dolphin: social relationships in a fission-fusion society. In: Mann J, Connor RC, Tyack PJ, Whitehead H, editors. Cetacean societies: field studies of whales and dolphins. Chicago: University of Chicago Press; 2000. p. 91–126.

Constable S, Parslow A, Dutton G, Rogers T, Hogg C. Urinary cortisol sampling: a non-invasive technique for examining cortisol concentrations in the Weddell seal, *Leptonychotes weddelli*. Zoo Biol. 2006;25:137–44.

Corkeron P, Rolland RM, Hunt KE, Kraus SD. A right whale pootree: classification trees of faecal hormones identify reproductive states in North Atlantic right whales (*Eubalaena glacialis*). Conserv Physiol. 2017;5:1–9.

Corkeron P, Hamilton P, Bannister J, Best P, Charlton C, Groch KR, Findlay K, Rowntree V, Vermeulen E, Pace RM III. The recovery of North Atlantic right whales, *Eubalaena glacialis*, has been constrained by human-caused mortality. R Soc Open Sci. 2018;5:180892–12.

Cornell LH, Asper ED, Antrim JE, Searles SS, Young WG, Goff T. Progress report: results of a long-range captive breeding program for the bottlenose dolphin, *Tursiops truncatus* and *Tursiops truncatus gilli*. Zoo Biol. 1987;6:41–53.

Costa DP. The secret lives of marine mammals: novel tools for studying their behaviour and biology at sea. Oceanography. 1993;6:120–8.

Cummins JM, Woodall PF. On mammalian sperm dimensions. J Reprod Fertil. 1985;75:153–75.

Danil K, Chivers SJ. Habitat based spatial and temporal variability in life history characteristics of female common dolphins (*Delphinus delphis*) in the eastern tropical Pacific. Mar Ecol Prog Ser. 2006;318:277–86.

Davidson AD, Boyer AG, Kim H, Pompa-Mansilla S, Hamilton MJ, Costa DP, et al. Drivers and hotspots of extinction risk in marine mammals. Proc Natl Acad Sci. 2012;109:3395–400.

Dawbin WH. Temporal segregation of humpback whales during migrations in southern hemisphere waters. Memoirs Queensland Museum. 1997;42:105–38.

de Bruyn PJN, Tosh CA, Bester MN, Cameron EZ, McIntyre T, Wilkinson IS. Sex at sea: alternative mating system in an extremely polygynous mammal. Anim Behav. 2011;82:445–51.

de Mello DMD, de Oliveira CA. Biological matrices for sampling free-ranging cetaceans and the implications of their use for reproductive endocrine monitoring. Mammal Rev. 2016;46:77–91.

DeMaster DP. Estimating the average age of first birth in marine mammals. Can J Fish Aquat Sci. 1981;38:237–9.

Derocher AE, Stirling I. Temporal variation in reproduction and body mass of polar bears in western Hudson Bay. Can J Zool. 1995;73:1657–65.

Desportes G, Saboureau M, Lacroix A. Reproductive maturity and seasonality of male long-finned pilot whales off the Faroe Islands. Rep Int Whaling Comm (Special Issue). 1993;14:233–62.

Desportes G, Saboureau M, Lacroix A. Growth-related changes in testicular mass and plasma testosterone concentrations in long-finned pilot whales, *Globicephala melas*. Reproduction. 1994;102:237–44.

Deutsch CJ, Reid JP, Bonde RK, Easton DE, Kochman HI, O'Shea TJ. Seasonal movements, migratory behavior, and site fidelity of West Indian manatees along the Atlantic coast of the United States. Wildl Monogr. 2003;151:1–77.

Dizon AE, Perrin WF, Akin PA. Stocks of dolphins (Stenella spp., and Delphinus delphis) in the eastern tropical Pacific: a phylogenetic classification. 1994. 20 pp.

Dover SR, van Bonn W. Flexible and rigid endoscopy in marine mammals. In: Dierauf LA, Gulland FMD, editors. CRC handbook of marine mammal medicine. 2nd ed. Boca Raton: CRC Press; 2001. p. 621–42.

Duffield DA, Odell DK, McBain JF, Andrews B. Killer whale (*Orcinus orca*) reproduction at SeaWorld. Zoo Biol. 1995;14:417–30.

Edds PL, Odell S, Tershy BR. Calls of a captive juvenile and free-ranging adult-calf pairs of Bryde's whales, *Balaenoptera edeni*. Mar Mamm Sci. 1993;9:269–84.

Eisert R, Oftedal OT, Barrell GK. Milk composition in the Weddell seal *Leptonychotes weddellii*: evidence for a functional role of milk carbohydrates in pinnipeds. Physiol Biochem Zool. 2013;86:159–75.

Ellis S, Franks DW, Nattrass S, Currie TE, Cant MA, Giles D, et al. Analyses of ovarian activity reveal repeated evolution of post-reproductive lifespans in toothed whales. Sci Rep. 2018;8:12833.

Essapian FS. Observations of abnormalities of parturition in captive bottlenose dolphins, *Tursiops truncatus*, and concurrent behavior of other porpoises. J Mammal. 1962;44:405–14.

Evans K, Hindell MA. The age structure and growth of female sperm whales (*Physeter macrocephalus*) in southern Australian waters. J Zool. 2004;263:237–50.

Everitt RD, Krogman BD. Sexual behavior of bowhead whales observed off the north coast of Alaska. Arctic. 1979;32:277–80.

Favaro L, Gnone G, Pessani D. Postnatal development of echolocation abilities in a bottlenose dolphin (*Tursiops truncatus*). Zoo Biol. 2013;32:210–5.

Fiori L, Doshi A, Martinez E, Orams MB, Breen B. The use of unmanned aerial systems in marine mammal research. Remote Sens. 2017;9:543.

Fitzpatrick JL, Almbro M, Gonzalez-Voyer A, Kolm N, Simmons LW. Male contest competition and the coevolution of weaponry and testes in pinnipeds. Evolution. 2012;66:3595–604.

Fleming AD, Yanagimachi R, Yanagimachi H. Spermatozoa of the Atlantic bottlenose dolphin, *Tursiops truncatus*. J Reprod Fertil. 1981;63:509–14.

Fleming AH, Kellar NM, Allen CD, Kurle CM. The utility of combining stable isotope and hormone analyses in marine megafauna research. Front Mar Sci. 2018;5:338.

Foote AD. Mortality rate acceleration and post-reproductive lifespan in matrilineal whale species. Biol Lett. 2008;4:189–91.

Ford MJ, Hanson MB, Hempelmann JA, Ayres KL, Emmons CK, Schorr GS, et al. Inferred paternity and male reproductive success in a killer whale (*Orcinus orca*) population. J Hered. 2011;102:537–53.

Fossi C, Marsili L. The use of non-destructive biomarkers in the study of marine mammals. Biomarkers. 1997;2:205–16.

Fowler CW. Density dependence as related to life history strategy. Ecology. 1981;62:602–10.

Fowler CW. Density dependence in cetacean populations. Rep Int Whaling Comm (Special Issue). 1984;6:373–9.

Fragalà S, Medica P, Grande F, Vazzana I, Fazio E. Evaluation of seasonal changes of serum and plasma estradiol-17β, progesterone and testosterone in dolphins (*Tursiops truncatus*) by chemiluminescence. Veterinary World. 2015;8:977–82.

Francis-Floyd R, White JR, Chen CL, Cardeilhac PT, Cichra CE. Serum progesterone and estradiol concentrations in captive manatess *Trichechus manatus*. J Aquat Anim Health. 1991;3:70–3.

Fujihira T, Kobayashi M, Hochi S, Hirabayashi M, Ishikawa H, Ohsumi S, Fukui Y. Developmental capacity of Antarctic minke whale (*Balaenoptera bonaerensis*) vitrified oocytes following *in vitro* maturation, and parthenogenetic activation or intracytoplasmic sperm injection. Zygote. 2006;14:89–95.

Fukui Y, Mogoe T, Jung YG, Terawaki Y, Miyamoto A, Ishikawa H, Fujise Y, Ohsumi S. Relationships among morphological status, steroid hormones, and post-thawing viability of frozen spermatozoa of male minke whales (*Balaenoptera acutorostrata*). Mar Mamm Sci. 1996;12:28–37.

Fukui Y, Iwayama H, Matsuoka T, Nagai H, Koma N, Mogoe T, Ishikawa H, Fujise Y, Hirabayashi M, Hochi S, Kato H, Ohsumi S. Attempt at intracytoplasmic sperm injection of *in vitro* matured oocyctes in common minke whales (*Balaenoptera acutorostrata*) captured during the Kushiro coast survey. J Reprod Dev. 2007;53:945–52.

Funasaka N, Yoshioka M, Suzuki M, Ueda K, Miyahara H, Uchida S. Seasonal difference of diurnal variations in serum melatonin, cortisol, testosterone, and rectal temperature in indo-Pacific bottlenose dolphins (*Tursiops aduncus*). Aquat Mamm. 2011;37:433–42.

Gardiner KJ, Boyd IL, Racey PA, Reijnders PJH, Thompson PM. Plasma progesterone concentrations measured using an enzyme-linked immunosorbent assay useful for diagnosing pregnancy in harbor seals (*Phoca vitulina*). Mar Mamm Sci. 1996;12:265–73.

Gardiner KJ, Boyd IL, Follett BK, Racey PA, Reijnders PJH. Changes in pituitary, ovarian, and testicular activity in harbour seals (*Phoca vitulina*) in relation to season and sexual maturity. Can J Zool. 1999;77:211–21.

Gero S, Whitehead H. Suckling behavior in sperm whale calves: observations and hypotheses. Mar Mamm Sci. 2007;23:398–413.

Gibbens J, Parry LJ, Arnould JPY. Influences on fecundity in Australian fur seals (*Arctocephalus pusillus doriferus*). J Mammal. 2010;91:510–8.

Gillett RM, Frasier TR, Rolland RM, White BN. Molecular identification of individual North Atlantic right whales (*Eubalaena glacialis*) using free-floating feces. Mar Mamm Sci. 2010;26:917–36.

Glockner-Ferrari DA, Ferrari MJ. Reproduction in the humpback whale (*Megaptera novaeangliae*) in Hawaiian waters, 1975-1988: the life history, reproductive rates and behavior of known individuals identified through surface and underwater photography. Rep Int Whaling Comm Special Issue. 1990;12:161–9.

Goldstein J, Reese E, Reif JS, Varela RA, McCulloch SD, Defran RH, Fair PA, Bossart GD. Hematologic, biochemical, and cytologic findings from apparently healthy Atlantic bottlenose dolphins (*Tursiops truncatus*) inhabiting the Indian River lagoon, Florida, USA. J Wildl Dis. 2006;42:447–54.

Green ML, Herzing DL, Baldwin JD. Noninvasive methodology for the sampling and extraction of DNA from freeranging Atlantic spotted dolphins (*Stenella frontalis*). Mol Ecol Notes. 2007;7:1287–92.

Hadley GL, Rotella JJ, Garrot RA, Nichols JD. Variation in probability of first reproduction of Weddell seals. J Anim Ecol. 2006;75:1058–70.

Halpern BS, Walbridge S, Selkoe KA, Kappel CV, Micheli F, D'Agrosa C, Bruno JF, Casey KS, Ebert C, Fox HE, et al. A global map of human impact on marine ecosystems. Science. 2008;319:948–52.

Hamilton PK, Knowlton AR, Marx MK. Right whales tell their own stories: the photo-identification catalog. In: Kraus SD, Rolland RM, editors. The urban whale. Cambridge: Harvard University Press; 2007. p. 75–104.

Hao YJ, Chen DQ, Zhao QZ, Wang D. Serum concentrations of gonadotropins and steroid hormones of *Neophocaena phocaenoides asiaeorientalis* in middle and lower regions of the Yangtze River. Theriogenology. 2007;67:673–80.

Harcourt RG, Turner E, Hall A, Waas JR, Hindell M. Effects of capture stress on free-ranging, reproductively active male Weddell seals. J Comp Physiol A. 2010;196:147–54.

Harrison RJ, Ridgway SH. Gonadal activity in some bottlenose dolphins (*Tursiops truncatus*). J Zool. 1971;165:355–66.

Hartman DS. Ecology and behavior of the manatee (*Trichechus manatus*) in Florida. Am Soc Mammal Special Publ. 1979;5:1–153.

Hayden M, Bhawal R, Escobedo J, Harmon C, O'Hara TM, Klein D, San-Francisco S, Zabet-Moghaddam M, Godard-Codding CAJ. Nanospray liquid chromatography/tandem mass spectrometry analysis of steroids from gray whale blubber. Rapid Commun Mass Spectrom. 2017;31:1088–94.

Hays GC, Ferreira LC, Sequeira AMM, Meekan MG, Duarte CM, Bailey H, Bailleul F, Bowen WD, Caley MJ, Costa DP, et al. Key questions in marine megafauna movement ecology. Trends Ecol Evol. 2016;31:463–75.

Hernandez P, Reynolds JE III, Marsh H, Marmontel M. Age and seasonality of spermatogenesis of Florida manatees. In: O'Shea TJ, Ackerman BB, Percival HF, editors. Population biology of the Florida manatee. Information and technology report 1: National Biological Service; 1995. p. 84–97.

Heyning JE. Functional morphology involved in intraspecific fighting of the beaked whale, *Mesoplodon carlhubbsi*. Can J Zool. 1984;62:1645–54.

Higgins DP, Noad MJ. Standardised protocols for the collection of biological samples from stranded cetaceans: Department of the Environment and Heritage; 2006.

Hildebrandt TB, Brown JL, Hermes R, Göritz F. Ultrasound analysis of reproductive function in wildlife species. Conserv Biol. 2003;8:166–82.

Hiwasa M, Suzuki Y, Watanabe H, Bhuiyan MMU, Matsuoka K, Fujise Y, Ishikawa H, Ohsumi S, Fukui Y. Effects of semen extenders and storage temperatures on characteristics of frozen-thawed Bryde's (*Balaenoptera edeni*) *whale spermatozoa*. J Reprod Dev. 2009;55:599–606.

Hoberecht LK, Vos DJ, VanBlaricom GR. A remote biopsy system used to sample Steller Sea lion (*Eumetopias jubatus*) blubber. Mar Mamm Sci. 2006;22:683–9.

Hogg C, Vickers E, Rogers T. Determination of testosterone in saliva and blow of bottlenose dolphins (*Tursiops truncatus*) using liquid chromatography-mass spectrometry. J Chromatogr B. 2005;814:339–46.

Hogg C, Rogers T, Shorter A, Barton K, Miller P, Nowacek D. Determination of steroid hormones in whale blow: it is possible. Mar Mamm Sci. 2009;25:605–18.

Hohn AA, Read AJ, Fernández S, Vidâl O, Findley LT. Life history of the vaquita, *Phocoena sinus* (Phocoenidae, Cetacea). J Zool (Lond). 1996;239:235–51.

Holt WV, Waller J, Moore A, Jepson PD, Deaville R, Bennett PM. Smooth muscle actin and vimentin as markers of testis development in the harbour porpoise (*Phocoena phocoena*). J Anat. 2004;205:201–11.

Hückstädt L, Antezana T. An observation of parturition in a stranded *Kogia breviceps*. Mar Mamm Sci. 2001;17:362–5.

Hückstädt LA, Holser RR, Tift MS, Costa DP. The extra burden of motherhood: reduced dive duration associated with pregnancy status in a deep-diving mammal, the northern elephant seal. Biol Lett. 2018;14:20170722.

Hunt KE, Trites AW, Wasser SK. Validation of a fecal glucocorticoid assay for Steller Sea lions (*Eumetopias jubatus*). Physiol Behav. 2004;80:595–601.

Hunt KE, Rolland RM, Kraus SD, Wasser SK. Analysis of fecal glucocorticoids in the North Atlantic right whale (*Eubalaena glacialis*). Gen Comp Endocrinol. 2006;148:260–72.

Hunt KE, Moore MJ, Rolland RM, Kellar NM, Hall AJ, Kershaw J, Raverty SA, Davis CE, Yeates LC, Fauquier DA, Rowles TK, Kraus SD. Overcoming the challenges of studying conservation physiology in large whales: a review of available methods. Conserv Physiol. 2013;1:1–24.

Hunt KE, Rolland RM, Kraus SD. Detection of steroid and thyroid hormones via immunoassay of North Atlantic right whale (*Eubalaena glacialis*) respiratory vapor. Mar Mamm Sci. 2014a;30:796–809.

Hunt KE, Stimmelmayr R, George C, Hanns C, Suydam R, Brower H, Rolland RM. Baleen hormones: a novel tool for retrospective assessment of stress and reproduction in bowhead whales (*Balaena mysticetus*). Conserv Physiol. 2014b;2:cou030.

Hunt KE, Rolland RM, Kraus SD. Conservation physiology of an uncatchable animal: the North Atlantic right whale (*Eubalaena glacialis*). Integr Comp Biol. 2015;55:577–86.

Hunt KE, Lysiak NS, Moore MJ, Rolland RM. Longitudinal progesterone profiles in baleen from female North Atlantic right whales (*Eubalaena glacialis*) match known calving history. Conserv Physiol. 2016;4:cow014–9.

Hunt KE, Lysiak NS, Robbins J, Moore MJ, Seton RE, Torres L, Buck CL. Multiple steroid and thyroid hormones detected in baleen from eight whale species. Conserv Physiol. 2017;5:162.

Hunt KE, Lysiak NSJ, Matthews CJD, Lowe C, Fernández Ajó A, Dillon D, Willing C, Heide-Jørgensen MP, Ferguson SH, Moore MJ, et al. Multi-year patterns in testosterone, cortisol and corticosterone in baleen from adult males of three whale species. Conserv Physiol. 2018;6:285–16.

IUCN 2018. The IUCN Red List of Threatened species. Version 2018–2. http://www.iucnredlist.org.

Iwayama H, Ishikawa H, Ohsumi S, Fukui Y. Attempt at *in vitro* maturation of minke whale (*Balaenoptera bonaerensis*) oocytes using a portable CO_2 incubator. J Reprod Dev. 2005;51:69–75.

Jameson RJ, Johnson AM. Reproductive characteristics of female sea otters. Mar Mamm Sci. 1993;9:156–67.

Jenssen TA, Lovern MB, Congdo JD. Field-testing the protandry-based mating system for the lizard, *Anolis carolinesis*: does the model organism have the right model? Behav Ecol Sociobiol. 2001;50:162–72.

Karamanlidis AA, Paravas V, Trillmich F, Dendrinos P. First observations of parturition and postpartum behavior in the Mediterranean monk seal (*Monachus monachus*) in the eastern Mediterranean. Aquat Mamm. 2010;36:27–32.

Karczmarski L, Thornton M, Cockroft VG. Description of selected behaviours of humpback dolphins, *Sousa chinensis*. Aquat Mamm. 1997;23:127–34.

Karpovich SA, Skinner JP, Kapronczai LA, Smith JA, Janz DM. Examination of relationships between stable isotopes and cortisol concentrations along the length of phocid whiskers. Mar Mamm Sci. 2018;81:67–21.

Kaschner K, Tittensor DP, Ready J, Gerrodette T, Worm B. Current and future patterns of global marine mammal biodiversity. PLoS One. 2011;6:e19653.

Kato H, Sakuramoto K. Age at sexual maturity of southern minke whales: a review and some additional analyses. Rep Int Whaling Comm. 1991;41:331–7.

Katsumata E. Study on reproduction of captive marine mammals. J Reprod Dev. 2010;56:1–8.

Katsumata E, Furuta C, Katsumata H, Watanabe G, Taya K. Basal body temperature method for detecting ovarian cycle in the captive beluga (*Delphinapterus leucas*). J Reprod Dev. 2006a; 52:59–63.

Katsumata E, Jaroenporn S, Katsumata H, Konno S, Maeda Y, Watanabe G, Taya K. Body temperature and circulating progesterone levels before and after parturition in killer whales (*Orcinus orca*). J Reprod Dev. 2006b;52:65–71.

Katsumata E, Ueda Y, Arai K, Katsumata H, Kishimoto M, Watanabe G, Taya K. Circulating testosterone and inhibin levels at different ages in the male beluga (*Delphinapterus leucas*). J Vet Med Sci. 2012;74:292–300.

Kellar NM, Trego ML, Marks CI, Dizon AE. Determining pregnancy from blubber in three species of delphinids. Mar Mamm Sci. 2006;22:1–16.

Kellar NM, Trego ML, Marks CI, Chivers SJ, Danil K, Archer FI. Blubber testosterone: a potential marker of male reproductive status in short beaked common dolphins. Mar Mamm Sci. 2009;25:507–22.

Kellar NM, Keliher J, Trego ML, Catelani KN, Hanns C, George J, Rosa C. Variation of bowhead whale progesterone concentrations across demographic groups and sample matrices. Endanger Species Res. 2013a;22:61–72.

Kellar NM, Trego ML, Chivers SJ, Archer FI. Pregnancy patterns of pantropical spotted dolphins (*Stenella attenuata*) in the eastern tropical Pacific determined from hormonal analysis of blubber biopsies and correlations with the purse-seine tuna fishery. Mar Biol. 2013b;160:3113–24.

Kellar NM, Trego ML, Chivers SJ, Archer FI. From progesterone in biopsies to estimates of pregnancy rates: large scale reproductive patterns of two sympatric species of common dolphin, *Delphinus* spp. off California, USA and Baja, Mexico. Bull South Calif Acad Sci. 2014;113:58–80.

Kellar NM, Speakman TR, Smith CR, Lane SM, Balmer BC, Trego ML, Catelani KN, Robbins MN, Allen CD, Wells RS, et al. Low reproductive success rates of common bottlenose dolphins *Tursiops truncatus* in the northern Gulf of Mexico following the deepwater horizon disaster (2010-2015). Endanger Species Res. 2017;33:143–58.

Keller KV. Training of the Atlantic bottlenose dolphins (*Tursiops truncatus*) for artificial insemination. Int Assoc Aquat Anim Med. 1986;14:22–4.

Kellogg Hunter M, Broderick D, Ovenden JR, Tucker KP, Bonde RK, McGuire PM, Lanyon JM. Characterization of highly informative cross-species microsatellite panels for the Australian dugong (*Dugong dugon*) and Florida manatee (*Trichechus manatus latirostris*) including five novel primers. Mol Ecol Resour. 2011;10:368–77.

Kinoshita K, Kiwata M, Kuwano R, Sato N, Tanaka T, Nagata M, Taira H, Kusunoki H. Temporal association of serum progesterone concentrations and vaginal cytology in walruses (*Odobenus rosmarus*). Theriolgenology. 2012;77:933–9.

Kita S, Yoshioka M, Kashiwagi M, Ogawa S, Tobayama T. Comparative external morphology of cetacean spermatozoa. Fish Sci. 2001;67:482–92.

Kjeld M, Alfredsson A, Ólafsson O, Tryland M, Christensen I, Stuen S, Árnason A. Changes in blood testosterone and progesterone concentrations of the North Atlantic minke whale (*Balaenoptera acutorostrata*) during the feeding season. Can J Fish Aquat Sci. 2004;61:230–7.

Kjeld M, Ólafsson O, Víkingsson GA, Sigurjónsson J. Sex hormones and reproductive status of the North Atlantic fin whales (*Balaenoptera physalus*) during the feeding season. Aquat Mamm. 2006;32:75–84.

Knowlton AR, Kraus SD, Kenney RD. Reproduction in North Atlantic right whales (*Eubalaena glacialis*). Can J Zool. 1994;72:1297–305.

Knowlton AR, Hamilton PK, Marx MK, Pettis HM, Kraus SD. Monitoring North Atlantic right whale *Eubalaena glacialis* entanglement rates: a 30 yr retrospective. Mar Ecol Prog Ser. 2012;466:293–302.

Knowlton AR, Robbins J, Landry S, McKenna HA, Kraus SD, Werner TB. Effects of fishing rope strength on the severity of large whale entanglements. Conserv Biol. 2016;30:318–28.

Kooyman GL, Castellini MA, Davis RW. Physiology of diving in marine mammals. Annu Rev Physiol. 1981;43:343–56.

Kraus SD, Rolland RM. The urban whale. Cambridge: Harvard University Press; 2007.

Kraus SD, Hamilton PK, Kenney RD, Knowlton AR, Slay CK. Reproductive parameters of the North Atlantic right whale. J Cetacean Res Manag. 2001;2:231–6.

Kraus SD, Kenney RD, Mayo CA, McLellan WA, Moore MJ, Nowacek DP. Recent scientific publications cast doubt on North Atlantic right whale future. Front Mar Sci. 2016;3:1401–3.

Kumar A, Kalra B, Patel A, McDavid L, Roudebush WE. Development of a second generation anti-Müllerian hormone (AMH) ELISA. J Immunol Methods. 2010;362:51–9.

Kwan D. Towards a sustainable indigenous fishery for dugongs in Torres Strait: a contribution of empirical data analysis and process. Unpublished PhD thesis, James Cook University; 2002. 282 pp.

Lacave G, Eggermont M, Verslycke T, Brook FM, Salbany A, Roque L, Kinoshita R. Prediction from ultrasonographic measurements of the expected delivery date in two species of bottlenose dolphin (*Tursiops truncatus* and *Tursiops aduncus*). Vet Rec. 2004;154:228–33.

Laist DW, Knowlton AR, Pendleton D. Effectiveness of mandatory vessel speed limits for protecting North Atlantic right whales. Endanger Species Res. 2014;23:133–47.

Lambrechts MM, Perret P, Maistre M, Blondel J. Do experiments with captive non-domesticated animals make sense without population field studies? A case study in blue tit's breeding time. Proc R Soc Lond B. 1999;266:1311–5.

Langtimm CA, O'Shea TJ, Pradel R, Beck CA. Estimates of annual survival probabilities for adult Florida manatees (*Trichechus manatus latirostris*). Ecology. 1998;79:981–97.

Langtimm CA, Beck CA, Edwards H, Fick-Child K, Ackerman BB, Barton S, Hartley WC. Survival estimates for Florida manatees from the photo-identification of individuals. Mar Mamm Sci. 2004;20:438–63.

Lanyon JM. Management of megafauna in estuaries and coastal waters: Moreton Bay as a case study. In: Wolanski E, Day JW, Elliott M, Ramachandran R, editors. Coasts and estuaries. Burlington: Elsevier; 2019. p. 87–101.

Lanyon JM, Burgess EA. Methods to examine reproductive biology in free-ranging, fully-marine mammals. In: Reproductive sciences in animal conservation. New York: Springer; 2014. p. 241–74.

Lanyon JM, Sneath HL, Kirkwood JM, Slade RW. Establishing a mark-recapture program for dugongs in Moreton Bay, south-East Queensland. Aust Mammal. 2002;24:51–6.

Lanyon JM, Smith K, Carrick FN. Reproductive steroids are detectable in the faeces of dugongs. Aust Zool. 2005;33:247–50.

Lanyon JM, Slade RW, Sneath HL, Broderick D, Kirkwood JM, Limpus D, Limpus CJ, Jessop T. A method for capturing dugongs (*Dugong dugon*) in open water. Aquat Mamm. 2006;32:196–201.

Lanyon JM, Sneath HL, Ovenden JR, Broderick D, Bonde RK. Sexing sirenians: validation of visual and molecular sex determination in both wild dugongs (*Dugong dugon*) and Florida manatees (*Trichechus manatus latirostris*). Aquat Mamm. 2009;35:187–92.

Lanyon JM, Sneath HL, Long T. Three skin sampling methods for molecular characterisation of free-ranging dugong (*Dugong dugon*) populations. Aquat Mamm. 2010a;36:298–306.

Lanyon JM, Sneath HL, Long T, Bonde RK. Physiological response of wild dugongs (*Dugong dugon*) to out-of-water sampling for health assessment. Aquat Mamm. 2010b;36:46–58.

Lanyon JM, Sneath HL, Long T. Evaluation of exertion and capture stress in serum of wild dugongs, *Dugong dugon*. J Zoo Wildl Med. 2012;43:20–32.

Larkin ILV, Gross TS, Reep RL. Use of faecal testosterone concentrations to monitor male Florida manatee (*Trichechus manatus latirostris*) reproductive status. Aquat Mamm. 2005;31:52–61.

Lasley BL, Kirkpatrick JF. Monitoring ovarian function in captive and free-ranging wildlife by means of urinary and fecal steroids. J Zoo Wildl Med. 1991;22:23–31.

Laws RM. Reproduction, growth, and age of southern fin whales. Discov Rep. 1961;31:327–486.

Lawson JW, Parsons JL, Craig SJ, Eddingotn JD, Kimmins WC. Use of electroejaculation to collect semen samples from wild seals. J Am Vet Med Assoc. 1996;209:1615–7.

Le Boeuf BJ. Pinniped mating systems on land, ice and in the water: emphasis on the Phocidae. In: Renouf D, editor. Behavior of pinnipeds. London: Chapman and Hall; 1991. p. 45–65.

Leatherwood S, Reeves RR. The bottlenose dolphin. San Diego: Elsevier; 2012.

Lee MM, Donahoe PK. Müllerian inhibiting substance: a gonadal hormone with multiple functions. Endocr Rev. 1993;14:152–64.

Lee E, Bhuiyan MMU, Watanabe H, Matsuoka K, Fujise Y, Ishikawa H, Fukui Y. Production of cloned sei whale (*Balaenoptera borealis*) embryos by interspecies somatic cell nuclear transfer using enucleated pig oocyctes. J Vet Sci. 2009;10:285–92.

Lefebvre KA, Bargu S, Kieckhefer T, Silver MW. From sanddads to blue whales: the pervasiveness of domoic acid. Toxicon. 2002;40:971–7.

Lockyer C. Review of baleen whale (Mysticeti) reproduction and implications for management. Rep Int Whaling Comm Special Issue. 1984;6:27–50.

Lockyer C, Martin AR. The sei whale off western Iceland: II age, growth, and reproduction. Rep Int Whaling Comm. 1983;33:465–76.

Lockyer C, Smellie CG. Assessment of reproductive status of female fin and sei whales taken off Iceland, from a histological examination of the uterine mucosa. Rep Int Whaling Comm. 1985;35:343–8.

Lotze HK, Flemming JM, Magera AM. Critical factors for the recovery of marine mammals. Conserv Biol. 2017;31:1301–11.

Lyamin O, Pryaslova J, Lance V, Siegel J. Continuous activity in cetaceans after birth. Nature. 2005;435:1177.

MacLeod CD. Intraspecific scarring in odontocete cetaceans: an indicator of male 'quality' in aggressive social interactions. J Zool. 1998;244:71–7.

Mann J, Connor RC, Barre LM, Heithaus MR. Female reproductive success in bottlenose dolphins (*Tursiops* sp.): life history, habitat, provisioning, and group-size effects. Behav Ecol. 2000a;11:210–9.

Mann J, Connor RC, Tyack PL, Whitehead H. Cetacean societies: field studies of dolphins and whales. Chicago: University of Chicago Press; 2000b. 448 pp.

Mansour AAH, McKay DW, Lien J, Orr JC, Banoub JH, Øien N, Stenson G. Determination of pregnancy status from blubber samples in minke whales (*Balaenoptera acutorostrata*). Mar Mamm Sci. 2002;18:112–20.

Marcoux M, Whitehead H, Rendell L. Sperm whale feeding variation by location, year, social group and clan: evidence from stable isotopes. Mar Ecol Prog Ser. 2007;333:309–14.

Marmontel M. Age and reproduction in female Florida manatees. In: O'Shea TJ, Ackerman BB, Percival HF, editors. Population biology of the Florida manatee. Information and technology report 1: National Biological Service; 1995. p. 98–119.

Marsh H, Kasuya T. Changes in the ovaries of the short-finned pilot whale, *Globicephala macrorhynchus*, with age and reproductive activity. Rep Int Whaling Comm Special Issue. 1984;6:331–5.

Marsh H, Kasuya T. Evidence for reproductive senescence in female cetaceans. Rep Int Whaling Comm Special Issue. 1986;8:57–74.

Marsh H, Kwan D. Temporal variability in the life history and reproductive biology of female dugongs in Torres Strait: the likely role of seagrass dieback. Cont Shelf Res. 2008;28:2152–9.

Marsh H, Heinsohn GE, Channells PW. Changes in the ovaries and uterus of the dugong, *Dugong dugon* (Sirenia: Dugongidae), with age and reproductive activity. Aust J Zool. 1984a;32:743–66.

Marsh H, Heinsohn GE, Glover TD. Changes in the male reproductive organs of the dugong, *Dugong dugon* (Sirenia: Dugondidae) with age and reproductive activity. Aust J Zool. 1984b;32:721–42.

Marsh H, Heinsohn GE, Marsh LM. Breeding cycle, life history and population dynamics of the dugong, *Dugon dugon* (Sirenia: Dugongidae). Aust J Zool. 1984c;32:767–88.

Marsh H, O'Shea TJ, Reynolds JE III. Ecology and conservation of the Sirenia. Dugongs and manatees. Cambridge: Cambridge University Press; 2011. 521 pp.

Mate BR, Gisiner R, Mobley J. Local and migratory movements of Hawaiian humpback whales tracked by satellite telemetry. Can J Zool. 1998;76:863–8.

Mate BR, Lagerquist BA, Urban-Ramirez J. A note on using satellite telemetry to document the use of San Ignacio lagoon by gray whales (*Eschrichtius robustus*) during their reproductive season. J Cetacean Res Manag. 2003;5:149–54.

Maxwell SM, Hazen EL, Bograd SJ, Halpern BS, Breed GA, Nickel B, Teutschel NM, Crowder LB, Benson S, Dutton PH, Bailey H. Cumulative human impacts on marine predators. Nature comm. 2013;4:2688.

McBride AF, Hebb DO. Behavior of the captive bottlenose dolphin *Tursiops truncatus*. J Comp Physiol Psychol. 1948;41:111–23.

McBride AF, Kritzler H. Observations on pregnancy, parturition, and postnatal behavior in the bottlenose dolphin. J Mammal. 1951;32:251–66.

McCauley DJ, Pinsky ML, Palumbi SR, Estes JA, Joyce FH, Warner RR. Marine defaunation: animal loss in the global ocean. Science. 2015;347:1255641.

McHale M, Broderick D, Ovenden JR, Lanyon JM. A multiplexed PCR assay for gender assignment in dugong (*Dugong dugon*) and West Indian manatee (*Trichechus manatus*). Mol Ecol Notes. 2008;8:669–70.

McIntyre T. Trends in tagging of marine mammals: a review of marine mammal biologging studies. Afr J Mar Sci. 2014;36:409–22.

McKenna S, Jarvis J, Sankey T, Reason C, Coles R, Rasheed M. Declines of seagrasses in a tropical harbour, North Queensland, Australia, are not the result of a single event. J Biosci. 2015;40:389–98.

McKenzie J, Parry LJ, Page B, Goldsworthy SD. Estimation of pregnancy rates and reproductive failure in New Zealand fur seals (*Arctocephalus forsteri*). J Mammal. 2005;86:1237–46.

Meager JJ, Limpus C. Mortality of inshore marine mammals in eastern Australia is predicted by freshwater discharge and air temperature. PLoS One. 2014;16;9(4):e94849.

Mellish JE, Calkins DG, Christen DR, Horning M, Rea LD, Atkinson SK. Temporary captivity as a research tool: comprehensive study of wild pinnipeds under controlled conditions. Aquat Mamm. 2006;32:58–65.

Mello D, Colosio A, Marcondes M, Viau P, Oliveira C. Feasibility of using humpback whale blubber to measure sex hormones. J Exp Mar Biol Ecol. 2017;486:32–41.

Mercado E III, deLong CM. Dolphin cognition: representations and processes in memory and perception. Int J Comp Psychol. 2010;23:344–78.

Meyer-Gutbrod EL, Greene CH. Uncertain recovery of the North Atlantic right whale in a changing ocean. Glob Chang Biol. 2018a;24:455–64.

Meyer-Gutbrod EL, Greene CH, Davies K. Marine species range shifts necessitate advanced policy planning: the case of the North Atlantic right whale. Oceanography. 2018b;31:1–6.

Mignucci-Giannoni AA, Toyos-Gonzalez GM, Perez-Padilla J, Rodriguez-Lopez MA, Overing J. Mass stranding of pygmy killer whales (*Feresa attenuata*) in the British Virgin Islands. J Mar Biol Assoc U K. 2000;80:759–60.

Miller DL, Dougherty MM, Decker SJ, Bossart GD. Ultrastructure of the spermatozoa from a Florida manatee (*Trichechus manatus latirostris*). Anat Histol Embryol. 2001;30:253–6.

Miller DL, Bossart GD, Nadji M, Tarpley R, Roberts B, O'Hara TM. A note on the possibility of identifying Leydig and Sertoli cells by immunohistochemistry in bowhead whales (*Balaena mysticetus*). J Cetacean Res Manag. 2002a;4:149–53.

Miller DL, Styer EL, Decker SJ, Robeck T. Ultrastructure of the spermatozoa from three odonocetes: a killer whale (*Orcinus orca*), a Pacific white-sided dolphin (*Lagenorhynchus obliquidens*) and a beluga (*Delphinapterus leucas*). Anat Histol Embryol. 2002b;31:158–68.

Miller CA, Best PB, Perryman WL, Baumgartner MF, Moore MJ. Body shape changes associated with reproductive status, nutritive condition and growth in right whales *Eubalaena glacialis* and *E. australis*. Mar Ecol Prog Ser. 2012;459:135–56.

Miller DL, Styer EL, Kita S, Menchaca M. The mature cetacean spermatozoon. In: Reproductive biology and phylogeny of cetacea: whales, porpoises and dolphins. Enfield: Science Publishers; 2016. p. 245.

Mingramm FMJ, Dunlop RA, Blyde D, Whitworth DJ, Keeley T. Evaluation of respiratory vapour and blubber samples for use in endocrine assessments of bottlenose dolphins (*Tursiops* spp.). Gen Comp Endocrinol. 2019;274:37–49.

Miyazaki N. Further analyses of reproduction in the striped dolphin, *Stenella coeruleoalba*, off the Pacifi c coast of Japan. Rep Int Whaling Comm (Special Issue). 1984;6:343–53.

Mogoe T, Fukui Y, Ishikawa H, Ohsumi S. Effects of diluent composition and temperature on motility and viability after liquid storage and precryopreservation of minke whale (*Balaenoptera acutorostrata*) spermatozoa. Mar Mamm Sci. 1998;14:854–60.

Montano GA, Kraemer DC, Love CC, Robeck TR, O'Brien JK. Evaluation of motility, membrane status and DNA integrity of frozen-thawed bottlenose dolphin (*Tursiops truncatus*) spermatozoa after sex-sorting and recryopreservation. Reproduction. 2012;143:799–813.

Montano GA, Robeck TR, Steinman KJ, O'Brien JK. Circulating anti-Müllerian hormone concentrations in relation to age and season in male and female beluga (*Delphinapterus leucas*). Reprod Fertil Dev. 2017;29:1642–52.

Muraco H, Clough P, Teets V, Arn D, Muraco M. Ovarian follicular dynamics during the luteinizing hormone surge in the bottlenose dolphin (*Tursiops truncatus*). Int J Comp Psychol. 2010;23:723–33.

Murphy S, Collet A, Rogan E. Mating strategy in the male common dolphin (*Delphinus delphis*): what gonadal development tells us. J Mammal. 2005;86:1247–58.

Myers MJ, Litz B, Atkinson S. The effects of age, sex, season and geographic region on circulating serum cortisol concentrations in threatened and endangered Steller Sea lions (*Eumetopias jubatus*). Gen Comp Endocrinol. 2010;165:72–7.

Newell CL, Cowles TJ. Unusual gray whale *Eschrichtius robustus* feeding in the summer of 2005 off the Central Oregon coast. Geophys Res Lett. 2006;33:L22S11.

Nichols C. Temporal change in dugong body condition and seagrass in Moreton Bay, a temperate environment. Unpublished Honours thesis, The University of Queensland; 2005. 55 pp.

Noad M, Cato DH. A combined acoustic and visual survey of humpback whales off Southeast Queensland. Memoirs Queensland Museum. 2001;47:507–23.

Noren DP, Mocklin JA. Review of cetacean biopsy techniques: factors contributing to successful sample collection and physiological and behavioral impacts. Mar Mamm Sci. 2012;28:154–99.

O'Brien JK, Robeck TR. Development of sperm sexing and associated assisted reproductive technology for sex pre-selection of captive bottlenose dolphins (*Tursiops truncatus*). Reprod Fertil Dev. 2006;18:319–29.

O'Brien JK, Robeck TR. Preservation of beluga (*Delphinapterus leucas*) spermatozoa using a trehalose-based cryodiluent and directional freezing technology. Reprod Fertil Dev. 2010;22:653–63.

O'Brien JK, Robeck TR. The relationship of maternal characteristics and circulating progesterone concentrations with reproductive outcome in the bottlenose dolphin (*Tursiops truncatus*) after artificial insemination, with and without ovulation induction, and natural breeding. Theriogenology. 2012a;78:469–82.

O'Brien JK, Robeck TR. The value of *ex situ* cetacean populations in understanding reproductive physiology and developing assisted reproductive technology for *ex situ* and *in situ* species management and conservation efforts. Int J Comp Psychol. 2012b;23:227–48.

O'Brien JK, Steinman KJ, Schmitt TL, Robeck TR. Semen collection, characterisation and artificial insemination in the beluga (*Delphinapterus leucas*) using liquid-stored spermatozoa. Reprod Fertil Dev. 2008;20:770–83.

O'Brien JK, Steinman KJ, Robeck TR. Application of sperm sorting and associated reproductive technology for wildlife management and conservation. Theriogenology. 2009;71:98–101.

Odell DK, Bossart GD, Lowe MT, Hopkins TD. Reproduction of the West Indian manatee in captivity. In: O'Shea TJ, Ackerman BB, Percival HF, editors. Population biology of the Florida manatee. Information and technology report 1: National Biological Service; 1995. p. 192–3.

Ohsumi S, Masaki Y. Biological parameters of the Antarctic minke whale at the virginal population level. J Fish Res Board Can. 1975;32:995–1004.

Oosthuizen WC, de Bruyn PJN, Bester MN. Unmarked individuals in mark-recapture studies: comparisons of marked and unmarked southern elephant seals at Marion Island. Austral Ecol. 2012;37:556–68.

Pace RM III, Corkeron PJ, Kraus SD. State-space mark-recapture estimates reveal a recent decline in abundance of North Atlantic right whales. Ecol Evol. 2017;7:8730–41.

Pallin LJ, Baker CS, Steel D, Kellar NM, Robbins J, Johnston DW, Nowacek DP, Read AJ, Friedlaender AS. High pregnancy rates in humpback whales (*Megaptera novaeangliae*) around the Western Antarctic peninsula, evidence of a rapidly growing population. R Soc Open Sci. 2018;5:180017–5.

Panin M, Gabai G, Ballarin C, Peruffo A, Cozzi B. Evidence of melatonin secretion in cetaceans: plasma concentration and extrapineal HIOMT-like presence in the bottlenose dolphin *Tursiops truncatus*. Gen Comp Endocrinol. 2012;177:238–45.

Parks SE, Brown MW, Conger LA, Hamilton PK, Knowlton AR, Kraus SD, Slay CK, Tyack PL. Occurrence, composition, and potential functions of North Atlantic right whale (*Eubalaena glacialis*) surface active groups. Mar Mamm Sci. 2007;23:868–87.

Parsons KM, Dallas JF, Claridge DE, Durban JW, Balcomb KC III, Thompson PM, Noble LR. Amplifying dolphin mitochondrial DNA from faecal plumes. Mol Ecol. 1999;8:1776–68.

Parsons KM, Durban JW, Claridge DE. Comparing two alternative methods for sampling small cetaceans for molecular analysis. Mar Mamm Sci. 2003;19:224–31.

Perez HIC. Reproductive anatomy and histology of the male Florida manatee (Trichechus manatus latirostris). Unpublished PhD dissertation, The University of Florida; 2015. 78 pp.

Pérez S, García-LÓpez A, De Stephanis R, Giménez J, García-Tiscar S, Verborgh P, Mancera JM, Martínez-Rodriguez G. Use of blubber levels of progesterone to determine pregnancy in free-ranging live cetaceans. Mar Biol. 2011;158:1677–80.

Perrin WF, Henderson JR. Growth and reproductive rates in two populations of spinner dolphins, *Stenella longirostris*, with different histories of exploitation. Rep Int Whaling Comm Special Issue. 1984;6:417–30.

Perrin WF, Reilly SB. Reproductive parameters of dolphins and small whales of the family delphinidae. Rep Int Whaling Comm Special Issue. 1984;6:97–133.

Perrin WF, Scott MD, Walker GJ, Cass VL. Review of geographical stocks of tropical dolphins (*Stenella* spp. and *Delphinus delphis*) in the eastern Pacific. NOAA technical report NMFS 28, US Department of Commerce, Washington; 1985.

Perry CT, Smithers SG, Kench PS, Pears B. Impacts of cyclone Yasi on nearshore, terrigenous sediment-dominated reefs of the central great barrier reef, Australia. Geomorphology. 2014;222:92–105.

Perryman WL, Donahue MA, Perkins PC, Reilly SB. Gray whale calf production 1994-2000: are observed fluctuations related to changes in seasonal ice cover? Mar Mamm Sci. 2002;18:121–44.

Pettis HM, Rolland RM, Hamilton PK, Knowlton AR, Burgess EA, Kraus SD. Body condition changes arising from natural factors and fishing gear entanglements in North Atlantic right whales *Eubalaena glacialis*. Endanger Species Res. 2017;32:237–49.

Pettis HM, Pace RM III, Hamilton PK. Annual report card, North Atlantic Right Whale Consortium. 2018. http://www.narwc.org.

Pietraszek J, Atkinson S. Concentrations of estrone sulfate and progesterone in plasma and saliva, vaginal cytology, and bioelectric impedance during the estrous cycle of the Hawaiian monk seal (*Monachus schauinslandi*). Mar Mamm Sci. 1994;10:430–41.

Pirotta V, Smith A, Ostrowski M, Russell D, Jonsen ID, Grech A, Harcourt R. An economical custom-built drone for assessing whale health. Front Mar Sci. 2017;4:217–2.

Piscitelli MA, Raverty SA, Lillie MA, Shadwick RE. A review of cetacean lung morphology and mechanics. J Morphol. 2013;274:1425–40.

Plön S, Bernard RTF. A review of spermatozoan morphology in Cetacea with new data for the genus *Kogia*. J Zool. 2006;2006:466–73.

Preen AR. Observations of mating behaviour in dugongs (*Dugong dugon*). Mar Mamm Sci. 1989;5:382.

Preen AR. Interactions between dugongs and seagrasses in a sub-tropical environment. Unpublished PhD thesis, James Cook University, Townsville, Queensland; 1992.

Rathbun GB, Reid JP, Bonde RK & Powell JA 1995 Reproduction in free-ranging Florida manatees. Pages 135–156 In O'Shea TJ, Ackerman BB & Percival HF (Eds.) Population biology of the Florida manatee. Information and technology report 1. National Biological Service

Read AJ. Estimation of body condition in harbour porpoises, *Phocoena phocoena*. Can J Zool. 1990;68:1962–6.

Read AJ, Hohn AA. Life in the fast lane: the life history of harbor porpoises from the Gulf of Maine. Mar Mamm Sci. 1995;11:423–40.

Read FL, Hohn AA, Lockyer CH. A review of age estimation methods in marine mammals with special reference to monodontids. NAMMCO Scientific Publications 10; 2018.

Reeves RR, Rolland RM, Clapham PJ. Causes of reproductive failure in North Atlantic right whales: new avenues of research. Woods Hole: National Marine Fisheries Service Northeast Region; 2001. Northeast Fisheries Science Center Reference Document 01-16.

Reyes JC, Mead JG, Van Waerebeek K. A new species of beaked whale *Mesoplodon peruvianus* sp. N (Cetacea: Ziphiidae) from Peru. Mar Mamm Sci. 1991;7:1–24.

Richard JT, Robeck TR, Osborn SD, Naples L, McDermott A, LaForge R, Romano TA, Sartini BL. Testosterone and progesterone concentrations in blow samples are biologically relevant in belugas (*Delphinapterus leucas*). Gen Comp Endocrinol. 2017;246:183–93.

Ridgway S, Carder D, Jeffries M, Todd M. Spontaneous human speech mimicry by a cetacean. Curr Biol. 2012;22:860–1.

Riekkola L, Zerbini AN, Andrews O, Andrews-Goff V, Baker CS, Chandler D, Childerhouse S, Clapham PJ, Dodémont R, Donnelly D, et al. Application of a multi-disciplinary approach to reveal population structure and Southern Ocean feeding grounds of humpback whales. Ecol Indic. 2018;89:455–65.

Robeck TR. Advances in the understanding and manipulation of bottlenose dolphin reproduction. In: Duffield DA, Robeck TR, editors. Proceedings bottlenose dolphin reproduction workshop, San Diego, June 1999; 2000. p. 109–31.

Robeck TR, Monfort S. Characterization of male killer whale (*Orcinus orca*) sexual maturation and reproductive seasonality. Theriogenology. 2006;66:242–50.

Robeck TR, O'Brien JK. Effect of cryopreservation methods and precryopreservation storage on bottlenose dolphin (*Tursiops truncatus*) spermatozoa. Biol Reprod. 2004;70:1340–8.

Robeck TR, Schneyer AL, McBain JF, Dalton LM, Walsh MT, Czekala NM, Kraemer DC. Analysis of urinary immunoreactive steroid metabolites and gonadotrophins for characterization of the estrous cycle, breeding period, and seasonal estrous activity of captive killer whales (*Orcinus orca*). Zoo Biol. 1993;12:173–87.

Robeck TR, Curry BE, McBain JF, Kraemer DC. Reproductive biology of the bottlenose dolphin (*Tursiops truncatus*) and the potential application of advanced reproductive technologies. J Zoo Wildl Med. 1994;25:321–36.

Robeck TR, McBain JF, Mathey S, Kraemer DC. Ultrasonographic evaluation of the effects of exogenous gonadotrophin on follicular recruitment and ovulation induction in the Atlantic bottlenose dolphin, *Tursiops truncatus*. J Zoo Wildl Med. 1998;29:6–13.

Robeck TR, Atkinson SKC, Brook F. Reproduction. In: Dierauf LA, Gulland FMD, editors. CRC handbook of marine mammal medicine. 2nd ed. Boca Raton: CRC Press; 2001. p. 193–236.

Robeck TR, Steinman KJ, Gearhart S, Reidarson TR, McBain JF, Monfort SL. Reproductive physiology and development of artificial insemination technology in killer whales (*Orcinus orca*). Biol Reprod. 2004;71:650–60.

Robeck TR, Monfort SL, Calle PP, Dunn JL, Jensen E, Boehm JR, Young S, Clark ST. Reproduction, growth and development in captive beluga (*Delphinapterus leucus*). Zoo Biol. 2005a;24:29–49.

Robeck TR, Steinman KJ, Yoshioka M, Jensen E, O'Brien JK, Katsumata E, Gili C, McBain JF, Sweeney J, Monfort SL. Estrous cycle characterization and artificial insemination using frozen-thawed spermatozoa in the bottlenose dolphin (*Tursiops truncatus*). Reproduction. 2005b;129:659–74.

Robeck TR, Steinman KJ, Ramirez K, Greenwell M, Van Bonn W, Yoshioka M, Katsumata E, Dalton L, Osborn S, O'Brien JK. Seasonality, estrous cycle characterization, estrus synchronization, semen cryopreservation and artificial insemination in the Pacific white-sided dolphin (*Lagenorhynchus obliquidens*). Reproduction. 2009;138:391–405.

Robeck TR, Steinman KJ, Montano GA, Katsumata E, Osborn S, Dalton L, Dunn JL, Schmidt TL, Reidarson TR, O'Brien JK. Deep intra-uterine artificial inseminations using cryopreserved spermatozoa in beluga (*Delphinapterus leucas*). Theriogenology. 2010;74:989–1001.

Robeck TR, Willis K, Scarpuzzi MR, O'Brien JK. Comparisons of life-history parameters between free-ranging and captive killer whale (*Orcinus orca*) populations for application toward species management. J Mammal. 2015;96:1055–70.

Robeck TR, Blum JL, Steinman KJ, Ratner JR, Bergfelt DR, O'Brien JK. Longitudinal profiles of relaxin and progestagens during pregnancy, pregnancy loss and false pregnancy in the killer whale (*Orcinus orca*). Gen Comp Endocrinol. 2018;267:98–108.

Robinson KJ, Twiss SD, Hazon N, Pomeroy PP. Maternal oxytocin is linked to close mother-infant proximity in grey seals (*Halichoerus grypus*). PLoS One. 2015;10:e0144577–17.

Rolland RM, Hunt KE, Kraus SD, Wasser SK. Assessing reproductive status of right whales (*Eubalaena glacialis*) using fecal hormone metabolites. Gen Comp Endocrinol. 2005;142:308–17.

Rolland RM, Hamilton PK, Kraus SD, Davenport B, Gillett RM, Wasser SK. Faecal sampling using detection dogs to study reproduction and health in North Atlantic right whales (*Eubalaena glacialis*). J Cetacean Res Manag. 2006;8:121–5.

Rolland RM, Hunt KE, Doucette GJ, Rickard LG, Wasser SK. The inner whale: hormones, biotoxins and parasites. In: Kraus SD, Rolland RM, editors. The urban whale: North Atlantic right whales at the crossroads. Cambridge: Harvard University Press; 2007. p. 232–72.

Rolland RM, Parks SE, Hunt KE, Castellote M, Corkeron PJ, Nowacek DP, Wasser SK, Kraus SD. Evidence that ship noise increases stress in right whales. Proc R Soc B Biol Sci. 2012;279:2363–8.

Rolland RM, Hunt KE, Kraus SD, Claridge DE, Dunn CA. Assessing stress responses in beaked and sperm whales in the Bahamas: Report to the Office of Naval Research, Marine Mammals and Biology Program; 2015.

Rolland RM, Schick RS, Pettis HM, Knowlton AR, Hamilton PK, Clark JS, Kraus SD. Health of North Atlantic right whales *Eubalaena glacialis* over three decades: from individual health to demographic and population health trends. Mar Ecol Prog Ser. 2016;542:265–82.

Rolland RM, McLellan WA, Moore MJ, Harms CA, Burgess EA, Hunt KE. Fecal glucocorticoids and anthropogenic injury and mortality in North Atlantic right whales *Eubalaena glacialis*. Endanger Species Res. 2017;34:417–29.

Rommel SA, Pabst DA, McLellan WA, Mead JG, Potter CW. Anatomical evidence for a countercurrent heat exchanger associated with dolphin testes. Anat Rec. 1992;232:150–6.

Rowntree VJ, Payne RS, Schell DM. Changing patterns of habitat use by southern right whales (*Eubalaena australis*) on their nursery ground at Península Valdés, Argentina, and in their long-range movements. J Cetacean Res Manag Special Issue. 2001;2:133–43.

Russell DJF, McConnell B, Thompson D, Duck C, Morris C, Harwood J, Matthiopoulos J. Uncovering the links between foraging and breeding regions in a highly mobile mammal. J Appl Ecol. 2013;50:499–509.

Sánchez-Calabuig MJ, García-Vázquez FA, Laguna-Barraza R, Barros-García C, García-Parraga D, Rizos D, Adan AG, Pérez-Gutíerrez JF. Bottlenose dolphin (*Tursiops truncatus*) spermatozoa: collection, cryopreservation, and heterologous *in vitro* fertilization. J Vis Exp. 2017;126:e55237.

Schaefer AM, Stavros H-CW, Bossart GD, Fair PA, Goldstein JD, Reif JS. Associations between mercury and hepatic, renal, endocrine, and hematological parameters in Atlantic bottlenose dolphins (*Tursiops truncatus*) along the eastern coast of Florida and South Carolina. Arch Environ Contam Toxicol. 2011;61:688–95.

Schipper J, Chanson JS, Chiozza F, et al. The status of the world's land and marine mammals: diversity, threat, and knowledge. Science. 2008;322:225–30.

Schmitt TL, St. Aubin DJ, Schaeffer AM, Dunn JL. Baseline, diurnal variations and stress-induced changes of stress hormones in three captive beluga whales, *Delphinapterus leucas*. Mar Mamm Sci. 2010;26:635–47.

Schroeder JP, Keller KV. Seasonality of serum testosterone levels and sperm density in *Tursiops truncatus*. J Exp Zool. 1989;249:316–21.

Schwabe C, Büllesbach EE, Heyn H, Toshioka M. Cetacean relaxin: isolation and sequence of relaxins from *Balaenoptera acutorostrata* and *Balaenoptera edeni*. J Biol Chem. 1989;264:940–3.

Schwarzenberger F, Möstl E, Palme R, Bamberg E. Faecal steroid analysis for non-invasive monitoring of reproductive status in farm, wild and zoo animals. Anim Reprod Sci. 1996;42:515–26.

Schwierzke WL. Assessment of fertility potential in bottlenose dolphins (*Tursiops truncates*): an ELISA-based biomarker analysis: Unpublished Master of Science thesis, University of South Florida; 2011. 89 pp.

Scott MD, Wells RS, Irvine AB. A long-term study of bottlenose dolphins on the west coast of Florida. In: Leatherwood SJ, Reeves RR, editors. The bottlenose dolphin. San Diego: Academic; 1990. p. 235–44.

Scott EM, Mann J, Watson-Capps JJ, Sargeant BL, Connor RC. Aggression in bottlenose dolphins: evidence for sexual coercion, male-male competition, and female tolerance through analysis of tooth-rake marks and behaviour. Behaviour. 2005;142:21–44.

Shane SH, Wells RS, Wursig B. Ecology, behavior and social organization of the bottlenose dolphin: a review. Mar Mamm Sci. 1986;2:34–63.
Simmonds MP, Isaac SJ. The impacts of climate change on marine mammals: early signs of significant problems. Oryx. 2007;41:19–26.
Smolker RA, Richards AF, Connor RC, Pepper JW. Sex differences in patterns of association among Indian Ocean bottlenose dolphins. Behaviour. 1992;123:38–69.
Sobtzick S, Cleguer C, Hagihara R, Marsh H. Distribution and abundance of dugong and large marine turtles in Moreton Bay, Hervey Bay and the southern Great Barrier Reef. A report to the Great Barrier Reef Marine Park Authority. Centre for Tropical Water & Aquatic Ecosystem Research (TropWATER) Publication 17/21, James Cook University, Townsville; 2017.
Stacey PJ, Baird RW. Birth of a 'resident' killer whale off Victoria, British Columbia, Canada. Mar Mamm Sci. 1997;13:504–8.
Steinman KJ, O'Brien JK, Monfort SL, Robeck TR. Characterization of the estrous cycle in female beluga (*Delphinapterus leucas*) using urinary endocrine monitoring and transabdominal ultrasound: evidence of facultative induced ovulation. Gen Comp Endocrinol. 2012;175:389–97.
Sumich J, Goff T, Perryman WL. Growth of two captive gray whale calves. Aquat Mamm. 2001;27:231–3.
Suzuki T, Mogoe T, Asada M, Miyamoto A, Tetsuka M, Ishikawa H, Ohsumi S, Fukui Y. Plasma and pituitary concentrations of gonadotropins (FSH and LH) in minke whale (*Balaenoptera acutorostrata*) during the feeding season. Theriogenology. 2001;55:1127–41.
Tavolga MC, Essapian FS. The behavior of the bottle-nosed dolphin (*Tursiops truncatus*): mating, pregnancy, parturition and mother-infant behavior. Zoologica. 1957;42:11–31.
Terasawa F, Yokoyama Y, Kitamure M. Rectal temperature before and after parturition in bottlenose dolphins. Zoo Biol. 1999;18:153–6.
Thewissen JGM, Heyning J. Embryogenesis and development in *Stenella attenuata* and other cetaceans. In: Miller D, editor. Reproductive biology and phylogeny of cetacea: whales, dolphin and porpoises. Enfield: Science Publishers; 2007. p. 307–29.
Thomas PO, Reeves RR, Brownell RL Jr. Status of the world's baleen whales. Mar Mamm Sci. 2015;32:682–734.
Thompson KV, Baker CS, van Helden A, Patel S, Millar C, Constantine R. The world's rarest whale. Curr Biol. 2012;22:905–6.
Tizzi R, Accorsi PA, Azzali M. Non-invasive multidisciplinary approach to the study of reproduction and calf development in bottlenose dolphin (*Tursiops truncatus*): the Rimini Delfinario experience. Int J Comp Psychol. 2010;23
Touma C, Palme R. Measuring fecal glucocorticoid metabolites in mammals and birds: the importance of validation. Ann N Y Acad Sci. 2005;1046:54–74.
Trana MR, Roth JD, Tomy GT, Anderson WG, Ferguson SH. Influence of sample degradation and tissue depth on blubber cortisol in beluga whales. J Exp Mar Biol Ecol. 2015;462:8–13.
Trego ML, Kellar NM, Danil K. Validation of blubber progesterone concentrations for pregnancy determination in three dolphin species and a porpoise. PLoS One. 2013;8:e69709–10.
Tremellen KP, Kolo M, Gilmore A, Laekamge DN. Anti-Müllerian hormone as a marker of ovarian reserve. Aust N Z J Obstet Gynecol. 2006;45:20–4.
Tripp KM, Verstegen JP, Deutsch CJ, Bonde RK, de Wit M, Manire CA, Gaspard J, Harr KE. Evaluation of adrenocortical function in Florida manatees (*Trichechus manatus latirostris*). Zoo Biol. 2010;29:1–15.
Trumble SJ, Robinson EM, Berman-Kowalewski M, Potter CW, Usenko S. Blue whale earplug reveals lifetime contaminant exposure and hormone profiles. Proc Natl Acad Sci U S A. 2013;110:16922–6.
Valenzuela-Molina M, Atkinson S, Mashburn K, Gendron D, Brownell RL Jr. Fecal steroid hormones reveal reproductive state in female blue whales sampled in the Gulf of California, Mexico. Gen Comp Endocrinol. 2018;261:127–35.
Van der Hoop JM, Moore MJ, Barco SG, Cole TVN, Daoust P-Y, Henry AG, McAlpine DF, McLellan WA, Wimmer T, Solow AR. Assessment of management to mitigate anthropogenic effects on large whales. Conserv Biol. 2013;27:121–33.

Van der Hoop J, Corkeron P, Moore M. Entanglement is a costly life-history stage in large whales. Ecol Evol. 2016;7:92–106.

Van der Schoot P. Studies on the fetal development of the gubernaclum in cetacea. Anat Rec. 1995;243:449–60.

Van Waerebeek K, Read AJ. Reproduction of dusky dolphins, *Lagenorhynchus obscurus,* from coastal Peru. J Mammal. 1994;75:1054–62.

Vergara V, Michaud R, Barrett-Lennard L. What can captive whales tell us about their wild counterparts? Identification, usage, and ontogeny of contact calls in belugas (*Delphinapterus leucas*). Int J Comp Psychol. 2010;23:278–309.

Villegas-Amtmann S, Atkinson S, Costa DP. Low synchrony in the breeding cycle of Galapagos Sea lions revealed by seasonal progesterone concentrations. J Mammal. 2009;90:1232–7.

Vu ET, Clark C, Catelani K, Kellar NM, Calambokidis J. Seasonal blubber testosterone concentrations of male humpback whales (*Megaptera novaeangliae*). Mar Mamm Sci. 2015;31:1258–64.

Wade PR, De Robertis A, Hough KR, Booth R, Kennedy A, LeDuc RG, Munger L, Napp J, Shelden K, Rankin S, et al. Rare detections of North Pacific right whales in the Gulf of Alaska, with observations of their potential prey. Endanger Species Res. 2011;13:99–109.

Wakai Y, Hasegawa K, Sakamoto S, Asano S, Watanabe G, Taya K. Annual changes of urinary progesterone and estradiol-17β of the dugong (*Dugong dugon*) in captivity. Zool Sci. 2002;19:679–82.

Walker L, Cornell L, Dahl K, Czekala N, Dargen C, Joseph B, Hsueh A, Lasley B. Urinary concentrations of ovarian steroid hormone metabolites and bioactive follicle-stimulating hormone in killer whales (*Orcinus orcus*) during ovarian cycles and pregnancy. Biol Reprod. 1988;39:1013.

Waples KA, Gales NJ. Evaluating and minimising social stress in the care of captive bottlenose dolphins (*Tursiops aduncus*). Zoo Biol. 2002;21:5–26.

Wasser SK, Lundin JI, Ayres K, Seely E, Giles D, Balcomb K, Hempelmann J, Parsons K, Booth R. Population growth is limited by nutritional impacts on pregnancy success in endangered southern resident killer whales (*Orcinus orca*). PLoS One. 2017;12:e0179824–2.

Watanabe H, Mogoe T, Asada M, Hayashi K, Fujise Y, Ishikawa H, Ohsumi S, Miyamoto A, Fukui Y. Relationship between serum sex hormone concentrations and histology of seminiferous tubules of captured baleen whales in the western North Pacific during the feeding season. J Reprod Dev. 2004;50:419–27.

Watanabe H, Tateno H, Kusakbe H, Matsuoka T, Kamiguchi Y, Fujise Y, Ishikawa H, Ohsumi S, Fukui Y. Fertilizability and chromosomal integrity of frozen-thawed Bryde's whale (*Balaenoptera edeni*) spermatozoa intracytoplasmically injected into mouse oocytes. Zygote. 2007;15:9–14.

Wells RS. Reproduction in wild bottlenose dolphins: overview of the patterns observed during a long-term study. In: Duffield DA, Robeck TR, editors. Proceedings bottlenose dolphin reproduction workshop, San Diego, June 1999; 2000. p. 57–74.

Wells RS. Dolphin social complexity: lessons from long-term study and life history. In: Animal social complexity: intelligence, culture, and individualized societies. Cambridge: Harvard University Press; 2003. p. 32–56.

Wells RS, Rhinehart HL, Hansen LJ, Sweeney JC, Townsend FI, Stone R, Casper DR, Scott MD, Hohn AA, Rowles TK. Bottlenose dolphins as marine ecosystem sentinels: developing a health monitoring system. EcoHealth. 2004;1:246–54.

Wells RS, Tornero V, Borrell A, Aguilar A, Rowles TK, Rhinehart HL, Hofmann S, Jarman WM, Hohn AA, Sweeney JC. Integrating potential life-history and reproductive success data to examine relationships with organochlorine compounds for bottlenose dolphins (*Tursiops truncatus*) in Sarasota Bay, Florida. Sci Total Environ. 2005;349:106–19.

West K, Atkinson S, Carmichael M, Sweeney J, Krames B, Krames J. Concentrations of progesterone in milk from bottlenose dolphins during different reproductive states. Gen Comp Endocrinol. 2000;117:218–24.

Wetzel DL, Reynolds JE III, Breuel KF, Roudebush WE. Validation of an inhibin-B ELISA for use in an endangered marine mammal, the West Indian manatee (*Trichechus manatus*). Biol Reprod (Suppl). 2009;215:97.

Whitehead H. The behaviour of mature male sperm whales on the Galapagos Islands breeding grounds. Can J Zool. 1993;71:689–99.

Whitehead H. Delayed competitive breeding in roving males. J Theor Biol. 1994;166:127–33.

Whitehead H, Mann J. Female reproductive strategies of cetaceans. In: Mann J, Connor RC, Tyack PL, Whitehead H, editors. Cetacean societies: field studies of dolphins and whales. Chicago: University of Chicago Press; 2000. p. 219–46.

Williamson P, Gales NJ, Lister S. Use of real-time B-mode ultrasound for pregnancy diagnosis and measurement of fetal growth in captive bottlenose dolphins (*Tursiops truncatus*). J Reprod Fertil. 1990;88:543–8.

Wilson RC, Reynolds JE III, Wetzel DL, Schwierzke-Wade L, Bonde RK, Breuel KF, Roudebush WE. Secretion of anti-Müllerian hormone in the Florida manatee *Trichechus manatus latirostris*, with implications for assessing conservation status. Endanger Species Res. 2011;14:107–12.

Worm B, Barbier EB, Beaumont N, Duffy JE, Folke C, Halpern BS, Jackson JBC, Lotze HK, Micheli F, Palumbi SR, et al. Impacts of biodiversity loss on ocean ecosystem services. Science. 2006;314:787–90.

Würsig B, Jefferson TA. Methods of photo-identification for small cetaceans. Rep Int Whaling Comm (Special Issue). 1990;12:43–52.

Yochem PK, Leatherwood S. Blue whale Balaenoptera musculus (Linnaeus, 1758). In: Ridgway SH, Harrison R, editors. The sirenians and baleen whales, Handbook of marine mammals, vol. 3. London: Academic Press; 1985. p. 193–240.

Yoshioka M, Aida K, Hanyu I. Annual changes in serum reproductive hormone levels in captive bottlenose dolphins. Bull Jap Soc Sci Fish. 1986;52:1939–46.

Yoshioka M, Okumura T, Aida K, Fujise Y. A proposed technique for quantifying muscle progesterone content in minke whales (*Balaenoptera acutorostrata*). Can J Zool. 1994;72:368–70.

Yuen QWH, Brook FM, Kinoshita RE, Ying MTC. Semen collection and ejaculate characteristics in the indo-Pacific bottlenose dolphin (*Tursiops truncatus*). J Androl. 2009;30:432–9.

Chapter 14
Amphibian Assisted Reproductive Technologies: Moving from Technology to Application

J. Clulow, R. Upton, V. L. Trudeau, and S. Clulow

Abstract Amphibians have experienced a catastrophic decline since the 1980s driven by disease, habitat loss, and impacts of invasive species and face ongoing threats from climate change. About 40% of extant amphibians are under threat of extinction and about 200 species have disappeared completely. Reproductive technologies and biobanking of cryopreserved materials offer technologies that could increase the efficiency and effectiveness of conservation programs involving management of captive breeding and wild populations through reduced costs, better genetic management and reduced risk of species extinctions. However, there are relatively few examples of applications of these technologies in practice in on-the-ground conservation programs, and no example that we know of where genetic diversity has been restored to a threatened amphibian species in captive breeding or in wild populations using cryopreserved genetic material. This gap in the application of technology to conservation programs needs to be addressed if assisted reproductive technologies (ARTs) and biobanking are to realise their potential in amphibian conservation. We review successful technologies including non-invasive gamete collection, IVF and sperm cryopreservation that work well enough to be applied to many current conservation programs. We consider new advances in technology (vitrification and laser warming) of cryopreservation of aquatic embryos of fish and some marine invertebrates that may help us to overcome factors limiting amphibian oocyte and embryo cryopreservation. Finally, we address two case studies

J. Clulow (✉) · R. Upton
School of Environmental and Life Sciences, University of Newcastle, Newcastle, NSW, Australia
e-mail: john.clulow@newcastle.edu.au; rose.upton@uon.edu.au

V. L. Trudeau
Department of Biology, University of Ottawa, Ottawa, ON, Canada
e-mail: trudeauv@uottawa.ca

S. Clulow
Department of Biological Sciences, Faculty of Science and Engineering, Macquarie University, Sydney, NSW, Australia
e-mail: simon.clulow@mq.edu.au

P. Comizzoli et al. (eds.), *Reproductive Sciences in Animal Conservation*, Advances in Experimental Medicine and Biology 1200,
https://doi.org/10.1007/978-3-030-23633-5_14

that illustrate the urgent need and the opportunity to implement immediately ARTs, cryopreservation and biobanking to amphibian conservation. These are (1) managing the biosecurity (disease risk) of the frogs of New Guinea which are currently free of chytridiomycosis, but are at high risk (2) the Sehuencas water frog of Bolivia, which until recently had only one known surviving male.

1 Introduction

Amphibian assisted reproductive technologies (ARTs) have been a focus of research for more than two decades (Clulow et al. 1999, 2014; Kouba and Vance 2009; Kouba et al. 2009, 2013; Clulow and Clulow 2016) since the emergence of the global amphibian crisis in the 1970s and 1980s. The scale of the decline is difficult to comprehend, even in an era of the sixth mass extinction (Ceballos et al. 2017). The numbers of amphibian species at risk of extinction, or extinct continues to grow. Bishop et al. (2012), based on earlier data from Stuart et al. (2004, 2008) recognised 34 amphibian species extinctions and 32% (1893 species) of species as threatened with extinction, when adjusted for data-deficient species this figure is about 40% of all described amphibian species. Using Bayesian approaches, Alroy (2015) estimated as many as 200 species to be extinct and Scheele et al. (2019) estimated 90 species extinctions from chtyridiomycosis alone, with 501 species in decline from chytridiomycosis (6.5% of all amphibian species), 124 of those having declined in abundance by >90%. Habitat modification is responsible for declines affecting 63% of amphibian species (Bishop et al. 2012; Chanson et al. 2008). The understanding of species-specific susceptibility to chytridiomycosis has improved remarkably in recent years (Berger et al. 1998; Longcore et al. 1999; Lips et al. 2006; Skerratt et al. 2007, 2010; Murray et al. 2010; Campbell et al. 2019) and there is evidence of limited, localised recovery in around 20% (60) of affected species (Scheele et al. 2019). Research into habitat manipulation, translocation, selection for resistance (Scheele et al. 2014) and altering environmental salinity (Clulow et al. 2018b; Klop-Toker et al. 2018) are all being actively pursued to mitigate the effects of chytridiomycosis, but none have successfully been deployed at scale in wild situations.

Sometime after the amphibian declines became apparent in the 1970–1980s, proposals for the use of assisted reproductive technologies and cryopreservation as conservation tools began to emerge (Clulow et al. 1999). Indeed, in the period since then, there have been many advances and improvements in the technologies at the laboratory level. The array of technologies have included gamete induction using hormones (Clulow et al. 2018a; Vu and Trudeau 2016), in vitro fertilization (IVF) (Kouba et al. 2009; Clulow et al. 2014), cell cryopreservation (Kouba et al. 2013; Mollard 2018; Zimkus et al. 2018; Ryder and Onuma 2018) and sperm cryopreservation (Kouba et al. 2009, 2013; Clulow et al. 2014; Browne et al. 2019) and even the use of non-conventional procedures for the manipulation of reproduction such as nuclear transfer (cloning) and androgenesis (Clulow et al. 2014). Of all these technologies, the most useful advances have been in the areas of non-invasive (non-lethal) gamete collection and sperm cryopreservation. These particular techniques support IVF with the generation of viable embryos from cryopreserved sperm. As a back-up to these procedures, the generation of cultures and cell lines (Kouba et al. 2013; Zimkus et al. 2018; Ryder and Onuma 2018) have provided another source of cells for cryopreservation and future use as technologies such stem cells, nuclear transfer and generation of germ cells from chimeras advance further towards utilisation in amphibians.

Nevertheless, despite the commitment and progress of a growing number of researchers there has been far less progress in converting the technologies into successful examples of applied case studies in real world conservation programs. While the black-footed ferret is the gold-standard example of restoration of genetic diversity from cryopreserved genomes (Howard et al. 2003, 2016; Santymire et al. 2014), there is no evidence of the use of biobanked, cryopreserved genomes having added to the outcomes in any on-the-ground amphibian conservation program (Clulow and Clulow 2016). The shortage of "proof of concept" examples of the usefulness of ART and cryobiology in amphibian conservation arguably result in a credibility gap between the promise of the technology and the realisation of conservation outcomes (a broader problem with the use of ART's in conservation (Monfort 2014)). It is important to continue the development of the underpinning technologies, given the promise and justification of this approach. Potential benefits range from reduced costs and greater efficiencies in breeding programs (Clulow et al. 2014; Clulow and Clulow 2016) to restored genetic diversity in wild populations (Frankham 2010, 2015), and reduced loss of genetic diversity in captive colonies (Williams and Hoffman 2009). In one example, reproductive technologies (induction of gamete release (Kouba and Vance 2009)) were required to breed endangered Wyoming toads, *Anaxyrus baxteri* in captivity, and aided a breeding program leading to the release of more than 100,000 offspring into the wild (Harding et al. 2016). In another example (Kouba et al. 2011), refrigerated (stored) sperm were transported between zoos, allowing IVF without the need to transport adults of the endangered gopher frog, *Rana sevosa*. More examples like these are required to build confidence and promote utilisation of reproductive technologies.

There are estimates that, globally, up to 950 amphibian species require captive breeding programs, but adequate resources for only around 50 species across zoos and other institutions (Bishop et al. 2012; Clulow et al. 2014; Browne et al. 2018), although Browne et al. (2018) have argued that the major expansion of private hobbyist networks could increase capacity. There is a broad recognition of the need to accelerate and progress the implementation of genome storage and biobanking for amphibian conservation (Kouba and Vance 2009; Kouba et al. 2013; Clulow et al. 2014; Clulow and Clulow 2016; Zimkus et al. 2018; Browne et al. 2018, 2019; Silla and Byrne 2019). A number of convincing proof-of-concept examples of the usefulness of ART and cryobiology in amphibian conservation would go a long way towards increasing the support of agencies and funding bodies towards further basic research in the area. This review, therefore, addresses three things: where the technology stands, where likely advances may occur that will further improve the power of the technology and finally, approaches that might be pursued to achieve early success in proof-of-concept applications in amphibian conservation programs.

2 Technologies up to Speed: IVF and Hormonal Induction of Gamete Release

2.1 *In Vitro Fertilisation (IVF)*

IVF in frogs (if not in caecilians and urodeles) has been available as a reproductive technique for many decades (for reviews see Kouba et al. 2009; Clulow et al. 2014). It has been well characterised for a few species, particularly the mainstays of laboratory studies of developmental biology (Rugh 1962; Hollinger and Corton 1980; McKinnell 1978), but IVF has also been partly characterised in non-laboratory species of various families e.g. Ranidae, *Pelophylax lessonae* (Uteshev et al. 2013), *Rana temporaria* (Shishova et al. 2011) and *P. kurtmuelleri* (Turani et al. 2015); Bufonidae, *Bufo baxteri* (Browne et al. 2006a), *Rhinella marina* (Browne et al. 1998); Myobatrachidae, *Limnodynastes tasmaniensis* (Edwards et al. 2004), *Pseudophryne corroboree* (Byrne and Silla 2010) and *P. guentherii* (Silla 2013); Pelodryadidae, *Litoria peronii* (Sherman et al. 2008); Ceratophryidae, *Lepidobatrachus laevis* and *L. llanensis* (Waggener and Carroll 1998); and a direct-developing member of the Eleutherodactylidae, *Eleutherodactylus coqui* (Toro and Michael 2004). Fertilisation in anurans is external, and hence avoids complex fertilisation media, mimicry of an internal oviduct or uterine environment and the need for implantation into a hormonally primed and synchronised uterus. The primary challenge is to have available simultaneously gametes of suitable number and quality.

Various studies have examined the problem of optimising the availability of viable gametes and the IVF outcome (Hollinger and Corton 1980; Edwards et al. 2004; Silla 2012). These studies act as useful guides for those considering IVF in less studied species, since high rates of fertilisation, cleavage and early development with few abnormal embryos are likely to produce larger proportions of larvae completing metamorphosis and developing into sexually mature adults. There are likely to be species differences in terms of optimal conditions for IVF, although some basic principles are likely to apply across most species that are externally fertilized in freshwater.

Some basic principles include:

1. Utilisation of low osmolality to activate sperm, since amphibian sperm, like fresh water fish sperm, are usually activated by an osmotic shock (osmolality reduced from isotonic to body fluids, around 220 mosm kg^{-1} to values closer to pond water, perhaps in the region of 20–40 mosm kg^{-1}). Motility (forward progression) is usually a function of the decrease in osmolality (Hollinger and Corton 1980; Browne et al. 1998; Edwards et al. 2004; Inoda and Morisawa 1987; Kouba et al. 2003; Fitzsimmons et al. 2007).
2. High ratios of sperm to eggs is advantageous during fertilisation. Various studies demonstrated higher fertilisation rates (essentially a sigmoid curve) with increasing sperm concentrations (Browne et al. 1998; Edwards et al. 2004).

3. Minimisation of the holding time of oocytes (however these are obtained prior to fertilisation). Where such curves have been determined, there is usually a strong negative correlation between oocyte holding time and fertilisation rate. It is not fully determined why this is the case, but may involve the activation of apoptotic pathways (Du Pasquier et al. 2011; Messaoud et al. 2015) in oocytes that fail to fertilise. Interestingly, even holding oocytes at reduced temperatures (e.g. at 4 °C which is associated with much extended storage times for amphibian sperm in vitro, see below) does not increase holding times for anuran oocytes in most species (Kouba and Vance 2009; Edwards et al. 2004; Wolf and Hedrick 1971; Browne et al. 2001). This was demonstrated by Browne et al. (2001) who showed that meiosis-II arrested oocytes of the toad *Rhinella marina* lost fertilising capacity within 2–4 h at 5–10 °C, but retained fertilising capacity at 15–25 °C for up to 8–10 h. This is the opposite outcome to holding spermatozoa in vitro, where sperm may be viable for only minutes to hours at room temperature, but survive for days to weeks under when refrigerated at 4 °C (with subsequent activation initiated by increasing temperature and reducing osmolality). There are clearly cellular and metabolic processes initiated in unfertilised amphibian oocytes that are not reversibly arrested simply by the obvious step of cooling gametes and reducing metabolic rates. This could be a fruitful area of investigation if processes such as apoptosis could be inhibited, then oocyte storage might be extended substantially (Du Pasquier et al. 2011; Messaoud et al. 2015). One recent paper on a temperate northern hemisphere frog (*Rana temporaria*) demonstrated extended periods of storage of oocytes either within or outside the oviducts (Uteshev et al. 2018). It is not clear if this unusual phenomenon is due to cold-adaptation of this species, in which case, it would be of interest to investigate it further to understand the mechanism.
4. Standard IVF hygiene procedures improve survival to later development stages. This may be as simple as removing non-progressing, abnormal embryos before they degenerate and support extensive microbial proliferation, or changing media on a regular basis until embryos hatch as swimming and feeding larvae.

2.2 Induction of Gamete Release

Viable gametes are required for fertilisation (unless there is some modification to procedures such as Intra-cytoplasmic Sperm Injection (Ishibashi et al. 2007a; Ishibashi et al. 2007b)). Collection is normally reliant on hormonal induction to manipulate and initiate final stages of oogenesis (ovulation) or spermatogenesis (spermiation) via pathways in the neuroendocrine hypothalamic-pituitary-gonadal (HPG) axis (Vu and Trudeau 2016). Surprisingly, the HPG axis is far less understood in amphibians than might be expected given the knowledge of the equivalent systems of fish and mammals (Clulow et al. 2014; Vu and Trudeau 2016). Indeed, when Vu and Trudeau (2016) reviewed the neuroendocrine release of gametes in amphibians, it was the first complete review of the topic since the 1980s (Moore 1987).

This was followed by another that reviewed the differential success of gamete induction between males and female frogs (Clulow et al. 2018a), a difference that is likely to arise from more complex processes in the neuroendocrine pathways of the HPG axis in females, as well as incompatibilities between exogenous hormones used for induction and endogenous receptors (ligand-receptor incompatibilities arising from non-conserved regions in active units of hormones). There are many gaps in knowledge that, if addressed, might provide more refined and effective protocols for the induction of the maturation of gametes and the synchronised, accurately timed release of these from the testes and ovaries.

Gametes released from the ovaries and testes of amphibians through hormonal induction are generally capable of fertilisation when mixed together in vitro (Kouba et al. 2009; Clulow et al. 2014), whether they were collected by oviposition (Kouba and Vance 2009; Kouba et al. 2009; Clulow et al. 2012, 2014; McKinnell et al. 1976), removed directly from oviducts, or collected in the urine of males (known as spermic urine (Uteshev et al. 2013; Shishova et al. 2011; McKinnell et al. 1976; Silla 2011)). In the case of males, if mature sperm are present in the testes, these can be collected by simple maceration of the testes into an isotonic Ringer medium, and then diluted immediately for activation in a hypotonic solution (and are capable of immediate fertilisation without undergoing the capacitation process occurring in mammals) (Browne et al. 1998). Removal of testes after euthanasia is not a procedure that is desirable with threatened species, nor is post-mortem collection of eggs from oviducts (Clulow et al. 2014), hence collection of gametes after release from the gonads by hormonal induction is the goal.

Given that gamete release has been reviewed in detail recently (Clulow et al. 2014, 2018a; Clulow and Clulow 2016; Vu and Trudeau 2016), this topic won't be reviewed extensively here. Rather, the focus will be on recognising that, even with techniques that are not yet fully optimised, there are nevertheless, protocols available that result in release of viable gametes capable of fertilisation in many species. There are two main targets for gamete induction protocols, the neurohormones of the hypothalamus and the pituitary (Vu and Trudeau 2016), although the effects of initiating hormonal activity in the hypothalamic-pituitary section of the HPG axis are ultimately mediated through effects on steroid synthesis and release at the gonad level, especially progestogens and androgens (Clulow et al. 2014, 2018a; Vu and Trudeau 2016). Importantly, there is a considerable degree of difference in the difficulty of designing protocols for the initiation of spermiation and the collection of sperm in urine in comparison to the induction of ovulation, with the former much easier than the latter (Clulow et al. 2018a).

Techniques for the collection of gametes in laboratory species, especially Pipidae *Xenopus laevis* and *Silurana tropicalis* (Clulow et al. 2012, 2014) and some Ranidae (McKinnell 1978) have allowed reproducible and highly controlled and timed collection of games for IVF with high fertilisation rates and developmental success. Indeed, the most utilised species (*X laevis*, *S. tropicalis*) are rare amongst amphibians in respect of the highly reliable induction of egg release in females in response to various doses of human chorionic gonadotropin (hCG) alone, with or without priming doses (reviewed by Clulow et al. 2012), generally without any requirement for the involvement of hypothalamic induction. The primary action of the luteinising hormone (LH) analogue hCG is through its activation of LH receptors, and ovulation can be induced with just this stimulus in these species, although it should be recognised that the hCG doses (on a body weight basis) are extraordinarily high (Kouba and Vance 2009; Clulow et al. 2012) in comparison to doses administered to mammalian species to induce ovulation. This reflects the large phylogenetic distance (non-conserved sequences) in active β unit peptides between the mammalian gonadotropin and the amphibian gonadotropin receptor (Clulow et al. 2014). Females of few other amphibian species are so responsive to hCG, and reliable induction of ovulation remains one of the key challenges of working with most amphibian species. Unusually, Clulow et al. (2012) were able to achieve reasonably high (greater than 50% ovulation rates) in one Australian myobatrachid, *Mixophyes fasciolatus*. However, for optimisation, this required priming over several days with another mammalian gonadotropin, pregnant mare serum gonadotropin (PMSG), which primarily has follicle stimulating hormone (FSH) activity. FSH is considered to be primarily involved in the maturation of follicles prior to the final ovulatory response occurring in response to an LH/progesterone spike (Clulow et al. 2014, 2018a). The report of Clulow et al. (Clulow et al. 2012) was the first publication to show in hCG responsive amphibian species, that the rate of ovulation is significantly increased by adding the PMSG priming step, indicating the benefit of promoting oocyte maturation prior to the ovulatory signal.

Most of the work focussing on the collection of sperm and eggs has used laboratory species, or non-laboratory species from the Americas and Europe (Clulow et al. 2014; Uteshev et al. 2013; Shishova et al. 2011; Browne et al. 2006a, b; Trudeau et al. 2013). More recently, considerable progress has been made in the collection of gametes from species of the three main Australian anuran families, the Myobatrachidae, Limnodynastidae and the Pelodryadidae (Clulow et al. 2018a; Silla et al. 2019). Amongst the Australian Pelodryadidae, in particular, Clulow et al. (2018a) highlighted a major difference in response of sexually mature males and gravid females to hormonal induction. All Pelodryadidae species tested (Clulow et al. 2018a) responded strongly in males to induction protocols targeting the pituitary (e.g. the hypothalamic decapeptide gonadotropin releasing hormone (GnRH), synthetic GnRH analogues (especially D-Ala^6-GnRH) and the gonads (hCG)). However, females of Pelodryadidae species tested were largely unresponsive to synthetic and mammalian hormones, but were responsive to pituitary extracts (source of endogenous hormones) from other Pelodryadidae species. The reasons for the different responses between the sexes within the Pelodryadidae have not been elucidated, but as discussed above may relate to more complex hormone/receptor interactions in the HPG axis of females. We suspect this is a much more general and widespread phenomenon across the anurans than is generally recognised, while perhaps not as pronounced as for the Australian Pelodryadidae. A similar, but less pronounced effect was seen across the Limnodynastidae and Myobatrachidae as well, in which gravid females were induced in some species with non-endogenous hormones, but with an efficiency substantially below the rate of induction of spermiation in sexually mature males (Clulow et al. 2018a). Our view (Clulow et al. 2018a;

Vu and Trudeau 2016) is that much more research needs to be focussed on understanding the neuroendocrinology of female follicular maturation and spawning if gamete induction for IVF and spawning is to be successfully initiated across threatened amphibian species in conservation programs.

Vu and Trudeau (2016) and Clulow et al. (2018a) outlined a range of avenues for research to improve the understanding of the neuroendocrinology and the effectiveness and efficiency of gamete release and spawning protocols. These included:

1. GnRH. GnRH agonists are applied extensively to ovulation induction protocols (Kouba et al. 2009; Clulow et al. 2014, 2018a; Vu and Trudeau 2016; Byrne and Silla 2010; Silla 2011, 2013; Waggener and Carroll 1998; Browne et al. 2006b; Trudeau et al. 2010, 2013; Roth et al. 2010), and the classical stimulatory effects of GnRH on FSH and LH secretion are assumed, but there is little direct data confirming this (largely because of a lack of gonadotropin assays (see below)). Although more than 20 GnRH sub-types are recognised across the vertebrates (Clulow et al. 2014; Vu and Trudeau 2016), it is likely that the primary, hypophysiotropic form is mammalian GnRH (GnRH-I), given its high levels in the amphibian brain and its location in the hypothalamus and the median eminence (Licht et al. 1994; Collin et al. 1995). The two main GnRH agonists that are highly active and used commonly in anurans (Vu and Trudeau 2016) are des-Gly^{10}-His(Bzl)^{6}-GnRH-ethylamide and des-Gly^{10}, D-Ala^{6}, Pro^{9}-GnRH-ethylamide; the high activity of these analogues is probably due to high affinity for the dominant GnRHR-1 receptor in the adenohypophysis (Seong et al. 2003; Wang et al. 2001). Nevertheless, these processes and activities are poorly understood at a molecular level, and there is a scope for further research to investigate GnRH analogues and ligand-receptor interactions (Wang et al. 2001; Acharjee et al. 2002), especially the role of GnRH-1 (Wang et al. 2001) which is the most abundant receptor sub-type in the amphibian pituitary. The combination of receptor-ligand dynamics studies in combination with appropriate gonadotropin immunoassays would be the most powerful approach to better achieve optimisation of protocols for gamete (especially oocyte) release, but these should be combined with the determination of downstream effects on ovarian maturation and ovulation. This approach should also be combined with understanding the interaction of dopamine antagonists (see below) and GnRH analogues, an approach that has been productive in the development of protocols in teleost fish (Hill et al. 2009; Peter et al. 1988) and is being applied increasingly in amphibians (Vu and Trudeau 2016; Trudeau et al. 2010; Salgado Costa et al. 2016; Vu et al. 2017).
2. Dopamine (DA). Until recently, a potentially important role of dopamine mediated through dopamine receptors in the pituitary has not been considered with respect to effects on the induction of gamete release in amphibians, despite a strong body of evidence for the inhibition of pituitary gonadotropin release by DA in teleosts (Trudeau 1997; Dufour et al. 2005; Yaron et al. 2003; Popesku et al. 2008). This mechanism is well defined in teleosts and has resulted in the development of successful product (OVAPRIM (Hill et al. 2009; Peter et al. 1988))

for spawning fish that utilises a combination of salmon GnRH and a DA type-2 receptor antagonist (domperidone). There is evidence that this mechanism is conserved in amphibians: DA antagonists have triggered ovulation (Sotowska-Brochocka et al. 1994), DA agonists have inhibited ovulation or reduced LH (Sotowska-Brochocka 1988; Sotowska-Brochocka and Licht 1992), and DA receptor agonists have reduced male vocalisation (Creighton et al. 2013; Creighton and Wilczynski 2014) and inhibited spawning (Vu and Trudeau 2016) in various species. Dopaminergic inhibitory effects in amphibians are likely to be mediated through DA D2 receptors in the pituitary (Nakano et al. 2010a, b). Nevertheless, the role of dopamine in regulating (inhibiting) amphibian gamete release and inhibiting spawning behaviours through control of GnRH, LH and FSH synthesis and secretion is poorly understood. In some amphibian species, there has been a strong effect of DA agonists and antagonists (Clulow et al. 2018a; Vu and Trudeau 2016; Trudeau et al. 2010, 2013; Vu et al. 2017), but not in every species examined to date e.g., the lack of response of the Australian tree frog, *Litoria fallax*, to DA antagonist metoclopramide (Clulow et al. 2018a). A better understanding of the role of DA offers the potential to improve gamete release and spawning protocols across species.

3. Sex steroids. A better understanding of the role of sex steroids would be obtained through increased use of steroid immunoassays during investigation of sub-threshold endocrine and physiological response during hormonal induction with GnRH and gonadotropins. This would greatly increase the prospects for improving induction protocols based on an understanding of the quantitative steroidal responses within the gonads. Progesterone is important in oocyte maturation and ovulation (Wright 1961; Masui 1967; Smith et al. 1968), but more recent studies have also demonstrated a role for androgens in oocyte maturation in frogs (Rasar and Hammes 2006; Evaul et al. 2007; Deng et al. 2009). Amplecting behaviours in females associated with spawning are subject to estradiol/progesterone regulation (Chakraborty and Burmeister 2009, 2015; Gordon and Gerhardt 2009; Wilczynski and Lynch 2011). Steroids are a critical component of the HPG axis leading to successful gamete maturation and release, and should be given a great deal of attention in research and protocol improvement and optimisation.
4. Purified or recombinant sources of gonadotropins from native species. One factor strongly limiting research on the development and optimisation of gamete release and spawning protocols is the current lack of endogenous sources of amphibian gonadotropins. These are needed for trials of the induction of ovulation, and for the development of immunoassays to quantify hypothalamic-pituitary FSH and LH responses to induction protocols. Current protocols utilising hCG rely on extremely high doses to achieve sufficient LH activity to induce ovulation (Kouba and Vance 2009; Clulow et al. 2012) and these can have side-effects such as oedema, or simply work at sub-optimal efficiency in terms of egg quality and yield. Replacing these with recombinant amphibian LH and FSH could represent a major advance in the efficiency and effectiveness of induction protocols. LH alone (presumably through a surge in progesterone) is sufficient in some species to induce ovulation (Kouba and Vance 2009;

Browne et al. 2006a, b; Clulow et al. 2012). However, FSH is critical to earlier stages of follicular development, and may play a role in the penultimate induction process. Sources of gonadotropins with higher specificity than currently available (especially mammalian PMSG, hCG) would be an advantage.

5. Other factors affecting the HPG axis and ovulation. Existing protocols for the induction of gamete release and spawning in non-laboratory species is set apart from the well-established laboratory protocols for species such as *Xenopus* and *Silurana* because of the variability of such protocols. Some are highly efficient and repeatable for example in the induction of sperm release by hCG (Clulow et al. 2018a) while others have varying levels of effectiveness and reproducibility in the timing of gamete release (e.g. oviposition after induction of female *Mixophyes fasciolatus* (Clulow et al. 2012)). Other factors and pathways that are poorly understood, or have been subject to limited research, may offer avenues to improve protocols. These include the role of stress hormones impinging on the HPG axis (especially through corticosterone (Davis et al. 2015; Narayan and Hero 2011; Narayan et al. 2011)); manipulation of light levels and photoperiod affecting melatonin activity in the pineal (Skinner and Robinson 1997; Vanecek 1999) which may in turn affect ovarian function (Horseman et al. 1978; Alonso-Bedate et al. 1990), possibly mediated through interactions with arginine vasotocin (Vu and Trudeau 2016). There are also a number of other factors that may impinge on the lower efficiency of ovulation protocols through effects on ovarian function including, but not limited to, gonadotropin inhibitory hormone, kisspeptin, pituitary adenylate cyclase-activating polypeptide and endocannabinoids (Vu and Trudeau 2016). It is not clear whether more detailed investigation of the roles of these hormones and signal transduction factors will provide major improvements on induction protocols, but it is important that they be investigated more closely.

In conclusion, there are many amphibians, including threatened species (especially anurans) for which protocols exist to induce either spawning or the release of gametes for IVF. These may already be efficient and effective enough to be used in conservation programs. There is a strong case that more use should be made of these protocols directly in captive breeding programs and genome storage/biobanking actions. Nevertheless, a greater focus on the underlying endocrinology of the amphibian HPG axis as it relates to gamete maturation and release could only further strengthen the prospects for the use of ARTs in amphibian conservation programs for threatened species.

3 Technologies up to Speed: Sperm Storage Above Zero Degrees and Sperm Cryopreservation

Storage of unfrozen sperm (refrigerated) and cryopreserved sperm are the two most successful and advanced genome storage technologies available for amphibians (Clulow et al. 2014; Clulow and Clulow 2016; Browne et al. 2019; Silla and Byrne 2019).

3.1 Sperm Storage Above 0 °C

Sperm can be stored under different conditions at temperatures above zero (usually 0–4 °C) and these are capable of motility and even fertilisation, days to weeks after removal from the body. Sperm can be stored under different conditions, depending on how they have been collected or removed from the body including storage as whole testes (Browne et al. 2001), testis macerates (Browne et al. 2001, 2002b, d; Keogh et al. 2017) and as spermic urine (Keogh et al. 2017) after hormonal induction of sperm release. Refrigerated sperm have been held for extended periods e.g. up to 30 days in *Rhinella marina* (Browne et al. 2001) and still achieved fertilisation. In at least one threatened species (*Lithobates sevosus*), stored sperm have been transported over large distances between zoos and used to generate offspring by IVF (Kouba et al. 2011, 2013). Browne et al. (2019) have listed a large number of amphibian species (more than 40) for which refrigerated sperm storage has been recorded following collection by various methods, with various endpoints from recovery of motile sperm to generation of offspring after IVF. This list included anurans and urodeles. This technology is clearly viable and available currently for application in conservation programs, even if its use would need to be refined and optimised on a species by species basis.

3.2 Cryopreserved Sperm

Amphibian sperm cryopreservation technology has advanced and been expanded to a broad range of taxa (Table 14.1) since first reports of the generation of embryos from frozen sperm (Shishova et al. 2011; Browne et al. 1998), that followed on from earlier reports on the recovery of sperm after cryopreservation (Clulow et al. 2014). Progress on sperm cryopreservation has been reviewed on a number of occasions (Kouba and Vance 2009; Kouba et al. 2013; Clulow et al. 2014; Clulow and Clulow 2016; Browne et al. 2019; Silla and Byrne 2019) and won't be reviewed in detail here.

In general terms, most amphibian sperm cryopreservation protocols have involved the cryopreservation of sperm inactivated in cryodiluents containing non-penetrating osmotica of high osmolality (isotonic to hypertonic media); high concentrations of penetrating cryoprotectants, especially dimethyl sulphoxide (DMSO) and dimethyl formamide; and slow cooling rates, less than 5–10 °C min^{-1} (Kouba and Vance 2009; Kouba et al. 2013; Clulow et al. 2014; Browne et al. 1998, 2002a, b, c, 2008, 2019; Uteshev et al. 2013; Shishova et al. 2011; Michael and Jones 2004; Mansour et al. 2009; Luo et al. 2007). In our hands, amphibian sperm appear to be very tolerant to extended exposure to penetrating cryoprotectants such as dimethyl sulphoxide (Fitzsimmons, Mahony, Clulow unpublished data), a fact that facilitates the handling of sperm during procedures. Vitrification or cryopreservation in dilute osmotic solutions has not been reported as the basis of viable

Table 14.1 Summary of sperm cryopreservation studies performed on amphibian species

Species	Protocol		Study endpoint	Reference
	CPA	Freeze/thaw		
Xenopus laevis	15% DMSO; 10% sucrose	F: −80 °C freezer; T: 37 °C	Fertilisation attempted; failed	Buchholz et al. (2004)
	5% DMSO;73 mM sucrose	F: 10 cm above LN_2 for 10 min; store at −196 °C; T: RT	Tadpoles	Mansour et al. (2009)
Xenopus laevis; *Silurana tropicalis*	20% EY; 0.4 M sucrose; 10 mM $NaHC0_3$; 2 mM PF	F: Esky stored at −80 °C; T: In the hand	Sperm motility	Morrow (2015) and Morrow et al. (2017)
	20% EY; 0.4 M sucrose; 10 mM $NaHC0_3$	F: Esky stored at −80 °C; T: 37 °C WB; 40 s	F1 generation from cryopreserved sperm/F2 generation from F1 animals (*Xenopus laevis*); Embryos (*Silurana tropicalis*)	Pearl et al. (2017)
	20% EY; 0.8 M sucrose; 20 mM $NaHC0_3$	F: Ramp −10 °C/min to −80 °C; T: 30 °C WB	Fertilisation	Sargent and Mohun (2005)
Anaxyrus americanus	20% EG; 75% FBS; 25% AM	F: Ramp −0.7 °C/min to −20 °C; Store at −80 °C; T: 25 °C WB	Sperm motility	Barton and Gutman (1972)
Anaxyrus americanus; *Anaxyrus fowleri*	15% DMSO; 10% sucrose	F: Store at −196 °C; T: NS	Sperm motility (urinic sperm)	Kouba and Vance (2009)
Anaxyrus fowleri	6% glycerol; TEST EY buffer			
	3.5% glycerol; AndroMed™ media			
Rhinella marina	10–15% DMSO; 10% sucrose	F: Hold at 2 °C for 10 min; −1 °C/min to −8 °C; −3 °C/min to −16 °C; −5 °C/min to −80 °C; Store at −196 °C; T: RT	Fertilisation	Browne et al. (1998, 2002a, c)
	10% DMSO	F: Hold at 4 °C for 10 min; Place in −20 °C freezer for 1 h; Store at −80 °C; T: 4 °C; 2 h	Embryos (tailbud)	Proaño and Pérez (2017)

(continued)

Table 14.1 (continued)

Species	Protocol		Study endpoint	Reference
	CPA	Freeze/thaw		
Atelopus zeteki	5–10% DMFA; 10% trehalose	F: 10 cm above LN_2 for 5 min; plunge in LN2; T:21 °C WB; 15 s	Sperm motility (urinic sperm)	Della Togna Nieto (2015)
Epidalea calamita	10% DMFA; 10% sucrose	F: NS; T: NS	Embryos (morula; urinic sperm)	Arregui et al. (2018)
Anaxyrus americanus; *Lithobates sylvaticus*; *Lithobates pipiens*	0.5 M DMSO; 50% FBS	F: Ramp −130 °C/min to −80 °C; store at −80 °C; T: NS	Sperm vitality	Beesley et al. (1998)
Anaxyrus boreas; *Lithobates sevosus*	10% DMFA; 10% trehalose	F: NS; T: 40 °C WB	Tadpoles (urinic sperm)	Langhorne et al. (2013)
	10% DMFA; 10% trehalose	F: NS; T: NS	F1 generation from urinic cryopreserved sperm/F2 generation from F1 animals	Vance et al. (2018)
Rana temporaria	–	–	Fertilisation	Kaurova et al. (1996, 1997)
	5% glycerol; 2.5% sucrose; 5% EY	F: 10 cm above LN2 for 7 min; plunge in LN_2; T: 22 °C; 40 s	Tadpoles	Mansour et al. (2010)
	12% DMFA; 10% sucrose	F: Hold at 4 °C for 10 min; 10 cm above LN2 for 5–7 min; plunge in LN_2; T: 40 °C WB; 10 s	Tadpoles (7 days old; urinic sperm)	Shishova et al. (2011)
Lithobates sylvaticus	1.5 M DMSO	F: Ramp −130 °C/min to −80 °C; Store at −80 °C; T: 30 °C WB	Sperm vitality	Mugnano et al. (1998)
Lithobates sylvaticus; *Lithobates pipiens*	150 mM glycerol	F: Ramp −0.2 °C/min to −8 °C; store at −8 °C; T: 23 °C WB; 20 min	Sperm motility/ vitality	Costanzo et al. (1998)
	None			

(continued)

Table 14.1 (continued)

Species	Protocol		Study endpoint	Reference
	CPA	Freeze/thaw		
Lithobates pipiens	Glycerol	F: 100 µl pellets on dry ice; store at −196 °CT: 21 °C	Sperm vitality (urinic sperm)	Hopkins and Herr (2007)
Pelophylax lessonae	24% DMFA; 20% sucrose	F: Hold at 4 °C for 10 min; Placed 10 cm over LN2 vapour for 15 min; Plunged into LN_2; T: 40 °C WB; 10 s	Tadpoles (urinic sperm)	Uteshev et al. (2013)
Rana catesbeiana	1–7% DMSO or MeOH; 1% lactose	F: 11.8 L dry shipper for 24 h; Store at −196 °C; T: 37 °C WB; 10 s	Sperm motility (urinic sperm)	Calado (2009)
Eleutherodactylus coqui	2 M glycerol/FBS	F: −80 °C freezer; store at −80 °C; T: 20 °C WB	Sperm vitality	Michael and Jones (2004)
Limnodynastes peronii	20% DMSO; 10% sucrose	F: Hold at 2 °C for 10 min; −1 °C/min to −8 °C; −3 °C/min to −16 °C; −5 °C/min to −80 °C; Store at −196 °C; T: RT	Sperm motility	Browne et al. (2002b)
Crinia signifera	15% DMSO; 10% sucrose			
Pseudophryne bibroni				
Philoria sp.				
Litoria brevipalmata				
Litoria latopalmata				
Litoria subglandulosa				
Litoria phyllocroa				
Litoria dentata				
Litoria lesueurii				
Litoria peronii				
Litoria nasuta				
Litoria fallax				
Litoria fallax	15% DMSO; 1% sucrose	F: Hold at 2 °C for 10 min; −1 °C/min to −8 °C; −3 °C/min to −16 °C; −3.4 °C/min to −80 °C; Store at −196 °C; T: RT	Sexually mature male (producing sperm)	Upton et al. (2018a)

(continued)

Table 14.1 (continued)

Species	Protocol		Study endpoint	Reference
	CPA	Freeze/thaw		
Litoria aurea	15% DMSO; 10% sucrose	F: Hold at 2 °C for 10 min; −1 °C/min to −8 °C; −3 °C/min to −16 °C; −3.4 °C/min to −80 °C; Store at −196 °C; T: RT	Juvenile frogs	Upton et al. (2018b)
Hypsiboas calcaratus; *Dendropsophus bifurcus*	10% DMSO	F: 45 μl aliquots in −80 °C freezer; Store at −80 °C; T: RT	Sperm vitality	Pers. com., Proano and Perez, in Browne et al. (2019)
"Frog"	2 M sucrose	F: Immersion in LOX (−183 °C); T: 20 °C PW	Sperm motility	Luyet and Hodapp (1938)
Cryptobranchus alleganiensis	NS	F: NS; T: NS	Larvae	Pers. com., McGinnity in Browne et al. (2019)
Cryptobranchus alleganiensis bishopi	2.5% DMSO; 10% sucrose	F: Hold at 4 °C for 1–4 days; −80 °C freezer; T: 35 °C WB; 10 s	Sperm motility	Unger et al. (2013)
Andrias davidianus	10% DMSO	F: Hold at 4 °C for 30 min; 0.05 mL pellets on copper mesh over LN_2 vapour; T: 26–28 °C WB	Sperm motility (urinic sperm)	Peng et al. (2011)
Ambystoma mexicanum	None	F: Ramp −7 to −300 °C/min; Store at −196 °C; T: 25 °C or 40 °C WB	Sperm motility	Figiel (2013)
Ambystoma tigrinum	5% DMSO; 0.5% w/v BSA	F: NS; T: NS	Embryos (early cleavage; urinic sperm)	Marcec et al. (2014)
	5% DMSO; 0.05% w/v BSA	F: NS; T: NS	F1 generation from urinic cryopreserved sperm/F2 generation from F1 animals	Vance et al. (2018)

Where multiple protocols were tested, data are presented for the protocol yielding the best results
DMSO dimethyl sulfoxide, *DMFA* dimethylformamide, *EG* ethylene glycol, *MeOH* methanol, *EY* egg yolk, *PF* pentoxyfylline, *FBS* foetal bovine serum, *BSA* bovine serum albumin, *AM* Alsever's medium, *PW* pond water, LN_2 liquid nitrogen, *LOX* liquid oxygen, *RT* room temperature, *NS* not specified

protocols (Browne et al. 1998). Nevertheless, in at least one urodele species (Figiel 2013), extremely rapid cooling rates did not reduce sperm recovery.

Amphibian cryopreservation protocols have largely been developed empirically and the reasons for the generally favourable responses to cryopreservation in high osmolality and slow cooling conditions have not been investigated in detail. It is assumed that slow cooling assists with dehydration of cells during freezing and perhaps membrane stabilisation. Anuran sperm are activated in solutions of low osmolality (hypotonic media) (Hollinger and Corton 1980; Browne et al. 1998; Edwards et al. 2004; Inoda and Morisawa 1987; Kouba et al. 2003; Fitzsimmons et al. 2007), so cryopreservation of sperm in low osmolar diluents would necessarily involve the cryopreservation of activated motile sperm that would probably be more hydrated (have higher cytoplasmic water content) than sperm in hypertonic media. Nevertheless, there may be some species, or groups of species, in which high diluent osmolality does not lead to optimal cryopreservation protocols (e.g. some pelodryadid species, Upton, Mahony, Clulow, and Clulow unpublished data), so there is clearly much scope for further refinement of sperm cryopreservation protocols within and between species, and a need for fundamental studies on cryobiological parameters in amphibians.

One important advance in cryopreservation of amphibian sperm has come through the procedure for cryopreserving urinic sperm (Shishova et al. 2011). This is an advance on earlier protocols that relied on cryopreservation of sperm from post-mortem testicular macerates (Kouba and Vance 2009; Browne et al. 1998, 2001, 2002a, b, c, d, 2019), and therefore were not suitable for use with endangered species. This combined procedure of hormonal induction of sperm release and cryopreservation has now been applied to a number of species with the production of embryos and larvae (Kouba et al. 2013; Clulow et al. 2014; Uteshev et al. 2013; Shishova et al. 2011; Vance et al. 2018), including two endangered species (*Rana sevosa* and *Anaryxus boreas boreas*) (Kouba et al. 2013; Clulow et al. 2014; Langhorne et al. 2013). Juvenile offspring have been reported for another endangered species, *Litoria aurea* from cryopreserved sperm (Upton et al. 2018b), however, the sperm were derived from testicular macerates.

Recently, further advances have been reported demonstrating the capacity of cryopreserved sperm to generate viable, reproductively competent (sexually mature) adult offspring from frozen sperm for several species including: (1) the dwarf green tree frog, *Litoria fallax* (Upton et al. 2018a), with completion of metamorphosis and generation of a sexually mature male (producing sperm); and (2) the production of F1/F2 generations of tiger salamanders (*Amystoma tigrinum*) (Vance et al. 2018) and frogs, *Xenopus laevis* (Pearl et al. 2017) from frozen sperm. Importantly, all these reports are proof of concept that there is no fundamental block to the production of multiple generations of amphibians produced from cryopreserved sperm. Proof of concept from these studies supports a more vigorous pursuit of the implementation of protocols involving the use of biobanked amphibian sperm. There are clear applications for the use of cryopreserved sperm in captive breeding and on-ground conservation programs. These applications could include the production of animals for translocation, reintroduction, and the restoration of genetic diversity and even genetic rescue of inbred wild populations.

4 Prospects for the Future Cryopreservation of Amphibian Embryos: Technological Advances in Aquatic Vertebrates and Invertebrates

4.1 Limitations of Current Technologies: Slow Cooling, Vitrification and Convective Thawing

Cryopreservation of oocytes and embryos is the holy grail of amphibian ARTs. Success in this area would obviate the need for other ARTs where the aim is to generate viable embryos from cryopreserved material, including such complex procedures as nuclear transfer (McKinnell 1978; Gurdon and Uehlinger 1966; Gurdon et al. 1975; Gurdon and Hopwood 2000; Gurdon and Byrne 2003; Di Berardino 1997) and androgenesis (Porter 1939; Gurdon 1960; Gillespie and Armstrong 1980; Parsons and Thorgaard 1984; Corley-Smith et al. 1996; Thorgaard et al. 2005). Clulow et al. (2014) reviewed the prospects for the development of female germline and embryo cryopreservation technologies in amphibians based on the limited amount of work on amphibian oocytes (Guenther et al. 2006; Kleinhans et al. 2006; Mazur and Kleinhans 2008) and embryos (Clulow et al. 2014) and the much broader body of research on fish oocyte and embryo cryopreservation which had been carried out over several decades (Harvey 1983; Hagedorn et al. 1996, 1997a, c, 1998; Zhang and Rawson 1996; Liu et al. 1999, 2001; Isayeva et al. 2004; Zhang et al. 2005; Hagedorn and Kleinhans 2011; Godoy et al. 2013). Some of the key problems and technical hurdles involved in the cryopreservation of fish embryos were identified and prospects for cryopreservation of amphibian oocytes considered in the light of a comparison of the structure and biology of the large diameters and volumes and high yolk content of eggs and embryos in both taxa. Notwithstanding differences in the structure and pattern of early embryonic development between fish and amphibians (Rugh 1962; Hagedorn et al. 1998; Wallace 1963; Balinsky 1975a, b), it was considered similarities in oocyte structure and cytoplasmic constituents between the two vertebrate classes (reflecting the common evolutionary origins of their reproductive and developmental biology as aquatic external fertilisers) indicate some similar technical barriers to oocyte and embryo cryopreservation for both groups. Recent, major advances in the technology of cryopreservation of fish and corals (Daly et al. 2018; Khosla et al. 2017), however, have resulted in a paradigm shift and greatly increased prospects for the cryopreservation of amphibian embryos. In this section, the structural and physico-chemical features of fish and amphibian oocytes and embryos are briefly summarised before considering the advances in technology that suggest there may be a way forward for amphibian female germline and embryo cryopreservation.

Primary problems with the cryopreservation of amphibian and fish oocytes appear to be derived from three major factors: size (diameter/volume), cytoplasmic yolk content and barriers to water and cryoprotectant permeability.

4.1.1 Size

Size per se is a major factor determining the prospects for oocytes and embryos to survive cryopreservation using conventional technologies (Mazur and Kleinhans 2008). The formation of intracellular ice crystals is likely to be the major cause of lethal cryoinjury during cryopreservation (Mazur 2004), either during the freezing and thawing process (slow cooling), or during thawing (after vitrification). The size of mammalian oocytes and embryos that have been successfully cryopreserved are much smaller (75 μm, mouse oocytes (Guenther et al. 2006; Mazur and Kleinhans 2008; Hagedorn and Kleinhans 2011); 110–130 μm, bovine oocytes (Otoi et al. 1997; Diez et al. 2012; Hwang and Hochi 2014); 150–200 μm, bovine blastulae (Hoelker et al. 2006; Leibo and Loskutoff 1993); 100 μm, human oocytes (Kuwayama et al. 2005; Rienzi et al. 2012; Argyle et al. 2016)) than the range of fish (e.g. zebrafish, 800 μm) and amphibian (e.g. *Xenopus* 1200 μm) oocytes from laboratory species in which cryopreservation has been attempted but not achieved (Table 14.2). Indeed, the size range of amphibian embryos and oocytes (Table 14.2) indicates that even the lower size limits are likely to present major challenges as structures to be frozen and thawed, with the larger sizes presenting a daunting, if not insurmountable challenge. The largest eggs and embryos tend to be associated with direct developers that require large yolk reserves to sustain larvae for extended periods without feeding (see also Sect. 5). The size of oocytes and ova in Table 14.2 range from 0.6 to 12.0 mm (Anura), 1.1 to 10.0 mm (Urodeles) and 1.0 to 42.0 mm (Gymnophora), but the list is not exhaustive of the potential upper limits in size (nor necessarily of the lower limits).

A major problem associated with large oocyte size is the relatively high temperatures at which intracellular ice crystals form, suggesting that these form as a result of ice nucleating events emanating from the surface of the oocyte (external ice acting as the nucleator through membrane pores or other discontinuities at the plasmalemma) and spreading inwards. Evidence for this comes from empirical data showing that intracellular ice formation in oocytes of fish (Liu et al. 2001; Hagedorn et al. 2004), amphibians (Guenther et al. 2006; Kleinhans et al. 2006; Mazur and Kleinhans 2008) and one marine invertebrate (Köseoglu et al. 2001) occurs at relatively high temperatures that are close to the temperature of extracellular ice formation. In the case of zebrafish and *Xenopus*, these temperatures are between −14 and −18 °C (zebrafish (Hagedorn et al. 2004)) and −8 to −10 °C (most *Xenopus* stage I and II oocytes and all mature stage V and VI oocytes (Guenther et al. 2006; Kleinhans et al. 2006)). In contrast, intracellular ice formation occurs at −41 °C in mouse oocytes (Mazur et al. 2005). Further evidence for this model of intracellular ice propagation comes from *Xenopus* oocytes in which direct observations have been made of ice formation originating at the periphery of the oocytes

(Guenther et al. 2006). Taken together, these data indicate that freezing the oocytes of amphibians (and fish) using slow cooling procedures is not a viable approach since lethal intracellular ice formation will necessarily happen at a stage before there is any chance of dehydration of the oocyte that would reduce ice formation, and with freezing occurring at the upper level of the melt to glassy ice temperature zone, the level of ice crystallization occurs very early in the cooling process and is irreversible and lethal.

4.1.2 Yolk

There is some evidence that yolk may be a factor negatively affecting cryopreservation outcomes in species with high oocyte yolk content. Evidence comes from the reduction in ice crystallization in zebrafish embryos during freezing when some of the yolk has been removed (Liu et al. 1999). It has been suggested that the high lipid content of yolk may interfere with cryopreservation or that yolk may limit water and cryoprotectant efflux during cryopreservation (Hagedorn et al. 1997c). The presence of a substantial yolk content is a common factor between the telolecithal eggs of fish and amphibians (Wallace 1985), although yolk content may be proportionally higher in amphibian eggs than in fish (Wallace 1963; Guenther et al. 2006; Hagedorn et al. 1997c; Lawson et al. 2013). This suggests that if yolk content is deleterious to cryopreservation, the contribution of yolk to failed cryopreservation in amphibians may be as large or larger in amphibian oocytes and embryos. Nevertheless, given that the ice flash (growth of ice) during freezing of amphibian oocytes proceeds from the periphery of the cell (Guenther et al. 2006), yolk is probably not the source of ice nucleation during cryopreservation. Vitrification (Hagedorn and Kleinhans 2011; Derakhshan et al. 2017) offers a mechanism to avoid ice crystallisation during cryopreservation in yolk containing oocytes or embryos of aquatic organisms, however, until a mechanism is available (see below) to prevent ice crystal formation during thawing, cryopreservation will still involve lethal effects of cryoinjury from intracellular ice crystals.

4.1.3 Barriers to Water and Cryoprotectant Permeability

Differences in the structure of early fish and amphibian embryos indicate models of water and cryoprotectant permeability may not be directly transferable between the two taxa, and comparisons should be treated with caution. In particular, fish develop a syncytium around a highly condensed yolk mass in early embryogenesis that is impermeable to water and cryoprotectants (Hagedorn et al. 1996, 1997b, 1998; Liu et al. 2001; Hagedorn and Kleinhans 2011). This is consequently a major problem during cryopreservation of fish embryos (Hagedorn et al. 1996, 1997a, 1998; Liu et al. 2001). The syncytium forms in fish but not in gastrulating amphibian embryos (Balinsky 1975a, b) and in amphibians the yolk is distributed through the cytoplasm of dividing embryonic cells, but is concentrated in cells of the

Table 14.2 Ovum diameters of amphibians

Order	Family/species	Ova diameter (mm)	Reproductive mode	Notes	References
	Homo sapiens	0.1		Routinely cryopreserved	Argyle et al. (2016) and Khalili et al. (2013)
	Mus musculus	0.01–0.02		Common laboratory model; cryopreservation achieved	Eroglu et al. (2009) and Zuccotti et al. (1994)
	Danio rerio	0.7		Fish model organism; cryopreservation success poor	Khosla et al. (2017), Guan et al. (2010), and Kimmel et al. (1995)
	Fungia scutaria	0.2–0.3		Recently embryos were successfully cryopreserved	Daly et al. (2018)
Anura	Bufonidae	0.6–3.5	Terrestrial/Arboreal	*Nimbaphrynoides occidentalis* has one of the smallest egg sizes of the anurans (low yolk content)	Duellman and Trueb (1994) and O'Brien et al. (2010)
	Dendrobatidae	1.6–3.5	Terrestrial/Arboreal		del Pino et al. (2007)
	Eleutherodactylidae	2.0–3.5	Direct Developers		Duellman and Trueb (1994) and del Pino et al. (2007)
	Hemiphractidae	3.0–12.0	Varied; Arboreal, Direct developers	*Gastrotheca cornuta* has the largest known eggs of the anurans	Duellman and Trueb (1994)
	Hylidae	0.9–2.0	Aquatic		Duellman and Trueb (1994)
	Limnodynastidae	1.0–1.9	Aquatic (foam nesters)		Duellman and Trueb (1994) and (Anstis 2017)
	Microhylidae	1.1–4.0	Aquatic		Duellman and Trueb (1994)
	Myobatrachidae	0.85–3.9	Aquatic (foam nesters)		Duellman and Trueb (1994) and Anstis 2017)
	Pelodryadidae	1.0–1.9	Aquatic		Anstis (2017) and Upton, Clulow, Mahony, Clulow (unpublished data)
	Pipidae	0.75–6.0	Aquatic		Duellman and Trueb (1994)
	Ranidae	1.0–3.0	Aquatic		Duellman and Trueb (1994)
	Xenopus laevis	1.2	Aquatic		Guenther et al. (2006)

Urodela	Ambystomatidae	1.1–6.0	Aquatic		Duellman and Trueb (1994)
	Plethodontidae	1.9–10.0	Direct Developers		Duellman and Trueb (1994) and Elinson and del Pino (2012)
	Plethodontidae	2.5–3.0	Aquatic		Duellman and Trueb (1994)
	Plethodontidae	2.5–3.0	Terrestrial/Arboreal		Duellman and Trueb (1994)
	Salamandridae	1.5–1.5	Aquatic		Duellman and Trueb (1994)
Gymnophiona	Various	1.0–42.0	Direct Developers	*Ichthyophis glutinosus* has the largest known eggs of all amphibians	Duellman and Trueb (1994) and Vitt and Caldwell (2013)

The largest of the amphibian ova are from the caecilian, *Ichthyophis glutinosus*, which is a direct developing amphibian, resulting in large yolky ovoid eggs at about 35 × 42 mm. Of the anurans, *Gastrotheca cornuta* (also a direct developer) has the largest eggs at 12 mm. Direct developers often have large yolky eggs beyond a 3 mm diameter and as large as 10 mm in the Urodela family Plethodontidae. *Nimbaphrynoides occidentalis*, *Silurana tropicalis* and *Hymenochirus boettgeri* have small eggs about 0.6–0.8 mm in diameter from the anuran families Bufonidae and Pipidae. Given that species from these families are often used as model organisms, these could provide useful starting points for testing cryopreservation of amphibian eggs. The Microhylidae and Pelodryadidae (See Sect. 5.1) have a range of egg sizes.

vegetal pole, which at least in theory makes the yolk compartment in amphibians more accessible to water and cryoprotectants. Nevertheless, amphibian oocytes do not express aquaporin channels in the plasmalemma (Kleinhans et al. 2006), and the permeability of the plasmalemma to water is extremely low (as low as in a pure lipid bilayer) (Zhang and Werkman 1991; Kleinhans et al. 2005). Taken together, these permeability barriers in both fish and amphibian oocytes, although located in different structural components of the embryo, point to a problem with water and cryoprotectant permeability during cryopreservation that must exacerbate the problem of size (diameter/volume and surface area to volume ratio) during cryopreservation of embryos.

4.2 Advances in the Cryopreservation of Aquatic Oocytes and Embryos Using Vitrification and Laser Warming: Implications for Amphibian Cryopreservation

As outlined in Clulow et al. (2014), and discussed above, the prospects for cryopreservation of amphibian embryos using slow cooling and standard warming procedures is extremely poor, with evidence that intracellular ice formation occurs at high temperatures during cooling of amphibian and other large aquatic oocytes. This prevents the reduced formation of ice that might occur with intracellular ice formation at lower temperatures, in a more dehydrated state, as found in mammalian oocytes. Cryoinjury during the slow cooling phase is likely to be even further compounded during the recrystallization of intracellular ice before thawing is complete. Little has been published on amphibian oocyte or embryo cryopreservation, and the limited amount of data available confounds freezing and thawing effects. Clulow et al. (2014), Fig. 14.4, show the disintegration of a vitrified Gosner Stage 20 embryo of a striped marsh frog (*Limnodynastes peronii*), vitrified in 2.5 or 5 M propylene glycol (PG) with rupture of many structures and dispersal of yolk contents post thaw. In that case, it was not possible to distinguish cryoinjury to the larva arising from cooling compared to thawing. Derakhshan et al. (Derakhshan et al. 2017) reported vitrification of blastulae and embryos of a bufonid, *Bufotes viridis* using 3 M PG and 2 M DMSO as the vitrifying medium, with better structural recovery, but no evidence of survival of embryos thawed by convective warming beyond neural fold. Once again, it was not possible to separate freezing versus thawing effects. Similarly, with fish survival of embryos after thawing using conventional thawing and warming protocols has not been demonstrated reproducibly and the structural damage is clearly evident (Khosla et al. 2017). However, in the case of zebrafish embryos (Khosla et al. 2017), there is now clear evidence that damage from conventional, convective thawing alone can be extreme and lethal, because for the first time, embryos were cryopreserved under conditions where convective thawing was replaced by ultrarapid laser warming, and those embryos recovered.

Recent advances in the cryopreservation of aquatic embryos including fish (Khosla et al. 2017) and coral larvae (Daly et al. 2018) indicate that cryopreservation of large, yolky embryos is possible. These recent studies, and the advances in understanding of the underlying theory (Bischof 2018; Khosla et al. 2018b) represent the most important advances in cryopreservation of aquatic oocytes and embryos in the more than three decades since the early work of Harvey (Harvey 1983) indicated the challenges associated with freezing such structures, subsequently confirmed in the many later studies attempting cryopreservation of zebrafish embryos (Harvey 1983; Hagedorn et al. 1996, 1997a, b, c, 1998; Zhang and Rawson 1996; Liu et al. 1999, 2001; Isayeva et al. 2004; Zhang et al. 2005; Hagedorn and Kleinhans 2011; Godoy et al. 2013). Success is dependent on a combination of vitrification and laser warming to achieve high rates of cooling during freezing and even higher rates of warming during thawing, which together eliminate formation of intracellular ice during both phases of cryopreservation.

4.2.1 Vitrification

Vitrification occurs when the rate of cooling in specimens is sufficiently high to reach the glass transition temperature (T_g) without ice crystallization occurring during specimen cooling from the melting temperature (T_m) to T_g through $\Delta T_{crystallization}$, the region between T_m and T_g where ice crystallization occurs (Khosla et al. 2017, 2018a; Fahy et al. 1984; Wowk et al. 2000; Mazur and Seki 2011; Peyridieu et al. 1996). The magnitude of $\Delta T_{crystallization}$ is not constant, but varies inversely with the concentration of cryoprotectant in the medium (Fahy et al. 1984; Khosla et al. 2018a). Mammalian embryos have been successfully cryopreserved by vitrification for many years due to their small size (see above) because they can accommodate lower rates of heating during thawing but still pass successfully through the T_g to T_m transition zone without forming ice crystals. Nevertheless, even mammalian embryo cryopreservation has been improved with technical modifications that have allowed cooling rates to be increased from 2500 °C min^{-1} in traditional straws to 22,500 °C min^{-1} in open pulled straws (Vajta et al. 1998) to potentially >69,000 °C min^{-1} in Cryotops (Jin et al. 2014; Mazur and Seki 2011) and >100,000 °C min^{-1} in quartz micro-capillaries (He et al. 2008). Under such conditions, cooling rates are not limited to oocytes or embryos with a diameter less than 100–150 μm (see Fig. 14.1). As well, higher concentrations of cryoprotectants (e.g. 6–8 M in some tissues and mammalian embryos (Fahy et al. 1984; Rall and Fahy 1985a, b)) allow glass transition states to be achieved at slower cooling rates (Fahy et al. 1984; Peyridieu et al. 1996) because the gap between T_g and T_m ($\Delta T_{crystallization}$) reduces greatly at higher cryoprotectant concentrations (due mainly to an exponential decline in T_m with a linear increase in cryoprotectant concentration (Wowk et al. 2000; Khosla et al. 2018a)). Under conditions of much higher cryoprotectant concentrations, the size of tissue slices that can be vitrified and thawed successfully is increased substantially (Peyridieu et al. 1996).

Vitrifying larger embryos is a much greater problem, however, when high cryoprotectant concentrations are not tolerated well by some aquatic embryos such as zebrafish (Hagedorn et al. 1998; Khosla et al. 2017). In this case, a double problem is encountered: lower cryoprotectant concentrations increase the gap between T_m and T_g, making ice crystallization more likely during cooling, and the rate of cooling achievable is reduced because the size (diameter/volume) of the aquatic oocytes and embryos is larger. Khosla et al. (2018a) achieved vitrification of droplets up to 2 μl using a cryoprotectant formulation of 2 M polyethylene glycol (PG) and 1 M trehalose in open pulled straws with known cooling rates of 20,000 °C min^{-1} (Vajta et al. 1998) (although not with 2 M or 3 M PG alone). 2 M PG alone is estimated to have a critical cooling rate (CCR) required to achieve the glass ice transition without ice formation of 290,000 °C min^{-1}, although that is theoretically reduced to a CCR of only 23,900 °C min^{-1} with 3 M PG (Liebermann and Tucker 2007). In smaller droplets of 1 μl, Khosla et al. (2018a) did achieve vitrification (visible as translucent droplets following vitrification) with 3 M PG alone in modified Cryotops, estimated in modelling to be achieving a cooling rate of 93,000 °C min^{-1} in 1 μl droplets. Taken together, these data suggest amphibian oocytes or other biomaterials up to 2 μl (D = 1560 μm), or even greater with higher concentrations of cryoprotectants, could be vitrified in LN_2 (Fig. 14.1) with appropriately designed cooling devices (such as modified Cryotops).

Khosla et al. (2017) overcame the problem of vitrifying zebrafish embryos (D = 800 μm) and achieved successful vitrification by reaching estimated cooling rates of greater than 50,000 °C min^{-1} (up to 90,000 °C min^{-1}) using modified

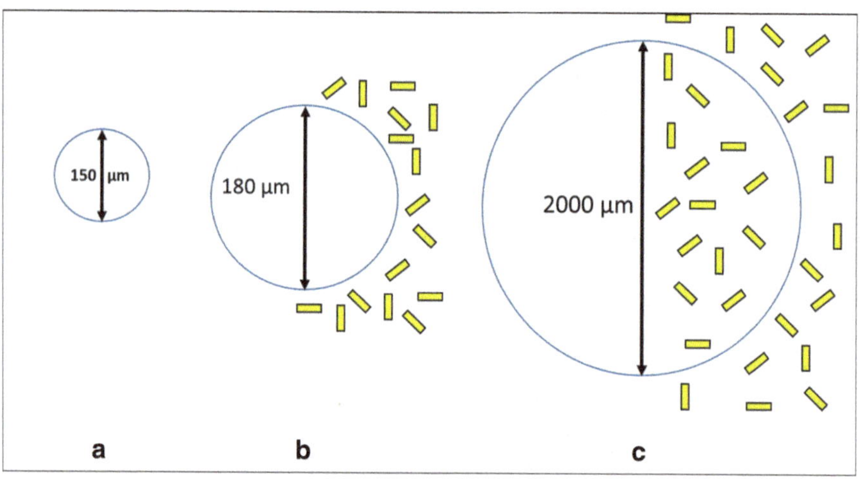

Fig. 14.1 Observed and predicted limits to oocyte and embryo size during various types of thawing post vitrification. (**a**) Maximum mammalian oocyte diameters cryopreserved by vitrification and convective warming (see text) (**b**) predicted limits for biomaterial volumes thawed by laser warming of external GNRs (Khosla et al. 2018a) (**c**) predicted limits for biomaterial volumes thawed by laser warming of internal and external GNRs (Khosla et al. 2018a)

Cryotop devices that increased the width of the blade from the standard 0.4–1.5 mm so that they could accommodate zebrafish embryos (D = 800 μm without chorion, >1 mm with chorion). This rate of cooling was sufficient to allow the use of intra-specimen cryoprotectant concentrations of ≈2 M PG achieved by micro-injection into the embryo prior to vitrification (Khosla et al. 2017). This was the first time that successful vitrification of a fish embryo was confirmed (confirmed with the observation of transparent embryos, but only confirmed as successful vitrification with minimal or no ice formation by laser warming of the specimen during thawing). Although the modified Cryotop approach could potentially be used to vitrify aquatic embryos >800 um in diameter, the achievable rate of specimen warming is a limiting factor (Khosla et al. 2017, 2018a, b) in determining the upper size limit (see below), especially given that required critical warming rates (CWR) are likely to be 10× greater than CCRs during thawing to avoid ice nucleation and crystal formation during the reverse $\Delta T_{crystallization}$ transition.

4.2.2 Laser Warming

Successful vitrification achieved by rapid cooling that transitions the specimen through the critical thermal zone of ice nucleation between T_m and T_g without initiating intracellular ice formation is only the first step in achieving cryopreservation of larger structures. Thawing specimens is a greater problem for larger structures because the CWR needs to be in the order of tenfold the CCR if ice nucleation events are to be avoided in the T_g to T_m transition range in the cells and media (Peyridieu et al. 1996). While this is achievable in small mammalian oocytes (Rall and Fahy 1985a; Rall and Fahy 1985b) and larger tissue slices (Fahy et al. 1984) where cooling and warming rates of only 10–100 °C min^{-1} can be tolerated because of very high 6–8 M cryoprotectant concentrations, thawing rates become rate limiting for larger structures such as oocytes and embryos of aquatic taxa. Where the size of the specimen reduces the convective rate of heat transfer (low surface area to volume), and cryoprotectant levels need to be reduced to avoid toxicity (Hagedorn et al. 1998; Rall and Fahy 1985b), high CCRs (Khosla et al. 2017) have been employed to achieve vitrification. Conventional conductive warming protocols simply cannot achieve heating rates that are sufficient to avoid ice crystallisation for larger structures. Zebrafish embryos that have been vitrified but thawed through conductive heating (estimated to only be in the range of 21,000 °C min^{-1}) show severe morphological and structural damage and have zero survival (Khosla et al. 2017). Estimated warming rates required to achieve rapid transition from the T_g past the T_m without formation of ice crystals in zebrafish embryos (allowing for embryo size and the concentration of cryoprotectants, ≈2 M PG, that the embryos can tolerate (Hagedorn et al. 1998; Rall and Fahy 1985b)) were in the order of 1.4×10^7 °C min^{-1} (Khosla et al. 2017). The only feasible technology that can achieve this rate of warming is short (millisecond) laser pulses (Daly et al. 2018; Khosla et al. 2017, 2018a, b) capable of achieving heating rates up to 2×10^7 °C min^{-1} in a controlled system that rapidly heats, but doesn't destroy, the specimen.

The use of laser warming to thaw vitrified embryos was pioneered by Mazur and colleagues who achieved ultrarapid warming of mouse oocytes (Jin et al. 2014) using extracellular India ink (carbon black) particles in the vitrification solution to absorb the laser beam and heat the oocytes sufficiently quickly to avoid intracellular ice formation during thawing. This process relied on external heating of the medium surrounding the oocyte (with emission of heat from the carbon particles absorbing the laser light) at rates of 10^7 °C min^{-1} with convective transfer to the oocytes. This process could achieve a sufficiently high rate of thaw within the mouse oocytes (relying on convective heat transfer), whose diameter was only in the order of 80 μm, to avoid intracellular ice formation. This approach was successful with small mammalian oocytes that were also capable of being thawed with conventional convective warming techniques (Fahy et al. 1984; Rall and Fahy 1985a, b). However, it was recognised (Daly et al. 2018; Khosla et al. 2017, 2018a) that the approach of only using laser heating of the external medium, and relying on convective heat transfer to thaw the specimen, would not achieve a sufficiently rapid internal heating rate to prevent ice crystal formation during the transition from T_g to above T_m for specimens above D = 180 μm i.e. CWRs would not be achieved in larger specimens.

This problem was resolved (Daly et al. 2018; Khosla et al. 2017, 2018a) with a major advance in the technology of non-destructive specimen warming through use of gold nanorods (GNRs) that could be introduced into the specimen by microinjection (Khosla et al. 2017). This allowed an ultrarapid laser pulse to generate heat simultaneously internally within the specimen and externally in the medium during thawing. This meant that, unlike Mazur's original innovation, convective heat transfer from the external medium was not relied upon to thaw the internal glassy ice with the limitations of reduced heat dissipation occurring as a function of diffusion distance.

Nevertheless, some smaller aquatic oocytes and embryos can be thawed using ultrarapid laser warming where GNRs are present only in the external but not the internal medium. Daly et al. (2018) successfully thawed the oocytes of the coral *Fungia scutaria* (D ≈ 100–200 μm) with external NGRs employing a vitrification medium of 3.5 M cryoprotectant solution (approx. 1.3 M PG, 0.7 M DMSO and 1 M trehalose) and specimen warming at 4,500,000 °C min^{-1}. They suggested that this approach might be adapted to other coral species (acroporids) with diameters ≈400–600 μm, although this limit may be uncertain given the suggested upper limit by Khosla et al. (2018a) of ≤180 μm for external GNRs (Fig. 14.1).

The advent of GNRs and their use to mediate ultrarapid warming of aquatic embryos and larvae represent a very recent advance in technology and protocols for cryopreservation of larger, yolky aquatic embryos and oocytes. Further research is needed to define the upper size limits of the structures that can be successfully cryopreserved. The nanorods themselves used to date are poly-ethylene coated GNRs approximately 110 nm long with a diameter around 20 nm (Khosla et al. 2018a) and have a peak energy absorption (and therefore heat generation) with infra-red laser wavelength at 1064 nm. Theoretically, warming rates of up to 1–2 × 10^7 °C min^{-1} are achievable, but the upper limits in energy generation when near-infrared laser light is converted to heat at the surface of the particles will be a function of the intracellular concentration of particles that can be tolerated by specimens.

In zebrafish embryos, injected NGRs appear to disperse evenly through the embryos (important for achieving consistent rates of heating across the embryos to avoid stress fractures during thawing), and particle concentrations around 1.2×10^{18} m^{-3} were deployed by micro-injection when thawing zebrafish embryos with laser warming (Khosla et al. 2017). These were well tolerated in embryos successfully recovered following vitrification and laser warming, with demonstrated survival of more than 10% of recovered embryos for 24 h or more. The morphology of recovered embryos was intact with development resumed and movement at 24+ hours. In comparison, vitrified embryos that were warmed by convective heating (Khosla et al. 2017) had an amorphous morphology and zero survival. Further refinement of the protocols is expected to achieve development of embryos and larvae through to the completion of development. The achievement by Khosla et al. (2017) represents the first reproducible recovery of cryopreserved frozen fish embryos. It is a major advance in protocols with important implications for prospects of similar success in amphibians. This is true irrespective of the stage of development reached after thawing in the initial experiments.

The current upper size limit of aquatic oocytes and embryos that can be recovered by vitrification and laser warming remains to be determined. If one accepts that limits for amphibians and fish embryos may be similar, it can be inferred that the limits for amphibians might at least reach those demonstrated for other aquatic embryos. The current limits of recovery by laser warming of vitrified specimens achieved in corals and zebrafish are in the order of 180 μm with external GNRs (Daly et al. 2018) and 800 μm (Khosla et al. 2017) with internal and external GNRs. However, larger diameter specimens might be successfully recovered (Khosla et al. 2018a) with higher levels of cryoprotectants, modified freezing devices, increased concentrations of GNRs and wider laser beams. Using modified cryotops and laser warming of GNR impregnated droplets, successful laser warming of biological specimens with diameters up to 180 μm for external only GNRs (warming rates 70,000–20,000,000 °C min^{-1}) to 2000 μm (warming rates 400,000–20,000,000 °C min^{-1}) with external and internal GNRs are predicted from modelling (Khosla et al. 2018a). Cryopreservation under these conditions is predicted to be achievable with cooling rates in the range of 66,000–210,000 °C min^{-1}, cryoprotectants propylene glycol ≥2.0 M, and sugar additives 0.1–1.0 M.

Potentially, even larger amphibian structures might be able to be cryopreserved if higher concentrations of cryoprotectants could be tolerated than considered by Khosla et al. (2018a, b). There has not been extensive investigation of the tolerance of amphibian embryos to cryoprotectants other than Derakhshan et al. (2017) who reported recovery of *B. viridis* embryos up to neural fold after vitrification, and reported that embryos could tolerate concentrations of PG up to 3 M (but not higher) for 15–30 min. Curphey, Mahony and Clulow (unpublished data; Fig. 14.2) investigated survival of unfrozen *Limnodynastes peronii* stage 20 embryos exposed to PG and found survival after exposure to at least 2.6 M (higher concentrations weren't tested), depending on the period of exposure and the temperature (embryos were more tolerant of higher concentrations for longer at 4 °C compared to room temperature), and so vitrification with total concentrations of cryoprotectant at 4 M or above may be feasible by combining PG and other cryoprotectants. Should vitrification and laser warming technologies be tested for their application to amphibians, optimising cryoprotectant concentrations would be an important component of research aimed at determining the upper size limits of oocytes and embryos that can be cryopreserved with this approach.

If an upper limit of 800–2000 µm, as proposed by Khosla et al. (2018a), is assumed to be feasible, the mature oocytes and embryos of many amphibian species might be recoverable through vitrification and laser warming (see Fig. 14.2, Table 14.1). However, as can be seen in a representative range of oocytes and embryo sizes recorded across a number of amphibian taxa (Table 14.2), species with smaller eggs would be candidates for laser warming, but species with larger diameters (often direct developers with much larger eggs carrying large yolk reserves to support extended developmental phases without feeding) would be clearly excluded from direct use of the technology. For these species, other approaches that have been outlined previously (Clulow et al. 2014) might still be required to achieve storage and recovery of viable genomes. These approaches might include the cryopreservation of early embryonic cells that are smaller structures than whole, intact oocytes or complete embryos, combined with reconstituting viable embryos through the generation of chimeras or use of nuclear transfer. This approach has not been investigated extensively for amphibians, although Uteshev et al. (Uteshev et al. 2002; Uteshev and Gakhova 2005) reported the generation of blastulae from vitrified early embryonic cells of *Bufo bufo* by nuclear transfer and Lawson et al. (2013) recovered live cells from dissociated and cryopreserved gastrulae and neuralas of the striped marsh frog, *Limnodynastes peronii*. Nevertheless, this approach has been extensively investigated in fish with successful cryopreservation of blastomeres from dissociated embryos (Harvey 1983; Calvi and Maisse 1998, 1999; Cardona-Costa and García-Ximénez 2007; Dash et al. 2008; Kusuda et al. 2002; Lin et al. 2009; Nilsson and Cloud 1993; Routray et al. 2010). As well, fish chimeras have been generated from unfrozen (Hong et al. 2012; Lin et al. 1992; Nilsson and Cloud 1992; Yamaha et al. 1997) and cryopreserved (Yasui et al. 2011; Kusuda et al. 2004) blastomeres and PGCs (Higaki et al. 2010), an approach that should be feasible for amphibians. Promordial germ cells (PGCs) or other gonadal stem cells (e.g. spermatogonial stem cells) may also be candidates for cryopreservation and regeneration of embryos through nuclear transfer or chimeras. For example, PGCs from later stage embryos of zebrafish have been isolated, cryopreserved and transferred into other zebrafish embryos in which the host embryo PGCs have been sterilised (Higaki et al. 2010). Similar approaches have been pursued with fish spermatogonial cells (Franěk et al. 2018; Lujić et al. 2018; Marinović et al. 2018a, b).

In conclusion, while the advent of vitrification combined with laser warming is the most promising advance in cryopreservation for several decades for larger aquatic embryos and oocytes, considerable work will be required to determine the full potential and limits of the approach for amphibians. Nevertheless, pursuit of research on this approach with amphibians is a matter of urgency.

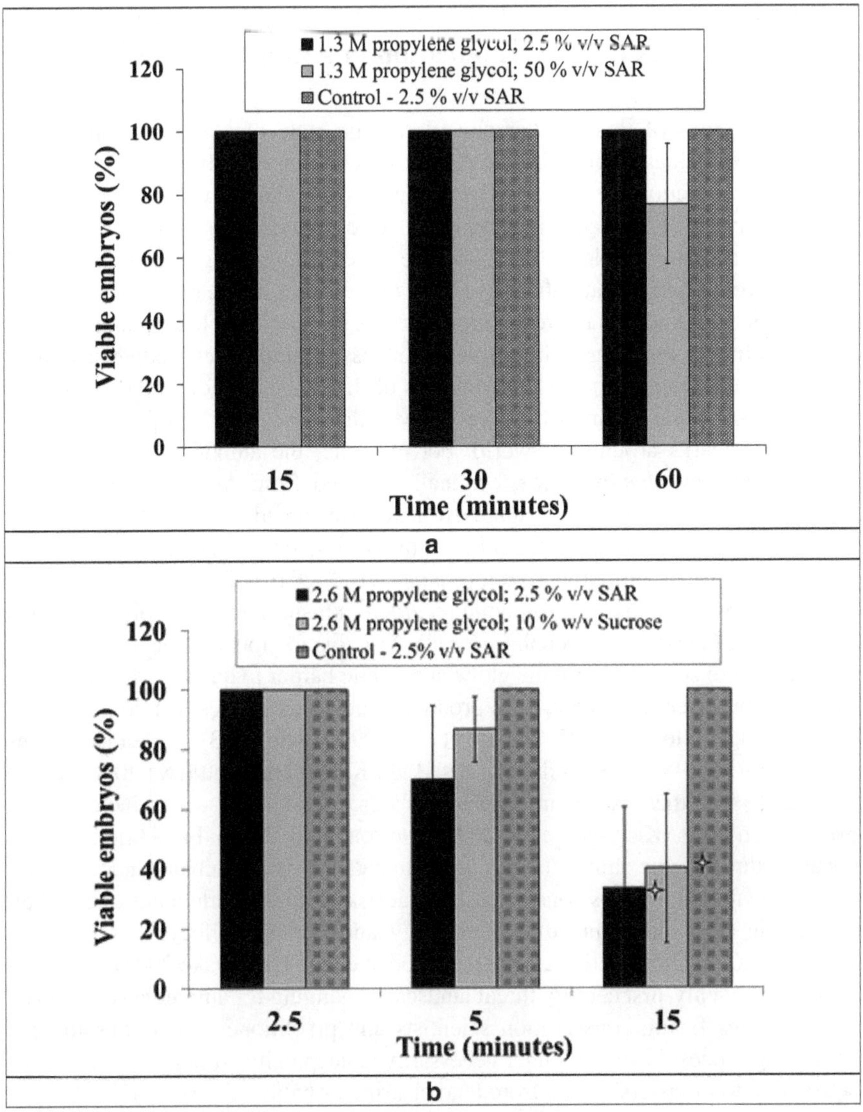

Fig. 14.2 Survival of stage 20 embryos of the striped marsh frog *Limnodynastes peronii* after exposure to two concentrations of propylene glycol (PG) for varying time periods in different media: *SAR* simplified amphibian ringer (Rugh 1962); Sucrose. (**a**) 1.3 M PG. (**b**) 2.6 M PG. Viability assessed at 24 h post cryoprotectant exposure. N = 3, mean ± s.e.m., 5–10 embryos per replicate. Diamond—Significant ($p < 0.05$) difference to control at same time interval. Curphey, Mahony, Clulow (unpublished data)

5 Putting the ARTs into Practice: Emerging Opportunities and Priorities for the Application of Amphibian ARTs

Previous sections of this chapter considered the state of amphibian ARTs and technology gaps that would advance ARTS if technological advances are realised. In the current section we consider two case studies where amphibian ARTs and biobanking could be applied to achieve improved conservation outcomes of amphibian species: the frogs of New Guinea and the water holding frog of Bolivia.

In the twenty-first century, the very nature of the threat landscape has changed for amphibians, with traditional threatening processes such as land clearing and overharvesting no longer the clear leading threats for loss of amphibian biodiversity. With globalisation that allows travel to most parts of the globe in under a day, emerging threats have changed to impact species in even the most intact and pristine forests and national parks around the world. For example, the amphibian chytrid fungi (*Batrachochytrium dendrobatidis*, originally derived from Asia (O'Hanlon et al. 2018) and the emerging *B. salamandrivorans*) have spread to all continents except Antarctica (Bower et al. 2017) resulting in the decline or extinction of 501 species, around 6.5% of the amphibian biota (Scheele et al. 2019). This qualifies it as the worst global wildlife disease pandemic recorded in history and arguably one that is still emerging (Wake and Vredenburg 2008). Also due to globalisation, invasive species continue to spread around the globe and, while harder to quantify the impact on amphibian biodiversity to date, have produced numerous species that compete with (Hecnar and M'Closkey 1997; Lawler et al. 1999; Moyle 1973), consume (Adams 2000; Bradford 1989; Goodsell and Kats 1999; Knapp and Matthews 2000; Knapp et al. 2007) or otherwise harm amphibians (Crossland et al. 2008) including by spreading chytrids (Kiesecker et al. 2001; Mazzoni et al. 2003). In addition to these threats, anthropogenic climate change is set to change the amphibian landscape permanently, with amphibians considered most at risk of all the vertebrates due to their ectothermic physiology, breeding phenology and complex lifecycle (Carey and Alexander 2003; Gibbs and Breisch 2001; Foden et al. 2013; Wake 2007).

With the twenty-first century threat landscape changing for amphibians, so too do the challenges facing conservation scientists and practitioners. Solutions to fight threatening processes are no longer necessarily conceptually 'simple'. For example, the solution to biodiversity loss from land clearing or habitat destruction is obvious: stop the land clearing and create habitat reserves such and national parks. Overharvesting can be addressed by education around sustainability and government bans or limits on protected species. But how do we address a threat such as a microscopic fungus that survives and persists indefinitely in the environment, an established invasive species or climate change, all of which pay no regard to national park boundaries? While some investigators are working on novel approaches to remove or eliminate these threats, for example, using environmental manipulation and environmental 'mismatch' to mitigate against emerging chytrids (Clulow et al. 2018b), these solutions are not likely to become available in the timeframe required to prevent further amphibian declines or extinctions without more immediate intervention.

It is in this space that genome storage and ARTs in combination with other *ex situ* conservation approaches are the technologies most likely to be of immediate conservation assistance and benefit for declining amphibians.

But where to start? With finite conservation resources and so many declining species it is imperative to select prioritised targets based on the level of threat, the urgency of action required, and the endemicity and evolutionary significance (uniqueness) whilst assessing and selecting projects that are likely to succeed and generate community interest and support (including financial) from the public and other stakeholders. In this light, there are some opportunities and actions occurring that we are currently aware of that are worth discussing and highlighting in the context of the potential for conservation value adding through application of ARTs, cryobiology and genome storage. Most of these, perhaps unsurprisingly, are focussed in areas considered global amphibian biodiversity hotspots—areas with disproportionately high amphibian species richness per land unit area. These are primarily in the tropics and include regions such as New Guinea, Madagascar and the Amazon Basin in South America. While these emerging opportunities/priorities all require genome storage and ARTs to address amphibian declines, each differs in aspects of the technological approaches that would be applied and the cases differ in the stage of the 'conservation cycle'. In one case, proactive action ahead of predictable threats and declines is indicated, in the other, the application of ARTs and genome storage is reactive long after severe declines have taken place.

5.1 Island of Opportunity: A Proactive Role for ARTs to Prevent Catastrophic Amphibian Declines in the Last Frog Frontier—New Guinea

New Guinea is unique among the world for amphibian fauna. It is the largest and highest tropical island on Earth and, while it contains no caudates (salamanders and newts) or caecilians (neither group crossed Wallace's Line), it is mega-diverse for frogs. Despite constituting only around 0.6% of the world's land surface, New Guinea is home to >6% of the world's frog species, a proportion that becomes even more impressive since it is likely this is a gross underestimate of New Guinea's contribution to global amphibian diversity due to many undiscovered and undescribed species (Bower et al. 2017, 2019; Austin et al. 2008) (Fig. 14.3). New Guinea is also unique for another reason: it is the last major landmass on Earth that is free of both species of amphibian chytrid fungi, making it one of the globe's most important remaining refuges for anuran amphibian fauna (Bower et al. 2017, 2019). While this situation is a welcome reprieve for global frog fauna, it is considered only a matter of when, not if the chytrid fungi will arrive due to the favourable environment for chytrid in New Guinea, increasing international traffic and lack of biosecurity in both Papua New Guinea and Indonesia (the nations covering the eastern and western halves of New Guinea respectively) (Bower et al. 2019). Recent predictions based

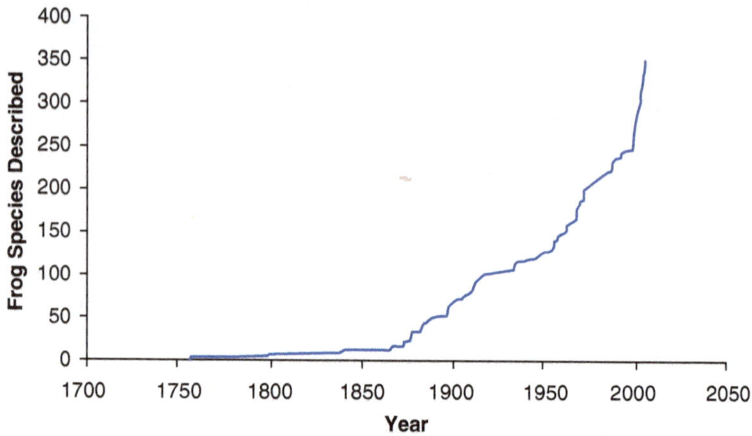

Fig. 14.3 The number of frog species described from New Guinea against time, reproduced from Austin et al. (2008). As at 2008, the number of species being described was still accelerating and showed no signs of plateau, with some estimates (Günther 2006) that only 50% of New Guinea's frog fauna is currently described

Table 14.3 Reproduced from Bower et al. (2019): projected declines of Papua New Guinean frogs based upon patterns of declines observed in Australia

Family	No. spp. in Australia	Percent declined in Australia	No. spp. in PNG	Predicted decline (n species) in PNG
Ceratobatrachidae	0		44	8[a]
Dicroglossidae	0		2	1[a]
Limnodynastidae	44	7	4	1
Microhylidae	24	0[a]	212	38[a]
Myobatrachidae	88	19	4	1
Pelodryadidae	88	22	102	22
Ranidae	1	0[a]	12	2
Total	245	Mean = 16	380	73

[a]Indicates where predictions for a particular family is based upon a mean decline of 16% calculated from the decline of the three main families in Australia collectively, due to being insufficient data, too few species or no species at all for that family in Australia to draw robust comparisons. PNG contains approximately 380 of the ca. 500 species in New Guinea. Extrapolating predicted declines in PNG to the rest of New Guinea suggests that up to ca. 100 species might be at risk

upon the Australian experience with chytrid suggest that as many as 73 species in PNG alone and perhaps up to 100 species across New Guinea might be impacted, although this prediction is untested and there could be fewer or more susceptible species (Bower et al. 2019) (Table 14.3).

Unfortunately, further unique traits of New Guinea anuran fauna make it one of the most prone of the global amphibian faunas to another major looming threat: anthropogenic climate change. The permeable skin, biphasic lifecycles (with one life stage typically aquatic or heavily reliant on constant moisture) and unshelled eggs of frogs make them extremely sensitive to even small changes in temperature and moisture (Carey and Alexander 2003; Donnelly and Crump 1998; Pounds and Crump 1994; Pounds et al. 1999). In New Guinea, the frog fauna is dominated by one particular family, the Microhylidae (Menzies 2006). Uniquely, the microhylids of New Guinea and Australia are all, as far as is known, direct-developing species in which the larval (tadpole) phase occurs entirely within the egg and hatchlings emerge as fully-formed metamorphosed froglets (Menzies 2006; Clulow and Swan 2018). Direct development is a specific adaptation that has the effect of eliminating the aquatic tadpole phase (presumably with the benefit of reduced predation by aquatic predators). In the case of New Guinea microhylids, this often results in the terrestrial or arboreal oviposition of the unprotected eggs, typically in nests amongst leaf litter or attached to leaves or branches in trees (Menzies 2006). The eggs typically avoid desiccation through high levels of humidity in lowland rainforest, moisture associated with high elevation cloud forests or being oviposited in soaks or damp soil. This puts this group at significant risk to even small amounts of drying and higher temperatures associated with climate change. Further, many New Guinea microhylids occur over extremely small ranges and at high elevations, with many being single (or present on just a few) mountain-top endemics (Menzies 2006). The tight physiological limits associated with living in narrow, high-elevational ranges (sometimes above the snow line) places the New Guinea frog fauna at unique risk to dehydration and physiological stress from changes in temperature and humidity (Araujo et al. 2006).

Since both of these major threats are predictable and seemingly unavoidable, it provides a unique opportunity among the world's amphibian fauna to invoke a rare, proactive and pre-emptive conservation approach involving ARTs to protect against catastrophic declines before they occur. As both threats are difficult to deal with *in situ* and take mitigating steps in the environment in the timeframe needed to avoid rapid declines and extinctions *in situ*, *ex situ* conservation has a major role to play, and ARTs could be front and centre in preventing the predicted catastrophic declines. In the first instance, we argue that facilities should be set up that are capable of establishing captive assurance colonies of frogs as required. As a part of this strategy, biobanking and storage of frozen lines of gametes and other cells in dedicated genome storage facilities should be initiated. Should the amphibian chytrids be detected in the country and in amphibian populations, susceptible or declining species could be brought into captivity at short notice with genetic diversity preserved and managed as an insurance against the extinctions seen in other parts of the world with the arrival of chytrids.

One such initiative is currently underway, with an agreement in place between Papua New Guinea government and various PNG and Australian zoos, museums and universities to establish an *ex situ* conservation facility at the existing Port Moresby Nature Park in PNG, including the capacity to biobank frozen gametes and cells (Banks et al. 2018). While this is promising, the program is still seeking funding and is in its infancy with regard to establishing the necessary infrastructure, expertise in-country and technologies required to deal with the predominately microhylid fauna. The challenges in biobanking include dealing with the large direct developing eggs and embryos of the dominant microhylid fauna, as well as the challenges of dealing with reproduction and direct development in captivity of this group. We are unaware of any successful attempts to store and reproduce microhylid frogs using genome storage and ARTs. Nevertheless, technologies to achieve successful restoration of adult frogs from cryopreserved gametes do exist for other groups of New Guinea frog fauna such as the Pelodryadidae (Clulow and Clulow 2016; Upton et al. 2018a).

An obvious starting point for the establishment of a working amphibian gene bank in PNG for the microhylids would be sperm storage (as discussed in Sect. 3), which is the most successful of the amphibian cryopreservation approaches and should be achievable for most species (Clulow and Clulow 2016). This would at least preserve the genetic diversity from the paternal line and could be coupled with establishing the technologies required to hormonally induce female microhylids to obtain fresh eggs (this should be achievable in many species including some from other New Guinea frog families (Clulow et al. 2018a)) and perform IVFs using frozen-thawed sperm (Upton et al. 2018a). This would require some investment in research and technology development as there have been few attempts (Michael et al. 2004; Michael and Jones 2004), that we are aware of, to produce offspring from direct-developing species from frozen-thawed sperm. Dealing with eggs and embryos on the other hand are likely to remain difficult as for other amphibians (Sect. 4) due to size and yolk content. This will be most challenging for microhylid frogs due to the fact that the eggs are much larger than average with extremely large yolks required for tadpoles to complete development and metamorphosis within the egg capsule without feeding (Table 14.2, Fig. 14.4 (Clulow and Swan 2018)).

Fig. 14.4 A male of the microhylid frog *Oreophryne oviprotector* from Papua New Guinea guards his direct developing embryos which were oviposited on the underside of a leaf. The large size of the eggs and embryos combined with large yolk content is typical of direct developing species throughout New Guinea and will likely make cryopreservation of eggs and embryos difficult or unachievable even with new technologies such as laser warming. Photograph: Stephen Richards

The emerging technologies that might overcome the cryopreservation block to egg and embryo cryopreservation for some amphibian species (Sect. 4) will be unlikely to be applicable to direct-developing species, with eggs and yolks typically many millimetres in diameter (Anstis 2017; Clulow and Swan 2018) (Table 14.2). Nevertheless, the establishment of a biobanking facility in New Guinea capable of, and dedicated to, storing sperm and other cells could be used in conjunction with ARTs to maintain genetic diversity and prevent catastrophic extinctions of New Guinea frogs, and avoid the catastrophic collapse of amphibian biodiversity that has occurred elsewhere with the arrival of chytridiomycosis. This would be a world-first example of a proactive conservation program using ARTs to prevent predictable species losses and declines. It would also facilitate research into more modern conservation approaches such as genetic rescue to manage and restore wild populations (Madsen et al. 1999). We would argue, therefore, that an amphibian ART and biobanking program for New Guinea should be considered a global conservation priority for the use of genome storage and ARTs.

5.2 *Reactive Conservation with ARTs: Restoring Amphibians on the Brink*

At the other end of the proactive-to-reactive conservation spectrum are the amphibians in systems that have already undergone massive declines to the point where they are not extinct, but few individuals remain. One such example is the Sehuencas water frog *Telmatobius yuracare* in Bolivia (Fig. 14.5). After repeated surveys of Bolivia's tropical cloud forests failed to locate any live individuals, for most of the past decade it was believed that a single male held at the Museo de Historia Natural Alcide d'Orbigny in Cochabamba—dubbed Romeo, the world's loneliest frog—was the last of its kind following massive declines in the species due to the chytrid fungus *B. dendrobatidis*. This trajectory was sadly reminiscent of another South American frog, the Panamanian Rabb's fringe-limbed tree frog *Ecnomiohyla rabborum*, which in 2016 went extinct (presumably) when the last male held at the Atlanta Botanical Gardens, dubbed 'Toughie', died (https://www.smithsonianmag.com/smart-news/adios-toughie-last-rabbs-fringe-limbed-tree-frog-dies-atlanta-180960671/). *Could assisted reproductive technologies have played a role in preventing the demise of 'Toughie' or rescued other amphibians on the brink such as Romeo?*

After learning about the single Sehuencas water frog remaining in captivity in Cochabamba, a collaboration was struck up in 2017 between Macquarie University scientists in Sydney (Simon Clulow), Global Wildlife Conservation and scientists at the Bolivian Amphibian Initiative project at the Museo de Historian Natural Alcide d'Orbigny in Cochabamba (including Arturo Muñoz and Teresa Camacho Badani), to attempt to rescue the species from the brink. It was recognised that, while never achieved before, advanced assisted reproductive techniques could potentially be

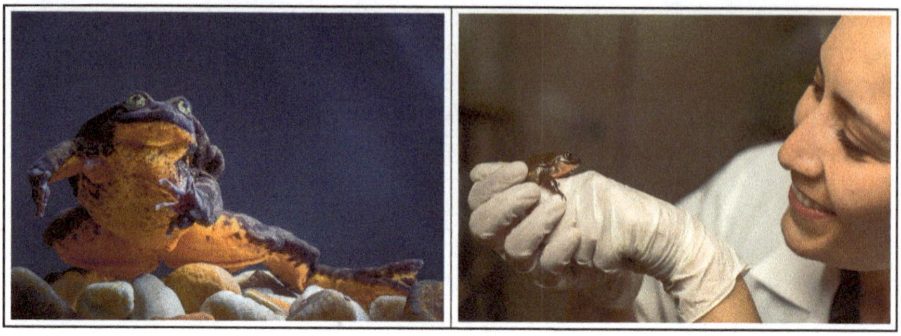

Fig. 14.5 Left image: 'Romeo, the loneliest frog in the world' was considered to be the last living Sehuencas water frog (*Telmatobius yuracare*) for most of the past decade. Right image: Dr. Teresa Camacho Badani holding 'Juliet', the first and only adult female Sehuencas water frog captured in more than 10 years. This rediscovery has created additional options for genome storage and ARTs to play a role in bringing the Sehuencas water frog back from the brink of extinction. Photographs: Robin Moore, Global Wildlife Conservation

used to generate offspring from Romeo through techniques such as androgenesis. Plans to freeze Romeo's sperm for insurance and use advanced assisted reproductive technologies to potentially rescue the species from the brink as a proof of concept of this approach became part of a publicity campaign undertaken as a collaboration between Global Wildlife Conservation and researchers at the Museo de Historia Natural Alcide d'Orbigny to raise funding for last-ditch surveys for more individuals of the water frog. The campaign involved setting Romeo up with a profile on an online dating website to find his 'Juliet', which gained international conservation publicity, raising enough money to conduct one last round of surveys in the Bolivian cloud forests. Amazingly, this last-ditch attempt to find more Sehuencas water frogs was successful, with Dr. Teresa Camacho Badani, a herpetologist at the museum and current caretaker of Romeo, and her team discovering 5 more Sehuencas water frogs in late 2018, remarkably including one adult female (Fig. 14.5). While attempts since then to breed 'Romeo and Juliet' in captivity have been so far unsuccessful, the discovery opens up real possibilities for ARTs to play a significant role in saving the species and others like it from joining Rabb's fringe-limbed tree frog in extinction. Some of the options and scenarios for the Sehuencas water frog are proposed here:

1. *Banking sperm or other cell lines for insurance.* While eggs and embryos are difficult to successfully freeze and recover (Sect. 4), freezing the sperm is an established technology (Sect. 3) and there is no reason to suspect that it will not be possible for *T. yuracare*. In the first instance, spermiation and release via the injection of hormones such as gonadotropins (Clulow and Clulow 2016; Clulow et al. 2018a) should be performed on Romeo and the other collected males with the aim of collecting and cryopreserving urinic sperm. In the first instance, these samples would be intended as a permanent insurance against the complete loss of the species, should something happen to the adults. The same could potentially be done with cell lines, using biopsies from the skin webbing and creating a cell

line from this material before freezing it down, even though establishing cell lines in amphibians can be challenging. A significant bank of sperm (and other cells) should be built up from these founding animals both for insurance purposes and to help manage potential issues with genetic diversity in the future, as demonstrated with the black-footed ferret when genetic diversity was restored after multiple generations of inbreeding and reduced heterozygosity (Howard et al. 2016).

2. *Hormonally inducing sperm and egg release and performing IVF with fresh gametes*. If 'Romeo and Juliet' continue to experience problems with natural breeding in captivity, which might be expected with Romeo's long period in captivity, ARTs have an obvious and significant role. The release of both fresh sperm from the male (as above) along with eggs from the female via induction of ovulation and oviposition (Sect. 2) would allow for attempts at IVF to produce fertilised embryos.
3. *Using frozen-thawed sperm to fertilize hormonally induced eggs from the live female should Romeo and the other males die, but the female persists*. If something was to happen to Romeo or the other newly discovered Sehuencas water frogs, the 'insurance' policy of storing their sperm could be called upon. While generally more difficult and with lower fertilisation rates than fresh sperm, the creation of viable embryos from frozen-thawed sperm, including frozen urinic sperm (Sect. 3) is feasible, noting that recent studies have shown that embryos created with frozen-thawed sperm are able to complete development and reproduce into the F1 and F2 generations (Sect. 3).
4. *Worst case scenario (death of all Sehuencas water frogs) example 1: use frozen-thawed sperm to create hybrid embryos with fresh eggs from a closely related species and back-cross*. If all male and female frogs died, under the strategy outlined here, there would be a store of sperm but no way of obtaining eggs from a live female and no frozen stored eggs (given technology to cryopreserve eggs is not available). In this scenario, there could be the possibility of IVF with frozen-thawed sperm as above, but using donor eggs from a closely related species to create hybrid embryos, and back crossing through multiple generations to genetically retrieve *T. yuracare*.
5. *Worst case scenario (death of all Sehuencas water frogs) example 2: use frozen-thawed sperm to create androgenotes*. This could function in much the same way as the original plans for Romeo prior to the discovery of Juliet and other water frogs, although it would employ frozen-thawed sperm. While no proof-of-concept example exists whereby androgenotes have been created to restore a species from the brink of extinction, this should be theoretically possible (sections above). How this would proceed would depend upon the sex-determining system of *T. yuracare*, which is currently unknown. Frogs have genetic sex determination which can be either male heterogametic (males XY, females XX) or female heterogametic (males ZZ, females ZW) (Nakamura 2009; Sarre et al. 2011). In the case of an XX, XY sex determination system, androgenotes could theoretically be produced by inseminating an enucleated donor egg from another species (e.g. through UV or physical enucleation) with a sperm cell from *T. yuracare* via IVF or ICSI. The removal of the female donor maternal pronucleus means that insemination

with the sperm would produce a haploid zygote initially, but this could be converted into a diploid zygote by blocking first cleavage through heat or cold shock before allowing development to continue as a diploid embryo (doubled haploid, see Clulow et al. (2014) and Sect. 1). The resulting offspring would either be XX females (if an X sperm is used for insemination) or YY males (if a Y sperm is used for insemination), with the outcome that normal XY males could potentially be generated by crossing YY males with XX females, resulting in both XX females and XY males to continue the species on through regular breeding.

A ZZ, ZW sex determination system would be more problematic. The creation of androgenotes using the above techniques (IVF or ICSI with donor eggs) would create 100% ZZ male offspring. Sex-reversal to produce phenotypic females but still genetic ZZ males (Sarre et al. 2011; Lambert et al. 2019) could allow the production of fresh eggs and sperm and permit breeding, but again, all resulting offspring would be ZZ and phenotypically male. To maintain the species into the future would require constant sex reversals or the implementation of the creation of ZW females through something like nuclear transfer or chimeras if female somatic or stem cells were frozen down prior to the death of all female frogs. This example highlights the importance of considering sex determination systems in the use of advanced assisted reproductive technologies.

The above case study and current proposed efforts to restore the Sehuencas water frog with the assistance of ARTs is an extreme example being considered for a species that is essentially functionally extinct (if only one surviving individual, or very few individuals persist), but one that is possible and necessary if species on the verge of complete extinction are to be retrieved (nothing can be lost by these last-ditch strategies). The reality with the current amphibian extinction crisis is that the declines and extinctions of amphibians observed around the world from chytridiomycosis and other causes are extreme. A proof-of-concept example such as the ambitious Sehuencas water frog program is needed to create confidence in using biotechnological approaches in conservation in the twenty-first century. This could lead to further public (philanthropic) and governmental funding to support and secure purpose-built facilities for genome storage and ARTs for threatened amphibians and other fauna in places such as Bolivia. This is a goal of the program with the water frog. Some success would help further existing efforts currently being made towards amphibian conservation around the world using ARTs to secure an overwhelming number of threatened amphibian species.

6 Conclusions

The previous sections have provided an overview of amphibian reproductive technologies that could be applied in conservation programs involving biobanking and the use of ARTs. We described technologies (gamete collection, IVF and sperm storage and cryopreservation) that work well enough to be implemented in existing conservation and biobanking programs. We recognise the value of ongoing research to refine and improve those basic protocols, and in particular, research that could lead to breakthroughs in oocyte and embryo cryopreservation through vitrification and laser warming. In support of our thesis that we should be actively implementing biobanking and ARTs we gave case studies of which we are familiar, or involved in, to demonstrate the urgent need and the opportunities to contribute to the preservation of amphibian biodiversity in mega-diverse regions that are currently under-resourced to implement the needed conservation strategies. Also critically missing are training programs for conservationists to learn and employ ART. The clarion call for cooperation between researchers, educators and conservationists is thus sounded.

Acknowledgements The authors acknowledge funding support from the Australian Research Council, the University of Newcastle and WWF (JC); Australian Government Post-Graduate Awards (RU); University of Ottawa Research Chair Program, Natural Sciences and Engineering Research Council of Canada (VLT); Macquarie University Research Fellowship (SC).

References

Acharjee S, Maiti K, Soh JM, Im W-B, Seong JY, Kwon HB. Differential desensitization and internalization of three different bullfrog gonadotropin-releasing hormone receptors. Mol Cells. 2002;14:101–7.

Adams MJ. Pond permanence and the effects of exotic vertebrates on anurans. Ecol Appl. 2000; 10:559–68.

Alonso-Bedate M, Carballada R, Delgado MJ. Effects of melatonin on gonadal steroids and glucose plasma levels in frogs (*Rana perezi* and *Rana temporaria*). J Pineal Res. 1990;8:79–89.

Alroy J. Current extinction rates of reptiles and amphibians. Proc Natl Acad Sci. 2015;112(42):13003–8.

Anstis M. Tadpoles and frogs of Australia. 2nd ed. Sydney: New Holland; 2017.

Araujo MB, et al. Climate warming and the decline of amphibians and reptiles in Europe. J Biogeogr. 2006;33:1712–28.

Argyle CE, Harper JC, Davies MC. Oocyte cryopreservation: where are we now? Hum Reprod Update. 2016;22:440–9.

Arregui L, Boveda P, Gosalvez J, Kouba AJ. Seasonal effect on sperm cryopreservation in *Epidalea calamita* (Anura: Bufonidae). Cryobiology. 2018;85:177.

Austin C, Hayden C, Bigilale I, Dahl C, Anaminiato J. Checklist and comments on the terrestrial amphibian and reptile fauna from Utai, Northwestern Papua New Guinea. Herpetol Rev. 2008;39(1):40–6.

Balinsky BI. Cleavage. In: An introduction to embryology. 4th ed. Philadelphia: W.B. Saunders; 1975a. p. 101–47.

Balinsky BI. Morphological aspects of gastrulation and primary organ formation. In: An Introduction to embryology. 4th ed. Philadelphia: W.B. Saunders; 1975b. p. 151–88.

Banks C, Clulow S, McGeorge M, Bower D. A future-proofing plan for Papua New Guinea frogs. AArk Newslett. 2018;42:16–7.

Barton HL, Gutman SL. Low temperature preservation of toad spermatozoa (Genus *Bufo*). Texas J Sci. 1972;23:363–70.

Beesley SG, Costanzo JP, Lee RE. Cryopreservation of spermatozoa from freeze-tolerant and -intolerant anurans. Cryobiology. 1998;37:155–62.

Berger L, Speare R, Daszak P, Green DE, Cunningham AA, Goggin CL, et al. Chytridiomycosis causes amphibian mortality associated with population declines in the rain forests of Australia and Central America. Proc Natl Acad Sci. 1998;95:9031–6.

Bischof J. Nanoparticle and metal form heating for improved cryopreservation. Cryobiology. 2018;85:164.

Bishop PJ, Angulo A, Lewis JP, Moore RD, Rabb GB, Moreno JG. The amphibian extinction crisis—what will it take to put the action into the Amphibian Conservation Action Plan? SAPIENS. 2012;(5.2). http://sapiens.revues.org/1406.

Bower DS, Lips K, Schwarzkopf L, Georges A, Clulow S. Amphibians on the brink. Science. 2017;357(6350):454–5.

Bower D, Lips K, Amepou Y, Richards S, Dahl C, Nagombi E, Supuma M, Dabek L, Alford R, Schwarzkopf L, Ziembicki M, Noro J, Hamidy A, Gillespie G, Berger L, Eisemberg C, Li Y, Liu X, Jennings C, Tjaturadi B, Peters A, Krockenberger A, Nason D, Kusrini M, Webb R, Skerratt L, Banks C, Mack A, Georges A, Clulow S. Island of opportunity: can New Guinea protect amphibians from a globally emerging pathogen? Front Ecol Environ. 2019; https://doi.org/10.1002/fee.2057.

Bradford DF. Allotopic distribution of native frogs and introduced fishes in high Sierra Nevada lakes of California: implication of the negative effect of fish introductions. Copeia. 1989;1989:775–8.

Browne RK, Clulow J, Mahony M, Clark A. Successful recovery of motility and fertility of cryopreserved cane toad (*Bufo marinus*) sperm. Cryobiology. 1998;37:339–45.

Browne RK, Clulow J, Mahony M. Short-term storage of cane toad (*Bufo marinus*) gametes. Reproduction. 2001;121:167–73.

Browne RK, Clulow J, Mahony M. The effect of saccharides on the post-thaw recovery of cane toad (*Bufo marinus*) spermatozoa. Cryo Letters. 2002a;23:121–8.

Browne RK, Clulow J, Mahony M. The short-term storage and cryopreservation of spermatozoa from Hylid and Myobatrachid frogs. CryoLetters. 2002b;23:129–36.

Browne RK, Clulow J, Mahony M. A comparison of sucrose, saline, and saline with egg-yolk diluents on the cryopreservation of cane toad (*Bufo marinus*) sperm. Cryobiology. 2002c;44:251–7.

Browne RK, Davis J, Pomering M, Clulow J. Storage of cane toad (Bufo marinus) sperm for 6 days at 0 °C with subsequent cyropreservation. Reprod Fertil Dev. 2002d;14:267–73.

Browne RK, Seratt J, Vance C, Kouba A. Hormonal priming, induction of ovulation and in-vitro fertilization of the endangered Wyoming toad (Bufo baxteri). Reprod Biol Endocrinol. 2006a;4:34.

Browne RK, Li H, Seratt J, Kouba A. Progesterone improves the number and quality of hormone induced Fowler toad (*Bufo fowleri*) oocytes. Reprod Biol Endocrinol. 2006b;4:3.

Browne R, Gaikhorst G, Vitali S, Roberts JD, Matson P. Exogenous hormones induce poor rates of oviposition in the anurans, *Litoria moorei* and *L. aurea*. Appl Herpetol. 2008;5(1):81–6.

Browne R, Janzen P, Bagaturov M, van Houte D. Amphibian keeper conservation breeding programs. J Zool Res. 2018;2:29–46.

Browne RK, et al. Sperm collection and storage for the sustainable management of amphibian biodiversity. Theriogenology. 2019;133:187–200.

Buchholz DR, Fu L, Shi YB. Cryopreservation of Xenopus transgenic lines. Mol Reprod Dev. 2004;67:65–9.

Byrne P, Silla A. Hormonal induction of gamete release, and in-vitro fertilisation, in the critically endangered Southern Corroboree Frog, *Pseudophryne corroboree*. Reprod Biol Endocrinol. 2010;8:144.

Calado LL. Coleta e criopreservação do sêmen de rã-touro. Universidade Federal de Viçosa; 2009.

Calvi SL, Maisse G. Cryopreservation of Rainbow Trout (Oncorhynchus mykiss) blastomeres: influence of embryo stage on postthaw survival rate. Cryobiology. 1998;36:255–62.

Calvi SL, Maisse G. Cryopreservation of carp (Cyprinus carpio) blastomeres. Aquat Living Resour. 1999;12:71–4.

Campbell L, Bower DS, Clulow S, Stockwell M, Clulow J, Mahony M. Interaction between temperature and sublethal infection with the amphibian chytrid fungus impacts a susceptible frog species. Sci Rep. 2019;9:83.
Cardona-Costa J, García-Ximénez F. Vitrification of zebrafish embryo blastomeres in microvolumes. CryoLetters. 2007;28:303–9.
Carey C, Alexander MA. Climate change and amphibian declines: is there a link? Divers Distrib. 2003;9:111–21.
Ceballos G, Ehrlich PR, Dirzo R. Biological annihilation via the ongoing sixth mass extinction signaled by vertebrate population losses and declines. Proc Natl Acad Sci. 2017;114:E6089–96.
Chakraborty M, Burmeister SS. Estradiol induces sexual behavior in female túngara frogs. Horm Behav. 2009;55:106–12.
Chakraborty M, Burmeister SS. Effects of estradiol on neural responses to social signals in female túngara frogs. J Exp Biol. 2015;218:3671–7.
Chanson J, et al. The State of the world's amphibians. In: Stuart SN, Hoffmann M, Chanson JS, Cox NA, Berridge RJ, Ramani P, Young BE, editors. Threatened amphibians of the world. Barcelona, Gland, and Arlington: Lynx Edicions, IUCN, and Conservation International; 2008. p. 33–52.
Clulow J, Clulow S. Cryopreservation and other assisted reproductive technologies for the conservation of threatened amphibians and reptiles: bringing the arts up to speed. Reprod Fertil Dev. 2016;28:1116–32.
Clulow S, Swan M. A complete guide to frogs of Australia. Sydney: Australian Geographic; 2018. 336 pp.
Clulow J, Mahony M, Browne R, Pomering M, Clark A. Applications of assisted reproductive technologies (ART) to endangered anuran amphibians. In: Campbell A, editor. Declines and disappearances of Australian frogs. Canberra: Environment Australia; 1999. p. 219–25.
Clulow J, Clulow S, Guo J, French AJ, Mahony MJ, Archer M. Optimisation of an oviposition protocol employing human chorionic and pregnant mare serum gonadotropins in the barred frog *Mixophyes fasciolatus* (Myobatrachidae). Reprod Biol Endocrinol. 2012;10(1):60.
Clulow J, Trudeau VL, Kouba AJ. Amphibian declines in the twenty-first century: why we need assisted reproductive technologies. In: Holt WV, et al., editors. Reproductive sciences in animal conservation, Advances in experimental medicine and biology 753. New York: Springer; 2014. p. 275–316. https://doi.org/10.1007/978-1-4939-0820-2_12.
Clulow J, Pomering M, Herbert D, Upton R, Calatayud N, Clulow S, Mahony M, Trudeau V. Differential success in obtaining gametes between male and female Australian temperate frogs by hormonal induction: a review. Gen Comp Endocrinol. 2018a;265:141–8.
Clulow S, Gould J, James H, Stockwell M, Clulow J, Mahony M. Elevated salinity blocks pathogen transmission and improves host survival from the global amphibian chytrid pandemic: implications for translocations. J Appl Ecol. 2018b;55:830–40.
Collin F, Chartrel N, Fasolo A, Conlon JM, Vandesande F, Vaudry H. Distribution of two molecular forms of gonadotropin-releasing hormone (GnRH) in the central nervous system of the frog Rana ridibunda. Brain Res. 1995;703:111–28.
Corley-Smith GE, Lim CJ, Brandhorst BP. Production of androgenetic zebrafish (*Danio rerio*). Genetics. 1996;142:1265–76.
Costanzo JP, Mugnano JA, Wehrheim HM, Lee RE. Osmotic and freezing tolerance in spermatozoa of freeze-tolerant and -intolerant frogs. Am J Physiol Regul Integr Comp Physiol. 1998;275:R713–9.
Creighton AE, Wilczynski W. Influence of dopamine D2-type receptors on motor behaviors in the green tree frog, *Hyla cinerea*. Physiol Behav. 2014;127:71–80.
Creighton A, Satterfield D, Chu J. Effects of dopamine agonists on calling behavior in the green tree frog, *Hyla cinerea*. Physiol Behav. 2013;116:54–9.
Crossland MR, Brown GP, Anstis M, Shilton CM, Shine R. Mass mortality of native anuran tadpoles in tropical Australia due to the invasive cane toad (Bufo marinus). Biol Conserv. 2008;141:2387–94.

Daly J, Zuchowicz N, Nuñez Lendo CI, Khosla K, Lager C, Henley M, Bischof J, Kleinhans FW, Lin C, Peters E, et al. Successful cryopreservation of coral larvae using vitrification and laser warming. Scientific Reports. 2018; https://doi.org/10.1038/s41598-018-34035-0.
Dash SN, Routray P, Dash C, Guru BC, Swain P, Sarangi N. Use of the nontoxic cryoprotectant trehalose enhances recovery and function of fish embryonic stem cells following cryogenic storage. Curr Stem Cell Res Ther. 2008;3:277–87.
Davis A, Abraham E, McEvoy E, Sonnenfeld S, Lewis C, Hubbard CS, Dolence EK, Rose JD, Coddington E. Corticosterone suppresses vasotocin-enhanced clasping behavior in male rough-skinned newts by novel mechanisms interfering with V1a receptor availability and receptor-mediated endocytosis. Horm Behav. 2015;69:39–49.
del Pino EM, Venegas-Ferrín M, Romero-Carvajal A, Montenegro-Larrea P, Sáenz-Ponce N, Moya IM, Alarcón I, Sudou N, Yamamoto S, Taira M. A comparative analysis of frog early development. Proc Natl Acad Sci U S A. 2007;104:11882–8.
Della Togna Nieto G. Structural and functional characterization of the Panamanian golden frog (*Atelopus zeteki*) spermatozoa—impact of medium osmolality and cryopreservation on motility and cell viability. Dissertation, University of Maryland. 2015.
Deng J, Carbajal L, Evaul K, Rasar M, Jamnongjit M, Hammes SR. Nongenomic steroid-triggered oocyte maturation: of mice and frogs. Steroids. 2009;74:595–601.
Derakhshan Z, Nokhbatolfoghahai M, Zahiri S. Cryopreservation of *Bufotes viridis* embryos by vitrification. Cryobiology. 2017;75:60–7.
Di Berardino MA. Genomic potential of differentiated cells. New York: Columbia University Press; 1997.
Diez C, Munoz M, Caamano J, Gomez E. Cryopreservation of the bovine oocyte: current status and perspectives. Reprod Domest Anim. 2012;47:76–83.
Donnelly MA, Crump ML. Potential effects of climate change on two neotropical amphibian assemblages. Clim Change. 1998;39:541–61.
Du Pasquier D, Dupré A, Jessus C. Unfertilized xenopus eggs die by bad-dependent apoptosis under the control of cdk1 and jnk. PLoS One. 2011;6:e23672.
Duellman WE, Trueb L. Biology of amphibians. Baltimore: Johns Hopkins University Press; 1994.
Dufour S, Weltzien FA, Sebert ME, Le Belle N, Vidal B, Vernier P, Pasqualini C. Dopaminergic inhibition of reproduction in teleost fishes: ecophysiological and evolutionary implications. Ann N Y Acad Sci. 2005;1040:9–21.
Edwards DL, Mahony MJ, Clulow J. Effect of sperm concentration, medium osmolality, and oocyte storage on artificial fertilisation success in a myobatrachid frog (*Limnodynastes tasmaniensis*). Reprod Fertil Dev. 2004;16:347–54.
Elinson RP, del Pino EM. Developmental diversity of amphibians. Wiley Interdiscip Rev Dev Biol. 2012;1:345–69.
Eroglu A, Bailey SE, Toner M, Toth TL. Successful cryopreservation of mouse oocytes by using low concentrations of trehalose and dimethylsulfoxide. Biol Reprod. 2009;80:70–8.
Evaul K, Jamnongjit M, Bhagavath B, Hammes SR. Testosterone and progesterone rapidly attenuate plasma membrane Gβγ-mediated signaling in *Xenopus laevis* oocytes by signaling through classical steroid receptors. Mol Endocrinol. 2007;21:186–96.
Fahy GM, MacFarlane D, Angell C, Meryman H. Vitrification as an approach to cryopreservation. Cryobiology. 1984;21:407–26.
Figiel CR. Cryopreservation of sperm from the axolotl *Ambystoma mexicanum*: implications for conservation. Herpetol Conserv Biol. 2013;8(3):748–55.
Fitzsimmons C, McLaughlin EA, Mahony MJ, Clulow J. Optimisation of handling, activation and assessment procedures for Bufo marinus spermatozoa. Reprod Fertil Dev. 2007;19(4):594–601.
Foden WB, Butchart SH, Stuart SN, Vié J-C, Akçakaya HR, Angulo A, DeVantier LM, Gutsche A, Turak E, Cao L. Identifying the world's most climate change vulnerable species: a systematic trait-based assessment of all birds, amphibians and corals. PLoS One. 2013;8:e65427.
Franěk R, Marinović Z, Lujić J, Urbányi B, Fučíková M, Kašpar V, Pšenička M, Horváth Á. Cryopreservation and transplantation of common carp spermatogonia. bioRxiv. 2018:429449.

Frankham R. Challenges and opportunities of genetic approaches to biological conservation. Biol Conserv. 2010;143(9):1919–27.
Frankham R. Genetic rescue of small inbred populations: meta-analysis reveals large and consistent benefits of gene flow. Mol Ecol. 2015;24(11):2610–8.
Gibbs JP, Breisch AR. Climate warming and calling phenology of frogs near Ithaca, New York, 1900-1999. Conserv Biol. 2001;15:1175–8.
Gillespie LL, Armstrong JB. Production of androgenetic diploid axolotls by suppression of first cleavage. J Exp Zool. 1980;213:423–5.
Godoy LC, Streit DP Jr, Zampolla T, Bos-Mikich A, Zhang T. A study on the vitrification of stage III zebrafish (*Danio rerio*) ovarian follicles. Cryobiology. 2013;67:347–54.
Goodsell JA, Kats LB. Effect of introduced mosquitofish on Pacific treefrogs and the role of alternative prey. Conserv Biol. 1999;13:921–4.
Gordon NM, Gerhardt HC. Hormonal modulation of phonotaxis and advertisement-call preferences in the gray treefrog (*Hyla versicolor*). Horm Behav. 2009;55:121–7.
Guan M, Rawson DM, Zhang T. Cryopreservation of zebrafish (*Danio rerio*) oocytes by vitrification. Cryo Letters. 2010;31:230–8.
Guenther JF, Seki S, Kleinhans FW, Edashige K, Roberts DM, Mazur P. Extra- and intra-cellular ice formation in stage I and II *Xenopus laevis* oocytes. Cryobiology. 2006;52:401–16.
Günther R. Derived reproductive modes in New Guinean anuran amphibians and description of a new species with paternal care in the genus *Callulops* (Microhylidae). J Zool. 2006;268:153–70.
Gurdon JB. The effects of ultraviolet irradiation on uncleaved eggs of *Xenopus laevis*. Q J Microsc Sci. 1960;101:299–311.
Gurdon J, Byrne J. The first half-century of nuclear transplantation. Proc Natl Acad Sci U S A. 2003;100:8048–52.
Gurdon JB, Hopwood N. The introduction of *Xenopus laevis* into developmental biology: of empire, pregnancy testing and ribosomal genes. Int J Dev Biol. 2000;44:43–50.
Gurdon JB, Uehlinger V. "Fertile" intestine nuclei. Nature. 1966;210:1240–1.
Gurdon JB, Laskey RA, Reeves OR. The developmental capacity of nuclei transplanted from keratinized skin cells of adult frogs. J Embryol Exp Morphol. 1975;34:93–112.
Hagedorn M, Kleinhans FW. Problems and prospects in cryopreservation of fish embryos. In: Tiersch TR, Green CC, editors. Cryopreservation in aquatic species. 2nd ed. Baton Rouge: World Aquaculture Society; 2011. p. 483–502.
Hagedorn M, Hsu EW, Pilatus U, Wildt DE, Rall WF, Blackband SJ. Magnetic resonance microscopy and spectroscopy reveal kinetics of cryoprotectant permeation in a multicompartmental biological system. Proc Natl Acad Sci U S A. 1996;93:7454–9.
Hagedorn M, Hsu E, Kleinhans FW, Wildt DE. New approaches for studying the permeability of fish embryos: toward successful cryopreservation. Cryobiology. 1997a;34:335–47.
Hagedorn M, Kleinhans FW, Freitas R, Liu J, Hsu EW, Wildt DE, et al. Water distribution and permeability of zebrafish embryos, *Brachydanio rerio*. J Exp Zool. 1997b;278:356–71.
Hagedorn M, Kleinhans FW, Wildt DE, Rall WF. Chill sensitivity and cryoprotectant permeability of dechorionated zebrafish embryos, *Brachydanio rerio*. Cryobiology. 1997c;34:251–63.
Hagedorn M, Kleinhans FW, Artemov D, Pilatus U. Characterization of a major permeability barrier in the zebrafish embryo. Biol Reprod. 1998;59:1240–50.
Hagedorn M, Peterson A, Mazur P, Kleinhans FW. High ice nucleation temperature of zebrafish embryos: slow-freezing is not an option. Cryobiology. 2004;49:181–9.
Harding G, Griffiths RA, Pavajeau L. Developments in amphibian captive breeding and reintroduction programs. Conserv Biol. 2016;30:340–9.
Harvey B. Cooling of embryonic cells, isolated blastoderms, and intact embryos of the zebra fish *Brachydanio rerio* to -196°C. Cryobiology. 1983;20:440–7.
He X, Park EY, Fowler A, Yarmush ML, Toner M. Vitrification by ultra-fast cooling at a low concentration of cryoprotectants in a quartz micro-capillary: a study using murine embryonic stem cells. Cryobiology. 2008;56:223–32.
Hecnar SJ, M'Closkey RT. Changes in the composition of a ranid frog community following bullfrog extinction. Am Midl Nat. 1997;137:145–50.

Higaki S, Eto Y, Kawakami Y, Yamaha E, Kagawa N, Kuwayama M, et al. Production of fertile zebrafish (*Danio rerio*) possessing germ cells (gametes) originated from primordial germ cells recovered from vitrified embryos. Reproduction. 2010;139:733–40.

Hill JE, Kilgore KH, Pouder DB, Powell JF, Watson CA, Yanong RP. Survey of ovaprim use as a spawning aid in ornamental fishes in the United States as administered through the University of Florida Tropical Aquaculture Laboratory. N Am J Aquac. 2009;71:206–9.

Hoelker M, Schmoll F, Schneider H, Rings F, Gilles M, Tesfaye D, Jennen D, Tholen E, Griese J, Schellander K. Bovine blastocyst diameter as a morphological tool to predict embryo cell counts, embryo sex, hatching ability and developmental characteristics after transfer to recipients. Reprod Fertil Dev. 2006;18:551–7.

Hollinger T, Corton G. Artificial fertilization of gametes from the South African clawed frog, Xenopus laevis. Gamete Res. 1980;3:45–7.

Hong N, Chen S, Ge R, Song J, Yi M, Hong Y. Interordinal chimera formation between medaka and zebrafish for analyzing stem cell differentiation. Stem Cells Dev. 2012;21:2333–41.

Hopkins BK, Herr C. 78 Cryopreservation of frog (*Rana pipiens*) sperm cells collected by nonlethal methods. Reprod Fertil Dev. 2007;20:120.

Horseman ND, Smith CA, Culley DD Jr. Effects of age and photoperiod on ovary size and condition in bullfrogs (*Rana catesbeiana*, Shaw) (Amphibia, Anura, Ranidae). J Herpetol. 1978; 12:287–90.

Howard JG, Marinari PE, Wildt DE. Black-footed ferret: model for assisted reproductive technologies contributing to in situ conservation. In: Holt WV, Pickard AR, Rodger JC, Wildt DE, editors. Reproductive science and integrated conservation. Cambridge: University Press; 2003. p. 249–66.

Howard JG, Lynch C, Santymire RM, Marinari PE, Wildt DE. Recovery of gene diversity using long-term cryopreserved spermatozoa and artificial insemination in the endangered black-footed ferret. Anim Conserv. 2016;19:102–11.

Hwang I-S, Hochi S. Recent progress in cryopreservation of bovine oocytes. Biomed Res Int. 2014;2014:570647.

Inoda T, Morisawa M. Effect of osmolality on the initiation of sperm motility in *Xenopus laevis*. Comp Biochem Physiol A Physiol. 1987;88(3):539–42.

Isayeva A, Zhang T, Rawson DM. Studies on chilling sensitivity of zebrafish (*Danio rerio*) oocytes. Cryobiology. 2004;49:114–22.

Ishibashi S, Kroll KL, Amaya E. Generation of transgenic *Xenopus laevis*: II. Sperm nuclei preparation. CSH Protoc. 2007a;2007:pdb.prot4839.

Ishibashi S, Kroll KL, Amaya E. Generation of transgenic *Xenopus laevis*: III. Sperm nuclear transplantation. Cold Spring Harb Protoc. 2007b;2007:pdb.prot4840.

Jin B, Kleinhans FW, Mazur P. Survivals of mouse oocytes approach 100% after vitrification in 3-fold diluted media and ultra-rapid warming by an IR laser pulse. Cryobiology. 2014; 68:419–30.

Kaurova S, Chekurova N, Melnikova E, Uteshev V, Gakhova E. Cryopreservation of frog Rana temporaria sperm without loss of fertilizing capacity. In: Paper presented at genetic resource conservation proc of XIV working meeting, Pushchino, 13–15 May 1996; 1996.

Kaurova S, Uteshev V, Chekurova N, Gakhova E. Cryopreservation of testis of frog Rana temporaria. Infusionsther Transfusionsmed. 1997;24:78–9.

Keogh LM, Byrne PG, Silla AJ. The effect of gentamicin on sperm motility and bacterial abundance during chilled sperm storage in the Booroolong frog. Gen Comp Endocrinol. 2017;243:51–9.

Khalili MA, A Nottola S, Shahedi A, Macchiarelli G. Contribution of human oocyte architecture to success of in vitro maturation technology. Iranian J Reprod Med. 2013;11:1–10.

Khosla K, Wang Y, Hagedorn M, Qin Z, Bischof J. Gold nanorod induced warming of embryos from the cryogenic state enhances viability. ACS Nano. 2017;11:7869–78.

Khosla K, Zhan L, Bhati A, Carley-Clopton A, Hagedorn M, Bischof J. Characterization of laser gold nanowarming: a platform for millimeter-scale cryopreservation. Langmuir. 2018a; https://doi.org/10.1021/acs.langmuir.8b03011.

Khosla K, Zhan L, Bhati A, Carley-Clopton A, Hagedorn M, Bischof J. Physical limits of laser gold nanowarming. Cryobiology. 2018b;85:161.

Kiesecker JM, Blaustein AR, Miller CL. Transfer of a pathogen from fish to amphibians. Conserv Biol. 2001;15:1064–70.

Kimmel CB, Ballard WW, Kimmel SR, Ullmann B, Schilling TF. Stages of embryonic development of the zebrafish. Dev Dyn. 1995;203:253–310.

Kleinhans FW, Guenther JF, Seki S, Edashige K, Roberts DM, Mazur P. Water P_f of *Xenopus* (AQP +/-) Stage I and II oocytes. Cryobiology. 2005;51:389.

Kleinhans FW, Guenther JF, Roberts DM, Mazur P. Analysis of intracellular ice nucleation in *Xenopus* oocytes by differential scanning calorimetry. Cryobiology. 2006;52:128–38.

Klop-Toker KL, Valdez JW, Stockwell MP, Edgar ME, Fardell L, Clulow S, Clulow J, Mahony MJ. Assessing host response to disease treatment: how chytrid-susceptible frogs react to increased water salinity. Wildl Res. 2018;44:648–59.

Knapp RA, Matthews KR. Non-native fish introductions and the decline of the mountain yellow-legged frog from within protected areas. Conserv Biol. 2000;14:1–12.

Knapp RA, Boiano DM, Vredenburg VT. Removal of nonnative fish results in population expansion of a declining amphibian (mountain yellow-legged frog, *Rana muscosa*). Biol Conserv. 2007;135(1):11–20.

Köseoglu M, Eroglu A, Toner M, Sadler KC. Starfish oocytes form intracellular ice at unusually high temperatures. Cryobiology. 2001;43:248–59.

Kouba AJ, Vance CK. Applied reproductive technologies and genetic resource banking for amphibian conservation. Reprod Fertil Dev. 2009;21:719–37.

Kouba AJ, Vance CK, Frommeyer MA, Roth TL. Structural and functional aspects of *Bufo americanus* spermatozoa: effects of inactivation and reactivation. J Exp Zool. 2003;295A:172–82.

Kouba AJ, Vance CK, Willis EL. Artificial fertilization for amphibian conservation: current knowledge and future considerations. Theriogenology. 2009;71:214–27.

Kouba A, Willis E, Vance C, Hasenstab S, Reichling S, Krebs J, Linhoff L, Snoza M, Langhorne C, Germano J. 116 development of assisted reproduction technologies for the endangered Mississippi gopher frog (*Rana sevosa*) and sperm transfer for in vitro fertilization. Reprod Fertil Dev. 2011;24:170.

Kouba AJ, Lloyd RE, Houck ML, Silla AJ, Calatayud N, Trudeau VL, et al. Emerging trends for biobanking amphibian genetic resources: the hope, reality and challenges for the next decade. Biol Conserv. 2013;164:10–21.

Kusuda S, Teranishi T, Koide N. Cryopreservation of chum salmon blastomeres by the straw method. Cryobiology. 2002;45:60–7.

Kusuda S, Teranishi T, Koide N, Nagai T, Arai K, Yamaha E. Pluripotency of cryopreserved blastomeres of the goldfish. J Exp Zool A Comp Exp Biol. 2004;301(2):131–8.

Kuwayama M, Vajta G, Kato O, Leibo SP. Highly efficient vitrification method for cryopreservation of human oocytes. Reprod Biomed Online. 2005;11:300–8.

Lambert MR, et al. Molecular evidence for sex reversal in wild populations of green frogs (*Rana clamitans*). PeerJ. 2019;7:e6449. https://doi.org/10.7717/peerj.6449.

Langhorne CJ, Calatayud NE, Kouba AJ, Feugang JM, Vance CK, Willard ST. 026 Cryoconservation: successful sperm cryopreservation and develop-mental outcomes using endangered North American amphibians. Cryobiology. 2013;67(3):405.

Lawler SP, Dritz D, Strange T, Holyoak M. Effects of introduced mosquitofish and bullfrogs on the threatened California red-legged frog. Conserv Biol. 1999;13:613–22.

Lawson B, Clulow S, Mahony MJ, Clulow J. Towards gene banking amphibian maternal germ lines: short-term incubation, cryoprotectant tolerance and cryopreservation of embryonic cells of the frog, Limnodynastes peronii. PLoS One. 2013;8(4):e60760.

Leibo S, Loskutoff N. Cryobiology of in vitro-derived bovine embryos. Theriogenology. 1993;39:81–94.

Licht P, Tsai P-S, Sotowska-Brochocka J. The nature and distribution of gonadotropin-releasing hormones in brains and plasma of ranid frogs. Gen Comp Endocrinol. 1994;94:186–98.

Liebermann J, Tucker MJ. Vitrifying and warming of human blastocysts using the Cryotop. In: Vitrification in assisted reproduction. London: CRC Press; 2007. p. 258–64.
Lin S, Long W, Chen J, Hopkins N. Production of germ-line chimeras in zebrafish by cell transplants from genetically pigmented to albino embryos. Proc Natl Acad Sci U S A. 1992;89:4519–23.
Lin C, Zhang T, Rawson DM. Cryopreservation of zebrafish (*Danio rerio*) blastomeres by controlled slow cooling. CryoLetters. 2009;30:132–41.
Lips KR, Brem F, Brenes R, Reeve JD, Alford RA, Voyles J, et al. Emerging infectious disease and the loss of biodiversity in a Neotropical amphibian community. Proc Natl Acad Sci U S A. 2006;103:3165–70.
Liu X-H, Zhang T, Rawson DM. The effect of partial removal of yolk on the chilling sensitivity of zebrafish (*Danio rerio*) embryos. Cryobiology. 1999;39:236–42.
Liu X-H, Zhang T, Rawson DM. Differential scanning calorimetry studies of intramembrane freezing and cryoprotectant penetration in zebrafish (*Danio rerio*) embryos. J Exp Zool. 2001;290:299–310.
Longcore JE, Pessier AP, Nichols DK. Batrachochytrium dendrobatidis gen. et sp. nov., a chytrid pathogenic to amphibians. Mycologia. 1999;91:219–27.
Lujić J, Franěk R, Marinović Z, Kašpar V, Pšenička M, Urbányi B, Horváth Á. Cryopreservation of common carp (*Cyprinus carpio*) spermatogonial stem cells. Cryobiology. 2018;85:158.
Luo T, Xu Y, Hoffman TL, Zhang T, Schilling T, Sargent TD. Inca: a novel p21-activated kinase-associated protein required for cranial neural crest development. Development. 2007;134(7):1279–89.
Luyet BJ, Hodapp EL. Revival of frog's spermatozoa vitrified in liquid air. Proc Soc Exp Biol Med. 1938;39:433–4.
Madsen T, Shine R, Olsson M, Wittzell H. Conservation biology: restoration of an inbred adder population. Nature. 1999;402:34–5.
Mansour N, Lahnsteiner F, Patzner RA. Optimization of the cryopreservation of African clawed frog (*Xenopus laevis*) sperm. Theriogenology. 2009;72:1221–8.
Mansour N, Lahnsteiner F, Patzner RA. Motility and cryopreservation of spermatozoa of European common frog, Rana temporaria. Theriogenology. 2010;74:724–32.
Marcec R, Langhorne C, Vance C, Kouba A, Willard S. C-1013: cryopreservation of spermic milt in the model species *Ambystoma tigrinum* (Tiger salamander) for application in endangered salamanders. Cryobiology. 2014;69:515.
Marinović Z, Li Q, Lujić J, Iwasaki Y, Csenki Z, Urbányi B, Horváth Á, Yoshizaki G. Testis cryopreservation and spermatogonia transplantation as a tool for zebrafish line reconstitution. Cryobiology. 2018a;85:145–6.
Marinović Z, Lujić J, Kása E, Csenki Z, Urbányi B, Horváth Á. Cryopreservation of zebrafish spermatogonia by whole testes needle immersed ultra-rapid cooling. J Vis Exp. 2018b;133:e56118.
Masui Y. Relative roles of the pituitary, follicle cells, and progesterone in the induction of oocyte maturation in *Rana pipiens*. J Exp Zool. 1967;166:365–75.
Mazur P. Principles of cryobiology. In: Fuller BJ, Lane N, Benson EE, editors. Life in the frozen state. Boca Rotan: CRC Press; 2004. p. 3–65.
Mazur P, Kleinhans FW. Relationship between intracellular ice formation in oocytes of the mouse and *Xenopus* and the physical state of the external medium—a revisit. Cryobiology. 2008;56:22–7.
Mazur P, Seki S. Survival of mouse oocytes after being cooled in a vitrification solution to −196 °C at 95 to 70,000 °C/min and warmed at 610 to 118,000 °C/min: a new paradigm for cryopreservation by vitrification. Cryobiology. 2011;62:1–7.
Mazur P, Seki S, Pinn IL, Kleinhans FW, Edashige K. Extra- and intracellular ice formation in mouse oocytes. Cryobiology. 2005;51:29–53.
Mazzoni R, Cunningham AA, Daszak P, Apolo A, Perdomo E, et al. Emerging pathogen of wild amphibians in frogs (*Rana catesbeiana*) farmed for international trade. Emerg Infect Dis. 2003;9:3–30.
McKinnell RG. Cloning: nuclear transplantation in amphibia. Minneapolis: University of Minnesota Press; 1978.

McKinnell RG, Picciaano DJ, Kreig RE. Fertilization and development of frog eggs after repeated spermiation induced by human chorionic gonadotrophin. Lab Anim Sci. 1976;26:932–5.
Menzies J. The frogs of New Guinea and the Solomon Islands. Sofia: Pensoft; 2006.
Messaoud NB, Yue J, Valent D, Katzarova I, López JM. Osmostress-induced apoptosis in *Xenopus* oocytes: role of stress protein kinases, calpains and smac/diablo. PLoS One. 2015;10:e0124482.
Michael SF, Jones C. Cryopreservation of spermatozoa of the terrestrial Puerto Rican frog, *Eleutherodactylus coqui*. Cryobiology. 2004;48:90–4.
Michael S, Buckley C, Toro E, Estrada A, Vincent S. Induced ovulation and egg deposition in the direct developing anuran *Eleutherodactylus coqui*. Reprod Biol Endocrinol. 2004;2:6.
Mollard R. Culture, cryobanking and passaging of karyotypically validated native Australian amphibian cells. Cryobiology. 2018;81:201–5.
Monfort SL. "Mayday Mayday Mayday", the Millenium Ark is sinking! In: Holt WV, et al., editors. Reproductive sciences in animal conservation, Advances in experimental medicine and biology 753. New York: Springer; 2014. p. 15–31. https://doi.org/10.1007/978-1-4939-0820-2_2.
Moore FL. Reproductive endocrinology of amphibians. In: Chester-Jones L, Ingleton PM, Phillips JG, editors. Fundamentals of comparative vertebrate endocrinology. New York: Springer; 1987. p. 207–21.
Morrow SG. The effects of DNA damage caused by sperm cryopreservation in Xenopus. University of Portsmouth; 2015.
Morrow S, Gosálvez J, López-Fernández C, Arroyo F, Holt WV, Guille MJ. Effects of freezing and activation on membrane quality and DNA damage in *Xenopus tropicalis* and *Xenopus laevis* spermatozoa. Reprod Fertil Dev. 2017;29:1556–66.
Moyle PB. Effects of introduced bullfrogs (*Rana catesbeiana*), on the native frogs of the San Joaquin Valley, California. Copeia. 1973;1:18–22.
Mugnano JA, Costanzo JP, Beesley SG, Lee RE. Evaluation of glycerol and dimethyl sulfoxide for the cryopreservation of spermatozoa from the wood frog (*Rana sylvatica*). CryoLetters. 1998;19:249–54.
Murray K, Retallick R, McDonald KR, Mendez D, Aplin K, Kirkpatrick P, et al. The distribution and host range of the pandemic disease chytridiomycosis in Australia spanning surveys from 1956 to 2007. Ecology. 2010;91:1557.
Nakamura M. Sex determination in amphibians. Semin Cell Dev Biol. 2009;20(3):271–82. https://doi.org/10.1016/j.semcdb.2008.10.003.
Nakano M, Hasunuma I, Okada R, Yamamoto K, Kikuyama S, Machida T, Kobayashi T. Molecular cloning of bullfrog D2 dopamine receptor cDNA: tissue distribution of three isoforms of D2 dopamine receptor mRNA. Gen Comp Endocrinol. 2010a;168:143–8.
Nakano M, Minagawa A, Hasunuma I, Okada R, Tonon M-C, Vaudry H, Yamamoto K, Kikuyama S, Machida T, Kobayashi T. D2 Dopamine receptor subtype mediates the inhibitory effect of dopamine on TRH-induced prolactin release from the bullfrog pituitary. Gen Comp Endocrinol. 2010b;168:287–92.
Narayan E, Hero J-M. Urinary corticosterone responses and haematological stress indicators in the endangered Fijian ground frog (*Platymantis vitiana*) during transportation and captivity. Aust J Zool. 2011;59:79–85.
Narayan EJ, Cockrem JF, Hero J-M. Urinary corticosterone metabolite responses to capture and captivity in the cane toad (*Rhinella marina*). Gen Comp Endocrinol. 2011;173:371–7.
Nilsson EE, Cloud JG. Rainbow trout chimeras produced by injection of blastomeres into recipient blastulae. Proc Natl Acad Sci U S A. 1992;89:9425–8.
Nilsson EE, Cloud JG. Cryopreservation of rainbow trout (*Oncorhynchus mykiss*) blastomeres. Aquat Living Resour. 1993;6:77–80.
O'Hanlon SJ, Rieux A, Farrer RA, Rosa GM, Waldman B, Bataille A, Kosch TA, Murray KA, Brankovics B, Fumagalli M. Recent Asian origin of chytrid fungi causing global amphibian declines. Science. 2018;360:621–7.
O'Brien ED, Salicioni AM, Cabada MO, Arranz SE. Vitellogenesis in Bufo arenarum: identification, characterization and immunolocalization of high molecular mass lipovitellin during oogenesis. Comp Biochem Physiol B Biochem Mol Biol. 2010;155:256–65.

Otoi T, Yamamoto K, Koyama N, Tachikawa S, Suzuki T. Bovine oocyte diameter in relation to developmental competence. Theriogenology. 1997;48:769–74.

Parsons JE, Thorgaard GH. Production of androgenetic diploid rainbow trout. J Hered. 1984;76:177–81.

Pearl E, Morrow S, Noble A, Lerebours A, Horb M, Guille M. An optimized method for cryogenic storage of Xenopus sperm to maximise the effectiveness of research using genetically altered frogs. Theriogenology. 2017;92:149–55.

Peng L-Y, Xiao Y-M, Liu Y. Effect of cryopreservation and short-term storage of Chinese giant salamander sperm. Acta Hydrobiol Sinica. 2011;35:325–31.

Peter RE, Lin H-R, Van Der Kraak G. Induced ovulation and spawning of cultured freshwater fish in China: advances in application of GnRH analogues and dopamine antagonists. Aquaculture. 1988;74:1–10.

Peyridieu J, Baudot A, Boutron P, Mazuer J, Odin J, Ray A, Chapelier E, Payen E, Descotes J. Critical cooling and warming rates to avoid ice crystallization in small pieces of mammalian organs permeated with cryoprotective agents. Cryobiology. 1996;33:436–46.

Popesku JT, Martyniuk CJ, Mennigen J, Xiong H, Zhang D, Xia X, Cossins AR, Trudeau VL. The goldfish (*Carassius auratus*) as a model for neuroendocrine signaling. Mol Cell Endocrinol. 2008;293:43–56.

Porter KR. Androgenetic development of the egg of *Rana pipiens*. Biol Bull. 1939;77:233–57.

Pounds JA, Crump ML. Amphibian declines and climate disturbance: the case of the golden toad and the harlequin frog. Conserv Biol. 1994;8:72–85.

Pounds JA, Fogden MPL, Campbell JH. Biological response to climate change on a tropical mountain. Nature. 1999;398:611–5.

Proaño B, Pérez OD. In vitro fertilizations with cryopreserved sperm of *Rhinella marina* (Anura: Bufonidae) in Ecuador. Amphibian Reptile Conserv. 2017;11:1–6.

Rall W, Fahy G. Vitrification: a new approach to embryo cryopreservation. Theriogenology. 1985a;23:220.

Rall WF, Fahy GM. Ice-free cryopreservation of mouse embryos at −196°C by vitrification. Nature. 1985b;313:573.

Rasar MA, Hammes SR. The physiology of the *Xenopus laevis* ovary. In: Liu XJ, editor. Xenopus protocols: cell biology and signal transduction. Totowa: Humana Press; 2006. p. 17–30.

Rienzi L, Balaban B, Ebner T, Mandelbaum J. The oocyte. Hum Reprod. 2012;27:i2–i21.

Roth TL, Szymanski DC, Keyster ED. Effects of age, weight, hormones, and hibernation on breeding success in boreal toads (*Bufo boreas boreas*). Theriogenology. 2010;73:501–11.

Routray P, Dash C, Dash SN, Tripathy S, Verma DK, Swain SK, et al. Cryopreservation of isolated blastomeres and embryonic stem-like cells of Leopard danio, *Brachydanio frankei*. Aquacult Res. 2010;41:579–89.

Rugh R. Experimental embryology techniques and procedures. Minneapolis: Burgess Publishing; 1962.

Ryder OA, Onuma M. Viable cell culture banking for biodiversity characterization and conservation. Annu Rev Anim Biosci. 2018;6:83–98.

Salgado Costa C, Trudeau VL, Ronco AE, Natale GS. Exploring antipredator mechanisms: new findings in ceratophryid tadpoles. J Herpetol. 2016;50:233–8.

Santymire R, Livieri T, Branvold-Faber H, Marinari P. The black-footed ferret: on the brink of recovery? In: Holt WV, Brown JL, Comizzoli P, editors. Reproductive sciences in animal conservation, vol. 753. New York: Springer; 2014. p. 119–34.

Sargent MG, Mohun TJ. Cryopreservation of sperm of *Xenopus laevis* and *Xenopus tropicalis*. Genesis. 2005;41:41–6.

Sarre SD, Ezaz T, Georges A. Transitions between sex-determining systems in reptiles and amphibians. Annu Rev Genomics Hum Genet. 2011;12(1):391–406. https://doi.org/10.1146/annurev-genom-082410-101518.

Scheele BC, Hunter DA, Grogan LF, Berger L, Kolby JE, McFadden MS, Marantelli G, Skerratt LF, Driscoll DA. Interventions for reducing extinction risk in chytridiomycosis-threatened amphibians. Conserv Biol. 2014;28:1195–205.

Scheele BC, Pasmans F, Skerratt LF, Berger L, Martel A, Beukema W, Acevedo AA, Burrowes PA, Carvalho T, Catenazzi A. Amphibian fungal panzootic causes catastrophic and ongoing loss of biodiversity. Science. 2019;363:1459–63.

Seong JY, Wang L, Oh DY, Yun O, Maiti K, Li JH, Soh JM, Choi HS, Kim K, Vaudry H. Ala/Thr201 in extracellular loop 2 and Leu/Phe290 in transmembrane domain 6 of type 1 frog gonadotropin-releasing hormone receptor confer differential ligand sensitivity and signal transduction. Endocrinology. 2003;144:454–66.

Sherman CD, Wapstra E, Uller T, Olsson M. Male and female effects on fertilization success and offspring viability in the Peron's tree frog, *Litoria peronii*. Austral Ecol. 2008;33:348–52.

Shishova NR, Uteshev VK, Kaurova SA, Browne RK, Gakhova EN. Cryopreservation of hormonally induced sperm for the conservation of threatened amphibians with *Rana temporaria* as a model research species. Theriogenology. 2011;75(2):220–32.

Silla A. Effect of priming injections of luteinizing hormone-releasing hormone on spermiation and ovulation in Gunther's toadlet, *Pseudophryne guentheri*. Reprod Biol Endocrinol. 2011;9:68.

Silla AJ. Artificial fertilisation in a terrestrial toadlet (*Pseudophryne guentherii*): effect of medium osmolality, sperm concentration and gamete storage. Reprod Fertil Dev. 2012; https://doi.org/10.1071/RD12223.

Silla AJ. Artificial fertilisation in a terrestrial toadlet (*Pseudophryne guentheri*): effect of medium osmolality, sperm concentration and gamete storage. Reprod Fertil Dev. 2013;25:1134–41.

Silla AJ, Byrne PG. The role of reproductive technologies in amphibian conservation breeding programs. Annu Rev Anim Biosci. 2019;7:499–519.

Silla AJ, McFadden MS, Byrne PG. Hormone-induced sperm-release in the critically endangered Booroolong frog (*Litoria booroolongensis*): effects of gonadotropin-releasing hormone and human chorionic gonadotropin. Conserv Physiol. 2019;7:coy080.

Skerratt LF, Berger L, Speare R, Cashins S, McDonald KR, Phillott AD, et al. Spread of chytridiomycosis has caused the rapid global decline and extinction of frogs. Ecohealth. 2007;4:125–34.

Skerratt LF, McDonald KR, Hines HB, Berger L, Mendez D, Phillott AD, et al. Application of the survey protocol for chytridiomycosis to Queensland, Australia. Dis Aquat Organ. 2010;92:117–29.

Skinner DC, Robinson JE. Luteinising hormone secretion from the perifused ovine pars tuberalis and pars distalis: effects of gonadotropin-releasing hormone and melatonin. Neuroendocrinology. 1997;66:263–70.

Smith LD, Ecker RE, Subtelny S. In vitro induction of physiological maturation in *Rana pipiens* oocytes removed from their ovarian follicles. Dev Biol. 1968;17:627–43.

Sotowska-Brochocka J. The stimulatory and inhibitory role of the hypothalamus in the regulation of ovulation in grass frog, *Rana temporaria*. Gen Comp Endocrinol. 1988;70:83–90.

Sotowska-Brochocka J, Licht P. Effect of infundibular lesions on GnRH and LH release in the frog, *Rana temporaria*, during hibernation. Gen Comp Endocrinol. 1992;85:43–54.

Sotowska-Brochocka J, Martyńska L, Licht P. Dopaminergic inhibition of gonadotropic release in hibernating frogs, *Rana temporaria*. Gen Comp Endocrinol. 1994;93:192–6.

Stuart SN, Chanson JS, Cox NA, Young BE, Rodrigues AS, Fischman DL, et al. Status and trends of amphibian declines and extinctions worldwide. Science. 2004;306:1783–6.

Stuart SN, Hoffmann M, Chanson JS, Cox NA, Berridge RJ, Ramani P, Young BE. Threatened amphibians of the world. Barcelona, Gland, and Arlington: Lynx Edicions, IUCN, and Conservation International; 2008.

Thorgaard G, Wheeler P, Fields R. Utilization of androgenesis for strain recovery from cryopreserved sperm. In: Tiersch TR, Green CC, editors. Cryopreservation in aquatic species. 2nd ed. Sorrento: World Aquaculture Society; 2005. p. 732–7.

Toro E, Michael SF. In vitro fertilization and artificial activation of eggs of the direct-developing anuran *Eleutherodactylus coqui*. Reprod Biol Endocrinol. 2004;2:60.

Trudeau VL. Neuroendocrine regulation of gonadotrophin II release and gonadal growth in the goldfish, *Carassius auratus*. Rev Reprod. 1997;2:55–68.

Trudeau VL, Somoza GM, Natale GS, Pauli B, Wignall J, Jackman P, Doe K, Schueler FW. Hormonal induction of spawning in 4 species of frogs by coinjection with a gonadotropin-releasing hormone agonist and a dopamine antagonist. Reprod Biol Endocrinol. 2010;8:36.

Trudeau VL, Schueler FW, Navarro-Martin L, Hamilton CK, Bulaeva E, Bennett A, Fletcher W, Taylor L. Efficient induction of spawning of Northern leopard frogs (*Lithobates pipiens*) during and outside the natural breeding season. Reprod Biol Endocrinol. 2013;11:14.

Turani B, Aliko V, Jona D. In vitro fertilization and maturation of Balkan water frog (*Pelophylax kurtmuelleri*, Gayda, 1940)—a case study in reproductive amphibian biotechnology. Int J Ecosyst Ecol Sci. 2015;5:557–60.

Unger S, Mathis A, Wilkinson R. A comparison of sperm health in declining and stable populations of Hellbenders (*Cryptobranchus alleganiensis alleganiensis* and *C.a. bishopi*). Am Midl Nat. 2013;170:382–92.

Upton R, Clulow S, Mahony M, Clulow J. Generation of a sexually mature individual of the amphibian *Litoria fallax* from cryopreserved testicular macerates: proof of capacity of cryopreserved sperm derived offspring to complete development. Conserv Physiol. 2018a;6(1):coy043. https://doi.org/10.1093/conphys/coy043.

Upton R, Clulow S, Seeto R, Wong L, Mahony M, Clulow J. Successful sperm cryopreservation and generated offspring of the endangered frog, *Litoria aurea*. Cryobiology. 2018b;85:148–9.

Uteshev V, Gakhova E. Gene cryobanks for conservation of endangered amphibian species. Russ J Herpetol. 2005;12(Suppl):233–4.

Uteshev VK, Melnikova EV, Kaurova SA, Nikitin VA, Gakhova EN, Karnaukhov VN. Fluorescent analysis of cryopreserved totipotent cells of amphibian embryos. Biofizika. 2002;47:539–45.

Uteshev VK, Shishova N, Kaurova SA, Manohkin AA, Gakhova EN. Collection and cryopreservation of hormonally induced sperm of pool frog (*Pelophylax lessonae*). Russ J Herpetol. 2013;20(2):105–9.

Uteshev VK, Gakhova EN, Kramarova LI, Shishova NV, Kaurova SA, Browne RK. Refrigerated storage of European common frog *Rana temporaria* oocytes. Cryobiology. 2018;83:56–9.

Vajta G, Holm P, Kuwayama M, Booth PJ, Jacobsen H, Greve T, Callesen H. Open pulled straw (ops) vitrification: a new way to reduce cryoinjuries of bovine ova and embryos. Mol Reprod Dev. 1998;51:53–8.

Vance CK, Julien A, Counsell K, Marcec R, Agcanas L, Tucker A, Kouba A. Amphibian art over the generations: frozen sperm offspring produce viable F2 generation. Cryobiology. 2018; 85:178.

Vanecek J. Inhibitory effect of melatonin on GnRH-induced LH release. Rev Reprod. 1999;4:67–72.

Vitt LJ, Caldwell JP. Herpetology: an introductory biology of amphibians and reptiles. San Diego: Academic Press; 2013.

Vu M, Trudeau VL. Neuroendocrine control of spawning in amphibians and its practical applications. Gen Comp Endocrinol. 2016;234:28–39.

Vu M, Weiler B, Trudeau VL. Time-and dose-related effects of a gonadotropin-releasing hormone agonist and dopamine antagonist on reproduction in the northern leopard frog (lithobates pipiens). Gen Comp Endocrinol. 2017;254:86–96.

Waggener WL, Carroll EJ. A method for hormonal induction of sperm release in anurans (eight species) and in vitro fertilization in *Lepidobatrachus* species. Dev Growth Differ. 1998;40:19–25.

Wake DB. Climate change implicated in amphibian and lizard declines. PNAS. 2007;104:8201–2.

Wake DB, Vredenburg VT. Are we in the midst of the sixth mass extinction? A view from the world of amphibians. Proc R Soc Lond Ser B Biol Sci. 2008;105:11466–73.

Wallace RA. Studies on amphibian yolk III. A resolution of yolk platelet components. Biochim Biophys Acta. 1963;74:494–504.

Wallace RA. Vitellogenesis and oocyte growth in nonmammalian vertebrates. In: Browder LW, editor. Oogenesis. Boston: Springer; 1985. p. 127–77.

Wang L, Bogerd J, Choi HS, Seong JY, Soh JM, Chun SY, Blomenröhr M, Troskie BE, Millar RP, Wen HY. Three distinct types of GnRH receptor characterized in the bullfrog. Proc Natl Acad Sci. 2001;98:361–6.

Wilczynski W, Lynch KS. Female sexual arousal in amphibians. Horm Behav. 2011;59:630–6.

Williams SE, Hoffman EA. Minimizing genetic adaptation in captive breeding programs: a review. Biol Conserv. 2009;142(11):2388–400.

Wolf DP, Hedrick JL. A molecular approach to fertilization: II. Viability and artificial fertilization of *Xenopus laevis* gametes. Dev Biol. 1971;25(3):348–59.

Wowk B, Leitl E, Rasch CM, Mesbah-Karimi N, Harris SB, Fahy GM. Vitrification enhancement by synthetic ice blocking agents. Cryobiology. 2000;40:228–36.

Wright P. Induction of ovulation in vitro in Rana pipiens with steroids. Gen Comp Endocrinol. 1961;1:20–3.

Yamaha E, Mizuno T, Hasebe Y, Yamazaki F. Chimeric fish produced by exchanging upper parts of blastoderms in goldfish blastulae. Fish Sci. 1997;63:514–9.

Yaron Z, Gur G, Melamed P, Rosenfeld H, Elizur A, Levavi-Sivan B. Regulation of fish gonadotropins. Int Rev Cytol. 2003;225:131–85.

Yasui GS, Fujimoto T, Sakao S, Yamaha E, Arai K. Production of loach (*Misgurnus anguillicaudatus*) germ-line chimera using transplantation of primordial germ cells isolated from cryopreserved blastomeres. J Anim Sci. 2011;89:2380–8.

Zhang T, Rawson DM. Feasibility studies on vitrification of intact zebrafish (*Brachydanio rerio*) embryos. Cryobiology. 1996;33:1–13.

Zhang RB, Werkman AS. Water and urea permeability properties of *Xenopus* oocytes: expression of mRNA from toad urinary bladder. Am J Physiol Cell Physiol. 1991;260:C26–34.

Zhang T, Isayeva A, Adams SL, Rawson DM. Studies on membrane permeability of zebrafish (*Danio rerio*) oocytes in the presence of different cryoprotectants. Cryobiology. 2005;50:285–93.

Zimkus B, Hassapakis C, Houck M. Integrating current methods for the preservation of amphibian genetic resources and viable tissues to achieve best practices for species conservation. Amphibian Reptile Conserv. 2018;12:1–27.

Zuccotti M, Piccinelli A, Marziliano N, Mascheretti S, Redi CA. Development and loss of the ability of mouse oolemma to fuse with spermatozoa. Zygote (Cambridge, England). 1994;2:333–9.

Chapter 15
Reproductive Science in Sharks and Rays

Linda M. Penfold and Jennifer T. Wyffels

Abstract Sharks and rays make up 96% of the class Chondrichthyes. They are among the most endangered of any taxa, threatened through habitat loss, overfishing and hunting for shark fin soup, traditional medicines or sport, and because many species are slow to mature and produce low numbers of offspring. Sharks and rays are ecologically and reproductively diverse, though basic knowledge of their reproductive physiology is lacking for many species. There has been a move towards non-lethal approaches of data collection in sharks and rays, especially with reproductive technologies such as ultrasound and hormone analysis. Additionally, technologies such as semen collection and artificial insemination are lending themselves to develop tools to manage small or closed populations, with cold-stored sperm being shipped between institutions to maximize genetic diversity in managed populations. The role of steroid hormones in elasmobranch reproduction appears broadly conserved, though heavily influenced by environmental cues, especially temperature. For this reason elasmobranchs are likely at risk of reproductive perturbations due to environmental changes such as ocean warming. Current reproductive technologies including computer assisted sperm assessments to study warming effects on sperm motility and intra-uterine satellite tags to determine birthing grounds will serve to generate data to mitigate anthropogenic changes that threaten the future of this vulnerable groups of fish.

Keywords Elasmobranch · Reproduction · Spermatozoa · Ultrasound · Endocrinology · Seasonality · Artificial insemination

L. M. Penfold (✉) · J. T. Wyffels
South-East Zoo Alliance for Reproduction & Conservation, Yulee, FL, USA
e-mail: Linda.Penfold@sezarc.com

P. Comizzoli et al. (eds.), *Reproductive Sciences in Animal Conservation*, Advances in Experimental Medicine and Biology 1200,
https://doi.org/10.1007/978-3-030-23633-5_15

1 Introduction

Chondrichthyans are one of the evolutionarily oldest and most ecologically diverse groups of animals in the world, found in rivers, lakes, estuaries and oceans (Compagno 2008; Rosa et al. 2010; Dulvy et al. 2014) and comprised of over 1250 species (Eschmeyer and Fong 2019). Sharks and rays are among the most endangered of taxa, with approximately 25% of the species threatened with extinction, and the conservation status of many others as yet unknown (Dulvy et al. 2014; Camhi 2008). Most elasmobranchs are slow to mature and produce low numbers of large offspring, rendering them particularly susceptible to extinction threats. The elasmobranch reproductive system is evolutionally and physiologically unique, yet compared to mammalians species, the physiology of reproduction for the vast majority of sharks and rays is still undescribed. Fertilization is exclusively internal, and relatively large sized offspring are born, in low numbers for many species, arguably rendering them more similar in reproduction to reptiles, birds and mammals than other fishes. Current findings have shown that elasmobranchs are either aseasonal and reproduce throughout the year, have a partially defined annual season with one or two peaks of offspring production or have defined annual, biennial or even triennial cycle (Wourms 1977; see frill shark, below). Reproductive modes in elasmobranchs are extraordinarily diverse and defined by the site of development (internal or external) and how the embryo receives nutrition during development (Fig. 15.1). Lecithotrophic reproductive modes include oviparity and yolk sac viviparity (Hamlett et al. 2005a). Matrotrophic modes of reproduction include mucoid and lipid histotrophy, placental viviparity, embryotrophy practiced by the tiger shark, *Galeocerdo cuvier*, oophagy and adelphophagy practiced by the sand tiger shark, *Carcharius taurus*, whereby the first embryo to hatch methodically kills and consumes any other developing embryos before continuing to feed on ovulated eggs (Conrath and Musick 2012; Maruska and Gelsleichter 2011; Castro et al. 2015). Generally, there is a common mode of reproduction for species within an order, though some species display multiple modes of nutrition throughout gestation (e.g. white shark, Sato et al. 2016). Nonetheless, while the general mode of reproduction is known for every order, it is still unknown for many species within the orders. The reproductive cycle of elasmobranchs includes a period of vitellogenesis and a gestation period that occur simultaneously or consecutively depending on the species (Castro 2009). Gestation lengths vary greatly from a few months (thorny stingray, *Dasyatis centroura*, Capapé 1993; marbled stingray, *Dasytis marmorata*, Capapé and Zaouali 1995) to up to 3.5 years (frilled shark, *Chlamydoselachus anguineus*, Tanaka et al. 1990).

Reproductive intervals (time between pregnancies) vary from a few days (Blackspotted smoothhound, *Mustelus punctulatus*, Saidï et al. 2009) to a year, such as the sand tiger shark (*C. taurus*), that undergoes a resting period of a year after a reproductively active year. Reproduction is generally aseasonal in deepwater, pelagic and tropical non-migratory species presumably because of relatively constant water temperatures; partially seasonal in a few prolonged oviparous species and seasonal

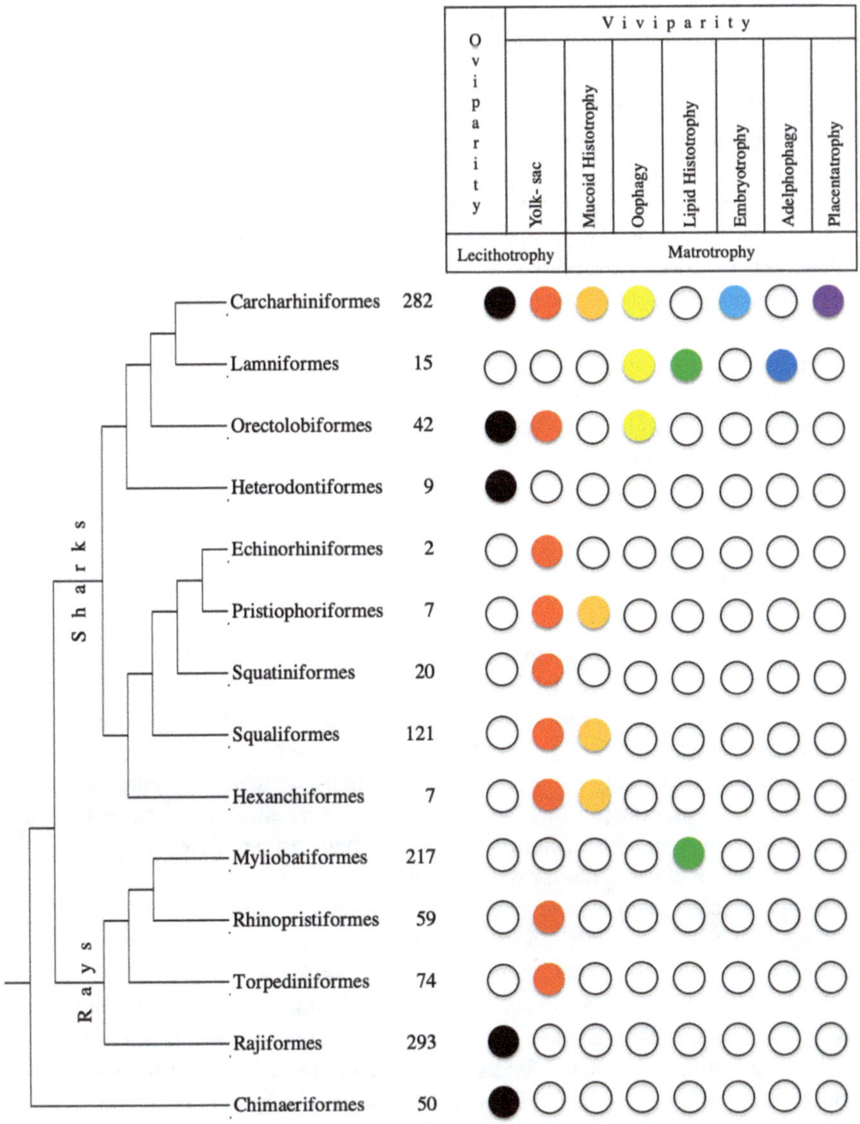

Fig. 15.1 Phylogenetic tree of Chondrichthyan orders and reproductive modes. Adapted from Buddle et al. (2018)

in coastal sub-tropical regions whereby seasonal migrations, thought to be linked to temperature and food availability see large numbers of animals moving along the coast in large aggregates (blacktip shark, *Carcharhinus limbatus*, sand tiger shark, *C. taurus*, cownose ray, *Rhinoptera bonasus*). But even within species, geographic location may affect reproductive parameters such as age and length at maturity (starspotted dogfish, *Mustelus manazo*, Yamaguchi et al. 2004) or the breeding periodicity where a tropical population may have a more protracted breeding season than the same species in a more temperate region (bonnethead shark, *Sphyrna tiburo*; Parsons and Grier 1992). There may also be regional variability in breeding periodicity, such as the finetooth sharks, *Carcharhinus isodon*, that demonstrates both a biennial cycle or an annual reproductive cycle in the Northern Gulf of Mexico (Driggers and Hoffmayer 2009), and the gummy shark, *Mustelus antarcticus*, demonstrates biennial or annual periodicity depending on whether the population is east or west of Kangaroo Island (Walker 2007). This is referred to as 'reproductive plasticity' and illustrates the complexity of elasmobranch reproductive strategies and challenges of classifying a species with a single reproductive strategy (Parsons and Grier 1992).

Parthenogenesis has been documented in reptiles, birds and more recently in several elasmobranch species. Although the majority of the cases were observed in aquaria (spotted eagle rays *Aetobatus narinari*, Harmon et al. 2016; swell *Cephaloscyllium ventriosum*, Feldheim et al. 2017a, b; bonnethead *S. tiburo*, Chapman et al. 2007; blacktip *Carcharinus limbatus*, Chapman et al. 2008; white-spotted bamboo, *Chiloscyllium plagiosum*, Feldheim et al. 2010 and zebra sharks, *Stegastoma fasciatum*, Dudgeon et al. 2017), there is a least one case noted in a wild smalltooth sawfish *Pristis pectinata* (Fields et al. 2015). The incidence of parthenogenesis in the wild is unknown, but may have conservation implications for small, isolated populations. Interestingly, several of the species in which parthenogenesis has been documented are coastal seasonal migrators (cownose ray, swell, bonnethead, blacktip sharks). These are generally considered to be more evolutionarily modern species, that congregate in large numbers, where finding a mate might be considered to be relatively straightforward. It remains to be seen whether parthenogenesis is an evolutionary conserved strategy from more ancient pelagic, deep water species, or a more recent evolutionary development. Irrespective, it is likely that other elasmobranchs will demonstrate this trait.

Many species demonstrate behavioral polyandry and multiple paternity. Polyandry has been directly observed in nurse sharks, *Ginglymostoma cirratum* (Carrier et al. 1994; Pratt and Carrier 2001), and confirmed through genetic analysis in the lemon shark, *Negaprion brevirostris* (Feldheim et al. 2001) and thornback ray, *Raja clavata* (Chevolot et al. 2007). The physical nature of elasmobranch mating involves biting the fins and flanks of the female, resulting sometimes in serious wounds that can threaten the female's health thus polyandry may be a risky strategy (Byrne and Avise 2012). Conversely, single paternity has been documented in species like the bonnethead shark (Chapman et al. 2004).

Another reproductive strategy that is common in some elasmobranchs is sperm storage in the female which has been documented in multiple species, including the dusky, *Carcharhinus obscurus*, blue, *Prionace glauca*, Atlantic sharpnose, *Rhizoprionodon terrenovae*, and scalloped hammerhead, *Sphyna lewini*, sharks, and is likely common in many elasmobranchs (Hamlett et al. 1998, 2005a, b; Pratt 1993). Interestingly, although sperm storage might be expected to be a useful post-copulation strategy to maximize paternity, studies have revealed that the wide range of polyandry in elasmobranch species indicates a relative inefficiency of this strategy.

Future studies investigating semen quality and sperm competition may generate information that could shed light on the practicality of this mechanism.

Sperm are quiescent until activated by exposure to an ionic medium change in the female tract and by dilution in various media in laboratory settings (Minamikawa and Morisawa 1996) whereon they display vigorous forward motility. Occasionally, reverse sperm motility is observed and has been described in banded houndshark, *Triakis scyllia* (Minamikawa and Morisawa 1996), white-spotted bamboo shark (*C. plagiosum*) and sand tiger shark, *C. taurus*. Reverse motility has also been reported in gastropods and fruitflies (Buckland-Nicks and Chia 1981; Köttgen et al. 2011; Shiba et al. 2014) where it has been suggested that it is another strategy for sperm competition. In elasmobranchs, reverse motility might be advantageous if sperm are stored, allowing sperm to undergo reorientation to allow better movement through the female tract.

The reproductive anatomy in the female consists of paired ovaries in ancient shark species (Hexanchiformes, Squaliformes), and a single functional right ovary in Heterodontiformes and the more evolutionarily modern, Orectolobiformes, Lamniformes, and Carcharhiniformes. Males have paired testes and fertilization is internal for all elasmobranchs. Deposition of semen inside the female is facilitated by the use of claspers, a modification of the median edge of the pelvic fin that lengthens and calcifies with sexual maturity. Maturity for males is determined by one or more of the following factors: total length in sharks and disc width in rays, the degree of calcification of the claspers, coiling of the vas deferens or epididymis and/or presence of semen in the ampullae or seminal vesicles. For females, shark length, ovarian follicle diameter and width of the uterus and nidamental gland are used to determine maturity in the absence of developing young or eggs. Reproductive hormone concentrations also have been used also to assess maturity (Awruch 2013; Sulikowski et al. 2006).

In spite of the notoriety of sharks and rays, little is known about their role in the environment. The impact of removal of apex predators (blacktip, dusky, great hammerhead, sandbar, tiger, great white etc.) on mesopredator (predominantly skates and rays) on prey population control indicates a critical role in maintaining ocean ecosystems (Myers et al. 2007). Disappearance of white sharks at Seal Island, False Bay, South Africa, coincided with appearance of allopatric apex predator, sevengill sharks (*Notorynchus cepedianus*) demonstrating a behavioral change associated with local and temporal species loss (Hammerschlag et al. 2019). Subtler impacts of disturbances in marine ecosystems, such as seasonal increases in certain species, may similarly have negative impacts on local species and (Dill et al. 2003; Heithaus 2001). Aquatic species in the wild are typically difficult to study and census accurately to generate data to determine their conservation status (McCully et al. 2013; Lauria et al. 2015), for example up to 30% of elasmobranch species in the Mediterranean are data deficient to even make a determination as to their relative abundance (Abdulla 2004) and combined with overfishing, entanglement in fishing gear, hunting for shark fin soup, traditional medicines or sport and habitat degradation, the future for many of these species is bleak.

Aquariums can play an important role in supporting the research of elasmobranch species, especially in understanding the basic reproductive biology, by providing animals in a managed environment for study and by having trained professional staff for animal handling, ultrasound, and biological sampling (Carrier et al. 2004). By studying the same individual over time the nuances of hormone changes can be correlated with, for example, observations of increased ova, uterus and nidamental or shell gland size for females and accumulation of semen for males that are otherwise 'averaged' for *in situ* populations that usually involve single timepoints for individuals. Many elasmobranch species, such as cownose rays, southern stingrays and bamboo sharks easily adapt to managed care and reproduce well. As such they may serve as model species, allowing intensive studying of the mechanisms of elasmobranch reproductive biology. Conversely, other species are not able to be held long-term in aquariums such as the great white shark, thought to be associated with their requirements of fast swimming of great distances in an open ocean (Ezcurra et al. 2012).

2 State of the Art

2.1 Non-lethal Methods of Reproductive Assessment

Historically, most reproductive data was collected using sacrificed animals, but now that increasing numbers of elasmobranch species are being identified as threatened or endangered, it is beneficial to limit euthanasia of animals when possible for study (Hammerschlag and Sulikowski 2011) and new approaches to studying elasmobranchs increasingly include non-lethal sampling methods. The move towards the development of non-lethal techniques, already widely employed in aquariums, such as ultrasound and short-term restraint for biological sampling has continued to advance the field of research in both wild and aquarium sharks and rays (Carrier et al. 2004). In wild bonnethead sharks (*S. tiburo*) ultrasound was proven useful by validation alongside dissection with 90.9% agreement to determine pregnancy and fecundity (Anderson et al. 2018). However, the accuracy of determining specific pregnancy stages depended on whether a curvilinear or transverse was probe was used. The broad usefulness of ultrasound has been demonstrated in studies that have characterized, diagnosed and monitored reproductive disease in southern stingrays (*Hypanus americanus*, Mylniczenko et al. 2019), studied folliculogenesis in the broadnose seven gill shark (*Notorhynchus cepedianus*; Daly et al. 2007), monitored fetal movement between uteri during gestation in tawny nurse sharks (*Nebrius ferrugineus*) (Tomita et al. 2018), assessed maturity in oviparous elasmobranchs (Whittamore et al. 2010), confirmed and defined biennial cycles and reproductive seasonality in female sand tiger sharks (Fig. 15.2) and the presence of semen within the ampullae of breeding sand tiger males (Fig. 15.3), and diagnosed pregnancy in wild sand tiger sharks (*C. taurus*; Christiansen, personal communication).

Best practices for ultrasounding rays and sharks now include a consistent and methodical use of the ultrasound probe to capture images of what can often be a

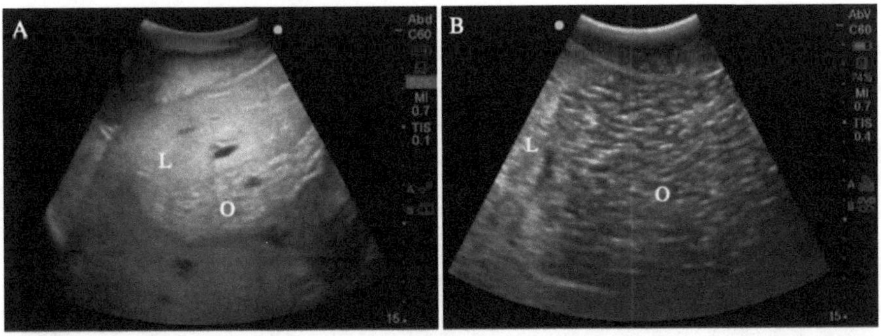

Fig. 15.2 Ultrasound images of the ovary (O) with surrounding liver (L) of sand tiger sharks, *Carcharias taurus*, for sharks (**a**) in a reproductive resting phase and (**b**) during active folliculogenesis

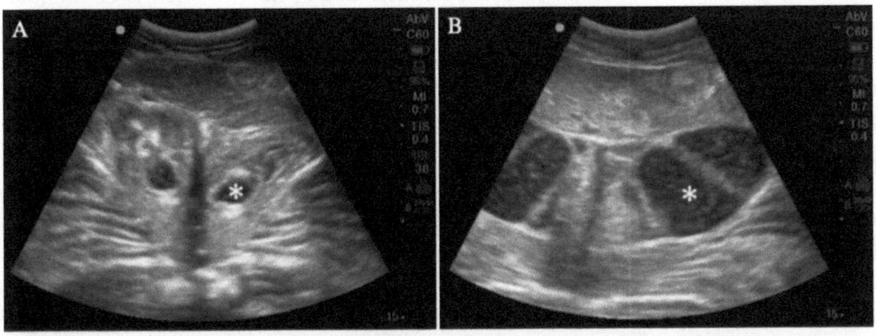

Fig. 15.3 Ultrasound images of the ampullae (∗) of sand tiger sharks, *Carcharias taurus*, for sharks (**a**) without semen and (**b**) with semen

large, active ovary in its entirety, along with complete views of the liver for potential body scoring. Also, the dorsoventral rotation of large sharks combined with the weight and size of the active ovary has resulted in gut torsion and/or ovarian artery tears resulting in death (Daly and Jones 2017). Where possible, reproductive exams on large gravid females might be accomplished by partially rolling the female to one side to access the ventral aspect of the shark, while keeping the female in the water to maintain buoyancy.

2.2 *Semen and Sperm*

Semen can be collected from elasmobranchs using a couple of techniques: The first is by applying manual pressure directed towards the distal end of the reproductive tract adjacent to the cloaca, appropriate for smaller species (cloudy catshark,

Scyliorhinus torazame; Masuda et al. 2003; ocellate river stingray *Potamotrygon motoro*, Dzyuba et al. 2019), and the second is by aspiration of semen from the ampullae by catheterization via the urinary papillae (Minamikawa and Morisawa 1996). Bamboo sharks grow to about 1 m, have teeth that are used for grasping and crushing, and can be handled fairly easily for semen collection. In contrast, a larger shark species such as the sand tiger shark, that reaches 2.5 m in length, and has extremely sharp teeth that point outwards, requires a skilled team of 5 or 6 individuals to restrain and monitor, to prevent risk of injury. Fortunately, as with many shark species, the sand tiger shark undergoes 'tonic immobility' when rotated dorsoventrally and stays in a trance-like condition until righted. Short-term (~20 min) restraint using tonic immobility is a useful technique for catheterization of the ampullae for semen collection (Fig. 15.4). While useful for handling males, caution is advised when using this technique on gravid females, There are multiple terms used to describe the reproductive anatomy of the elasmobranch reproductive tract, the epididymis is also referred to as the extra-testicular ducts, which is continuous with the ductus deferens, also described as the vas deferens. The ductus deferens continues to the seminal vesicles, also described as ampullae. The Leydig gland releases fluid into the both the epididymis and ductus deferens, and the ampullae open into a single urogenital sinus that opens into the cloaca via a urogenital papilla. Individual sperm move through the epididymis to form aggregates that are laterally aligned with the heads embedded in matrix, known as spermatozeugmata; (white-spotted bamboo sharks *C. plagiosum* Fig. 15.5) or completely encased within matrix to form ovoid shaped spermatophores (Pratt and Tanaka 1994; Hamlett et al. 2002).

The form, structure, and post-testicular development of elasmobranch spermatozoa has been well described in the Port Jackson shark (*Heterodontus portusjacksoni*; Jones et al. 1984) and sand tiger shark with the basic structure composed of a helical shaped head terminating in a small acrosome, with a midpiece and flagellum, and a cytoplasmic sleeve over the midpiece region (Tanaka et al. 1995; Fig. 15.6). The helical shaped head differs in length and numbers of gyres between species and is unrelated to the size of the species (Table 15.1).

The use and validation of combinations of fluorescent stains like SYBR-14® and propidium iodide (LIVE-DEAD stain, Molecular Probes) are useful in determining the plasma membrane integrity of elasmobranch sperm (Fig. 15.7) and the use of the lectin *Pisum sativum* (PSA) confirmed the presence of acrosin within acrosomal membranes at the tip of the sperm head (Fig. 15.8). Elasmobranch sperm has proven to be remarkably tolerant of cold storage and young have been produced in the brown banded bamboo shark *C. punctatum* after 8 h cold storage (Daly and Jones 2017) and 24 h cold storage (Lance Adams, personal communication) following overnight shipping. This demonstrates a practical approach to genetically managing isolated populations by moving gametes rather than translocating animals and is potentially highly useful for locally philopatric species.

Although semen quality is generally good in the aseasonal bamboo shark (*C. punctatum, C. plagiosum*) in aquaria, with high numbers of motile, sperm, that is not the case for the seasonally reproductive sand tiger shark (*C. taurus*). Semen and plasma collected quarterly from aquarium sand tiger sharks was found to have sperm

Fig. 15.4 Assessing the reproductive status of *in situ* female sand tiger sharks, *Carcharias taurus*, using ultrasound (**a**) boatside and (**b**) using a livewell and (**c**) collecting semen from a sand tiger shark in tonic immobilization

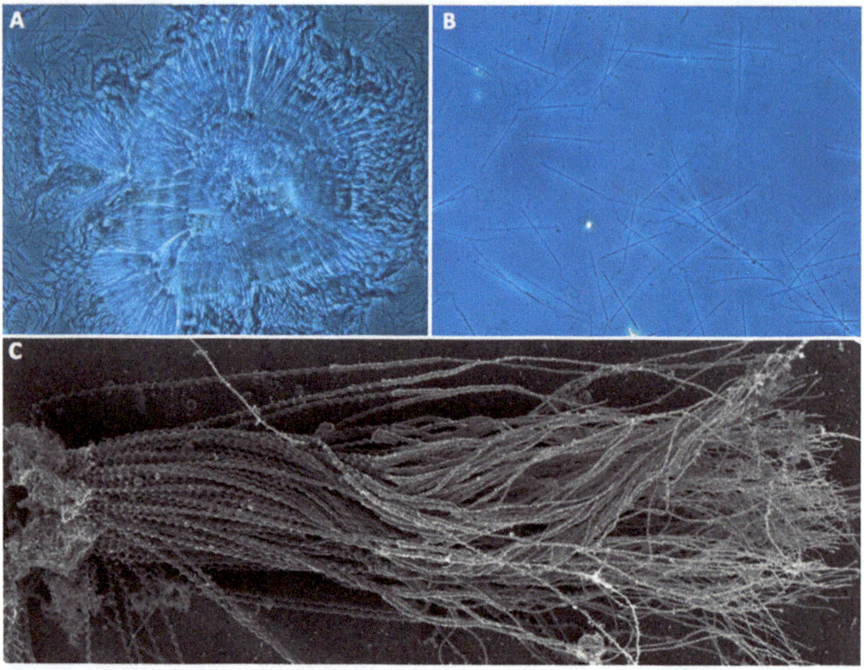

Fig. 15.5 *Chiloscyllium plagiosum* spermatozeugmata (**a**, **c**) and free spermatozoa (**b**)

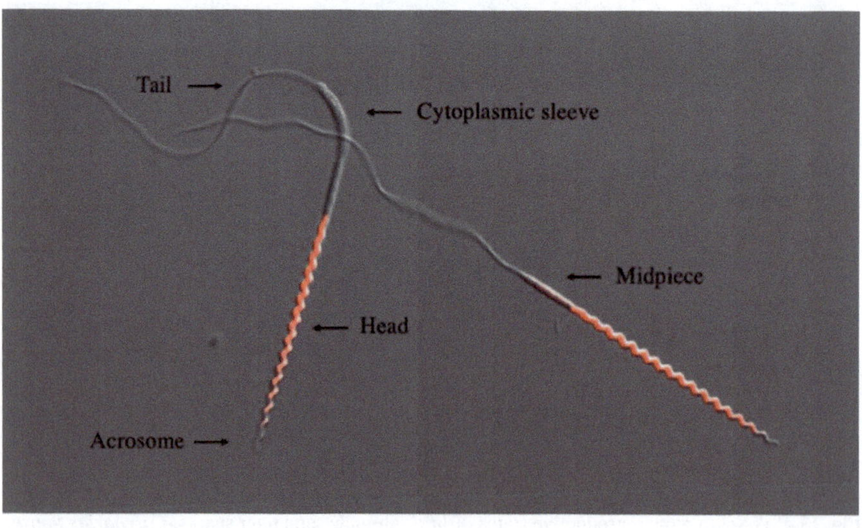

Fig. 15.6 Sand tiger shark, *Carcharias taurus*, sperm

Table 15.1 Elasmobranch sperm morphometrics

Species	Common name	TL/DW (cm)	N=	Length (μm)				Gyres
				Total	Head	Midpiece	Flagellum	Head and midpiece
Chiloscyllium punctatun	Whitespotted bamboo shark	65	4	154.7 ± 0.5	43.3 ± 0.3	17.4 ± 0.3	94 ± 0.5	28.7 ± 0.4
Chiloscyllium plagiosum	Brownbanded bamboo shark	65	5	178.3 ± 0.8	57.7 ± 0.2	19.9 ± 0.2	100.7 ± 0.6	34.8 ± 0.1
Hypanus americanus	Southern stingray	51	4	126.6 ± 0.3	41.5 ± 0.2	14.3 ± 0.7	70.8 ± 0.3	6.4 ± 0.04
Carcharias taurus	Sand tiger shark	190	5	110 ± 0.3	36.5 ± 0.2	8.9 ± 0.1	64.6 ± 0.2	20.2 ± 0.1

TL total length or DW disc width (cm); 10 sperm per fish measured and data presented as mean ± SEM

Fig. 15.7 Plasma membrane intact (green) and plasma membrane damaged (red) Brownbanded bamboo shark, *Chiloscyllium punctatum*, sperm stained with SYBR® 14 and propidium iodide

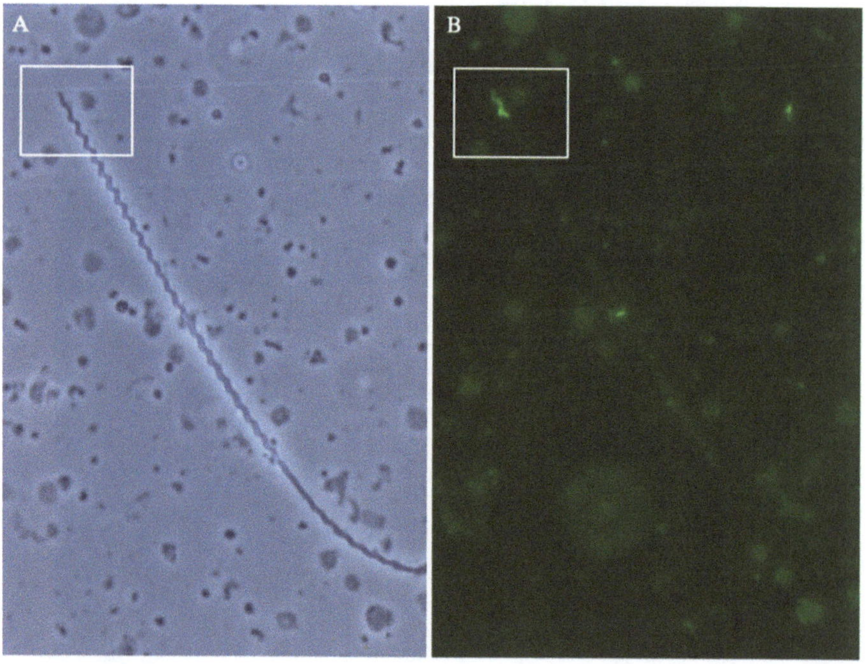

Fig. 15.8 Whitespotted bamboo shark, *Chiloscyllium plagiosum*, phase contrast (**a**) and fluorescent (**b**) micrographs of a sperm head with the acrosome stained with *Pisum sativum*

Table 15.2 Breeding season sperm parameters and blood plasma testosterone for aquarium and *in situ* sand tiger sharks, *Carcharias taurus*

	Aquarium			In situ		
	N=	Mean ± SEM	Range	N=	Mean ± SEM	Range
Total sperm (10^6)	12	292 ± 109	16–120	17	329 ± 77	2–988
Normal morphology (%)	10	78 ± 4.7	42–96	17	96 ± 1.3	87–100
Plasma membrane intact (%)	11	31 ± 9.2	0–74	15	68 ± 8	10–95
Motile (%)	12	24 ± 7.2	0–60	15	62 ± 7.7	10–95
Progressive motile (%)	12	3.7 ± 2.2	0–26	12	20 ± 3.9	0–50
Plasma testosterone (ng/ml)	12	10 ± 6.9	0.2–85	15	35 ± 7.6	8.6–103

with lower progressive motility and membrane integrity, and lower testosterone, respectively, compared to semen from wild sharks and independent from seasonal changes (Table 15.2). Thus, although the seasonal timing of sperm production was similar to that of wild sharks, semen quality of semen during peak breeding season remained inferior to that of wild counterparts. Current hypotheses include a lack of appropriate environmental cues and/or nutritional deficiencies that result in suboptimal sperm production and is a subject of ongoing investigation. Computer assisted sperm assessment (CASA) provides objective assessments of sperm motility parameters. CASA was used for the first time in an elasmobranch species, in the ocellate river stingray *P. motoro*, to study changes in sperm motility parameters that were likened to sperm capacitation and confirmed post-testicular sperm maturation (Dzyuba et al. 2019).

2.3 *Artificial Insemination*

Published accounts of artificial insemination in elasmobranchs report success in cloudy catshark, *Scyliorhinus torazame* (Masuda et al. 2003), white-spotted bamboo sharks, *Chilosycillium plagiosum* (Masuda et al. 2005) and clearnose skate *Raja eglanteria* (Luer et al. 2007). The technique also has been successful in brown banded bamboo sharks (*Chiloscyllium punctatum*; Melbourne SEALIFE aquarium, Australia and Lance Adams, personal communication) and zebra sharks (Lance Adams, personal communication). Much like avian species, oviductal or intrauterine insemination results in superior fertilization rates compared with cloacal insemination (Blanco et al. 2009; Daly and Jones 2017). In brief, insemination is achieved non-surgically by inserting a catheter through the cervix and delivering raw or extended semen. Large species that can be restrained or placed into tonic immobility can be inseminated without anesthesia (sand tiger shark, *C. taurus*). Smaller species, like the white-spotted bamboo sharks (*C. plagiosum*), can be anesthetized for catheter insertion.

The resilience to chilling and longevity of elasmobranch sperm lends itself well to artificial insemination. Offspring have been produced in the white-spotted bamboo shark (*C. plagiosum*) following oviductal insemination with fresh and 24 h cold stored (4 °C) semen, shipped overnight within the US (Lance Adams, personal communication). The ability to produce offspring following AI with cold-stored semen demonstrates a useful application for managing single sex groups in aquaria, or for moving genes between isolated populations to maintain genetic diversity. The recent discovery that several elasmobranch species, including bamboo sharks, can undergo parthenogenesis has required the concomitant use of genetic analysis to confirm paternity, which to date has proven useful and accurate (Kevin Feldheim, unpublished data) and in the case of the white-spotted bamboo shark (*C. plagiosum*) confirmed the paternity of the offspring.

2.4 Endocrinology

Elasmobranchs have been broadly used as research models to study the evolution of hormonal mechanisms and endocrine regulation (Callard et al. 1989; Baker et al. 2015), and excellent comprehensive reviews of the hormones governing reproduction in elasmobranchs are found in Maruska and Gelsleichter (2011), Gelsleichter and Evans (2012) and Awruch (2013). However, the precise role and mechanism of action of reproductive hormones is unclear for many elasmobranch species. The recognition that hormone analysis could confirm and eventually replace gross morphological changes (Sulikowski et al. 2007) set the path for increased non-lethal sampling and better understanding of seasonal profiles (Awruch et al. 2008) and habitat utilization (Sulikowski et al. 2016), but there is still a general lack of information for many elasmobranch species.

Reproductive steroids are generally analyzed from blood products, usually plasma or serum and historically has been completed using blood plasma or serum and radioimmunoassay (RIA; Sulikowski et al. 2007) or more recently with enzyme-linked antibodies (EIA) (Nozu et al. 2018; Mylniczenko et al. 2019) or chemiluminescent-linked antibodies (CLIA; Shields 2018). Blood is collected from the brachial or tail vein in rays and from the caudal vein, pectoral fin radial vein or artery or dorsal cutaneous vein in larger elasmobranchs (Ueda et al. 2017). Blood sampling is usually accomplished from the animal under restraint or anesthesia but in some species (whale shark *Rhincodon typus*, sand tiger shark *C. taurus*), blood has been collected without restraint while swimming beside the animal using two syringes connected by a 3-way stopcock and extension set allowing blood collection to start in one syringe and then switching the stopcock to allow a fresh uncontaminated blood sample to be collected in the second syringe (Ueda et al. 2017). There is little information on fecal monitoring in elasmobranchs though it has been used with success in the epaulette shark (*Hemiscyllium ocellatum*) to measure corticosterone by high performance liquid chromatography (HPLC; Karsten and Turner 2003).

Steroid hormones also have been measured in yolk (McComb et al. 2005), uterine fluid and follicles (Fasano et al. 1992) and muscle (Prohaska et al. 2013).

Unlike terrestrial mammals where feces and urine can be collected non-invasively for hormone profiling, this type of sampling is challenging for elasmobranchs and better suited for aquarium animals, although there is some debate as to whether a muscle biopsy, gathered without removing the animal form the water is more or less stressful than blood sampling which usually requires removal from the water (Prohaska et al. 2013). Unlike terrestrial mammals where muscle biopsy would be considered a more invasive procedure, rapid wound healing combined with regionally thickened skin designed for biting and holding during mating in many elasmobranchs suggests this may indeed be an alternative for occasional sampling.

Development of EIAs using serum and plasma in some stingray species (southern stingray *Hypanus americanus*, cownose ray *R. bonasus*) has proven challenging (Cayman Adams, James Gelsleichter, personal communication) and might be attributed to sex hormone binding proteins that interfere with the assay and have a high affinity for steroid hormones (Idler and Freeman 1969) and the presence of low molecular weight albumin-like plasma proteins that aggregate or dissociate depending on urea concentrations as described for the common stingray *Dasyatis pastonaca* (Andreeva and Fedorov 2010) and the cownose ray *R. bonasus* (Cray et al. 2015). As a result, serum and plasma benefits from some type of pretreatment such as ether extraction or addition of a dissociation reagent. Conversely, steroid analysis in the sand tiger shark appears to be without similar complications (Cayman Adams, personal communication).

Antibody-independent techniques such as liquid chromatography coupled with tandem mass spectrometry (LC-MS/MS) overcome the need for hormone-specific antibodies and potential cross-reactivity by measuring ionized hormones directly and increasingly are considered a gold standard. LC-MS/MS is useful to validate more feasible and inexpensive antibody-based assays suited for routine endocrine monitoring. For example, testosterone was detected in plasma from cownose rays using RIA and found to increase in early gestation (Sheldon et al. 2018) but testosterone *per se* was not detected using LC-MS/MS (Table 15.3; Erin Legacki, unpublished data)

Table 15.3 Testosterone concentration (ng/ml) range for elasmobranchs using EIA and LC-MS/MS

				Females					
	Males			Non-pregnant			Pregnant		
Species	N=	LC-MS/MS	EIA	N=	LC-MS/MS	EIA	N=	LC-MS/MS	EIA
Carcharias taurus	34(65)	0.3–61.6	0.1–103.4	16(50)	0–33.1	0–72.3	4(4)	0.8–1.5	5.4–65.9
Rhinoptera bonasus	5(5)	16.8–99.6		9(15)	0		7(13)	0	
Carcharhinus melanopterus	3(9)	3.7–434.1		3(40)	0–19.1		2(3)	0–1.2	

LC-MS/MS liquid chromatography tandem mass spectrometry, *EIA* enzyme-linked immunoassay N= animals (samples)

suggesting another cross-reacting androgen was detected in the RIA. Concentrations of testosterone in male and female sand tiger sharks using EIA were twice or more as high as measured by LC-MS/MS (Table 15.3). Reasons for the discrepancy in measurements from female sand tiger sharks are not understood but may include interfering proteins and/or cross-reacting hormones for this species. Among elasmobranchs, the wide range of reproductive modes makes generalization of hormonal control of the reproductive system challenging and species-specific endocrine patterns should be expected. For example, when measured using LC-MS/MS, testosterone was nearly four times higher for male blacktip (*Carcharhinus melanopterus*) compared to male sand tiger or male cownose rays (Table 15.3; Erin Legacki, unpublished data).

Androgens and estrogens seem to be the most important steroid hormones and are correlated with seasonal reproductive changes such as folliculogenesis and testicular recrudescence and reproductive behaviors (Henningsen et al. 2008; Tricas et al. 2000; Parsons and Grier 1992; Heupel et al. 1999). The regulation of vitellogenesis by estradiol 17ß has been established in several vertebrate species and seasonal changes likely upregulate estrogen receptors and estrogen stimulates the hepatic cells to synthesize and secrete vitellogenin. Testosterone is broadly implicated in the onset of puberty, as well as testis development and spermatogenesis. Appropriate environmental cues may be drivers of the hypothalamus-pituitary-gonadal (HPG) axis, and testosterone has been shown to decrease in captive round stingrays maintained at 25 °C but increases to concentrations similar to their wild counterparts at temperatures of 18–20 °C (Mull et al. 2010). Similarly, testosterone has been shown to be higher in captive male sand tiger sharks that receive natural environmental cues, in the form of seasonal water temperature changes and natural light cycles, compared to those that do not. Semen quality (higher percent motility) also is better in wild sand tiger sharks, suggesting an environmentally driven reproductive cascade that has management implications for institutions hoping to breed sand tiger sharks. In contrast, reproductive hormones in the seasonally breeding sandbar (*Carcharhinus plumbaeus*) and lemon (*Negaprion brevirostris*) sharks, were found to be similar to those of their wild counterparts (Rasmussen and Murro 1992). In the bonnethead shark (*S. tiburo*) highest concentrations of dihydrotestosterone were observed during mating and preovulation periods, preovulatory rises in testosterone and estradiol were measured and highest concentrations of progesterone were measured following ovulation (Manire et al. 1995). Longitudinal monitoring of male sand tiger sharks showed that male sand tiger sharks have a hierarchy and that dominant males have higher testosterone. Seasonal increases in testosterone also have been observed in the spring in one female, in agreement with the onset of breeding season (Henningsen et al. 2008). Seasonal changes in testosterone were also observed in male cownose rays while increasing estradiol concentrations were seen throughout gestation in female cownose rays (Sheldon et al. 2018). Thyroid hormones, triiodothyronine (T3) and thyroxine (T4) are highly conserved among vertebrates and cycles of thyroid activity associated with reproductive seasonality have been observed in elasmobranchs (McComb et al. 2005; Volkoff et al. 1999; Woodhead 1966). Differences between *in situ* and aquarium sand tiger sharks

(*C. taurus*) for T3 and T4 have been observed (Cayman Adams, personal communication), with higher T3 and lower T4 than *in situ* males that may contribute to the general poorer reproductive response in aquarium sharks.

3 Influence of Environmental Change

Ocean warming, together with ocean acidification caused by atmospheric CO_2 entering the water and lowering pH, is expected to drive profound changes to elasmobranch diversity and biogeography. Empirical studies based on current and projected ocean warming and synergistic effects of acidification, have shown that in embryos of the little skate, *Leucoraja erinacea*, increased metabolic costs, decreased body condition (that puts young at risk of predation through earlier feeding behavior), and decreased survival are correlated with increased temperatures and acidification (Di Santo 2015). For some shark species, adults may be more thermally and pH tolerant, while juveniles are more susceptible to environmental changes as demonstrated in the brown banded bamboo shark *C. punctatum* where hatchlings had lower body condition and increased mortality rates after 30 days, and adults were unaffected (Rosa and Seibel 2008; Rosa et al. 2014). Temperature has been directly correlated with embryonic growth rate in the dogfish, *Scyliorhinus canicula* (Harris 1952) and it has been suggested that elasmobranchs select warmer waters to increase embryo growth to enhance reproductive success (tiger shark *Galeocerdo cuvier*, Sulikowski et al. 2016, Atlantic stingray *Hypanus sabinus*, Wallman and Bennett 2006; round stingray *Urobatis halleri*, Mull et al. 2010; Jirik and Lowe 2012). Elasmobranchs that are more thermally tolerant and adaptable are likely to outcompete species that are not, especially for philopatric species.

Additionally, some sharks employ thermal gradients possibly to exploit prey availability, though whether the movement is physiologically based to take advantage of thermal conditions or whether it coincides with prey abundance and availability is yet unknown. Examples include the Australian Tiger shark, *G. cuvier*, and US sand tiger shark, *C. carcharinus*. Nonetheless, the effects of large sharks on local prey populations has been well described (Heithaus 2001), and potential failure of apex predators to undergo thermal location changes may negatively impact local prey populations.

Though climate change and ocean warming is attributed to human impacts, some anthropogenic change could be considered useful for example, man-made reefs and shipwrecks off the coast of North Carolina provide habitat for diverse fishes, invertebrates and algae and are popular locations to observe aggregations of sand tiger sharks. The annual congregation of large numbers of sand tiger sharks at these wrecks and recent discovery of site fidelity (Paxton et al. 2019), demonstrates rapid adjustment to a changing environment, although the specific reasons as to when and why sand tiger sharks congregate there (e.g. food resources, mating grounds) is an area of active research.

4 Priorities for the Future

A better understanding of the role of environmental cues on elasmobranch reproductive biology will be important to fully appreciate how environmental change might impact the future of stingrays and sharks. There could be an increased role for aquariums in studying elasmobranch species, leveraging skills in animal handling, husbandry and veterinary medicine, not to mention that added advantage of public engagement and education for conservation. Aquarium and field-based research complement each other and integration will be beneficial and necessary to understand reproduction for many species.

The effects of ocean warming on stingray and shark physiology is a potential area for future research for reproductive biologists. Sperm motility is temperature dependent and in teleost studies, the percentage of motile sperm and the total duration of forward movement decreased over time with increased temperature (Alava and Cosson 2005). The implication of potential increased sperm motility caused by ocean warming on stingray and shark fertilization is not known, but an expected risk might include a decreased window of opportunity for fertilization. CASA is one mechanism to quantitatively collect data addressing the question as to whether elasmobranch sperm motility will be affected by ocean warming. Basic studies documenting sperm motility and viability parameters will provide immediate indications as to the potential impact of warming temperatures on shark and ray sperm physiology. Annual documentation of average sperm parameters for wild elasmobranch species displaying site fidelity might allow long-term monitoring of reproductive health.

Cold tolerance and longevity of elasmobranch sperm and success in artificial inseminations indicates there may be practical value in the future for managing fragmented populations by moving genes from one area to another via AI to maintain gene diversity. Cryopreservation of elasmobranch sperm is not widely practiced and barriers such as membrane phase transitions at critical temperatures is currently hindering development of protocols for freezing shark sperm (Linda Penfold, unpublished results; Jon Daly, unpublished results).

Emerging technology such as ‘birthing tags’ (J. A. Sulikowski and Lotek wireless, unpublished), intrauterine satellite tags designed to be inserted into the uterus of a pregnant female that are expelled when young are born, is an emerging tool hoped to shed light on birthing grounds for species such as the tiger shark *G. cuvier* and sand tiger shark *C. taurus* for which the fundamental question ‘where do the females give birth’, is still unknown. This exciting technology is not without concerns as to the impact on the female and young, so this an example of where an aquarium shark that can be carefully monitored over time, might serve as a model for its wild counterparts. Emerging technologies may often require integration of several non-lethal techniques such as ultrasound and birth tags, to effectively monitor reproduction. The continued study of anthropogenic effects on stingrays and sharks will hopefully help to secure these populations for the future, maintaining a critical balance in ocean ecosystem health.

Well documented migrations (cownose ray *R. bonasus*; Schwartz 1990; great white shark *C. carcharius*; Bonfil et al. 2005), together with observations of site fidelity (sand tiger shark, *C. taurus*; Paxton et al. 2019) suggest these migrations may be resource driven to find food and mates and/or influenced by natal philopatry (lemon shark *Negaprion brevisrostris*, Feldheim et al. 2014; smalltooth sawfish *P. pectinata* Feldheim et al. 2017a, b) and may be similarly environmentally driven by temperature or photoperiod (Smith and Merriner 1986; Blaylock 1993; Luer et al. 2007).

Acknowledgements There are numerous elasmobranch scientists who have been working in this field for far longer than us and we are grateful to them all for their work that has influenced our studies. We are also grateful to Jim Gelsleichter and James Sulikowski for their thoughtful edits and additions to this manuscript. We are especially indebted to the combined efforts of the dedicated staff of AZA institutions that have funded and supported elasmobranch research, especially the Sand Tiger Shark Consortium [North Carolina Aquariums, Florida Aquarium, Georgia Aquarium and affiliate Marineland Dolphin Adventure, Ripley's Aquariums, New York Aquarium, Norwalk Aquarium, Aquarium of the Pacific, Adventure Aquarium, National Aquarium, Shark Reef at Mandalay Bay, Jenkinson's Aquarium, Minnesota Zoo and Aquariums and Omaha's Henry Doorly Zoo]. We are grateful for *in situ* samples collected through collaboration and support from North Carolina Aquariums, Ripley's Aquariums, the National Aquarium and the Cooperative Atlantic States Pupping and Nursery Survey (COASTSPAN) and Cooperative Shark Tagging Program (CSTP) administered by the Apex Predators Program (APP). We especially thank the Morris Animal Foundation and SeaWorld/Busch Gardens Conservation Fund for critical funding to support this research.

References

Abdulla A. (2004). Status and conservation of sharks in the Mediterranean Sea. IUCN Technical paper, 7 p.

Alava SMH, Cosson J. Sperm motility in fishes. I Effects of temperature and pH: a review. Cell Biol Int. 2005;29:101–10.

Anderson B, Belcher C, Slack J, Gelsleichter J. Evaluation of the use of portable ultrasonography to determine pregnancy status and fecundity in bonnethead shark (*Sphyrna tiburo*). Fish Biol. 2018;93:1163–70.

Andreeva AM, Fedorov RA. Features of the organization of low molecular weight proteins from the blood and tissue fluid of the common stingray *Dasyatis pastinaca* (Chondroichthyes: Trygonidae). Russ J Mar Biol. 2010;36:469–72.

Awruch C. Reproductive endocrinology in chondrichthyans: the present and the future. Gen Comp Endocrinol. 2013;192:60–70.

Awruch CA, Frusher SD, Pankhurst NW, Stevens JD. Non-lethal assessment of reproductive characteristics for management and conservation of sharks. Mar Ecol Prog Ser. 2008;355:277–85.

Baker ME, Nelson DR, Studer RA. Origin of the response to adrenal and sex steroids: roles of promiscuity and co0evolution of enzymes and steroid receptors. J Steroid Biochem Mol Biol. 2015;151:12–24.

Blanco JM, Wildt DE, Höfle U, Voelker W, Donoghue AM. Implementing artificial insemination as an effective tool for *ex situ* conservation of endangered avian species. Theriogenology. 2009;71:200–13.

Blaylock RA. Distribution and abundance of the cownose ray *Rhinoptera bonasus*, in the lower Chesapeake Bay. Estuaries. 1993;16:255–63.

Bonfil R, Meÿer M, Scholl MC, Johnson R, O'Brien S, Oosthuizen H, Swanson S, Kotze D, Paterson M. Transoceanic migration, spatial dynamics, and population linkages of white sharks. Science. 2005;310:100–3.

Buckland-Nicks JA, Chia FS. Locomotion of the filiform sperm of Littorina (Gastropoda, Prosobranchia). Cell Tissue Res. 1981;219:27–39.

Buddle AL, Van Dyke JU, Thonpson MB, Simpfendorfer CA, Whittington CM. Evolution of placentotrophy: using viviparous sharks as a model to understand vertebrate placental evolution. Mar Freshw Res. 2018:1–17.

Byrne RJ, Avise JC. Genetic mating system of the brown smoothhound shark (Mustelus henlei), including a literature review of multiple paternity in other elasmobranch species. Mar Biol. 2012;159:749–56.

Callard IP, Klosterman LL, Sorbera LA, Fileti LA, Reese JC. Endocrine regulation of reproduction in elasmobranchs: archetype for terrestrial vertebrates. J Exp Zool Suppl. 1989;2:12–22.

Camhi MD. Conservation status of pelagic elasmobranchs. In: Camhi MD, Pikitch EK, Babcock EA, editors. Sharks of the open ocean: biology, fisheries and conservation: Blackwell; 2008. https://doi.org/10.1002/9781444302516.

Capapé C. New data on the reproductive biology of the thorny stingray, Dasytis centroura, from off the Tunisian coasts. Environ Biol Fish. 1993;38:73–80.

Capapé C, Zaouali J. Reproductive biology of the marbled stingray, *Dasytis marmorata*, in Tunisian waters. J Aquaric Aquat Sci. 1995;7:108–19.

Carrier JC, Pratt HL, Martin LK. Group Reproductive Behaviors in Free-Living Nurse Sharks, Ginglymostoma cirratum. Copeia 1994;3:646.

Carrier JC, Murru FL, Walsh MT, Pratt HL. Assessing reproductive potential and gestation in nurse sharks (*Ginglymostoma cirratum*) using ultrasonography and endoscopy: an example of bridging the gap between field research and captive studies. Zoo Biol. 2004;22:179–87.

Castro JI, Sato K, Bodine AB. A novel mode of embryonic nutrition in the tiger shark, *Galerocerdo cuvier*. Mar Biol Res. 2015;12:200–5.

Chapman DD, Prodohl PA, Gelsleichter J, Manire CA, Shivji MS. Predominance of genetic monogamy by females in a hammerhead shark, *Sphyrna tiburo*: implications for shark conservation. Mol Ecol. 2004;13:1965–74.

Chapman DD, Shivji MS, Louis E, Sommer J, Fletcher H, Prodöhl PA. Virgin birth in a hammerhead shark. Biol Lett. 2007;3:425–7.

Chapman DD, Firchau B, Shivji MS. Parthenogenesis in a large-bodied requiem shark, the blacktip Carcharhinus limbatus. J Fish Biol. 2008;73:1473–7.

Chevolot M, Ellis JR, Rijnsdorp AD, Stam WT, Olsen JL. Multiple paternity analysis in the thornback ray *Raja clavata*. J Hered. 2007;98:712–5.

Compagno LJV. Pelagic elasmobranch diversity. In: Camhi MD, Pikitch EK, Babcock EA, editors. Sharks of the open ocean: biology, fisheries and conservation. Oxford: Blackwell; 2008. p. 14–22.

Conrath CL, Musick JA. Reproductive biology of elasmobranchs. In: Carrier JC, Musick JA, Hiethaus MR, editors. Biology of sharks and their relatives. Boca Raton: CRC Press; 2012. p. 291–312.

Cray C, Rodriguez M, Field C, McDermott A, Leppert L, Clauss T, Bossart GD. Protein and cholesterol electrophoresis of plasma samples from captive cownose ray *Rhinoptera bonasus*. J Vet Diagn Invest. 2015;27:688–95.

Castro JI (2009). Observations on the reproductive cycles of some viviparous North American sharks. Aqua, Int J Ichthy. 2009;15:205-222.

Daly J, Jones R. The use of reproductive technologies in breeding programs for elasmobranchs in aquaria. In: The Elasmobranch Husbandry Manual II: recent advances in the care of sharks, rays and their relatives. Columbus: Ohio Biological Survey; 2017. p. 363–74.

Daly J, Gunn I, Kirby N, Jones R, Galloway D. Ultrasound examination and behavior scoring of captive broadnose sevengill sharks, *Notorhynchus cepedianus*. Zoo Biol. 2007;26:331–43.

Dill LM, Heithaus MR, Walters CJ. Behaviorally mediated indirect interactions in marine communities and their conservation implications. Ecology. 2003;85:1151–7.

Driggers WB III, Hoffmayer ER. Variability in the reproductive cycle of finetooth sharks, *Charcharhinus isodon*, in the Northern Gulf of Mexico. Copeia. 2009;2009:390–3.

Dudgeon CL, Coulton L, Bone R, Ovenden JR, Thomas S. Switch from sexual to parthenogenetic reproduction in a zebra shark. Sci Rep. 2017;7:40537.

Dulvy NK, Davidson LNK, Kyne PM, Simpfendorfer CA, Harrison LR, Jk C, Fordham SV. Ghosts of the coast: global extinction risk and conservation of sawfishes. Aquat Conserv Mar Freshwat Ecosyst. 2014;26:134–53. https://doi.org/10.1002/aqc.2525.

Dzyuba V, Ninhaus-Silveira A, Kahanec M, Veríssimo-Silveira R, Rodina M, Holt WV, Dzyuba B. Sperm motility in ocellate river stingrays: evidence for post-testicular sperm maturation and capacitation in Chondrichthyes. J Zool. 2019;307:9–16.

Eschmeyer WN, Fong JD. Species by family/subfamily. 2019. http://researcharchive.calacademy.org/research/ichthyology/catalog/SpeciesByFamily.asp. Accessed 24 Feb 2019.

Ezcurra JM, Lowe CL, Mollet HF, Ferry LA, O'Sullivan JB. Captive feeding and growth of young-of-the-year white sharks, Charcharodon charcharias, at the Monterey Bay Aquarium. In: Domeier ML, editor. Global perspectives on the biology and life history of the white shark. Boca Raton: CRC Press; 2012.

Fasano S, D'Antonio M, Pierantoni R, Chieffi G. Plasma and follicular tissue steroid levels in the elasmobranch fish, *Torpedo marmorata*. Gen Comp Endocrinol. 1992;85:327–33.

Feldheim KA, Gruber S, Ashley MV. Multiple paternity of a lemon shark litter (Chondrichthyes: Carcharhinidae). Copeia. 2001;3:81–6.

Feldheim KA, Chapman DD, Sweet D, Fitzpatrick S, Prodöhl PA, Shivji MS, Snowden B. Shark virgin birth produces multiple viable offspring. J Hered. 2010;101:374–7.

Feldheim KA, Gruber S, Di Battista J, Babcock E, Kessel S, Hendry A, Pikitch E, Ashley MV, Chapman DD. Two decades of genetic profiling yields first evidence of natal philopatry and long-term fidelity to parturition sites in sharks. Mol Ecol. 2014;23:110–7.

Feldheim KA, Clews A, Henningsen A, Todorov L, McDermott C, Meyers M, Bradley J, Pulver A, Anderson E, Marshall A. Multiple births by a captive swellshark *Cephaloscyllium ventriosum* via facultative parthenogenesis. J Fish Biol. 2017a;90:1047–53.

Feldheim KA, Fields AT, Chapman DD, Scharer RM, Poulakis GR. Insights into reproduction and behavior of the smalltooth sawfish *Pristis pectinata*. Endanger Species Res. 2017b;34:463–71.

Fields AT, Feldheim KA, Poulakis GR, Chapman DD. Facultative parthenogenesis in a critically endangered wild vertebrate. Curr Biol. 2015;25:R439–47.

Gelsleichter J, Evans AN. Hormonal regulation of elasmobranch physiology. In: Carrier JC, Musick JA, Heithaus MR, editors. Biology of sharks and their relatives. Boca Raton: CRC Group, Taylor and Francis; 2012.

Hamlett WC, Knight DP, Koob TJ, Jezior M, Luong T, et al. Survey of oviducal gland structure and function in elasmobranchs. J Exp Zool. 1998;282:399–420.

Hamlett WC, Reardon M, Clark J, Walker TI. Ultrastructure of sperm storage and male genital ducts in the elephant fish, *Callorhynchus milli*. J Exp Zool. 2002;292:111–28.

Hamlett WC, Kormanik G, Storrie M, Stevens B, Walker TI. Chondrichthyan parity, lecithotropy and matrotrophy. In: Hamlett WC, editor. Reproductive biology and phylogeny of Chondrichthyes. Enfield: Science Publishers; 2005a. p. 395–434.

Hamlett WC, Knight DP, Periera FTV, Steele J, Sever DM. Oviducal glands in Chondrichthyans. In: Hamlett WC, editor. Reproductive biology and phylogeny of Chondrichthyes. Enfield: Science Publishers; 2005b. p. 301–35.

Hammerschlag N, Sulikowski J. Killing for conservation: the need for alternatives for lethal sampling of apex predatory sharks. Endanger Species Res. 2011;14:135–40.

Hammerschlag N, Williams L, Fallows M, Fallows C. Disappearance of white sharks leads to the novel emergence of an allopatric apex predator, the sevengill shark. Sci Rep. 2019;9:1–6.

Harmon TS, Kamerman TY, Corwin AL, Sellas AB. Consecutive parthenogenetic births in a spotted eagle ray Aetobatus narinari. J Fish Biol. 2016;88:741–5.

Harris JE. A note of the breeding season, sex ratio and embryo development of the dogfish, *Scyliorhinus canicula*. J Mar Biol Assoc UK. 1952;31:269–74.

Heithaus MR. The biology of tiger sharks, *Galeocerdo cuvier*, in Shark Bay, Western Australia: sex ratio, size distribution, diet, and seasonal changes in catch rates. Environ Biol Fish. 2001;61:25–36.

Henningsen AD, Murru FL, Rasmussen LEL, Whitaker BR, Violetta GC. Serum levels of reproductive steroid hormones in captive sand tiger sharks, *Carcharias taurus* (Rafinesque), and comments on their relation to sexual conflicts. Fish Physiol Biochem. 2008;34:437–46.

Heupel MR, Whittier JM, Bennett MB. Plasma steroid hormone profiles and reproductive biology of the epaulette shark, *Hemiscyllium ocellatum*. J Exp Zool. 1999;594:586–94.

Idler DR, Freeman HC. Sex hormone binding proteins I. Binding of steroids by serum of an elasmobranch (*Raja radiate*). Gen Comp Endocrinol. 1969;13:75–82.

Jirik JE, Lowe CG. An elasmobranch maternity ward: female round stingrays Urobatis halleri use warm, restored estuarine habitat during gestation. J Fish Biol. 2012;80:1227–45.

Jones RC, Jones N, Djakiew D. Luminal composition and maturation of spermatozoa in the male genital ducts of the Port Jackson shark, *Heterodontus portusjacksoni*. J Exp Zool. 1984;230:417–26.

Karsten AH, Turner JW. Fecal corticosterone assessment in the epaulette shark (*Hemiscyllium ocellatum*). J Exp Zool. 2003;299A:188–96.

Köttgen M, Hofherr A, Li W, Chu K, Cook S, Montell C, Watnick T. Drosophila sperm swim backwards in the female reproductive tract and are activated via TRPP2 ion channels. PLoS One. 2011;6:1–8.

Lauria V, gristina M, Attrill MJ, Fiorentino F, Garofalo G. Predictive habitat suitability models to aid conservation of elasmobranch diversity in the central Mediterranean Sea. Sci Rep. 2015;5:13245.

Luer CA, Walsh CJ, Bodine AB, Wyffels J. Normal embryonic development in the clearnose skate, *Raja eglanteria*, with experimental observations on artificial insemination. Environ Biol Fish. 2007;80:239–55.

Manire CA, Rasmussen LEL, Hess DL, Heuter RE. Serum steroid hormones and the reproductive cycle in the female bonnethead shark (*Sphyrna tiburo*). Gen Comp Endocrinol. 1995;97:366–76.

Maruska KP, Gelsleichter J. Hormones and reproduction in Chondrichthyan fishes. In: Norris DO, Lopez KH, editors. Hormones and reproduction of vertebrates, Vol 1—fishes. San Diego: Academic Press; 2011. p. 209–31.

Masuda M, Izawa Y, Kametsuta S, Ikuta H, Isogai T. Artificial insemination of the cloudy catshark. J Jpn Assoc Zool Gardens Aquariums. 2003;44:39–43.

Masuda M, Izawa Y, Kametsuta S, Ikuta H, Isogai T. Artificial insemination of the white-spotted bamboo shark, *Chiloscyllium plagiosum*. J Jpn Assoc Zool Gardens Aquariums. 2005;46:91–6.

McComb DM, Gelsleichter J, Manire CA, Brinn R, Brown CL. Comparative thyroid hormone concentration in maternal serum and yolk of the bonnethead shark (*Sphyrna tiburo*) from two sites along the coast of Florida. Gen Comp Endocrinol. 2005;144:167–73.

Minamikawa S, Morisawa M. Acquisition, initiation and maintenance of sperm motility in the shark, *Triakis scyllia*. Comp Biochem Physiol. 1996;113A:387–92.

Mull CG, Lowe CG, Young KA. Seasonal reproduction of female round stingrays (Urobatis halleri): steroid hormone profiles and assessing reproductive state. Gen Comp Endocrinol. 2010;166:379–87.

Myers RA, Baum JK, Shepherd TD, Powers SP, Peterson CH. Cascading effects of the loss of predatory sharks from a coastal ocean. Science. 2007;315:1846–50.

Mylniczenko ND, Sumigama S, Wyffels JT, Wheaton CJ, Guttridge TL, DiRocco S, Penfold LM. Am J Vet Res. 2019. In Press.

McCully SR, Scott F, Ellis JR, Pilling GM. Productivity and susceptibility analysis: Application and suitability for data poor assessment of elasmobranchs in northern European seas. Collect. Vol. Sci. Pap. ICCAT. 2013; 69:1679-1698.

Nozu R, Murakumo K, Yano N, Furuyama R, Matsumoto R, Yanagisawa M, Sato K. Changes in sex steroid hormone levels reflect the reproductive status of captive female zebra sharks (*Stegostoma fasciatum*). Gen Comp Endocrinol. 2018;265:174–9.

Parsons GR, Grier HJ. Seasonal changes in shark testicular structure and spermatogenesis. J Exp Zool. 1992;261:173–84.

Paxton AB, Blair E, Blawas C, Fatzinger MH, Marens M, Holmberg J, Kingen C, Houppermans T, Keusenkothen M, McCord J, Silliman BR, Penfold LM. Citizen science reveals female sand tiger sharks (*Carcharias taurus*) exhibit signs of site fidelity on shipwrecks. Ecology. 2019:e02687.

Pratt HL Jr. The storage of spermatozoa in the oviducal glands of western North Atlantic sharks. Environ Biol Fish. 1993;38:139–49.

Pratt HL Jr, Tanaka S. Sperm storage in male elasmobranchs: a description and survey. J Morphol. 1994;219:297–308.

Pratt HL, Carrier JC. A review of elasmobranch reproductive behavior with a case study on the nurse shark, Ginglymostoma cirratum. Environ. Biol. Fishes. 2001:60;157-188

Prohaska BK, Tsang PCW, Driggers WB III, Hoffmaer ER, Wheeler CR, Brown AC, Sulikowski JA. Assessing reproductive status in elasmobranch fishes using steroid hormones extracted from skeletal muscle tissue. Conserv Physiol. 2013;1:1–13.

Rasmussen LEL, Murro FL. Long-term studies of serum concentrations of reproductively related steroid hormones in individual captive carcharhinids. Aust J Mar Freshwat Res. 1992;43:273–81.

Rosa R, Seibel BA. Synergistic effects of climate-related variables suggest future physiological impairment in a top predator. Proc Natl Acad Sci. 2008;105:20776–80.

Rosa RS, Charvet-Almeida P, Quijada CCD. Biology of the south American Potamotrygonid stingrays. In: Carrier JC, Musick JA, Heithaus MR, editors. Sharks and their relatives II. Boca Raton: CRC Press; 2010.

Rosa R, Baptista M, Lopes VM, Pegado MR, Ricardo Paula J, Trubenbach K, Leal MC, Calado, R, Repolho T. Early-life exposure to climate change impairs tropical shark survival. Proceedings of the Royal Society B: Biol. Sci. 2014;281(1793):20141738–20141738.

Saidï B, Nejmeddine Bradeï M, Bouaïn A. Reproductive biology and diet of *Mustelus punctulatus* from the Gulf of Gabès, central Mediterranean Sea. Sci Mar. 2009;73:249–58.

Santo D. Ocean acidification exacerbates the impacts of global warming on embryonic little skate, *Leucoraja erinacea*. J Exp Mar Biol Ecol. 2015;463:72–8.

Sato K, Nakamura M, Tomita T, Toda M, Miyamoto K, Nozu R. How great white sharks nourish their embryos to a large size: evidence of lipid histotrophy in lamnoid shark reproduction. Biol Open. 2016;5:1211–5. https://doi.org/10.1242/bio.017939.

Schwartz FJ. Mass migratory congregations and movements of several species of cownose rays, Genus Rhinoptera: a world-wide review. J Elisha Mitchell Sci Soc. 1990;106:10–3.

Sheldon JD, Allender MC, George RH, Bulman F, Abney K. Reproductive hormone patterns in male and female cownose rays (Rhinoptera bonasus) in an aquarium setting and correlation to ultrasonographic staging. J Zoo Wildl Med. 2018;49:638–47.

Shiba K, Shibata D, Inaba K. Autonomous changes in the swimming direction of sperm in the gastropod *Strombus luhuanus*. J Exp Biol. 2014;21:986–96.

Shields C. Reproductive biology of the tiger shark in the Western Atlantic Ocean. UNF graduate theses and dissertations. 2018. https://digitalcommons.unf.edu/etd/818.

Smith JW, Merriner JV. Observations on the reproductive biology of the cownose ray, *Rhinoptera bonasus*, in the lower Chesapeake Bay. Fish Bull. 1986;84:871–7.

Sulikowski JA, Kneebone J, Elzey S, Jurek J, Howell WH, Tsang PCW. Using the composite variables of reproductive morphology, histology and steroid hormones to determine age and size at sexual maturity for the thorny skate *Amblyraja radiata* in the western Gulf of Maine. J Fish Biol. 2006;69:1449–65.

Sulikowski JA, Driggers WB III, Ingram GW Jr, Kneebone J, Fergusen DE, Tsang PCW. Profiling plasma steroid hormones: a non-lethal approach for the study of skate reproductive biology and its potential use in conservation management. Environ Biol Fish. 2007;80:285–92.

Sulikowski JA, Wheeler CR, Gallagher AJ, Prohaska BK, Langan JA, Hammerschlag N. Seasonal and life-stage variation in the reproductive ecology of a marine apex predator, the tiger shark, *Galeocerdo cuvier*, at a protected female-dominated site. Aquat Biol. 2016;24:175–84.

Tanaka S, Shiobara Y, Hioli S, Abe H, Nishi G, Yano K, Suzuki K. The reproductive biology of the frilled shark, *Chlamydoselachus anguineus* from Suruga Bay, Japan. Jpn J Ichthyol. 1990;37:273–91.

Tanaka S, Kurokawa H, Masako H. Comparative morphology of the sperm in chondrichthyan fishes. Mémoires Du Muséum National d'Histoire Naturelle. 1995;166:321–32.

Tomita T, Murakumo K, Ueda K, Ashida H, Furuyama R. Locomotion is not a privilege after birth: ultrasound images of viviparous shark embryos swimming from one uterus to the other. Ethology. 2018;00:1–5.

Tricas C, Maruska KP, Rasmussen LEL. Annual cycles of steroid hormone production, gonad development, and reproductive behavior in the Atlantic stingray. Gen Comp Endocrinol. 2000;118:209–25.

Ueda K, Yanagisawa M, Murakumo K, Matsumoto Y, Sato K, Uchida S. Physical examination, blood sampling, and sedation of large elasmobranchs. In: The Elasmobranch Husbandry Manual II: recent advances in the care of sharks, rays and their relatives, vol. 26; 2017. p. 255–62.

Volkoff H, Wourms JP, Amesbury E, Snelson FF. Structure of the thyroid gland, serum thyroid hormones, and the reproductive cycle of the Atlantic stingray, *Dasyatis sabina*. J Exp Zool. 1999;284:505–16.

Walker TI. Spatial and temporal variation in the reproductive biology of gummy shark *Mustelus antarticus* (Chondrichthyes:Triakidae) harvested off southern Australia. Mar Freshw Res. 2007;58:67–97.

Wallman HL, Bennett WA. Effects of parturition and feeding on thermal preference of Atlantic stingray *Dasyatis sabina* Lesueur. Environ Biol Fish. 2006;75:259–67.

Whittamore JM, Bloomer C, Hanna GM, McCarthy ID. Evaluating ultrasonography as a non-lethal method for the assessment of maturity in oviparous elasmobranchs. Mar Biol. 2010;157:2613–24.

Woodhead AD. Thyroid activity in the ovo-vivparous elasmobranch *Squalus acanthias*. J Zool. 1966;148:238–75.

Wourms JP. Reproduction and development of chondrichthyan fishes. Am Zool. 1977;17:379–410.

Yamaguchi A, Tanuichi T, Shimizu M. Geographic variations in the reproductive parameters of the starspotted dogfish, *Mustelus manazo*, from five localities in Japan and in Taiwan. Environ Biol Fish. 2004;57:221–33.

Chapter 16
Cryopreservation as a Tool for Reef Restoration: 2019

Mary Hagedorn, Rebecca Spindler, and Jonathan Daly

Abstract Throughout the world coral reefs are being degraded at unprecedented rates. Locally, reefs are damaged by pollution, nutrient overload and sedimentation from out-dated land-use, fishing and mining practices. Globally, increased greenhouse gases are warming and acidifying oceans, making corals more susceptible to stress, bleaching and newly emerging diseases. The coupling of climate change impacts and local anthropogenic stressors has caused a widespread and well-recognized reef crisis. While the establishment and enforcement of marine protected areas and preventing the acceleration of climate change are essential to management of these stressors, the inexorable impacts of climate change will continue to cause declines in genetic diversity and population viability. Gamete cryopreservation has already acted as an effective insurance policy to maintain the genetic diversity of many wildlife species, and has now begun to be explored and applied to coral conservation. Cryopreservation can act to preserve reef biodiversity and genetic diversity. To date, we have had a great deal of success with cryopreserving sperm from ~30 coral species of coral species. Moreover, we are creating the basic science to freeze and thaw coral larvae that can soon be used to help secure and restore reefs. Building on these successes, we have established genetic banks using frozen samples and use those samples to help mitigate threats to the Great Barrier Reef and other areas.

Keywords Coral · Reef · Cryobiology · Cryobanking · Assisted reproduction · Invertebrate

M. Hagedorn (✉) · J. Daly
Smithsonian Conservation Biology Institute, Smithsonian Institution, Washington, DC, USA

Hawaii Institute of Marine Biology, University of Hawaii, Kaneohe, HI, USA
e-mail: hagedornm@si.edu

R. Spindler
Bush Heritage Australia, Melbourne, VIC, Australia

P. Comizzoli et al. (eds.), *Reproductive Sciences in Animal Conservation*, Advances in Experimental Medicine and Biology 1200,
https://doi.org/10.1007/978-3-030-23633-5_16

1 Introduction

Coral reefs are natural superstructures created by hard coral species that provide economic, ecological and social services. In Australia alone, the Great Barrier Reef directly supports 39,000 jobs and contributes an estimated $6.4 billion to the national economy each year (O'Mahony et al. 2017). Globally, the combined value of marine tourism, fishing industry, ecological services and pharmaceutical development is estimated at $375 billion each year (Costanza et al. 1997).

Coral reefs provide invaluable ecological services, for example acting as nursery grounds for marine fish and invertebrates, providing natural storm barriers for coastlines, promoting mangroves and sea grass beds, facilitating nitrogen fixation and carbon/calcium regulation, waste assimilation and providing potential sources for undiscovered pharmaceuticals. These ecosystem services, in turn have significant social impact by providing food security, supporting recreation, supporting spiritual and cultural practices, and aesthetic values (Moberg and Folke 1999; Cesar 2000; O'Mahony et al. 2017).

Despite their value, reefs are being degraded at unprecedented rates. Locally, reefs are damaged by pollution, nutrient overload and sedimentation from inappropriate land-use, fishing and mining practices. Globally, increased greenhouse gases are warming and acidifying oceans, making corals more susceptible to stress, bleaching and newly emerging diseases. The coupling of climate change impacts and local anthropogenic stressors has caused a widespread and well-recognized reef crisis (Glynn and D'Crox 1990; Glynn 1996; Hoegh-Guldberg 1999; Goreau et al. 2000; Hughes et al. 2003, 2017a; Shearer et al. 2009; Anthony 2016). Urgent and effective conservation action is now required to address this widespread crisis facing coral reefs. To emphasize this point, the Chair of the IUCN's Species Survival Commission, Simon Stuart, lists corals as one of the planet's three major species extinction crises. Most importantly, the proportion of corals threatened with extinction has increased dramatically in recent decades, exceeding most terrestrial groups (IUCN 2019).

The full impact of the potential loss of coral reefs on our global ecosystems and economies is still unknown. This is because we do not understand enough about the inter-relationships between reefs and other marine ecosystems. Coral may be a keystone ecosystem within the marine environment. For example, coral reefs only encompass 0.2% of the Earth's surface, but over one quarter of all marine life lives on a coral reef at some point in their life cycle. Moreover, they are some of the oldest and most diverse ecosystems on our planet. Approximately 50% of all the Earth's oxygen is produced in our oceans by green algae, if reefs failed would it affect this most important element for life on Earth? It is fair to assume that the web of ocean life is closely tied together, and the reef failures around the world will impact most aspects of ocean life, such as the availability of seafood. According to Cassandra de Young of the UN Food and Agricultural Organization, with over 1 billion people on the planet already hungry, the disappearance of seafood may cause great instability in food security around the world.

Of all the reefs, Caribbean reefs are suffering the most severe declines, and their fate may predict the future of the corals throughout the world. For example, *Acropora palmata* (Elkhorn coral) and *Acropora cervicornis* (Staghorn coral), both critical Caribbean reef-building species, have declined 80–99% from their historical population levels (Bruckner 2002; Bellwood et al. 2004), resulting in a loss of structure and function of reefs throughout the Caribbean (Buddemeier and Ware 2003; Gardner et al. 2003). Pandolfi et al. (2003) have identified similar patterns of ecosystems degradation and species loss in all the oceans, including the Great Barrier Reef (GBRMPA 2017). Ultimately, if changes in the world's climate become too severe, no management actions will be able to climate-proof the Great Barrier Reef ecosystem (Great Barrier Reef Outlook Report 2009). This prediction was borne out in 2016 and 2017 when back-to-back bleaching events were recorded on the Great Barrier Reef for the first time, resulting in coral bleaching across 91% of surveyed reefs (Hughes et al. 2017b) and the loss of 29% of shallow-water coral cover in 2016 alone (GBRMPA 2017).

The solutions to coral conservation must be innovative and multivariant, and incorporate a sound knowledge of the complex coral biology. Each coral individual (holobiont) consists of a complex of coral cells, algal symbionts (for many coral species) and bacteria. All of these cells play an important role in the growth, development and health of mature coral colonies. This complex relationship requires that each element of these living systems be adequately preserved as part of a global coral conservation program. Cryopreservation of gametes and small coral fragments would achieve these goals and this is currently under development, as part of a multi-disciplinary recovery process.

Critical to any restoration effort or conservation strategy for coral is a consideration of how much genetic diversity remains in our wild populations. However, this diversity is not well-described for most coral species (Shearer et al. 2009). These authors suggest that "coral restoration strategies using 10–35 randomly selected local donor colonies would retain at least 50–90% of the genetic diversity of the original population". But some populations, such as *Pocillopora damicornis*, that fragment easily, suffer from inbreeding (Combosch and Vollmer 2011). So, clearly

Box 16.1 Species in Worldwide Coral Bank
Caribbean (3)

Acropora cervicornis
Acropora palmata
Orbicella faveolata

Hawaii (3)

Lobactis scutaria
Montipora capitate

Pocillopora meandrina

French Polynesia (4)

Acropora globiceps
Acropora hyacinthus
Acropora nasuta
Acropora pulchra

Great Barrier Reef (21)

Acropora aspera
Acropora austera
Acropora cytherea
Acropora florida
Acropora hyacinthus.
Acropora loripes
Acropora millepora
Acropora muricata
Acropora nobilis
Acropora sarmentosa
Acropora spathulate
Acropora tenuis
Acropora valida
Astrea curta
Coelastrea aspera
Cyphastrea micropthalma
Dipsastraea matthai
Echinopora lamellose
Goneastrea retiformis
Platygyra daedalea
Platygyra lamellina

conservation and restoration strategies must be carefully tailored to the populations they will serve.

Cryopreservation has already acted as an effective insurance policy to maintain the genetic diversity of many wildlife species (Wolf et al. 2001; Wildt et al. 2010; Combosch and Vollmer 2011). We have had a great deal of success with cryopreserving sperm and coral embryos (Hagedorn et al. 2012; Daly et al. 2018). Building on this success, we have now begun to establish genetic banks using frozen samples, to help offset these threats to the Great Barrier Reef and other areas and have applied these techniques to 31 coral species encompassing 12 genera worldwide (Box 16.1).

Cryopreservation has been seen as a solution for the catastrophic event and would certainly provide an opportunity in the long term for large scale restoration. However, the resources conserved using this technology could also be used in the short term to facilitate movement of genes across geographic barriers and the acceleration of adaptation to increasing temperatures and acidic ocean conditions while avoiding the overarching loss of genetic diversity. The medium term opportunity to restore species in small areas is also facilitated by a stock of material that can be used to create viable corals, shoring up reef function and overarching integrity of the system (Hagedorn et al. 2017).

Cells and embryos that are cryopreserved and banked properly can retain viability for years, or even centuries, without DNA damage. The greatest challenge facing this critical conservation effort is the ability to scale up the effort to conserve diversity before it is diminished further. Time and resources are required to train individuals and form satellite banks globally. The successful implementation of this worldwide capacity building would create an insurance population for reefs, securing their biodiversity and helping to maintain or bolster their related economies.

2 State of the Art

2.1 *The Fundamentals of Cryobiology*

Box 16.2 Cryopreservation Primer

1. *Slow Freezing Cryopreservation:*

   Uses extracellular ice to dehydrate cells, slowly dehydrating and freezing cells over minutes to hours.

   Advantage: amenable to most cells.

   Disadvantage: some cells are damaged by a slow reduction in temperature.
2. *Vitrification:*

   Uses high concentrations of cryoprotectants and ultrafast freezing temperatures to form a glass instead of ice.

   Advantage: good for chill sensitive cells or organisms.

   Disadvantage: solutions can be toxic, thawing must use very fast warming temperatures to prevent ice formation.

Cryopreservation (the storage of cells under cold conditions) is an extremely effective conservation tool for maintaining genetic diversity. In this approach (see Box 16.2), cells are frozen in sugar-like compounds, called cryoprotectants, to approximately −80 °C and then placed into liquid nitrogen (−196 °C) where they can remain frozen, but alive, for decades in a genetic bank. Most technological innovations in the field of germplasm cryopreservation arose from a sound understanding of the mechanisms of cryodamage and cryoprotection (Mazur 1970,

1984). Successful cryopreservation of cells, germplasm and tissue must address intrinsic biophysical properties (e.g., water and cryoprotectant permeability, osmotic tolerance limits, intracellular ice nucleation, etc.) to maximize survival (Rall 1993). A similar systematic approach is vital to improving post-thaw survival of coral and its associated organisms.

Cryopreservation (the storage of cells under cold conditions) is an extremely effective conservation tool for maintaining genetic diversity. In this approach (see Box 16.2), cells are frozen in sugar-like compounds, called cryoprotectants, to approximately −80 °C and then placed into liquid nitrogen (−196 °C) where they can remain frozen, but alive, for decades in a genetic bank. Most technological innovations in the field of germplasm cryopreservation arose from a sound understanding of the mechanisms of cryodamage and cryoprotection (Mazur 1970, 1997; Gao and Critser 2000). Successful cryopreservation of cells, germplasm and tissue must address intrinsic biophysical properties (e.g., water and cryoprotectant permeability, osmotic tolerance limits, intracellular ice nucleation, etc.) to maximize survival. A similar systematic approach is vital to improving post-thaw survival of coral and its associated organisms.

Conventional cryopreservation of many types of cells relies upon cryoprotectants and slow freezing to dehydrate and shrink the cell. Cryoprotectants that enter the cell, such as dimethyl sulfoxide, propylene glycol, or glycerol, are effective, yet their mechanisms of action are not completely understood. They depress the freezing point of solutions in and around the cells and may directly alter membrane bilayers or interact with bound proteins on the external cell surface (Hammerstedt et al. 1990). Too little entering the cell before cooling reduces effectiveness and may lead to damaging intracellular ice formation (Taylor et al. 1974); too much entering the cell causes osmotic swelling and rupture during thawing and dilution (Levin and Miller 1981). Often, these procedures must be tailored for each type of cell, based upon a thorough understanding of its properties.

Preventing intracellular ice formation is essential to successful cryopreservation. During slow cooling, extracellular fluid freezes before intracellular fluid, pulling pure water out of the cell, leading to osmotic dehydration of the cells as they supercool. If ~90% of the intracellular water can be removed before lethal intracellular ice forms, then many cells will survive thawing and dilution (Mazur 1984).

However, certain cells can be damaged by the slow-freezing process because a sudden reduction in temperature can cause cold shock injury (or chilling sensitivity), often resulting in severe membrane damage. It is common in many mammalian sperm cells, but particularly in aquatic oocytes, embryos and larvae (Hagedorn et al. 2006b). Vitrification, whereby cell water is converted to a glass rather than undergoing a damaging phase transition to ice, may prove to be a more viable technique for aquatic cells. Vitrification entails the use of: (1) highly concentrated cryoprotectants (typically 5–6 M), which cause dehydration before cooling; and (2) rapid cooling of the cell suspension, forming a transparent glass-state. Vitrification permits rapid cryopreservation with improved survival in some cells (Rall and Fahy 1985).

2.2 *Current Status of the Cryobiology of Reef Organisms*

Storage of important coral and related cells through cryopreservation will profoundly advance basic research in embryology, genetics, systematics and molecular biology, as well as enhance management strategies for reef restoration. Although cryopreservation is a proven method for long-term maintenance of genetic material, current protocols for coral and associated organisms are not fully developed, and so the associated programs that could employ these important genetic resources have not reached their full potential. An important point, however, is that once the material is frozen, a great deal of research can be done to determine how it might best be used in the future. In the past 14 years, we have characterised some of the fundamental cryobiology for coral sperm, larvae and associated symbionts (Hagedorn et al. 2006a, b, 2010, 2012, 2017; Hagedorn and Carter 2015; Daly et al. 2018).

Germplasm Cryopreservation (Successful)

The sperm from 31 different coral species including 12 genera (Box 16.1) has been successfully cryopreserved, using the same standardised cryopreservation protocol and preserved in banks around the world. The general cryopreservation method for coral germplasm has been described in detail in (Hagedorn et al. 2012). Briefly, the sperm are collected and held in a concentrated form (approximately 2×10^9 cells/ml). Sub-samples are diluted either 1:100 or 1:1000 in filtered seawater, counted with a hemocytometer and their motility assessed on a phase microscope approximately 30–45 min after collection. This standardized process is important because some acroporid species only reach full motility 20–30 min after they have been released from their bundle (Hagedorn et al. unpublished data). Sperm samples with 50% motility or greater are either pooled across males or kept singly and prepared for cryopreservation. The sample is diluted 1:1 with 20% dimethyl sulfoxide in filtered seawater and aliquots (1 ml) are loaded into 1.8 ml cryovials held at 26–28 °C. After a 10 min exposure to the cryoprotectant, the cells in the vials are frozen at 20 °C/min, quenched in liquid nitrogen, and then placed into a dry shipper for transport to permanent storage. A single sample from each freezing trial can then be thawed to examine post-thaw motility and fertilization success with fresh eggs.

Assessment of Sperm Viability and Use of the Frozen Bank (Successful)

Frozen-thawed sperm have been used to fertilize conspecific eggs released during the same spawn and from successive spawns (Hagedorn et al. 2012, 2017). While variability remains across species and even within individuals on different nights of the same spawn these sperm have reached fertilization success of 60% (Hagedorn et al. 2012). From 2013 to 2015, tens of thousands of *A. tenuis* and *A. millepora* larvae were generated with (1) sperm collected immediately after spawning; (2) this same sperm frozen for 1 h and then thawed, and; (3) sperm that been frozen for 1 year and then thawed. These larvae produced from frozen/thawed *A. tenuis* and *A. millepora* sperm developed, settled and assimilated symbionts over an 8 week-period (Hagedorn et al. 2017) (Fig. 16.1). These studies will help guide future usage of these invaluable frozen resources, because one day they may be needed to help expand and diversify shrinking coral populations worldwide. These longer-duration small-scaled experiments suggested that larger-scale grow outs would be possible to examine the effects of cryopreservation on the growth and maturity of coral over several years.

Recent data from yet unpublished findings (Hagedorn et al. 2018; Daly et al. 2019a in prep) suggest that cryopreservation could be a powerful tool for assisted gene flow

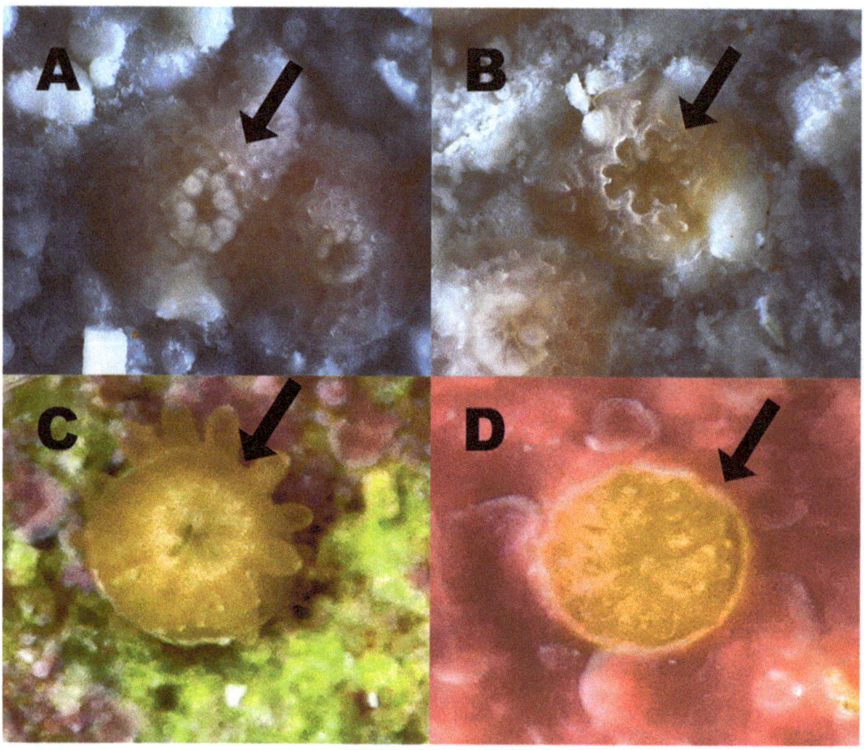

Fig. 16.1 Coral settlers from *A. tenuis* (**a** and **b**) and *A. millepora* (**c** and **d**) demonstrated similar settlement size (~2 mm in diameter), complexity of their septa, tentacle development and calcareous skeleton, regardless of whether they were fertilized with fresh (**a**, **b**) or cryopreserved (**c**, **d**) sperm. The *A. millepora* images (arrows) were taken at approximately 4 weeks post-fertilization and the settlers had acquired their *Symbiodinium*. The images of the *A. tenuous* settlers (arrows) were taken at 2 weeks post-fertilization and had not yet acquired their symbionts. Bar = 1 mm. This figure was modified from Hagedorn et al. (2017)

experiments. In collaboration with international partners, Hagedorn et al. (2018) used cryopreserved sperm from *Acropora palmata* collected from around the Caribbean and produced trans-regional crosses that were separated by hundreds of miles. These crosses are settled and are now growing to maturity in two closed systems in Florida. Similar experiments were conducted on the Great Barrier Reef in 2018 where cryopreserved *Acropora tenuis* sperm was crossed with fresh eggs from the Northern Great Barrier Reef, and cryopreserved sperm from the Northern GBR was crossed with Central Great Barrier Reef eggs (Daly et al. 2019a in prep). Until recently, assisted gene migration for corals was executed by moving colonies from one area of the reef to another (Dixon et al. 2015), however, these latest experiments demonstrated the efficacy of using cryopreserved sperm for gene flow experiments in the future.

Coral Larval Cryopreservation (Successful)

Cryopreservation of coral larvae had never been achieved due to their large size and complex structure, which cause lethal ice damage upon warming using normal convective methods (tens of degrees C/min). However, recent studies (Daly et al. 2018) created an advanced laser-warming process that overcame this lethal ice formation. Additionally, more recent work has shown that cryopreserved larvae can be successfully infected with Symbiodiniaceae and settled, indicating that cryopreserved larvae retain competence post-thaw (Daly et al. 2019b in prep). Successful cryopreservation of coral larvae and proving that they are competent post-thaw is a major step forward towards maintaining the biodiversity of coral worldwide, prevent species extinctions and being able to use these methodologies for large-scale restoration processes.

Symbiodiniaceae Cryopreservation (Successful)

The dinoflagellate algae in the family Symbiodiniaceae (often referred to as symbionts) live within some coral cells and produce energy-rich compounds in exchange for the carbon substrates needed for photosynthesis. The cryobiology of Symbiodiniaceae isolated from three species of Hawaiian coral has been described (Hagedorn et al. 2010). The different subtypes studied demonstrated remarkable similarities in their morphology, sensitivity to cryoprotectants and permeability characteristics; however, they differed greatly in their sensitivity to hypo- and hyperosmotic challenges and sensitivity to chilling, suggesting that standard slow freezing cryopreservation may not work well for Symbiodiniaceae. Following these cryobiological descriptions, a method for the successful cryopreservation of Symbiodiniaceae was developed and used successfully in Symbiodiniaceae isolated from three species of corals in Hawaii (Hagedorn and Carter 2015).

Ice Recrystallization Inhibitors (Cryo-Studies Underway in Hawaii)

Recent advances in cryobiology (Briard et al. 2016) may transform the field of cryobiology. These authors have created and tested a suite of low molecular mass bioengineered compounds called ice recrystallization inhibitors (IRIs). These compounds block the growth of ice-crystals during warming, vastly improving post-thaw viability. This work has indicated that the use of IRIs may increase the efficiency of cryopreservation, reducing the concentration of cryoprotectant required and thereby reducing toxicity to tissues and cells, thus increasing the range of samples that can be banked. Additionally, use of IRIs has also shown promise in preliminary studies with coral larvae vitrification and laser-warming, increasing both the rate of recovery and quality of larvae produced post-thaw using vitrification and laser warming protocols in larvae from *Fungia scutaria* Daly, personal comm. These experiments will greatly reduce toxicity in our banks and help us to bank complex tissues, such as larvae and coral fragments, more effectively.

Coral Fragment Cryopreservation (Cryo-Studies Underway in Hawaii)

Although there is some information on the cryosensitivities of coral fragments (Hagedorn et al. 2013), the size of these fragments would be prohibitive for most banking. Instead, we would like to take advantage of new microfragmentation technology (Page et al. 2018) that might be able to be warmed at ultra-rapid rates. If this became a common and robust process, preservation could happen within many ecosystems around the world throughout the year. This process would preserve the entire holobiont and would substantially secure reef biodiversity.

3 Influence of Environmental Change

3.1 Sperm Motility and Cryopreservation, the Ability to Create New Larvae

Warming events can trigger coral bleaching either locally or globally, and the frequency of mass bleaching events is increasing (Hughes et al. 2018). The 2007 report from the Intergovernmental Panel on Climate Change (IPCC 2007) predicted that, during the period 2012–2040, coral reefs will experience more frequent bleaching events closer together in time, and identified these bleaching events as the greatest threat to the world's reef systems. Bleaching profoundly affects reproduction in coral often for more than one season (Cox and Ward 2002; Ward et al. 2002). Moreover, stressful long-term effects of warming can lead to the reduced reproduction in coral that visibly bleach, as well as on those that do not (Levitan et al. 2014). So, even though corals were not conspicuously losing their *Symbiodinium*, the long-term stress was found to disrupt reproduction. But how do these bleaching events affect specific reproductive traits? Hagedorn et al. (2016) has long-term reproductive measurements for two species of coral in Kaneohe Bay, Hawaii, namely *Fungia scutaria* and *Montipora capitata*. They examined spawning periods, egg morphometry, sperm concentration, fresh and cryopreserved sperm motility exposed to different concentrations of dimethyl sulfoxide, time of first cleavage, larval survival with fresh and cryopreserved spermatozoa, infection success and settlement success over a period of 10 years that encompassed a bleaching event in 2014–2015. Many of these reproductive parameters were reduced in 2015, especially sperm motility. Once the reduced-motility spermatozoa from 2015 were cryopreserved, there was a steep decline in post-thaw viability and this prevented any substantive further use of these samples in reproduction for conservation benefit. Even though the bleaching event has ended, mean sperm motility has not returned to their pre-bleach levels for either species as late as 2017 (Fig. 16.2). This suggests that this one stressful event had long-term effects or that local impacts, such as sedimentation, pollution, etc. are adding sufficient stress to the populations to prevent rapid recovery.

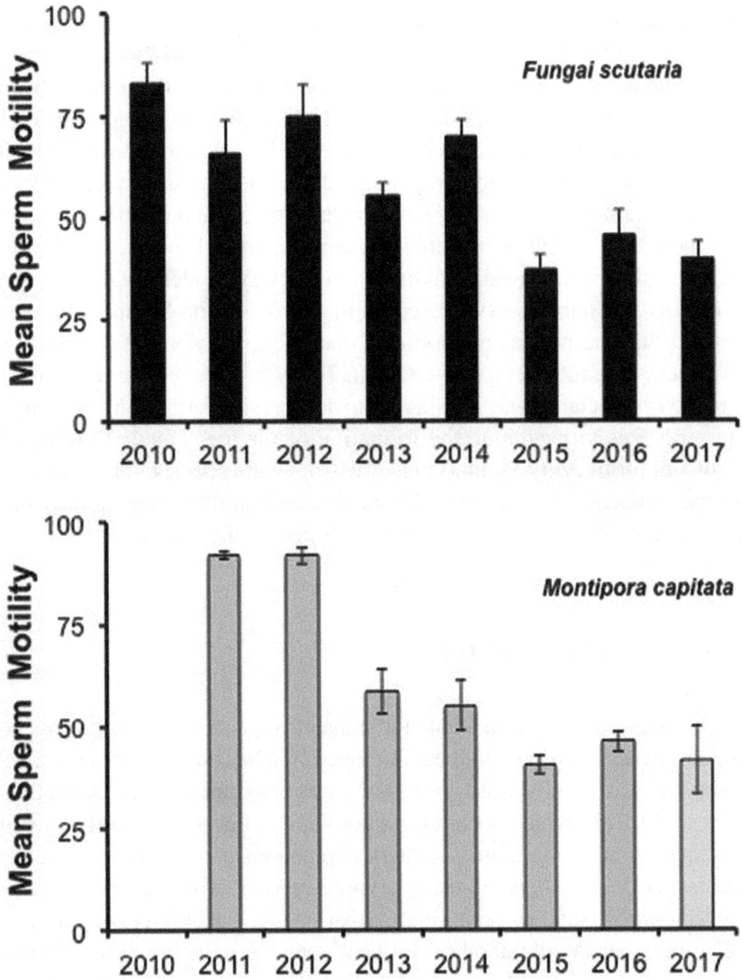

Fig. 16.2 Long-term changes in sperm motility in Kaneohe Bay. Note the lack of recovery in both species after the 2015 bleaching event. Figure was modified from Hagedorn et al. (2016)

Worldwide, as bleaching events become more frequent, the ability to bank and conserve coral sperm *ex situ* may be significantly reduced. Thus, it is imperative that while genetic diversity is still high in these populations, intensive efforts are made to bank sperm from coral species during non-bleaching periods.

3.2 Climate Change and Alterations in Symbiodiniaceae

When the coral holobiont begins to stress, it can lose its life-giving symbionts and bleach. One of the effects of the bleaching process is that the thylakoid membranes in the photosynthetic apparatus of the symbionts become damaged by the warming (Tchernov et al. 2004), thus making them inefficient or unable to undertake photosynthesis. Prior to the bleaching event in Hawaii in 2014 and 2015, Hagedorn and Carter (2015) devised a cryopreservation process that included encasing the symbionts in agar and using cryotops to freeze and thaw them. This cryopreservation process worked very well in the winter and spring but less well in the summer and fall, demonstrating a seasonal variability, perhaps due to increased fat within the symbionts during the summer months. However, recent studies in our lab have identified a decline in post-thaw viability in symbionts isolated from Hawaiian corals after the 2015 bleaching event in Hawaii. Studies are underway to identify the cause of this decline and to re-establish cryopreservation protocols for Symbiodiniaceae. We hypothesized that the new adaptation that made the symbionts more heat tolerant made them less tolerant to cryopreservation. Today, we are re-examining the sensitivities to cryoprotectants and sensitivity to ice crystal formation by using laser-warming processes to overcome the formation of the ice-crystals to re-establish a robust protocol for freezing Symbiodiniaceae from coral species. It is clear that cryopreservation protocols, must continually keep pace with the changing physiology of coral as they adapt to new environmental conditions in the ocean.

4 Priorities for the Future

Cryopreservation will help to facilitate the transfer of genetic material across populations to boost genetic diversity in areas damaged by bleaching, storms and predation. Cryopreserved sperm can be used to fertilize fresh eggs and produce larvae that settle (Hagedorn et al. 2017), which opens up the possibility of genetic crosses between individuals from northern, central, and southern populations of wide-ranging species (e.g. *Acropora tenuis* or *A. millepora*) using cryopreserved sperm. Preservation of sperm will enable reproduction between species that spawn on different nights to create hybrid offspring that would not otherwise be possible. It will also allow the storage of genetic material from individual coral with desirable traits for selective breeding across different nights or even different years. New and existing cryopreservation technologies will be used to bank sperm samples during spawning with an emphasis on increasing the diversity of species and populations that have been sampled to date. The use of coral produced from sexual reproduction (e.g., using cryopreserved sperm) for assisted gene flow also provides greater genetic diversity compared to using clonal fragments from mature colonies (Hagedorn et al. 2018).

The development and application of new technologies for low-temperature storage of a range of sample types will permit large-scale, long-term security of coral reef biodiversity, and maximize the diversity and utility of samples available to reef adaptation research and recovery efforts now and in the future. The development of vitrification, laser warming and high throughput freezing of coral larvae and cryopreservation of coral fragments will expand the diversity of samples and genetics that can be transferred between populations if required, and enable the low-cost storage of mature coral polyps from thousands of individuals. Development of high-throughput processing (defined as producing 1000s of larvae per minute) and refinement of warming and recovery parameters will enable the production of large batches of larvae for proof-of-concept experiments on Symbiodiniaceae infection and larval settlement, and storage of larvae from selectively-bred corals will provide a source of material for experimentation and for grow-out and *in-situ* planting of resilient corals for reef restoration. It is hoped that application of vitrification and laser warming technologies in a range of species will permit strategic biobanking of coral larvae to begin in coming years.

Cryopreservation of bioengineered Symbiodiniaceae will provide a low-cost way to secure valuable research strains and natural diversity from the reef. Modification of the cryopreservation protocol to account for effects of warming stress on post-thaw viability and for application to cultured Symbiodiniaceae will be required. The ability of these Symbiodiniaceae to infect will be assessed in coral larvae when they are available during the spawning period, and potentially using larvae that have been cryopreserved and laser-warmed.

A critical element of maintaining biodiversity in biorepositories, is the long-term commitment to maintaining these assets. Today, there are three main sites in the world where coral cells are stored long-term, in the U.S. at the Smithsonian Institution at the Hawaii Institute of Marine Biology, the U.S. Department of Agriculture's Animal Germplasm Program and in Australia at the Taronga Western Plains Zoo. While the banks already established contain billions of cells, we have been able to collect in highly controlled environments and capture many cells from a modest number of individuals of each species. In the wild, the recruitment rates are curtailed by predation, wave motion, lack of available settling sites, etc. Thus, restoring all of these cells to the source reef would overwhelm the gene pool and possibly result in a less adaptive reef, post restoration. For this reason, future use of this frozen material, will more than likely, include three separate but complementary streams:

1. A proportion of cells remain banked for future generations, cloistered in a long-term repository for large scale restoration;
2. A larger proportion could be used to effect local restoration efforts in response to a specific event such as bleaching, dredging, disease, silting etc. when paired with effective use of land- or sea-based nurseries. The threat must be mitigated

before cells will be thawed and the resources allocated to the growth and care of developing coral colonies. If local threats can be mitigated, there is great hope that we can restore the function of these much needed global drivers of ecosystem health; and

3. The remaining cells will be used to advance our understanding of biology (i.e., in systematics, genetic, development and disease and pharmaceutical explorations). For example, these cells will provide source material for innovative work by our colleagues on the exploration, selection and nurturing of coral most likely to maintain resilience to warmer and more acidic conditions.

In combination with other much needed restoration and threat mitigation efforts, the potential outcomes of these genome resource banks would: (1) preserve gene diversity; (2) prevent extinctions; (3) store the entire genome, including as yet unknown but critically valuable epigenetic factors; (4) create opportunities for diversifying shrinking populations by avoiding natural losses in heterozygosity due to genetic drift; and (5) advance the science of coral biology. In particular, coral developmental biology which on the Great Barrier Reef for example has been limited to a period of only a few days per year during the brief annual spawning event.

In order for the coral cryobank to fulfill its potential, there is an immediate need to capture a comprehensive representation of genetic diversity in each selected species. Our progress to date has been promising and, ultimately, this is easily accomplished as only 3–4 genotypes from a reef are needed to capture much of the allelic diversity (Baums 2018). This makes the number of individuals needed to be stored from a species fairly tractible. Furthermore, the bank must expand in terms of morphological diversity, functional roles and geographic location. We will include priority species (as defined by the worlds coral experts) to ensure ecosystem function and system resilience are maximised by the species maintained in the bank.

In addition to these biological benefits, however, their maybe cultural benefits to these resource banks, as well. For example, both Western Australia and the Great Barrier Reef encompass World Heritage Sites important to the history and culture of the Aboriginal and Torres Strait Islander groups. Some of these sites, such as Shark Bay in Western Australia have evidence of continual occupation by Aboriginal groups for over 30,000 years, and it could be longer. The loss of these biomes to the social and cultural fabric within Australia would be incalculable. Traditional local knowledge must be incorporated in decision-making and prioritisation of corals, as well as understanding reef function and cultural roles.

Overall, the early success in coral banking is promising and should provide some avenues to further research and protect the reefs of the world. Both aspects are fundamentally impacted by the level of cooperation and collaboration, and the now rapidly growing interest in the value and potential uses of the cells in these banks. Looking to the future, cryobanks have the potential to contribute to many aspects of fundamental and applied coral science, but may also provide an avenue for local organisations to expand coral nurseries to include sexually reproduced corals. This opportunity provides expanded income for these groups and maintains the skills required for rapid generation of many corals for local re-seeding of reefs.

The potential of these banks is therefore relevant across the economic, cultural and ecological spectra.

Summary of Priorities for the Future:

1. Train more professionals in all aspects of coral cryopreservation and standard operating protocols for banking and transportation of all types of coral cells;
2. Build and sustain biorepositories in critical areas of the world that could act as regional nodes if local biorepositories were not feasible. Ideally, you would have one regional node in each ocean;
3. Create a list of priority reefs and species that are tagged and genetically identified to ensure diversity;
4. Create teams that can collect and bank various types of coral material at these priority sites; and,
5. Continue to create new and faster research protocols to cryopreserve large quantities of high-quality coral material.

References

Anthony KRN. Coral reefs under climate change and ocean acidification: challenges and opportunities for management and policy. Annu Rev Environ Resour. 2016;41:59–81.

Baums IB. How to maximize future adaptive potential of restored coral populations. In: Reef Futures, Key Largo; 2018.

Bellwood DR, Hughes TP, Folke C, Nystrom M. Confronting the coral reef crisis. Nature. 2004;429:827–33.

Briard JG, Poisson JS, Turner TR, Capicciotti CJ, Acker JP, Ben RN. Small molecule ice recrystallization inhibitors mitigate red blood cell lysis during freezing, transient warming and thawing. Sci Rep. 2016;6:23619.

Bruckner AW (2002) Proceedings of the Caribbean Acropora workshop: potential application of the U.S. endangered species act as a conservation strategy.

Buddemeier RW, Ware JR. Coral reef decline in the Caribbean. Science. 2003;302:391–3. author reply 391–393.

Cesar HLP. Coral reefs: their functions, threats and economic value. In: Cesar HLP, editor. Collected essays on the economics of coral reefs. Kalmar: CORDIO, University of Kalmar; 2000. p. 14–39.

Combosch DJ, Vollmer SV. Population genetics of an ecosystem-defining reef coral *Pocillopora damicornis* in the tropical eastern Pacific. PLoS One. 2011;6:e21200.

Cox EF, Ward S. Impact of elevated ammonium on reproduction in two Hawaiian scleractinian corals with different life history patterns. Mar Pollut Bull. 2002;44:1230–5.

Daly J, Zuchowicz N, Nunez Lendo CI, Khosla K, Lager C, Henley EM, Bischof J, Kleinhans FW, Lin C, Peters EC, Hagedorn M. Successful cryopreservation of coral larvae using vitrification and laser warming. Sci Rep. 2018;8:15714.

Daly J, Zuchowicz N, Hagedorn M. Proof of competence in laser-warmed coral larvae. 2019a (in prep).

Daly J, Zuchowicz N, Hobbs R, O'Brien J, Bay LK, Hagedorn M. Cryopreservation can assist gene flow on the Great Barrier Reef. 2019b (in prep).

Dixon GB, Davies SW, Aglyamova GV, Meyer E, Bay LK, Matz MV. Genomic determinants of coral heat tolerance across latitudes. Science. 2015;348:1460–2.

Gao D, Critser JK. Mechanisms of cryoinjury in living cells. ILAR J. 2000;41:187–96.

Gardner TA, Cote IM, Gill JA, Grant A, Watkinson AR. Long-term region-wide declines in Caribbean corals. Science. 2003;301:958–60.

GBRMPA. Final report: 2016 coral bleaching event on the Great Barrier Reef, Townsville. 2017.

Glynn PW. Coral reef bleaching: facts, hypotheses and implications. Glob Chang Biol. 1996;2: 495–509.

Glynn PW, D'Crox L. Experimental evidence for high temperature stress as the cause of El Niño-coincident coral mortality. Coral Reefs. 1990;8:181–91.

Goreau TJ, Hayes RL, McClanahan T. Conservation of coral reefs after the 1998 global bleaching event. Conserv Biol. 2000;14:5–15.

Great BarrierReef Outlook Report. 2009. http://elibrary.gbrmpa.gov.au/jspui/bitstream/11017/429/1/Great-Barrier-Reef-outlook-report-2009-in-brief.pdf

Hagedorn M, Carter VL. Seasonal preservation success of the marine dinoflagellate coral symbiont, *Symbiodinium sp*. PLoS One. 2015;10:e0136358.

Hagedorn M, Carter VL, Steyn RA, Krupp D, Leong JA, Lang RP, Tiersch TR. Preliminary studies of sperm cryopreservation in the mushroom coral, *Fungia scutaria*. Cryobiology. 2006a; 52:454–8.

Hagedorn M, Pan R, Cox EF, Hollingsworth L, Krupp D, Lewis TD, Leong JC, Mazur P, Rall WF, MacFarlane DR, Fahy G, Kleinhans FW. Coral larvae conservation: physiology and reproduction. Cryobiology. 2006b;52:33–47.

Hagedorn M, Carter VL, Leong JC, Kleinhans FW. Physiology and cryosensitivity of coral endosymbiotic algae (*Symbiodinium*). Cryobiology. 2010;60:147–58.

Hagedorn M, Carter V, Martorana K, Paresa MK, Acker J, Baums IB, Borneman E, Brittsan M, Byers M, Henley M, Laterveer M, Leong JA, McCarthy M, Meyers S, Nelson BD, Petersen D, Tiersch T, Uribe RC, Woods E, Wildt D. Preserving and using germplasm and dissociated embryonic cells for conserving Caribbean and Pacific coral. PLoS One. 2012;7:e33354.

Hagedorn M, Farrell A, Carter VL. Cryobiology of coral fragments. Cryobiology. 2013;66:17–23.

Hagedorn M, Carter VL, Lager C, Camperio Ciani JF, Dygert AN, Schleiger RD, Henley EM. Bleaching effects on coral reproduction. Reprod Fertil Dev. 2016;28:1061–71.

Hagedorn M, Carter VL, Henley EM, van Oppen MJH, Hobbs R, Spindler RE. Producing coral offspring with cryopreserved sperm: a tool for coral reef restoration. Sci Rep. 2017;7:14432.

Hagedorn M, Page CA, Oneill K, Flores DM, Tichy L, Chamberland VF, Lager C, Zuchowicz N, Lohr K, Blackburn H, Vardi T, Moore J, Moore T, Vermeij MJA, Marhaver KL. Successful demonstration of assisted gene flow in the threatened coral *Acropora palmata* across genetically-isolated Caribbean populations using cryopreserved sperm. bioRxiv. 2018:492447.

Hammerstedt RH, Graham JK, Nolan JP. Cryopreservation of mammalian sperm: what we ask them to survive. J Androl. 1990;111:73–88.

Hoegh-Guldberg O. Climate change, coral bleaching and the future of the world's coral reefs. Mar Freshw Res. 1999;50:839–66.

Hughes TP, Baird AH, Bellwood DR, Card M, Connolly SR, Folke C, Grosberg R, Hoegh-Guldberg O, Jackson JBC, Kleypas J, Lough JM, Marshall P, Nystrom M, Palumbi SR, Pandolfi JM, Rosen B, Roughgarden J. Climate change, human impacts, and the resilience of coral reefs. Science. 2003;301:929–33.

Hughes TP, Barnes ML, Bellwood DR, Cinner JE, Cumming GS, Jackson JBC, Kleypas J, van de Leemput IA, Lough JM, Morrison TH, Palumbi SR, van Nes EH, Scheffer M. Coral reefs in the Anthropocene. Nature. 2017a;546:82–90.

Hughes TP, Kerry JT, Álvarez-Noriega M, Álvarez-Romero JG, Anderson KD, Baird AH, Babcock RC, Beger M, Bellwood DR, Berkelmans R, Bridge TC, Butler IR, Byrne M, Cantin NE, Comeau S, Connolly SR, Cumming GS, Dalton SJ, Diaz-Pulido G, Eakin CM, Figueira WF, Gilmour JP, Harrison HB, Heron SF, Hoey AS, Hobbs J-PA, Hoogenboom MO, Kennedy EV, Kuo C-Y, Lough JM, Lowe RJ, Liu G, McCulloch MT, Malcolm HA, McWilliam MJ, Pandolfi JM, Pears RJ, Pratchett MS, Schoepf V, Simpson T, Skirving WJ, Sommer B, Torda G, Wachenfeld DR, Willis BL, Wilson SK. Global warming and recurrent mass bleaching of corals. Nature. 2017b;543:373–7.

Hughes TP, Anderson KD, Connolly SR, Heron SF, Kerry JT, Lough JM, Baird AH, Baum JK, Berumen ML, Bridge TC, Claar DC, Eakin CM, Gilmour JP, Graham NAJ, Harrison H, Hobbs J-PA, Hoey AS, Hoogenboom M, Lowe RJ, McCulloch MT, Pandolfi JM, Pratchett M, Schoepf V, Torda G, Wilson SK. Spatial and temporal patterns of mass bleaching of corals in the Anthropocene. Coral Reefs. 2018;359:80–3.

IPCC. Contribution of Working Group I to the fourth assessment report of the Intergovernmental Panel on Climate Change. Cambridge; 2007.

IUCN. Coral reefs. Gland: International Union for Conservation of Nature; 2019.

Levin RL, Miller TW. An optimum method for the introduction or removal of permeable cryoprotectants: isolated cells. Cryobiology. 1981;18:32–48.

Levitan DR, Boudreau W, Jara J, Knowlton N. Long-term reduced spawning in *Orbicella* coral species due to temperature stress. Mar Ecol Prog Ser. 2014;515:1–10.

Mazur P. Cryobiology: the freezing of biological systems. Science. 1970;168:939–49.

Mazur P. Freezing of living cells: mechanisms and implications. Am J Phys. 1984;247(3. Pt 1):C125–42.

Mazur P. The role of intracellular freezing in the death of cells cooled at supraoptimal rates. Cryobiology. 1997;14:251–72.

Moberg F, Folke C. Ecological goods and services of coral reef ecosystems. Ecol Econ. 1999; 29:215–33.

O'Mahony J, Simes R, Redhill D, Heaton K, Atkinson C, Hayward E, Nguyen M. At what price? The economic, social and icon value of the Great Barrier Reef: Deloitte Access Economics; 2017.

Page C, Muller E, Vaughan D. Microfragmenting for the successful restoration of slow growing massive corals. Ecol Eng. 2018;123:86–94.

Pandolfi JM, Bradbury RH, Sala E, Hughes TP, Bjorndal KA, Cooke RG, McArdle D, McClenachan L, Newman MJ, Paredes G, Warner RR, Jackson JB. Global trajectories of the long-term decline of coral reef ecosystems. Science. 2003;301:955–8.

Rall WF. Advances in the cryopreservation of embryos and prospects for the application to the conservation of salmonid fishes. In: Thorgaard GH, Cloud JG, editors. Genetic conservation of salmonid fishes. New York: Plenum Press; 1993. p. 137–58.

Rall WF, Fahy GM. Ice-free cryopreservation of mouse embryos at −196°C by vitrification. Nature. 1985;313:573–5.

Shearer TL, Porto I, Zubillaga AL. Restoration of coral populations in light of genetic diversity estimates. Coral Reefs. 2009;28:727–33.

Taylor R, Adams GD, Boardman CF, Wallis RG. Cryoprotection—permeant vs nonpermeant additives. Cryobiology. 1974;11:430–8.

Tchernov D, Gorbunov MY, de Vargas C, Narayan Yadav S, Milligan AJ, Haggblom M, Falkowski PG. Membrane lipids of symbiotic algae are diagnostic of sensitivity to thermal bleaching in corals. Proc Natl Acad Sci U S A. 2004;101:13531–5.

Ward S, Harrison P, Hoegh-Guldberg O. Coral bleaching reduces reproduction of scleractinian corals and increases susceptibility to future stress. Proceedings of the ninth international coral reef symposium, Bali, 23–27 Oct 2000. 2002;2:1123–8.

Wildt DE, Comizzoli P, Pukazhenthi B, Songsasen N. Lessons from biodiversity—the value of nontraditional species to advance reproductive science, conservation, and human health. Mol Reprod Dev. 2010;77:397–409.

Wolf KN, Wildt DE, Vargas A, Marinari PE, Ottinger MA, Howard JG. Reproductive inefficiency in male black-footed ferrets (*Mustela nigripes*). Zoo Biol. 2001;19:517–28.

Chapter 17
Fertility Control in Wildlife: Review of Current Status, Including Novel and Future Technologies

Cheryl Asa and Anneke Moresco

Abstract Wildlife conservation requires varying degrees of management of endangered species as well as management of their potential predators and competitors. Conservation also depends on ameliorating human-wildlife conflict, especially where there is a threat to the safety of people and of their pets and livestock. In addition, sustainability of wildlife populations can be affected by feral domestic animals or invasive species, that require removal or control. With the increasing concern for animal welfare, non-lethal methods of limiting population size and distribution, such as fertility control, are gaining favor. Breeding programs in zoos depend on highly effective and selective contraception to manage sustainable insurance populations. This review covers fertility control methods currently in use, those that have not lived up to past promises, and others that are under development and present hope for addressing remaining challenges.

Keywords Contraception · Immunocontraception · Population management · Endangered species

1 Introduction

The interface of wildlife with human activities has produced conflict for centuries. Predators present a potential threat to the safety of people and their livestock, and herbivores compete for valuable resources. The responses have varied over the years from accommodation to lethal methods. Innovative approaches to resolving these conflicts often require compromise among the various stakeholders. A common approach to controlling or reducing wildlife numbers has been through culling,

C. Asa
Saint Louis Zoo, Saint Louis, MO, USA

A. Moresco (✉)
Animal Welfare and Research Department, Denver Zoo, Denver, CO, USA
e-mail: amoresco@denverzoo.org

P. Comizzoli et al. (eds.), *Reproductive Sciences in Animal Conservation*, Advances in Experimental Medicine and Biology 1200,
https://doi.org/10.1007/978-3-030-23633-5_17

which produces an immediate but seldom a long-lasting effect. Among the possible outcomes is compensatory reproduction, due to reduced competition among remaining conspecifics for territory and access to food, which allows numbers to quickly rebound (e.g., coyotes, *Canis latrans*: Knowlton et al. 1999; introduced rabbits, *Oryctolagus cuniculus*: Twigg and Kent Williams 1999). This increased reproductive rate, as well as disruption in social structure, especially from non-targeted culling of carnivores, can also lead to higher predation on livestock, since parents with young to feed may be more likely to focus on prey such as livestock that are easier to catch (Sacks et al. 1999). Even in cases where culling successfully targets offending predators, the open habitat niche created may function as a sink for surrounding populations and thus not actually decrease the density of the target population in the area (e.g., European badger, *Meles meles*: Pope et al. 2007; African lions, *Panthera leo*: Woodroffe and Frank 2005; American mink, *Neovison vison*: Oliver et al. 2016). In cases where culling of predators (e.g., gray wolves, *C. lupus*) has led to their extirpation, relative overpopulation of their prey species (white-tailed deer, *Odocoileus virginianis*) or competitors (coyotes) has resulted.

Although not a panacea, fertility control presents a potential compromise. Unlike most lethal methods, it avoids boom and bust population cycles and favors more stable social structures, which can reduce predation and human contact (Bromley and Gese 2001a; Shuster et al. 2018). As an alternative to lethal management of wildlife populations, fertility control has become a favored option, due to a shift in public attitudes away from acceptance of lethal approaches (Oogjes 1997; Reiter et al. 1999). Contraception can be part of conservation in many ways (Cope et al. 2018). Both permanent sterilization and reversible methods have been used to limit population size in overabundant species and to reduce human/animal conflict (Massei and Cowan 2014).

2 Wildlife Fertility Control and Conservation

Successful wildlife conservation requires safe, adequate habitat, as well as successful reproduction in populations large enough to sustain genetic health. The idea of limiting reproduction seems counter-intuitive, but it may be needed to manage the local number of individuals, even in endangered species, within protected-area boundaries. Native, free-ranging wildlife may also face competition or predation from invasive or feral domestic species that might be best controlled by reducing their reproductive rates. Feral and introduced animals, such as cats, dogs, donkeys, horses, and pigs, may prey on or compete with endangered wildlife, as well as destroy habitat when present in large numbers. However, lethal control of charismatic feral species, especially horses, cats and dogs, is unacceptable to most people, and hunting is not allowed in areas with high human density where, for example, white-tailed deer may be a particular problem. Fertility control presents an alternative management strategy in situations such as these.

Wildlife under managed care, especially when part of endangered species breeding programs, are also part of the conservation equation. The IUCN

(International Union for Conservation of Nature) Conservation Planning Specialist Group recently introduced the One Plan Approach for formally incorporating zoo wildlife programs into field conservation efforts. This process integrates wild and captive population management for endangered species recovery. Although not all captive endangered species populations will be used for reintroduction or supplementation of wild populations, most are managed genetically and demographically so that they can be candidates for return to the wild. A critical component of their management involves reproductive control to allow maintenance of social and family groups while achieving genetic management to prevent inbreeding and equalize founder representation.

Contraception has been used in zoos since the mid-1970s, with almost 45,000 records of use in the joint database held by the AZA Reproductive Management Center (RMC) and the EAZA (European Association of Zoos and Aquariums) Group for Zoo Animal Contraception (EGZAC). More recently, contraception has also become an important part of managing great apes and other wildlife in sanctuaries and rehabilitation centers, where there is a need to prevent reproduction while under human care, as well as a need to maintain fertility for potential release back into the wild.

3 State of the Art: Contraceptive Methods

Fertility control approaches have been designed to interfere with the reproductive process at various points (Fig. 17.1). Most target the hormonal cascade that ultimately stimulates ovulation in females and sperm production in males, while others are directed at the gametes, that is, eggs and sperm, with the intent of preventing fertilization. This review covers the methods that are currently in use, as well as those that did not live up to their early promise and those that offer hope for improved efficacy, safety or application in the future.

3.1 Steroid Hormones

3.1.1 Progestins

Progesterone is a naturally occurring steroid hormone, produced by the ovary after ovulation. In mammals, it is responsible for the maintenance of pregnancy. Progestin-based contraceptives are marketed in various formulations, such as pills, slow-release depot-injections and implants. However, not all products are available in all countries, and they may be marketed under different brand names in different regions of the world. Progestins have effectively been used for decades as contraceptives in females of many species, because progestins can prevent reproduction via negative feedback on the hypothalamus and pituitary, suppressing release of

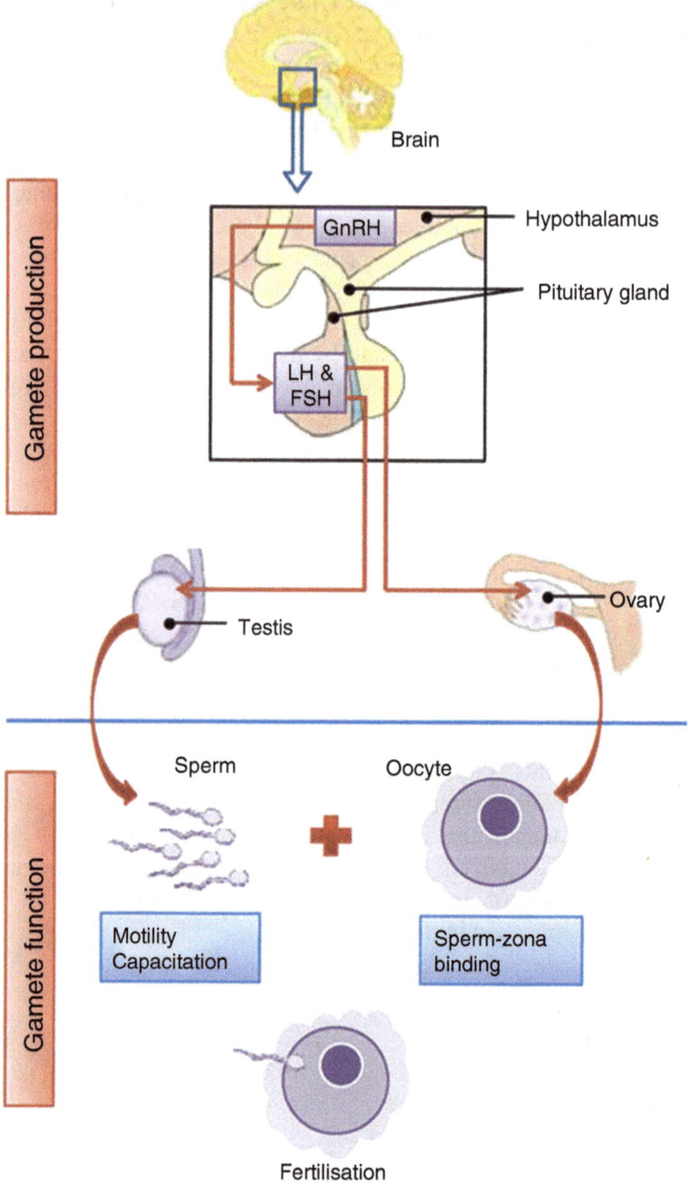

Fig. 17.1 Targets for reproductive inhibition by contraceptives. From Garside et al. (2014)

follicle-stimulating hormone (FSH) and luteinizing hormone (LH), which in turn prevents stimulation of follicle growth and ovulation. However, sufficient follicle growth and estradiol production (Kazensky et al. 1998; Beijerink et al. 2007; Moresco et al. 2009) may continue to stimulate estrous behavior, which may be seen as an advantage or disadvantage, depending on the situation. Other contraceptive effects of progestins include impeding the movement of sperm and eggs to the site of fertilization and interfering with implantation if fertilization might occur (Attardi 1984; Brache et al. 1985). Products that have been used in wildlife include melengestrol acetate (MGA), medroxyprogesterone acetate (MPA), megestrol acetate (MA), levonorgestrel (LNG), altrenogest, etonogestrel, and norgestomet. An obstacle to using progestins more widely in free-ranging animals is the need for slow-release or depot formulations for longer efficacy. This, along with the ability of progestins to be effective if ingested, introduces the potential for secondary exposure by predators or scavengers. See Table 17.1 for wildlife species treated with progestin-based contraceptives. Also see Wallace et al. (2016) for tabular summaries of RMC database records for use of progestin-only products in primates.

Megestrol acetate (MA) was previously approved in the U.S. for contraception in dogs but not cats. However, it is currently only available through compounding pharmacies in pill form, since the brands (Ovaban® in the U.S. and Ovarid® in Europe) became unavailable. MA was also previously available in the U.S. as the compounded product FeralStat®, specifically for preventing reproduction in feral cat colonies. However, MA as Megace® (Bristol-Myers Squibb) marketed for humans in the U.S., is sometimes used off-label for other species. MA has both anti-androgenic and anti-estrogenic effects, but is a cortisol agonist, which has made it useful for treating some dermatologic conditions in cats (Middleton et al. 1987), but caution is needed to prevent more profound immune suppression.

Medroxyprogesterone acetate (MPA) is most commonly used in its injectable, slow-release formulation (Depo-Provera®: Pfizer). Of the synthetic progestins currently available, MPA is the most androgenic (Labrie et al. 1987) and immunosuppressive (Hapgood et al. 2004). Experience with zoo animals has shown that the duration of action can vary by species and often results in shorter inter-dosing intervals than those recommended for women (RMC/EGZAC Database).

Etonorgestrel, levonorgestrel and norgestomet are progestin components of contraceptive products formulated as either pills or implants (numerous formulations and brands) and marketed for women. Non-human application has been mostly to primates, especially apes. Altrenogest, marketed most commonly as Regu-Mate® (Merck Animal Health), is an oral formulation developed originally for domestic horses but sometimes used for contraception in giraffe and marine mammals trained for hand-feeding in zoos and aquariums.

Most side-effects of progestin contraception are associated with the natural response of tissues to endogenous progesterone but occur at an unnatural time, duration or intensity. More serious side effects are evident at higher doses and longer-term administration, especially in carnivores, such as canids, felids and mustelids, including increased endometrial gland secretory activity and mammary gland proliferation. When reproductive tissues respond in a normal manner to an abnormally strong stimulus (or of extended duration) without the natural events that would follow, such as pregnancy, birth and lactation, proliferative lesions of the uterus and mammary glands can develop (hyperplasia or neoplasia). Additionally, synthetic progestins have an affinity for receptors other than those for progesterone,

Table 17.1 Progestins used to control reproduction in wildlife populations

Order	Species	Product	Reference
Artiodactyla	Deer, black-tailed (*Odocoileus hemionus*)	Norgestomet ballistic implants	Jacobsen et al. (1995)
	Deer, white-tailed (*O. virginianus*)	Levonorgestrel implant	Plotka and Seal (1989)
	Deer, white-tailed	MGA implant	Plotka and Seal (1989)
	Deer, white-tailed	MGA oral	Roughton (1979)
	Deer, white-tailed	Norgestomet ballistic implants	DeNicola et al. (1997)
	Giraffe (*Giraffa camelopardalis*)	DMPA	Patton et al. (2005)
	Hippopotamus (*Hippopotamus amphimbius*)	DMPA-high failure rate	Patton et al. (2007)
	Zoo bovids	MGA oral (pelleted)	Raphael et al. (2003)
Carnivora	Cat, domestic (*Felis catus*)	Levonorgestrel	Baldwin et al. (1994)
	Cat, domestic	Megestrol acetate	Remfry (1978)
	Lions (*Panthera leo*)	MGA implant	Orford et al. (1988)
	Tigers (*P. tigris*)	MGA implant	Chuei et al. (2007)
	Various zoo felids, canids and ursids	MGA implant	DeMatteo (2005)
Diprodontia	Kangaroo, eastern gray (*Macropus giganteus*)	Levonorgestrel implant; 2 × 70 mg	Nave et al. (2002)
	Koala (*Phascolarctos cinereus*)	Etonogestrel	Hynes et al. (2011)
	Koala (*P. cinereus*)	Levonorgestrel	Middleton et al. (2003)
	Wallaby, tammar (*Notamacropus eugenii*)	Levonorgestrel implant; 2 × 70 mg	Nave et al. (2002)
Perissodactyla	Horse, domestic (*Equus caballus*)	Regumate® DMPA no effect	Gee et al. (2009)
	Zoo tapir spp	DMPA	Patton et al. (2007)
	Zoo tapir spp	MGA implant	Patton et al. (2007)
Primata	Macaques, barbary (*Macaca sylvanus*)	Implanon®	Maijer and Semple (2016)
	Tamarin, cotton top (*Saguinus oedipus*)	Levonorgestrel	Wheaton et al. (2011)
	Various primate species	Various progestins	Wallace et al. (2016)

MGA melengestrol acetate, *DMPA* depo-medroxyprogesterone acetate

such as androgen and glucocorticoid receptors (Duncan et al. 1964; Fekete and Szeberényi 1965; Kloosterboer et al. 1988). Which receptor and the degree of affinity vary with the specific progestin (Sitruk-Ware 2008). The specific receptor with which it cross-reacts determines the range of possible side effects (Sloan and Oliver 1975; Selman et al. 1997). For example, binding androgen receptors can stimulate physical or behavioral masculinization (Wilkins 1960; Asa et al. 2007), and binding to glucocorticoid receptors can disrupt glucose dynamics and suppress immune function (Middleton et al. 1987).

Progestin treatment is also commonly associated with increased appetite and weight gain (Romatowski 1989), whereas reports of hair loss or discoloration are infrequent (Evans and Sutton 1989). In domestic dogs and cats, growth hormone is locally produced in the mammary gland as a natural response to progesterone (Selman et al. 1994; Mol et al. 1996). Suppression of the immune system and altered glucose metabolism which can be associated with progestational-type diabetes mellitus have also been reported. This effect has also reported anecdotally for primates and some carnivores (Reproductive Health Surveillance Program and RMC/EGZAC Databases). However, no systematic study has been conducted to document that glucose intolerance and bouts of diabetes are caused by contraceptive use, as opposed to exacerbation of a pre-existing condition.

In zoos, reproductive lesions such as endometrial hyperplasia (EH), hydrometra, and uterine infections (e.g., pyometra) have been observed in non-treated animals (Munson et al. 2005), but prevalence tends to be higher in progestin-treated animals. A study that focused specifically on canids (Asa et al. 2014) found higher association of EH and pyometra with progestin contraception, but interestingly that association was also true for females separated from males and not allowed to breed. A subsequent review (Penfold et al. 2014) found that trend for higher prevalence of uterine pathology in non-breeding females across a number of other taxa. Thus, separation to prevent reproduction did not protect those females from uterine changes that may impair fertility. In contrast, in free-ranging lions implanted with MGA, removal of implants allowed reversal, but a high mortality rate was reported (Orford et al. 1988). The authors did not explain the causes of mortality, but mentioned that pyometra was not found. However, those lions were treated for only 22 months; studies by Munson and colleagues (Munson et al. 2005) in various felid species showed that longer treatment increased the likelihood of uterine pathology.

One effect of progestin treatment is that folliculogenesis is not completely suppressed. In one published study, altrenogest-implanted white-tailed deer each had one large abnormal Graafian follicle but no pregnancies (Matschke 1977a, b). Similarly, folliculogenesis and even ovulation were also documented in felids and canids while exposed to MGA (Kazensky et al. 1998; Moresco et al. 2009).

Oral MGA administration to pregnant white-tailed deer was not associated with interrupted pregnancies, changes in gestation length, stillbirth, or parturition difficulty (dystocia) (Roughton 1979). However, in another white-tailed deer study, MGA implants did not interrupt pregnancy but did interfere with parturition (Plotka and Seal 1989). Perhaps these contrasting results of effect on parturition in deer may have been related to dose (e.g., in rat: Duncan et al. 1964), but circulating levels were not measured, and differing routes of delivery prevent direct comparison of dosage administered. MGA was found to delay parturition in mink (Jarosz and Dukelow 1985) and laboratory rats (Duncan et al. 1964), suggesting carnivores and rodents may also be subject to this effect. According to reports to the RMC/EGZAC Database, primates may be an exception, in that progestin-treated females have given birth normally. Nevertheless, caution is warranted, and particularly in non-primate species progestins are not recommended in pregnant animals. For seasonally breeding species, treatment should occur or begin well in advance of the breeding season. For others, and ultrasound exam or pregnancy test should precede treatment.

The various progestin contraceptive formulations have different durations of action and therefore different re-dosing intervals, which could be an advantage or disadvantage in managing free-ranging animals. Melengestrol acetate implants are effective a minimum of 2 years but require surgical placement. Depot injection formulations (MPA) last approximately 1.5–3 months, although duration can vary by species. Norgestomet implants have been delivered remotely into the muscle in biobullets (Jacobsen et al. 1995; DeNicola et al. 1997), but that product was never commercially available. Daily administration of pre-mixed feed for ungulates (MGA feed) and oral contraceptive pills are available but unlikely to be useful for free-ranging individuals. When treating free-ranging populations, there is the potential to affect non-target species, because steroids remain systemically bioactive after administration, and predators or scavengers can be unintentionally affected by secondary exposure. However, the potential for steroid hormone-treated animals to enter the human food chain (e.g., hunted white-tailed deer) is an even more important factor preventing such application to free-ranging wildlife.

3.1.2 Estrogens

Synthetic estrogens are more effective than progestins at preventing follicle growth, but they also have been associated with more serious side effects, although those vary by species. Early in the development of contraception in wildlife, diethylstilbestrol (DES) was used with variable success in coyotes (Balser 1964) red foxes, *Vulpes vulpes* (Linhart and Enders 1964), white-tailed deer (Matschke 1977a) and black-tailed prairie dogs, *Cynomys ludovicianus* (Garrett and Franklin 1983) among others. However, the main challenges were that efficacy was heavily dependent on precisely timed administration, and in some species pregnancy was disrupted but females rebred. Additionally, in the trials, the drug was administered in bait, which can be ingested by non-target animals (Balser 1964). Chronic exposure to estrogens (exogenous or endogenous) has been associated with bone marrow suppression and potentially life-threatening pancytopenia in domestic dogs and ferrets, and is not considered a good option for reproductive control.

3.1.3 Estrogen and Progestin Combinations

When combined with a progestin, lower doses of estrogens are able to suppress follicle growth, while reducing the potential deleterious effects of estrogens. Combination pills are used to manage reproduction at rehabilitation facilities for orangutans (F. Sulistyo, pers. comm), gorillas, chimpanzees and bonobos (M. McDonald, pers. comm). In many cases, the animals will be returned to the wild, so fertility needs to be maintained without adding to the burden of the center. Combination implants also have been used in female feral horses (Eagle et al. 1992).

Foaling rates were lower than for controls, but due to loss of implants when placed in the neck region, implants were placed intraperitoneally, an approach associated with more serious consequences if infection occurs. The risk of secondary exposure through predation or scavenging remains.

3.2 GnRH Agonists: Deslorelin and Leuprolide

GnRH agonists achieve contraception by reversibly suppressing the reproductive endocrine system, preventing release of pituitary follicle-stimulating hormone (FSH) and luteinizing hormone (LH) and hence the gonadal hormones (estradiol and progesterone in females and testosterone in males). The observed effects are similar to those following either ovariectomy in females or castration in males, but can be reversible. GnRH agonists first stimulate the reproductive system, which can result in estrus and ovulation in females or temporary enhancement of testosterone and semen production in males. Then, down-regulation follows the initial stimulation. The stimulatory phase can be prevented in females by administration of a synthetic progestin for 1 week before and 1 week after implant placement (Wright et al. 2001). The effects of GnRH agonists on gestation have not been characterized in detail for most wildlife species. Therefore, they are not recommended for use during pregnancy because they can cause spontaneous abortion or prevent mammary development necessary for lactation. They also may prevent initiation of lactation by inhibiting progesterone secretion, but effects on established lactation are less likely.

Although GnRH agonists can also be an effective contraceptive in males, they are used more commonly in females, because monitoring efficacy by suppression of estrous behavior or cyclic gonadal steroids in feces is usually easier than ensuring continued absence of sperm in males. In free-ranging wildlife, females are often targeted because any remaining untreated, fertile males could still impregnate females, especially true for polygynous species. When used in males, following down-regulation of testosterone, disappearance of sperm from the ejaculate may take an additional 6 weeks, as is the case following vasectomy. In seasonally breeding species, it should be easier to suppress the onset of spermatogenesis than to interrupt ongoing sperm production. Onset begins at least 2 months before the first appearance of mature sperm, so treatment should be initiated at least 2 months before the anticipated onset of breeding.

The GnRH agonist used most widely as a contraceptive in captive wildlife is deslorelin acetate (Suprelorin®: Virbac Animal Health) in slow-release, lipid-based implants (see Table 17.2; Wallace et al. 2016). Two formulations are available, designed to be effective either 6 (4.7 mg) or 12 (9.4 mg) months, but those are only minimum durations. Leuprolide acetate (Lupron Depot®: AbbVie) is commercially available as a long-acting injection for treatment of endometriosis, prostate cancer and precocious adolescence in humans in three doses, with expected durations of 1, 3, 4 and 6 months. The major drawback of these products is that time to reversal

Table 17.2 GnRH agonists used to control reproduction in wildlife populations

Order	Species	Product	Reference
Artiodactyl	Deer, black-tailed (*Odocoileus hemionus*)	Leuprolide	Baker et al. (2004)
	Elk (*Cervus elpahus*)	Leuprolide	Baker et al. (2002)
	Giraffe (*Giraffa camelopardalis*)	Suprelorin® 9.4 mg	Patton et al. (2006)
	Hog, red river (*Potamochoerus porcus*)	Deslorelin and leuprolide	Goblet et al. (2019)
	Muntjac, Chinese (*Muntiacus reevesi*)	Deslorelin	RMC
	Warthog (*Phacochoerus africanus*)	Deslorelin	Goblet et al. (2019)
Carnivora	African painted dog (*Lycaon pictus*)	Deslorelin 6 mg	Bertschinger et al. (2001)
	Cat, domestic (Felis catus)	Deslorelin 6.0 mg, and Suprelorin® 4.7 mg	Goericke-Pesch et al. (2011) and Munson et al. (2001)
	Cheetah (*Acinonyx jubatus*)	Deslorelin 6 mg	Bertschinger et al. (2001)
	Ferret, domestic (*Mustela putorius furo*)	Deslorelin	Prohaczik et al. (2003)
	Fox, fennec (*Vulpes zerda*)	Deslorelin	RMC
	Jaguar (*Panthera onca*)	Deslorelin	RMC
	Leopard (*P. pardus*)	Deslorelin 6 mg	Bertschinger et al. (2001)
	Lion (*P. leo*)	Suprelorin® 4.7 mg or 9.4 mg	Bertschinger et al. (2001)
	Meerkat, slender-tailed (*Suricata suricatta*)	Deslorelin	RMC
	Mongoose, banded (*Mungos mungo*)	Deslorelin	RMC
	Otter, Asian small-clawed (*Aonyx cinerea*)	Deslorelin	RMC
	Otter, giant (*Pteronura brasiliensis*)	Deslorelin	RMC
	Otter, North American river (*Lutra canadensis*)	Deslorelin	RMC
	Otter, sea (*Enhydra lutris*)	Suprelorin® 4.7 mg or 9.4 mg	Larson et al. (2013)
	Pinniped species	Lupron	Calle (2005)
	Red panda (*Ailurus fulgens*)	Suprelorin® 4.7 mg	Koeppel et al. (2014)
	Tiger (*P. tigris*)	Suprelorin® 4.7 mg or 9.4 mg	Bertschinger et al. (2008)
	Wolf, maned (*Chrysocyon brachyurus*)	Deslorelin	RMC
	Wolf, Mexican (*Canis lupus bailey*)	Deslorelin	RMC
	Wolf, red (*C. rufus*)	Deslorelin	RMC

(continued)

Table 17.2 (continued)

Order	Species	Product	Reference
Chiroptera	Flying fox, Indian (*Pteropus giganteus*)	Deslorelin	RMC
	Flying fox, large (*P. vampyrus*)	Deslorelin	RMC
	Flying fox, variable (*P. hypomelanus*)	Suprelorin® 4.7 mg	Metrione et al. (2008); RMC
Columbiformes	Pigeon (*Columba livia*)	Suprelorin® 4.7 mg	Cowan et al. (2014)
Dasyuromorphia	Tasmanian devil (*Sarcophilus harrisii*)	Suprelorin® 4.7 mg	Cope et al. (2018)
Diprotodontia	Brushtail possum (*Trichosurus vulpecula*)	Suprelorin® 4.7 mg	Lohr et al. (2009)
	Kangaroo, eastern grey (*Macropus giganteus*)	Suprelorin® 9.4 mg	Wilson et al. (2013)
	Tammar wallaby (*Notamacropus eugenii*)	Suprelorin® 5 mg	Herbert et al. (2005)
	Wallaby, red-flanked rock (*Petrogale lateralis*)	Suprelorin® 4.7 mg and 9.4 mg	Willers et al. (2015)
Galliformes	Quail, Japanese (*Coturnix coturnix japonica*)	Suprelorin® 4.7 mg	Petritz et al. (2013)
Hyracoidea	Hyrax, rock (*Procavia capensis*)	Suprelorin®	Raines and Fried (2016)
Primata	Baboon, hamadryas (*Papio hamadryas*)	Deslorelin	RMC
	Colobus, eastern black and white (*Colobus guereza*)	Deslorelin	RMC
	Lemur, black and white ruffed (*Varecia variegata*)	Deslorelin	RMC
	Lemur, red ruffed (*V. rubra*)	Deslorelin	RMC
	Lemur, ring-tailed (*Lemur catta*)	Deslorelin	RMC
	Macaque, Japanese (*Macaca fiscata*)	Deslorelin	RMC
	Mandrill (*Mandrillus sphinx*)	Leuprolide	Okeson et al. (2016)
	Marmoset, Goeldi's (*Callimico goeldii*)	Deslorelin	RMC
	Marmoset, pygmy (*Cebuella pygmaea*)	Deslorelin	RMC
	Monkey, black handed spider (*Ateles xxx*)	Deslorelin	RMC
	Monkey, black headed spider (*A. fusciceps fusciceps*)	Deslorelin	RMC
	Monkey, black howler (*Alouatta caraya*)	Deslorelin	RMC
	Monkey, squirrel (*Saimiri scuireus*)	Deslorelin	RMC
	Saki, white-faced (*Pithecia pithecia*)	Deslorelin	RMC
	Tamarin, cotton-top (*Saguinus oedipus*)	Deslorelin	RMC
	Tamarin, golden lion (*Leontopithecus rosalia*)	Deslorelin	RMC
	Tamarin, golden-headed (*L. chrysomelas*)	Deslorelin	RMC

(continued)

Table 17.2 (continued)

Order	Species	Product	Reference
Proboscidae	Elephant Asian (*Elephas maximus*)	Lupron Depot® 3 mo, 22.5 mg	de Oliveira et al. (2004)
Psittaciformes	Cockatiels (*Nymphicus hollandicus*)	Leuprolide	Millam and Finney (1994)
Squamata	Bearded dragon (*Pogona vitticeps*)	Suprelorin® 4.75 mg	Rowland (2011)
Stringiformes	Owl, elf (*Micrathene whitneyi*)	Leuprolide, suspect anaphylactic reaction	Stringer et al. (2011)

Data from the Reproductive Management Center (RMC) Database are included for species with 10 or more bouts available, but the complete list of species treated with deslorelin is more extensive

cannot be predicted. However, removal of the Suprelorin® implants, facilitated by placement that makes them easier to locate and retrieve, seems to shorten reversal time (Cowl et al. 2018). In some cases, ultrasound can be used to aid locating implants (Moresco et al. 2014). A disadvantage of the depot injection (Lupron Depot®) is that it cannot be removed to shorten the duration of efficacy.

Main adverse effects of the GnRH agonists are similar to those of surgical gonadectomy: weight gain and loss of secondary sexual characteristics, particularly in males such as lions that may lose their manes. The absence of sexual behavior can also profoundly affect social organization and cohesion. In a colony of macaques where the males were castrated in an attempt to curb reproduction, massive social disruption resulted in more pronounced human-animal conflicts (Malaivijitnond and Hamada 2008).

In free-ranging animals, treatment of pregnant animals may be inevitable, as diagnosing a pregnancy prior to treatment may not always be possible. GnRH agonists did not disrupt pregnancy, even in the diapause stage, of tammar wallabies, *Macropus eugenii*, and domestic ferrets (Herbert et al. 2004; Wilson et al. 2013) but was believed responsible for early pregnancy loss in gray wolves, *Canis lupus* (W. Jöchle report to the RMC/EGZAC Database). Evidence of disruption of lactation is sparse and not unanimous, with some case reports suggesting administration of GnRH agonists negatively affected lactation (e.g., African painted dogs and Mexican gray wolves, *C. l. baileyi*: report to RMC/EGZAC Database), while others suggest GnRH did not affect lactation or maternal care (Wilson et al. 2013). No experimental studies have been carried out specifically with the goal of assessing safety in pregnant animals and the effect on lactation. Therefore, the differences in these reports may be due to species differences or the small number of individuals.

GnRH agonists have been used in captive and free-ranging lions, cheetahs, leopards and African painted dogs (Bertschinger et al. 2001). Contraception was successful in females for about 1 year and longer in males of some species. Social hierarchy of cheetahs was specifically noted to be unaffected. No notes on social dynamics were reported for treated African painted dogs, leopards and lions. Leuprolide implants have been used in captive elk, *Cervus elaphus nelsoni*, as a pilot study for proposed application to free-ranging elk (Baker et al. 2002, 2005). Pregnancy was successfully prevented, no adverse effects (physiologic or behavior) were observed, and females returned to fertility the following year. See Table 17.2 for a list of wildlife species treated with GnRH agonists.

Because GnRH agonists are peptide hormones and would most likely be denatured in the gastro-intestinal tract of an animal exposed to them by predation or scavenging, this class is unlikely to affect the next trophic level. Disadvantages to field application would more likely be due to the short duration of efficacy (1–3 years) and to the need for capture to administer the implant. Analyses of zoo canid populations showed them to be at increased risk of pyometra if the stimulation phase following administration of the GnRH agonist Suprelorin® was not prevented by the administration of oral progestin (Asa et al. 2014). However, field administration of a progestin prior to GnRH agonist administration is not feasible for free-ranging carnivores. To date no other serious adverse effects have been documented in the species under managed care (RHSP, RMC/EGZAC Databases). Similarly, no deleterious effects were seen in either captive or free-ranging lions and tigers treated with Suprelorin (Bertschinger et al. 2008).

3.3 *Immunocontraception*

Immunocontraceptives are the most extensively tested and used class of contraceptives for free-ranging wildlife. They interfere with reproduction by provoking an immune reaction, thus generating antibodies to some aspect of the reproductive process. The two most studied vaccines for wildlife contraception are directed against the zona pellucida (ZP), the outer membrane of the mammalian oocyte that must be penetrated by sperm to achieve fertilization, and against GnRH, the hypothalamic peptide that controls the hormonal cascade culminating in ovulation in females and sperm production in males. The efficacy, duration, and side effects can vary according to the active component of the vaccine, the delivery system, dose and adjuvant and by species, age, gender and individual (Miller et al. 2008; Kirkpatrick et al. 2011). Differing results are also a function of the formulations, that is, whether they are native, purified or recombinant ZP vaccines (Miller et al. 2009; Kirkpatrick et al. 2011; Bechert et al. 2013).

3.3.1 Zona Pellucida Vaccines: ZonaStat-H, ZonaStat-D, pZP-22, SpayVac, Recombinant ZP

Anti-zona pellucida vaccines prevent conception by blocking sperm binding and penetration of the zona pellucida membrane of the oocyte. However, Bechert et al. (2013) documented impaired ovarian function in mares vaccinated with a porcine ZP (pZP) vaccine, using transrectal ultrasound and steroid hormone concentrations. These findings were confirmed in a subsequent study with pony mares (Joonè et al.

2017a), suggesting that additional mechanisms of action for pZP vaccines are likely, at least in horses. The most commonly used formulation of pZP (Liu et al. 1989) is registered with EPA for feral horses and burros (ZonaStat-H) and for white-tailed deer (ZonaStat-D). Because pZP is only weakly immunogenic, it is emulsified with an adjuvant (originally with Freund's Complete Adjuvant and since 2001, with Modified Freund's Adjuvant in the primer and Freund's Incomplete Adjuvant in the boosters). ZonaStat-H and -D are most effective when the primer is followed by a first booster within 1 month, then additional boosters every 8–12 months, depending on whether the species is a seasonal breeder. There are also species differences in duration of suppression. A notable exception to the need to only re-treat annually was found in Himalayan tahr (*Hemitragus jemlahicus*), mouflon (*Ovis orientalis*), Dall sheep (*O. dalli*), big horn sheep (*O. canadensis*), and domestic goats (*Capra hircus*). Suppression in these species extended for 2–3 years, suggesting that members of the Capridae and Ovidae families may have more prolonged responses (Lyda et al. 2013). PZP-22 uses a ZonaStat injection combined with controlled-release polymer pellets, rates of contraception in free-ranging mares were 94%, 86%, 68%, and 52% for 1–4 years post-treatment (Turner et al. 2007). However, in a subsequent study with mares, contraceptive efficacy dropped to 79% and 38% for 1–2 years post-treatment (Rutberg et al. 2017).

Another pZP product (SpayVac®, SpayVac-for-Wildlife, Inc.) encapsulates the pZP vaccine in liposomes to extend contraception efficacy (Brown et al. 1997) and uses either Modified Freund's or AdjuVac™ as the adjuvant. SpayVac® has achieved multi-year contraceptive efficacy with a single injection in a variety of species, including 10 years in grey seals, 4 years in horses, and over 3 years in deer (see Bechert and Fraker 2018 for review). A single injection of SpayVac® resulted in 7+ years of consistently elevated pZP-antibody titers in six captive African elephants (Bechert and Fraker 2016), but reversibility has not been confirmed.

Various pZP formulations have been used extensively in ungulates (see Table 17.3). A relatively small number of carnivore species has also been treated. This method was also tried in feral cats, but efficacy was poor and severe adverse effects have been documented in various species of felids (Munson et al. 2005; Fischer et al. 2018).

Safety of the pZP vaccine for non-target animals is a clear advantage of this method. One aspect that may be considered an advantage in free-ranging wildlife is that, in contrast to animals under managed care, reduction in litter size is also a good result, as it still curbs reproduction but does not eliminate the genetic contribution from the treated individual. Such a reduction, without complete contraception, has been documented in coyotes (Miller et al. 2006) and in white-tailed deer (Miller et al. 2009).

Treatment of coyotes with pZP did not interfere with the estrous cycle, number of ova produced and ovulated, or whelping dates (Miller et al. 2006). In contrast, there are numerous reports of ovarian dysfunction in other species, especially after prolonged pZP treatment (reviewed in Joonè et al. 2017b). Possible explanations for these changes include contamination with non-ZP ovarian proteins, the glycosylation of the ZP molecule, direct effects of Freund's adjuvants, and cytotoxic t-cells

Table 17.3 Earliest or most complete reports of immunocontraceptives used to control reproduction in wildlife populations

Order	Species	Product	Reference
pZP			
Artiodactyla	Bighorn sheep (*Ovis canadensis*)	pZP emulsified	Frank et al. (2005)
	Bison (*Bison bison*)	pZP emulsified	Frank et al. (2005)
	Deer, fallow (*Dama dama*)		Fraker et al. (2002)
	Deer, white-tailed (*Odocoileus virginianus*)	Microencapsulated	McShea et al. (1997)
	Deer, white-tailed	pZP emulsified	McShea et al. (1997)
	Elk, tule (*Cervus elaphus nannodes*)	pZP emulsified	Shideler et al. (2002)
	Gerenuk (*Litocranius wallerii*)	pZP emulsified	Frank et al. (2005)
	Giraffe (*Giraffa* spp)	pZP emulsified	Frank et al. (2005)
	Himalayan Tahr (*Hemitragus jemlahicus*)	pZP emulsified	Frank et al. (2005)
	Ibex (*Capra ibex*)	pZP emulsified	Frank et al. (2005)
	Mountain goat (*Oreamnos americanus*)	pZP emulsified	Frank et al. (2005)
	Reindeer (*Rangifer tarandus*)	pZP emulsified	Frank et al. (2005)
	Sambar deer (*Cervus unicolor*)	pZP emulsified	Frank et al. (2005)
	Sika deer (*C. Nippon*)	pZP emulsified	Frank et al. (2005)
	Wapiti (*C. elaphus*)	pZP emulsified	Frank et al. (2005)
Carnivora	Bear, Asian black (*Selenarctos thibetanus*)	pZP emulsified	Frank et al. (2005)
	Bear, black (*Ursus americanus*)	pZP emulsified	Frank et al. (2005)
	Bear, brown (*U. arctos*)	pZP emulsified	Frank et al. (2005)
	Bear, sun (*Helarctos malayanus*)	pZP emulsified	Frank et al. (2005)
	Coyote (*Canis latrans*)	pZP-Dunbar and Liu	Miller et al. (2006)
	Coyote (*C. latrans*)	pZP-SpayVac™	Miller et al. (2006)
Otariidae	California sea lion (*Zalophus californian*)	pZP emulsified	Frank et al. (2005)
Perissodactyla	Burro, feral (*Equus asinus*)	pZP emulsified	Turner et al. (1996)
	Horse, Przewalski (*E ferus przewalski*)	pZP emulsified	Frank et al. (2005)
	Horse, feral (*E. caballus*)	pZP emulsified	Kirkpatrick et al. (1992)
	Zebra (*E. quagga burchellii*, *E. grevyi* and *E. zebra*)	pZP emulsified	Frank et al. (2005)
Primata	Baboon (*Papio anubis*)	ZP1 and ZP3	Dunbar et al. (1989)
	Macaque, cynomolgus (*Macaca fascicularis*)	pZP with alhydrogel (adjuvant)	Gulyas et al. (1983)
	Monkey, squirrel (*Saimiri sciureus*)	ZP3	Sacco et al. (1983)
	Eulemur spp.	pZP	Wallace et al. (2016)
	Otolemur spp.	pZP	Wallace et al. (2016)
	Aotus spp.	pZP	Wallace et al. (2016)
	Callimico goeldii	pZP	Wallace et al. (2016)
	Cercopithecus spp.	pZP	Wallace et al. (2016)
	Colobus spp.	pZP	Wallace et al. (2016)
	Erythrocebus patas	pZP	Wallace et al. (2016)
	Pan spp	pZP	Wallace et al. (2016)
	Pongo pygmaeus	pZP	Wallace et al. (2016)

(continued)

Table 17.3 (continued)

Order	Species	Product	Reference
Proboscidea	Elephant, African (*Loxodonta africana*)	pZP emulsified	Fayrer-Hosken et al. (1999)
GnRH			
Artiodactyla	Deer, black-tailed (*Odocoileus hemionus*)	GonaCon™	Perry et al. (2006)
	Deer, fallow (*Dama dama*)	GonaCon™	Gionfriddo et al. (2008)
	Deer, rusa (*Rusa timorensis*)	Improvac®	Phraluk et al. (2015)
	Deer, white-tailed (*Odocoileus virginianus*)	GnRH (LA Miller)	Curtis et al. (2002)
	Deer, white-tailed	GonaCon™	Killian et al. (2008)
	Elk (*Cervus elaphus*)	GonaCon™	Killian et al. (2009)
	Cattle, feral (Bos taurus)	GonaCon™	Massei et al. (2015)
	Goat, domestic (*Capra hircus*)	Improvac®	Bishop et al. (2016)
	Swine, domestic (*Sus scrofa*)	Improvact®	Dunshea et al. (1993)
	Swine, feral (*S. scrofa*)	GonaCon™	Massei et al. (2008)
Carnivora	Cat (*Felis catus*)	GonaCon™	Levy et al. (2011)
Marsupialia	Kangaroo, eastern gray (*M. acropus giganteus*)	GonaCon™	Snape (2012)
	Wallaby, tammar (*M eugenii*)	GonaCon™	Snape (2012)
Perissodactyla	Rhinoceros, greater one-horned (*Rhinoceros unicornis*)	Improvac®	Hermes et al. (2016)
	Rhinoceros, southern white (*Ceratotherium simum simum*)	Improvac®	Hermes et al. (2016)
	Horse, domestic (*Equus caballus*)	Improvac®	Schulman et al. (2013)
Proboscidea	Elephant, African (*Loxodonta Africana*)	Improvac®	Lueders et al. (2017)
	Elephant, Asian (*Elephas maximas*)	Improvac®	Lueders et al. (2014)
Rodentia	Prairie dog, black-tailed (*Cynomys ludovicianus*)	GonaCon™	Yoder and Miller (2010)
	Squirrel, California ground (*Spermophilus beecheyi*)	GonaCon™	Nash et al. (2004)
	Squirrel, eastern grey (*Sciurus carolinensis*)	GonaCon™	Pai et al. (2011)

Only species with N ≥ 10 from Frank et al. (2005) are included here. The section on GnRH immuno-contraception includes species for which males were contracepted, as well as females. In some cases, success was measured by hormone suppression, histology and/or titers

that mediate the immune response (Joonè et al. 2017b). These ovarian changes do not decrease contraceptive efficacy but do interfere with ovulatory cycles and the expression of estrous behavior.

In felids, treatment with pZP led to severe reactions at the injection site, lameness and hypercalcemia and was unsuccessful in achieving contraception (Munson et al. 2005). Adverse effects include the occurrence of abscesses at the injection sites (more prevalent with darting than hand injections). However, in ungulates these abscesses do not seem to have severe or long-term clinical effects. In feral horses, only 5–6% of doses delivered by dart resulted in abscesses and most were resolved by 90 days (Roelle and Ransom 2009). Recent success with a non-Freund's adjuvant in domestic horses may address the concern about possible injection-site reactions (Nolan et al. 2019). The success of initial trials with a recombinant ZP vaccine in mares after an immune stimulant Poly(I:C) was incorporated into the formulation (Nolan et al. 2018) has also been encouraging.

Hypothetical adverse effects on the population include the possibility that immunocontraception would select against animals with the best immune system (Nettles 1997), thereby putting the population at a higher risk for infectious disease. However, to date that has not been observed, even after long-term treatment of some populations. In fact, pZP-managed feral horses on Assateague Island are documented to be in better body condition and to live longer than non-treated females (Kirkpatrick and Turner 2007). Because pZP is a protein, it does not persist in the environment or bio-accumulate in the food chain, so presents no threat to non-target species.

3.3.2 Anti-Gonadotropin-Releasing Hormone (GnRH) Vaccines: GonaCon™, Improvac®, Improvest®

As described for GnRH agonists, effectively interrupting the action of GnRH can prevent reproduction in both males and females, making it an attractive target. Antibodies generated against GnRH can neutralize GnRH molecules in the vasculature between the hypothalamus and the anterior pituitary (Fagerstone et al. 2010), which eliminates stimulation of gonadotropin release. However, the small GnRH molecule is not immunogenic (Herbert and Trigg 2005), but efficacy can be improved by coupling to another molecule, such as keyhole limpet hemacyanin (KLH), as has been reported for GonaCon™ (Miller et al. 2003).

Addition of an adjuvant is also necessary to promote an adequate immune response. The most effective adjuvant identified over time has been Freund's Complete Adjuvant (FCA), but because it causes treated animals to subsequently test positive to the tuberculosis skin test and can cause injection-site reactions, FDA does not allow FCA as the adjuvant in wildlife fertility-control agents (Eisemann et al. 2006). To address this problem, the USDA National Wildlife Research Center (NWRC) developed AdjuVac™, which consists of a modified USDA-approved vaccine for Johne's disease (Mycopar: Ft. Dodge Animal Health) containing inactivated *Mycobacterium avium*. GonaCon™, coupled with KLH and AdjuVac, has been successful in limiting reproduction in numerous species in research trials (see Table 17.3), but at present it is only registered for use in white-tailed deer. GonaCon™ also has been tested in domestic dogs and cats in anticipation of its possible use in feral animals, as part of a strategy to include contraception in an effort to combat spread of infectious diseases such as rabies (Vargas-Pino et al. 2013).

Another anti-GnRH vaccine, marketed as Improvac® in Europe and South Africa and as Improvest® in the U.S. (Zoetis Animal Health), was developed to prevent "boar taint" in domestic swine but is being used in some captive wildlife, especially ungulates. Similar Zoetis products are marketed in Australia for horses (Equity®) and in South America for cattle (Bopriva®). Because the product compositions are considered proprietary, the types of adjuvant and presence of any other coupled molecules for increasing antigenicity are unknown. Improvac and Improvest are less useful for use in free-ranging wildlife due to the need for frequent booster injections.

Abscesses and granulomas can occur following GnRH vaccination injections, as with pZP (Powers et al. 2011; Gionfriddo et al. 2008, 2011). Female elk treated with GonaCon showed pre-copulatory behaviors throughout the normal breeding season, suggesting that ovarian hormone suppression may not be complete (Powers et al. 2011). Treatment of pregnant females with GnRH vaccines has been reported to not affect pregnancy in elk (Powers et al. 2011), feral pigs (Killian et al. 2003), bison (Miller et al. 2004), and dogs (Vargas-Pino et al. 2013), but may have affected pregnancy success in white-tailed deer (Miller et al. 2000).

Similar to surgical castration, vaccination with GonaCon™ negatively affected antler development and testes morphology in white-tailed deer (Miller et al. 2000). Also similar to gonadectomy, body-condition scores in treated adult males were higher than in non-treated individuals (Gionfriddo et al. 2011). Most studies have not attempted to evaluate reversibility after long-term administration of GnRH vaccines, but reversibility after short-term administration appears to be good in elk (Powers et al. 2011), mares (Killian et al. 2006), and white-tailed deer (Miller et al. 2008).

Safety trials involving gross and histological analyses of organs have not yet been conducted for GnRH vaccines such as GonaCon™. GnRH vaccines may have long-term side effects (Kirkpatrick et al. 2011), because GnRH receptors are located in a variety of tissues, including central nervous system, liver, kidney, adrenal gland, spleen urinary tract and heart (Skinner et al. 2009). However, effects in non-reproductive tissues have not been reported for either the anti-GnRH vaccines or GnRH agonists. Because administration is by dart or by hand-injection, accidental primary administration to non-target animals should be rare. As with pZP vaccine, the GnRH vaccine has a protein-based antigen and is not a threat to non-target species via consumption of treated animals.

3.4 Chemical Methods

3.4.1 Nicarbazin

As a compound approved by FDA to treat coccidiosis in broiler chickens, nicarbazin was found to reduce egg hatchability by increasing permeability of the membrane separating the yolk from albumin. There are no other known side effects. It is cleared within 48 h, so requires regular ingestion to be effective, but consequently is readily reversible (Fagerstone et al. 2010). Nicarbazin has been shown to be effective in Canada geese, *Branta canadensis*; mallard ducks, *Anas platyrhynchos* (Yoder et al. 2005); and pigeons, *Columba livia* (Avery et al. 2008a). It is now EPA registered for Canada geese as OvoControl®G and for pigeons as OvoControl®P (Innolytics). Species-specific delivery of these oral products is accomplished with feeders designed especially for the target species, and game cameras can be used to track numbers of birds visiting the feeders.

3.4.2 DiazaCon

First registered as Ornitrol for pigeons in the late 1960s, this cholesterol mimic (22,25-diazacholesterol) inhibits biosynthesis of pregnenolone, the precursor to the reproductive steroid hormones, e.g., testosterone in males and estrogens and progesterone in females, leaving the treated animal infertile (Fagerstone et al. 2010). Registration was cancelled in 1993, perhaps related to challenges to ensuring sufficiently regular ingestion in this year-round breeder. However, it persists in the body for several months and high doses can produce serious, even fatal, outcomes (Lofts et al. 1968), likely related to the need for cholesterol as a precursor for other steroid hormones that regulate metabolism and kidney function. It has been tested more recently, as the product DiazaCon™, in other species, including monk parakeets (*Myiopsitta monarchus*), Japanese quail (*Coturnix japonica*), ring-necked doves (*Streptopelia capicola*), brown-headed cowbirds (*Molothrus ater*), gray squirrels (*Sciurus carolinensis*) and prairie dogs (*Cynomys ludovicianus*) (Avery et al. 2008b; Yoder et al. 2011; Fagerstone et al. 2010; www.aphis.usda.gov/aphis/ourfocus/wildlifedamage/programs/nwrc/research-areas/sa_reproductive_control/ct_diazacon). Although originally targeting birds, the biosynthetic pathway for sex steroid hormones in mammals is similar, so the product should work in those species as well.

With all orally delivered methods, one of the most important concerns is that non-target species may consume the contraceptive. Effects of oral chemical contraception are exerted largely only while the chemical is in the body. Nicarbazin is eliminated relatively rapidly, so continuous feeding is needed. Adverse effects have not been reported for nicarbazin.

3.4.3 Zinc Gluconate

Zinc gluconate is administered as an intratesticular injection that inhibits development of germ cells at several points, interfering with spermatogenesis, and affects androgen metabolism (Kellokumpu and Rajaniemi 1981; Hidiroglou and Knipfel 1984). First approved by FDA as Neutersol®, zinc gluconate neutralized with arginine was developed to sterilize domestic dogs. It was subsequently acquired by a different company (Ark Sciences) and marketed as Esterisol™ in the U.S. and as Zeuterin™ in several Latin American countries and in Turkey, in hopes of using it to control populations of stray dogs and cats. However, although apparently successful in permanently sterilizing male dogs and cats with a single intra-testicular injection (Oliveira et al. 2007; Fagundes et al. 2014), it is no longer commercially available.

3.4.4 Calcium Chloride

Calcium chlorate dihydrate, when injected directly into the testis, causes degeneration of the seminiferous tubules and supporting structures, thus eliminating sperm production (Jana and Samanta 2006). Efficacy depends on the volume, concentration and vehicle used. This method has been successful in limited trials with domestic dogs (Jana and Samanta 2007), cats (Jana and Samanta 2011), donkeys (Ibrahim et al. 2016) and goats (Jana et al. 2005). The resulting tissue reaction causes testicular swelling, which subsides in three to 4 weeks. Care must be taken to prevent leakage on or under the scrotal skin or local necrosis can occur around the injection site (Kutzler 2015). Calcium chloride is inexpensive compared to other methods, but concern continues about the possible discomfort associated with injection, particularly if injection is not limited to tissue within the testicular capsule.

3.5 *Surgical Sterilization*

The major feature of surgical approaches is that they are permanent. This can be seen as an advantage in some populations, since re-treatment of an individual is never needed. But in others, reversible methods are preferred as a failsafe in case of a population crash due to disease or other factors.

Castration is a relatively simple procedure for most species, but is more difficult and carries more risk in species with undescended testes, such as elephants, pinnipeds and cetaceans. Even in species with descended but ascrotal testes, such as hippopotamus, *Hippopotamus amphibius*, the procedure remains challenging. However, castration results include loss of secondary sex characteristics (e.g., lion's mane) and disruption of seasonal cycles (e.g., deer antler growth and shedding). Especially for polygynous species, castration is unlikely to have much if any effect on population-wide reproductive rates, since castrated males will be out-competed for access to females by intact males. Even for monogamous species, the resulting absence of courtship and mating behavior may result in dissolution of pair bonds, freeing the female to bond with an intact male.

If a permanent method is desired, though, vasectomy is a possible alternative. All species-typical male behaviors and secondary sex characteristics are spared, so the only resulting change expected is the absence of pregnancy and offspring. Vasectomy was effective at reducing birth rates in feral horses without changing social and band dynamics (Eagle et al. 1993; Asa 1999). In studies with vasectomized coyotes, predation rates on sheep were lower but territories and pair-bonds were maintained (Bromley and Gese 2001a, b). Pair bonds were also maintained in vasectomized, free-ranging gray wolves (Spence et al. 1999). Vasectomy of free-ranging male hamadryas baboons in Saudi Arabia resulted in stable social groups even though no females reproduced during the 4 years of monitoring (Biquand et al. 1994). As expected, no behavioral changes were noted in free-ranging vasectomized compared to castrated Eastern gray kangaroos, *Macropus giganteus*, on an Australian golf course (Tribe et al. 2014). Despite the difficulty of the surgery, especially under field conditions, vasectomies were performed on 45 African elephants (Marais et al. 2013). Neither behavioral nor population data were reported, but complications related to the surgery occurred in several males and two died. However, vasectomy is generally not recommended for species with induced ovulation (e.g. felids and mustelids), as females would be exposed to repeated, extended periods of endogenous progesterone and its deleterious effects on the endometrium (see section on progestin 3.1.1 contraception) during the luteal phase that follows mating with sterile males.

Removal of the ovaries (ovariectomy) or ovaries plus uterus (ovariohysterectomy) would of course prevent reproduction in females but entails abdominal surgery or laparoscopy, which would entail significant risk under field conditions. However, this approach can be practical in captivity for some species, e.g., carnivores such as African painted dogs with a high risk of pyometra (a potentially fatal uterine infection). Unilateral hysterectomy (in which one uterine horn is removed) or tying one tube has been proposed for litter size reduction in lions (Miller et al. 2013), although these techniques have not been sufficiently studied to warrant recommendation. Parallel to vasectomy, tubal transection or ligation allows females to remain hormonally intact. Although this procedure also requires surgery in the field (with its accompanying costs, difficulties and risks), it can be accomplished with laparoscopy, which is less invasive than laparotomy. This procedure has been used in white-tailed deer (MacLean et al. 2006) and coyotes (Bromley and Gese 2001a, b). However, these procedures are likely not widely used because they are not cost effective.

3.6 *Methods That Have Not Worked and Why*

3.6.1 Bisdiamine

The bisdiamine WIN 18446 first showed promise as a contraceptive pill for men and then for other mammals. Although it reversibly inhibits spermatogenesis without affecting testosterone levels (Beyler et al. 1961), it also inhibits activity of alcohol dehydrogenase, the liver enzyme that detoxifies alcohol (Heller et al. 1961). This is a drawback to marketing for human use but should pose no problem for captive animals. It was shown to be safe and effective in rhesus macaques (*Macaca mulatta*), several rodent species, domestic cats (Munson et al. 2004), domestic dogs (Drobeck and Coulston 1962) and gray wolves (*Canis lupus*: Asa et al. 1996). However, the large, daily dose required made the cost of commercial development and marketing for wildlife prohibitive. Additionally, bisdiamine has teratogenic effects (Binder 1985), so ingestion by pregnant females in mixed-sex groups would be problematic.

3.6.2 Indenopyridine

During testing for its antihistamine properties, the indenopyridine Sandoz 20-438 was also found to disrupt spermatogenesis in rats, mice and dogs without affecting testosterone levels (Hodel and Suter 1978; Matter et al. 1979). Unfortunately, in pilot tests with domestic cats it caused diarrhea that was sometimes blood-tinged (C. Asa, D. Kunze, and P. Fail, unpublished), indicating it was not an acceptable contraceptive, at least for felids.

4 Promising New Approaches

4.1 BMP-15 and GDF-9

Immunization against bone morphogenetic protein-15 (BMP-15) or growth differentiation factor-9 (GDF-9) was shown to interfere with follicle development and depress ovulation rate in sheep and cattle (Juengel et al. 2002, 2009). A study to evaluate the contraceptive efficacy of these vaccines in domestic mares found that immunization against BMP-15 resulted in decreased follicle size and ovulation rate, whereas immunization against GDF-9 caused abnormal follicle growth and small ovulatory follicle size (Davis et al. 2018). The authors hypothesize that the effect of the GDF-9 vaccine might deplete oocyte reserves resulting in infertility. In female white-tailed deer (Eckery et al. 2014), GDF-9 vaccination did not affect fertility in the first year of treatment, but females were infertile in years 2 and 3. In contrast, BMP-15 vaccination resulted in higher fecundity than seen in controls, revealing a species difference in response. Research continues with these vaccines to improve efficacy and understand appropriate timing of treatment.

4.2 Cytotoxin Targeting Gonadal Cells

A new technique uses a lipid-based nanoparticle coupled with a cytotoxin (saponin). This nanoparticle complex is carried by an antibody against the anti-Mullerian hormone II (AMHII) receptor on gonadal cells of either males or females (Meadows et al. 2019). The complex can be guided by the antibody to receptors for AMHII on specific cells necessary for development of sperm or eggs. These cells also produce sex steroid hormones that support reproductive behaviors, so the outcomes of destruction of these cells by the attached cytotoxin are similar to those following gonadectomy. It remains to be seen whether the effect is permanent, though. Trials with rats as a model have shown apoptosis of these cells within 24 h of injection. Research is extending to cats, dogs, pigs and chickens. In earlier research in dogs treated with GnRH conjugated to a cytotoxin targeting pituitary gonadotrophs (Ball et al. 2006), suppression of reproduction was not permanent.

4.3 Vectored Gene Delivery

Permanent sterilization via vectored gene delivery has been proposed by Hay et al. (2018), with a single injection that would stimulate expression and secretion of monoclonal antibodies or other proteins that can inhibit fertility. Possible targets include zona pellucida, GnRH, and AMH, as pivotal points in the reproductive process. Successful proof of concept has been shown in mice by targeting ZP and GnRH (Li et al. 2015) and anti-Mullerian hormone (Kano et al. 2017). Although research in this methodology is in early stages but holds promise for application in wildlife.

4.4 Adjudin

Adjudin, an indazole-carboxylic acid like lonidamine that is used for chemotherapy, causes germ cell loss from seminiferous tubule epithelium by disrupting cell adhesion between Sertoli and germ cells, so developing sperm are sloughed. However, it does not affect spermatogonia or testosterone, so is expected to be reversible with no effect on libido (Mok et al. 2011). Because it has low oral availability, the high dose needed for contraception causes serious side effects in muscle and liver. However, coupled with mutant FSH, it binds directly to Sertoli cells, so a lower dose is effective on the testis target. Adjudin-FSH can be injected or given as an implant or gel (Cheng and Mruk 2010).

4.5 Indenopyridines

First tested as anti-histamines, indenopyridines also showed anti-spermatogenic effects via germ cell loss from the seminiferous tubules. Efficacy was first demonstrated in rats (Hild et al. 2001a) and extended to cynomolgus macaques, *M. fascicularis* (Hild et al. 2007) and horses (Pozor et al. 2013). The effects of indenopyridines can vary by species. Treatment of rats was not reversible without prior administration of a GnRH agonist (Hild et al. 2001b), and treatment of cats resulted in bloody diarrhea (Asa et al. unpublished, see Sect. 3.6 above). Some of these differences may also be due to the specific action of the various indenopyridine forms.

4.6 Possible Improvements to Drug Delivery

A review of drug delivery technologies revealed some that are promising for development of injectable contraceptives with longer action (Halpern et al. 2015). Examples include existing polymer systems, such as poly-lactic-co-glycolic acid and poly(epsilon-caprolactone). Advances in the manufacture of microspheres could also benefit the development of a longer-acting injectable contraceptives. In addition, novel polymers and materials that the review identified might have advantages over existing polymer systems. Unfortunately, products with new materials could take much longer to gain regulatory approval.

Long et al. (2018) have proposed using 3D printing to produce customized drug delivery devices. For example, a biodegradable projectile might contain different doses of a contraceptive using fused deposition modelling 3D printing. They used three concentrations of progesterone with polylactic acid prepared as a 1.75 mm filament by hot melt extrusion. Progesterone was successfully incorporated in the polymer, and an interaction between the contraceptive drug and the polymer was confirmed. Penetration assessment showed that the drug-loaded projectiles had sufficient kinetic energy to penetrate thin and medium-thickness skin, demonstrating the feasibility of using fused deposition modelling 3D printing for customizing drug delivery devices.

5 Application and Implications

Despite decades of research, challenges remain. These vary by species and local circumstance but include species-specific targeted delivery and effective identification of treated animals to avoid duplicate dosing or to deliver needed boosters. An advantage of injectable methods is the possibility of remote darting, but hand-injection is more reliable, since darts may not reach the animal or may not discharge their contents completely resulting in an inadequate dose. However, hand-injection requires capture and restraint, which is much more labor-intensive, more expensive, and causes more stress and social disruption. Ideally, a product would be long-lasting to reduce the number of times animals have to be treated. Work to develop long-acting agents continues, but most require repeated treatment every 1–3 years, and ZonaStat® is most effective when a booster is given within a few months of initial inoculation. Even GonaCon™, which seems to be effective longer than ZonaStat®, has lower efficacy in the second and third years. Oral agents administered in baits can be made species-specific via specially designed feeders or incorporation in a bait that non-target species would not want or be able to consume. For example, the Boar-Operated System (Campbell et al. 2011) delivers bait specifically to wild boar and feral pigs, and automatic feeders deliver OvoControl® to pigeons. Deployment strategy, that is, whether baits are aggregated or dispersed, can also affect success (European badger: Cagnacci et al. 2007). A disadvantage of most orally active products is that they must be consumed repeatedly, sometimes daily, to be effective.

Implants are often preferable for captive wildlife, but are less practical for field application since they require capture for insertion. However, implants containing the synthetic progestin levonoregestrel have been used successfully for long-term fertility control in wallabies and eastern gray kangaroos in Australia (Nave et al. 2002).

In contrast to the larger silastic implants used for MGA delivery, the very small size of the implants containing GnRH agonists (e.g., Suprelorin®) or commercially available protestagen implants (e.g., Implanon® or Nexplanon®) makes insertion roughly comparable to hand-injection, and while they may require restraint, they do not require anesthesia.

For overabundant populations, especially those that are increasing, fertility control may not be sufficient for reducing numbers (Hone 1992). A combination of initial population reduction by culling (Hobbs et al. 2000) and/or removal (feral swine: Pepin et al. 2017; feral horses: Garrott and Oli 2013) would shorten the time necessary to reach target population size. Once populations are at target size, fertility control is more likely to be adequate for maintaining that level. However, for long-lived species, island populations, and those with slow turnover, fertility control alone may decrease population size but take a longer period of time. In general, it is more practical to use fertility control to manage longer-lived species, especially if permanent methods are used (Hone 1999; Hobbs et al. 2000). As a special case, feral cat populations are especially difficult to control, as the influx is not only due to immigration from neighboring populations but also to people abandoning cats that add to these populations. The latter factor is an important component of the low success of trap-neuter-release programs in populations that are not isolated (Miller et al. 2014).

Selection among available methods will depend to a great extent on practicalities such as delivery and duration of efficacy, but in some cases potential effects on behavior as well as welfare and safety may be important. When populations are limited by preventing reproduction, females that are not subject to the energy demands of gestation and lactation show improved condition and extended lifespan (Kirkpatrick and Turner 2007; Gionfriddo et al. 2011). In a study where both male and female coyotes were sterilized (vasectomy and tubal ligation, respectively), sterilized coyotes lived longer (Bromley and Gese 2001b), indicating this effect extends beyond feral horses. This increase in life-span and body condition may be appealing to the public, but it also requires adjustments in population models regarding the target number of individuals to treat, given the longer lifespan of treated females.

Methods that inhibit sex steroid production will also result in an inhibition of sexual behavior. In species with year-round social units based on male/female bonds, e.g., canids such as wolves, coyotes, and feral horses, this could result in social disruption and population instability. However, in many species, absence of sexual behavior could have the benefit of reducing disease transmission by reducing contact rates (Ramsey 2007). In some cases, contraception has been used specifically to control disease by limiting reproduction, for example, to control brucellosis in bison (Miller et al. 2004) and rabies in feral dogs (Vargas-Pino et al. 2013). For some species, e.g., red foxes, sterilization can be combined with rabies vaccination programs to achieve multiple goals (Taylor et al. 2017). A disadvantage, though, of continuing ovulatory cycles in sterilized polygynous species is potential extension of the breeding season, prolonging the period of male/male competition, increasing the risks of injury and of exhaustion going into winter (Miller et al. 2000; Nuñez et al. 2010).

6 Regulatory Considerations

Acquiring regulatory approval for use of sterilizing agents and contraceptive products in wildlife and invasive species has been challenging. In the U.S., application of contraceptive products to free-ranging wildlife and feral animals, including those opportunistically using zoo property (e.g., raccoons, rodents, geese and pigeons), typically requires approval from the Environmental Protection Agency (EPA), whereas the Food and Drug Administration (FDA) oversees treatment of captive species in zoos and sanctuaries (see Eisemann et al. 2006).

The registration process can be prohibitively expensive for products intended for wildlife application, since the market, and thus potential income, is very limited compared to products for pets and agricultural species. A somewhat less expensive option has been established in FDA for Minor Use Minor Species (MUMS), but no products for wildlife contraception have yet been approved using this process. However, contraceptive vaccines that do not currently have EPA or FDA registration approval can be used under EPA Experimental Use Permits or FDA Investigational New Animal Drug Exemptions.

Counterparts to the FDA in other countries include the European Medicines Agency (EMA); the Veterinary Medicines Directorate, part of the Department of Environment, Food and Rural Affairs (DEFRA) in the UK; Health Canada-Veterinary Drugs Directorate; Australian Pesticide and Veterinary Medicine Authority. These agencies have different requirements for approval of animal drugs, most evident in the availability of products in Europe and Australia not approved in the U.S., e.g., the GnRH agonist contraceptive Suprelorin® (Virbac Animal Health). Suprelorin-F® (Virbac Animal Health), however, is commercially available in the U.S. as an index listed product, labeled specifically for domestic pet ferrets, for treatment of adrenal gland cortical disease. Index listing is a provisional category, which does not allow off-label application to other species for other purposes such as wildlife contraception. A "listed" product such as Suprelorin-F can be pulled from the market if any adverse events are reported, whereas more extensive evidence of harm is required before removal of approved products.

EPA registration requires data on product chemistry, residue chemistry, toxicology, environmental effects and fate, product performance and worker protection. For oral delivery, further information may be required on hazards to non-target species and on environmental effects and fate, especially if bait stations are not used. Requirements may be less stringent for injectable products, since risk of accidental non-target species treatment is minimal and potential environmental effects are lower. Following national EPA approval, these products must also meet approval by a state before use within its jurisdiction (Eisemann et al. 2013). The potential impact of fertility control on hunting opportunities is the major concern for most states.

Unfortunately, contraceptives and sterilizing agents registered with the EPA must be categorized as pesticides. This carries the requirement that treatment must be administered by Certified Pesticide Applicators, which necessitates hiring these specialists or entails wildlife managers acquiring this training and certification. This designation also has caused much confusion and misunderstanding among the public. Although fertility control products are intended in most cases to be used in place of lethal methods, the designation of those products as pesticides suggests that they are lethal, generating protest among people who would otherwise be supportive. Permanent sterilization of wildlife is not permitted in many countries, so government regulations should be consulted prior to choosing a method.

References

Asa CS. Male reproductive success in free-ranging feral horses. Behav Ecol Sociobiol. 1999;47:89–93.

Asa CS, Zaneveld LJD, Munson L, Callahan M, Byers AP. Efficacy, safety, and efficacy of a bisdiamine male-directed oral contraceptive in gray wolves (*Canis lupus*). J Zoo Wildl Med. 1996;27:501–6.

Asa CS, Porton IJ, Junge R. Reproductive cycles and contraception of black lemurs (*Eulemur macaco macaco*) with depot medroxyprogesterone acetate during the breeding season. Zoo Biol. 2007;26:289–98.

Asa CS, Bauman KL, Devery S, Zordan M, Camilo GR, Boutelle S, Moresco A. Factors associated with uterine endometrial hyperplasia and pyometra in wild canids: implications for fertility. Zoo Biol. 2014;33:8–19.

Attardi B. Progesterone modulation of the luteinizing hormone surge: regulation of hypothalamic and pituitary progestin receptors. Endocrinology. 1984;115:2113–22.

Avery ML, Keacher KL, Tillman EA. Nicarbazin bait reduces reproduction by pigeons (*Columba livia*). Wildl Res. 2008a;35:80–5.

Avery ML, Yoder CA, Tillman EA. Diazacon inhibits reproduction in invasive monk parakeet populations. J Wildl Manag. 2008b;72:1449–52.

Baker DL, Wild MA, Conner MM, Ravivarapu HB, Dunn RL, Nett TM. Effects of GnRH agonist (leuprolide) on reproduction and behaviour in female wapiti (*Cervus elaphus nelsoni*). Reprod Suppl. 2002;60:155–67.

Baker DL, Wild MA, Connor MM, Ravivarapu HB, Dunn RL, Nett TM. Gonadotropin-releasing hormone agonist: a new approach to reversible contraception in female deer. J Wildl Dis. 2004;40:713–24.

Baker DL, Wild MA, Hussain MD, Dunn RL, Nett TM. Evaluation of remotely delivered leuprolide acetate as a contraceptive agent in female elk (*Cervus elaphus nelsoni*). J Wildl Dis. 2005;41:758–67.

Baldwin CJ, Peter AT, Bosu W, Dubielzig RR. The contraceptive effects of levonorgestrel in the domestic cat. Lab Anim Sci. 1994;44:261–9.

Ball BA, Sabeur K, Nett T, Liu IKM. Effects of a GnRH cytotoxin on reproductive function in peripubertal male dogs. Theriogenology. 2006;66:766–74.

Balser DS. Management of predator populations with antifertility agents. J Wildl Manag. 1964;28:352–8.

Bechert U, Fraker MA. The response of African elephants to a single-dose of SpayVac, a pZP contraceptive vaccine, over a 7-year period. Pachyderm. 2016;57:97–108.

Bechert U, Fraker MA. Twenty years of SpayVac research: potential implications for regulating feral horse and burro populations in the United States. Hum Wildl Interact. 2018;12:117–20.

Bechert U, Bartell J, Kutzler M, Menino A, Bildfell R, Anderson M, Fraker M. Effects of two porcine zona pellucida immunocontraceptive vaccines on ovarian activity in horses. J Wildl Manag. 2013;77:1386–400.

Beijerink NJ, Buijtels JJ, Okkens AC, Kooistra HS, Dieleman SJ. Basal and GnRH-induced secretion of FSH and LH in anestrous versus ovariectomized bitches. Theriogenology. 2007;67:1039–45.

Bertschinger HJ, Asa CS, Calle PP, Long JA, Bauman K, DeMatteo K, Jochle W, Trigg TE, Human A. Control of reproduction and sex related behaviour in exotic wild carnivores with the GnRH analogue deslorelin: preliminary observations. J Reprod Fertil Suppl. 2001;57:275–83.

Bertschinger H, de Barros Vaz Guimarães MA, Trigg T, Human A. The use of deslorelin implants for the long-term contraception of lionesses and tigers. Wildl Res. 2008;35:525–30.

Beyler AL, Poots GO, Coulston F, Surrey AR. The selective testicular effects of certain bis(dichloroacetyl)diamines. Endocrinology. 1961;69:819–33.

Binder M. The teratogenic effects of a bis (dichloroacetyl) diamine on hamster embryos. Aortic arch anomalies and the pathogenesis of the DiGeorge syndrome. Am J Pathol. 1985;118:179.

Biquand S, Boug A, Biquand-Guyot V, Gautier J-P. Management of commensal baboons in Saudi Arabia. Revue d'Ecologie. 1994;49:213–22.

Bishop CC, Fleming PA, Barnes AL, Collins T, Miller DW. Immunisation against gonadotrophin-releasing hormone (GnRH) reduces agonistic behaviours in male rangeland goats. Anim Prod Sci. 2016;56:1882–7.

Brache V, Faundes A, Johansson E. Anovulation, inadequate luteal phase and poor sperm penetration in cervical mucus during prolonged use of Norplant implants. Contraception. 1985;31:261–73.

Bromley C, Gese EM. Surgical sterilization as a method of reducing coyote predation on domestic sheep. J Wildl Manag. 2001a;65:510–9.

Bromley D, Gese EM. Effects of sterilization on territory fidelity and maintenance, pair bonds, and survival rates of free-ranging coyotes. Can J Zool. 2001b;79:386–92.

Brown RG, Bowen WD, Eddington JD, Kimmins WC, Mezei M, Parsons JL, Pohajdak B. Evidence for a long-lasting single administration contraceptive vaccine in wild grey seals. J Reprod Immunol. 1997;35:43–51.

Cagnacci F, Massei G, Cowan DP, Walker N, Delahay RJ. Effects of bait type and deployment strategy on uptake by free-living badgers. Wildl Res. 2007;34:454–60.

Calle PP. Contraception in pinnipeds and cetaceans. In: Asa CS, Porton IJ, editors. Wildlife contraception: issues, methods and applications. Baltimore: Johns Hopkins University Press; 2005. p. 168–76.

Campbell TA, Long DB, Massei G. Efficacy of the boar-operated-system to deliver baits to feral swine. Prev Vet Med. 2011;98:243–9.

Cheng CY, Mruk DD. New frontiers in nonhormonal male contraception. Contraception. 2010;82:476–82.

Chuei JY, Asa CS, Hall-Woods M, Ballou J, Traylor-Holzer K. Restoration of reproductive potential after expiration or removal of melengestrol acetate contraceptive implants in tigers (*Panthera tigris*). Zoo Biol. 2007;26:275–88.

Cope HR, Hogg CJ, White PJ, Herbert CA. A role for selective contraception of individuals in conservation. Conserv Biol. 2018;32:546–58.

Cowan ML, Martin GB, Monks DJ, Johnston SD, Doneley RJT, Blackberry MA. Inhibition of the reproductive system by deslorelin in male and female pigeons (*Columba livia*). J Avian Med Surg. 2014;28:102–8.

Cowl VB, Walker SL, Feltrer Rambaud Y. Assessing the efficacy of deslorelin acetate implants (Suprelorin) in alternative placement sites. J Zoo Wildl Med. 2018;49:1–8.

Curtis P, Pooler R, Richmond M, Miller L, Mattfeld G, Quimby F. Comparative effects of GnRH and porcine zona pellucida (PZP) immunocontraceptive vaccines for controlling reproduction in white-tailed deer (*Odocoileus virginianus*). Reprod Suppl. 2002;60:131–41.

de Oliveira CA, West GD, Houck R, Leblanc M. Control of musth in an Asian elephant bull (*Elephas maximus*) using leuprolide acetate. J Zoo Wildl Med. 2004;35:70–6.

Davis KA, Klohonatz KM, Mora DS, Twenter HM, Graham PE, Pinedo P, Eckery DC, Bruemmer JE. Effects of immunization against bone morphogenetic protein-15 and growth differentiation factor-9 on ovarian function in mares. Anim Reprod Sci. 2018;192:69–77.

DeMatteo KE. Contraception in carnivores. In: Asa CS, Porton IJ, editors. Wildlife contraception: issues, methods, and applications. Baltimore: Johns Hopkins University Press; 2005. p. 105–18.

DeNicola AJ, Kesler DJ, Swihart RK. Dose determination and efficacy of remotely delivered norgestomet implants on contraception of white-tailed deer. Zoo Biol. 1997;16:31–8.
Drobeck HP, Coulston F. Inhibition and recovery of spermatogenesis in rats, monkeys and dogs medicated with bis(dichloroacetyl)diamines. Exp Mol Pathol. 1962;1:251–74.
Dunbar BS, Lo C, Powell J, Stevens VC. Use of a synthetic peptide adjuvant for the immunization of baboons with denatured and deglycosylated pig zona pellucida glycoproteins. Fertil Steril. 1989;52:311–8.
Duncan GL, Lyster SC, Hendrix JW, Clark JJ, Webster HD. Biologic effects of melengestrol acetate. Fertil Steril. 1964;15:419–32.
Dunshea F, King R, Campbell R, Sainz R, Kim Y. Growth performance, carcass characteristics and meat quality of group-penned surgically castrated, immunocastrated (Improvac®) and entire male pigs and individually penned entire male pigs. J Anim Sci. 1993;71:2913–30.
Eagle TC, Plotka ED, Siniff DB, Tester JR. Efficacy of chemical contraception in feral mares. Wildl Soc Bull. 1992;20:211–6.
Eagle T, Asa CS, Garrott R, Siniff DV. Efficacy of dominant male contraception to reduce reproduction in feral horses. Wildl Soc Bull. 1993;21:116–21.
Eckery DC, Miller LA, Killian GJ, DeNicola AJ. Effect of vaccination against GDF9 and BMP15 on fertility and ovarian function in white-tailed deer. In: Timm RM, O'Brien JM, editors. Proceedings 26th vertebrate pest conference. Davis: University of California-Davis; 2014. p. 391–5.
Eisemann JD, Fagerstone KA, O'Hare JR. Wildlife contraceptives: a regulatory hot potato. In: Timm RM, O'Brien JM, editors. Proc 22nd vertebrate pest conference. Berkeley: University of California-Davis; 2006. p. 63–6.
Eisemann JD, O'Hare JR, Fagerstone KA. State-level approaches to managing the use of contraceptive in wildlife in the United States. J Zoo Wildl Med. 2013;44:S47–51.
Evans JM, Sutton DJ. The use of hormones, especially progestagens, to control oestrus in bitches. J Reprod Fertil Suppl. 1989;39:163–73.
Fagerstone KA, Miller LA, Killian G, Yoder CA. Review of issues concerning the use of reproductive inhibitors, with particular emphasis on resolving human-wildlife conflicts in North America. Integr Zool. 2010;5:15–30.
Fagundes AKF, Oliveira ECS, Tenorio BM, Melo CCS, Nery LTB, Santos FAB, Alves LC, Douglas RH, Silva VA. Injection of a chemical castration agent, zinc gluconate, into the testes of cats results in the impairment of spermatogenesis: a potentially irreversible contraceptive approach for this species? Theriogenology. 2014;81(2):230–6.
Fayrer-Hosken RA, Bertschinger HJ, Kirkpatrick JF, Grobler D, Lamberski N, Honneyman G, Ulrich T. Contraceptive potential of the porcine zona pellucida vaccine in the African elephant (*loxodonta africana*). Theriogenology. 1999;52(5):835–46.
Fekete G, Szeberényi S. Data on the mechanism of adrenal suppression by medroxyprogesterone acetate. Steroids. 1965;6:159–66.
Fischer A, Benka VA, Briggs J, Driancourt M-A, Maki J, Mora DS, Morris K, Myers KA, Rhodes L, Vansandt LM, Weedon GR, Wolf J, Levy JK. Effectiveness of GonaCon as an immunocontraceptive in colony-housed cats. J Feline Med Surg. 2018;20:786–92.
Fraker MA, Brown RG, Grant GE, Kerr JA, Pohajdak B. Long-lasting, single dose, immunocontraception of feral fallow deer in British Columbia. J Wildl Manag. 2002;66:1141–7.
Frank KM, Lyda RO, Kirkpatrick JF. Immunocontraception of captive exotic species. IV. Species differences in response to the porcine zona pellucida vaccine, timing of booster inoculations and procedural failures. Zoo Biol. 2005;24:349–58.
Garrett MG, Franklin WL. Diethylstilbestrol as a temporary chemosterilant to control black-tailed prairie dog populations. J Range Manag. 1983;36:753–6.
Garrott RA, Oli MK. A critical crossroad for BLM's wild horse program. Science. 2013;341:847–8.
Garside D, Gebril A, Alsaadi M, Ferro VA. Fertility control in wildlife: review of current status, including novel and future technologies. In: Holt WV, Brown JL, Comizzoli P, editors. Reproductive sciences in animal conservation: advances in experimental medicine and biology. New York: Springer; 2014. p. 467–88.

Gee EK, DeLuca C, Stylski JL, McCue PM. Efficacy of medroxyprogesterone acetate in suppression of estrus in cycling mares. J Equine Vet Sci. 2009;29:140–5.
Gionfriddo JP, Gates NB, DeNicola AJ, Fagerstone KA, Miller LA. Field test of GonaCon™ immunocontraceptive vaccine in free-ranging female fallow deer. In: Timm RM, Modon MB, editors. Proc 23rd vertebrate pest conference. Davis: University of California-Davis; 2008.
Gionfriddo JP, DeNicola AJ, Miller LA, Fagerstone KA. Health effects of GnRH immunocontraception of wild white-tailed deer in New Jersey. Wildl Soc Bull. 2011;35:149–60.
Goblet CC, Moresco A, Garner MM, Agnew DW, Newell-Fugate AE. Retrospective characterization of reproductive tract lesions in relation to age, parity, and contraception in captive suidae and tayassuidae. Theriogenology. 2019;127:137–44.
Goericke-Pesch S, Georgiev P, Antonov A, Albouy M, Wehrend A. Clinical efficacy of a GnRH-agonist implant containing 4.7mg deslorelin, Suprelorin, regarding suppression of reproductive function in tomcats. Theriogenology. 2011;75:803–10.
Gulyas BJ, Gwatkin RBL, Yuan LC. Active immunization of cynomolgus monkeys (*Macaca fascicularis*) with porcine zonae pellucidae. Gamete Res. 1983;4:299–307.
Halpern V, Stalter RM, Owen DH, Dorflinger LJ, Lendvay A, Rademacher KH. Towards the development of a longer-acting injectable contraceptive: past research and current trends. Contraception. 2015;92:3–9.
Hapgood JP, Koubovec D, Louw A, Africander D. Not all progestins are the same: implications for usage. Trends Pharmacol Sci. 2004;25:554–7.
Hay BA, Li J, Guo M. Vectored gene delivery for lifetime animal contraception: overview and hurdles to implementation. Theriogenology. 2018;112:63–74.
Heller CG, Moore DJ, Paulsen CA. Suppression of spermatogenesis and chronic toxicity in men by a new series of bis (dichloroacetyl) diamines. Toxicol Appl Pharmacol. 1961;3:1–11.
Herbert CA, Trigg TE. Applications of GnRH in the control and management of fertility in female animals. Anim Reprod Sci. 2005;88:141–53.
Herbert CA, Trigg T, Cooper DW. Effect of deslorelin implants on follicular development, parturition and post-partum oestrus in the tammar wallaby (*Macropus eugenii*). Reproduction. 2004;127:265–73.
Herbert CA, Trigg TE, Renfree MB, Shaw G, Eckery DC, Cooper DW. Long-term effects of deslorelin implants on reproduction in the female tammar wallaby (*Macropus eugenii*). Reproduction. 2005;129:361–9.
Hermes R, Schwarzenberger F, Göritz F, Oh S, Fernandes T, Bernardino R, Leclerc A, Greunz E, Mathew A, Forsyth S, Saragusty J, Hildebrandt TB. Ovarian down regulation by GnRF vaccination decreases reproductive tract tumour size in female white and greater one-horned rhinoceroses. PLoS One. 2016;11:e0157963.
Hidiroglou M, Knipfel JE. Zinc in mammalian sperm: a review. J Dairy Sci. 1984;67:1147–56.
Hild SA, Meistrich ML, Blye RP, Reel JR. Lupron depot prevention of anti-spermatogenic/antifertility activity of the indenopyridine, CDB-4022, in the rat. Biol Reprod. 2001a;65:165–72.
Hild SA, Reel JR, Larner JM, Blye RP. Disruption of spermatogenesis and Sertoli cell structure and function by the indenopyridine CDB-4022 in rats. Reproduction. 2001b;65:1771–9.
Hild SA, Marshall GR, Attardi BJ, Hess RA, Schlatt S, Simorangkir DR, Ramaswamy S, Koduri S, Reel JR, Plant TM. Development of l-CDB-4022 as a nonsteroidal male oral contraceptive: induction and recovery from severe oligospermia in the adult male cynomolgus monkey (*Macaca fascicularis*). Endocrinology. 2007;148:1784–96.
Hobbs NT, Bowden DC, Baker DL. Effects of fertility control on populations of ungulates: general, stage-structured models. J Wildl Manag. 2000;64:473–91.
Hodel C, Suter K. Reversible inhibition of spermatogenesis with an indenopyridine (20-438). Arch Toxicol Suppl. 1978;1:323–6.
Hone J. Rate of increase and fertility control. J Appl Ecol. 1992;29:695–8.
Hone J. On rate of increase (r): patterns of variation in Australian mammals and the implications for wildlife management. J Appl Ecol. 1999;36:709–18.
Hynes EF, Handasyde KA, Shaw G, Renfree MB. The effects of gestagen implants on the behaviour of free-ranging female koalas. Appl Anim Behav Sci. 2011;134:209–16.

Ibrahim A, Ali MM, Abou-Khalil NS, Ali MF. Evaluation of chemical castration with calcium chloride versus surgical castration in donkeys: testosterone as an endpoint marker. BMC Vet Res. 2016;12:46.

Jacobsen NK, Jessup DA, Kessler DJ. Contraception of captive black-tailed deer by remotely delivered norgestomet ballistic implants. Wildl Soc Bull. 1995;23:718–22.

Jana K, Samanta PK. Evaluation of single intratesticular injection of calcium chloride for nonsurgical sterilization in adult albino rats. Contraception. 2006;73:289–300.

Jana K, Samanta PK. Sterilization of male stray dogs with a single intratesticular injection of calcium chloride: a dose-dependent study. Contraception. 2007;75:390–400.

Jana K, Samanta PK. Clinical evaluation of non-surgical sterilization of male cats with single intratesticular injection of calcium chloride. BMC Vet Res. 2011;7:39.

Jana K, Samanta P, Ghosh D. Evaluation of single intratesticular injection of calcium chloride for nonsurgical sterilization of male Black Bengal goats (*Capra hircus*): a dose-dependent study. Anim Reprod Sci. 2005;86:89–108.

Jarosz SJ, Dukelow WR. Effect of progesterone and medroxyprogesterone acetate on pregnancy length and litter size in mink. Lab Anim Sci. 1985;35:156–8.

Joonè CJ, Bertschinger HJ, Gupta SK, Fosgate GT, Arukha AP, Minhas V, Dieterman E, Schulman ML. Ovarian function and pregnancy outcome in pony mares following immunocontraception with native and recombinant porcine zona pellucida vaccines. Equine Vet J. 2017a;49:189–95.

Joonè CJ, Schulman ML, Bertschinger HJ. Ovarian dysfunction associated with zona pellucida–based immunocontraceptive vaccines. Theriogenology. 2017b;89:329–37.

Juengel JL, Hudson NL, Heath DA, Smith P, Reader KL, Lawrence SB, O'Connell AR, Laitinen MPE, Cranfield M, Groome NP, Ritvos O, McNatty KP. Growth differentiation factor 9 and bone morphogenetic protein 15 are essential for ovarian follicular development in sheep1. Biol Reprod. 2002;67:1777–89.

Juengel JL, Hudson NL, Berg M, Hamel K, Smith P, Lawrence SB, Whiting L, McNatty KP. Effects of active immunization against growth differentiation factor 9 and/or bone morphogenetic protein 15 on ovarian function in cattle. Reproduction. 2009;138:107.

Kano M, Sosulski AE, Zhang L, Saatcioglu HD, Wang D, Nagykery N, Sabatini ME, Gao G, Donahoe PK, Pépin D. AMH/MIS as a contraceptive that protects the ovarian reserve during chemotherapy. PNAS. 2017;114:E1688–97.

Kazensky CA, Munson L, Seal US. The effects of melengestrol acetate on the ovaries of captive wild felids. J Zoo Wildl Med. 1998;29:1–5.

Kellokumpu S, Rajaniemi H. Effect of zinc on the uptake of human chorionic gonadotropin (hCG) in rat testis and testosterone response in vivo. Biol Reprod. 1981;24:298.

Killian G, Miller L, Rhyan J, Dees T, Perry D, Doten H. Evaluation of GnRH contraceptive vaccine in captive feral swine in Florida. In: Fagerstone KA, Witmer GW, editors. Proc 10th wildlife damage management conf; 2003. p. 128–33.

Killian G, Diehl NK, Miller LA, Rhyan J, Thain D. Long-term efficacy of three contraceptive approaches for population control of wild horses. Proc Vertebr Pest Conf. 2006;22.

Killian G, Wagner D, Fagerstone K, Miller L. Long-term efficacy and reproductive behavior associated with GonaCon(R) use in white-tailed deer (*Odocoileus virginianus*). In: Timm RM, editor. Proc 23rd vertebrate pest conference. Davis: University of California-Davis; 2008. p. 240–3.

Killian G, Kreeger TJ, Rhyan J, Fagerstone K, Miller L. Observations on the use of GonaCon™ in captive female elk (*Cervus elaphus*). J Wildl Dis. 2009;45:184–8.

Kirkpatrick JF, Liu IMK, Turner JW, Naugle R, Keiper R. Long-term effects of porcine zonae pellucidae immunocontraception on ovarian function in feral horses (*Equus caballus*). Reproduction. 1992;94(2):437–44.

Kirkpatrick JF, Turner A. Immunocontraception and increased longevity in equids. Zoo Biol. 2007;26:237–44.

Kirkpatrick JF, Lyda R, Frank KM. Contraceptive vaccines for wildlife: a review. Am J Reprod Immunol. 2011;66:40–50.

Kloosterboer HJ, Vonk-Noordegraaf CA, Turpijn EW. Selectivity in progesterone and androgen receptor binding of progestagens used in oral contraceptives. Contraception. 1988;38:325–32.

Knowlton FF, Gese EM, Jaeger MM. Coyote depredation control: an interface between biology and management. J Range Manag. 1999;52:398–412.

Koeppel KN, Barrows M, Visser K. The use of the gonadotropin-releasing hormone analog deslorelin for short-term contraception in red pandas (*Ailurus fulgens*). Theriogenology. 2014;81:220–4.

Kutzler MA. Intratesticular and intraepididymal injections to sterilize male cats: from calcium chloride to zinc gluconate and beyond. J Feline Med Surg. 2015;17:772–6.

Labrie C, Cusan L, Plante M, Lapointe S, Labrie F. Analysis of the androgenic activity of synthetic "progestins" currently used for the treatment of prostate cancer. J Steroid Biochem. 1987;28:379–84.

Larson S, Belting T, Rifenbury K, Fisher G, Boutelle SM. Preliminary findings of fecal gonadal hormone concentrations in six captive sea otters (*Enhydra lutris*) after deslorelin implantation. Zoo Biol. 2013;32:307–15.

Levy JK, Friary JA, Miller LA, Tucker SJ, Fagerstone KA. Long-term fertility control in female cats with GonaCon™, a GnRH immunocontraceptive. Theriogenology. 2011;76:1517–25.

Li J, Olvera AI, Akbari OS, Moradian A, Sweredoski MJ, Hess S, Hay BA. Vectored antibody gene delivery mediates long-term contraception. Curr Biol. 2015;25:R820–2.

Linhart SB, Enders RK. Some effects of diethylstilbestrol on reproduction in captive red foxes. J Wildl Manag. 1964;28:358–63.

Liu IKM, Bernoco M, Feldman M. Contraception in mares heteroimmunized with pig zona pellucidae. J Reprod Fertil. 1989;85:19–28.

Lofts B, Murton RK, Thearle RJP. The effects of 22,25-diazacholesterol dihydrochloride on the pigeon testis and on reproductive behaviour. Reproduction. 1968;15(1):145–8.

Lohr CA, Mills H, Robertson H, Bencini R. Deslorelin implants control fertility in urban brushtail possums (*Trichosurus vulpecula*) without negatively influencing their body-condition index. Wildl Res. 2009;36:324–32.

Long J, Nand AV, Ray S, Mayhew S, White D, Bunt CR, Seyfoddin A. Development of customised 3D printed biodegradable projectile for administrating extended-release contraceptive to wildlife. Int J Pharm. 2018;548:349–56.

Lueders I, Hildebrandt TB, Gray C, Botha S, Rich P, Niemuller C. Supression of testicular function in a male Asian elephant (*Elephas maximus*) treated with gonadotropin-releasing hormone vaccines. J Zoo Wildl Med. 2014;45:611–9.

Lueders I, Young D, Maree L, Van der Horst G, Luther I, Botha S, Tindall B, Fosgate G, Ganswindt A, Bertschinger HJ. Effects of GnRH vaccination in wild and captive African elephant bulls (*Loxodonta africana*) on reproductive organs and semen quality. PLoS One. 2017;12:e0178270.

Lyda RO, Frank KM, Wallace R, Lamberski N, Kirkpatrick JF. Immunocontraception of captive exotic species: V. Prolonged antibody titers in Dall sheep (*Ovis dalli dalli*) and domestic goats (*Capra hircus*) immunized with porcine zona pellucida. J Zoo Wildl Med. 2013;44:S21–5.

MacLean RA, Mathews NE, Grove DM, Frank ES, Paul-Murphy J. Surgical technique for tubal ligation in white-tailed deer (*Odocoileus virginianus*). J Zoo Wildl Med. 2006;37:354–61.

Maijer AM, Semple S. Investigating potential effects of the contraceptive Implanon on the behavior of free-ranging adult female Barbary macaques. J Appl Anim Welf Sci. 2016;19:16–23.

Malaivijitnond S, Hamada Y. Current situation and status of long-tailed macaques (*Macaca fascicularis*) in Thailand. Trop Nat Hist. 2008;8:185–204.

Marais HJ, Hendrickson DA, Stetter M, Zuba JR, Penning M, Siegal-Willott J, Hardy C. Laparoscopic vasectomy in African savannah elephant (*Loxodonta africana*); surgical technique and results. J Zoo Wildl Med. 2013;44:S18–20.

Massei G, Cowan D. Fertility control to mitigate human–wildlife conflicts: a review. Wildl Res. 2014;41:1–21.

Massei G, Cowan DP, Coats J, Gladwell F, Lane JE, Miller LA. Effect of the GnRH vaccine GonaCon on the fertility, physiology and behaviour of wild boar. Wildl Res. 2008;35:540–7.

Massei G, Koon K-K, Benton S, Brown R, Gomm M, Orahood DS, Pietravalle S, Eckery DC. Immunocontraception for managing feral cattle in Hong Kong. PLoS One. 2015;10:e0121598.

Matschke GH. Fertility control in white-tailed deer by steroid implants. J Wildl Manag. 1977a;41:731–5.

Matschke GH. Micro-encapsulated diethylstilbestrol as an oral contraceptive in white-tailed deer. J Wildl Manag. 1977b;41:87–91.

Matter BE, Jaeger I, Suter W, Tsuchimoto T, Deysenroth H. Actions of an antispermatogenic but non-mutagenic indenopyridine derivative in mice and Salmonella typhimurium. Mutat Res. 1979;66:113–27.

McShea WJ, Monfort SL, Hakim S, Kirkpatrick J, Liu I, Turner JW Jr, Chassy L, Munson L. The effect of immunocontraception on the behavior and reproduction of white-tailed deer. J Wildl Manag. 1997;61:560–9.

Meadows KL, Li Y, Xu Q, Ayres S. A novel non-surgical method to reduce fertility using the rat as a model. Clin Theriogenol. 2019;11:31–41.

Metrione LC, Verstegen JP, Heard DJ, Leblanc D, Walsh AL, Penfold LM. Preliminary evaluation of deslorelin, a GnRH agonist for contraception of the captive variable flying fox *Pteropus hypomelanus*. Contraception. 2008;78:336–45.

Middleton DJ, Watson A, Howe C, Caterson I. Suppression of cortisol responses to exogenous adrenocorticotrophic hormone, and the occurrence of side effects attributable to glucocorticoid excess, in cats during therapy with megestrol acetate and prednisolone. C J Vet Res. 1987;51:60–5.

Middleton DR, Walters B, Menkhorst P, Wright P. Fertility control in the koala, *Phascolarctos cinereus*: the impact of slow-release implants containing levonorgestrel or oestradiol on the production of pouch young. Wildl Res. 2003;30:207–12.

Millam JR, Finney HL. Leuprolide acetate can reversibly prevent egg laying in cockatiels (*Nymphicus hollandicus*). Zoo Biol. 1994;13:149–55.

Miller LA, Bynum K, Zemlick D. PZP immunocontraception in coyotes: a multi-year study with three vaccine formulations. In: Timm RM, O'Brien JM, editors. Proc 22nd vertebrate pest conf. Davis: University of California-Davis; 2006. p. 88–95.

Miller LA, Johns BE, Killian GJ. Immunocontraception of white-tailed deer with GnRH vaccine. Am J Reprod Immunol. 2000;44:266–74.

Miller L, Rhyan J, Killian G. Evaluation of GnRH contraceptive vaccine using domestic swine as a model for feral hogs. In: Proc 10th wildlife damage management conference; 2003.

Miller LA, Rhyan JC, Drew M. Contraception of bison by GnRH vaccine: a possible means of decreasing transmission of brucellosis in bison. J Wildl Dis. 2004;40:725–30.

Miller L, Killian G, Thain D, Diehl NK, Rhyan J. Four-year contraception rates of mares treated with single-injection porcine zona pellucida and GnRH vaccines and intrauterine devices. Wildl Res. 2008;35:531–9.

Miller LA, Fagerstone KA, Wagner DC, Killian GJ. Factors contributing to the success of a single-shot, multiyear PZP immunocontraceptive vaccine for white-tailed deer. Hum Wildl Confl. 2009;3:103–15.

Miller S, Bissett C, Burger A, Courtenay B, Dickerson T, Druce D, Ferreira S, Funston P, Hofmeyr D, Kilian P. Management of reintroduced lions in small, fenced reserves in South Africa: an assessment and guidelines. Afr J Wildl Res. 2013;43:138–55.

Miller PS, Boone JD, Briggs JR, Lawler DF, Levy JK, Nutter FB, Slater M, Zawistowski S, Allen BL. Simulating free-roaming cat population management options in open demographic environments. PLoS One. 2014;9(11):e113553.

Mok K-W, Mruk DD, Lie PP, Lui W-Y, Cheng CY. Adjudin, a potential male contraceptive, exerts its effects locally in the seminiferous epithelium of mammalian testes. Reproduction. 2011;141:571–80.

Mol JA, van Garderen E, Rutteman GR, Rijnberk A. New insights in the molecular mechanism of progestin-induced proliferation of mammary epithelium: induction of the local biosynthesis of

growth hormone (GH) in the mammary glands of dogs, cats and humans. J Steroid Biochem Mol Biol. 1996;57:67–71.
Moresco A, Munson L, Gardner IA. Naturally occurring and melengestrol acetate-associated reproductive tract lesions in zoo canids. Vet Pathol. 2009;46:1117–28.
Moresco A, Dadone L, Arble J, Klaphake E, Agnew DW. Location and removal of deslorelin acetate implants in female African lions (*Panthera leo*). J Zoo Wildl Med. 2014;45:397–401.
Munson L, Bauman JE, Asa CS, Jochle W, Trigg TE. Efficacy of the GnRH analogue deslorelin for suppression of oestrus cycle in cats. J Reprod Fertil Suppl. 2001;57:269–73.
Munson L, Chassy LM, Asa C. Efficacy, safety, and reversibility of bisdiamine as a male contraceptive in cats. Theriogenology. 2004;62:81–92.
Munson L, Moresco A, Calle PP. Adverse effects of contraceptives. In: Asa CS, Porton IJ, editors. Wildlife contraception: issues, methods, and application. Baltimore: Johns Hopkins University Press; 2005. p. 66–82.
Nash PB, James DK, Hui LT, Miller LA. Fertility control of California ground squirrels using GnRH immunocontraception. In: Timm RM, Gorenzel WP, editors. 21st vertebrate pest conference. Davis: University of California-Davis; 2004. p. 274–8.
Nave CD, Coulson G, Poiani A, Shaw G, Renfree MB. Fertility control in the eastern grey kangaroo using levonorgestrel implants. J Wildl Manag. 2002;66:470–7.
Nettles VF. Potential consequences and problems with wildlife contraceptives. Reprod Fertil Dev. 1997;9:137–44.
Nolan MB, Bertschinger HJ, Roth R, Crampton M, Martins IS, Fosgate GT, Stout TA, Schulman ML. Ovarian function following immunocontraceptive vaccination of mares using native porcine and recombinant zona pellucida vaccines formulated with a non-Freund's adjuvant and anti-GnRH vaccines. Theriogenology. 2018;120:111–6.
Nolan MB, Schulman ML, Botha AE, Human A-M, Roth R, Crampton MC, Bertschinger HJ. Serum antibody immunoreactivity and safety of native porcine and recombinant zona pellucida vaccines formulated with a non-Freund's adjuvant in horses. Vaccine. 2019;37:1299–306.
Nuñez CMV, Adelman JS, Rubenstein DI. Immunocontraception in wild horses (*Equus caballus*) extends reproductive cycling beyond the normal breeding season. PLoS One. 2010;5:e13635.
Okeson DM, Higbie CT, Mylniczenko ND, Haynes A, Bennett S, Klocke E, Carpenter JW. Management of endometriosis in two captive mandrills (*Mandrillus sphinx*). J Zoo Wildl Med. 2016;47:614–7.
Oliveira ECS, Moura MR, Silva VA, Peixoto CA, Saraiva KLA, de Sá MJC, Douglas RH, de Pinho Marques A. Intratesticular injection of a zinc-based solution as a contraceptive for dogs. Theriogenology. 2007;68(2):137–45.
Oliver M, Piertney S, Zalewski A, Melero Y, Lambin X. The compensatory potential of increased immigration following intensive American mink population control is diluted by male-biased dispersal. Biol Invasions. 2016;18:3047–61.
Oogjes G. Ethical aspects and dilemmas of fertility control of unwanted wildlife: an animal welfarist's perspective. Reprod Fertil Dev. 1997;9:163–8.
Orford H, Perrin M, Berry H. Contraception, reproduction and demography of free-ranging Etosha lions (*Panthera leo*). J Zool. 1988;216:717–33.
Pai M, Bruner R, Schlafer DH, Yarrow GK, Yoder CA, Miller LA. Immunocontraception in eastern gray squirrels (*Sciurus carolinensis*): morphologic changes in reproductive organs. J Zoo Wildl Med. 2011;42:718–22.
Patton ML, Jochle W, Penfold LM. Contraception in ungulates. In: Asa CS, Porton IJ, editors. Wildlife contraception: issues, methods and application. Baltimore: Johns Hopkins University Press; 2005. p. 149–67.
Patton ML, Bashaw MJ, del Castillo SM, Jochle W, Lamberski N, Rieches R, Bercovitch FB. Long-term suppression of fertility in female giraffe using the GnRH agonist deslorelin as a long-acting implant. Theriogenology. 2006;66:431–8.
Patton ML, Jochle W, Penfold LM. Review of contraception in ungulate species. Zoo Biol. 2007;26:311–26.

Penfold LM, Powell D, Traylor-Holzer K, Asa CS. "Use it or lose it": characterization, implications, and mitigation of female infertility in captive wildlife. Zoo Biol. 2014;33(1):20–8.
Pepin KM, Davis AJ, Cunningham FL, VerCauteren KC, Eckery DC. Potential effects of incorporating fertility control into typical culling regimes in wild pig populations. PLoS One. 2017;12:e0183441.
Perry KR, Arjo WM, Bynum KS, Miller LA. GnRH single-injection immunocontraception of black-tailed deer. In: Timm RM, O'Brien JM, editors. Proc 22nd vertebrate pest conference. Davis: University of California-Davis; 2006. p. 72–7.
Petritz OA, Guzman DS-M, Paul-Murphy J, Fecteau K, Mete A, Kass PH, Hawkins MG. Evaluation of the efficacy and safety of single administration of 4.7-mg deslorelin acetate implants on egg production and plasma sex hormones in Japanese quail (*Coturnix coturnix japonica*). Am J Vet Res. 2013;74:316–23.
Phraluk O, Wajjwalku W, Siriaroonrat B, Booddee O, Thongtip N. Effects of immunization against gonadotropin releasing hormone on reproductive functions in male rusa deer (*Rusa timorensis*). Thai J Vet Med. 2015;45:1.
Plotka ED, Seal US. Fertility control in female white-tailed deer. J Wildl Dis. 1989;25:643–6.
Pope LC, Butlin RK, Wilson GJ, Woodroffe R, Erven K, Conyers CM, Franklin T, Delahay RJ, Cheeseman CL, Burke T. Genetic evidence that culling increases badger movement: implications for the spread of bovine tuberculosis. Mol Ecol. 2007;16:4919–29.
Powers JG, Baker DL, Davis TL, Conner MM, Lothridge AH, Nett TM. Effects of gonadotropin-releasing hormone immunization on reproductive function and behavior in captive female Rocky Mountain elk (*Cervus elaphus nelsoni*). Biol Reprod. 2011;85:1152–60.
Pozor MA, Macpherson ML, McDonnell SM, Nollin M, Roser JF, Love C, Runyon S, Thomas BF, Troedsson MH. Indenopyride derivative RTI-4587-073(l): a candidate for male contraception in stallions. Theriogenology. 2013;80(9):1006–16.
Prohaczik A, Kulcsar M, Trigg T, Huszenicza G. Treatments suppressing ovarian activity in ferrets (*Mustela putorius furo*). Reprod Domest Anim. 2003;38:331.
Raines JA, Fried JJ. Use of deslorelin acetate implants to control aggression in a multi-male group of rock hyrax (*Procavia capensis*). Zoo Biol. 2016;35:201–4.
Ramsey D. Effects of fertility control on behavior and disease transmission in brushtail possums. J Wildl Manag. 2007;71:109–16.
Raphael BL, Kalk P, Thomas P, Calle PP, Doherty JG, Cook RA. Use of melengestrol acetate in feed for contraception in herds of captive ungulates. Zoo Biol. 2003;22:455–63.
Reiter DK, Brunson MW, Schmidt RH. Public attitudes toward wildlife damage management and policy. Wildl Soc Bull. 1999;27:746–58.
Remfry J. Control of feral cat populations by long term administration of megestrol acetate. Vet Rec. 1978;28:403–4.
Roelle JE, Ransom J. Injection-site reactions in wild horses (*Equus caballus*) receiving an immunocontraceptive vaccine. US Geological Survey Scientific Investigations Report: 2009-5038; 2009. 15 pp.
Romatowski J. Use of megestrol acetate in cats. J Am Vet Med Assoc. 1989;194:700–2.
Roughton RD. Effects of oral melengestrol acetate on reproduction in captive white-tailed deer. J Wildl Manag. 1979;43:428–36.
Rowland MN. Use of a deslorelin implant to control aggression in a male bearded dragon (*Pogona vitticeps*). Vet Rec. 2011;169:127.
Rutberg A, Grams K, Turner JW Jr, Hopkins H. Contraceptive efficacy of priming and boosting doses of controlled-release PZP in wild horses. Wildl Res. 2017;44:174–81.
Sacco AG, Subramanian MG, Yurewicz EC, DeMayo FJ, Dukelow WR. Heteroimmunization of squirrel monkeys (Saimiri sciureus) with a purified porcine zona antigen (PPZA): immune response and biologic activity of antiserum. Fertil Steril. 1983;39:350–8.
Sacks BN, Jaeger MM, Jennifer CCN, McCullough DR. Territoriality and breeding status of coyotes relative to sheep predation. J Wildl Manag. 1999;63:593–605.
Schulman ML, Botha AE, Muenscher SB, Annandale CH, Guthrie AJ, Bertschinger HJ. Reversibility of the effects of GnRH-vaccination used to suppress reproductive function in mares. Equine Vet J. 2013;45:111–3.

Selman PJ, Mol JA, Rutteman GR, Rijnberk A. Progestin treatment in the dog. I. Effects on growth hormone, insulin-like growth factor I and glucose homeostasis. Eur J Endocr. 1994;131:413–21.
Selman PJ, Mol JA, Rutteman GR, van Garderen E, van den Ingh TS, Rijnberk A. Effects of progestin administration on the hypothalamic-pituitary-adrenal axis and glucose homeostasis in dogs. J Reprod Fertil Suppl. 1997;51:345–54.
Shideler SE, Stoops MA, Gee NA, Howell JA, Lasley BL. Use of porcine zona pellucida (PZP) vaccine as a contraceptive agent in free-ranging tule elk (*Cervus elaphus nannodes*). J Reprod Fertil Suppl. 2002;60:169–76.
Shuster SM, Pyzyna B, Mayer LP, Dyer CA. The opportunity for sexual selection and the evolution of non-responsiveness to pesticides, sterility inducers and contraceptives. Heliyon. 2018;4:e00943.
Sitruk-Ware R. Reprint of pharmacological profile of progestins. Maturitas. 2008;61:151–7.
Skinner DC, Albertson AJ, Navratil A, Smith A, Mignot M, Talbott H, Scanlan-Blake N. Effects of gonadotropin-releasing hormone outside the hypothalamis-pituitary-reproductive axis. J Neuroendocrinol. 2009;21:282–92.
Sloan JM, Oliver IM. Progestogen-induced diabetes in the dog. Diabetes. 1975;24:337–44.
Snape MA. Reproductive and behavioural effects of a GnRH-targeted immunocontraceptive vaccine in macropodids. Canberra: Australian National University; 2012. p. 167.
Spence CE, Kenyon JE, Smith DR, Hayes RD, Baer AM. Surgical sterilization of free-ranging wolves. Can Vet J. 1999;40:118–21.
Stringer EM, De Voe RS, Loomis MR. Suspected anaphylaxis to leuprolide acetate depot in two elf owls (*Micrathene whitneyi*). J Zoo Wildl Med. 2011;42:166–8.
Taylor LH, Wallace RM, Balaram D, Lindenmayer JM, Eckery DC, Mutonono-Watkiss B, Parravani E, Nel LH. The role of dog population management in rabies elimination—a review of current approaches and future opportunities. Front Vet Sci. 2017;4 https://doi.org/10.3389/fvets.2017.00109.
Tribe A, Hanger J, McDonald I, Loader J, Nottidge B, McKee J, Phillips C. A reproductive management program for an urban population of eastern grey kangaroos (*Macropus giganteus*). Animals. 2014;4:562–82.
Turner JW Jr, Liu IKM, Kirkpatrick JF. Remotely delivered immuno-contraception in free-roaming feral burros (*Equus asinus*). J Reprod Fertil. 1996;107:31–5.
Turner JW, Liu IKM, Flanagan DR, Rutberg AT, Kirkpatrick JF. Immunocontraception in wild horses: one inoculation provides two years of infertility. J Wildl Manag. 2007;71:662–7.
Twigg L, Kent Williams C. Fertility control of overabundant species; can it work for feral rabbits? Ecol Lett. 1999;2:281–5.
Vargas-Pino F, Gutierrez-Cedillo V, Vanales-Vargas EJ, Gress-Ortega LR, Miller LA, Rupprecht CE, et al. Concomitant administration of GonaCon™ and rabies vaccine in female dogs (*Canis familiaris*) in Mexico. Vaccine. 2013;31:4442–7.
Wallace PY, Asa CS, Agnew M, Cheyne SM. A review of population control methods in captive-housed primates. Anim Welf. 2016;25:7–20.
Wheaton CJ, Savage A, Shukla A, Neiffer DL, Qu W, Sun Y, Lasley B. The use of long acting subcutaneous levonorgestrel (LNG) gel depot as an effective contraceptive option for cotton-top tamarins (*Saguinus oedipus*). Zoo Biol. 2011;30:498–522.
Wilkins L. Masculinization of female fetus due to use of orally given progestins. J Am Vet Med Assoc. 1960;172:1028–32.
Willers N, Martin G, Matson P, Mawson P, Morris K, Bencini R. Finding the balance: fertility control for the management of fragmented populations of a threatened rock-wallaby species. Animals. 2015;5:0414.
Wilson ME, Coulson G, Shaw G, Renfree MB. Deslorelin implants in free-ranging female eastern grey kangaroos (*Macropus giganteus*): mechanism of action and contraceptive efficacy. Wildl Res. 2013;40:403–12.
Woodroffe R, Frank LG. Lethal control of African lions (*Panthera leo*): local and regional population impacts. Anim Conserv. 2005;8:91–8.

Wright PJ, Verstegen JP, Onclin K, Jochle W, Armour AF, Martin GB, Trigg TE. Suppression of the oestrous responses of bitches to the GnRH analogue deslorelin by progestin. J Reprod Fertil Suppl. 2001;57:263–8.
Yoder CA, Miller LA. Effect of GonaCon™ vaccine on black-tailed prairie dogs: immune response and health effects. Vaccine. 2010;29:233–9.
Yoder C, Bynum K, Miller L. Development of Diazacon as an avian contraceptive. In: Wildlife damage management. Lincoln: University of Nebraska; 2005. p. 190–201.
Yoder CA, Mayle BA, Furcolow CA, Cowan DP, Fagerstone KA. Feeding of grey squirrels (*Sciurus carolinensis*) with the contraceptive agent DiazaCon™: effect on cholesterol, hematology, and blood chemistry. Integr Zool. 2011;6:409–19.

Chapter 18
From the Ivory Tower to Reality! Conclusions of the New Edition

Pierre Comizzoli, Janine L. Brown, and William V. Holt

Abstract While many of the traditional scientific disciplines have developed over centuries, animal conservation is a relative newcomer. It relies on multiple specialties with different levels of expertise that, eventually, generate vast amounts of data. More specifically, conservation physiology is an emerging area that can be defined as '*an integrative scientific discipline applying physiological concepts, tools, and knowledge to characterizing biological diversity and its ecological implications; understanding and predicting how organisms, populations, and ecosystems respond to environmental change and stressors; and solving conservation problems across the broad range of taxa, including microbes, plants, and animals*' (Cooke et al. 2013). Reproductive biology is more focused, given that it mainly deals with the physiology underlying the production of gametes, embryos, and offspring, and the many associated processes that control these events. However, it is integrated into the different components of conservation physiology. In bringing together the various contributors for this book, the editors' purpose was to provide readers with a new perspective about the complexity behind reproduction and the role it plays in species conservation. Chapters highlight the diversity of reproductive mechanisms across taxa, and provide insight into how they may have evolved, and likely will continue to evolve in a changing environment. To conservation physiologists, the hope is that this information will be applied to sustain populations in both natural habitats and managed facilities. Ultimately, a major goal is to forecast and mitigate negative impacts of environmental change or anthropogenic pressures on animal fitness, which will only follow once we have acquired a solid understanding of reproductive processes.

P. Comizzoli (✉)
Smithsonian Conservation Biology Institute, National Zoological Park,
Washington, DC, USA
e-mail: Comizzolip@si.edu

J. L. Brown
Smithsonian Conservation Biology Institute, National Zoological Park,
Front Royal, VA, USA

W. V. Holt
Academic Unit of Reproductive and Developmental Medicine, The University of Sheffield,
Sheffield, UK

P. Comizzoli et al. (eds.), *Reproductive Sciences in Animal Conservation*,
Advances in Experimental Medicine and Biology 1200,
https://doi.org/10.1007/978-3-030-23633-5_18

1 The Growing and Evolving Role of Reproductive Sciences in a Changing Planet

While many of the traditional scientific disciplines have developed over centuries, animal conservation is a relative newcomer. It relies on multiple specialties with different levels of expertise that, eventually, generate vast amounts of data. More specifically, conservation physiology is an emerging area that can be defined as '*an integrative scientific discipline applying physiological concepts, tools, and knowledge to characterizing biological diversity and its ecological implications; understanding and predicting how organisms, populations, and ecosystems respond to environmental change and stressors; and solving conservation problems across the broad range of taxa, including microbes, plants, and animals*' (Cooke et al. 2013). Reproductive biology is more focused, given that it mainly deals with the physiology underlying the production of gametes, embryos, and offspring, and the many associated processes that control these events. However, it is integrated into the different components of conservation physiology. In bringing together the various contributors for this book, the editors' purpose was to provide readers with a new perspective about the complexity behind reproduction and the role it plays in species conservation. Chapters highlight the diversity of reproductive mechanisms across taxa, and provide insight into how they may have evolved, and likely will continue to evolve in a changing environment. To conservation physiologists, the hope is that this information will be applied to sustain populations in both natural habitats and managed facilities. Ultimately, a major goal is to forecast and mitigate negative impacts of environmental change or anthropogenic pressures on animal fitness, which will only follow once we have acquired a solid understanding of reproductive processes.

Nowadays, scientists are conscious that we live in a time when climate factors are changing rapidly, and wildlife habitats are being modified dramatically by humans. This has been emphasized by the recent Intergovernmental Science-Policy Platform on Biodiversity and Ecosystem Services (IPBES) in Paris, France (www.ipbes.net/news/Media-Release-Global-Assessment). Wild species will either adapt to new conditions, or eventually succumb to extinction. A variety of environmental changes affect wildlife in a multitude of ways, and species are starting to respond. For example, to avoid the consequences of global warming, animals are changing their preferred latitudes or migration patterns to more suitable habitats: i.e., fish moving to cooler oceanic regions; amphibians and reptiles preferring higher elevations where the air is cooler. Yet, these are only temporary and deceptively simple solutions to allay threats to their survival, as reproduction and health are highly dependent on seasonality, the availability of suitable diets, and the ability to withstand the higher prevalence of diseases—all of which will continue to change if the climate crisis persists. Acidification of the oceans is another important stressor faced by marine wildlife, with potentially devastating impacts upon coral reefs and their attendant fish populations. Some effects persist in subsequent generations, affecting transgenerational programming in a way that will impact fitness for decades to come. As if these examples were not enough, exposure to man-made chemicals in the environment can disrupt reproduction on many levels (from conception to birth) over multiple generations. Reproductive science therefore plays a key role in species conservation to understand and mitigate those multi-factorial phenomena.

2 Moving Toward Transdisciplinary Approaches

In the present edition, it is clear that the boundaries between different scientific disciplines are becoming ever more blurred. It is no longer easy or even possible to remain focused on a highly specialized topic in reproduction or conservation, without having at least some understanding of allied subjects. These interactions emerge from different chapters, emphasising how "reproductive science" is more than a topic of interest to people who only want to breed animals. The opening chapter by Manlik presents examples that are highly relevant to many species. The conclusion that for several slow-growing animal populations, reproduction is more important than survival is significant because it goes against dogma and emphasizes the importance of reproduction when devising conservation strategies. Proper modelling approaches that take into account the role of reproduction in population dynamics are therefore needed.

The chapters by Edwards et al. and Comizzoli are also connected to other chapters. They illustrate the need to integrate reproductive and welfare sciences, as they are inexorably linked, both *ex situ* and *in situ*. Threats to species survival include intrinsic (perceived threats, disease) and extrinsic (pollutants, noise, resource limitation, predator pressure, pathogens, temperature, salinity) stressors, all of which can challenge an organism's homeostasis and ability to survive. How animals respond to these stressors is further driven by temperament and coping skills; i.e., resilience, which likely are determined by genetics. We can no longer rely on single factor analyses, but need to use holistic approaches to examine multiple biological systems to determine the wear and tear on the body and how it impacts reproductive fitness, a pervasive theme throughout this book.

Both Acevedo-Whitehouse and Lanyon show that inbreeding is a problem for marine mammals. Acevedo-Whitehouse states that captive breeding must be carefully planned to avoid inbreeding, while Lanyon suggests that cryopreserved gametes, etc. could be useful in this context. Interestingly, Lanyon, like Manlik, mentions that rate of offspring production is the most important predictor of risk.

Most wildlife research has been conducted in larger, more charismatic, terrestrial mammalian species, so although we touch on some of these—e.g., Brown's update on elephants and Martin-Wintle's chapter on giant pandas—we also endeavoured to highlight advances in less-studied taxa. There is much in common between Mayer's chapter on fish, Hagedorn's chapter about corals, Penfold's chapter on sharks and rays, and Clulow's chapter on amphibians. They all recognize the lessons learned from early studies in mammals and how technologies are now developed in non-mammalian taxa. The same is true for understudied mammals, such as the marsupials described in Rodger's and Johnston's chapters. Rodger takes a broad view of marsupial conservation and suggests that the only realistic way to ensure that marsupial genetics are conserved for the future is to collect and preserve genetic samples "now" in the expectation that they will eventually be usable. In some ways this is consistent with Johnston's chapter, which shows that despite huge efforts to develop ART methods for breeding only one marsupial (koala), it is still not possible to use frozen semen and traditional artificial insemination methods. It is clear that a new way forward is needed.

While the need for more fundamental knowledge is still essential, the contribution of new technologies will help to make progress faster, and be especially important to mitigating threats to a multitude of species throughout the biosphere. What the future holds for our most fragile populations (e.g., see chapters on marine mammals and coral) cannot be predicted, but we may be moving to a time when specialized species are no longer able to adapt, leaving primarily generalists that thrive under a variety of natural and anthropogenic conditions. This could have a devastating effect on the planet's biodiversity, as many specialists fill niches important to maintaining the equilibrium of ecosystems (e.g., coral). While the integration of reproductive biotechnologies into conservation programs is far from complete, we are moving toward transdisciplinary approaches (learning from each other and building stronger joint efforts) that take a holistic approach to species conservation and management, and understanding what the greatest threats are.

3 Looking Forward

The last decade has seen enormous progress in many aspects of sciences and technologies relevant to reproduction. The optimization of new tools and approaches spanning from biomarkers to bioinformatics have helped to accelerate the exploration and understanding of basic mechanisms (gametogenesis, fertilization, embryo development, implantation, and gestation) and how they can be influenced by different factors (seasons, aging, pathologies, or environmental changes) (Comizzoli and Holt 2019). In the next decade, more fundamental research will help to decipher complex biological mechanisms in wild animal species. For instance, a better understanding about enhancement or suppression of fertility through the central nervous system (via GnIH or kisspeptins) is expected. Likewise, we predict that more information on final gamete maturation in the epididymis and the oviduct (through exosomes) will be available. Assisted Reproductive Techniques (ART) and biobanking no doubt will continue to be developed and systematically implemented to change the paradigm in conservation programs (from pachyderms to amphibians and reptiles). This will be associated with the development of alternative methods to produce gametes and store germplasm at room temperature. Integration of new research tools will be the key to success in conservation programs; for example, new biomarkers and non-invasive detection and monitoring of reproductive status using 'omics' and bioinformatics (Comizzoli and Holt 2019).

While further use of existing approaches is critical, innovations will be central to saving species. However, it must be done methodically, with basic and applied research coming first to directly inform breeding programs or to develop ARTs and biobanking. Choosing priority species will remain a major point of discussion, and contention, as it depends on many factors that are often not scientific. Ideally, we should study reproductive biology in common species (not only the rarest ones) in each taxonomic group, and apply that knowledge to more endangered counterparts. Finally, we need to clearly designate target/flagship species in large taxa like fish and amphibians, while being particularly cognizant of the need to maintain a diversity of invertebrates for planetary health. New directions in wildlife research should not only enrich our understanding of basic reproduction, but also add to our knowledge of the molecular mechanisms and principles in cellular and organismal biology that accompany it.

The role that synthetic biology (or more specifically the production of genetically engineered organisms) can play in species conservation still needs to be defined, as is a need to 'separate the good from the scary'. For instance, based on a survey of life science student attitudes, perception of synthetic biology application for biodiversity conservation was positive, and the technique was perceived as being beneficial (Kemal 2019). However, it also was considered risky, and in general student attitude toward risk in tackling conservation problems was somewhat cautious. Nevertheless, there was a significant correlation between support for new technology development for biodiversity conservation with the belief in promises that synthetic biology will benefit conservation science. The respondents were optimistic that the application would not decrease people's interest in biodiversity conservation. So, there are aspects of genetical engineering that could potentially enhance reproduction and fertility of wild populations, but we need to be cautious about misuse of reproductive technologies, and keep welfare as well as ethical priorities in mind.

4 Old and New Challenges

Reproductive studies will always remain a high priority in the giant conservation puzzle. However, we have to be realistic about what our challenges are and make sure that we address them early on. While we can expect tremendous progress in the study of reproduction of wild species using new technologies, we need to keep in mind that conservation means supporting whole organisms and reproductive fitness of entire populations. Genomic and epigenomic information will only have value when solid phenotypical data are integrated to them. Similar to the discipline of 'precision medicine' currently developed in humans, it is time to create and develop a concept of 'precision conservation' or 'precision conservation breeding'—a more customized/tailored approach to optimize the potential of each individual in a population.

Future challenges are not only related to the application of new technologies to wild species preservation. As pointed out earlier, it is urgent to study species while population numbers are still high, so we can understand their reproductive processes and be prepared to take action if needed. Furthermore, for many reasons that often are not scientifically justified, access to live animals and creation of conservation breeding facilities are still difficult in regions that are rich in biodiversity (South-East Asia, South America). This has to be addressed by promoting the benefits of *ex situ* conservation to decision makers and broader audiences, and avoiding the consideration of zoos as mere amusement parks.

Interestingly, success is not always guaranteed by money or the promise of money (i.e., Mickelson prize https://www.michelsonprizeandgrants.org/ shows how difficult it still is to develop a permanent contraceptive for male and female dogs and cats). Another set of challenges in reproductive science as applied to threatened species is the diffusion of knowledge. Data generated in studies of reproduction in wild species are often not easily publishable in peer-reviewed journals because of the limited number of observations that can be made over long time periods. The current emphasis by journals on statistical power analysis, while not unreasonable, means that the required number of observations on separate individuals are frequently unattainable due to many constraints. These include difficult working environments (field studies, dangerous species, or stress-sensitive species), small population sizes, and poor funding. In some cases, the entire population of a threatened species may be less than 10 individuals. Does this mean that research conducted on that species is not worthwhile? Without an efficient way to share those valuable data, too many studies remain inaccessible and unknown to the scientific community. Instead of creating new ways of sharing data on wildlife reproductive biology, existing peer-reviewed journals should be persuaded to accept and publish studies that are perhaps more descriptive than in traditional species, but nevertheless have a high conservation value.

It also is highly recommended to foster communication and coordination between experts in reproduction with physiological and conservation backgrounds. This could be achieved by making sure that scientific societies that specialize in cell and organismal biology make room for presentations about species conservation. This also is valid for the creation of funding opportunities bridging multiple fields to tackle the lack of knowledge in biodiversity, while recognizing the critical role of comparative studies. Unfortunately, the current international community of reproductive physiologists and veterinarians specialized in wild species preservation is very small. It therefore is urgent to attract and train more students in our field through clearly identified tracks at the university or in veterinary/medical schools.

In sum, success will require collective efforts to identify extant limitations and fundamental gaps in knowledge, both intellectual and practical, and joint efforts to secure long-overdue improvements. Animal and plant conservation is now studied, understood, and enhanced through multiple disciplines (from bio-economy to molecular technologies) that have integrated new tools and are generating considerable amount of data. Science is therefore conducted in a different way to draw conclusions and define solutions. Biotechnologies are increasingly contributing to biological knowledge at different levels (molecular, cellular, organismal, individual, population, or entire habitats). However, there is an urgent need to bridge the gap between those new technologies and traditional approaches through transdisciplinary projects, proper training of students, and general public as well as decision maker awareness.

References

Comizzoli P, Holt WV. Breakthroughs and new horizons in reproductive biology of rare and endangered animal species. Biol Reprod. 2019; https://doi.org/10.1093/biolre/ioz031.

Cooke SJ, Sack L, Franklin CE, Farrell AP, Beardall J, Wikelski M, Chown SL. What is conservation physiology? Perspectives on an increasingly integrated and essential science. Conserv Physiol. 2013;1:cot001.

Kemal RA. Short communication: perception of synthetic biology application for biodiversity conservation among life science students in Institut Teknologi Bandung, Indonesia. Nusant Biosci. 2019;10:36–40.

Index

P. Comizzoli et al. (eds.), *Reproductive Sciences in Animal Conservation*, Advances in Experimental Medicine and Biology 1200,
https://doi.org/10.1007/978-3-030-23633-5

FSC
www.fsc.org
MIX
Papier | Fördert
gute Waldnutzung
FSC® C083411

Kolibri 360 GmbH
Ferdinand-Jühlke-Straße 7
99095 Erfurt, Deutschland
produktsicherheit@kolibri360.de